10 0351526 8
KU-030-317

UNIVERSITY OF NOTTINGHAM
WITHDRAWN
FROM THE LIBRARY
NOT TO
BE TAKEN
OUT OF
THE
LIBRARY

Compendium of Organic Synthetic Methods

Compendium of Organic Synthetic Methods

Volume 11

MICHAEL B. SMITH

DEPARTMENT OF CHEMISTRY
THE UNIVERSITY OF CONNECTICUT
STORRS, CONNECTICUT

A JOHN WILEY & SONS, INC., PUBLICATION

Cover illustration was adapted from "Disconnect By the Numbers: A Beginner's Guide to Synthesis" by M. B. Smith. *Journal of Chemical Education,* **1990,** 67, 848–856.

Copyright © 2003 by John Wiley & Sons, Inc. All rights reserved.

Published by John Wiley & Sons, Inc., Hoboken, New Jersey.
Published simultaneously in Canada.

No part of this publication may be reproduced, stored in a retrieval system or transmitted in any form or by any means, electronic, mechanical, photocopying, recording, scanning or otherwise, except as permitted under Section 107 or 108 of the 1976 United States Copyright Act, without either the prior written permission of the Publisher, or authorization through payment of the appropriate per-copy fee to the Copyright Clearance Center, Inc., 222 Rosewood Drive, Danvers, MA 01923, (978) 750-8400, fax (978) 750-4470, or on the web at www.copyright.com. Requests to the Publisher for permission should be addressed to the Permissions Department, John Wiley & Sons, Inc., 111 River Street, Hoboken, NJ 07030, (201) 748-6011, fax (201) 748-6008, e-mail: permreq@wiley.com.

Limit of Liability/Disclaimer of Warranty: While the publisher and author have used their best efforts in preparing this book, they make no representation or warranties with respect to the accuracy or completeness of the contents of this book and specifically disclaim any implied warranties of merchantability or fitness for a particular purpose. No warranty may be created or extended by sales representatives or written sales materials. The advice and strategies contained herein may not be suitable for your situation. You should consult with a professional where appropriate. Neither the publisher nor author shall be liable for any loss of profit or any other commercial damages, including but not limited to special, incidental, consequential, or other damages.

For general information on our other products and services please contact our Customer Care Department within the U.S. at 877-762-2974, outside the U.S. at 317-572-3993 or fax 317-572-4002.

Wiley also publishes its books in a variety of electronic formats. Some content that appears in print, however, may not be available in electronic format.

Library of Congress Cataloging Card Number: 71-162800

ISBN 0-471-25965-9

Printed in the United States of America.

10 9 8 7 6 5 4 3 2 1

UNIVERSITY OF NOTTINGHAM
WITHDRAWN
FROM THE LIBRARY
UNIVERSITY LIBRARY NOTTINGHAM

CONTENTS

PREFACE

Since the original volume in this series by Ian and Shuyen Harrison, the goal of the *Compendium of Organic Synthetic Methods* was to facilitate the search for functional group transformations in the original literature of Organic Chemistry. In Volume 2, difunctional compounds were added and this compilation was continued by Louis Hegedus and Leroy Wade for Volume 3 of the series. Wade became the author for Volume 4 and continued with Volume 5. I began editing the series with Volume 6, where I introduced an author index for the first time and added a new chapter (Chapter 15, Oxides). Volume 7 introduced Sections 378 (Oxides - Alkynes) through Section 390 (Oxides - Oxides). This volume introduces Section 74G (Cyclobutanations). The *Compendium* is a handy desktop reference that will remain a valuable tool to the working Organic chemist, allowing a "quick check" of the literature. It also allows one to "browse" for new reactions and transformations that may be of interest. The number of articles that comprise Organic literature is very large, and the *Compendium* is a focused and highly representative review of the literature and is offered in that context.

Compendium of Organic Synthetic Methods, Volume 11 contains both functional group transformations and carbon-carbon bond forming reactions from the literature appearing in the years 1999, 2000 and 2001. The classification schemes used for Volumes 6–10 have been continued, but one new section was added. Section 74G (Cyclobutanations) describes methods that produce cyclobutane rings. As in the past, difunctional compounds appear in Chapter 16. The experienced user of the *Compendium* will require no special instructions for the use of Volume 11. Author citations and the Author Index have been continued as in Volumes 6–10.

Every effort has been made to keep the manuscript error free. Where there are errors, I take full responsibility. If there are questions or comments, the reader is encouraged to contact me directly at the address, phone, fax, or Email addresses given below.

As I have throughout my writing career, I thank my wife Sarah and my son Steven, who have shown unfailing patience and devotion during this work. Steven Smith also prepared many of the drawings used in this volume, and I thank him for that contribution. I also thank Dr. Darla Henderson and Amy Romano, the editors for this volume.

Michael B. Smith

Department of Chemistry
University of Connecticut
55 N. Eagleville Road
Storrs, Connecticut 06269-3060

Voice phone: (860) 486-2881
Fax: (860) 486-2981
Email: smith@nucleus.chem.uconn.edu
Homepage: http://orgchem.chem.uconn.edu/home/mbs-home.html

Storrs, Connecticut
April, 2003

ABBREVIATIONS

Ac	Acetyl	(structure: acetyl, CH_3)
acac	Acetylacetonate	
AIBN	*azo-bis*-isobutyronitrile	
aq.	Aqueous	
(structure: 9-BBN boryl)	9-Borabicyclo[3.3.1]nonylboryl	
9-BBN	9-Borabicyclo[3.3.1]nonane	
BER	Borohydride exchange resin	
BINAP	*2R,3S*-2,2'-*bis*-(diphenylphosphino)-1,1'-binaphthyl	
Bmim	1-butyl-3-methylimidazolium	
Bn	benzyl	
Bz	benzoyl	
BOC	*t*-Butoxycarbonyl	(structure: O*t*-Bu)
bpy (Bipy)	2,2'-Bipyridyl	
Bu	*n*-Butyl	$-CH_2CH_2CH_2CH_3$
CAM	Carboxamidomethyl	
CAN	Ceric ammonium nitrate	$(NH)_2Ce(NO_3)_6$
c-	cyclo-	
cat.	Catalytic	
Cbz	Carbobenzyloxy	(structure: OCH_2Ph)
Chirald	2S,3R-(+)-4-dimethylamino-1,2-diphenyl-3-methylbutan-2-ol	
COD	1,5-Cyclooctadienyl	
COT	1,3,5-cyclooctatrienyl	
Cp	Cyclopentadienyl	
CSA	Camphorsulfonic acid	
CTAB	cetyltrimethylammonium bromide	$C_{16}H_{33}NMe_3^+Br^-$
Cy (c-C_6H_{11})	Cyclohexyl	(structure: cyclohexyl)
°C	Temperature in Degrees Centigrade	
DABCO	1,4-Diazabicylco[2.2.2]octane	
dba	dibenzylidene acetone	
DBE	1,2-Dibromoethane	$BrCH_2CH_2Br$
DBN	1,8-Diazabicyclo[5.4.0]undec-7-ene	
DBU	1,5-Diazabicyclo[4.3.0]non-5-ene	
DCC	1,3-Dicyclohexylcarbodiimide	c-C_6H_{11}-N=C=N-c-C_6H_{11}
DCE	1,2-Dichloroethane	$ClCH_2CH_2Cl$
DCM	dichloromethane	CH_2Cl_2
DDQ	2,3-Dichloro-5,6-dicyano-1,4-benzoquinone	

% de	% Diasteromeric excess	
DEA	Diethylamine	$HN(CH_2CH_3)_2$
DEAD	Diethylazodicarboxylate	$EtO_2C\text{-}N{=}NCO_2Et$
Dibal-H	Diisobutylaluminum hydride	$(Me_2CHCH_2)_2AlH$
Diphos (dppe)	1,2-*bis*-(Diphenylphosphino)ethane	$Ph_2PCH_2CH_2PPh_2$
Diphos-4 (dppb)	1,4-*bis*-(Diphenylphosphino)butane	$Ph_2P(CH_2)_4PPh_2$
DMAP	4-Dimethylaminopyridine	
DMA	Dimethylacetamide	
DME	Dimethoxyethane	$MeOCH_2CH_2OMe$
DMF	*N,N'*-Dimethylformamide	O H N(CH$_3$)$_2$
dmp	*bis*-[1,3-Di(*p*-methoxyphenyl)-1,3-propanedionato]	
dpm	dipivaloylmethanato	
dppb	1,4-*bis*-(Diphenylphosphino)butane	$Ph_2P(CH_2)_4PPh_2$
dppe	1,2-*bis*-(Diphenylphosphino)ethane	$Ph_2PCH_2CH_2PPh_2$
dppf	*bis*-(Diphenylphosphino)ferrocene	
dppp	1,3-*bis*-(Diphenylphosphino)propane	$Ph_2P(CH_2)_3PPh_2$
dvb	Divinylbenzene	
e$^-$	Electrolysis	
% ee	% Enantiomeric excess	
EE	1-Ethoxyethoxy	EtO(Me)CHO-
Et	Ethyl	$\text{-}CH_2CH_3$
EDA	Ethylenediamine	$H_2NCH_2CH_2NH_2$
EDTA	Ethylenediaminetetraacetic acid	
Emim	1-ethyl-3-methylimidazolium	
FMN	Flavin mononucleotide	
fod	*tris*-(6,6,7,7,8,8,8)-Heptafluoro-2,2-dimethyl-3,5-octanedionate	
Fp	Cyclopentadienyl-*bis*-carbonyl iron	
FVP	Flash Vacuum Pyrolysis	
Grubbs' catalyst	$Cl_2(PCy_3)_2Ru{=}CHPh$	
Grubbs' catalyst II	Mes N N Mes Cl Ru = CHPh Cl PCy$_3$	
h	hour (hours)	
hν	Irradiation with light	
1,5-HD	1,5-Hexadienyl	
HMPA	Hexamethylphosphoramide	$(Me_2N)_3P{=}O$
HMPT	Hexamethylphosphorus triamide	$(Me_2N)_3P$
iPr	Isopropyl	$\text{-}CH(CH_3)_2$
LICA (LIPCA)	Lithium cyclohexylisopropylamide	
LDA	Lithium diisopropylamide	$LiN(iPr)_2$
LHMDS	Lithium hexamethyl disilazide	$LiN(SiMe_3)_2$
LTMP	Lithium 2,2,6,6-tetramethylpiperidide	
MABR	Methylaluminum *bis*-(4-bromo-2,6-di-*tert*-butylphenoxide)	

MAD	*bis*-(2,6-di-*t*-butyl-4-methylphenoxy)methyl aluminum	
mCPBA	*meta*-Chloroperoxybenzoic acid	
Me	Methyl	$-CH_3$
MEM	β-Methoxyethoxymethyl	$MeOCH_2CH_2OCH_2$-
Mes	Mesityl	2,4,6-tri-Me-C_6H_2
MOM	Methoxymethyl	$MeOCH_2$-
Ms	Methanesulfonyl	CH_3SO_2-
MS	Molecular Sieves (3Å or 4Å)	
MTM	Methylthiomethyl	CH_3SCH_2-
NAD	Nicotinamide adenine dinucleotide	
NADP	Sodium triphosphopyridine nucleotide	
Napth	Naphthyl ($C_{10}H_8$)	
NBD	Norbornadiene	
NBS	*N*-Bromosuccinimide	
NCS	*N*-Chlorosuccinimide	
NIS	*N*-Iodosuccinimide	
Ni(R)	Raney nickel	
NMP	*N*-Methyl-2-pyrrolidinone	
Oxone	2 $KHSO_5 \cdot KHSO_4 \cdot K_2SO_4$	
	Polymeric backbone	
PCC	Pyridinium chlorochromate	
PDC	Pyridinium dichromate	
PEG	Polyethylene glycol	
Ph	Phenyl	
PhH	Benzene	
PhMe	Toluene	
Phth	Phthaloyl	
pic	2-Pyridinecarboxylate	
Pip	Piperidino	N
PMP	4-methoxyphenyl	
Pr	*n*-Propyl	$-CH_2CH_2CH_3$
Py	Pyridine	N
quant.	Quantitative yield	
Red-Al	$[(MeOCH_2CH_2O)_2AlH_2]Na$	
sBu	*sec*-Butyl	$CH_3CH_2CH(CH_3)$
sBuLi	*sec*-Butyllithium	$CH_3CH_2CH(Li)CH_3$
Siamyl	Diisoamyl	$(CH_3)_2CHCH(CH_3)$-
TADDOL	α,α,α',α'-tetraaryl-4,5-dimethoxy-1,3-dioxolane	
TASF	*tris*-(Diethylamino)sulfonium difluorotrimethyl silicate	
TBAB	Tetrabutylammonium bromide	*n*-$Bu_4N^+Br^-$
TBAF	Tetrabutylammonium fluoride	*n*-$Bu_4N^+F^-$
TBAI	Tetrabutylammonium iodide	*n*-$Bu_4N^+I^-$
TBDMS	*t*-Butyldimethylsilyl	*t*-$BuMe_2Si$
TBHP (*t*-BuOOH)	*t*-Butylhydroperoxide	Me_3COOH

t-Bu	*tert*-Butyl	$-C(CH_3)_3$
TEBA	Triethylbenzylammonium	$PhCH_2(Et)_3N^+$
TEMPO	Tetramethylpiperdinyloxy free radical	
TFA	Trifluoroacetic acid	CF_3COOH
TFAA	Trifluoroacetic anhydride	$(CF_3CO)_2O$
Tf (OTf)	Triflate	$-SO_2CF_3$ $(-OSO_2CF_3)$
THF	Tetrahydrofuran	
THP	Tetrahydropyran	
TMEDA	Tetramethylethylenediamine	$Me_2NCH_2CH_2NMe_2$
TMG	1,1,3,3-Tetramethylguanidine	
TMS	Trimethylsilyl	$-Si(CH_3)_3$
TMP	2,2,6,6-Tetramethylpiperidine	
TPAP	tetra-*n*-Propylammonium perruthenate	
Tol	Tolyl	$4-CH_3-C_6H_4$
Tr	Trityl	$-CPh_3$
TRIS	Triisopropylphenylsulfonyl	
Ts(Tos)	Tosyl = *p*-Toluenesulfonyl	$4-MeC_6H_4SO_2$
X_c	Chiral auxiliary	

INDEX, MONOFUNCTIONAL COMPOUNDS

Sections—**heavy type**
Pages—light type

FROM ↓ / PREPARATION OF →	Alkynes	Carboxylic acid derivatives	Alcohols, phenols	Aldehydes	Alkyls, methylenes, aryls	Amides	Amines	Esters	Ethers, epxoides	Halides, sulfonates	Hydrides (RH)	Ketones	Nitriles	Alkenes	Oxides
Alkynes	**1** 1	**16** 9		**46** 90	**61** 115	**76** 226	**91** 264	**106** 310	**121** 335			**166** 386		**196** 425	**211** 470
Carboxylic acid derivatives		**17** 9	**32** 18	**47** 90	**62** 120	**77** 226		**107** 310		**137** 366	**152** 377	**167** 387			**212** 470
Alcohols, phenols		**18** 10	**33** 19	**48** 91	**63** 120	**78** 230	**93** 265	**108** 314	**123** 335	**138** 367	**153** 378	**168** 389		**198** 439	**213** 471
Aldehydes	**4** 4	**19** 10	**34** 19	**49** 98	**64** 121	**79** 239	**94** 265	**109** 317	**124** 339	**139** 368		**169** 393	**184** 417	**199** 440	**214** 471
Alkyls, methylenes, aryls		**20** 11	**35** 42	**50** 98	**65** 122			**110** 319		**140** 369	**155** 378		**185** 419	**200** 443	**215** 472
Amides		**21** 12		**51** 98		**81** 232	**96** 268	**111** 320	**126** 343			**171** 395	**186** 419		**216** 472
Amines			**37** 42	**52** 99	**67** 125	**82** 246	**97** 271		**127** 343		**157** 378	**172** 395	**187** 420	**202** 444	**217** 473
Esters		**23** 12	**38** 43		**68** 125	**83** 251	**98** 288	**113** 321	**128** 344	**143** 369	**158** 379	**173** 397		**203** 444	
Ethers, epxoides		**24** 13	**39** 45	**54** 99	**69** 131	**84** 253		**114** 325	**129** 345	**144** 370	**159** 380	**174** 398		**204** 445	**219** 474
Halides, sulfonates	**10** 5	**25** 13	**40** 49	**55** 102	**70** 132	**85** 254	**100** 289	**115** 327	**130** 347	**145** 370	**160** 381	**175** 399	**190** 421	**205** 446	**220** 477
Hydrides (RH)	**11** 7	**26** 14	**41** 50	**56** 103	**71** 153	**86** 255	**101** 294	**116** 328		**146** 370		**176** 401			**221** 477
Ketones	**12** 7	**27** 14	**42** 52	**57** 103	**72** 155	**87** 255	**102** 295	**117** 330	**132** 348		**162** 383	**177** 405		**207** 450	
Nitriles	**13** 7				**73** 156	**88** 257	**103** 297					**178** 409	**193** 422	**208** 451	
Alkenes		**29** 15	**44** 70	**59** 103	**74** 156	**89** 258	**104** 298	**119** 332	**134** 349	**149** 375		**179** 409	**194** 423	**209** 452	**224** 480
Miscellaneous	**15** 8	**30** 16	**45** 73	**60** 105	**75** 221	**90** 259	**105** 298	**120** 333	**135** 363	**150** 376	**165** 385	**180** 410	**195** 423	**210** 468	**225** 481

PROTECTION

	Sect.	Pg.
Carboxylic acids	30A	17
Alcohols, phenols	45A	73
Aldehydes	60A	107
Amides	90A	263
Amines	105A	307
Ketones	180A	415

Blanks in the table correspond to sections for which no additional examples were found in the literature

INDEX, DIFUNCTIONAL COMPOUNDS

Sections—**heavy type**
Pages—light type

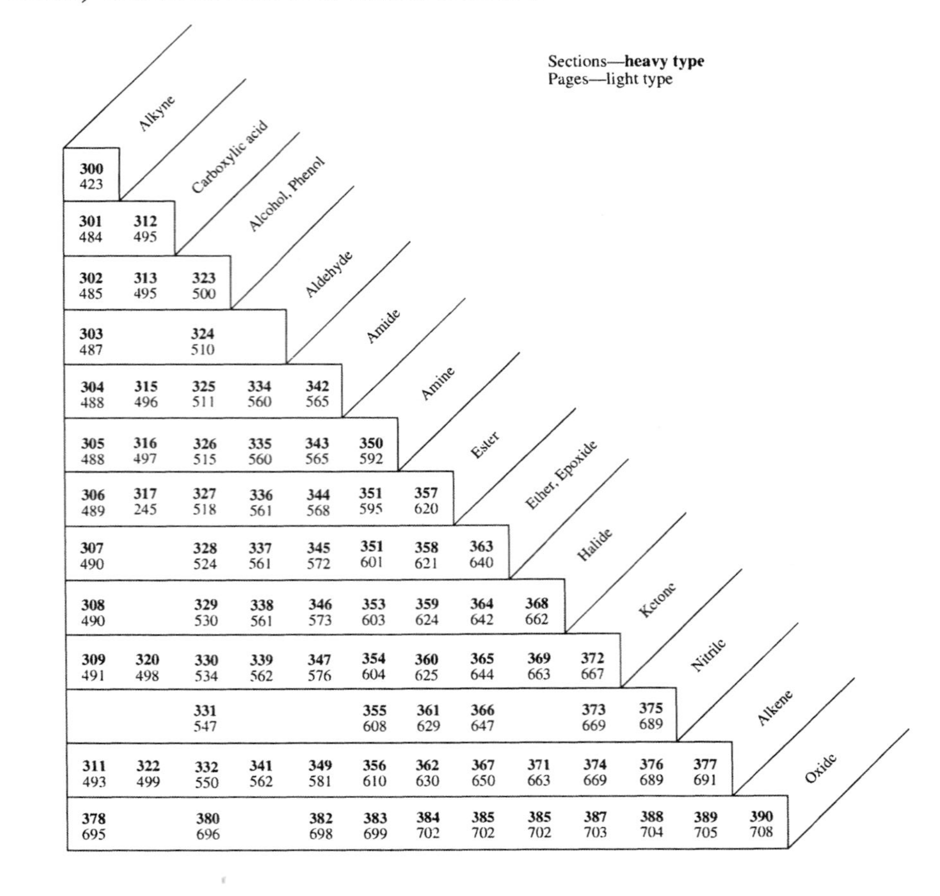

Blanks in the table correspond to sections for which no additional examples were found in the literature.

INTRODUCTION

Relationship between Volume 11 and Previous Volumes. *Compendium of Organic Synthetic Methods, Volume 11* presents 2781 examples of published reactions for the preparation of monofunctional compounds, updating the 13,050 in Volumes 1–10. Volume 11 contains 1212 examples of reactions that prepare difunctional compounds with various functional groups. Reviews have long been a feature of this series, and Volume 11 adds 41 pertinent reviews in the various sections.

Chapters 1–14 continue as in Volumes 1–10, as does Chapter 15, introduced in Volume 6. Difunctional compounds appear in Chapter 16, as in Volumes 6–10. The sections on oxides as part of difunctional compounds, introduced in Volume 7, continues in Chapter 16 of Volumes 8–11 with Sections 378 (Oxides-Alkynes) through Section 390 (Oxides-Oxides).

Following Chapter 16 is a complete alphabetical listing of all authors (last name, initials). The authors for each citation appear <u>below</u> the reaction. The principal author is indicated by <u>underlining</u> (i.e., Kwon, T.W.; <u>Smith, M.B.</u>), as done previously in Volumes 7–10.

Classification and Organization of Reactions Forming Monofunctional Compounds. Chemical transformations are classified according to the reacting functional group of the starting material and the functional group formed. Those reactions that give products with the same functional group form a chapter. The reactions in each chapter are further classified into sections on the basis of the functional group of the starting material. Within each section, reactions are loosely arranged in descending order of year cited (2001-1999), although an effort has been made to put similar reactions together when possible. Review articles are collected at the end of each appropriate section.

The classification is unaffected by allylic, vinylic, or acetylenic unsaturation appearing in both starting material and product, or by increases or decreases in the length of carbon chains; for example, the reactions *t*-BuOH → *t*-BuCOOH, $PhCH_2OH$ → PhCOOH, and $PhCH=CHCH_2OH$ → PhCH=CHCOOH would all be considered as preparations of carboxylic acids from alcohols. Conjugate reduction and alkylation of unsaturated ketones, aldehydes, esters, acids, and nitriles have been placed in Sections 74D and 74E (Alkyls from Alkenes), respectively.

The terms hydrides, alkyls, and aryls classify compounds containing reacting hydrogens, alkyl groups, and aryl groups, respectively; for example, RCH_2-H → RCH_2COOH (carboxylic acids from hydrides), RMe → RCOOH (carboxylic acids from alkyls), RPh → RCOOH (carboxylic acids from aryls). Note the distinction between R_2CO → R_2CH_2 (methylenes from ketones) and RCOR' → RH (hydrides from ketones). Alkylations involving additions across double bonds are found in Section 74 (alkyls, methylenes, and aryls from alkenes).

The following examples illustrate the classification of some potentially confusing cases:

RCH=CHCOOH	→	$RCH{=}CH_2$	Hydrides from carboxylic acids
$RCH{=}CH_2$	→	RCH=CHCOOH	Carboxylic acids from hydrides
ArH	→	ArCOOH	Carboxylic acids from hydrides
ArH	→	ArOAc	Esters from hydrides
RCHO	→	RH	Hydrides from aldehydes
RCH=CHCHO	→	$RCH{=}CH_2$	Hydrides from aldehydes
RCHO	→	RCH_3	Alkyls from aldehydes
R_2CH_2	→	R_2CO	Ketones from methylenes
RCH_2COR	→	R_2CHCOR	Ketones from ketones
$RCH{=}CH_2$	→	RCH_2CH_3	Alkyls from alkenes (Hydrogenation of Alkenes)
RBr + HC≡CH	→	RC≡CR	Acetylenes from halides; also acetylenes from acetylenes
ROH + RCOOH	→	RCOOR	Esters from alcohols; also esters from carboxylic acids
RCH=CHCHO	→	RCH_2CH_2CHO	Alkyls from alkenes (Conjugate Reduction)
RCH=CHCN	→	RCH_2CH_2CN	Alkyls from alkenes (Conjugate Reduction)

How to Use the Book to Locate Examples of the Preparation of Protection of Monofunctional Compounds. Examples of the preparation of one functional group from another are found in the monofunctional index on p xiii, which lists the corresponding section and page. Sections that contain examples of the reactions of a functional group are found in the horizontal rows of this index. Section 1 gives examples of the reactions of acetylenes that form new acetylenes; Section 16 gives reactions of acetylenes that form carboxylic acids; and Section 31 gives reactions of acetylenes that form alcohols.

Examples of alkylation, dealkylation, homologation, isomerization, and transposition are found in Sections 1, 17, 33, and so on, lying close to a diagonal of the index. These sections correspond to such topics as the preparation of acetylenes from acetylenes; carboxylic acids from carboxylic acids; and alcohols, thiols, and phenols from alcohols, thiols, and phenols. Alkylations that involve conjugate additions across a double bond are found in Section 74E (Alkyls, Methylenes, and Aryls from Alkenes).

Examples of name reactions can be found by first considering the nature of the starting material and product. The Wittig reaction, for instance, is in Section 199 (Alkenes from Aldehydes) and Section 207 (Alkenes from Ketones). The aldol condensation can be found in the chapters on difunctional compounds in Section 324 (Alcohol, Thiol-Aldehyde) and in Section 330

(Alcohol, Thiol-Ketone). Examples of the synthetically important alkene metathesis reaction are mostly found in Section 209 (Alkenes from Alkenes).

Examples of the protection of acetylenes, carboxylic acids, alcohols, phenols, aldehydes, amides, amines, esters, ketones, and alkenes are also presented. Sections (designated with an A: 15A, 30A, etc.) are labeled "protecting group: reactions" and are located at the end of pertinent chapters.

Some pairs of functional groups such as alcohol, ester; carboxylic acid, ester; amine, amide; and carboxylic acid, amide can be interconverted by simple reactions. When a member of these groups is the desired product or starting material, the other member should also be consulted in the text.

The original literature must be used to determine the generality of reactions, although this is occasionally stated in the citation. This is only done in cases where such generality is stated clearly in the original citation. A reaction given in this book for a primary aliphatic substrate may also be applicable to tertiary or aromatic compounds. This book provides very limited experimental conditions or precautions and the reader is referred to the original literature before attempting a reaction. **In no instance should a citation in this book be taken as a complete experimental procedure. Failure to refer to the original literature prior to beginning laboratory work could be hazardous.** The original papers usually yield a further set of references to previous work. Papers that appear after those publications can usually be found by consulting *Chemical Abstracts* and the *Science Citation Index.*

Classification and Organization of Reactions Forming Difunctional Compounds. This chapter considers all possible difunctional compounds formed from the groups acetylene, carboxylic acid, alcohol, thiol, aldehyde, amide, amine, ester, ether, epoxide, thioether, halide, ketone, nitrile, and alkene. Reactions that form difunctional compounds are classified into sections on the basis of two functional groups in the product that are pertinent to the reaction. The relative positions of the groups do not affect the classification. Thus preparations of 1,2-amino-alcohols, 1,3-amino-alcohols, and 1,4-amino-alcohols are included in a single section (Section 326, Alcohol-Amine). Difunctional compounds that have an oxide as the second group are found in the appropriate section (Sections 278–290). The nitroketone product of oxidation of a nitroalcohol is found in Section 386 (Ketone-Oxide). Conversion of an oxide (such as nitro or a sulfone moiety) to another functional group is generally found in the "Miscellaneous" section of the sections concerning monofunctional compounds. Conversion of a nitroalkane to an amine, for example, is found in Section 105 (Amines from Miscellaneous Compounds). The following examples illustrate applications of this classification system:

Difunctional Product	*Section Title*
$RC{\equiv}C{-}C{\equiv}CR$	Acetylene-Acetylene
RCH(OH)COOH	Carboxylic acid-Alcohol
RCH=CHOMe	Ether-Alkene
$RCHF_2$	Halide-Halide
$RCH(Br)CH_2F$	Halide-Halide
$RCH(OAc)CH_2OH$	Alcohol-Ester
$RCH(OH)CO_2Me$	Alcohol-Ester
$RCH{=}CHCH_2CO_2Me$	Ester-Alkene
RCH=CHOAc	Ester-Alkene
$RCH(OMe)CH_2SO_2CH_2CH_2OH$	Alcohol-Ether
$RSO_2CH_2CH_2OH$	Alcohol-Oxide

How to Use the Book to Locate Examples of the Preparation of Difunctional Compounds. The difunctional index on p xiv gives the section and page corresponding to each difunctional product. Thus Section 327 (Alcohol, Thiol-Ester) contains examples of the preparation of hydroxyesters; Section 323 (Alcohol, Thiol-Alcohol, Thiol) contains examples of the preparation of diols.

Some preparations of alkene and acetylenic compounds from alkene and acetylenic starting materials can, in principle, be classified in either the monofunctional or difunctional sections; for example, the transformation RCH=CHBr $\rightarrow$ RCH=CHCOOH could be considered as preparing carboxylic acids from halides (Section 25, monofunctional compounds) or preparing a carboxylic acid-alkene (Section 322, difunctional compounds). The choice usually depends on the focus of the particular paper where this reaction was found. In such cases both sections should be consulted.

Reactions applicable to both aldehyde and ketone starting materials are in many cases illustrated by an example that uses only one of them. Likewise, many citations for reactions found in the Aldehyde-X sections, will include examples that could be placed in the Ketone-X section. Again, the choice is dictated by the paper where the reaction was found.

Many literature preparations of difunctional compounds are extensions of the methods applicable to monofunctional compounds. As an example, the reaction RCl $\rightarrow$ ROH might be used for the preparation of diols from an appropriate dichloro compound. Such methods are difficult to categorize and may be found in either the monofunctional or difunctional sections, depending on the focus of the original paper.

The user should bear in mind that the pairs of functional groups alcohol, ester; carboxylic acids, ester; amine, amide; and carboxylic acid, amide can be interconverted by simple reactions. Compounds of the type $RCH(OAc)CH_2OAc$ (ester-ester) would thus be of interest to anyone preparing the diol $RCH(OH)CH_2OH$ (alcohol-alcohol).

Sources of Literature Citations. I thought it would be useful for a reader of this *Compendium* to see those journals that contain the most new synthetic methodology. The accompanying graph shows that *Tetrahedron Letters, Organic Letters,* and *Journal of Organic Chemistry* account for roughly 46% of all the citations in Volume 11. In past volumes, *Tetrahedron Letters* and *Journal of Organic Chemistry* have accounted for 50–60% of all citations. The advent of the new journal *Organic Letters,* the consolidation of several European journals into the *European Journal of Organic Chemistry,* and the apparent expansion of synthetic methodology relative to the last two volumes probably accounts for the new relationship of the three cited journals. This book was not edited to favor one journal, section or type of

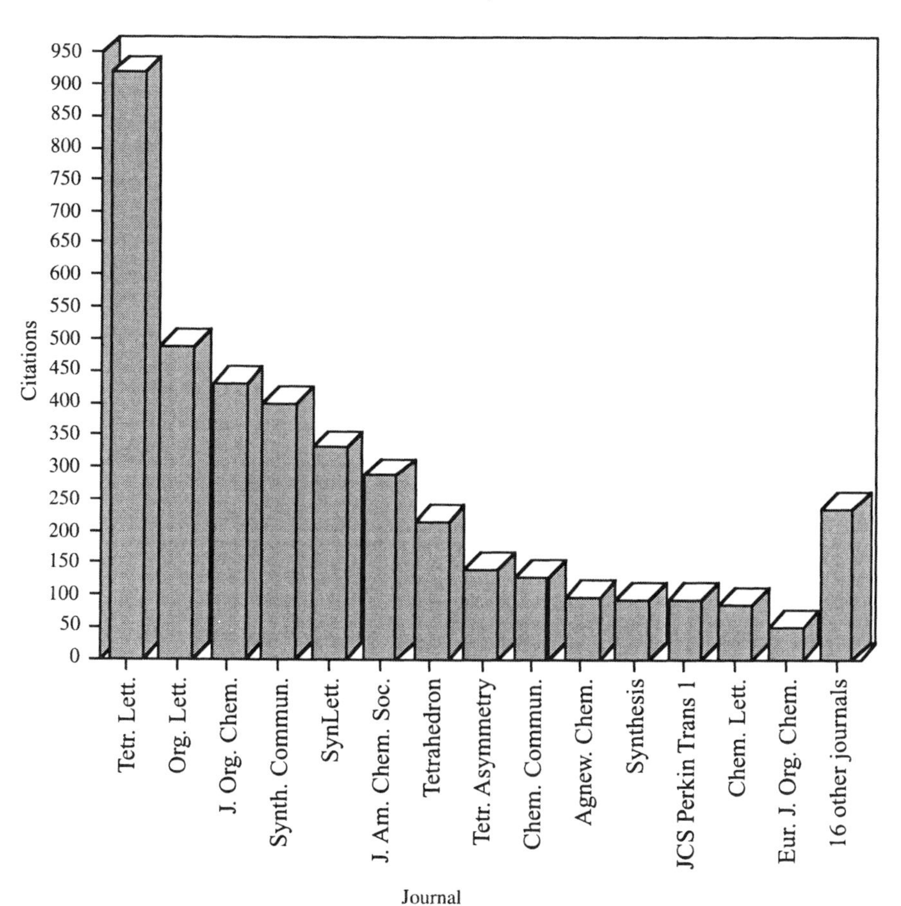

article over another. Undoubtedly, my own personal preferences are part of the selection, but I believe that this compilation is an accurate representation of new synthetic methods that appear in the literature for this period. Therefore, I believe the accompanying graph reflects those journals where new synthetic methodology is located. I should point out that the category "16 other journals" includes: *Helvetica Chimica Acta, Journal of Heterocyclic Chemistry, Pure and Applied Chemistry, Journal of Chemical Research (S), Monatshefte für Chemie, Russian Journal of Organic Chemistry, Acta Chemica Scandinavica, Israel Journal of Chemistry, Chemistry, A European Journal, Canadian Journal of Chemistry, Chemistry Letters, Chemical Reviews, Bulletin of the Chemical Society of Japan, Organic Preparations and Procedures International, Accounts of Chemical Research, Heterocycles,* and the *Australian Journal of Chemistry.* In addition, 6 more journals were examined but no references were recorded.

Compendium of Organic Synthetic Methods

CHAPTER 1
PREPARATION OF ALKYNES

SECTION 1: ALKYNES FROM ALKYNES

Br

, CuI , NEt_3

Ph—≡

10% $PdCl_2(PPh_3)_2$, toluene
80°C , 1 d

Ph—≡

45%

+ 20% 1,4-diphenyl-
1,3-butadiyne

Chow, H.-F.; Wan, C.-W.; Low, K.-H.; Yeung, Y.-Y. *J. Org. Chem.*, ***2001***, *66*, 1910.

1. BuLi , KO*t*-Bu , THF
hexane , –78°C → –25°C

2. 2 eq MeI , –43°C → rt

86%

Kowalik, J.; Tolbert, L.M. *J. Org. Chem.*, ***2001***, *66*, 3229.

$Pb(OAc)_3$

Ph—≡

5% $Pd_2(dba)_3$•$CHCl_3$, 10% CuI , rt
2 eq NaOMe , MeOH/MeCN , rt , 3 h

Ph—≡

74%

Kang, S.-K.; Ryu, H.-C.; Lee, S.-H. *Synth. Commun.*, ***2001***, *31*, 1059.

Ph—≡ → ($Ph_2I^+BF_4^-$, DME/H_2O, rt; $NaHCO_3$, 30 min) → Ph—≡—Ph 85%

Kang, S.-K.; Yoon, S.-K.; Kim, Y.-M. *Org. Lett.*, ***2001***, *3*, 2697.

Ph—≡—$SiMe_3$ → ($Ph_3Sb(OAc)_2$, 10% CuI, 5 h; 5% $Pd_2(dba)_3$•$CHCl_3$, MeCN, 50°C) → Ph—≡—Ph 81%

Kang, S.-K.; Ryu, H.-C.; Hong, Y.-T. *J. Chem. Soc., Perkin Trans. 1*, ***2001***, 736.

PhSe—≡—$CH(OEt)_2$ → (CF_3SiMe_3, Bu_4NF; THF, 0°C) → Me_3Si—≡—$CH(OEt)_2$ 88%

Yoshimatsu, M.; Kuribayashi, M. *J. Chem. Soc., Perkin Trans. 1*, ***2001***, 1256.

MeO_2C—≡ → (1. LDA, THF; 2. $ZnBr_2$, THF, −78°C; 3. PhI, cat $Pd(PPh_3)_4$, −78°C → 23°C, 3 h) → MeO_2C—≡—Ph 98%

Anastasia, L.; Negishi, E.-i *Org. Lett.*, ***2001***, *3*, 3111.

C_6H_{13}—≡ → (PhI, 2 eq TBAF, 2% CuI; 1% $PdCl_2(PPh_3)_2$, THF, rt) → C_6H_{13}—≡—Ph 93%

Mori, A.; Shimada, T.; Kondo, T.; Sekiguchi, A. *Synlett*, ***2001***, 649.

Ph—≡ → (Et_3SiH, 1% $Ir_4(CO)_{12}$; 12% PPh_3) → Ph—≡—$SiEt_3$ 96%

Shimizu, R.; Fuchikami, T. *Tetrahedron Lett.*, ***2000***, *41*, 907.

Ph—≡ → (PhBr, 3% $Pd(PhCN)_2Cl_2$, 6% $P(t\text{-}Bu)_3$, 2% CuI; 1.2 eq $i\text{-}Pr_2NH$, rt, dioxane) → Ph—≡—Ph 94%

Hundertmark, T.; Littke, A.F.; Buchwald, S.L.; Fu, G.C. *Org. Lett.*, ***2000***, *2*, 1729.

Ph—≡ → (I—C_6H_4—OMe, THF, 60°C; 0.5% $Pd_2(dba)_3$, 2% PPh_3, TBAF, 6 h) → Ph—≡—C_6H_4—OMe 83%

Mori, A.; Kawashima, J.; Shimada, T.; Suguro, M.; Hirabayashi, K.; Nishihara, Y. *Org. Lett.*, ***2000***, *2*, 2935.

$C_7H_{15}H_2C$—≡ —[PhI, neat, microwaves; Pd–CuI–PhPh$_3$/KF–Al_2O_3]→ $C_7H_{15}H_2C$—≡—Ph 94%

Kabalka, G.W.; Wang, L.; Namboodiri, V.; Pagni, R.M. *Tetrahedron Lett.*, ***2000***, *41*, 5151.

C_6H_{13}—≡ —[BuLi, B(O*i*-Pr)$_3$, DME-THF; *p*-Tol–Br, 3% Pd(PPh$_3$)$_4$]→ C_6H_{13}—≡—*p*-Tol 98%

Kang, S.; Jang, T.-S.; Keum, G.; Kang, S.B.; Han, S.-Y.; Kim, Y. *Org. Lett.*, ***2000***, *2*, 3615.

TMS-CF_3, cat CsF, THF; 0.5 h

Ts—N ... TMS quant

Ishizaki, M.; Hoshino, O. *Tetrahedron*, ***2000***, *56*, 8657, 8661.

Br—C6H4—Ac; 0.5% Pd$_2$(dba)$_3$, NEt$_3$, THF; 0.5% P(*t*-Bu)$_3$, rt, 20 h

Ph—≡ → Ph—≡—C6H4—Ac >99%

Böhm, V.P.W.; Herrmann, W.A. *Eur. J. Org. Chem.*, ***2000***, 3679.

MeO; MeO$_2$C—C6H4—I; 2.5% Pd$_2$(dba)$_3$, 20% CuI; 20% 2,6-di-*t*-Bu-phenol; DMF, *i*-Pr$_2$NEt, –20°C; MeO; MeO$_2$C

Nakamura, K.; Okubo, H.; Yamaguchi, M. *Synlett*, ***1999***, 549.

K_2FeO_4–Al_2O_3, pentane; rt, 2 h 52%

Caddick, S.; Murtaugh, L.; Weacing, R. *Tetrahedron Lett.*, ***1999***, *40*, 3655

6% $(t\text{-BuO})_3W{\equiv}CCt\text{-Bu}$
toluene, 80°C

69%

Fürstner, A.; Guth, O.; Rumbo, A.; Seidel, G. *J. Am. Chem. Soc.*, ***1999***, *121*, 11108.

SECTION 2: ALKYNES FROM ACID DERIVATIVES

NO ADDITIONAL EXAMPLES

SECTION 3: ALKYNES FROM ALCOHOLS AND THIOLS

NO ADDITIONAL EXAMPLES

SECTION 4: ALKYNES FROM ALDEHYDES

CHO

1. CCl_3COOH, $CCl_3COO^-Na^+$
DMF, 35°C
2. TsCl, TEA/DABCO, DCM, 25°C
3. 4.5 eq MeLi, THF, –10°C
4. H_3O^+

95x91x94%

Wang, Z.; Campagna, S.; Yang, K.; Xu, G.; Pierce, M.E.; Fortunak, J.M.; Confalone, P.N. *J. Org. Chem.*, ***2000***, *65*, 1889.

Br CHO

1. $Ph_3\overset{+}{P}CHBr_2\ \overset{-}{Br}$, THF
t-BuOK, rt
2. t-BuOK

Br C≡CH

66%

Michel, P.; Gennet, D.; Rassat, A. *Tetrahedron Lett.*, ***1999***, *40*, 8575.

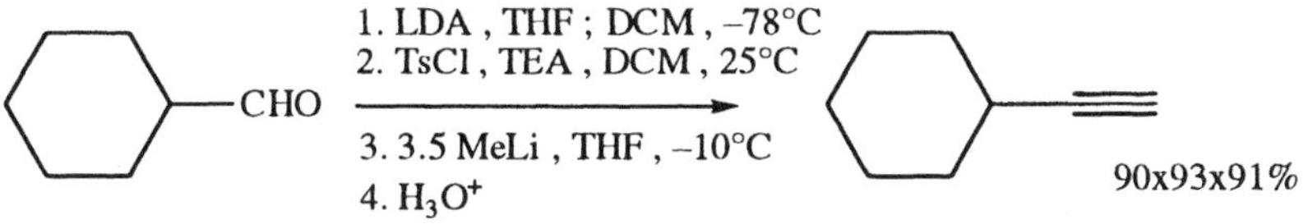

1. LDA, THF; DCM, –78°C
2. TsCl, TEA, DCM, 25°C
3. 3.5 MeLi, THF, –10°C
4. H_3O^+

CHO → 90x93x91%

Wang, Z.; Yin, J.; Campagna, S.; Pesti, J.A.; Fortunak, J.M. *J. Org. Chem.*, ***1999***, *64*, 6918.

SECTION 5: ALKYNES FROM ALKYLS, METHYLENES AND ARYLS

NO ADDITIONAL EXAMPLES

SECTION 6: ALKYNES FROM AMIDES

NO ADDITIONAL EXAMPLES

SECTION 7: ALKYNES FROM AMINES

NO ADDITIONAL EXAMPLES

SECTION 8: ALKYNES FROM ESTERS

NO ADDITIONAL EXAMPLES

SECTION 9: ALKYNES FROM ETHERS, EPOXIDES AND THIOETHERS

NO ADDITIONAL EXAMPLES

SECTION 10: ALKYNES FROM HALIDES AND SULFONATES

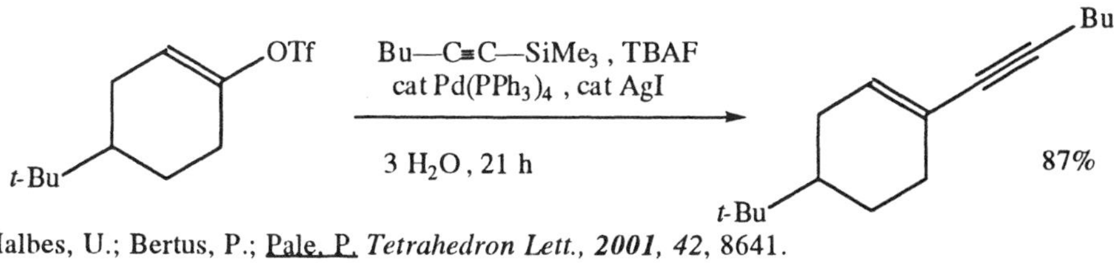

OTf, t-Bu

Bu—C≡C—$SiMe_3$, TBAF
cat $Pd(PPh_3)_4$, cat AgI
3 H_2O, 21 h

Bu, t-Bu, 87%

Halbes, U.; Bertus, P.; Pale, P. *Tetrahedron Lett.*, ***2001***, *42*, 8641.

Ph—≡—SiMe$_3$ → (PhI, Pd-CuI-PPh$_3$, neat; KF-Al$_2$O$_3$, microwaves, 2.5 min) → Ph—≡—Ph 90%

Kabalka, G.W.; Wang, L.; Pagni, R.M. *Tetrahedron, **2001**, 57*, 8017.

2-Iodoaniline (NH$_2$, I) → (≡—SiMe$_3$, Pd(PPh$_3$)$_2$Cl$_2$; Et$_2$NH, DMF, microwaves 120°C, 5 min) → 2-(trimethylsilylethynyl)aniline (NH$_2$, —≡—SiMe$_3$)

Erdélyi, M.; Gogoll, A. *J. Org. Chem., **2001**, 66,* 4165.

Bu$_3$Sn—≡ → (1. PhI, cat Pd(PPh$_3$)$_4$; 2. Merrifield resin, NaI) → Ph—≡ 93%

Lipshutz, B.H.; Blomgren, P.A. *Org. Lett., **2001**, 3*, 1869.

Ph—≡—SiMe$_2$OH → (PhI, THF, TBAF, 60°C; 5% Pd(PPh$_3$)$_4$, 2.5 h) → Ph—≡—Ph 97%

Chang, S.; Yang, S.H.; Lee, P.H. *Tetrahedron Lett., **2001**, 42*, 4833.

Ph—≡—SiMe$_3$ → (I—C$_6$H$_4$—OMe; 0.5 AgCO$_3$, 5% Pd(OAc)$_2$/PPh$_3$, Bu$_4$NCl, THF, 65°C, 2 h) → Ph—≡—C$_6$H$_4$—OMe 92%

Koseki, Y.; Omino, K.; Anzai, S.; Nagasaka, T. *Tetrahedron Lett., **2000**, 41*, 2377.

PhI → (Li[Br—≡—B(OHPh)$_3$]; 5% Pd(PPh$_3$)$_4$, 5% CuI, 15h, DMF) → Ph—≡—Br 94%

Oh, C.H.; Jung, S.H. *Tetrahedron Lett., **2000**, 41*, 8513.

REVIEW:

"Formation of Acetylenes by Ring–Opening of 1,1,2-Trihalocyclopropanes,"
Sydnes, L.K. *Eur. J. Org. Chem., **2000***, 3511.

SECTION 11: ALKYNES FROM HYDRIDES

For examples of the reaction RC≡CH → RC≡C-C≡CR[1], see section 300 (Alkyne-Alkyne).

NO ADDITIONAL EXAMPLES

SECTION 12: ALKYNES FROM KETONES

1. NaH ; Tf_2O
2. TFA
3. K_2CO_3 , acetone , 4 h reflux

$Ph(H_2C)_4$ ... CO_2t-Bu → $Ph(H_2C)_4$—≡ 72x68%

Fleming, I.; Ramarao, C. *Chem. Commun.*, ***1999***, 1113.

SECTION 13: ALKYNES FROM NITRILES

Ph—CO—CN → (Br—CH₂—C≡C—X; In , aq THF) → Ph—CO—C(X)=C=CH₂ + Ph—CO—CH₂—C≡C—X

X = H 69% –
X = Me – 73%

Yoo, B.-W.; Lee, S.-J.; Choi, K.-H.; Keum, S.-R.; Ko, J.-J.; Choi, K.-I.; Kim, J.-H. *Tetrahedron Lett.*, ***2001***, *42*, 7287.

SECTION 14: ALKYNES FROM ALKENES

NO ADDITIONAL EXAMPLES

SECTION 15: ALKYNES FROM MISCELLANEOUS COMPOUNDS

Bu—≡—SMe → (2-MeO-C$_6$H$_4$B(OH)$_2$, $Pd_2(dba)_3$; Cu(I)-thiophene-2-carboxylate, THF, 50°C) → 2-MeO-C$_6$H$_4$—≡—Bu 68%

Grasa, G.A.; Nolan, S.P. *Org. Lett.*, ***2001***, *3*, 119.

$PhCH_2SO_2Ph$ → (1. BuLi, THF; 2. PhCHO; 3. Ac_2O; 4. LDA, THF, –78°C → rt) → Ph—≡—Ph 74%

Orita, A.; Yoshioka, N.; Struwe, P.; Braier, A.; Beckmann, A.; Otera, J. *Chem. Eur. J.*, ***1999***, *5*, 1355.

SECTION 15A: PROTECTION OF ALKYNES

NO ADDITIONAL EXAMPLES

CHAPTER 2

PREPARATION OF ACID DERIVATIVES

SECTION 16: ACID DERIVATIVES FROM ALKYNES

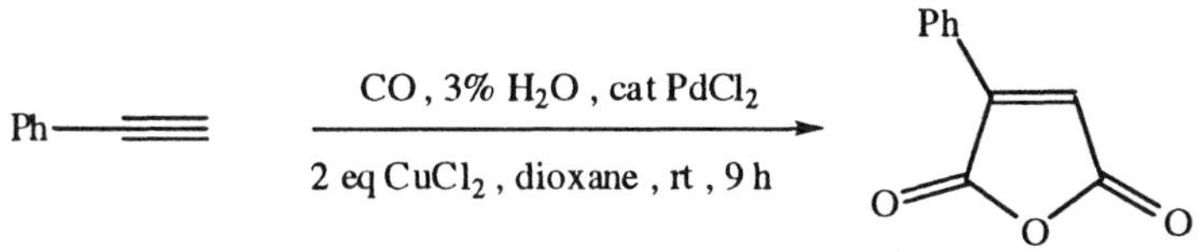

Ph—≡ → (CO, 3% H_2O, cat $PdCl_2$; 2 eq $CuCl_2$, dioxane, rt, 9 h) 98%

Li, J.; Li, G.; Jiang, H.; Chen, M. *Tetrahedron Lett.*, ***2001***, *42*, 6923.

SECTION 17: ACID DERIVATIVES FROM ACID DERIVATIVES

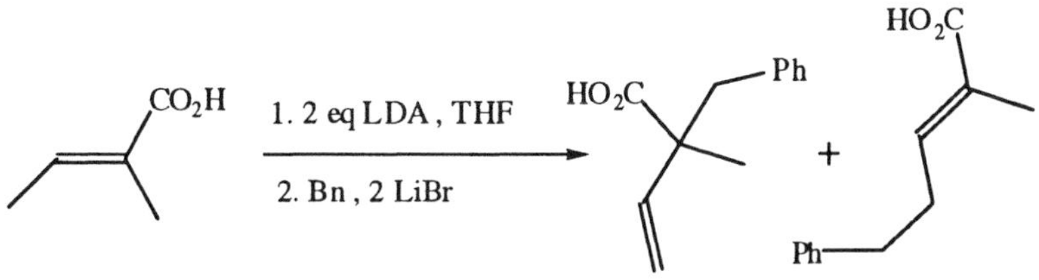

CO_2H → (1. 2 eq LDA, THF; 2. Bn, 2 LiBr) HO_2C ... Ph + HO_2C ... Ph

(84 : 16) 85%

in hexane + 12-c-4 (30 : 70) 74%

Brun, E.M.; Gil, S.; Parra, M. *Synlett.*, ***2001***, 156.

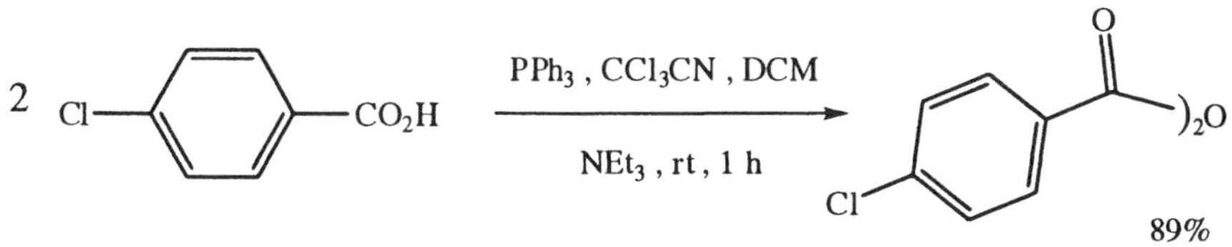

2 Cl—C₆H₄—CO_2H → (PPh_3, CCl_3CN, DCM; NEt_3, rt, 1 h) (Cl—C₆H₄—C(O)—$)_2O$ 89%

Kim, J.; Jang, D.O. *Synth. Commun.*, ***2001***, *31*, 395.

CO_2H (1-naphthoic acid) —— 1. 2.2 eq *sec*-BuLi, THF, −78°C; 2. MeI; 3. H_2O ——> (Me, CO_2H dihydronaphthalene product) 80% (90:10)

Mortier, J.; Vaultier, M.; Plunian, B.; Sinbandhit, S. *Can. J. Chem.* ***1999***, *77*, 98.

SECTION 18: ACID DERIVATIVES FROM ALCOHOLS AND THIOLS

$C_5H_{11}CH_2OH$ —— Na_2WO_4•2 H_2O, Bu_4NHSO_4 / 30% aq H_2O_2, microwaves, 20 min ——> $C_5H_{11}CO_2H$ 75%

Bogdal, D.; Łukasiewicz, M. *Synlett*, ***2000***, 143.

$PhCH_2OH$ —— 1. TEMPO, MeCN, 35°C sodium phosphate buffer / 2. $NaClO_2$, dilute bleach ——> $PhCO_2H$ 98%

Zhao, M.; Li, J.; Mano, E.; Song, Z.; Tschaen, D.M.; Grabowski, E.J.J.; Reider, P.J. *J. Org. Chem.*, ***1999***, *64*, 2564.

Ph(CH$_2$)$_3$OH —— $NaIO_4$, $RuCl_3$•H_2O, EtOAc / MeCN/H_2O ——> Ph(CH$_2$)$_2$CO$_2$H 93%

Prashad, M.; Lu, Y.; Kim, H.-Y.; Hu, B.; Repic, O.; Blacklock, T.J. *Synth. Commun.*, ***1999***, *29*, 2937.

SECTION 19: ACID DERIVATIVES FROM ALDEHYDES

PhCHO —— urea–H_2O_2, HCOOH / rt, 1.5 h ——> $PhCO_2H$ 96%

Balicki, R. *Synth. Commun.*, ***2001***, *31*, 2195.

PhCHO —— N-bromophthalimide–Hg(OAc)$_2$ ——> PhCOOH 94%

Anjum, A.; Srinivas, P. *Chem. Lett.*, ***2001***, 900.

PhCHO —— urea–H_2O_2, aq NaOH, 1 h ——> $PhCO_2H$ 94%

Heaney, H.; Newbold, A.J. *Tetrahedron Lett.*, ***2001***, *42*, 6607.

O_2N–C₆H₄–CHO (meta) $\xrightarrow[\text{reflux, 5 h}]{H_2O_2\text{–}SeO_2\text{, THF}}$ O_2N–C₆H₄–CO_2H (meta) 99%

Wójtowicz, H.; Brząszcz, M.; Kloc, K.; Młochowski, J. *Tetrahedron,* ***2001***, *57*, 9743.

$Ph(CH_2)_2CHO \xrightarrow[90°C\text{, 2 h}]{30\%\ H_2O_2\text{, [Me(}n\text{-}C_8H_{17})_3N]HSO_4} Ph(CH_2)_2CO_2H$ 73%

Sato, K.; Hyodo, M.; Takagi, J.; Aoki, M.; Noyori, R. *Tetrahedron Lett.*, ***2000***, *41*, 1439.

PhCHO $\xrightarrow[\text{octane, reflux}]{\text{bromate exchange resin}}$ PhCOOH 88%

Chetri, A.B.; Kalita, B.; Das, P.J. *Synth. Commun.*, ***2000***, *30*, 3317.

PhCHO $\xrightarrow[\text{reflux, 2.5 h}]{H_2O_2\text{, cat }SeO_2\text{, THF}}$ PhCOOH 97%

Młochowski, J. *Synth. Commun.*, ***2000***, *30*, 4425.

PhCHO $\xrightarrow{O_2\text{, }Al_2O_3\text{, microwaves}}$ $PhCO_2H$ 58%

Reddy, D.S.; Reddy, P.P.; Reddy, P.S.N. *Synth. Commun.*, ***1999***, *29*, 2949.

SECTION 20: ACID DERIVATIVES FROM ALKYLS, METHYLENES AND ARYLS

$CH_4 \xrightarrow[CF_3COOH]{CO\text{, VO(acac)}_2\text{, }K_2S_2O_8} CH_3COOH$ 93%

Taniguchi, Y.; Hayashida, T.; Shibasaki, H.; Piao, D.; Kitamura, T.; Yamaji, T.; Fujiwara, Y. *Org. Lett.*, ***1999***, *1*, 557.

$Ph\text{—}CH_3 \xrightarrow[\text{MeCN, rt, 1 h}]{0.5\ CrO_3\text{, 3.5 eq }H_5O_6} Ph\text{—}COOH$ 85%

Yamazaki, S. *Org. Lett.*, ***1999***, *1*, 2129.

O_2N–C₆H₄–Me (para) $\xrightarrow{\text{NaOCl, MeCN, rt, 1 h}}$ O_2N–C₆H₄–CO_2H (para) 98%

Yamazaki, S. *Synth. Commun.*, ***1999***, *29* , 2211.

SECTION 21: ACID DERIVATIVES FROM AMIDES

$PhCONHNMe_2$ —— 2 eq PhI(OH)OTs , H_2O ——→ $PhCO_2H$ 91%

Wuts, P.G.M.; Goble, M.P. *Org. Lett.*, ***2000***, *2*, 2139.

PhC(O)NH_2 + phthalic anhydride —— 250°C , 4 atm / 30 min ——→ PhC(O)OH 99% + phthalimide

Chemat, F. *Tetrahedron Lett.*, ***2000***, *41*, 3855.

SECTION 22: ACID DERIVATIVES FROM AMINES

NO ADDITIONAL EXAMPLES

SECTION 23: ACID DERIVATIVES FROM ESTERS

2-Cl-5-($NHCO_2i$-Pr)-C_6H_3-CO_2i-Pr —— $AlCl_3$, $MeNO_2$ / 10°C , 1 d ——→ 2-Cl-5-($NHCO_2i$-Pr)-C_6H_3-CO_2H 92%

Chee, G.-L. *Synlett*, ***2001***, 1593.

Cl-C_6H_4-CO_2t-Bu —— SiO_2 , toluene / reflux , 7 h ——→ Cl-C_6H_4-CO_2H 74%

Jackson, R.W. *Tetrahedron Lett.*, ***2001***, *42*, 5163.

Bu—≡ —— 1. BuLi , THF / 2. $C_{10}H_{21}I$, reflux , 8 h ——→ Bu—≡—$C_{10}H_{21}$ 75%

Buck, M.; Chong, J.M. *Tetrahedron Lett.*, ***2001***, *42*, 5825.

$$PhCO_2t\text{-}Bu \xrightarrow[2.\ H_2O\ ,\ 2\ h]{1.\ ZnBr_2\ ,\ DCM\ ,\ 1\ d} PhCO_2H \quad 86\%$$

Wu, Y.-q.; Limburg, D.C.; Wilkinson, D.E.; Vaal, M.J.; Hamilton, G.S. *Tetrahedron Lett.*, ***2000***, *41*, 2847.

$$PhCH_2CO_2Me \xrightarrow[\text{microwaves}]{SiO_2\ ,\ InCl_3\ ,\ H_2O} PhCH_2CO_2H \quad 92\%$$

Ranu, B.C.; Dutta, P.; Sarkar, A. *Synth. Commun.*, ***2000***, *30*, 4167.

$$PhCO_2Me \xrightarrow[190^\circ C\ ,\ 10\ min]{PhSH\ ,\ K_2CO_3\ ,\ NMP} PhCO_2H \quad \text{quant}$$

Sharma, L.; Nayak, M.K.; Chakraborti, A.K. *Tetrahedron*, ***1999***, *55*, 9595.

Other reactions useful for the hydrolysis of esters may be found in Section 30A (Protection of Carboxylic Acids).

SECTION 24: ACID DERIVATIVES FROM ETHERS, EPOXIDES AND THIOETHERS

Ph-epoxide $\xrightarrow[\text{PhCH(OH)CO}_2\text{H}\ ,\ 80^\circ C\ ,\ 4.5\ h]{Bi(II)\ mandelate\ ,\ O_2\ ,\ DMSO}$ $PhCO_2H$ 58%

Coin, C.; Le Boisselier, V.; Favier, I.; Postsel, M.; Duñach, E. *Eur. J. Org. Chem.*, ***2001***, 735.

SECTION 25: ACID DERIVATIVES FROM HALIDES AND SULFONATES

$$PhCH_2Cl \xrightarrow[90^\circ C]{30\%\ H_2O_2\ ,\ cat\ Na_2WO_4\cdot 2\ H_2O \quad Me(C_8H_{17})_3N^+\ HSO_4^-\ ,\ MS\ 4\text{Å}} PhCOOH \quad 83\%$$

Shi, M.; Feng, Y.-S. *J. Org. Chem.*, ***2001***, *66*, 3235.

4-*t*-Bu-C$_6$H$_4$-CH$_2$Br $\xrightarrow{\text{mesoporous silica , hv} \quad \text{acetone , 20 h}}$ 4-*t*-Bu-C$_6$H$_4$-CO$_2$H 86%

Itoh, A.; Kodama, T.; Inagaki, S.; Masaki, Y. *Org. Lett.*, ***2000***, *2*, 2455.

Ph–CH_2Br → Ph–CH_2–CO_2H 93%

10% $[RhCl(cod)]_2$, 1 atm CO; 5% KI, HCOOEt, 60°C, 18 h

Giroux, A.; Nadeau, C.; Han, Y. *Tetrahedron Lett.*, ***2000***, *41*, 7601.

SECTION 26: ACID DERIVATIVES FROM HYDRIDES

cyclohexane → cyclohexane–SO_3H

SO_2/O_2, cat $VO(acac)_2$, 1 d

46% conversion

Ishii, Y.; Matsunaka, K.; Sakaguchi, S. *J. Am. Chem. Soc.*, ***2000***, *122*, 7390.

30 atm CO, $VO(acac)_2$; TFA, $K_2S_2O_8$, 70°C

CO_2H + CO_2H + CF_3 O O

(0.34 : 0.1 : 0.24) 58%

Asadullah, M.; Taniguchi, Y.; Kitamura, T.; Fujiwara, Y. *Tetrahedron Lett.*, ***1999***, *40*, 8867.

SECTION 27: ACID DERIVATIVES FROM KETONES

PhCOCH$_2$COPh → PhCOOH 97%

aq Oxone, $NaHCO_3$; MeCN, pH <5

Ashford, S.W.; Grega, K.C. *J. Org. Chem.*, ***2001***, *66*, 1523.

PhCOMe → $PhCO_2H$ 75%

[hydroxy (2,4-dinitrobenzene-sulfonyloxy)iodo] benzene; MeCN, reflux, 1 h

Lee, J.C.; Choi, J.-H.; Lee, Y.C. *Synlett*, ***2001***, 1563.

PhCOCH$_3$ → PhCOOH 47%

KOH/DMF, 68°C, 12 h

Žabjek, A.; Petrič, A. *Tetrahedron Lett.*, ***1999***, *40*, 6077.

PhCOCH$_3$ → PhCOOH 71%

Re_2O_7, 70% TBHP, 5 h

Gurunath, S.; Sudalai, A. *Synlett*, ***1999***, 559.

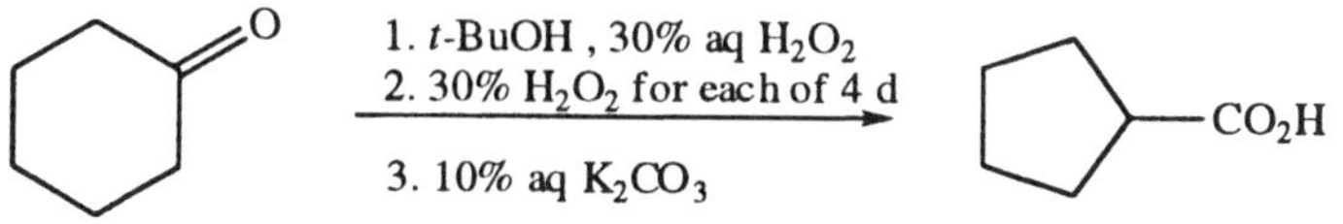

1. t-BuOH , 30% aq H_2O_2
2. 30% H_2O_2 for each of 4 d
3. 10% aq K_2CO_3

CO_2H 60%

Giurg, M.; Mlochowski, J. *Synth. Commun.*, ***1999***, *29*, 2281.

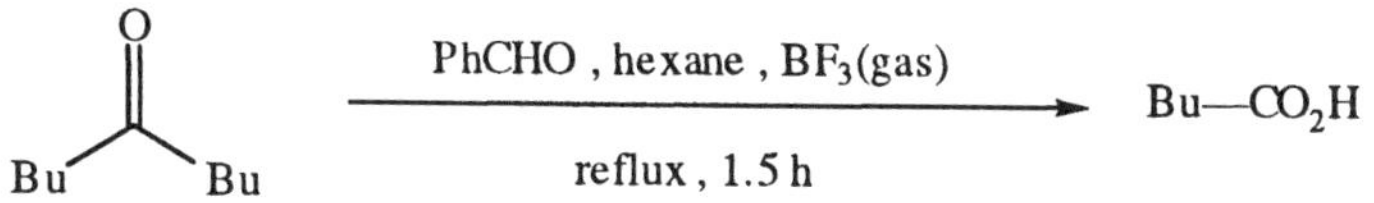

O

Bu Bu

PhCHO , hexane , BF_3(gas)

reflux , 1.5 h

Bu—CO_2H 67%

Kabalka, G.W.; Li, N.-S.; Tejedor, D.; Malladi, R.R.; Gao, X.; Trotman, S. *Synth. Commun.*, ***1999***, *29*, 2783.

SECTION 28: ACID DERIVATIVES FROM NITRILES

NO ADDITIONAL EXAMPLES

SECTION 29: ACID DERIVATIVES FROM ALKENES

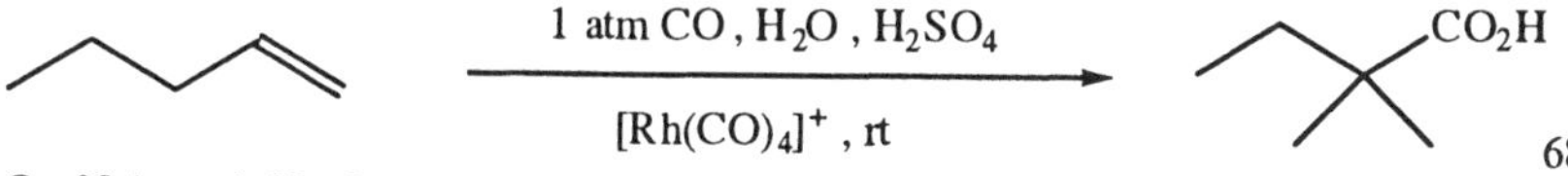

1 atm CO , H_2O , H_2SO_4

$[Rh(CO)_4]^+$, rt

CO_2H 68%

Xu, Q.; Nakatani, H.; Souma, Y. *J. Org. Chem.*, ***2000***, *65*, 1540.

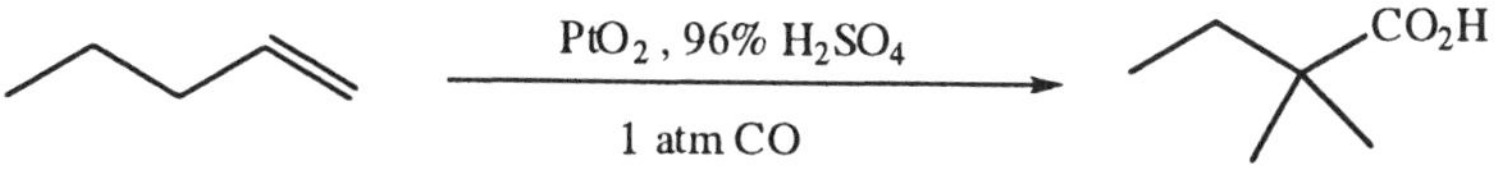

PtO_2 , 96% H_2SO_4

1 atm CO

CO_2H 70%

Xu, Q.; Fujiwara, M.; Tanaka, M.; Souma, Y. *J. Org. Chem.*, ***2000***, *65*, 8105.

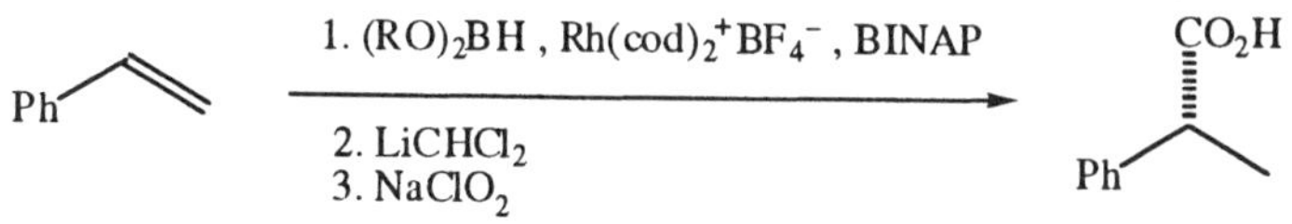

Ph

1. $(RO)_2BH$, $Rh(cod)_2^+BF_4^-$, BINAP
2. $LiCHCl_2$
3. $NaClO_2$

CO_2H

Ph

99x87% (93% ee, *R*)

Chen, A.; Ren, L.; Crudden, C.M. *J. Org. Chem.*, ***1999***, *64*, 9704.

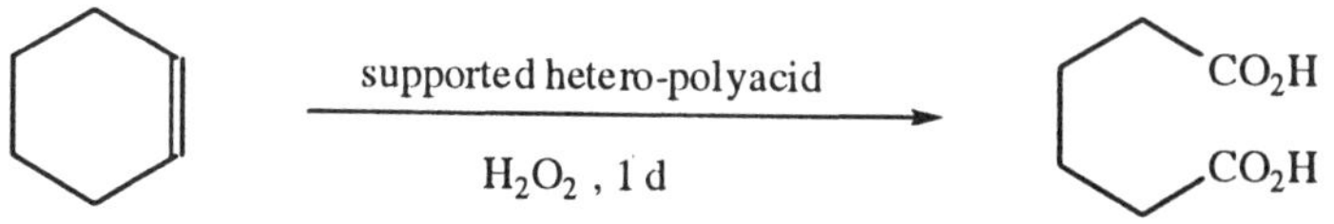

supported hetero-polyacid

H_2O_2 , 1 d

CO_2H

CO_2H 90%

Brooks, C.D.; Huang, L.-c.; McCarron, M.; Johnstone, R.A.W. *Chem. Commun.*, ***1999***, 37.

Ph

$PdCl_2(PPh_3)_2$, CO/H_2O

TsOH , LiCl

Ph CO_2H

91%

Seayad, A.; Jayasree, S.; Chaudhari, R.V. *Org. Lett.*, ***1999***, *1*, 459.

REVIEWS:

"Palladium-Catalysed Reppe Carbonylation," Kiss, G. *Chem. Rev.*, ***2001***, *101*, 3435.

SECTION 30: ACID DERIVATIVES FROM MISCELLANEOUS COMPOUNDS

$SnBu_3$

t-BuOH , 50°C

CHO + CO_2H

$VO(OCH_2CF_3)Cl_2$/36 h	–	100%
$VO(OCH_2CF_3)Cl_2$/3 h	60%	19%

Hirao, T.; Morimoto, C.; Takada, T.; Sakurai, H. *Tetrahedron Lett.*, ***2001***, *42*, 1961.

Se)$_2$ Se)$_2$ n

PhCH=N–OH → $PhCO_2H$ 99%

30% H_2O_2 , THF , reflux , 10 h

Giurg, M.; Said, S.B.; Syper, L.; Mlochowski, J. *Synth. Commun.*, ***2001***, *31*, 3151.

REVIEWS:

"Carboxylic Acids and Esters," Franklin, A.S. *J. Chem. Soc., Perkin Trans. 1*, ***1999***, 3537

SECTION 30A: PROTECTION OF CARBOXYLIC ACID DERIVATIVES

TMS O O NaH, DMF, rt O OH 92%

Serrano-Wu, M.H.; Regueiro-Ren, A.; St. Laurent, D.R.; Carroll, T.M.; Balasubramanian, B.N. *Tetrahedron Lett.*, ***2001***, *42*, 8593.

PhCOOH —(H N Ni-Pr OPMB; THF, rt)→ $PhCO_2PMB$ quant

Wang, M.F.; Golding, B.T.; Potter, G.A. *Synth. Commun.*, ***2000***, *30*, 4197.

Other reactions useful for the protection of carboxylic acids are included in Section 107 (Esters from Carboxylic Acids and Acid Halides) and Section 23 (Carboxylic Acids from Esters).

CHAPTER 3

PREPARATION OF ALCOHOLS

SECTION 31: ALCOHOLS AND THIOLS FROM ALKYNES

NO ADDITIONAL EXAMPLES

SECTION 32: ALCOHOLS AND THIOLS FROM ACID DERIVATIVES

$CH_3CH_2C(=O)CO_2H$ —[MeCHO / phenylpyruvate decarboxylase]→ $CH_3CH_2C(=O)CH(OH)CH_3$ 76% (87% ee)

Guo, Z.; Goswami, A.; Nandur, V.B.; Patel, R.N. *Tetrahedron Asymm.*, ***2001***, *12*, 571.

$BuCO_2H$ —[$BnEt_3N^+BF_4^-$, reflux, DCM]→ $BuCH_2OH$ 70%

Narasimhan, S.; Swarnalakshmi, S.; Balakumar, R. *Synth. Commun.*, ***2000***, *30*, 941.

$PhSO_2Cl$ —[3.5 eq Zn + Me_2SiCl_2, 3 eq DMA / 1.2 DCE, 75°C, 1.5 h]→ PhSH 95%

DMA = dimethylacetamide

Uchiro, H.; Kobayashi, S. *Tetrahedron Lett.*, ***1999***, *40*, 3179.

$Ph(CH_2)_2CO_2H$ —[1. 2,4,6-trichloro-1,3,5-triazine (Cl, N, N, Cl, N, Cl), DME, rt; N-methylmorpholine, 3 h; 2. $NaBH_4$, H_2O, 0°C]→ $Ph(CH_2)_3OH$ 92%

Falorni, M.; Porcheddu, A.; Taddei, M. *Tetrahedron Lett.*, ***1999***, *40*, 4395.

SECTION 33: ALCOHOLS AND THIOLS FROM ALCOHOL AND THIOLS

NHBoc / OH (diol substrate) → 1. $MeSO_2Cl$, DCM, 25° 2. $LiAlH_4$, THF → NHBoc / OH 47%

Benedetti, F.; Norbedo, S. *Tetrahedron Lett.*, ***2000***, *41*, 10071.

$C_6H_{13}CH_2CH_2OH$ → TFAA, MeCN; hydroslfide-supported polymer resin → $C_6H_{13}CH_2CH_2SH$ 85%

Bandgar, B.P.; Sadavarte, V.S.; Uppalla, L.S. *Chem. Lett.*, ***2000***, 1304.

SECTION 34: ALCOHOLS AND THIOLS FROM ALDEHYDES

The following reaction types are included in this section:
A. Reductions of Aldehydes to Alcohols
B. Alkylation of Aldehydes, forming Alcohols.

Coupling of Aldehydes to form Diols is found in Section 323 (Alcohol-Alcohol).

$PhCH_2(OMe)_2$ → 1. CF_3SO_3H–THF 2. *t*-$BuNH_2$–BH_3 → $PhCH_2OH$ 83%

Lukin, K.A.; Yang, C.X.; Bellettini, J.R.; Narayanan, B.A.; Leanna, M.R.; Rasmussen, M. *Synlett*, ***1999***, 59.

SECTION 34A: REDUCTIONS OF ALDEHYDES TO ALCOHOLS

PhCHO → $Ph_3PBu^+BH_4^-$, DCM, rt → $PhCH_2OH$ 94%

Hajipour, A.R.; Mallakpour, S.E. *Synth. Commun.*, ***2001***, *31*, 1177.

PhCHO → $PhB(OH)_2$, Bu_3SnH, DCM; rt, 4 h → $PhCH_2OH$ 88% (15% without $PhB(OH)_2$)

Yu, H.; Wang, B. *Synth. Commun.* ***2001***, *31*, 2719.

Cyclopentane-CHO → (3 eq HCO_2Na, H_2O; 340°C-350°C) → cyclopentyl-CH_2OH 58%

Bryson, T.A.; Jennings, J.M.; Gibson, J.M. *Tetrahedron Lett.*, ***2000***, *41*, 3523.

PhCHO → ($Mg–FeCl_3$, aq DMF; rt, 2.5 h) → $PhCH_2OH$ 92%

Swami, S.S.; Desai, D.G.; Bhosale, D.G. *Synth. Commun.*, ***2000***, *30*, 3097.

PhCHO → ([emim] PF_6, BBu_3, 16 h; ionic liquid medium) → $PhCH_2OH$ quant

Kabalka, G.W.; Malladi, R.R. *Chem. Commun.*, ***2000***, 2191.

PhCHO → ($NaBH_4$, SiO_2, rt, 10 min) → $PhCH_2OH$ 96%

Liu, W.-y.; Xu, Q.-h.; Ma, Y.-x. *Org. Prep. Proceed. Int.*, ***2000***, *32*, 596.

$PhCH_2CH_2CHO$ → (Bu_3SnH, MeOH; reflux) → $PhCH_2CH_2CH_2OH$ quant

Kamiura, K.; Wada, M. *Tetrahedron Lett.*, ***1999***, *40*, 9059.

PhCHO → (*i*-PrOH, 225°C, 1 d) → $PhCH_2OH$ 92%

Bagnell, L.; Strauss, C.R. *Chem. Commun.*, ***1999***, 287.

PhCHO → (HCHO, NaOH, 25 sec; microwaves) → $PhCH_2OH$ 90%

Thakuria, J.A.; Baruah, M.; Sandhu, J.S. *Chem. Lett.*, ***1999***, 995.

PhCHO → (Zn, $BnNEt_3Cl$, MeOH; reflux, 4 h) → $PhCH_2OH$ 90%

Kardile, G.B.; Desai, D.G.; Swami, S.S. *Synth. Commun.*, ***1999***, *29*, 2129.

PhCHO → (1. $SnCl_2•2H_2O$, Mg, THF; 2. H_2O) → $PhCH_2OH$ 90%

Bordoloi, M.; Sharma, R.P.; Chakraborty, V. *Synth. Commun.*, ***1999***, *29*, 2501.

SECTION 34B: ALKYLATION OF ALDEHYDES, FORMING ALCOHOLS

ASYMMETRIC ALKYLATIONS

PhCHO → [$Ti(Oi\text{-}Pr)_4$, THF, $AlEt_3$; chiral sulfonated amino alcohol ligand, 0°C] → PhCH(OH)Et 98% (96% ee)

You, J.-S.; Hsieh, S.-H.; Gau, H.-M. *Chem. Commun.*, ***2001***, 1546.

OTIPS aldehyde → [1. allyl$SiMe_3$, cat TiF_4, 0°C, *S*-BINOL, DCM; 2. aq HF, MeCN] → OTIPS OH homoallylic alcohol

93% (84% ee)

Bode, J.W.; Gauthier Jr., D.R.; Carreira, E.M. *Chem. Commun.*, ***2001***, 2560.

PhCHO → [Et_2Zn, bis-amino-alcohol, 0°C; toluene] → PhCH(OH)Et 84% (88% ee)

Ooi, T.; Saito, A.; Maruoka, K. *Chem. Lett.*, ***2001***, 1108.

Cl–C6H4–CHO → [0.65 Ph_2Zn, Et_2Zn, 10°C; toluene, chiral Re catalyst] → Cl–C6H4–CH(OH)Ph

97% (98% ee)

Bolm, C.; Kesselgruber, M.; Hermanns, N.; Hildebrand, J.P.; Raabe, G. *Angew. Chem. Int. Ed.*, ***2001***, *40*, 1488.

PhCHO → [Et_2Zn, toluene-hexane, rt; 5% 2-azanorbornyl-oxazolidine catalyst] → PhCH(OH)Et 72% (77% ee)

Nakano, H.; Okuyama, Y.; Iwasa, K.; Hongo, H. *Heterocycles*, ***2001***, *54*, 411.

PhCHO → [allyl$SnBu_3$; chiral phosphine–AgOTf, THF] → PhCH(OH)CH2CH=CH2

88% (96% ee)

Yanagisawa, A.; Nakashima, H.; Nakatsuka, Y.; Ishiba, A.; Yamamoto, H. *Bull. Chem. Soc. Jpn.*, ***2001***, *74*, 1129.

PhCHO → (Et_2Zn, hexane, toluene, rt; 15% chiral amino-alcohol, 1 d) → 1-phenylpropan-1-ol (OH, Ph) 97% (72% ee)

Liu, D.-X.; Zhang, L.-C.; Wang, Q.; Da, C.-S.; Xin, Z.-Q.; Wang, R.; Choi, M.C.K.; Chan, A.S.C. *Org. Lett.*, ***2001***, *3*, 2733.

PhCHO → (2 eq Et_2Zn, PhMe/hexane, 0°C; 2% bis(oxazoline) ligand, 45 h) → Ph–CH(OH)–Et 96% (93% ee)

Schinnerl, M.; Seitz, M.; Kaiser, A.; Reiser, O. *Org. Lett.*, ***2001***, *3*, 4259.

PhCHO → (*i*-Pr menthol-derived homoallylic alcohol (OH); 10% TSA•H_2O, 20°C, 20 h) → Ph–CH(OH)–CH=CH–CH₃ 48% (>99% ee, *S*)

Nokami, J.; Ohga, M.; Nakamoto, H.; Matusbara, T.; Hussain, I.; Kataoka, K. *J. Am. Chem. Soc.*, ***2001***, *123*, 9168.

4-nitrobenzaldehyde (CHO, NO_2) → (acetone; 20% *L*-proline, DMSO) → 4-hydroxy-4-(4-nitrophenyl)butan-2-one (O, OH, NO_2) 68% (76% ee)

Sakthivel, K.; Notz, W.; Bui, T.; Barbas III, C.F. *J. Am. Chem. Soc.*, ***2001***, *123*, 5260.

PhCHO → ($ZnEt_2$, norephedrine thiol ligand; toluene) → Ph–CH(OH)–Et 98% (99% ee, *R*)

Jimeno, C.; Moyano, A.; Pericàs, M.A.; Riera, A. *Synlett*, ***2001***, 1155.

PhCHO → (allyl-$SnBu_3$, CH_2Cl_2, 0°C; Ti(IV) catalyst, 2h) → Ph–CH(OH)–CH₂CH=CH₂ 95% (99% ee, *S*)

Kii, S.; Maruoka, K. *Tetrahedron Lett.*, ***2001***, *42*, 1935.

PhCHO → (2 eq Et_2Zn; chiral amino-alcohol) → Ph–CH(OH)–Et 99% (92% ee, *S*)

Hermsen, P.J.; Cremers, J.G.O.; Thijs, L.; Zwanenburg, B. *Tetrahedron Lett.* ***2001***, *42*, 4243.

PhCHO → ($CH_2=CHCH_2SnBu_3$, $SiCl_4$, 6 h; 5% chiral phosphoramide, –78°C) → product (OH)

85% (79% ee)

Denmark, S.E.; Wynn, T. *J. Am. Chem. Soc.*, ***2001***, *123*, 6199.

PhCHO → (Et_2Zn, toluene, 0°C; cat bis-β-amino-alcohol) → Ph–CH(OH)–Et

92% (95% ee *S*)

Xu, Q.; Wang, H.; Pan, X.; Chan, A.S.C.; Yang, T.-k. *Tetrahedron Lett.*, ***2001***, *42*, 6171.

PhCHO → (Ti—chiral diol, hexane; Et_2Zn, 0°C → rt) → Ph–CH(OH)–Et

97% (91% ee)

Yang, X.-w.; Shen, J.-h.; Da, C.-s.; Wang, H.-s.; Su, W.; Liu, D.-x.; Wang, R.; Choi, M.C.K.; Chan, A.S.C. *Tetrahedron Lett.*, ***2001***, *42*, 6573.

PhCHO → (Et_2Zn, chiral amino–alcohol; hexane, 0°C) → Ph–CH(OH)–Et

97% ee, R

Ohga, T.; Umeda, S.; Kawanami, Y. *Tetrahedron*, ***2001***, *57*, 4825.

PhCHO → (Et_2Zn, 10% $(C_6F_{13}CH_2CH_2)_3C$–OH; toluene–hexane, rt, 20 h) → Ph–CH(OH)–Et

90% (83% ee, *R*)

Nakamura, Y.; Takeuchi, S.; Okumura, K.; Ohgo, Y. *Tetrahedron*, ***2001***, *57*, 5565.

Cl–C_6H_4–CHO → (Et_2Zn, toluene, 0°C; 15% chiral pyrrolidinyl-methanol derivative) → Cl–C_6H_4–CH(OH)–Et

96% (93% ee, *SS*)

Zhao, G.; Li, X.-G.; Wang, X.-R. *Tetrahedron Asymm.*, ***2001***, *12*, 399.

PhCHO → (Br–CH_2CH=CH–OAc, Zn, aq NH_4Cl; 22°C) → Ph–CH(OH)–CH(OAc)–CH=CH_2

90% (65:35 *syn:anti*)

Lombardo, M.; Girotti, R.; Morganti, S.; Trombini, C. *Chem. Commun.*, ***2001***, 2310.

PhCHO —[Et_2Zn, 5% sulfinyl ferrocene ligand / toluene, rt]→ PhCH(OH)Et 80% (88% ee)

Priego, J.; Mancheño, O.G.; Cabrera, S.; Carretero, J.C. *Chem. Commun.*, ***2001***, 2026.

PhCHO —[Et_2Zn, toluene, hexane / 20% chiral quinoline derivative, 0°C → rt]→ PhCH(OH)Et 96% (83% ee, *R*)

Xu, Q.; Wang, G.; Pan, X.; Chan, A.S.C. *Tetrahedron Asymm.*, ***2001***, *12*, 381.

PhCHO —[allyl bromide, Sb, 0.5M HCl, H_2O / 16 h]→ PhCH(OH)CH₂CH=CH₂ quant

Li, L.-H.; Chan, T.H. *Can. J. Chem.*, ***2001***, *79*, 1536.

PhCHO —[Et_2Zn, toluene, 0°C, 78 h / cat iron tricarbonyl-amino alcohols]→ PhCH(OH)Et 67% (89% ee, *S*)

Okamoto, K.; Kimachi, T.; Ibuka, T.; Takemoto, Y. *Tetrahedron Asymm.*, ***2001***, *12*, 463.

PhCHO —[Et_2Zn, $Ti(Oi\text{-}Pr)_4$, toluene / chiral amine, MS 4Å, –30°C]→ PhCH(OH)Et quant (59% ee)

Lake, F.; Moberg, C. *Tetrahedron Asymm.*, ***2001***, *12*, 755.

PhCHO —[1. $ClCH_2C{\equiv}CH$, Mn, TMSCl, 10% Cr(salen), MeCN / 2. H^+, THF]→ PhCH(OH)CH₂C≡CH 58% (56% ee)

Bandini, M.; Cozzi, P.G.; Mechiorre, P.; Tino, R.; Umani-Ronchi, A. *Tetrahedron Asymm.*, ***2001***, *12*, 1063.

PhCHO —[2 eq $ZnEt_2$, toluene, rt / cat binaphthylazepine ligand]→ PhCH(OH)Et 99% (87% ee, *S*)

Superchi, S.; Mecca, T.; Giorgio, E.; Rosini, C. *Tetrahedron Asymm.*, ***2001***, *12*, 1235.

PhCHO —[2.2 eq Et_2Zn, toluene, –30°C / chiral amino-alcohol ligand]→ PhCH(OH)Et 99% (64% ee)

Cobb, A.J.A.; Marson, C.M. *Tetrahedron Asymm.*, ***2001***, *12*, 1547.

PhCHO → [2.2 eq Et_2Zn, toluene, 0°C, 7 h; dendritic BINOL-Ti catalyst] → PhCH(OH)Et 99% (84% ee)

Fan, Q.-H.; Liu, G.-H.; Chen, X.-M.; Deng, G.-J.; Chan, A.S.C. *Tetrahedron Asymm.*, ***2001***, *12*, 1559.

PhCHO → [Et_2Zn, toluene, 43 h; 10% chiral ferrocenyl-amino-alcohol ligand] → PhCH(OH)Et quant (83% ee)

Bastin, S.; Agbossou-Niedercorn, F.; Brucard, J.; Pélinski, L. *Tetrahedron Asymm.* ***2001***, *12*, 2395.

PhCHO → [1. Et_2Zn, β-carboline ester ligand, toluene, rt; 2. H_3O^+] → PhCH(OH)Et 74% (39% ee)

Zhu, H.J.; Zhao, B.T.; Zuo, G.-Y.; Pittman Jr., C.U.; Dai, W.M.; Hao, X.J. *Tetrahedron Asymm.* ***2001***, *12*, 2613.

PhCHO → [Et_2Zn, Ti(O*i*-Pr)$_4$, DCM, rt, 1 h; chiral sulfonated amino-alcohol] → PhCH(OH)Et quant (96% ee, *R*)

You, J.-S.; Shao, M.-Y.; Gau, H.-M. *Tetrahedron Asymm.*, ***2001***, *12*, 2971.

2-pyridyl-CHO → [DCM, 20°C, 19 h; $PhNMe_2$, chiral Al complex] → (2-pyridyl)CH(OH)(4-NMe_2-phenyl) 63% (43% ee)

Gothelf, A.S.; Hansen, T.; Jørgensen, K.A. *J. Chem. Soc., Perkin Trans. 1*, ***2001***, 854.

PhCHO → [Et_2Zn, neat; chiral amino-alcohol] → PhCH(OH)Et 99% (87% ee)

Sato, I.; Saito, T.; Soai, K. *Chem. Commun.*, ***2000***, 2471.

PhCHO → [2 eq Et_2Zn, toluene, rt, 1 d; 5% chiral amine bicyclo[2.2.1]heptane alcohol] → PhCH(OH)Et 99% (94% ee)

Hanya, N.; Mino, T.; Sakamoto, M.; Fujita, T. *Tetrahedron Lett.*, ***2000***, *41*, 4587.

PhCHO → (Et_2Zn, 5% chiral piperidine, hexane, 0°C) → 1-phenylpropan-1-ol (OH; Et, Ph) 88% (24% ee, *S*)

Shi, M.; Jiang, J.-K.; Feng, Y.-S. *Tetrahedron Asymm.*, ***2000***, *11*, 4923.

PhCHO → (allyl-$SnBu_3$, aq EtOH, –40°C; (*S*)-tolyl-BINAP, $AgNO_3$, 14 h) → 1-phenylbut-3-en-1-ol (Ph, OH)

93% (79% ee)

Loh, T.-P.; Zhou, J.-R. *Tetrahedron Lett.*, ***2000***, *41*, 5261.

4-Cl-C_6H_4-CHO → (20% bis-naphthyl, hexane; Ti(O*i*-Pr)$_4$, 45°C; perfluoro(methyldecalin)) → 4-Cl-C_6H_4-CH(OH)Et 99% (54% ee)

Tian, Y.; Chan, K.S. *Tetrahedron Lett.*, ***2000***, *41*, 8813.

Ph–C≡C–CHO → (1. allyl-dialkoxytitanium complex, ether, –78°C; 2. H_2O) → Ph–C≡C–CH(OH)–CH_2CH=CH_2

53% (96% ee)

Bouzbouz, S.; Pradaux, F.; Cossy, J.; Ferroud, C.; Falguiéres, A. *Tetrahedron Lett.*, ***2000***, *41*, 8877.

MeCHO → (1) $PhMe_2Si$–CH=CH–CH_2–B(Ipc)$_2$; 2) H_2O_2, pH 6 buffer; 3) Na_2SO_3, 0°) → Me–CH(OH)–CH($SiMe_2Ph$)–CH=CH_2 80% (93% ee)

Roush, W.R.; Pinchuk, A.N.; Micalizio, G.C. *Tetrahedron Lett.*, ***2000***, *41*, 9413.

PhCHO → (Et_2Zn, toluene, rt, 12 h; chiral carbohydrate derivative) → Ph–CH(OH)–Et 83% (90% ee)

Cho, B.T.; Chun, Y.S.; Yang, W.K. *Tetrahedron Asymm.*, ***2000***, *11*, 2149.

4-Cl-C_6H_4-CHO → (1. $ZnPh_2/ZnEt_2$, 5% chiral ferrocenyl catalyst; 2. workup) → 4-Cl-C_6H_4-CH(OH)Ph

92% (95% ee)

Bolm, C.; Hermanns, N.; Hildebrand, J.P.; Muñiz, K. *Angew. Chem. Int. Ed.*, ***2000***, *39*, 3465.

PhHO → [$AlEt_3$, THF, 12 h; Ti(IV)–TADDOL ate complex] → PhCH(OH)CH$_2$CH$_3$ 94% (84% ee)

Lu, J.-F.; You, J.-S.; Gau, H.-M. *Tetrahedron Asymm.*, ***2000***, *11*, 2531.

$C_6H_{13}CHO$ → [*i*-Pr_2Zn, a morpholino-xylofuranose; toluene, 0°C, 8 h] → C_6H_{13}CH(OH)*i*-Pr 88% (96% ee)

Yang, W.K.; Cho, B.T. *Tetrahedron Asymm.*, ***2000***, *11*, 2947.

PhCHO → [2 eq Et_2Zn, 5% chiral amino-alcohol; toluene, rt] → PhCH(OH)Et 99% (93% ee)

Hanyu, N.; Aoki, T.; Mino, T.; Sakamoto, M.; Fujita, T. *Tetrahedron Asymm.*, ***2000***, *11*, 2971.

PhCHO → [Et_2Zn, 6% hydroxy-benzylamine; toluene, rt] → PhCH(OH)Et 93% (89% ee)

Palmieri, G. *Tetrahedron Asymm.*, ***2000***, *11*, 3361.

PhCHO → [Et_2Zn, cat chiral amino-alcohol; toluene/hexane, 5 h] → PhCH(OH)Et 92% (77% ee)

Wu, Y.; Yun, H.; Wu, Y.; Ding, K.; Zhou, Y. *Tetrahedron Asymm.*, ***2000***, *11*, 3543.

PhCHO → [polymer-bound BINOL catalyst; Et_2Zn, Ti(O*i*-Pr)$_4$, DCM, 0°C] → PhCH(OH)CH$_2$CH$_3$ 92% (95% ee)

Lipshutz, B.H.; Shin, Y.-J. *Tetrahedron Lett.*, ***2000***, *41*, 9515.

4-Cl-C_6H_4-CHO → [Et_2Zn, toluene/hexane, 5 h; cat amino-alcohol] → 4-Cl-C_6H_4-CH(OH)Et

97% (91.5% ee)

Yun, H..; Wu, Y.; Wu, Y.; Ding, K.; Zhou, Y. *Tetrahedron Lett.*, ***2000***, *41*, 10263.

PhCHO → [2 eq Et_2Zn, cat chiral amino-alcohol; toluene, 2 h] → PhCH(OH)Et 99% (93% ee)

Hanyu, N.; Aoki, T.; Mino, T.; Sakamoto, M.; Fujita, T. *Tetrahedron Asymm.*, ***2000***, *11*, 4127.

4-$CF_3C_6H_4CHO$ → (allyl-$SnBu_3$, $Ti(Oi\text{-}Pr)_4$; binaphthyl, toluene, 25°C, 20 h) → 4-$CF_3C_6H_4CH(OH)CH_2CH=CH_2$ 94% (97% ee)

Doucet, H.; Santelli, M. *Tetrahedron Asymm.*, ***2000***, *11*, 4163.

PhCHO → (Et_2Zn, rt, 1 d; cat bridged chiral amino-alcohol) → PhCH(OH)Et 84% (96% ee)

Arroyo, N.; Haslinger, U.; Mereiter, K.; Widhalm, M. *Tetrahedron Asymm.*, ***2000***, *11*, 4207.

PhCHO → (1. Et_2Zn, cat thiazolidine derivative; 2. H^+) → PhCH(OH)Et 99% (89% ee)

Meng, Q.; Li, Y.; He, Y.; Guan, Y. *Tetrahedron Asymm.*, ***2000***, *11*, 4255.

PhCHO → (Et_2Zn, BINOL catalyst, DCM; $Ti(Oi\text{-}Pr)_4$, –70°C, hexane) → PhCH(OH)Et 96% (91% ee)

Shen, X.; Guo, H.; Ding, K. *Tetrahedron Asymm.*, ***2000***, *11*, 4321.

PhCHO → (Et_2Zn, chiral amino-alcohol; toluene, 18°C, 20 min) → PhCH(OH)Et 98% (97% ee, *S*)

Palco, M.R.; Cabeza, I.; Sardina, F.J. *J. Org. Chem.*, ***2000***, *65*, 2108.

AcO–CH_2–C≡C–CH(OMs)Me → (2.5% $Pd(OAc)_2$, 2.5% PPh_3, cyclohexane carboxaldehyde; 3 eq Et_2Zn, THF, –20°C) → AcO–CH_2–C≡C–CH(Me)–CH(OH)–C_6H_{11} 75% (95 : 5 *anti:syn*) 90% ee

Marshall, J.A.; Adams, N.D. *J. Org. Chem.*, ***1999***, *64*, 5201.

PhCHO → (allyl bromide, $Ce(OTf)_4$•xH_2O; chiral ligand) → PhCH(OH)$CH_2CH=CH_2$ 90% (92% ee, *R*)

Loh, T.-P.; Zhou, J.-R. *Tetrahedron Lett.*, ***1999***, *40*, 9115.

PhCHO → (Et_2Zn, toluene, 0°C, 3 h; N-pyrrolidinyl triphenyl amino alcohol, HO) → Ph–CH(OH)–Et

quant (97% ee, *S*)

Reddy, K.S.; Solá, L.; Moyano, A.; Pericàs, M.A.; Riera, A. *J. Org. Chem.*, ***1999***, *64*, 3969.

PhCHO → (allyl-$Si(OMe)_3$, MeOH, −20°C; cat *R*-Tolyl-BINAP-AgF, 4 h) → Ph–CH(OH)–$CH_2CH=CH_2$ 80% (94% ee)

Yanagisawa, A.; Kageyama, H.; Nakatsuka, Y.; Asakawa, K.; Matsumoto, Y.; Yamamoto, H. *Angew. Chem., Int. Ed.*, ***1999***, *38*, 3701.

PhCHO → (Et_2Zn, 2% chiral amino-alcohol; 0°C, toluene, 5 h) → Ph–CH(OH)–Et 88% (83% ee, *R*)

Cho, B.T.; Chun, Y.S. *Synth. Commun.*, ***1999***, *29*, 521.

MeO–C_6H_4–CHO → (Ph_2Zn, toluene, −30°C, 1 d; cat (chiral diaryl carbonate + $ZnEt_2$)) → MeO–C_6H_4–CH(OH)–Ph

84% (93% ee)

Huang W.-S.; Pu, L. *J. Org. Chem.*, ***1999***, *64*, 4222.

PhCHO → (2 eq Et_2Zn, toluene/DCM, rt; polymer bound aziridinylmethanol) → Ph–CH(OH)–Et 91% (96% ee, *S*)

ten Holte, P.; Wijgergans, J.-P.; Thijs, L.; Zwanenburg, B. *Org. Lett.*, ***1999***, *1*, 1095.

PhCHO → (allyl bromide, THF/hexane; In, cinchonine) → Ph–CH(OH)–$CH_2CH=CH_2$ 73% (72% ee, *S*)

Loh, T.-P.; Zhou, J.-R.; Yin, Z. *Org. Lett.*, ***1999***, *1*, 1855.

PhCHO → (2 eq Et_2-Zn; 5% N-Trt aziridinyl diphenylmethanol (Ph_2C–OH)) → Ph–CH(OH)–Et 90% (99% ee)

Gadhwal, S.; Baruah, M.; Sandhu, J.S. *Synlett*, ***1999***, 1573.

PhCHO → ($CH_2=CHCH_2SnBu_3$, DCM, rt; 5% (*S,S*-(Phebox)$RhCl_2(H_2O)$) → PhCH(OH)CH$_2$CH=CH$_2$

88% (61% ee)

Motoyama, Y.; Narusawa, H.; Nishiyama, H. *Chem. Commun.*, ***1999***, 131.

$(pinyl)_2B$–CH$_2$C(=CH$_2$)CH$_2$–B$(pinyl)_2$ → (PhCHO) → PhCH(OH)CH$_2$C(=CH$_2$)CH$_2$CH(OH)Ph

(84 : 16 dr) 55%
>95%ee

Barrett, A.G.M.; Braddock, D.C.; de Koning, P.D. *Chem. Commun.*, ***1999***, 459.

PhCHO → (Et_2Zn, hexane, 0°C → rt; 10% chiral pyrrolidine) → PhCH(OH)Et 97% (97% ee)

Yang, X.; Shen, J.; Da, C.; Wang, R.; Choi, M.C.K.; Yang, L.; Wong, K.-y. *Tetrahedron Asymm.*, ***1999***, *10*, 133.

PhCHO → (chiral amino alcohol–2 BuLi; THF, –78°C, 1 h) → PhCH(OH)Bu 17% (71% ee)

Schön, M.; Naef, R. *Tetrahedron Asymm.*, ***1999***, *10*, 169.

PhCHO → (BuLi, chiral bis-amide, –116°C; DEE/DMM) → PhCH(OH)Bu 72% ee

Arvidsson, P.I.; Davidsson, Ö; Hilmersson, G. *Tetrahedron Asymm.*, ***1999***, *10*, 527.

PhCHO → (N-Bn pyrrolidine, Et_2Zn, hexane, 0°C; 5% chiral azetidine) → PhCH(OH)Et 99% (92% ee)

Shi, M.; Jiang, J.-K. *Tetrahedron Asymm.*, ***1999***, *10*, 1673.

PhCHO → (Et_2Zn, 0°C, 30 h; chiral disulfide ligand) → PhCH(OH)Et 81% (>99% ee)

Braga, A.L.; Appelt, H.R.; Schneider, P.H.; Sliveira, C.C.; Wessjohann, L.A. *Tetrahedron Asymm.*, ***1999***, *10*, 1737.

PhCHO —(2 eq Et_2Zn, 2d, toluene, 0°C; 5% chiral pyridyl alcohol)→ PhCH(OH)Et 82% (90% ee)

Kang, J.; Kim, H.Y.; Kim, J.H. *Tetrahedron Asymm.*, ***1999***, *10*, 2523.

PhCHO —(Et_2Zn, toluene, 0°C; chiral 2,2'-bipyridyl diol)→ PhCH(OH)Et 86% (95% ee)

Kwong, H.-L.; Lee, W.-S. *Tetrahedron Asymm.*, ***1999***, *10*, 3791.

PhCHO —(Et_2Zn, toluene, –30°C; 10% camphor-derived 1,4-amino-alcohol)→ PhCH(OH)Et 92% (78% ee)

Knollmüller, M.; Ferencic, M.; Gärtner, P. *Tetrahedron Asymm.*, ***1999***, *10*, 3969.

PhCHO —(2.2 eq Et_2Zn, toluene, 0°C; cat Cinchona alkaloid)→ PhCH(OH)Et 91% (88% ee)

Lee, S.H.; Im, D.S.; Cheong, C.S.; Chung, B.Y. *Heterocycles*, ***1999***, *51*, 1913.

Cyclohexyl-CHO —(10 eq allyl-$SiCl_3$, HMPA; chiral amide, HMPA, 7 d, DCM, –78°C)→ cyclohexyl-CH(OH)CH$_2$CH=CH$_2$ 89% (76% ee)

Iseki, K.; Mizuno, S.; Kuroki, Y.; Kobayashi, Y. *Tetrahedron*, ***1999***, *55*, 977.

PhCHO —(Et_2Zn, toluene, 0°C, 70h; 2% polymer-supported ephedrine derivative)→ PhCH(OH)CH$_2$CH$_3$ 85% (83% ee)

Sung, D.W.L.; Hodge, P.; Stratford, P.W. *J. Chem. Soc., Perkin Trans. 1*, ***1999***, 1463.

NON-ASYMMETRIC ALKYLATIONS

3-Chloropyridine —(1. 3 eq BuLi–LiDMAE, hexane, –60°C, 1 h; 2. *t*-BuCHO, THF, –60°C, 1 h)→ 3-chloro-2-(CH(OH)*t*-Bu)pyridine 69%

Choppin, S.; Gros, P.; Fort, Y. *Eur. J. Org. Chem.*, ***2001***, 603.

PhCHO → ($SnBu_3$ allyltin; cat $Bu_4NBr–PbI_2$, H_2O) → Ph(OH) products + Ph(OH) products

Z-allyltin gives 93:7 *syn:anti*

(90 : 10) 80%

Shibata, I.; Yoshimura, N.; Yabu, M.; Baba, A. *Eur. J. Org. Chem.*, ***2001***, 3207.

$C_7H_{15}CHO$ → (1. I–CH=CH–CH$_2$Br, In; 2. PhI, cat $Pd(PPh_3)_4$, LiCl) → C_7H_{15}, OH, Ph 43%

Hirashita, T.; Yamamura, H.; Kawai, M.; Araki, S. *Chem. Commun.*, ***2001***, 387.

Ph–S(=O), OH → (PhCHO, aq THF, 70°C, 72 h; 3 eq $SnCl_2$, cat $Pd(PhCN)_2Cl_2$) → Ph–S(=O), OH, Ph 65%

Márquez, F.; Llebaria, A.; Delgado, A. *Tetrahedron Asymm.* ***2001***, *12*, 1625.

Ph, CO_2Et, O, O → (1. $HO_2C–C_6H_4–SO_2N_3$, NEt_3, MeCN, 25°C; 2. HO–CH$_2$CH$_2$–OH; 3. *p*-TsOH, PhH, 80°C) → O, O, CO_2Et, Ph 22%

Soai, K.; Konishi, T.; Shibata, T. *Heterocycles*, ***1999***, *51*, 1421.

Cl–C$_6$H$_4$–CHO → (allyl Br, β-SnO; cat $PtCl_2(PPh_3)_2$, aq THF) → HO, Cl 96%

Sinha, P.; Roy, S. *Chem. Commun.*, ***2001***, 1798.

PhCHO → (polymer-supported B–*p*-Tol, DME, H_2O; 3% $Rh(acac)(CO)_2$, 3% dppf, 80°C) → Ph–CH(OH)–*p*-Tol 62%

Pourbaix, C.; Carreuaux, F.; Carboni, B. *Org. Lett.*, ***2001***, *3*, 803.

PhCHO → 1. Cl/\=/\Cl , Mn , TMSCl; 10% Cr-salen , 10% salen , MeCN; 2. H^+ , THF → 40% + 5%

Bandini, M.; Cozzi, P.G.; Melchiorre, P.; Morganti, S.; Umani-Ronchi, A. *Org. Lett.*, ***2001***, *3*, 1153.

PhCHO → PhBr , $NiBr_2$•bipy; DMF , e^-, Fe/Cr/Ni anode → Ph–CH(OH)–Ph 65%

Durandetti, M.; Nédélec, J.-Y.; Périchon, J. *Org. Lett.*, ***2001***, *3*, 2073.

MeO_2C ... CHO → Et_2Zn , cat $Ni(acac)_2$, rt; Zn , hexane/THF , 15 min → OH 72%

Shibata, K.; Kimura, M.; Shimuzu, M.; Tamaru, Y. *Org. Lett.*, ***2001***, *3*, 2181.

CHO, O, N, PMP → $SnCl_4$, –78°C → O, OH, H, H, N, PMP 60%

Alcaide, B.; Pardo, C.; Rodríguez-Ranera, C.; Rodríguez-Vicente, A. *Org. Lett.*, ***2001***, *3*, 4205.

OHC, TBSO → OTBS, OMe; $B(C_6F_5)_3$, –78C°; CH_2Cl_2 , ether → TBSO, TBSO, CO_2Me 81%

vinylogous Mukaiyama aldol

Chirstmann, M.; Kalesse, M. *Tetrahedron Lett.*, ***2001***, *42*, 1269.

Ph–CH₂CH₂CH₂–CCl_3 → (PhCHO, $CrCl_2$; rt, THF) → Ph–CH₂CH₂CH=C(Cl)–CH(OH)Ph 91%

Barma, D.K.; Baati, R.; Valleix, A.; Mioskowski, C.; Falck, J.R. *Org. Lett.*, ***2001***, *3*, 4237.

1. Li, DTBB, *t*-BuCHO, THF, –78°C
2. phosphate buffer (pH 7), –78°C → 20°C
3. MeOH, *p*TsOH, 20°C, overnight

OH, *t*-Bu, CO_2Me 58%

Pastor, I.M.; Yus, M. *Tetrahedron Lett.*, ***2001***, *42*, 1029.

PhCHO → (allyl-$SiMe_3$, $Sc(OTf)_3$; CH_2Cl_2, rt, 5 h) → OMe, Ph 90%

Yadav, J.S.; Subba Reddy, B.V.; Srihari, P. *Synlett*, ***2001***, 673.

CHO, $SiMe_2Ph$ → (1. TMSOTf, CH_2Cl_2, –78°C; 2. aq NaOH) → HO, Me

81% (83:17 *cis:trans*)

Suginome, M.; Iwanami, T.; Yamamoto, A.; Ito, Y. *Synlett*, ***2001***, 1042.

Indole (N–H) → (EtO_2CCHO, 1% $Pd(MeCN)_2Cl_2$; DCM, rt, 18 h) → HO, CO_2Et, N–H 96%

Hao, J.; Taktak, S.; Aikawa, K.; Yusa, Y.; Hatano, M.; Mikami, K. *Synlett*, ***2001***, 1443.

OH → (PhCHO, $SnCl_2$, NCS; CH_2Cl_2, –60°C, 39 h) → OH, Ph 74%

Masuyama, Y.; Saeki, K.; Horiguchi, S.; Kurusu, Y. *Synlett*, ***2001***, 1802.

1. Cp_2ZrCl_2, BuLi
2. PhCHO

OBn → OH, Ph

96% (86:14 *anti:syn*)

Bertus, P.; CherouVrier, F.; Szymoniak, J. *Tetrahedron Lett.*, ***2001***, *42*, 1677.

OAc, NHBoc

2 eq InI, 5% $Pd(PPh_3)_4$, THF
PhCHO, rt

Ph, NHBoc OH

85%

Takemoto, Y.; Anzai, M.; Yanada, R.; Fujii, N.; Ohno, H.; Ibuka, T. *Tetrahedron Lett.*, ***2001***, *42*, 1725.

PhCHO

1. OTIPS, OEt , NHTs, CO_2H
BH_3•THF, CH_2Cl_2, −78°C
2. $(EtO)_2POCH_2CO_2Et$, 2 eq NaH
THF, −78°C → rt

OTIPS, Ph, CO_2Et

87x59%

Hillier, M.C.; Meyers, A.I. *Tetrahedron Lett.*, ***2001***, *42*, 5145.

Ph, CHO

Bu_3Sn
$NbCl_5$, ether, −78°C

OH, Ph + OH, Ph

(3 : 1) 63%

Andrade, C.K.Z.; Azevedo, N.R. *Tetrahedron Lett.*, ***2001***, *42*, 6473.

PhCHO

$(c\text{-}C_6H_{11})_2BCl$, hexane, rt
2,6-lutidine, 12 h

OH, Ph, $c\text{-}C_6H_{11}$

75%

Kabalka, G.W.; Wu, Z.; Ju, Y. *Tetrahedron Lett.*, ***2001***, *57*, 1663.

Bu, CHO

PhBr, cat $Pd(OAc)_2/PPh_3$
Cs_2CO_3, *o*-xylene, 3 h
160°C

Bu, Ph, Ph, CHO

29%

Terao, Y.; Kametani, Y.; Wakui, H.; Satoh, T.; Miura, M.; Nomura, M. *Tetrahedron*, ***2001***, *57*, 5967.

CHO; Br; In, H_2O; OH; 90%

Loh, T.-P.; Tan, K.-T.; Yang, J.-Y.; Xiang, C.-L. *Tetrahedron Lett.*, ***2001***, *42*, 8701.

PhCHO; S, S, $SiMe_3$; TBAF, DMF; S, S, OH, Ph; 79%

Degl'Innocenti, A.; Capperucci, A.; Nocentini, T. *Tetrahedron Lett.*, ***2001***, *42*, 4557.

PhCHO; Br, Mn*; THF-hexane; OH, Ph; 82%

Kakiya, H.; Nishimae, S.; Shinokubo, H.; Oshima, K. *Tetrahedron*, ***2001***, *57*, 8807.

OLi, Li; 1. PhCHO, acetophenone, 2 eq $TiCl(Oi\text{-}Pr)_3$, 20°C; 2. aq NH_4Cl; OH, OH, Ph; 60%

Pastor, I.M.; Yus, M. *Tetrahedron*, ***2001***, *57*, 2365.

PhCHO; Et_2Zn, hexane, 0°C; chiral leucine derivative; OH, Ph, Et; 91% (97% ee)

Kawanami, Y.; Mitsuie, T.; Miki, M.; Sakamoto, T.; Nishitani, K. *Tetrahedron*, ***2000***, *56*, 175.

PhCHO; 1. $)_2BCl$, NEt_3; 2. H_2O; OH, Ph; 75%

Kabalka, G.W.; Wu, Z.; Trotman, S.E.; Gao, X. *Org. Lett.*, ***2000***, *2*, 255.

Me; 2 eq $PhB(OH)_2$, DME/H_2O, 50°C; $Rh(acac)(coe)_2$, $P(t\text{-}Bu)_3$; OH, Ph; 99%

Ueda, M.; Miyaura, N. *J. Org. Chem.*, ***2000***, *65*, 4450.

Cl, Cl, C_3H_7, Cl → PhCHO, DMF, $CrCl_2$ → OH, Cl, Ph, C_4H_9, Cl

73% (56:44)

Baati, R.; Gouverneur, V.; Mioskowski, C. *J. Org. Chem.*, ***2000***, *65*, 1235.

OH → 5% $Pd(OAc)_2$, PhOH, 10% PPh_3, rt / PhCHO, 2 eq BEt_3, hexane/THF, 27 h → OH, Ph

78%

Kimura, M.; Tomizawa, T.; Horino, Y.; Tanaka, S.; Tamaru, Y. *Tetrahedron Lett.*, ***2000***, *41*, 3627.

PhCHO → $SiCl_3$, DMF, 0.2 h / 5% AgOTf → OH, Ph 92%

Fürstner, A.; Voigtländer, D. *Synthesis*, ***2000***, 959.

CHO, O_2N → BF_3^- K^+ / BF_3•OEt_2, –78°C, 15 min → OH, O_2N 93%

Batey, R.A.; Thadani, A.N.; Smil, D.V.; Lough, A.J. *Synthesis*, ***2000***, 990.

MeO, Br → PhCHO, Ni(dppe)Br_2/Zn / THF, 75°C → MeO, OH, Ph

91%

Keck, G.E.; Wager, C.A. *Org. Lett.*, ***2000***, *2*, 2307.

PhCHO → Br, In, DMF, rt → OH, Ph 78% (92:8 *syn:anti*)

Khan, F.A.; Prabhudas, B. *Tetrahedron*, ***2000***, *56*, 7595.

PhCHO → (allyl bromide, Sb/KF/H_2O) → PhCH(OH)CH$_2$CH=CH$_2$ quant

Li, L.-H.; Chan, T.H. *Tetrahedron Lett.*, ***2000***, *41*, 5009.

PhCHO → (allyl bromide, Cd, aq DMF) → PhCH(OH)CH$_2$CH=CH$_2$ 51%

Zheng, Y.; Bao, W.; Zhang, Y. *Synth. Commun.*, ***2000***, *30*, 3517.

CH$_2$=CHCH$_2$OH → (1. TMSCl–NaI, MeCN, rt; 2. Bi°; 3. PhCHO) → PhCH(OH)CH$_2$CH=CH$_2$ 68%

Miyoshi, N.; Nishio, M.; Murakami, S.; Furuma, T.; Wada, M. *Bull. Chem. Soc. Jpn.*, ***2000***, *73*, 689.

4-MeOC$_6$H$_4$CHO → (1. PhCH=CHCH$_2$Br, DMF, NiBr$_2$, CrCl$_3$, 50°C, TMSCl, 6 h; 2. H_2O) → 4-MeOC$_6$H$_4$CH(OH)CH=CHPh 96%

Kuroboshi, M.; Tanaka, M.; Kishimoto, S.; Goto, K.; Mochizuki, M.; Tanaka, H. *Tetrahedron Lett.*, ***2000***, *41*, 81.

CH$_3$CH$_2$CH$_2$CHO → (BrCH$_2$CO$_2$Et, In, THF, rt, 3 h, ultrasound) → CH$_3$CH$_2$CH$_2$CH(OH)CH$_2$CO$_2$Et 95%

Lee, P.H.; Bang, K.; Lee, K.; Sung, S.-y.; Chang, S. *Synth. Commun.*, ***2001***, *31*, 3781.

PhCHO → (allyl-SiMe$_3$, YbCl$_3$, MeNO$_2$) → PhCH(OH)CH$_2$CH=CH$_2$ 62%

Fang, X.; Watkin, J.G.; Warner, B.P. *Tetrahedron Lett.*, ***2000***, *41*, 447.

PhCHO → (allyl-SnBu$_3$, 20% ClCPh$_3$, Me$_3$SiCl, CH$_2$Cl$_2$, rt) → PhCH(OH)CH$_2$CH=CH$_2$ 65%

Chen, C.-T.; Chao, S.-D. *J. Org. Chem.*, ***1999***, *64*, 1090.

1. 2 eq BuLi
2. PhCHO

NH*t*-Bu

NH*t*-Bu

Ph

OH

77%

Stanetty, P.; Emerschitz, T. *Synth. Commun.*, ***2001***, *31*, 961.

PhCHO

, Mg/I_2

aq. 0.1 NH_4Cl

Zhang, W.-C.; Li, C.-J. *J. Org. Chem.*, ***1999***, *64*, 3230.

PhCHO

$(allyl)_2SnBr_2$, H_2O

NHPh

Ph

95%

Chan, T.H.; Yang, Y.; Li, C.J. *J. Org. Chem.*, ***1999***, *64*, 4452.

MeO—C6H4—$\overset{+}{I}$Ph $\overset{-}{BF_4}$

PhCHO , $CrCl_2$

DMF , 25°C , 5 h

Ph_2CHOH 60% + 20%

Chen, D.-W.; Ochiai, M. *J. Org. Chem.*, ***1999***, *64*, 6804.

Br

OAc

0.1% $NiCl_2$, $CrCl_2$

Ph CHO , DMF , rt

Ph OAc

OH

93%

Taylor, R.E.; Ciavarri, J.P. *Org. Lett.*, ***1999***, *1*, 467.

Ph CHO

Bu_3Sn , DCM , –78°C

10% (4-Br-2,6-di-*t*-Bu-phenoxy)$_2$AlMe

Ph

Ph

75%

Marx, A.; Yamamoto, H. *Synlett*, ***1999***, 584.

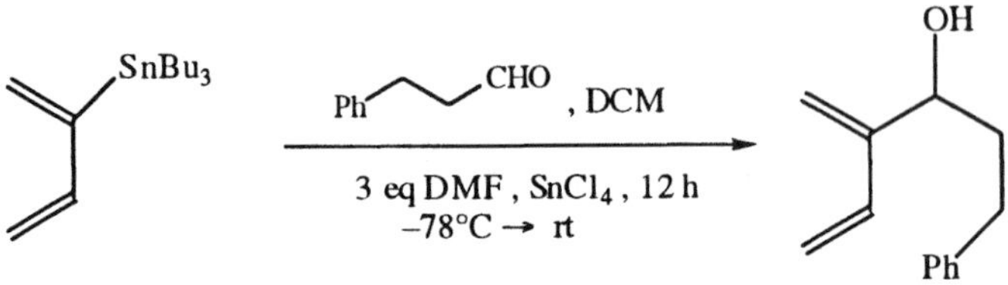

$SnBu_3$ — PhCH$_2$CH$_2$CHO, DCM; 3 eq DMF, $SnCl_4$, 12 h, –78°C → rt — OH, Ph — 88%

Luo, M.; Iwabuchi, Y.; Hatakeyama, S. *Synlett*, ***1999***, 1109.

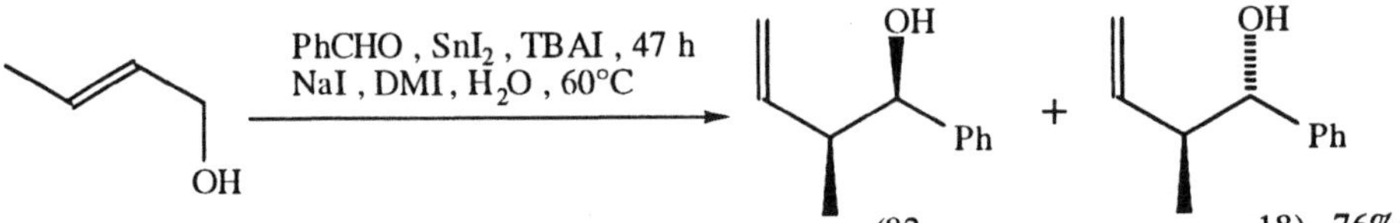

OH — PhCHO, SnI_2, TBAI, 47 h; NaI, DMI, H_2O, 60°C — OH, Ph + OH, Ph (82 : 18) 76%

Masuyama, Y.; Ito, T.; Tachi, K.; Ito, A.; Kurusu, Y. *Chem. Commun.*, ***1999***, 126.

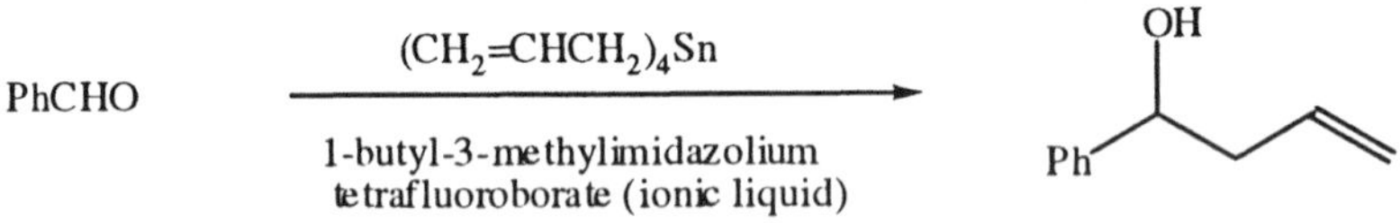

PhCHO — $(CH_2{=}CHCH_2)_4Sn$; 1-butyl-3-methylimidazolium tetrafluoroborate (ionic liquid) — OH, Ph — 79%

Gordon, C.M.; McCluskey, A. *Chem. Commun.*, ***1999***, 1431.

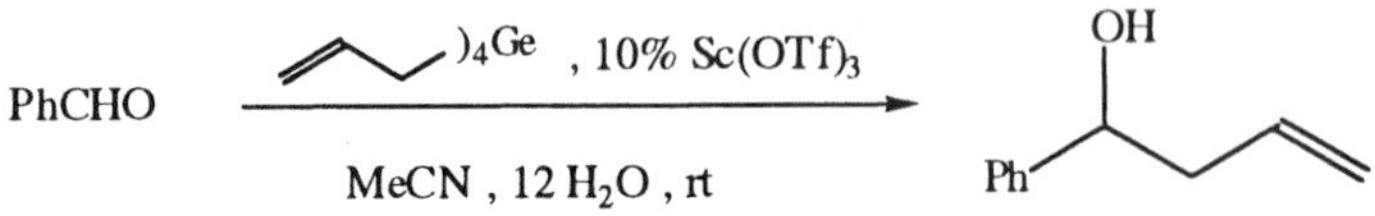

PhCHO — $)_4Ge$, 10% $Sc(OTf)_3$; MeCN, 12 H_2O, rt — OH, Ph — 81%

Akiyama, T.; Iwai, J.; Sugano, M. *Tetrahedron*, ***1999***, *55*, 7499.

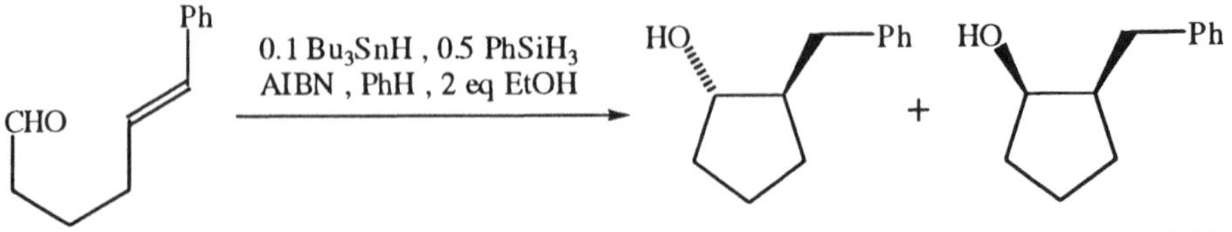

Ph, CHO — 0.1 Bu_3SnH, 0.5 $PhSiH_3$; AIBN, PhH, 2 eq EtOH — HO, Ph + HO, Ph (1.2 : 1) 85%

Hays, D.S.; Fu, G.C. *Tetrahedron*, ***1999***, *55*, 8815.

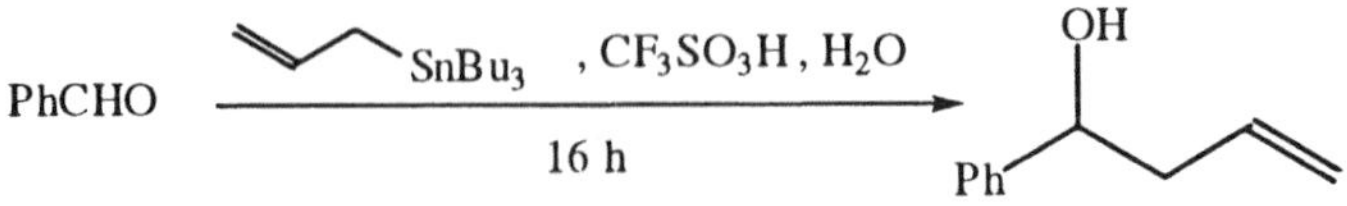

PhCHO — $SnBu_3$, CF_3SO_3H, H_2O; 16 h — OH, Ph — 70%

Loh, T.-P.; Xu, J. *Tetrahedron Lett.*, ***1999***, *40*, 2431.

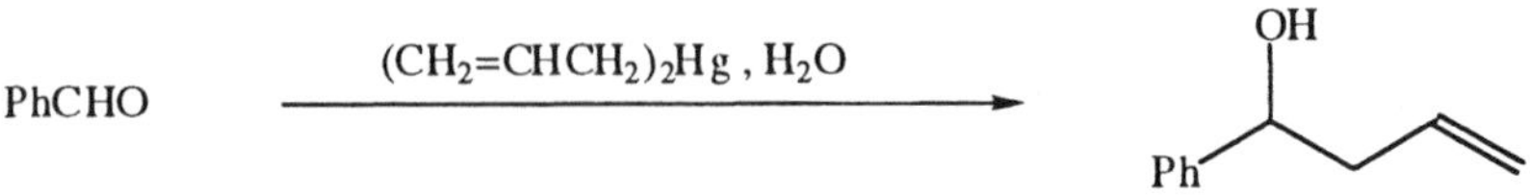

PhCHO — $(CH_2{=}CHCH_2)_2Hg$, H_2O — OH, Ph — 80%

Chan, T.H.; Yang, Y. *Tetrahedron Lett.*, ***1999***, *40*, 3863.

PhCHO , AgOTf, CH_2Cl_2
0°C , 15 min

NMe, NMe, O

SiCl$_3$ → OH, Ph

85%

Chataigner, I.; Piarulli, U.; Gennari, C. *Tetrahedron Lett.*, ***1999***, *40*, 3633.

C_5H_{11}–CHO

1.2 Bu_3Sn ... Br

Zn/CeCl$_3$, PhH–THF
ultrasound , 2 h

OH, C_5H_{11}, $SnBu_3$

73% (77:23 *E:Z*)

Lee, A.S.-Y.; Wu, C.-W. *Tetrahedron*, ***1999***, *55*, 12531.

$C_7H_{15}CHO$

Me, $BF_3^-K^+$

BF_3•OEt_2 , CH_2Cl_2
-78°C

OH, C_7H_{15}, Me

74% (>98:2 dr)

Batey, R.A.; Thadani, A.N.; Smil, D.V. *Tetrahedron Lett.*, ***1999***, *40*, 4289.

OHC, I

Bu_3SnH , BEt_3 , O_2

toluene , -78°C → 0°C

HO

87%

Devin, P.; Fensterbacnk, L.; Malacria, M. *Tetrahedron Lett.*, ***1999***, *40*, 5511.

PhCHO , BEt_3 , hexane/THF , rt

[Ni(acac)$_2$] , 4 h

OH, Ph

90% (1:15 *syn:anti*)

Kimura, M.; Fujimatsu, H.; Ezoe, A.; Shibata, K.; Shimizu, M.; Matsumoto, S.; Tamaru, Y. *Angew. Chem., Int. Ed.*, ***1999***, *38*, 397.

2 ... CO_2Me

PhCHO , 3.3 eq ATPH
2.3 eq LTMP , –78°C

toluene/THF

ATPH = tri(2,6-diphenylphenoxy)aluminum

OH, Ph, CO_2Me

97%

Saito, S.; Shiozawa, M.; Yamamoto, H. *Angew. Chem. Int. Ed.*, ***1999***, *38*, 1769.

PhCHO → 1. allyl bromide, cat In, Mn/TMSCl, H_2O, 1 d; 2. H_2O → Ph–CH(OH)–CH$_2$–CH=CH$_2$ 71%

Augé, J.; Lubin-Germain, N.; Thiaw-Woaye, A. *Tetrahedron Lett.*, ***1999***, *40*, 9245.

PhCHO → 1. crotyl bromide, Ga, KI, THF, LiCl, reflux; 2. H_2O → Ph–CH(OH)–CH(CH$_3$)–CH=CH$_2$ 94% (54:46 *syn:anti*)

Han, Y.; Chi, Z.; Huang, Y.-Z. *Synth. Commun.*, ***1999***, *29*, 1287.

EtCHO → 1,1-bis(trimethylsilyl)allyl (CH$_2$=CH–CH(SiMe$_3$)$_2$), 5 min, $TiCl_4$, DCM, –60°C → Me$_3$Si–CH=CH–CH$_2$–CH(OH)–Et 54%

Princet, B.; Anselme, G.; Pornet, J. *Synth. Commun.*, ***1999***, *29*, 3329.

REVIEWS:

"Catalytic Asymmetric Organic Zinc Addition to Carbonyl Compounds", Pu, L.; Yu, H.-B. *Chem. Rev.*, ***2001***, *101*, 757.

SECTION 35: ALCOHOLS AND THIOLS FROM ALKYLS, METHYLENES AND ARYLS

No examples of the reaction $RR^1 \rightarrow ROH$ (R^1 = alkyl, aryl, etc.) occur in the literature. For reactions of the type $RH \rightarrow ROH$ (R = alkyl or aryl) see Section 41 (Alcohols and Phenols from Hydrides).

SECTION 36: ALCOHOLS AND THIOLS FROM AMIDES

NO ADDITIONAL EXAMPLES

SECTION 37: ALCOHOLS AND THIOLS FROM AMINES

Me–CH=CH–CH=N–c-C_6H_{11} → 1. BuLi, THF, –45°C; 2. H^+; 3. $NaBH_4$, MeOH → Me–CH(Bu)–CH$_2$–CH$_2$–OH 74%

Tomioka, K.; Shieoya, Y.; Nagaoka, Y.; Yamada, K.-i. *J. Org. Chem.*, ***2001***, *66*, 7051.

Ph ... NH_2 → diethylene glycol, KOH / 210°C, 12 h → Ph ... OH 76%

Rahman, S.M.A.; Ohno, H.; Tanaka, T. *Tetrahedron Lett.*, ***2001***, *42*, 8007.

PhCH=NPh → 2 eq In, H_2O / 3 eq allyl bromide → PhCH(OH)CH₂CH=CH₂ 85%

Vilaivan, T.; Winotapan, C.; Shinada, T.; Ohfune, Y. *Tetrahedron Lett.*, ***2001***, *42*, 9073.

SECTION 38: ALCOHOLS AND THIOLS FROM ESTERS

$PhCO_2Et$ → $NaBH_4$–LiCl, 5 min / microwaves → $PhCH_2OH$ 75%

Feng, J.-C.; Liu, B.; Dai, L.; Yang, X.-L.; Tu, S.-J. *Synth. Commun.*, ***2001***, *31*, 1875.

2-phenylphenyl N-propylcarbamate (O, NHPr, Ph) → 1. BuLi, TMEDA 2. TMSCl / 3. 2M HCl → $SiMe_3$, OH, Ph 97%

Kauch, M.; Hoppe, D. *Can. J. Chem.*, ***2001***, *79*, 1736.

PhOAc → sulfurated borohydride exchange resin / MeOH, 25°C, 10 min → PhOH 94%

Bandgar, B.P.; Kamble, V.T. *J. Chem. Res. (S)*, ***2001***, 54.

CO_2Me → MeTi(O*i*-Pr)$_3$, 4 eq EtMgBr / THF, 0°C → Et, OH 67%

Quan, L.G.; Cha, J.K. *Org. Lett.*, ***2001***, *3*, 2745.

PhSAc → $TiCl_4$/Zn, CH_2Cl_2 / 0°C, rt → PhSH 87%

Jin, C.K.; Jeong, H.J.; Kim, M.K.; Kim, J.Y.; Yoon, Y.-J.; Lee, S.-G. *Synlett*, ***2001***, 1956.

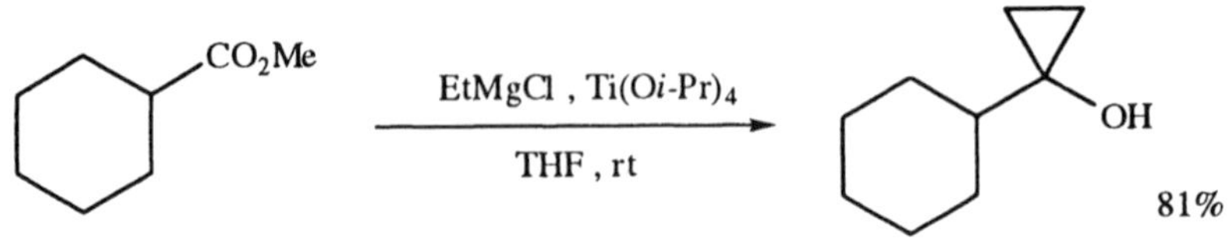

CO_2Me — EtMgCl, Ti(O*i*-Pr)$_4$ / THF, rt — OH 81%

Lee, J.C.; Sung, M.J.; Cha, J.K. *Tetrahedron Lett.*, ***2001***, *42*, 2059.

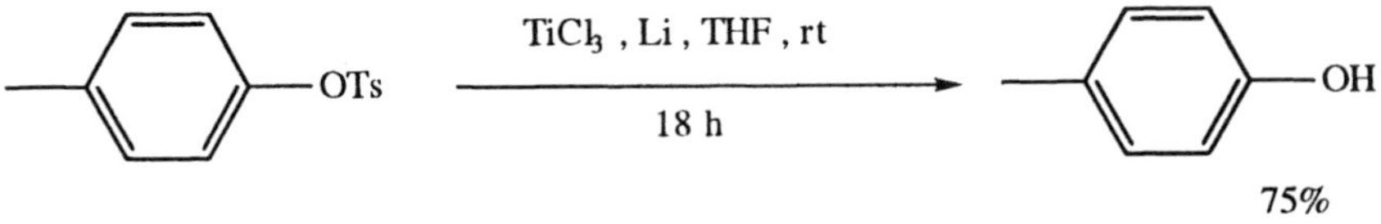

OTs — $TiCl_3$, Li, THF, rt / 18 h — OH 75%

Nayak, S.K. *Synthesis*, ***2000***, 1575.

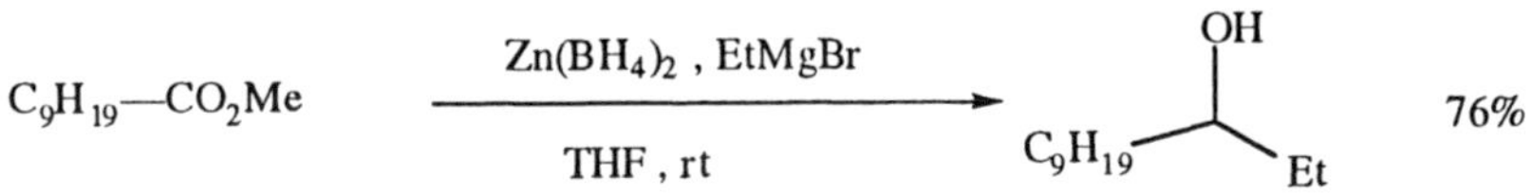

C_9H_{19}—CO_2Me — $Zn(BH_4)_2$, EtMgBr / THF, rt — OH, C_9H_{19}, Et 76%

Balasubramanian, S.; Nair, M.G. *Synth. Commun.*, ***2000***, *30*, 313.

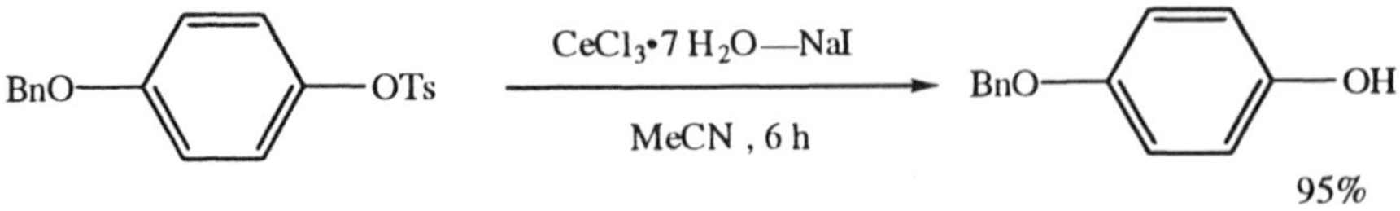

BnO— —OTs — $CeCl_3$•7 H_2O—NaI / MeCN, 6 h — BnO— —OH 95%

Reddy, G.S.; Mohan, G.H.; Iyengar, D.S. *Synth. Commun.*, ***2000***, *30*, 3829.

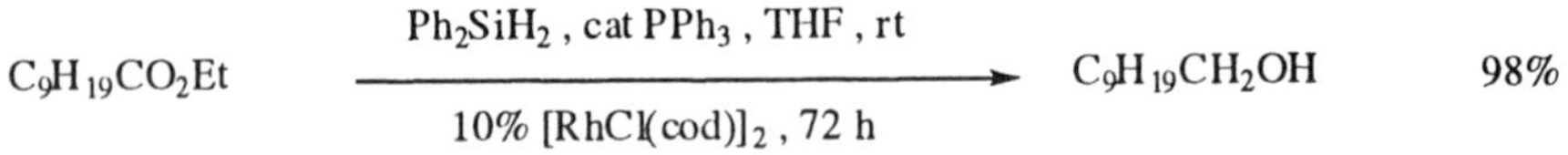

$C_9H_{19}CO_2Et$ — Ph_2SiH_2, cat PPh_3, THF, rt / 10% $[RhCl(cod)]_2$, 72 h — $C_9H_{19}CH_2OH$ 98%

Ohta, T.; Kamiya, M.; Kusui, K.; Michibata, T.; Nobutomo, M.; Furukawa, I. *Tetrahedron Lett.*, ***1999***, *40*, 6963.

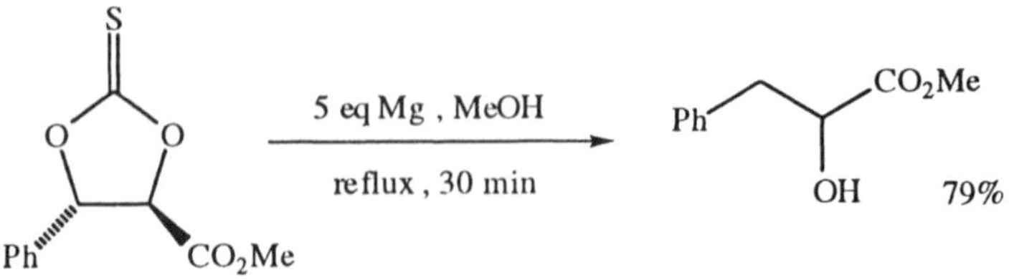

S, O, O, Ph, CO_2Me — 5 eq Mg, MeOH / reflux, 30 min — Ph, CO_2Me, OH 79%

Rho, H.-S.; Ko, B.-S. *Synth. Commun.*, ***1999***, *29*, 2875.

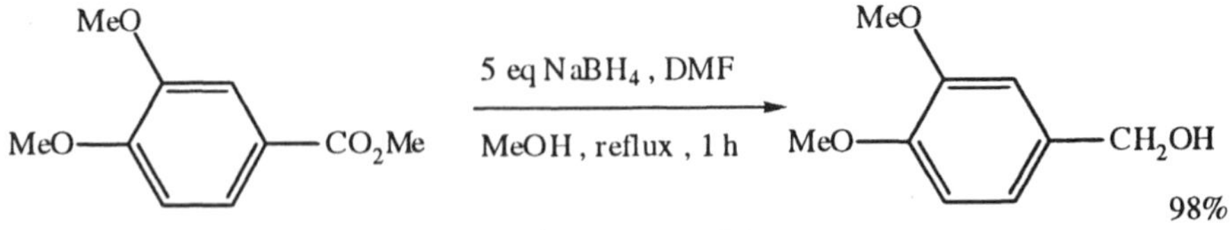

MeO, MeO— —CO_2Me — 5 eq $NaBH_4$, DMF / MeOH, reflux, 1 h — MeO, MeO— —CH_2OH 98%

Zanka, A.; Ohmori, H.; Okamoto, T. *Synlett*, ***1999***, 1636.

SECTION 39: ALCOHOLS AND THIOLS FROM ETHERS, EPOXIDES AND THIOETHERS

O

Et_3Al, 5% PPh_3, toluene

rt, 1 d

OH

Et

99%

Schneider, C.; Brauner, J. *Eur. J. Org. Chem.*, ***2001***, 4445.

O

HO_2C

NaN_3, H_2O, $AlCl_3$

pH 4, 30°C, 3.5 h

OH

HO_2C

N_3

99%

Fringuelli, F.; Pizzo, F.; Vaccaro, L. *Tetrahedron Lett.*, ***2001***, *42*, 1131.

O

I_2, CH_2Cl_2, 10°C, 1 h

OH

92%

Vatèle, J.-M. *Synlett*, ***2001***, 1989.

O

Ph

Br

In, THF, rt, 3 h

OH

Ph

+

OH

Ph

(90 : 10) 90%

Yadav, J.S.; Anjaneyulu, S.; Ahmed, Md.M.; Subba Reddy, B.V. *Tetrahedron Lett.*, ***2001***, *42*, 2557.

O

1. 4 eq Me_3Al, 1 eq H_2O
DCM, –20°C, 1 h

2. H_3O^+

OH

Wipf, P.; Ribe, S. *Org. Lett.*, ***2001***, *3*, 1503.

O

2 eq $AlEt_3$, 1 d
imidazolium salt catalyst

Et

OH

76%

Zhou, H.; Campbell, E.J.; Nguyen, S.T. *Org. Lett.*, ***2001***, *3*, 2229.

1. Li, 5% DTBB, THF
2. $ZnBr_2$, THF
3. 5% $Pd(PPh_3)_4$, THF, 60°C
4. HCl–H_2O

OH
Ph
77%

Yus, M.; Gomis, J. *Tetrahedron Lett.*, ***2001***, *42*, 5721.

O
Ph
NaN_3, rt, 6 h
phase transfer ammonium resin
N_3
Ph
OH
+
OH
Ph
N_3
(91 9) 95%

Tamami, B.; Mahdavi, H. *Tetrahedron Lett.*, ***2001***, *42*, 8721.

OH
O
OBn
1. BuLi, DCM
2. 3 eq Me_3Al, 0°C → rt
OH OH
OBn
96% (>97:<3 1,3-:1,2-diol)

Sasaki, M.; Tanino, K.; Miyashita, M. *Org. Lett.*, ***2001***, *3*, 1765.

O
O
C_5H_{11}
2.5 eq BuLi, THF, 1 h
–78°C → rt
HO
OH
C_5H_{11}
90%

Hodgson, D.M.; Stent, M.A.H.; Wilson, F.X. *Org. Lett.*, ***2001***, *3*, 3401.

O
Et_3Al, 5% $P(NMe_2)_3$
toluene, rt, 1 d
OH
99%

Schneider, C.; Brauner, J. *Tetrahedron Lett.*, ***2000***, *41*, 3043.

Me
O
Me
CO_2Bn
5% $Pd(PPh_3)_4$, $Me_2NH–BH_3$
3 eq AcOH, DCM, rt, 15 min
Me
H
OH
Me
CO_2Bn
92%

David, H.; Dupuis, L.; Guillerez, M.-G.; Guibe, F. *Tetrahedron Lett.*, ***2000***, *41*, 3335.

Bt–CH2–O–CH(Ph)Ph → 1. 2.2 eq BuLi; 2. H_2O → Ph2CH–CH(OH)Bu 40%

Katritzky, A.R.; Fang, Y. *Heterocycles*, **2000**, *53*, 1783.

Cyclohexene oxide → 1. Me_3SiN_3, Bu_4NCl, neat; 2. *p*-TsOH, MeOH → 2-azidocyclohexanol (N_3, OH) 89%

Schneider, C. *Synlett*, **2000**, 1840.

C_9H_{19} vinyl epoxide → allyl alcohol (OH), BF_3•OEt_2, DCM → C_9H_{19}, OH product 75%

Prestat, G.; Baylon, C.; Heck, M.-P.; Mioskowski, C. *Tetrahedron Lett.*, **2000**, *41*, 3829.

Resin-bound O-allyl salicylate → solid phase, DMF, 4 min; microwaves → 3-allyl-2-hydroxybenzoic acid (HO, O, OH) 92%

Kumar, H.M.S.; Anjaneyulu, S.; Reddy, B.V.S.; Yadav, J.C. *Synlett*, **2000**, 1129.

CO_2Me, Bu, O, Et epoxide → 5% $Pd_2(dba)_3CHCl_3$, 5% PPh_3, HCOOH-NEt_3; DMF, rt, 1 h → MeO_2C, Bu, Et, HO + MeO_2C, Bu, Et, HO

(95 5) 84%

Noguchi, Y.; Yamada, T.; Uchiro, H.; Kobayashi, S. *Tetrahedron Lett.*, **2000**, *41*, 7493, 7499.

1. $TMSN_3$, Cr-salen catalyst [emim] SbF_6, 20°C
2. CSA, MeOH

75% (87% ee)

Song, C.E.; Oh, C.R.; Roh, E.J.; Choo, D.J. *Chem. Commun.*, ***2000***, 1743.

NaN_3, 0.2 CAN
t-BuOH

(90 : 10) 93%

Iranpoor, N.; Kazemi, F. *Synth. Commun.*, ***1999***, *29*, 561.

2% $TiCl_3(OTf)$, MeOH
0°C, 5 min

95% (89% ee)

Iranpoor, N.; Zeynizadeh, B. *Synth. Commun.*, ***1999***, *29*, 1017.

Ph—OMe → Ph—OH 98%

1. 2 eq $AlCl_3$, 2 eq NaI
2. 5% aq $Na_2S_2O_3$

Ghiaci, M.; Asghari, J. *Synth. Commun.*, ***1999***, *29*, 973.

Li naphthalenide, THF
−78°C → rt, 10 min

78%

Jankowska, R.; Mhehe, G.-L.; Liu, H.-J. *Chem. Commun.*, ***1999***, 1581.

1. 2 eq *i*-PrSNa, DMF 160°C, 4 h
2. aq HCl

93%

Pinchart, A.; Dallaire, C.; Van Bierbeek, A.; Gingras, M. *Tetrahedron Lett.*, ***1999***, *40*, 5479.

1. $Hg(OAc)_2$, THF/H_2O
2. $NaBH_4$/aq K_2CO_3

62%

Crouch, E.D.; Mehlmann, J.F.; Herb, B.R.; Mitten, J.V.; Dai, G. *Synthesis*, ***1999***, 559.

1. MeLi, rt THF, 16 h
2. H_2O

75% + 21%

Maleczka Jr., R.E.; Geng, F. *Org. Lett.*, ***1999***, *1*, 1115.

Additional examples of ether cleavages may be found in Section 45A (Protection of Alcohols and Thiols).

SECTION 40: ALCOHOLS AND THIOLS FROM HALIDES AND SULFONATES

1. Mn*, THF, rt
2. PhCHO, THF

other electrophiles can be used

72%

Kim, S.-H.; Rieke, R.D. *Tetrahedron Lett.*, ***1999***, *40*, 4931.

1. BEt_3, DCC, DCE −50°C, syringe pump
2. MeOH ; Me_2S

82%

Kihara, N.; Ollivier, C.; Renaud, P. *Org. Lett.*, ***1999***, *1*, 1419.

PhBr

1. Bu_3MgLi, THF, 0°C, 0.5 h
2. EtCHO

87%

Inoue, A.; Kitagawa, K.; Shinokubo, H.; Oshima, K. *J. Org. Chem.*, ***2001***, *66*, 4333.

1. FSM-16, NaI, hv, acetone, rt, 1 d
2. FSM-16, CsI, hv, *i*-Pr_2O, 12 h

86x81%

Itoh, A.; Kodama, T.; Inagaki, S.; Masaki, Y. *Org. Lett.*, ***2001***, *3*, 2653.

SECTION 41: ALCOHOLS AND THIOLS FROM HYDRIDES

HO

N—CO_2Bn *Sphingomonas sp.* HXN-200 / 1 h → N—CO_2Bn

90% conversion

Li, Z.; Feiten, H.-J.; Chang, D.; Duetz, W.A.; van Beilen, J.B.; Witholt, B.
J. Org. Chem., ***2001***, *66*, 8424.

OH cultured cells of *Gocsypium hirsutum* → HO ... OH

72%

Hamada, H.; Tanaka, T.; Furuya, T.; Takahata, H.; Nemoto, H.
Tetrahedron Lett., ***2001***, *42*, 909.

O O, acetone / 0°C, 5 min → OH + OH OH

79% 21%

Curci, R.; D'Accolti, L.; Fusco, C. *Tetrahedron Lett.*, ***2001***, *42*, 7087.

Ph Ce(OTf)$_4$, MeCN, rt → Ph(C=O) Ph(CHOH)

	O	OH
+ 1.5% H_2O	9.5%	85.7%
+ 16.4% H_2O	99.7%	–

Laali, K.K.; Herbert, M.; Cushnyr, B.; Bhatt, A.; Terrano, D.
J. Chem. Soc., Perkin Trans. 1, ***2001***, 578.

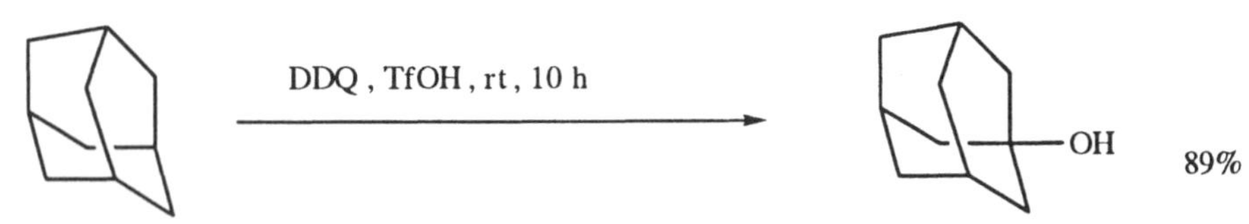

Tanemura, K.; Suzuki, T.; Nishida, Y.; Satsumabayashi, K.; Horaguchi, T.
J. Chem. Soc., Perkin Trans. 1, ***2001***, 3230.

$Cu(NO_3)\cdot 3\ H_2O$, MeCN
phosphate buffer , 50°C
30% H_2O_2 , 5 h

96%

Nasreen, A.; Adapa, S.R. *Org. Prep. Proceed. Int.*, ***2000***, *32*, 287.

B. megaterium

63% conversion (74% ee , ***R***)

Adam, W.; Kukacs, Z.; Harmsen, D.; Saha-Möller, C.R.; Schreier, P. *J. Org. Chem.*, ***2000***, *65*, 878.

Sphingomonas sp. HXN-200
2% glucose , 5 h

80%

Chang, D.; Witholt, B.; Li, Z. *Org. Lett.*, ***2000***, *2*, 3949.

Bacillus megaterium
O_2 , 30°C , 17 h

79%

Adam, W.; Lukacs, Z.; Saha-Möller, C.R.; Weckerle, B.; Schreirer, P. *Eur. J. Org. Chem.*, ***2000***, 2923.

Pseudomonas oleovorans GPol

62%

Li, Z.; Feiten, H.-J.; van Beilen, J.B.; Duetz, W.; Witholt, B. *Tetahedron Asymm.*, ***1999***, *10*, 1323.

O_2 , 10% NHPI , 2% Bu_4NBr
$PhCF_3$, H_2O

60% (73% conversion + 23% diol)

Matsunaka, K.; Iwahama, T.; Sakaguchi, S.; Ishii, Y. *Tetrahedron Lett.*, ***1999***, *40*, 2165.

O_2 , 5% $Mn(OAc)_2$•4 H_2O

DCM , 23°C , 14 h

89%

Christoffers, J. *J. Org. Chem.*, ***1999***, *64*, 7668.

Mn-salen catalyst , PhIO

PhCl , rt

58% (19% ee)

Punniyamurthy, T.; Katsuki, T. *Tetrahedron*, ***1999***, *55*, 9439.

1. ClCOSCl , NBu_3
2. $AlCl_3$
3. NaOH
4. H_3O^+

65x70%

Yoshida, Y.; Ogura, M.; Tanabe, Y. *Heterocycles*, ***1999***, *50*, 681.

SECTION 42: ALCOHOLS AND THIOLS FROM KETONES

The following reaction types are included in this section:
A. Reductions of Ketones to Alcohols
B. Alkylations of Ketones, forming Alcohols

BuOH , 5% $RuCl_2(PPh_3)_3$

KOH , dioxane , 40 h

80%

Cho, C.S.; Kim, B.T.; Kim, T.-J.; Shim, S.C. *J. Org. Chem.*, ***2001***, *66*, 9020.

Coupling of ketones to give diols is found in Section 323 (Alcohol → Alcohol).

SECTION 42A: REDUCTION OF KETONES TO ALCOHOLS

ASYMMETRIC REDUCTION

1. (–)-Ipc_2BH, THF, 2 d
2. NaOH, H_2O_2

Et / CO_2H → OH / Et / CO_2H

82% (95% ee, *R*)

Ramachandran, P.V.; Brown, H.C.; Pitre, S. *Org. Lett.*, ***2001***, *3*, 17.

bakers yeast, pet ether/H_2O, rt, 1 d

Ph → OH / Ph

62% conversion (90% ee, *S*)

Liu, X.; Zhu, T.-S.; Sun, P.-D.; Xu, J.-H. *Synth. Commun.*, ***2001***, *31*, 1521.

RuCl(cymene)$_2$, HCOOH, NEt_3
dendritic ligand, 20 h

Ph → OH / Ph

99% (97% ee)

Chen, Y.-C.; Wu, T.-F.; Deng, J.-G.; Liu, H.; Jiang, Y.-Z.; Choi, M.C.K.; Chan, A.S.C. *Chem. Commun.*, ***2001***, 1488.

3% CuCl, 3% NaO*t*-Bu, 3.5 h
3% (*R*)-3,5-xyl-MeOBIPHEP
0.34 PMHS, 0.5 M toluene
–50°C

Ph → OH / Ph

95% (95% ee)

Lipshutz, B.H.; Noson, K.; Chrisman, W. *J. Am. Chem. Soc.*, ***2001***, *123*, 12917.

Ru(*S*-Tol-P-Phos)PhH)Cl_2
H_2, 80°C, 2 h

CO_2Bn → OH / CO_2Bn

quant (98% ee)

Wu, J.; Chen, H.; Zhou, Z.-Y.; Yueng, C.H.; Chan, A.S.C. *Synlett*, ***2001***, 1050.

cat [Rh(cod)(O_2CCF_3)$_2$], toluene
50°C, η^6-arenechromium diphosphines

CO_2Et → OH / CO_2Et

90% (72% ee, *R*)

Pasquier, C.; Pélinski, L.; Brocard, J.; Mortreux, A.; Agbossou-Niedercorn, F. *Tetrahedron Lett.*, ***2001***, *42*, 2809.

PhC(O)CH$_3$ → 1. *i*-PrOH , *t*-BuOK , H_2O , rt, aminosulfonamide ligand; 2. $[Cp^*IrCl_2]_2$, 40°C , 2 h; 3. add ketone , aq *i*-PrOH, *t*-BuOK , 1 d → PhCH(OH)CH$_3$ 97% conversion (97% ee , *R*)

Thorpe, T.; Blacker, J.; Brown, S.M.; Bubert, C.; Crosby, J.; Fitzjohn, S.; Muxworthy, J.P.; Williams, J.M.J. *Tetrahedron Lett.*, ***2001***, *42*, 4041.

SmI_2/H_2O , THF , –78°C, 2 h

92% (75% de , *S*)

Fukuzawa, S.-i.; Miura, M.; Matsuzawa, H. *Tetrahedron Lett.*, ***2001***, *42*, 4167.

Bakers yeast, liquid petroleum gas

CO_2Et 74% (95% ee)

Johns, M.K.; Smallridge, A.J.; Trewhella, M.A. *Tetrahedron Lett.*, ***2001***, *42*, 4261.

1. proline ligand , DCE , 2% Ru complex; 2. HCOOH , NEt_3

96% (79% ee)

Rhyoo, H.Y.; Yoon, Y.-A.; Park, H.-J.; Chung, Y.K. *Tetrahedron Lett.*, ***2001***, *42*, 5045.

[bmim] PF_5 , H_2O , MeOH, immobilized Bakers yeast

35%

Howarth, J.; James, P.; Dai, J. *Tetrahedron Lett.*, ***2001***, *42*, 7517.

$RuCl_3$/*S*-Tol-BINAP, H_2 (4 bar) , MeOH , 1 d

99% ee

Madec, J.; Pfister, X.; Phansavath, P.; Ratovelomanana-Vidal, V.; Genêt, J.-P. *Tetrahedron*, ***2001***, *57*, 2563.

Rhizopus arrhizus

33% (69% ee, *S*)

Salvi, N.A.; Chattopadhyay, S. *Tetrahedron*, ***2001***, *57*, 2833.

Bakers yeast , 30°C

H_2O , 1 d

54% (75% ee)

Attolini, M.; Bourguir, F.; Iacazio, G.; Peiffer, G.; Maffei, M. *Tetrahedron*, ***2001***, *57*, 537.

BH_3•SMe_2 , 10% chiral Al alkoxide

toluene , 40°C , 10 min

96% (73% ee)

Fu, I.-P.; Uang, B.-J. *Tetrahedron Asymm.*, ***2001***, *12*, 45.

Bakers yeast/sucrose , rt

phosphate buffer , pH 6.5

18 h

92% (100% ee , ***R***)

Yadav, J.S.; Reddy, P.T.; Nanda, S.; Rao, A.B. *Tetrahedron Asymm.*, ***2001***, *12*, 63.

Bakers yeast

48% (>96% ee , *S*)

Weui, Z.-L.; Li, Z.-Y.; Lin, G.-Q. *Tetrahedron Asymm.*, ***2001***, *12*, 225.

i-PrOH , 20°C , 4 h

3% [(NHCONH*i*-Pr, NH_2 cyclohexane) /0.5 $RuCL_2$(*p*-cymene)]$_2$

75% (63% ee, *R*)

Bied, C.; Moreau, J.J.E.; Man, M.W.C. *Tetrahedron Asymm.*, ***2001***, *12*, 329.

i-PrNEtPh•BH_3 , THF , 25°C

98% (>99% ee ,*S*)

Cho, B.T.; Kim, D.J. *Tetrahedron Asymm.* ***2001***, *12*, 2043.

Ph–C(=O)–CH$_2$–SO$_2$Ph → *Curvularia lunata*, 10 h → Ph–CH(OH)–CH$_2$–SO$_2$Ph

90% (87% ee, *S*)

Gotor, V.; Rebolledo, F.; Liz, R. *Tetrahedron Asymm.*, ***2001***, *12*, 513.

Cl–CH$_2$–C(=O)–CH$_2$–CO$_2$Et → carbonyl reductase S1 → Cl–CH$_2$–CH(OH)–CH$_2$–CO$_2$Et

quant (>99% ee, *S*)

Yasohara, Y.; Kizaki, N.; Hasegawa, J.; Wada, M.; Kataoka, M.; Shimizu, S. *Tetrahedron Asymm.* ***2001***, *12*, 1713.

Ph–C(=O)–CH$_3$ → BH_3•SMe_2, squaric acid ligand / toluene, 50°C → Ph–CH(OH)–CH$_3$

95% (95% ee, *R*)

Zhang, J.; Zhou, H.-B.; Lü, S.-M.; Luo, M.-M.; Xie, R.-G.; Choi, M.C.K.; Zhou, Z.-Y.; Chan, A.S.C.; Yang, T.-K. *Tetrahedron Asymm.* ***2001***, *12*, 1907.

Ph–C(=O)–CH$_3$ → Ph_2SiH_2, cat [Rh)cod)Cl]$_2$, neat / cyclic ketimine ligand → Ph–CH(OH)–CH$_3$

86% (54% ee)

v. Keyserlingk, N.G.; Martens, J. *Tetrahedron Asymm.* ***2001***, *12*, 2213.

Ph–C(=O)–CH$_3$ → BH_3•oxazaborolidine, amino-alcohol ligand / THF → Ph–CH(OH)–CH$_3$

92% (74% ee, *S*)

Jiang, B.; Feng, Y.; Hang, J.-F. *Tetrahedron Asymm.*, ***2001***, *12*, 2323.

Ph–C(=O)–CH$_3$ → 1% chiral Ru complex, *i*-PrOH / microwaves, 9 min → Ph–CH(OH)–CH$_3$

90% (82% ee, *R*)

Lutsenko, S.; Moberg, C. *Tetrahedron Asymm.* ***2001***, *12*, 2529.

Ph–C(=O)–CH$_2$OH → *Yamadazyma farinosa* IFO 10896 / 1 d → Ph–CH(OH)–CH$_2$OH

91% (>99% ee, *S*)

Tsujigami, T.; Sugai, T.; Ohta, H. *Tetrahedron Asymm.* ***2001***, *12*, 2543.

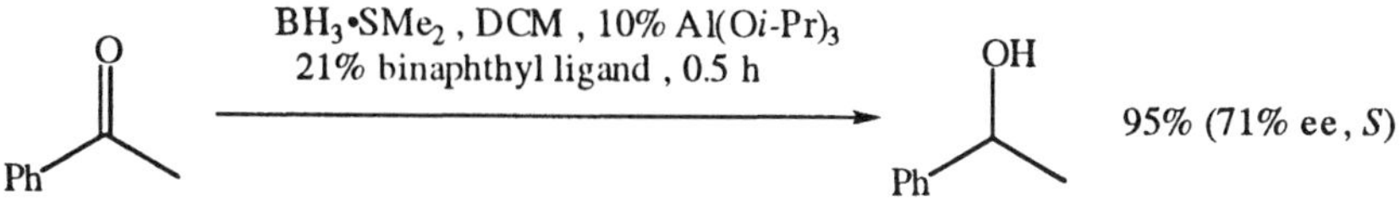

PhC(O)CH₃ → PhCH(OH)CH₃

BH_3•SMe_2, DCM, 10% Al(O*i*-Pr)$_3$
21% binaphthyl ligand, 0.5 h

95% (71% ee, *S*)

Lin, Y.-M.; Fu, I.-P.; Uang, B.-J. *Tetrahedron Asymm.* **2001**, *12*, 3217.

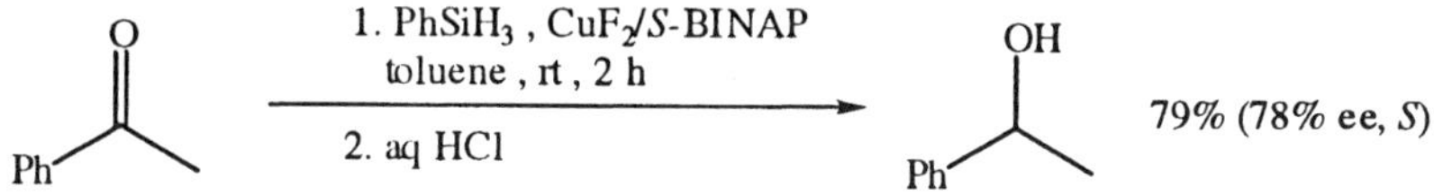

1. $PhSiH_3$, CuF_2/*S*-BINAP
toluene, rt, 2 h
2. aq HCl

79% (78% ee, *S*)

Sirol, S.; Courmarcel, J.; Mostefai, N.; Riant, O. *Org. Lett.* **2001**, *3*, 4111.

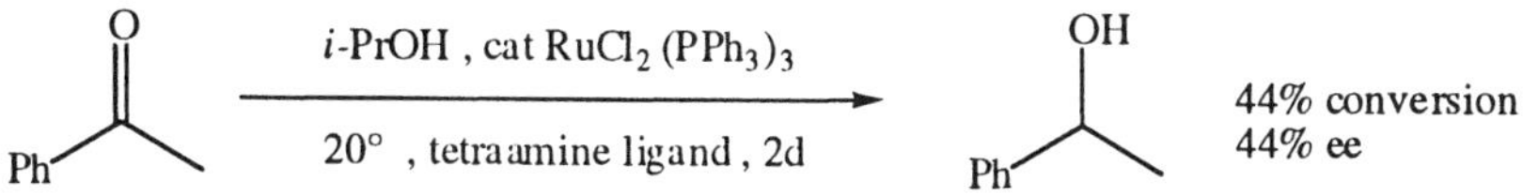

i-PrOH, cat $RuCl_2(PPh_3)_3$
20°, tetraamine ligand, 2d

44% conversion
44% ee

Marson, C.M.; Schwarz, I. *Tetrahedron Lett.*, **2000**, *41*, 8995.

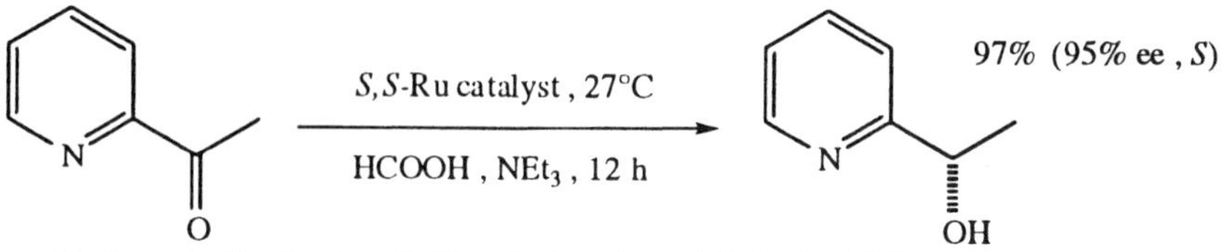

S,S-Ru catalyst, 27°C
HCOOH, NEt_3, 12 h

97% (95% ee, *S*)

Okano, K.; Murata, K.; Ikariya, T. *Tetrahedron Lett.*, **2000**, *41*, 9277.

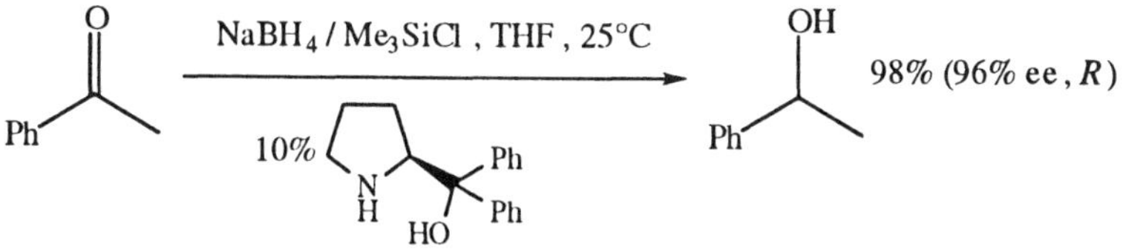

$NaBH_4$ / Me_3SiCl, THF, 25°C
10% (pyrrolidine diphenylmethanol: N H, Ph, Ph, HO)

98% (96% ee, *R*)

Jiang, B.; Feng, Y.; Zheng, J. *Tetrahedron Lett.*, **2000**, *41*, 10281.

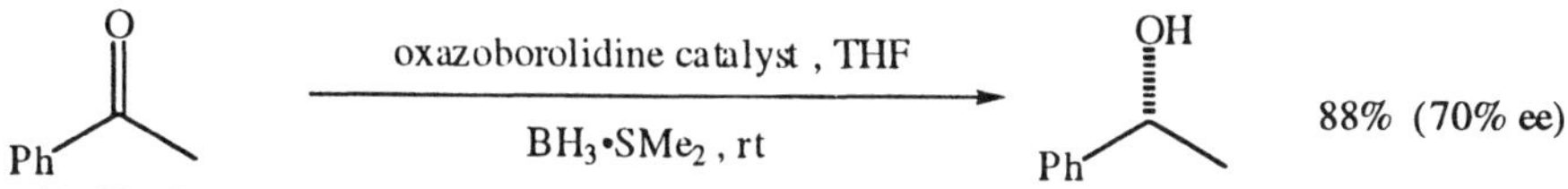

oxazoborolidine catalyst, THF
BH_3•SMe_2, rt

88% (70% ee)

Santhi, V.; Rao, J.M. *Synth. Commun.*, **2000**, *30*, 4329.

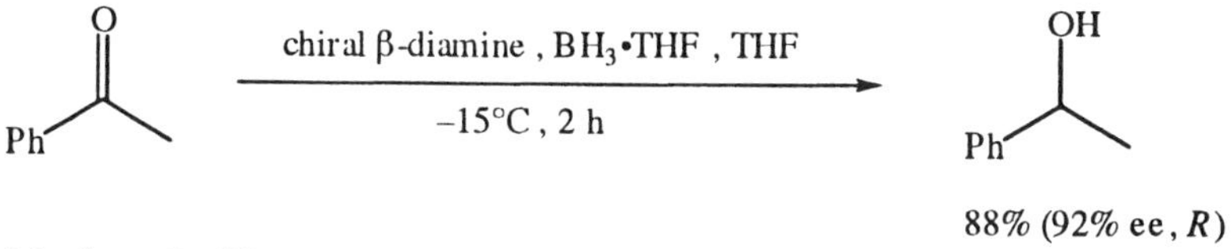

chiral β-diamine, BH_3•THF, THF
−15°C, 2 h

88% (92% ee, *R*)

Asami, M.; Sato, S.; Watanabe, H. *Chem. Lett.*, **2000**, 990.

cat $[Rh(cod)_2]$ BF_4 , Ph_2SiH_2

chiral bis-phosphine , THF , –40°C

HO

90% (91% ee)

Kuwano, R.; Sawamura, M.; Shirai, J.; Takahashi, M.; Ito, Y.
Bull. Chem. Soc. Jpn., ***2000***, *73*, 485.

O

Ph

BH_3•THF , proline , THF

rt , 10 min

OH

Ph

quant (85% ee)

Teodorović, A.V.; Joksović, M.D.; Konstantinović, S.K.; Mojsilović; Mihailović, M.L.
Monat. Chem., ***2000***, *131*, 91.

O

Ph

N–OMe

Bakers yeast , 1 d

OH

Ph

N–OMe

75% (97% ee)

Kreutz, O.C.; Segura, R.C.M.; Rodrigues, J.A.R.; Moran, P.J.S.
Tetrahedron Asymm., ***2000***, *11*, 2107.

O

Ph

Geotrichum candidum immobilized resting cells

supercritical CO_2

OH

Ph

51% (>99% ee , *S*)

Matsuda, T.; Harada, T.; Nakamura, K. *Chem. Commun.*, ***2000***, 1367.

O

Ph

Ru catalyst , H_2 , *i*-PrOH

t-BuOK

OH

Ph

98% ee

Burk, M.J.; Hems, W.; Herzberg, D.; Malan, C.; Zanotti-Gerosla, A. *Org. Lett.*, ***2000***, *2*, 4173.

O

Ph

aryl oxazaphospholidine oxide

BH_3 , THF

OH

Ph

95% (32% ee)

Brunel, J.M.; Legrand, O.; Buono, G. *Eur. J. Org. Chem.*, ***2000***, 3313.

O

Ph

CN

Curvularia lunata CECT 21130

phosphate buffer , pH 6 , MeCN

OH

Ph

CN

55% (96% ee , *S*)

Dehli, J.R.; Gotor, V. *Tetrahedron Asymm.*, ***2000***, *11*, 3693.

Synechococcus sp PCC7942

9 d

51% (98% ee, *S*)

Nakamura, K.; Yamanaka, R.; Tohi, K.; Hamada, H. *Tetrahedron Lett.*, ***2000***, ***41***, 6799.

chiral perfluoro salen ligand, 70°C
$[Ir(cod)Cl]_2$, *i*-PrOH, D-100, KOH

2 d

91% (56% ee)

Maillard, D.; Nguefack, C.; Pozzi, G.; Quici, S.; Valadé, B.; Sinou, D. *Tetrahedron Asymm.*, ***2000***, ***11***, 2887.

chiral oxazaboroline, $BH_3 \cdot SMe_2$

THF, rt, 2 h

94% (93% ee)

Santhi, V.; Rao, J.M. *Tetrahedron Asymm.*, ***2000***, ***11***, 3553.

H_2 (40 bar), cat diamine-BINAP

MeOH, 50°C, 14 h

quant (99% ee)

ter Halle, R.; Colasson, B.; Schulz, E.; Spagnol, M.; Lemaire, M. *Tetrahedron Lett.*, ***2000***, ***41***, 643.

0.1

2 eq BH_3, THF

>95% (84% ee)

Jones, S.; Atherton, J.C.C. *Tetrahedron Asymm.*, ***2000***, ***11***, 4543.

Geotrichum candidum IFO 4597

Amberlite XAD-7

83% (94% ee, *S*)

Nakamura, K.; Fujii, M.; Ida, Y. *J. Chem. Soc., Perkin Trans. 1*, ***2000***, 3205.

0.1 chiral diamine , BH_3•THF

THF , rt

88% (87% ee)

Sato, S.; Watanabe, H.; Asami, M. *Tetrahedron Asymm.*, ***2000***, *11*, 4329.

40 bar H_2 , *i*-PrOH/*t*-BuOK

chiral polymer catalyst
50°C , 18 h

quant (68% ee)

ter Halle, R.; Schulz, E.; Spagnol, M.; Lemaire, M. *Synlett*, ***2000***, 680.

Cr(*L*-His)$^+$, DMF-H_2O

rt , pH 6.5

33% (36% ee, *S*)

Patonay, T.; Hajdu, C.; Jekõ, J.; Lévai, A.; Micske, K.; Zucchi, C.
Tetrahedron Lett., ***1999***, *40*, 1373.

2 eq $LiBH_4/NiI_2$, THF

94% ee

Molvinger, K.; Lopez, M.; Court, J. *Tetrahedron Lett.*, ***1999***, *40*, 8375.

CpRhCl(chiral diamine) , 12 h

i-PrOH , *t*-BuOK , 30°C

92% (94% ee, *R*)

Murata, K.; Ikariya, T. *J. Org. Chem.*, ***1999***, *64*, 2186.

10,11-dihydrocinchonidine , AcOH
1% Pd/Al_2O_3 , 17°C , 5.8 bar H_2

91% ee , *R*

LeBlond, C.; Wang, J.; Liu, J.; Andrews, A.T.; Sun, Y.-K.
J. Am. Chem. Soc., ***1999***, *121*, 4920.

1%

BH_3•SMe_2 , THF , rt

>99% (96% ee)

Puijaner, C.; Vidal-Ferran, A.; Moyano, A.; Pericàs, M.A.; Riera, A.
J. Org. Chem., ***1999***, *64*, 7902.

bakers yeast, petro 40-60; 4°C, 7 d

59%

Florey, P.; Smallridge, A.J.; Ten, A.; Trewhella, M.A. *Org. Lett.*, ***1999***, *1*, 1879.

cinchona modified Pt catalyst; Al_2O_3, H_2, 60 bar, 25°C; AcOH, MeOH

96% ee

Studer, M.; Burkhardt, S.; Blaser, H.-U. *Chem. Commun.*, ***1999***, 1727.

$BH_3 \cdot SMe_2$, THF, rt; 10% dendritic chiral catalyst

80% (89% ee)

Bolm, C.; Derrien, N.; Seger, A. *Chem. Commun.*, ***1999***, 2087.

1. DIP-Cl, NEt_3, THF; 2. aq OH; aq H^+

DIP-Cl = B-chloro diisopinocampheylborane

87% (98% ee)

Wang, Z.; Zhao, C.; Pierce, M.E.; Fortunak, J.M. *Tetrahedron Asymm.*, ***1999***, *10*, 225.

$BH_3 \cdot SMe_2$, toluene, 25°C

92% (73% ee, *S*)

Yang, T.-K.; Lee, D.-S. *Tetrahedron Asymm.*, ***1999***, *10*, 405.

H_2, 5% $Ir(cod)_2 BF_4$; chiral diamine

quant (72% ee)

Tommasino. M.L.; Thomazeau, C.; Touchard, F.; Lemaire, M. *Tetrahedron Asymm.*, ***1999***, *10*, 1813.

10% chiral pyrrolidine alcohol; 12% $Al(OEt)_3$, BH_3–SMe_2

quant (97.5% ee)

Yanagi, T.; Kikuchi, K.; Takeuchi, H.; Ishikawa, T.; Nishimura, T.; Kamijo, T. *Chem. Lett.*, ***1999***, 1203.

20%, BH_3–SMe_2

THF, 50°C

91% ee

Li, X.; Yeung, C.-h.; Chan, A.S.C.; Yang, T.-K. *Tetrahedron Asymm.*, ***1999***, *10*, 759.

chiral Ga complex

95% 991% ee)

Ford, A.; Woodward, S. *Angew. Chem., Int. Ed.*, ***1999***, *38*, 335.

1. 0.1

2. $PhNEt_2$•BH_3, THF, 25°C

93% (95% ee)

Cho, B.T.; Chun, Y.S. *J. Chem. Soc., Perkin Trans. 1*, ***1999***, 2095.

BH_3•OEt_2, LiH, THF

5%

quant (95% ee)

Ford, A.; Woodward, S. *Synth. Commun.*, ***1999***, *29*, 189.

, BH_3•THF

toluene, –20°C, 1 h

99% (95:5 *R*:*S*)

Calmes, M.; Escale, F. *Synth. Commun.*, ***1999***, *29*, 1341.

YKER-1 enzyme
NADPH

92% (>99% ee, *S*)

Kawai, Y.; Hida, K.; Tsujimoto, M.; Kondo, S.-i.; Kitano, K.; Nakamura, K.; Ohno, A. *Bull. Chem. Soc. Jpn.*, ***1999***, *72*, 99.

NO_2 O → (LiBH$_4$, THF, 20°C; chiral Lewis acid) → NO_2 OH

56% ee

Nozaki, K.; Kobori, K.; Uemura, T.; Tsutsumi, T.; Takaya, H.; Hiyama, T. *Bull. Chem. Soc. Jpn.*, ***1999***, *72*, 1109.

NON-ASYMMETRIC REDUCTION

Ph(C=O)Ph → ($NiCl_2$, $NaBH_4$; THF, rt, 15 min) → Ph(CHOH)Ph 88%

Khurana, J.M.; Chauhan, S. *Synth. Commun.*, ***2001***, *31*, 3485.

cyclohexanone → (aq $TiCl_3$, NH_3, MeOH; 0°C) → cyclohexanol 68%

Clerici, A.; Pastori, N.; Porta, O. *Eur. J. Org. Chem.*, ***2001***, 2235.

PhC(=O)CH$_3$ → ($B_{10}H_{14}$, $CeCl_3$•7 H_2O; MeOH, 50°C, 4 h) → PhCH(OH)CH$_3$ 97%

Bae, W.; Lee, S.H.; Jung, Y.J.; Yoon, C.-O.M.; Yoon, C.M. *Tetrahedron Lett.*, ***2001***, *42*, 2137.

t-Bu-cyclohexanone → ($PhSiH_3$, Mn(dpm)$_3$; DCM, *i*-PrOH, O_2, rt) → *t*-Bu-cyclohexanol

98% (14:1 *trans:cis*)

Magnus, P.; Fielding, M.R. *Tetrahedron Lett.*, ***2001***, *42*, 6633.

PhC(=O)CH$_3$ → (0.2% $RuCl_2(PPh_3)_3$, 2.4% NaOH, 0.4 Yb(OTf)$_3$, *i*-PrOH, reflux; 1 h) → PhCH(OH)CH$_3$ 74%

Matsunaga, H.; Yoshioka, N.; Kunieda, T. *Tetrahedron Lett.*, ***2001***, *42*, 8857.

Ph(C=O)Ph → (H_2, 10% Pd/C–ethylenediamine; MeOH, 3 h) → Ph(CHOH)Ph 96%

Hattori, K.; Sajiki, H.; Hirota, K. *Tetrahedron*, ***2001***, *57*, 4781.

LiAlH(O*t*-Bu)$_3$, THF

−78°C → 18°C

80%

Liu, L.T.; Huang, H.-L.; Wang, C.-L.J. *Tetrahedron Lett.*, ***2001***, *42*, 1329.

Merulius tremellosus

61%

Hage, A.; Petra, D.G.I.; Field, J.A.; Schipper, D.; Wignberg, J.B.P.A.; Kamer, P.C.J.; Reek, J.N.H.; van Leeuven, P.W.N.M.; Wever, R.; Schoemaker, H.E. *Tetrahedron Asymm.*, ***2001***, *12*, 1025.

BH_3•THF , THF , reflux

81% (>99.5% *trans*)

Cha, J.S.; Moon, S.J.; Park, J.H. *J. Org. Chem.*, ***2001***, *66*, 7514.

1. [InH_3{$P(C_6H_{11})_3$}] , toluene

−78°C → rt , 15 h

2. aq H^+

quant

Abernathy, C.D.; Cole, M.L.; Davies, A.J.; Jones, C. *Tetrahedron Lett.*, ***2000***, *41*, 7567.

Na_2CO_3 , 135°C , 12 h

5% $Pd(OAc)_2$, 2 eq KOAc

DMF

69%

Quan, L.G.; Lamrani, M.; Yamamoto, Y. *J. Am. Chem. Soc.*, ***2000***, *122*, 4827.

macrocyclic Ni catalyst

KOH , *i*-PrOH , 82°C

78%

Phukan, P.; Sudalai, A. *Synth. Commun.*, ***2000***, *30*, 2401.

TiI_4, 0°C → rt

MeCN

94%

Hayakawa, R.; Sahara, T.; Shimizu, M. *Tetrahedron Lett.*, ***2000***, *41*, 7939.

$Ti(Oi\text{-}Pr)_4$

K-Selectride, THF
-100°C, 15 min

(>100 : 1) 91%

Lawson, E.C.; Zhang, H.–C.; Maryanoff, B.E. *Tetrahedron Lett.*, ***1999***, *40*, 593.

1.5 eq Cl_3SiH, CH_2Cl_2, 12 h

N–CHO, 0°C → rt

90%

Iwasaki, F.; Onomura, O.; Mishima, K.; Maki, T.; Matsumura, Y. *Tetrahedron Lett.*, ***1999***, *40*, 7507.

3 eq SmI_2, 2 eq H_2O

0°C, 1 h

96% (83:17 *anti:syn*)

Keck, G.E.; Wager, C.A.; Sell, T.; Wager, T.T. *J. Org. Chem.*, ***1999***, *64*, 2172.

$K(Trip)_2BHBn$, THF

Trip = triisopropylphenyl

85% (>99% cis)

Smith, K.; El-Hiti, G.A.; Hou, D.; DeBoos, G.A. *J. Chem. Soc., Perkin Trans. 1*, ***1999***, 2807.

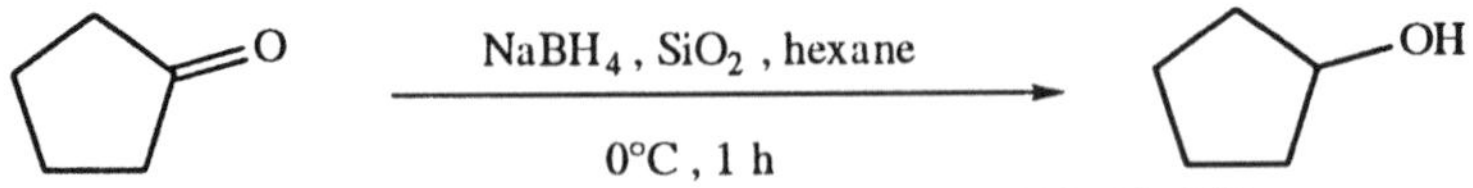
O
$NaBH_4$, SiO_2, hexane
0°C, 1 h
OH

95%

Yakabe, S.; Hirano, M.; Morimoto, T. *Synth. Commun.*, ***1999***, *29*, 295.

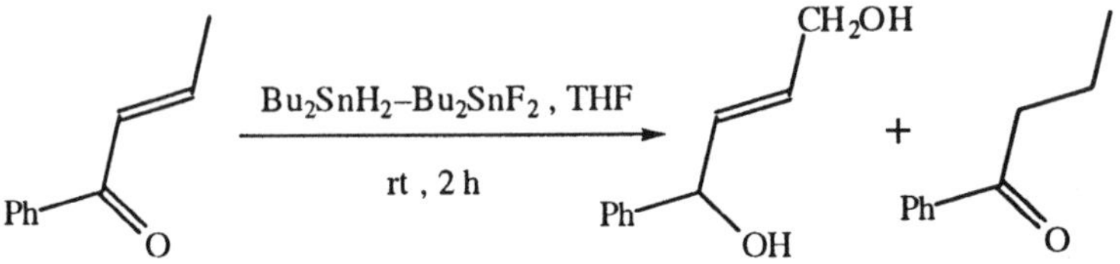
Ph
O
Bu_2SnH_2–Bu_2SnF_2, THF
rt, 2 h
CH_2OH
Ph
OH
+
Ph
O

+ HMPA	89%	0%
+ LiI	0%	83%

Moriuchi-Kawakami, T.; Matsuda, H.; Shibata, I.; Miyatake, M.; Suwa, T.; Baba, A. *Bull. Chem. Soc. Jpn.*, ***1999***, *72*, 465.

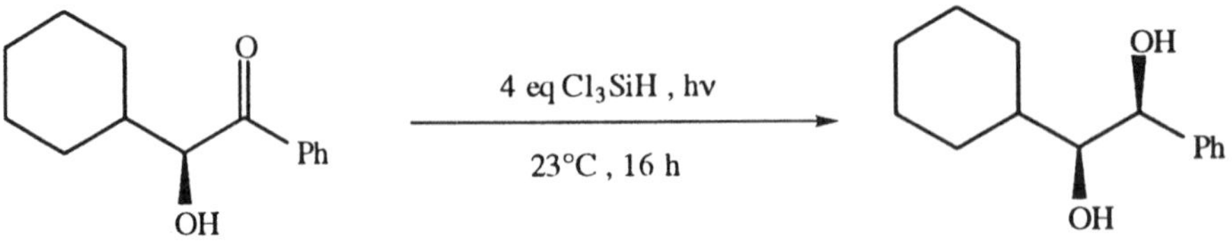
O
Ph
OH
4 eq Cl_3SiH, hν
23°C, 16 h
OH
Ph
OH

94% (63:1 syn:anti)

Enholm, E.J.; Schulte II, J.P. *J. Org. Chem.*, ***1999***, *64*, 2610.

REVIEWS:

"Enantioselective Reductions by Chirally Modified Alumino- and Borohydrides", Daverio, P.; Zanda, M. *Tetrahedron Asymm.* ***2001***, *12*, 2225.

SECTION 42B: ALKYLATION OF KETONES, FORMING ALCOHOLS

Aldol reactions are listed in Section 330 (Ketone-Alcohol)

ASYMMETRIC ALKYLATION

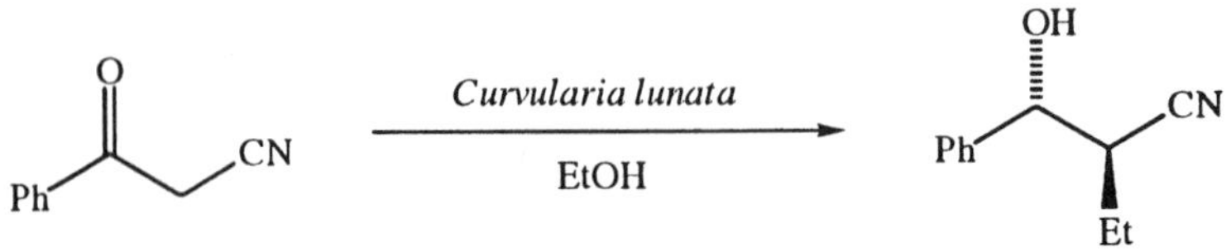
O
Ph
CN
Curvularia lunata
EtOH
OH
Ph
CN
Et

69% (98% ee)

Dehli, J.R.; Gotor, V. *Tetrahedron Asymm.* ***2001***, *12*, 1485.

$IrCl(cod)_2$–chiral amino-alcohol sulfoxide

$HCOOH/NEt_3$, 60°C , H_2

99% (65% ee)

Petra, D.G.I.; Kamer, P.C.J.; Spek, A.L.; Shoemaker, H.E.; van Leeuven, P.W.N.M. *J. Org. Chem.*, ***2000***, *65*, 3010.

0.25 $[RuCl_2(p\text{-cymene})]_2$, H_2 , 0.7 h

2% chiral amino-alcohol
2.5% *i*-PrOK , *i*-PrOH

97% (94% ee)

Alonso, D.A.; Nordin, S.J.M.; Roth, P.; Tarnai, T.; Andersson, P.G. *J. Org. Chem.*, ***2000***, *65*, 3116.

$(CH_2{=}CHCH_2)_4Sn$, DCM , rt

R-Binal-Ti

70% (55% ee)

Casolari, S.; D'Addario, D.; Tagliavini, E. *Org. Lett.*, ***1999***, *1*, 1061.

NON-ASYMMETRIC ALKYLATION

TMSCl , THF , In

\+ 0.1 CuI (– 60%
NO TMSCl , NO CuI (63% –))

Lee, P.H.; Ahn, H.; Lee, K.; Sung, S.-y.; Kim, S. *Tetrahedron Lett.*, ***2001***, *42*, 37.

Br , Sm , THF

rt , 20 min

80%

Basu, M.K.; Banik, B.K. *Tetrahedron Lett.*, ***2001***, *42*, 187.

1. $[LiC{\equiv}CCH_2Li]$, ether/TMEDA
2. $(CH_2O)_n$
3. aq NH_4Cl

96%

Cabezas, J.A.; Pereira, A.R.; Amey, A. *Tetrahedron Lett.*, ***2001***, *42*, 6819.

O

Ph

$Sn(CH_2CH=CH_2)_4$, DCE

10% $Cu(OTf)_2$, 10 h

HO

Ph

72%

Kamble, R.M.; Singh, V.K. *Tetrahedron Lett.*, ***2001***, *42*, 7525.

O

Ph

MgCl , THF , rt

$Cl_2Ti(OPh)_2$

Ph

OH

Me

63%

Quan, L.G.; Cha, J.K. *Tetrahedron Lett.*, ***2001***, *42*, 8567.

O

N—Bn

SO_2Ph

SmI_2 , acetone , THF , rt

O

N—Bn

OH

85%

Yoda, H.; Ujihara, Y.; Takabe, K. *Tetrahedron Lett.*, ***2001***, *42*, 9225.

O

Ph

1. HC≡CCH_2Br , NaI , Dy
THF , $HgCl_2$, −14°C

2. H_2O

OH

Ph

91%

Li, Z.; Jia, Y.; Zhou, J. *Synth. Commun.*, ***2000***, *30*, 2515.

OH

Ph

O_2 , pivaldehyde , MeCN , rt
Co salen complex , 8 h

O

Ph

75%

Fernández, I.; Pedro, J.R.; Roselló, A.L.; Ruiz, R.; Castro, I.; Ottenwaelder, X.; Jouyrnaux, Y. *Eur. J. Org. Chem.*, ***2001***, 1235.

O O

Me_2N

1. $TiCl_4$
2. MeMgCl–$CeCl_3$, ether

3. H_3O^+

O OH

Me_2N

85% (>98% de)

Bartoli, G.; Bosco, M.; Marcantoni, E.; Massaccesi, M.; Rinaldi, S.; Sambri, L. *Tetrahedron Lett.*, ***2001***, *42*, 6093.

SmI_2 , cat NiI_2 , THF

O OH I O O 94%

Molander, G.A.; Le Huérou, Y.; Brown, G.A. *J. Org. Chem.*, ***2001***, *66*, 4511.

1. Me CO_2Me (O) $TiCl_4$, DCM −78°C , 15 min 2. −78°C , 15 min $SnBr_3$

OMe H OH CO_2Me OMe + OH CO_2Me O 63% 20%

Ghosh, A.K.; Kawahama, R.; Wink, D. *Tetrahedron Lett.*, ***2000***, *41*, 8425.

Ph I 1. $TmI_2(DME)_x$, DME , <1 min 2. 4-*t*-butylcyclohexanone 3. aq NH_4Cl

Ph HO *t*-Bu 99% (79:21 *ax:eq*)

Wvans, W.J.; Allen, N.T. *J. Am. Chem. Soc.*, ***2000***, *122*, 2118.

O e⁻ (Hg cathode) , 0.1 MTBA•BF_4 DMF-pyridine , DMQ•2 BF_4

HO 96%

DDQ•2 BF_4 = N,N'-dimethylquinium tetrafluoroborate

Yadav, A.K.; Singh, A. *Synlett*, ***2000***, 1199.

Ph O $Sn(CH_2CH=CH_2)_4$, DCM 10% $Zn(OTf)_2$, 10% $PhNMe_2$ rt , 1d

Ph OH quant

Hamasaki, R.; Chounan, Y.; Horino, H.; Yamamoto, Y. *Tetrahedron Lett.*, ***2000***, *41*, 9883.

O BuTe Ph BuLi , THF , -78°C

Ph OH 64%

Dabdoub, M.J.; Jacob, R.G.; Ferreira, J.T.B.; Dabdoub, V.B.; Marques, F. de.A. *Tetrahedron Lett.*, ***1999***, *40*, 7159.

1. allyl iodide, 0.2 SmI_2, THF
mischmetall, 20°C
2. H_3O^+

C_5H_{11} → HO, C_5H_{11} 90%

mischmetall = alloy of [La/Ce/Nd/Pr/Sm]

Hélion, F.; Namy, J.-L. *J. Org. Chem.*, ***1999***, *64*, 2944.

Br
In, H_2O, rt, overnight
Ph → OH, Ph 90%

Chan, T.H.; Yang, Y. *J. Am. Chem. Soc.*, ***1999***, *121*, 3228.

REVIEWS

"Arylation with Organolead and Organobismuth Reagents,' Elliott, G.I.; Konopelski, J.P. *Tetrahedron*, ***2001***, *57*, 5683.

"Catalytic Asymmetric Organic Zinc Addition to Carbonyl Compounds,' Pu, L.; Yu, H.-B. *Chem. Rev.*, ***2001***, *101*, 757.

SECTION 43: ALCOHOLS AND THIOLS FROM NITRILES

NO ADDITIONAL EXAMPLES

SECTION 44: ALCOHOLS AND THIOLS FROM ALKENES

1. $BH_3 \cdot SMe_2$, THF, 0°C → rt
2. NEt_3, catechol, 0°C, 0°C → rt
3. O_2, rt, 12 h

Ph → Ph, OH + OH, Ph

(9 : 1) 69%

Cadot, C.; Dalko, P.I.; Cossy, J. *Tetrahedron Lett.*, ***2001***, *42*, 1661.

1. H, B, MeN, NMe, toluene, 18 h
1.6% $Cp_2 \cdot Sm{-}THF$
2. NaOH, H_2O_2

→ OH 86%

Molander, G.A.; Pfeiffer, D. *Org. Lett.*, ***2001***, *3*, 361.

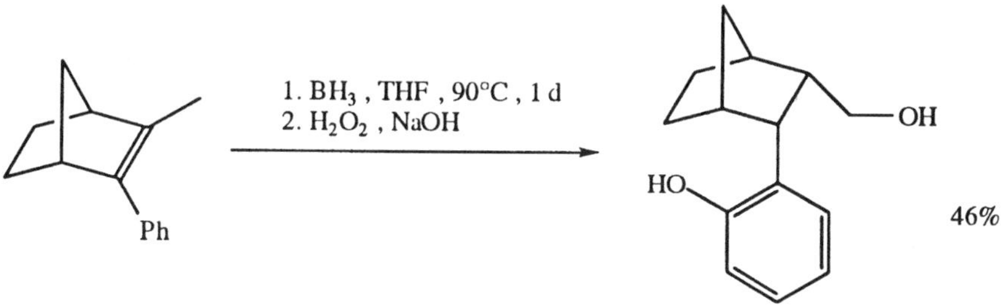

1. BH_3, THF, 90°C, 1 d
2. H_2O_2, NaOH

Ph

HO OH

46%

Varela, J.J.; Peña, D.; Goldfuss, B.; Polborn, K.; Knochel, P. *Org. Lett.*, ***2001***, *3*, 2395.

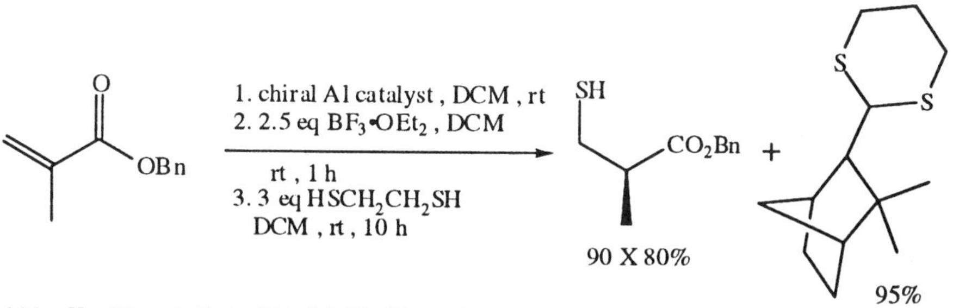

O

OBn

1. chiral Al catalyst, DCM, rt
2. 2.5 eq BF_3•OEt_2, DCM
rt, 1 h
3. 3 eq $HSCH_2CH_2SH$
DCM, rt, 10 h

SH

CO_2Bn +

S S

90 X 80%

95%

Nishide, K.; Ohsugi, S.-i.; Shiraki, H.; Tamakita, H.; Node, M. *Org. Lett.*, ***2001***, *3*, 3121.

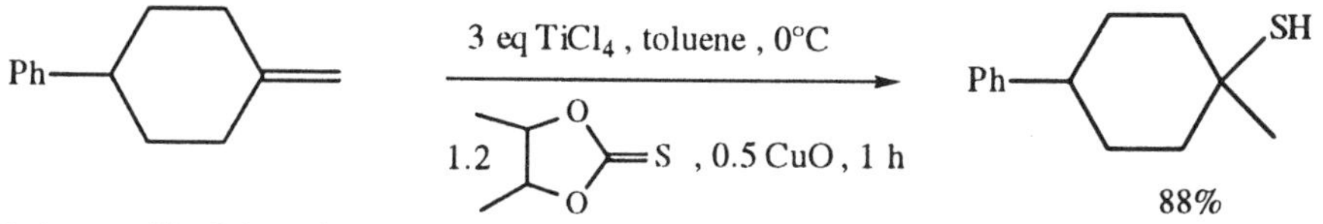

Ph

3 eq $TiCl_4$, toluene, 0°C

1.2 O O =S, 0.5 CuO, 1 h

Ph SH

88%

Mukaiyama, T.; Saitoh, T.; Jona, H. *Chem. Lett.*, ***2001***, 638.

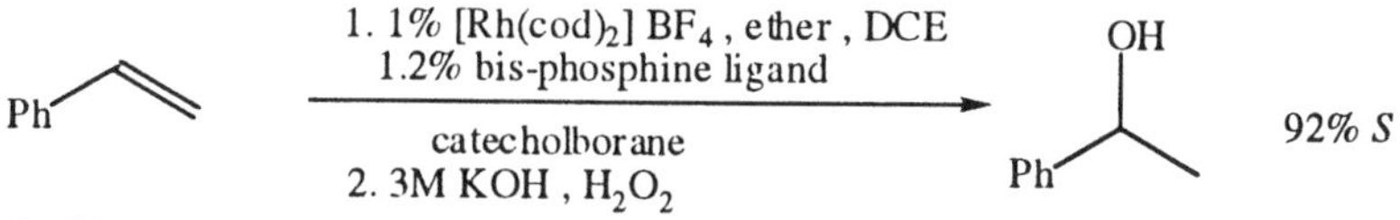

Ph

1. 1% $[Rh(cod)_2]$ BF_4, ether, DCE
1.2% bis-phosphine ligand
catecholborane
2. 3M KOH, H_2O_2

OH

Ph

92% *S*

Demay, S.; Volant, F.; Knochel, P. *Angew. Chem. Int. Ed.*, ***2001***, *40*, 1235.

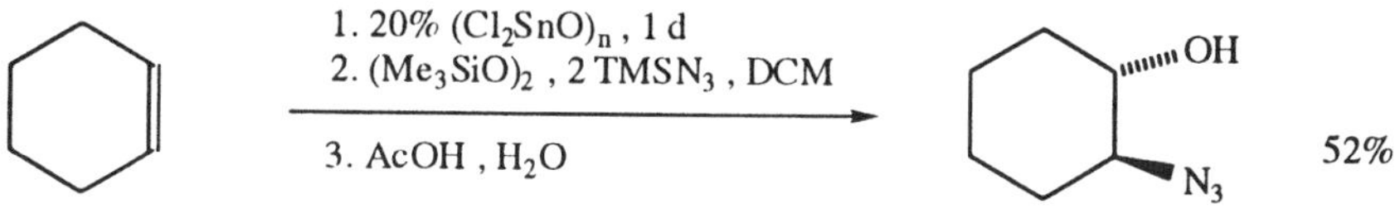

1. 20% $(Cl_2SnO)_n$, 1 d
2. $(Me_3SiO)_2$, 2 $TMSN_3$, DCM
3. AcOH, H_2O

OH

N_3

52%

Sakurada, I.; Yamasaki, S.; Kanai, M.; Shibasaki, M. *Tetrahedron Lett.*, ***2000***, *41*, 2415.

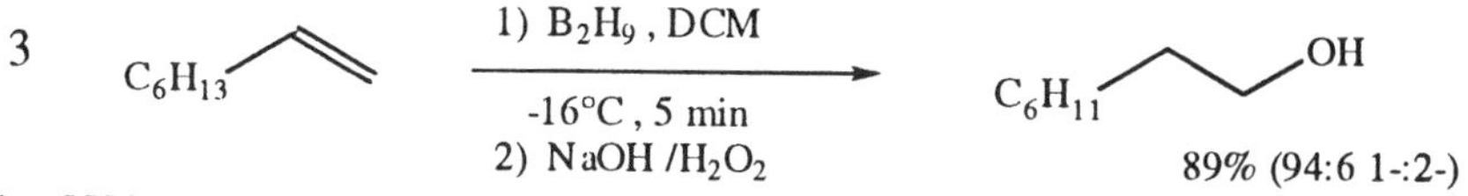

3 C_6H_{13}

1) B_2H_9, DCM
-16°C, 5 min
2) NaOH / H_2O_2

C_6H_{11} OH

89% (94:6 1-:2-)

Kantha, J.V.B.; Brown, H.C. *Tetrahedron Lett.*, ***2000***, *41*, 9361.

1. catecholborane (O,O-BH), 100°C
2. Me_3SiCHN_2, THF, reflux
3. H_2O_2, NaOH
4. Bu_4NF, THF

Ph–CH=CH$_2$ → Ph–CH$_2$CH$_2$CH$_2$–OH

60% overall

Goddard, J.-P.; LeGall, T.; Mioskowski, C. *Org. Lett.*, ***2000***, *2*, 1455.

1. 5% [1,10-phenanthrolinePdMeCl] $HSiMe_2OSiMe_3$, DCM, –20°C
2. KF, $MeCO_3H$, rt, 2 d

MeO_2C / MeO_2C → MeO_2C / MeO_2C ... OH

75%

Pei, T.; Widenhoefer, R.A. *Org. Lett.*, ***2000***, *1*, 1469.

cat $[Cp(TMS)_2Y(\mu\text{-Me})]_2$
$PhSiH_3$, cyclohexane
rt, 6 h

N → N ... OH

90% (98:2 ds)

Molander, G.A.; Schmitt, M.H. *J. Org. Chem.*, ***2000***, *65*, 3767.

1. THF, $NaBH_4$
2. BF_3•OEt_2, THF, 0°C
3. H_2O, rt, 2 h
4. Oxone, H_2O, 0°C → rt

OH

50%

Ripin, D.H.B.; Cai, W.; Brenek, S.J. *Tetrahedron Lett.*, ***2000***, *41*, 5817.

1. $HSiMe_2OTBDPS$, DCM –20°C, 12 h
2. TBAF, KF, $KHCO_3$, THF 50% H_2O_2, rt, 3 d

MeO_2C / MeO_2C → MeO_2C / MeO_2C ... OH, Me

99 x 48% (>50 : de) 90% ee

Pei, T.; Widenhoefer, R.A. *Tetrahedron Lett.*, ***2000***, *41*, 7597.

1. allyl-BBn_2, hexanes 0°C, 2 h
2. H_2O_2, NaOH

OH

93%

Frantz, D.E.; Singleton, D.A. *Org. Lett.*, ***1999***, *1*, 485.

SECTION 45: ALCOHOLS AND THIOLS FROM MISCELLANEOUS COMPOUNDS

4-$CH_3C_6H_4B(OH)_2$ —[30% H_2O_2, H_2O]→ 4-$CH_3C_6H_4OH$ 85%

Simon, J.; Salzbrunn, S.; Prakash, G.K.S.; Petasis, N.A.; Olah, G.A. *J. Org. Chem.*, ***2001***, *66*, 633.

BnS–SBn —[In, NH_4Cl, EtOH, reflux, 2 h]→ BnSH 97%

Reddy, G.V.S.; Rao, G.V.; Iyengar, D.S. *Synth. Commun.*, ***2000***, *30*, 859.

PhSSPh —[$ZrCl_4$, $NaBH_4$, THF]→ PhSH 95%

Chary, K.P.; Rajaram, S.; Iyengar, D.S. *Synth. Commun.*, ***2000***, *30*, 3905.

Ph, OMe, OMe —[Bakers yeast, sucrose/H_2O; EtOH, 30°C, 2 d]→ Ph, OH 82% (>98% ee)

Ferraboschi, P.; Reza-Elahi, S.; Verza, E.; Santaniello, E. *Tetrahedron Asymm.*, ***1999***, *10*, 2639.

SECTION 45A: PROTECTION OF ALCOHOLS AND THIOLS

HO —[Me$_2$Si(Cl), OAc / $LiNMe_2$; THF]→ O, $SiMe_2$, HO 80%

83% .01M MeCN, hv (254 nm), 30 min

Pirrung, M.C.; Fallon, L.; Zhu, J.; Lee, Y.R. *J. Am. Chem. Soc.*, ***2001***, *123*, 3638.

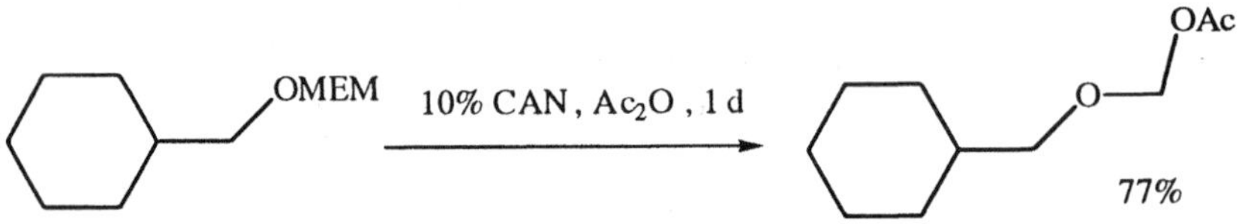

OMEM → (10% CAN, Ac_2O, 1 d) → OAc 77%

Tanemura, K.; Suzuki, T.; Nishida, Y.; Satsumabayashi, K.; Horaguchi, T. *Chem. Lett.*, ***2001***, 1012.

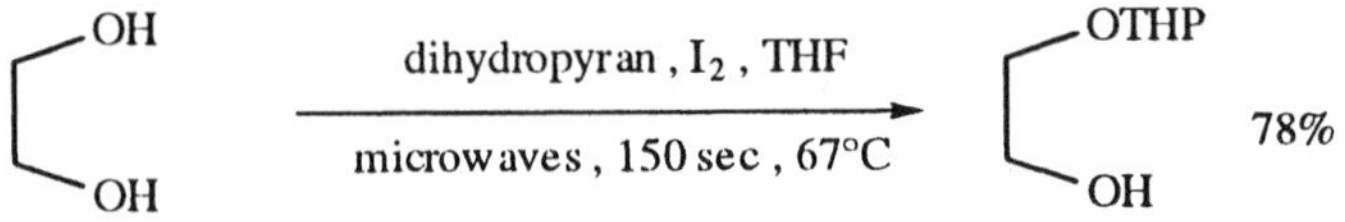

OH OH → (dihydropyran, I_2, THF; microwaves, 150 sec, 67°C) → OTHP OH 78%

Deka, N.; Sarma, J.C. *J. Org. Chem.*, ***2001***, *61*, 1947.

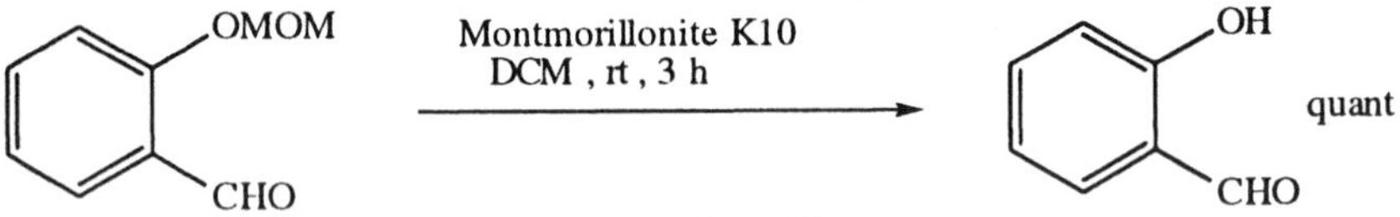

OMOM CHO → (Montmorillonite K10, DCM, rt, 3 h) → OH CHO quant

Deville, J.P.; Behar, V. *J. Org. Chem.*, ***2001***, *66*, 4097.

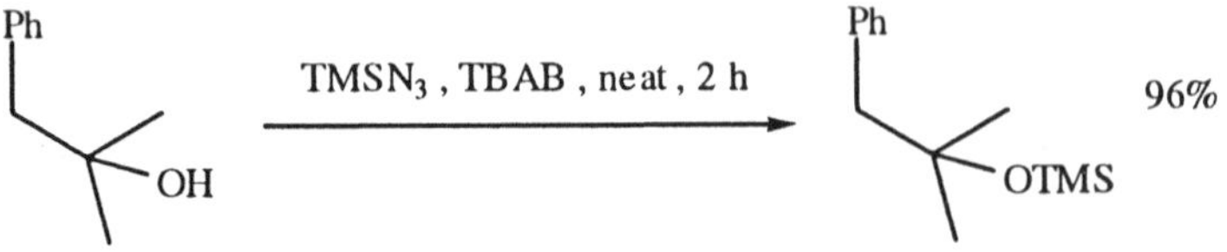

Ph OH → ($TMSN_3$, TBAB, neat, 2 h) → Ph OTMS 96%

Amantini, D.; Fringuelli, F.; Pizzo, F.; Vaccaro, L. *J. Org. Chem.*, ***2001***, *66*, 6734.

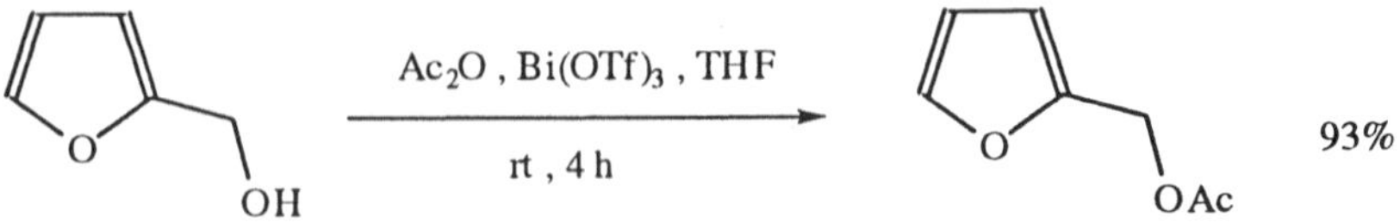

O OH → (Ac_2O, $Bi(OTf)_3$, THF; rt, 4 h) → O OAc 93%

Orita, A.; Tanahashi, C.; Kakuda, A.; Otera, J. *J. Org. Chem.*, ***2001***, *66*, 8926.

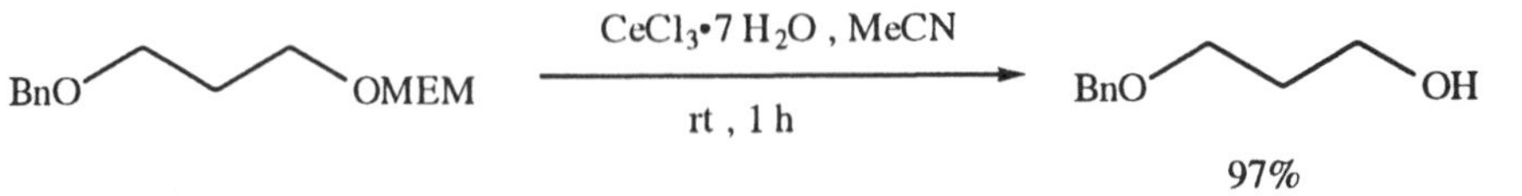

BnO OMEM → ($CeCl_3$•7 H_2O, MeCN; rt, 1 h) → BnO OH 97%

Sabitha, G.; Babu, R.S.; Rajkumar, M.; Srividya, R.; Yadav, J.S. *Org. Lett.*, ***2001***, *3*, 1149.

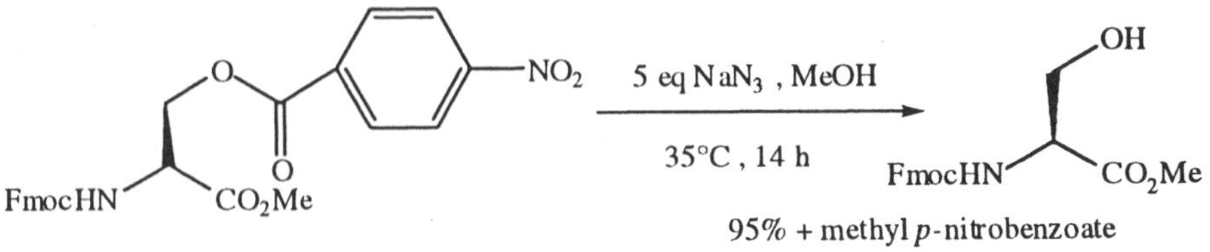

FmocHN CO_2Me O O NO_2 → (5 eq NaN_3, MeOH; 35°C, 14 h) → OH FmocHN CO_2Me 95% + methyl *p*-nitrobenzoate

Gómez-Vidal, J.A.; Forrester, M.T.; Silverman, R.B. *Org. Lett.*, ***2001***, *3*, 2477.

OH → ODPM 87%

10% $Yb(OTf)_3$ or $FeCl_3$

Ph_2CHOH , CH_2Cl_2 , 30min

Ph Ph

Sharma, G.V.M.; Prasad, T.R.; Mahalingam, A.K. *Tetrahedron Lett.*, ***2001***, *42*, 759.

Ph—OAc → Ph—OH 96%

kaolinilic clay , MeOH

25°C , 20 min

Bandgar, B.P.; Uppalla, L.S.; Sagar, A.D.; Sadavarte, V.S. *Tetrahedron Lett.*, ***2001***, *42*, 1163.

e^-

O O Ph → O OH + OH OH

62% 5%

Solis-oba, A.; Hudlicky, T.; Koroniak, L.; Frey, D. *Tetrahedron Lett.*, ***2001***, *42*, 1241.

OTs OTBDPS → OTs OH 94%

Br_2 , MeOH , reflux , 1 h

Barros, M.T.; Maycock, C.D.; Thomassigny, C. *Synlett*, ***2001***, 1146.

OH OH → OH OTHP 82%

DHP , silica chloride , rt

30 min

Ravindranath, N.; Ramesh, C.; Das, B. *Synlett*, ***2001***, 1777.

Br—C6H4—OTBDMS → Br—C6H4—OH 92%

KF/Al_2O_3 , DME

4 h

Blass, B.E.; Harris, C.L.; Portlock, D.E. *Tetrahedron Lett.*, ***2001***, *42*, 1611.

$PhCH_2OH$ → $PhCH_2OTHP$ 97%

DHP , $K_5CoW_{12}O_{40}$•3 H_2O , 5 min

quant

MeOH , $K_5CoW_{12}O_{40}$•3 H_2O , 1 h , rt

Habibi, M.H.; Tangestaninejad, S.; Mohammadpoor-Baltork, I.; Mirkhani, V.; Yadollahi, B. *Tetrahedron Lett.*, ***2001***, *42*, 2851.

21 eq $Py(HF)_x$, MeCN; 20 min , rt — 81%

Watanabe, Y.; Kiyosawa, Y.; Tatsukawa, A.; Hayashi, M. *Tetrahedron Lett.*, ***2001***, *42*, 4641.

Ac_2O , 5% $Cu(OTf)_2$, rt , 2.5 h — 75%

$ROSiR_3$ is converted to ROAc

Chandra, K.L.; Saravanan, P.; Singh, V.K. *Tetrahedron Lett.*, ***2001***, *42*, 5309.

NaH , THF , 0°C , 1 d — 65%

DDA , H_2O/CH_2Cl_2 , rt , 5 h — 74%

Sharma, G.V.M.; Rakesh *Tetrahedron Lett.*, ***2001***, *42*, 5571.

In , aq NH_4Cl , MeOH; 30 min — 98%

Valluri, M.; Mineno, T.; Hindupur, R.M.; Avery, M.A. *Tetrahedron Lett.*, ***2001***, *42*, 7153.

DHP , DCM , Bu_4NBr_3 , rt — 87%

Bu_4NBr_3 , MeOH , rt — 95%

Naik, S.; Gopinath, R.; Patel, B.K. *Tetrahedron Lett.*, ***2001***, *42*, 7679.

DCM , rt , 1 h — 88%

hv , aq EtOH — 91%

Loudwig, S.; Goeldner, M. *Tetrahedron Lett.*, ***2001***, *42*, 7957.

$EtO_2C-C_6H_4-OTHP \xrightarrow[EtOH]{H_2\text{ , Pd/C}} EtO_2C-C_6H_4-OH$ quant

Kaisalo, L.H.; Hase, T.A. *Tetrahedron Lett.*, ***2001***, *42*, 7699.

$PhCH_2OMOM \xrightarrow[2.5\text{ h}]{CBr_4\text{, }i\text{-PrOH, reflux}} PhCH_2OH$ 91%

Lee, A.S.-Y.; Hum, Y.-J.; Chu, S.-F. *Tetrahedron*, ***2001***, *57*, 2121.

$Ph(CH_2)_3OCH_2CH=CH_2 \xrightarrow[ZnCl_2/Pd(PPh_3)_4\text{ , THF , rt}]{\text{polymethylhydrosiloxane}} Ph(CH_2)_3OH$ 94%

Chandrasekhar, S.; Reddy, Ch.R.; Rao, R.J. *Tetrahedron*, ***2001***, *57*, 3435.

$PhCH_2OH \xrightarrow[\text{[bmim] }BF_4]{\text{DHP , TPP , HBr}} PhCH_2OTHP$ 96%

Branco, L.C.; Afonso, C.A.M. *Tetrahedron*, ***2001***, *57*, 4405.

$BuOCPh_3 \xrightarrow[\text{rt , 2 h}]{0.1\text{ }Ce(OTf)_4\text{ , wet MeCN}} BuOH$ 87%

Khalafi-Nezhad, A.; Alamdari, R.F. *Tetrahedron*, ***2001***, *57*, 6805.

$PhCH_2OH \xrightarrow[\text{4 min}]{\text{AcCl , neat , }KF-Al_2O_3} PhCH_2OAc$ 95%

Yadav, V.K.; Babu, K.G.; Mittal, M. *Tetrahedron*, ***2001***, *57*, 7047.

$PhCH_2OH \xrightarrow[\text{EtOAc , reflux , 3 h}]{5\%\text{ }K_5CoW_{12}O_{40}\bullet 3\text{ }H_2O} PhCH_2OAc$ 94%

Habibi, M.H.; Tangestaninejad, S.; Mirkhani, V.; Yadollahi, B. *Tetrahedron*, ***2001***, *57*, 8333.

$AcO-C_6H_4-OAc \xrightarrow[100°C\text{ , 3 h}]{1.6\text{ eq }K_2CO_3\text{ , NMP}} AcO-C_6H_4-OH$ 92%

Chakraborti, A.K.; Sharma, L.; Sharma, U. *Tetrahedron*, ***2001***, *57*, 9343.

$Ph(CH_2)_3OTBS \xrightarrow[\text{2. 10\% }Et_3N\text{ , }-78°C\text{ , 5 min}]{\text{1. THPOAc , 5\% TBS-OTf, EtCN , }-78°C\text{ , 15 min}} Ph(CH_2)_3OTHP$ 96%

Suzuki, T.; Oriyama, T. *Synthesis*, ***2001***, 555.

Ph–CH₂CH₂–OH → (1.5 eq Ac_2O, MeCN, 25 min; 0.1 $Bi(OTf)_3 \cdot x\,H_2O$) → Ph–CH₂CH₂–OAc 88%

Carrigan, M.D.; Freiberg, D.A.; Smith, R.C.; Zerth, H.M.; Mohan, R.S. *Synthesis*, ***2001***, 2091.

OBn, OBn, O, BnO, BnO, O (allyl glycoside) → (3 eq NMO, OsO_4, aq dioxane; 2 eq aq $NaIO_4$, 60°C, 18 h) → OBn, OBn, O, BnO, BnO, OH

60%

Kitov, P.I.; Bundle, D.R. *Org. Lett.*, ***2001***, *3*, 2835.

Ph—OH → (AcOH, P_2O_5, SiO_2, rt; solid state, 6 h) → Ph—OAc 60%

Eshghi, H.; Rafei, M.; Karimi, M.H. *Synth. Commun.*, ***2001***, *31*, 771.

$C_6H_{13}OH$ → ($(TMS)_2NH$, microwaves, $LiClO_4$) → $C_6H_{13}OTMS$ 83%

82% ← (microwaves, clay, H_2O)

Bandgar, B.P.; Kasture, S.P. *Monat. Chem.*, ***2001***, *132*, 1101.

BnO–(CH₂)₆–OTr → ($InBr_3$, aq MeCN; heat, 2.5 h) → BnO–(CH₂)₆–OH

90%

Yadav, J.S.; Reddy, B.V.S.; Srinivas, R.; Maiti, A. *J. Chem. Res. (S)*, ***2001***, 528.

$PhCH_2OTMS$ → ($BiCl_3$, MeOH, rt, 2 min) → $PhCH_2OH$ 97%

Firouzabadi, H.; Mohammadpoor-Baltork, I.; Kolagar, S. *Synth. Commun.*, ***2001***, *31*, 905.

$PhCH_2OH$ → (Ac_2O, microwaves, 8 min) → $PhCH_2OAc$ 80%

Bandgar, B.P.; Kasture, S.P.; Kamble, V.T. *Synth. Commun.* ***2001***, *31*, 2255.

OH (tetralol) → (TBDMSCl, DCM, rt; chiral quinidine base) → OTBDMS 50% (39% ee)

Isobe, T.; Fukuda, K.; Araki, Y.; Ishikawa, T. *Chem. Commun.*, ***2001***, 243.

$PhCH_2OTHP$ —— $Sc(OTf)_3$, MeCN/MeOH, rt ——> $PhCH_2OH$ 84%

also with ROMOM

Oriyama, T.; Watahiki, T.; Kobayashi, Y.; Hirano, H.; Suzuki, T.
Synth. Commun., ***2001***, ***31***, 2305.

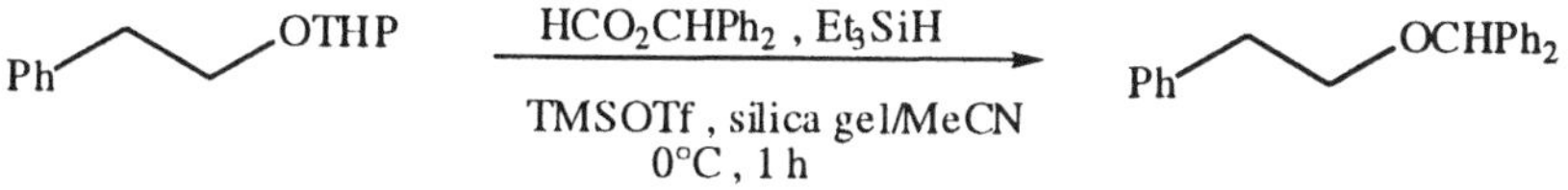

Ph⁀⁀OTHP —— HCO_2CHPh_2 , Et_3SiH / TMSOTf , silica gel/MeCN / 0°C , 1 h ——> Ph⁀⁀$OCHPh_2$ 90%

Suzuki, T.; Kobayashi, K.; Noda, K.; Oriyama, T. *Synth. Commun.* ***2001***, ***31***, 2761.

$PhCH_2OTHP$ —— In , I_2 , EtOAc , reflux , 15 h ——> $PhCH_2OAc$ 80%

also with RMOM

Ranu, B.C.; Hajara, A. *J. Chem. Soc., Perkin Trans. 1*, ***2001***, 355.
Ranu, B.C.; Hajra, A. *J. Chem. Soc., Perkin Trans. 1*, ***2001***, 2262.

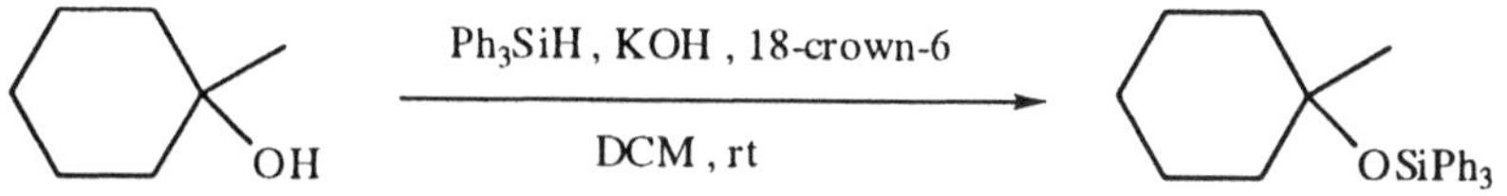

OH —— Ph_3SiH , KOH , 18-crown-6 / DCM , rt ——> $OSiPh_3$ 87%

Le Bideau, F.; Coradin, T.; Hénique, J.; Samuel, E. *Chem. Commun.*, ***2001***, 1408.

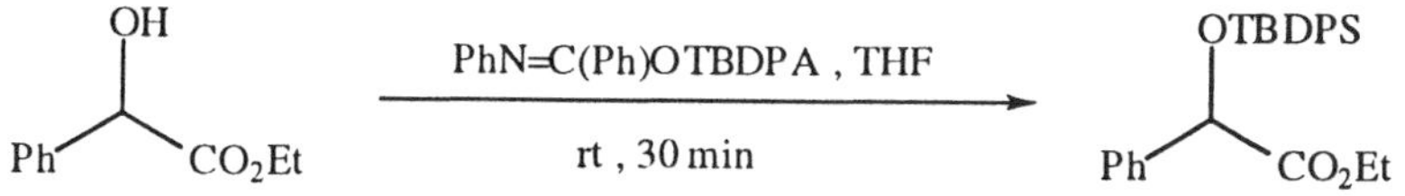

OH, Ph, CO_2Et —— PhN=C(Ph)OTBDPA , THF / rt , 30 min ——> OTBDPS, Ph, CO_2Et 99%

Misaki, T.; Kurihara, M.; Tanabe, Y. *Chem. Commun.*, ***2001***, 2478.

$PhCH_2OTHP$ —— NH_4^+ CrO_3Cl^- , Montmorillonite K10 / microwaves , 20 sec ——> PhCHO 94%

Heravi, M.M.; Hekmatshoar, R.; Beheshtiha, Y.S.; Ghassemzadeh, M.
Monat. Chem., ***2001***, ***132***, 651.

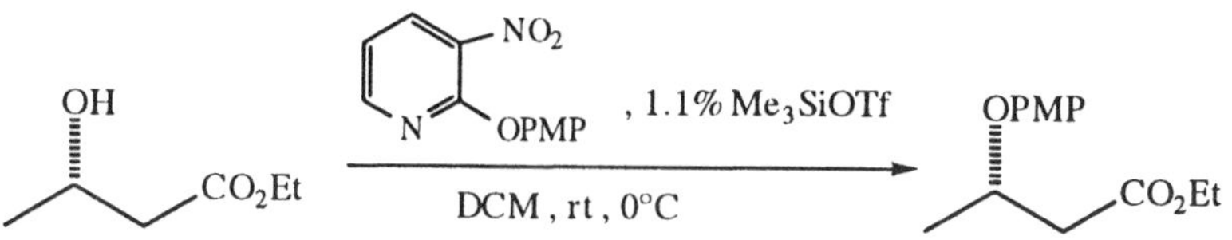

OH, CO_2Et —— NO_2, N, OPMP , 1.1% Me_3SiOTf / DCM , rt , 0°C ——> OPMP, CO_2Et 91%

Nakano, M.; Kikuchi, W.; Matsuo, J.-i.; Mukaiyama, T. *Chem. Lett.*, ***2001***, 424.

$PhCH_2OAc$ —— 10% P(2,4,6-trimethoxyphenyl)$_3$ / MeOH, 50°C , 3 h ——> $PhCH_2OH$ 90%

Yoshimoto, K.; Kawabata, H.; Nakamichi, N.; Hayashi, M. *Chem. Lett.*, ***2001***, 934.

Cl~~~OH → (1.2 eq BnOMs, 10% $LiB(C_6F_5)_4$, LiOTf, 1.6 eq MgO, rt, 1 d; cyclohexane-DCM) → Cl~~~OBn 97%

Nakano, M.; Matsuo, J.-i.; Mukaiyama, T. *Chem. Lett.*, ***2000***, 1352.

$PhCH_2OSiMe_3$ → (Oxone, MeCN, reflux, 5 h) → $PhCH_2OH$ 99%

Mohammadpoor-Baltork, I.; Amini, M.K.; Farshidipoor, S. *Bull. Chem. Soc. Jpn.*, ***2000***, *73*, 2775.

Ph~~C(Me)₂–OH → (1.2 eq [2-methylene-butyl]TBS; 2% Sc(OTF)₃, EtCN, rt, 30m) → Ph~~C(Me)₂–OTBS 88%

Suzuki, T.; Watahiki, T.; Oriyama, T. *Tetrahedron Lett.*, ***2000***, *41*, 8903.

Ph~~OH → (EDC, DMAP, DCM; BoBOH) → Ph~~O–C(=O)~~~OBn 96%

→ (H_2, 10% Pd / C, EtOAc) 90%

Clark, M.A.; Ganem, B. *Tetrahedron Lett.*, ***2000***, *41*, 9523.

HO~~~OH → (40% DHP, 5M aq $NaHSO_4$; toluene, 30°C, 50 min) → THPO~~~OH 93%

Nishiguchi, T.; Hayakawa, S.; Hirasaka, Y.; Saitoh, M. *Tetrahedron Lett.*, ***2000***, *41*, 9843.

Menthyl–OPMB → ($Mn(OAc)_3$-DDQ; CH_2Cl_2, rt, 8.5h) → Menthyl–OH 75%

Sharma, G.V.M.; Lavanya, B.; Mahalingam, A.K.; Krishna, P.R. *Tetrahedron Lett.*, ***2000***, *41*, 10323.

OH / OH → OBz / OH

1. 1% Me_2SnCl_2 , THF
2 eq K_2CO_3
2. PhCOCl

>99%

Iwasaki, F.; Maki, T.; Onomura, O.; Nakashima, W.; Matsumura, Y. *J. Org. Chem.*, ***2000***, *65*, 996.

Ph OTBDMS → Ph OH

DMSO, 80°C, 36 h

0.3 P—NMe, MeN MeN, N

90%

Yu, Z.; Verkade, J.G. *J. Org. Chem.*, ***2000***, *65*, 2065.

OH → OBn

BnOH, $BiBr_3$, CCl_4
rt, 96 h

77%

Boyer, B.; Keramane, E.-M.; Roque, J.-P.; Pavia, A.A. *Tetrahedron Lett.*, ***2000***, *41*, 2891.

Ph OH → Ph O O

$CrCl_2$, THF, CCl_4
rt

90%

Baati, R.; Valleix, A.; Mioskowski, C.; Barma, D.K.; Falck, J.R. *Org. Lett.*, ***2000***, *2*, 485.

Ph OH → Ph OAc

Ac_2O, Nafion-H, rt, 3 h

99%

Kumareswaran, R.; Pachamuthu, K.; Vankar, Y.D. *Synlett*, ***2000***, 1652.

C_5H_{11} O → C_5H_{11} OH

1. *t*-BuLi, pentane, –78°C → rt
2. H_3O^+

92%

Bailey, W.F.; England, M.D.; Mealy, M.J.; Thogsornkleeb, C.; Teng, L. *Org. Lett.*, ***2000***, *2*, 489.

$OSiPh_2SiMe_3$ / Br → OH / Br

1. Bu_3SnH, AIBN, PhH
0.05M
2. MeLi

55%

Studer, A.; Bossart, M.; Vasella, T. *Org. Lett.*, ***2000***, *2*, 985.

Ph~~~OBn → (3% $Sc(NTf_2)_3$, anisole; 100°C, 2.5 h) → Ph~~~OH 92%

Ishihara, K.; Hiraiwa, Y.; Yamamoto, H. *Synlett*, **2000**, 80.

Ph~~OBn → (MeOH, 30°C, 19 h; $[Cl(Bu)_2Sn(\mu\text{-}OH)]_2$ Cl(Bu)$_2$Sn(OH)$_2$Sn(Bu)$_2$Cl) → Ph~~OH 93%

Orita, A.; Sakamoto, K.; Hamada, Y.; Otera, J. *Synlett*, **2000**, 140.

PhO~allyl → (Et_2NH, $Pd(OAc)_2$, TPPTS; sec-(2,6-di-O-methyl)-β-cyclodextrin) → PhOH (+ allyldiethylamine) quant

Widehem, R.; Lacroix, T.; Bricout, H.; Monflier, E. *Synlett*, **2000**, 722.

Ph–C(Me)$_2$–OH → (3 eq Mg, 3 eq Me_3SiCl; DMF, rt) → Ph–C(Me)$_2$–$OSiMe_3$ 99%

Nishiguchi, I.; Kita, Y.; Watanabe, M.; Ishino, Y.; Ohno, T.; Maekawa, H. *Synlett*, **2000**, 1025.

PhC(O)CH$_2$N$_3$ → (10% oxazaborolidine (Ph, Ph), $BH_3{\bullet}SMe_2$; toluene–THF, 40°C) → PhCH(OH)CH$_2$N$_3$

92% (100% ee)

Yadav, J.S.; Reddy, P.T.; Hashim, S.R. *Synlett*, **2000**, 1049.

Ph~~~OTHP → (20% $TiCl_4$, DCM, Ac_2O; 0°C → rt, 6 h) → Ph~~~OAc 78%

Chandrasekhar, S.; Ramachandar, T.; Reddy, M.V.; Takhi, M. *J. Org. Chem.*, **2000**, *65*, 4729.

$PhCH_2OH$ → (dihydropyran, cat I_2, DCM, 30 min) → $PhCH_2$-OTHP 90%

83% ← (MeOH, I_2, heat, 3 h)

Kumar, H.M.S.; Reddy, B.V.S.; Reddy, E.J.; Yadav, J.S. *Chem. Lett.*, **2000**, 857.

Ph-CH$_2$CH$_2$-OPMB $\xrightarrow{\text{CBr}_4\text{, MeOH, reflux, 6 h}}$ Ph-CH$_2$CH$_2$-OH 87%

Yadav, J.S.; Reddy, B.V.S. *Chem. Lett.*, ***2000***, 566.

HO-CH(Ph)-CH(Ph)-OH $\xrightarrow[\text{0.5 DMAP, rt}]{\equiv\text{-CO}_2t\text{-Bu , MeCN}}$ 4,5-diphenyl-1,3-dioxolane-2-CH$_2$CO$_2t$-Bu 97%

(reverse: 5 eq pyrrolidine, THF, BuLi; rt, 16h)

Ariza, X.; Costa, A.M.; Faja, M.; Pineda, O.; Vilarrasa, J. *Org. Lett.*, ***2000***, *2*, 2809.

Ph-(CH$_2$)$_5$-OH $\xrightarrow[\text{DPQ, DCM, rt, 20 min}]{p\text{-Tol-CH}_2\text{OCPh}_3\text{, MS 4Å}}$ Ph-(CH$_2$)$_5$-OCPh$_3$ 96%

Sharma, G.V.M.; Mahalingam, A.K.; Prasad, T.R. *Synlett*, ***2000***, 1479.

TBSO-C$_6$H$_4$-CH$_2$OTBS $\xrightarrow[\text{20 min}]{\text{BiBr}_3\text{, MeCN, rt}}$ TBSO-C$_6$H$_4$-CH$_2$OH 85%

Bahwa, J.S.; Vivelo, J.; Slade, J.; Repič, O.; Blacklock, T. *Tetrahedron Lett.*, ***2000***, *41*, 6021.

C$_7$H$_{15}$CH$_2$CH$_2$OH $\xrightarrow[\text{THF, 60°C, add over 6 h}]{\text{3 eq BrCl}_3\text{, 2,4,6-collidine}}$ C$_7$H$_{15}$CH$_2$CH$_2$O-(2-tetrahydrofuranyl) 88%

Barks, J.M.; Gilbert, B.C.; Parsons, A.F.; Upeandran, B. *Tetrahedron Lett.*, ***2000***, *41*, 6249.

PhCH$_2$OH $\xrightarrow{\text{TIPSCl, imidazole, microwaves}}$ PhCH$_2$OTIPS 96%

Khalafi-Nezhad, A.; Alamdari, R.F.; Zekri, N. *Tetrahedron*, ***2000***, *56*, 7503.

C$_{10}$H$_{21}$OSEM $\xrightarrow{\text{10 eq MgBr}_2\text{/ether/MeNO}_2}$ C$_{10}$H$_{21}$OH 99%

Vakalopoulos, A.; Hoffmann, H.M.R. *Org. Lett.*, ***2000***, *2*, 1447.

Me$_3$CCH$_2$OH $\xrightarrow[\text{Ac}_2\text{O, DCM, rt, 15 h}]{\text{Zr - sulfophenyl phosphonate}}$ Me$_3$CCH$_2$OAc 74%

Curini, M.; Epifano, F.; Marcotullio, M.C.; Rosati, O.; Rossi, M. *Synth. Commun.*, ***2000***, *30*, 1319.

PMBO–(cyclohexane-1,3-dithiolane) —— $SnCl_4$, PhSH, DCM ——> HO–(cyclohexane-1,3-dithiolane) 92%

Yu, W.; Su, M.; Gao, X.; Yang, Z.; Jin, Z. *Tetrahedron Lett.*, ***2000***, *41*, 4015.

allyl-OH —— DHP, DCM, $ZrCl_4$ ——> allyl-OTHP 94%
(reverse: MeOH, $ZrCl_4$) 96%

Rezai, N.; Meybodi, F.A.; Salehi, P. *Synth. Commun.*, ***2000***, *30*, 1799.

$PhCH_2OSiMe_3$ —— air, Co or Mn salts ——> $PhCH_2OH$ 80%

Hashemi, M.M.; Kalantari, F. *Synth. Commun.*, ***2000***, *30*, 1857.

$PhCH_2OH$ —— AcOH, La Y-zeolite; 116°C, 8 h ——> $PhCH_2OAc$ 99%

Narender, N.; Srinivasu, P.; Kulkarni, S.J.; Raghavan, K.V. *Synth. Commun.*, ***2000***, *30*, 1887.

PhCH(OTHP)CH₃ —— $CuCl_2$•2 H_2O, MeOH; rt, 1 h ——> PhCH(OH)CH₃ 85%

Davis, K.J.; Bhalerao, U.T.; Rao, B.V. *Synth. Commun.*, ***2000***, *30*, 2301.

cyclohexanol —— 5% $MgBr_2$, Ac_2O, DCM, 3 h ——> cyclohexyl-OAc 65%

Pansare, S.V.; Malusare, M.G.; Rai, A.N. *Synth. Commun.*, ***2000***, *30*, 2587.

2-naphthyl-O-allyl —— $NaCNBH_3$, MeCN; TMSCl ——> 2-naphthol-OH 98%

Rao, G.V.; Reddy, D.S.; Mohan, G.H.; Iyengar, D.S. *Synth. Commun.*, ***2000***, *30*, 3565.

$PhCH_2OH$ —— DHP, LiBr, DCM ——> $PhCH_2OTHP$ 88%
(reverse: LiBr, MeOH, heat) 89%

Reddy, M.A.; Reddy, L.R.; Bhanumathi, N.; Rao, K.R. *Synth. Commun.*, ***2000***, *30*, 4323.

$PhCH_2OTHP$ —($CeCl_3$•7 H_2O, MeOH; 30 min)→ $PhCH_2OH$ 90%

Reddy, G.S.; Neelakantan, P.; Iyengar, D.S. *Synth. Commun.*, ***2000***, *30*, 4107.

4-(TBDMSO)C6H4CO2Me (OTBDMS, MeO_2C) —($Zr(KPO_4)_2$, aq acetone; 60°C, 7 h)→ 4-HOC6H4CO2Me (OH, MeO_2C) 79%

Curini, M.; Epifano, F.; Marcotullio, M.C.; Rosati, O.; Rossi, M.; Tsadjout, A. *Synth. Commun.*, ***2000***, *30*, 3181.

$PhCH_2OH$ —(DHP, cat I_2, microwaves, 7 min)→ $PhCH_2OTHP$ 91%

$PhCH_2OTHP$ —(MeOH, cat I_2, microwaves, 10 min)→ $PhCH_2OH$ 84%

Deka, N.; Sarma, J.C. *Synth. Commun.*, ***2000***, *30*, 4435.

4-methylphenyl allyl ether —(MeCN, NaI, Me_3SiCl; 2 h)→ 4-methylphenol (OH) 98%

Kamal, A.; Laxman, E.; Rao, N.V. *Tetrahedron Lett.*, ***1999***, *40*, 371.

t-$BuMe_3SiO$-C6H4-CH_2OSiMe_3t-Bu —(0.1 NaI, 0.1 M MeCN, 0.2 TMSCl/H_2O)→ t-$BuMe_3SiO$-C6H4-CH_2OH 86%

Grieco, P.A.; Markworth, C.J. *Tetrahedron Lett.*, ***1999***, *40*, 665.

$PhCH_2CH_2OTHP$ —(CBr_4, MeOH, reflux, 3 h)→ $PhCH_2CH_2OH$ 91%

Lee, A.S.-Y.; Su, F.-Y.; Liao, Y.-C. *Tetrahedron Lett.*, ***1999***, *40*, 1323.

1-benzyl-1-(OTBDMS)cyclohexane (Ph) —($Zn(BF_4)_2$, H_2O; 12 h)→ 1-benzylcyclohexanol (OH, Ph) 87%

Ranu, B.C.; Jana, U.; Majee, A. *Tetrahedron Lett.*, ***1999***, *40*, 1985.

PhCH(OH)CH$_3$ → (Ac_2O, CH_2Cl_2, 30 min; 2.5% $Cu(OTf)_2$) → PhCH(OAc)CH$_3$ 92%

Saravanan, P.; Singh, V.K. *Tetrahedron Lett.*, ***1999***, *40*, 2611.

$C_5H_{11}CH_2CH_2OTHP$ → (3% CAN, 70°C, MeCN; pH 8 borate buffer) → $C_5H_{11}CH_2CH_2OH$ 94%

Markó, I.E.; Ates, A.; Augustyns, B.; Gautier, A.; Quesnel, Y.; Turet, L.; Wiaux, M. *Tetrahedron Lett.*, ***1999***, *40*, 5613.

$PhOCH_2CH=CH_2$ → (e^-, $PdCl_2$–bipy, DMF, Bu_4NBF_4) → PhOH 75%

Franco, D.; Panyella, D.; Rocamora, M.; Gomez, M.; Clinet, J.C.; Muller, G.; Duñach, E. *Tetrahedron Lett.*, ***1999***, *40*, 5688.

2-Allyloxynaphthalene → ($CeCl_3{\bullet}7H_2O$, MeCN; NaI, reflux, 6 h) → 2-naphthol 90%

Thomas, R.M.; Reddy, G.S.; Iyengar, D.S. *Tetrahedron Lett.*, ***1999***, *40*, 7293.

$PhCH_2CH_2CH(OH)CH_3$ → (BzCl, TMEDA, MS 4Å; CH_2Cl_2, –78°C, 20 min) → $PhCH_2CH_2CH(OBz)CH_3$ 97%

Sano, T.; Ohashi, K.; Oriyama, T. *Synthesis*, ***1999***, 1141.

4-NC–C_6H_4–$OCH_2CH=CH_2$ → (electro-generated Ni, AcONa; DMF, 18 h, rt) → 4-NC–C_6H_4–OH 94%

Yasuhara, A.; Kasano, A.; Sakamoto, T. *J. Org. Chem.*, ***1999***, *64*, 4211.

$PhCH_2OH$ → (cat $B(C_6F_5)_3$, Ph_3SiH, toluene; rt, 20 h) → $PhCH_2OSiPh_3$ 93%

Blackwell, J.M.; Foster, K.L.; Beck, V.H.; Piers, W.E. *J. Org. Chem.*, ***1999***, *64*, 4887.

PhOAc → (PhSH, 5% K_2CO_3, NMP; reflux, 15 min) → PhOH 96%

Chakraborti, A.K.; Nayak, M.K.; Sharma, L. *J. Org. Chem.*, ***1999*** *64*, 8027.

OH; MeO-C6H4-CH2OH; 5% $Yb(OTf)_3$, DCM , rt , 4 h; Ph; OMPM; Ph; 80%

Sharma, G.V.M.; Mahalingam, A.K. *J. Org. Chem.*, ***1999***, *64*, 8943.

Ph–CH=CH–CH2OH; CH_2=CHOAc , THF , 17 h; $BnN=P(NMe_2)_3$; Ph–CH=CH–CH2OAc; 99%

Ilankumaran, P.; Verkade, J.G. *J. Org. Chem.*, ***1999***, *64*, 9063.

$BnO-(CH_2)_5-OTBDMS$ —Oxone , 58% aq MeOH; 3 h→ $BnO-(CH_2)_5-OH$ 90%

Sabitha, G.; Syamala, M.; Yadav, J.S. *Org. Lett.*, ***1999***, *1*, 1701.

$PhCH_2OTHP$ —K-10 Montmorillonite clay; MeOH , rt , 1 h→ $PhCH_2OH$ 91%

Tankguchi, T.; Kadota, K.; El Azab, A.S.; Ogasawara, K. *Synlett*, ***1999***, 1200.

Ph(CH2)5OH —AcOH , cat $FeCl_3$, 1 h→ Ph(CH2)5OAc 99%

Sharma, G.V.M.; Mahalingam, A.K.; Nagarajan, M.; Ilangovan, A.; Radhakrishna, P. *Synlett*, ***1999***, 1200.

Ph(CH2)2–O–CH2–C6H4–NO_2 —In , NH_4Cl , aq EtOH→ Ph(CH2)2OH 87%

Moody, C.J.; Pitts, M.R. *Synlett*, ***1999***, 1575.

$PhCH_2OH$ —0.1% $In(OTf)_3$, MeCN , rt; Ac_2O→ $PhCH_2OAc$ 97%

Chauhan, K.K.; Frost, C.G.; Love, I.; Waite, D. *Synlett*, ***1999***, 1743.

Ph(CH2)5OAc —*i*-PrOH , reflux , 12 h; 10% $Yb(OTf)_3$→ Ph(CH2)5OH 95%

Sharma, G.V.M.; Ilangovan, A. *Synlett*, ***1999***, 1963.

$C_8H_{17}CH_2CH_2OH$ (90%) — $TMS_3Si{-}Cl$, DMAP → ; ← hv, MeOH — $C_8H_{17}CH_2CH_2OSiTMS_3$ (85%)

Brook, M.A.; Balduzzi, S.; Mohamed, M.; Gottardo, C. *Tetrahedron,* ***1999****, 55,* 10027.

$C_8H_{17}{-}OTHP$ —Montmorillonite K10, MeOH; 50°C→ $C_8H_{17}{-}OH$

Li, T.-S.; Zhang, Z.-H.; Jin, T.-S. *Synth. Commun.* ***1999****, 29,* 181.

Cyclohexyl-CH₂OH —cat α-$Zr(O_3PMe)_{1.2}(O_3PC_6H_4SO_3H)_{0.8}$; DCM, TMS_2NH, rt→ Cyclohexyl-CH₂OTMS 95%

Curini, M.; Epifano, F.; Marcotullio, M.C.; Rosati, O.; Costantino, U. *Synth. Commun.,* ***1999****, 29,* 541.

$PhCH_2OH$ —DHP, H_2SO_4, SiO_2, 1 min→ $PhCH_2OTHP$ 92%

← H_2SO_4, SiO_2, MeOH, reflux, 15 min (90%)

Heravi, M.M.; Ajami, D.; Ghassemzadeh, M. *Synth. Commun.,* ***1999****, 29,* 1013.

4-Methylbenzyl alcohol (OH) —DHP, $SnCl_2{\bullet}2\,H_2O$→ 4-methylbenzyl-OTHP 95%

Davis, K.J.; Bhalerao, U.T.; Rao, B.V. *Synth. Commun.* ***1999****, 29,* 1679.

$PhCH_2OTMS$ —$TiCl_4$, EtOAc, 0.85 h→ $PhCH_2OAc$ 89%

Iranpoor, N.; Zeynizadeh, B. *Synth. Commun.,* ***1999****, 29,* 2123.

$Ph_2I^+\ BF_4^-$ —$PhBF_3K$, 5% $Pd(OAc)_2$, DME; 60°C, 1 h→ Ph—Ph 99%

Sabitha, G.; Reddy, B.V.S.; Srividya, R.; Yadav, J.S. *Synth. Commun.,* ***1999****, 29,* 2311.

$PhCH_2OTHP$ —Clayan, NH_4NO_3, 3 min; microwaves→ $PhCH_2OH$ 90%

Meshram, H.M.; Sumitra, G.; Raddy, G.S.; Ganesh, Y.S.S.; Yadav, J.S. *Synth. Commun.,* ***1999****, 29,* 2807.

$C_8H_{17}{-}OCPhAr_2$ —MeOH, DCM, $LiBF_4$; MeCN→ $C_8H_{17}{-}OH$ 94%

Ar = *p*-methoxyphenyl

Chen, A.; Zheng, Y.; Zhou, X. *Synth. Commun.,* ***1999****, 29,* 3421.

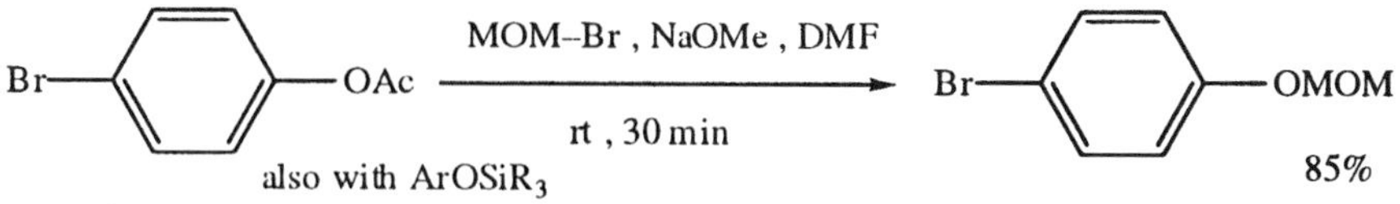

Br–C₆H₄–OAc → (MOM–Br, NaOMe, DMF; rt, 30 min) → Br–C₆H₄–OMOM 85%

also with $ArOSiR_3$

Oriyama, T.; Noda, K.; Sugawara, S. *Synth. Commun.*, ***1999***, *29*, 2217.

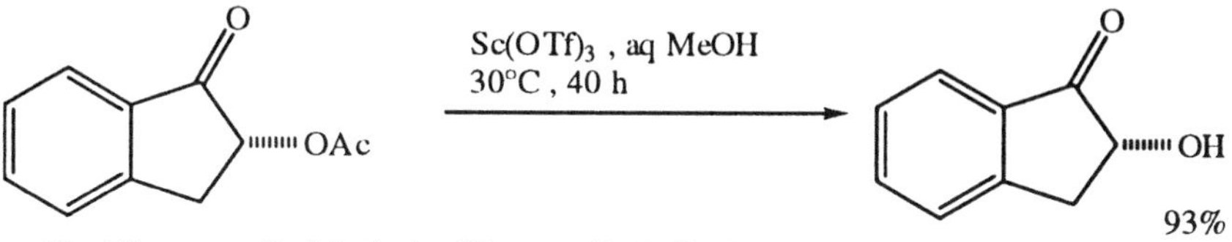

$Sc(OTf)_3$, aq MeOH
30°C, 40 h

OAc → OH 93%

Kajiro, H.; Mitamura, S.; Mori, A.; Hiyama, T. *Bull. Chem. Soc. Jpn.*, ***1999***, *72*, 1553.

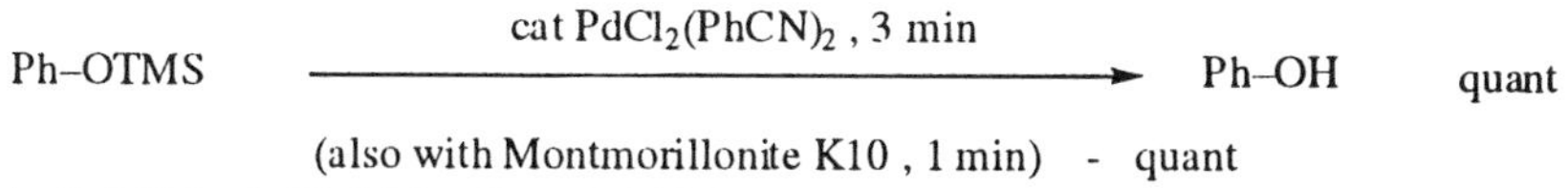

Ph–OTMS → (cat $PdCl_2(PhCN)_2$, 3 min) → Ph–OH quant

(also with Montmorillonite K10, 1 min) - quant

Mojtahedi, M.M.; Saidi, M.R.; Heravi, M.M.; Bolourtchian, M.
Monat. Chem., ***1999***, *130*, 1175.

CHAPTER 4
PREPARATION OF ALDEHYDES

SECTION 46: ALDEHYDES FROM ALKYNES

C_6H_{13}—≡ → (2% RuCpCl(dppm), H_2O, 12 h; *i*-PrOH, sealed tube, 100°C) → $C_6H_{13}CH_2CHO$ 93%

Suzuki, T.; Tokunaga, M.; Watatsuki, Y. *Org. Lett.*, ***2001***, *3*, 735.

Ph—≡ → (1. 5% [Ru(η^5-C_9H_7)(PPh_3)Cl], H_2O *i*-PrOH, 90°C, 2 d; 2. aq $C_{12}H_{25}SO_4^-Na^+$; 3. aq. $C_{16}H_{38}NMe_3^+Br^-$) → Ph⁀CHO 71% + PhC(O)CH₃ 17%

Alvarez, P.; Bassetti, M.; Gimeno, J.; Mancini, G. *Tetrahedron Lett.*, ***2001***, *42*, 8467.

SECTION 47: ALDEHYDES FROM ACID DERIVATIVES

PhC(O)Cl → (cat $InCl_3$, Bu_3SnH, toluene; 0.2 PPh_3, −30°C, 2 h) → PhC(O)H 97%

Inoue, K.; Yasuda, M.; Shibata, I.; Baba, A. *Tetrahedron Lett.*, ***2000***, *41*, 113.

BuC(O)Cl → (1. AlH_3; 2. PCC) → BuC(O)H 95%

Cha, J.S.; Kim, J.M.; Chun, J.H.; Kwon, O.O.; Kwon, S.Y.; Han, S.W. *Org. Prep. Proceed. Int*, ***1999***, *31*, 204.

$PhCO_2H$ → (1. $NaBH_4$; 2. Me_2SO_4; 3. PCC) → PhCHO 82%

Cha, J.S.; Lee, D.Y.; Kim, J.M. *Org. Prep. Proceed. Int.*, ***1999***, *31*, 694.

SECTION 48: ALDEHYDES FROM ALCOHOLS AND THIOLS

$PhCH_2OH$ —[Pd(II)-hydrotalcite, toluene, air / 0.2 Py, 65°C, 3 h]→ PhCHO 98%

Kakiuchi, N.; Maeda, Y.; Nishimura, T.; Uemura, S. *J. Org. Chem.*, **2001**, *66*, 6620.

$PhCH_2OH$ —[DMSO, THF, –30°C / 2,4,6-trichloro-1,3,5-triazine]→ PhCHO 91%

De Luca, L.; Giacomelli, G.; Porcheddu, A. *J. Org. Chem.* **2001**, *66*, 7907.

$PhCH_2OH$ —[polymer–Linker–TEMPO (N–O•)]→ PhCHO 75%

Fey, T.; Fischer, H.; Bachmann, S.; ALbert, K.; Bolm, C. *J. Org. Chem.* **2001**, *66*, 8154.

$C_6H_{13}CH_2CH_2CH_2OH$ —[1% $H_5[PMo_{10}V_2O_{40}]$, TEMPO / O_2 (2 atm), acetone, 100°C]→ $C_6H_{13}CH_2CH_2CHO$ 98%

Ben-Daniel, R.; Alsters, P.; Neumann, R *J. Org. Chem.*, **2001**, *66*, 8650.

1-heptanol —[trichloroisocyanuric acid / 0.1 TEMPO, DCM, rt]→ heptanal (CHO) 96%

De Luca, L.; Giacomelli, G.; Porchedda, A. *Org. Lett.*, **2001**, *3*, 3041.

$C_{10}H_{21}OH$ —[$[(ANS\text{-}O)_2Bi\text{—}O]_2$, CH_2Cl_2 / rt, 5 min]→ $C_9H_{19}CHO$ 96%

Matano, Y.; Nomura, H. *J. Am. Chem. Soc.*, **2001**, *123*, 6443.

$C_5H_{11}CH_2OH \xrightarrow[Bu_4NIO_4\text{, toluene, }60°C\text{ 1 d}]{\text{Schlenk tube, MS 4Å}} C_5H_{11}CHO$ quant

Friedrich, H.B.; Khan, F.; Singh, N.; van Staden, M. *Synlett*, ***2001***, 869.

$PhCH_2OH \xrightarrow[\text{MS 4Å, }CH_2Cl_2\text{, rt}]{\text{polymer—}C_6H_4\text{—}CH_2\text{–X, 20\% TPAP}} PhCHO$ 93%

X = NMO

Brown, D.S.; Kerr, W.J.; Lindsay, D.M.; Pike, K.G.; Ratcliffe, P.D. *Synlett*, ***2001***, 1257.

$PhCH_2OH \xrightarrow[\text{cyclohexane, 65°C}]{\text{phosphate buffered }SiO_2\text{–supported }KMnO_4} PhCHO$ 99%

Yakemoto, T.; Yasuda, K.; Ley, S.V. *Synlett*, ***2001***, 1555.

$PhCH_2OH \xrightarrow[\text{reflux, 1.5 h}]{PhCH_2PPh_3^+\ IO_4^-\text{, }AlCl_3/MeCN} PhCHO$ 78%

Hajipour, A.R.; Mallakpour, S.E.; Samimi, H.A. *Synlett*, ***2001***, 1735.

$C_6H_{13}CH_2OH \xrightarrow[\text{AcOH, cat }Co(NO_3)_2\text{, 40°C, 6 h}]{\text{TEMPO, }O_2\text{, cat }Mn(NO_3)_2} C_6H_{13}CHO$ 97%

Cecchetto, A.; Fontana, F.; Minisci, F.; Recupero, F. *Tetrahedron Lett.*, ***2001***, *42*, 6651.

$O_2N\text{–}C_6H_4\text{–}CH_2OH \xrightarrow{\text{2\% Ru complex, rt; hν, air, toluene, 1 d}} O_2N\text{–}C_6H_4\text{–}CHO$ quant

Miyata, A.; Murakami, M.; Irie, R.; Katsuki, T. *Tetrahedron Lett.*, ***2001***, *42*, 7067.

$PhCH_2OH \xrightarrow[\text{pH 4.5, 1 d}]{\text{TEMPO, laccase, }H_2O} PhCHO$ 92%

laccase = multi-copper oxidase

Fabbrini, M.; Galli, C.; Gentili, P.; Macchitella, D. *Tetrahedron Lett.*, ***2001***, *42*, 7551.

$PhCH_2OH \xrightarrow[\text{15 min}]{\text{PCC, no solvent, rt}} PhCHO$ 96%

Salehi, P.; Firouzabadi, H.; Farrokhi, A.; Gholizadeh, M. *Synthesis*, ***2001***, 2273.

$PhCH_2OH \xrightarrow[\text{reflux, 2.5 h}]{\text{4 eq }Ce(NO_3)_3BrO_3\text{, MeCN}} PhCHO$ 92%

Shirini, F.; Tajik, H.; Aliakbar, A.; Akbar, A. *Synth. Commun.*, ***2001***, *31*, 767.

$PhCH_2OH$ —(γ-picolinium chlorochromate; DCM, rt)→ PhCHO 75%

Khodaei, M.M.; Salehi, P.; Goudarzi, M. *Synth. Commun.*, ***2001***, *31*, 1253.

$PhCH_2OH$ —(SiO_2, Jones reagent, DCM)→ PhCHO 85%

Ali, M.H.; Wiggin, C.J. *Synth. Commun.*, ***2001***, *31*, 1389.

$PhCH_2OH$ —($KMnO_4$, $ZrOCl_2•H_2O$, ether; rt, 3.5 h)→ PhCHO 95%

Firouzabadi, H.; Fakoorpour, M.; Hazarkhani, H. *Synth. Commun.*, ***2001***, *31*, 3859.

Ph–CH2CH2–OH —(1.5 eq Ph(Cl)S=N*t*-Bu, DCM; 5 eq ZnO, 0°C, 30 min)→ Ph–CH2–CHO 91%

Matsuo, J.-i.; Kitagawa, H.; Iida, D.; Mukaiyama, T. *Chem. Lett.*, ***2001***, 150.

Ph–CH=CH–CH2–OH —(I_2, hv, *i*-Pr_2O, 12 h)→ Ph–CH=CH–CHO 95%

Itoh, A.; Kodama, T.; Masaki, Y. *Chem. Lett.*, ***2001***, 686.

C_5H_{11}–CH=CH–CH2–OH —(5% PhSNH*t*-Bu, NCS; 10 eq K_2CO_3, MS 4Å; DCM, 0°C, 1 h)→ C_5H_{11}–CH=CH–CHO >99%

Mukaiyama, T.; Matsuo, J.-i.; Iida, D.; Kitagwa, H. *Chem. Lett.*, ***2001***, 846.

$PhCH_2OH$ —(zeolite OMS-2, air)→ PhCHO 85%

Son, Y.-C.; Makwana, V.D.; Howell, A.R.; Suib, S.L. *Angew. Chem. Int. Ed.*, ***2001***, *40*, 4280.

$PhCH_2OH$ —($(NH_4)_2Cr_2O_7$, HIO_3, wet SiO_2; 30 min)→ PhCHO 87%

Shirini, F.; Zolfigol, M.A.; Azadbar, M.R. *Russ. J. Org. Chem.*, ***2001***, *37*, 1600.

$PhCH_2OH$ —(calcium hypochlorite, moist alumina; microwaves, 1 min)→ PhCHO 98%

Mojtahedi, M.M.; Saidi, M.R.; Bolourtchian, M.; Shirzi, J.S. *Monat. Chem.*, ***2001***, *132*, 655.

$PhCH_2OH$ —[$(NH_4)_2S_2O_8$, $AgNO_3$, hexane / Montmorillonite, 60°C, 25 h]→ PhCHO 99%

Hirano, M.; Kojima, K.; Yakabe, S.; Morimoto, T. *J. Chem. Res. (S)*, ***2001***, 274.

$PhCH_2OH$ —[$(NH_4)_2Cr_2O_7$, $ZrCl_4$, 0.05 h / wet SiO_2, solvent free]→ PhCHO 92%

Shirini, F.; ali Zolfigol, M.; Pourhabib, A. *J. Chem. Res. (S)*, ***2001***, 476.

$PhCH_2OH$ —[$Br^+(collidine)_2\ PF_6^-$, DCM, 2.5 h]→ PhCHO 75%

Rousseau, G.; Robin, S. *Tetrahedron Lett.*, ***2000***, ***41***, 8881.

$C_{11}H_{23}CH_2OH$ —[cat $Pd(OAc)_2$, toluene, 80°C, perfluorodecalin, MS 3Å, 10 h / O_2, fluoroalkyl acetal pyiridine ligand]→ $C_{11}H_{23}CHO$ 76%

Nishimura, T.; Maeda, Y.; Kakiuchi, N.; Uemura, S.
J. Chem. Soc., Perkin Trans. 1, ***2000***, 4301.

$PhCH_2OH$ —[5% $[RuCl_2(p\text{-cymene})_2$, O_2 / Cs_2CO_3, 100°C, toluene]→ Ph-CHO 91%

Lee, M.; Chang, S. *Tetrahedron Lett.*, ***2000***, ***41***, 7507.

$PhCH_2OH$ —[1.5 eq Ph(Cl)S=N*t*-Bu, 2 eq DBU / DCM, –78°C, 30 min]→ PhCHO 98%

Mukaiyama, T.; Matsuo, J.-i.; Yanagisawa, M. *Chem. Lett.*, ***2000***, 1072.

$C_7H_{15}CH_2OH$ —[cat Bu_4NBr, H_2O_2, DCM, 1 d / cat $[Ru_3O(OAc)_6(MeOH)_3]$ OAc]→ $C_7H_{15}CHO$ 67%

Wynne, J.H.; Lloyd, C.T.; Witsil, D.R.; Mushrush, G.W.; Stalick, W.M.
Org. Prep. Proceed. Int., ***2000***, ***32***, 588.

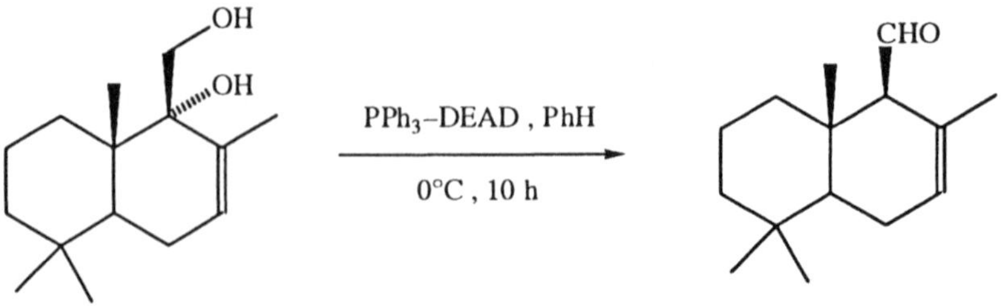

OH, OH —[PPh_3–DEAD, PhH / 0°C, 10 h]→ CHO 75%

Barrero, A.F.; Alvarez-Manzaneda, E.J.; Chahboun, R. *Tetrahedron Lett.*, ***2000***, ***41***, 1959.

PhCHO —[TEMPO, Oxone, rt / DCM, Bu_4NBr]→ $PhCH_2OH$ 90%

Bolm, C.; Magnus, A.S.; Hildebrand, J. *Org. Lett.*, ***2000***, *2*, 1173.

Ph–CH=CH–CH_2OH —[$NiBiO_3$, AcOH, 7 h / reflux, H_2O]→ Ph–CH=CH–CHO 46%

Banik, B.K.; Ghatak, A.; Ventraman, M.S.; Becker, I.F. *Synth. Commun.*, ***2000***, *30*, 2701.

$PhCH_2OH$ —[bis(1-benzyl-3,5,7-triaza-1-azoniatricyclo[3.3.1.$1^{3,7}$]decane peroxydisulfate, MeCN]→ PhCHO 98%

Minghu W.; Guichun, Z.; Zuxing, C. *Synth. Commun.*, ***2000***, *30*, 3127.

$PhCH_2OH$ —[$BnPPh_3^+ CrO_3{\bullet}HCl$, MeCN / reflux, 25 min]→ PhCHO 85%

Hajipour, A.R.; Mallakpour, S.E.; Backnejad, H. *Synth. Commun.*, ***2000***, *30*, 3855.

$PhCH_2OH$ —[CrO_3, 70% TBHP / microwaves, 10 min]→ PhCHO 87%

Singh, J.; Sharma, M.; Chhibber, M.; Kaur, J.; Kad, G.L. *Synth. Commun.*, ***2000***, *30*, 3941.

$PhCH_2OH$ —[poly[vinyl(pyridiniumfluorochromate)] / cyclohexane, rt, 15 min]→ PhCHO 95%

Srinivasan, R.; Balasubramanian, K. *Synth. Commun.*, ***2000***, *30*, 4397.

2-Bromobenzyl alcohol (OH, Br) —[2% perfluoroalkyl substituted bipyridine, $CuBr{\bullet}SMe_2$ / 3.5% TEMPO, 90°C, O_2, C_8F_{18}/PhCl]→ 2-bromobenzaldehyde (CHO, Br) 96%

Betzemeier, B.; Cavazzini, M.; Quici, S.; Knochel, P. *Tetrahedron Lett.*, ***2000***, *41*, 4343.

$PhCH_2OH$ —[HNO_3, 10% $Yb(OTf)_3$, DCM / reflux, 30 min]→ PhCHO 91%

Barrett, A.G.M.; Braddock, D.C.; McKinnell, R.M.; Waller, F.J. *Synlett*, ***1999***, 1489.

$C_3H_7CH_2OH$ —[$NaNO_2$–Ac_2O, 25°C / <1 min]→ C_3H_7CHO 60%

Bandgar, B.P.; Sadavarte, V.S.; Uppalla, L.S. *J. Chem. Soc., Perkin Trans. 1*, ***2000***, 3559.

$PhCH_2OH$ —[(N-benzyl-DABCO)$_2$ $Cr_2O_7^{-2}$; microwaves, 3 min]→ $PhCHO$ 78%

Hahipour, A.R.; Mallakpour, S.E.; Khoee, S. *Synlett*, ***2000***, 740.

$PhCH_2OH$ —[Th(III)-polypyridine complex]→ $PhCHO$

Kölle, U.; Fränzl, H. *Monat. Chem.*, ***2000***, ***131***, 1321.

Ph–CH=CH–CH$_2$OH —[Ru-Co-Al hydrotalcite, O_2; toluene, 60°C, 40 min]→ Ph–CH=CH–CHO 89%

Matsushita, T.; Ebitani, K.; Kaneda, K. *Chem. Commun.*, ***1999***, 265.

$PhCH_2OH$ —[zeolite HZSM-5-supported Fe(III) nitrate; microwaves, 10 sec]→ $PhCHO$ 99%

Heravi, M.M.; Ajami, D.; Aghapoor, K.; Ghassemzadeh, M. *Chem. Commun.*, ***1999***, 833.

$PhCH_2OH$ —[silica-supported TEMPO, DCM; aq KBr, aq NaOCl, $NaHCO_3$ (pH 9.1)]→ $PhCHO$ 75%

Bolm, C.; Fey, T. *Chem. Commun.*, ***1999***, 1795.

$PhCH_2OH$ —[$KMnO_4$/alumina, no solvent; 30 sec]→ $PhCHO$ 94%

Hajipour, A.R.; Mallakpour, S.E.; Imanzadeh, G. *Chem. Lett.*, ***1999***, 99.

3-furylmethanol (–OH) —[SiO_2, CH_2Cl_2; AcHN-(tetramethylpiperidine N=O)]→ 3-furaldehyde (CHO) 89%

Kernag, C.A.; Bobbitt, J.M.; McGrath, D.V. *Tetrahedron Lett.*, ***1999***, *40*, 1635.

$C_6H_{13}CH_2CH_2OH$ —[1% OsO_4, 1.5% CuCl, O_2 (1 atm); 5% Py, MS 4Å, toluene, 100°C]→ $C_6H_{13}CH_2CHO$ 32%

better yields with benzylic alcohols

Coleman, K.; Coppe, M.; Thomas, C.; Osborn, J.A. *Tetrahedron Lett.*, ***1999***, *40*, 3723.

$$PhCH_2OH \xrightarrow[O_2\text{ , reflux , 2 h}]{\text{2 eq CuCl-phen/PhH , 2 eq }K_2CO_3} PhCHO \quad 86\%$$

Markó, I.E.; Giles, P.R.; Tsukazaki, M.; Chellé-Regnaut, I.; Gautier, A.; Brown, S.M.; Urch, C.J. *J. Org. Chem.*, ***1999***, *64*, 2433.

$$PhCH_2OH \xrightarrow[\text{toluene , 80°C , }O_2]{5\%\ Pd()Ac)_2\text{ , Py , MS 3Å}} PhCHO \quad \text{quant}$$

Nishimura, T.; Onoue, T.; Ohe, K.; Uemura, S. *J. Org. Chem.*, ***1999***, *64*, 6750.

$$C_6H_{13}CH(OH)CH_2OH \xrightarrow[PhCF_3\text{ , 1 atm }O_2\text{ , 60°C , 15 h}]{4\%\ Ru(PPh_3)_2Cl_2/C_{KG}} C_6H_{13}CHO \quad 52\%$$

C_{KG} = Kurane coal GLC

Takezawa, E.; Sakaguchi, S.; Ishii, Y. *Org. Lett.*, ***1999***, *1*, 713.

$$PhCH_2OH \xrightarrow[\text{microwaves , 10 sec}]{TMSOTMS\text{ , }CrO_3\text{ , neat}} PhCHO \quad 98\%$$

Heravi, M.M.; Ajami, D.; Tabar-Heydar, K. *Synth. Commun.*, ***1999***, *29*, 163.

$$PhCH_2OH \xrightarrow[\text{neat , microwaves}]{CrO_3\text{–HY zeolite , 10 sec}} PhCHO \quad 98\%$$

Mirza-Aghayan, M.; Heravi, M.M. *Synth. Commun.*, ***1999***, *29*, 785.

$$PhCH_2OH \xrightarrow[MnSO_3\bullet 5\ H_2O\text{ , }KMnO_4\text{ , 80°C}]{\text{neutral }Al_2O_3\text{ , 40\% NaOH , }H_2O} PhCHO \quad 70\%$$

Stavrescu, R.; Kimura, T.; Fujita, M.; Vinatoru, M.; Ando, T. *Synth. Commun.*, ***1999***, *29*, 1719.

$$CH_3(CH_2)_4OH \xrightarrow{\text{quinolinium fluorochromate, }Al_2O_3\text{ , hexane , rt , 2 h}} CH_3(CH_2)_3CHO \quad 57\%$$

Rajkumar, G.A.; Arabindoo, B.; Murugesan, V. *Synth. Commun.*, ***1999***, *29*, 2105.

$$PhCH_2OH \xrightarrow[\text{DCM , 1 h}]{(\text{3-carboxypyridinium})\ CrO_3Cl^-\text{ , }Al_2O_3} PhCHO \quad 92\%$$

Heravi, M.M.; Kiakoojori, R.; Mirza-Aghayan, M.; Tabar-Hydar, K.; Bolourtchian, M. *Monat. Chem.*, ***1999***, *130*, 481.

$$PhCH_2OH \xrightarrow{NH_4^+ \ CrO_3Cl^- \text{ , Montmorillonite K10}} PhCHO \quad 92\%$$

Heravi, M.M.; Kiakojoori, R.; Tabar-Hydar, K. *Monat. Chem.*, ***1999***, *130*, 581.

SECTION 49: ALDEHYDES FROM ALDEHYDES

Conjugate reductions and Michael Alkylations of conjugated aldehydes are listed in Section 74 (Alkyls from Alkenes).

$$PhCH_2CHO \xrightarrow{(BnPPh_3)_2^+ \ Cr_2O_7^{-2} \text{ , MeCN , reflux}} PhCHO \quad \text{quant}$$

Hajipour, A.R.; Mohammadpoor-Baltork, I.; Niknam, K.
Org. Prep. Proceed. Int. ***1999***, *31*, 335.

Related Methods: Aldehydes from Ketones (Section 57)
Ketones from Ketones (Section 177)
Also via: Alkenyl aldehydes (Section 341)

SECTION 50: ALDEHYDES FROM ALKYLS, METHYLENES AND ARYLS

S S Me Ph — hv (Pyrex) , O_2/CCl_4, 1 h → S S CHO Ph 45%

Wan, Y.; Barnhurst, L.A.; Kutateladze, A.G. *Org. Lett.*, ***1999***, *1*, 937.

SECTION 51: ALDEHYDES FROM AMIDES

Ph NEt$_2$ (O) — $Cp_2Zr(H)Cl$, THF , rt , 15 min → Ph H (O)

96%

White, J.M.; Tunoori, A.R.; Georg, G.I. *J. Am. Chem. Soc.*, ***2000***, *122*, 11995.

SECTION 52: ALDEHYDES FROM AMINES

$PhCH_2NH_2$ → PhCHO 98x98%
1. MsCl , NEt_3 , DCM , 0°C
2. 2 eq Ph(Cl)S=N*t*-Bu , DCM
2 eq DBU , –78°C
3. 1M HCl , DCM-ether

Matsuo, J.-i.; Kawana, A.; Fukuda, Y.; Mukaiyama, T. *Chem. Lett.*, ***2001***, 712.

$PhCHN{-}NMe_2$ → PhCHO 95%
TMSCl , NaI , 1% H_2O
MeCN , 2 min

Kamal, A.; Ramana, K.V.; Arifuddin, M. *Chem. Lett.*, ***2000***, 827.

Related Methods: Ketones from Amines (Section 172)

SECTION 53: ALDEHYDES FROM ESTERS

NO ADDITIONAL EXAMPLES

SECTION 54: ALDEHYDES FROM ETHERS, EPOXIDES AND THIOETHERS

OMe → OH
$NiCl_2$, Zn , *p*-xylene
reflux , 3 d
97%

Maeyama, K.; Kobayashi, M.; Yonezawa, N. *Synth. Commun.*, ***2001***, *31*, 869.

$PhCH_2OTMS$ → PhCHO 90%
$BiCl_3$, $BnPPh_3$ HSO_5
DCM , microwaves , 3 min

Hajipour, A.R.; Mallakpour, S.E.; Balork, I.M.; Adibi, H. *Synth. Commun.* ***2001***, *31*, 1625.

$PhCH_2OSiMe_3$ → PhCHO 98%
zeolite HM-5 supported $Fe(NO_3)_3$
microwaves , 20 sec

Heravi, M.M.; Ajami, D.; Ghassemzadeh, M.; Tabar-Hydar, K. *Synth. Commun.*, ***2001***, *31*, 2097.

$PhCH_2OTHP$ —— zeolite HZSM-5 , $Fe(NO_3)_3$ / DCM , reflux , 6 h ——→ PhCHO 91%

Mohajerani, B.; Heravi, M.M.; Ajami, D. *Monat. Chem.*, ***2001***, *132*, 871.

$PhCH_2OSiMe_3$ —— K_2FeO_4 , clay , MeCN , 3 h ——→ PhCHO 95%

Tajbakhsh, M.; Heravi, M.M.; Habibzadeh, S.; Ghassemzadeh, M. *J. Chem.Res. (S)*, ***2001***, 39.

O

C_3H_7 SPh —— 3 eq Me_3Al , DCM , –30°C / 10 min ——→ OH C_3H_7 SPh Me 96%

Sasaki, M.; Tanino, K.; Miyashita, M. *J. Org. Chem.*, ***2001***, *66*, 5388.

Ph O Ph —— $Bi(OTf)_3$•x H_2O, DCM / 20 min ——→ Ph Ph CHO 92%

Bhatia, K.A.; Eash, K.J.; Loenard, N.M.; Oswald, M.C.; Mohan, R.S. *Tetrahedron Lett.*, ***2001***, *42*, 8129.

Ph Ph O —— $BiOClO_4$•xH_2O, CH_2Cl_2 / 25 min ——→ Ph Ph CHO 90%

Anderson, A.M.; Blazek, J.M.; Garg, P.; Payne, B.J.; Mohan, R.S. *Tetrahedron Lett.*, ***2000***, *41*, 1527.

O —— 2.2 eq Li/THF , reflux / cat biphenyl , 30 min ——→ OH 97%

Rama, K.; Pasha, M.A. *Tetrahedron Lett.*, ***2000***, *41*, 1073.

O —— 1. 2 eq Li , THF / 2. H_2O ——→ HO >95%

Azzena, U.; Demartis, S.; Pilo, L.; Pivas, E. *Tetrahedron*, ***2000***, *56*, 8375.

Ph-epoxide (O) → 1. $VO(OEt)Cl_2$, EtOH , rt , 5h; 2. aq NaCl , HCl → $PhCH_2CHO$ 90%

Martínez, F.; del Campo, C.; Llama, E.F. *J. Chem. Soc., Perkin Trans. 1*, ***2000***, 1749.

OTES / OTES / C_5H_{11} → DMSO, $(COCl)_2$; CH_2Cl_2 , NEt_3 , –70°C → OHC / OTES / C_5H_{11} 68%

Rodríguez, A.; Nomen, M.; Spur, B.W.; Godfroid, J.J. *Tetrahedron Lett.*, ***1999***, *40*, 5161.

C_8H_{17}-epoxide (O) → Fe(tpp)OTf , dioxane ; reflux → $C_8H_{17}CH_2CHO$ 96%

Suda, K.; Baba, K.; Nakajima, S.-i.; Takanami, T. *Tetrahedron Lett.*, ***1999***, *40*, 7243.

$PhCH_2OSiMe_3$ → wet alumina supported CrO_3 ; neat , 30 sec → PhCHO 90%

Heravi, M.M.; Ajami, D.; Ghassemzadeh, M. *Synthesis*, ***1999***, 393.
Heravi, M.M.; Ajami, D.; Ghassemzadeh, M. *Synth. Commun.*, ***1999***, *29*, 781.

$PhCH_2OSiMe_3$ → CrO_3 , TMSOTMS , SiO_2 ; DCM , 25°C , 25 min → PhCHO 95%

Heravi, M.M.; Ajami, D.; Tabar-Heydar, K. *Synth. Commun.*, ***1999***, *29*, 1009.

$PhCH_2OTHP$ → $NaBrO_3/AlCl_3$, MeCN ; 0.65 h , reflux → PhCHO 95%

Mohammadpoor-Baltork, I.; Nourozi, A.R. *Synthesis*, ***1999***, 487.

$PhCH_2CH=CHOTMS$ → PCWP–H_2O_2 , rt ; CH_2Cl_2 , 16 h → $PhCH_2CHO$

PCWP = peroxotungstophosphate

Sakaguchi, S.; Yamamoto, Y.; Sugiomoto, T.; Yamamoto, H.; Ishii, Y. *J. Org. Chem.*, ***1999***, *64*, 5954.

$PhOCH_2CH=CH_2$ → $ZrCl_4/NaBH_4$, THF ; 0°C → rt → PhOH 95%

Chary, K.P.; Mohan, G.H.; Iyengar, D.S. *Chem. Lett.*, ***1999***, 1223.

MeCN , 60°C , 5 h

Ot-Bu

NO ,

O N O

OH

CHO

72%

Eikawa, M.; Sakaguchi, S.; Ishii, Y. *J. Org. Chem.*, ***1999***, *64*, 4676.

$PhCH_2OTMS$ → ($FeNO_3$–Montmorillonite K10, microwaves) → PhCHO 90%

Mojtahedi, M.M.; Saidi, M.R.; Boloúrtchian, M.; Heravi, M.M. *Synth. Commun.*, ***1999***, *29*, 3283.

O

Ph CN , SmI_2

Br

Ph

THF , rt , 1 h

O O

Ph Ph

82%

Firouzabadi, H.; Etemadi, S.; Karimi, B.; Jarrahpour, A.A. *Synth. Commun.*, ***1999***, *29*, 4333.

Related Methods: Ketones from Ethers and Epoxides (Section 174)

SECTION 55: ALDEHYDES FROM HALIDES AND SULFONATES

Br Br

1. Bu_3MgLi , 0°C , toluene-THF

2. DMF , 0°C , 30 min

3. aq citric acid

Br CHO

84%

Iida, T.; Wada, T.; Tomimoto, K.; Mase, T. *Tetrahedron Lett.*, ***2001***, *42*, 4841.

Cl

DMSO , ZnO , NaBr

140°C , 14 h

CHO

58%

Guo, Z.; Sawyer, R.; Prakash, I. *Synth. Commun.*, ***2001***, *31*, 667.

Guo, Z.; Sawyer, R.; Prakash, I. *Synth. Commun.*, ***2001***, *31*, 3395.

Br

Ph

DMSO , $NaHCO_3$

rt , 2 h

CHO

Ph

70%

Ravichandran, S. *Synth. Commun.*, ***2001***, *31*, 2185.

SECTION 56: ALDEHYDES FROM HYDRIDES

OH OH

$(CH_2O)_n$, $MgCl_2$–NEt_3 → CHO

THF , 1.5 h 83%

Hòfsløkksen, N.U.; Skattebøl, L. *Acta Chem. Scand.* ***1999***, *53*, 258.

SECTION 57: ALDEHYDES FROM KETONES

Cl— POCl$_3$, DMF , SiO$_2$ → Cl— CHO

O microwaves , 2 min Cl 79%

Paul, S.; Gupta, M.; Gupta, R. *Synlett*, ***2000***, 1115.

SECTION 58: ALDEHYDES FROM NITRILES

NO ADDITIONAL EXAMPLES

SECTION 59: ALDEHYDES FROM ALKENES

P O 2.5% Rh(acac)(CO)$_2$, MeCN → P O CHO

50 psi H_2/CO , 65°C

92% (81:19 ***anti:syn***)

Krauss, I.J.; Wang, C.C.-Y.; Leighton, J.L. *J. Am. Chem. Soc.*, ***2001***, *123*, 11514.

PhCH=CH$_2$ —0.5% aq RuCl$_3$(H$_2$O)$_2$, 1.5 eq Oxone / 4.7 eq NaHCO$_3$, H$_2$O→ PhCHO 73%

Yang, D.; Zhang, C. *J. Org. Chem.*, ***2001***, *66*, 4814.

5% V catalyst

5 eq PhSH, 1 d

CHO

58%

Baucherel, X.; Uziel, J.; Jugé, S. *J. Org. Chem.* ***2001***, *66*, 4504.

CO/H_2, toluene, 180°C

$[Rh(acac)(CO)_2]$
chiral carbohydrate phosphine

CHO

Ph

58% conversion
50% ee, *S*

Diéguez, M.; Pàmies, O.; Net, G.; Ruiz, A.; Claver, C. *Tetrahedron Asymm.*, ***2001***, *12*, 651.

$Rh(acac)(CO)_2$, CO/H_2

chiral furanoside phosphite–phosphoramidate

CHO

Ph

68% (35% ee, *S*)

Diéguez, M.; Ruiz, A.; Claver, C. *Tetrahedron Asymm.* ***2001***, *12*, 2827.

furanoside phosphine-phosphite ligand
$Rh(acac)(CO)_2$, CO/H_2

CHO

Ph

55% (49% ee, *S*)

Pàmies, O.; Net, G.; Ruiz, A.; Claver, C. *Tetrahedron Asymm*, ***2001***, *12*, 3441.

CO/H_2, chiral diphosphine

40°C, 46 h

CHO

Ph

92% (47% ee, *R*)

Hegedüs, C.; Madrász, J.; Gulyás, H.; Szöllõsy, Á.; Bakos, J.
Tetrahedron Asymm. ***2001***, *12*, 2867.

CO/H_2, Pt catalyst, $SnCl_2$

chiral diphosphite ligand, 23°C
20 h

Ph CHO + Ph CHO

(85 15) 71%
86% ee

Bakos, J.; Cserépi-Szűcs, S.; Gömöry, Á.; Hegedüs, C.; Markó, L.; Szöllõsy, Á.
Can. J. Chem., ***2001***, *79*, 725.

cat $Rh(acac)(CO)_2$, NaPHOS

CO/H_2 (10 bar), 120°C

CHO

Klein, H.; Hackstell, R.; Wiese, K.-D.; Borgmann, C.; Beller, M.
Angew. Chem. Int. Ed., ***2001***, *40*, 3408.

$KMnO_4 \cdot CuSO_4 \cdot 5\,H_2O$

DCM , rt , 3 h

OHC CHO

65%

Göksu, S.; Altundaş, R. *Synth. Commun.*, ***2000***, *30*, 1615.

Rh(CO)$_2$(dpm) , CO/H_2 , 120°C

bis-phosphine complex , 1 h

C_5H_{11} ... C_5H_{11} ... CHO

90%

van der Veen, L.A.; Kamer, P.C.J.; van Leeuven, P.W.N.M. *Angew. Chem. Int. Ed.*, ***1999***, *38*, 336.

REVIEWS:

"Recent Advances on Chemo-, Regio- and Stereoselective Hydroformylation," Breit, B.; Seiche, W. *Synthesis*, ***2001***, 1.

Related Methods: Ketones from Alkenes (Section 179)

SECTION 60: ALDEHYDES FROM MISCELLANEOUS COMPOUNDS

SnBu$_3$ — *t*-BuOH , 50°C → CHO + CO_2H

	CHO	CO_2H
VO(OCH_2CF_3)Cl_2/36 h	–	100%
VO(OCH_2CF_3)Cl_2/3 h	60%	19%

Hirao, T.; Morimoto, C.; Takada, T.; Sakurai, H. *Tetrahedron Lett.*, ***2001***, *42*, 1961.

PhCH=N–OH —CuCl/Kieselguhr , O_2; DCM , 20 min→ PhCHO 96%

Hashemi, M.M.; Beni, Y.A. *Synth. Commun.*, ***2001***, *31*, 295.

PhCH=N–OH —quinolium dichromate; MeCN , reflux , 30 min→ PhCHO 97%

Sadeghi, M.M.; Mohammadpoor-Baltork, I.; Azarm, M.; Mazidi, M.R. *Synth. Commun.*, ***2001***, *31*, 435.

Ph–CH=CH–CH(OAc)(OAc) → (CBr_4, MeCN, reflux, 5 h) → Ph–CH=CH–CHO 81%

Ramalingam, T.; Srinivas, R.; Reddy, B.V.S.; Yadav, J.S. *Synth. Commun.*, ***2001***, *31*, 1091.

PhCH=N–OH → ($PyH^+ CrO_3F^-$, 3% H_2O_2, acetone, –10°C, 20 min) → PhCHO 83%

Ganguly, N.C.; Sukai, A.K.; De, S.; De, P. *Synth. Commun.* ***2001***, *31*, 1607.

PhCH=N–OH → ($BnPPh_3 HSO_4$, $BiCl_3$, MeCN, reflux, 50 min) → PhCHO 94%

Hajipour, A.R.; Mallakpour, S.E.; Baltork, I.M.; Adibi, H. *Synth. Commun.*, ***2001***, *31*, 3401.

PhCH=N–NHTs → (DDQ, DCM, H_2O, 2 h) → PhCHO 85%

Chandrasekhar, S.; Reddy, Ch.R.; Reddy, M.V. *Chem. Lett.*, ***2000***, 430.

$PhCH=N–NMe_2$ → ($[Ni(en)_3]_2S_2O_3$, $CHCl_3$, 5 min) → PhCHO 92%

Kamal, A.; Arifuddin, M.; Rao, M.V. *Synlett*, ***2000***, 1482.

2,4-(MeO)(OMe)-C6H3–CH=N–OH → (quinolium fluorochromate, MeCN) → 2,4-(MeO)(OMe)-C6H3–CHO 80%

Bose, D.S.; Narasaiah, A.V. *Synth. Commun.*, ***2000***, *30*, 1153.

CO_2Me–(CH2)4–CH=N-NHCONH$_2$ → ($(Bu_4N)_2S_2O_8$, DCE) → CO_2Me–(CH2)4–CHO 91%

Chen, F.-E.; Liu, J.-P.; Fu, H.; Peng, Z.-Z.; Shao, L.-Y. *Synth. Commun.*, ***2000***, *30*, 2295.

PhCH=N—OH → (Dess–Martin periodinane [AcO, OAc, I–OAc], wet CH_2Cl_2, 5°C, 20 min) → PhCHO 94%

Chaudhari, S.S.; Akamanchi, K.G. *Synthesis*, ***1999***, 760.

MeO–C₆H₃(OMe)–CH=N–OH → (peroxymonosulfate–SiO_2, microwaves, 1.5 min) → 2,4-(MeO)₂C₆H₃CHO 85%

Bose, D.S.; Narsaiah, A.V.; Lakshminarayana, V. *Synth. Commun.*, ***2000***, *30*, 3121.

PhCH=N–OH → ($Cu(NO_3)_2$•SiO_2, 5 min, microwaves) → PhCHO 95%

Ghiaci, M.; Asghari, J. *Synth. Commun.*, ***2000***, *30*, 3865.

2,4-Cl₂C₆H₃CH=N–OH → (DCM, rt, 40 min; Dess–Martin periodinane [AcO, OAc, OAc]) → 2,4-Cl₂C₆H₃CHO 80%

Bose, D.S.; Narsaiah, A.V. *Synth.Commun.*, ***1999***, *29*, 937.

2,4-(MeO)₂C₆H₃CH=N–OH → ($(Bu_4N)_2S_2O_8$, DCE, reflux, 1 h) → 2,4-(MeO)₂C₆H₃CHO 92%

Chen, F.; Liu, A.; Yan, Q.; Liu, M.; Zhang, D.; Dhao, L. *Synth. Commun.*, ***1999***, *29*, 1049.

$PhCH(OAc)_2$ → ($BiCl_3$, $CHCl_3$, reflux, 10 min) → PhCHO 90%

Mohammadpoor-Baltork, I.; Aliyan, H. *Synth. Commun.*, ***1999***, *29*, 2731.

PhCH=N–OH → (I_2, MeCN, reflux) → PhCHO 96%

Yadav, J.S.; Sasmal, P.K.; Chand, P.K. *Synth. Commun.*, ***1999***, *29*, 3667.

SECTION 60A: PROTECTION OF ALDEHYDES

2-Phenyl-1,3-dioxolane (Ph–) → (K_2FeO_4, Montmorillonite K10, DCM, reflux, 10 min) → PhCHO 95%

Heravi, M.M.; Tajbakhsh, M.; Habibzadeh, S.; Ghassemzadeh, M. *Monat. Chem.*, ***2001***, *132*, 985.

PhCH=NOH —(N-Me-piperidinium CrO$_3$Cp , Al$_2$O$_3$; dioxane , reflux , 30 min)→ PhCHO 92%

Tajbakhsh, M.; Heravi, M.M.; Mohanazadeh, F.; Sarabi, S.; Ghassemzadeh, M. *Monat. Chem.*, ***2001***, *132*, 1229.

PhCHO —(HS(CH$_2$)$_3$SH , CdI$_2$, 75 sec; microwaves)→ 2-Ph-1,3-dithiane 85%

Laskar, D.D.; Prajapati, D.; Sandhu, J.S. *J. Chem. Res. (S)*, ***2001***, 313.

PhCHO —(0.1% Bi(OTf)$_3$•x H$_2$O , MeCN; 3 eq Ac$_2$O)→ PhCH(OAc)$_2$ 91%

Carrigan, M.D.; Eash, K.J.; Oswald, M.C.; Mohan, R.S. *Tetrahedron Lett.*, ***2001***, *42*, 8133.

PhCHO —(HSCH$_2$CH$_2$SH , I$_2$, Al$_2$O$_3$; 10 min)→ 2-Ph-1,3-dithiolane 95%

Deka, N.; Sarma, J.C. *Chem. Lett.*, ***2001***, 794.

C$_{11}$H$_{23}$CH(SEt)$_2$ —(cat N$_2$O$_5$, H$_2$O$_2$, NH$_4$Br , 5 h; DCM/H$_2$O , 1.75 h)→ C$_{11}$H$_{23}$CHO 80%

Mondal, E.; Bose, G.; Sahu, P.R.; Khan, A.T. *Chem. Lett.*, ***2001***, 1158.

PhCHO —(HSCH$_2$CH$_2$OH , 210 min , DCE; cat [Pt(dppb)(μ-OH)]$_2$ (BF$_4$)$_2$, MgSO$_4$, H$_2$O)→ 2-Ph-1,3-oxathiolane 90%

Battaglia, L.; Pinna, F.; Strukul, G. *Can. J. Chem.*, ***2001***, *79*, 621.

PhCH$_2$CH$_2$CH(OBn)CH$_3$ —(3 eq CrCl$_2$, 4 eq LiI , aq EtOAc; 75°C , 12 h)→ PhCH$_2$CH$_2$CH(OH)CH$_3$ 92%

Falck, J.R.; Barma, D.K.; Baati, P.; Mioskowski, C. *Angew. Chem. Int. Ed.*, ***2001***, *40*, 1281.

PhCHO —(HS(CH$_2$)$_3$SH , cat LiOTf , 0.1 h; solvent free)→ 2-Ph-1,3-dithiane 99%

Firouzabadi, H.; Eslami, S.; Karimi, B. *Bull. Chem. Soc. Jpn.*, ***2001***, *74*, 2401.

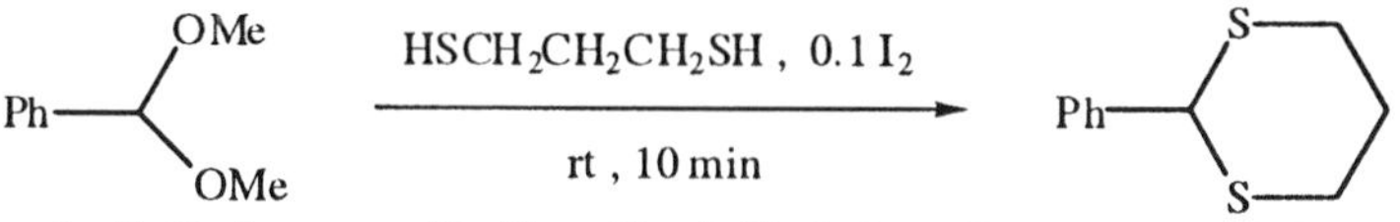

Ph–CH(OMe)(OMe) —[$HSCH_2CH_2CH_2SH$, 0.1 I_2 / rt , 10 min]→ 2-Ph-1,3-dithiane 94%

Firouzabadi, H.; Iranpoor, N.; Hazarkhani, H. *J. Org. Chem.*, ***2001***, *66*, 7527.

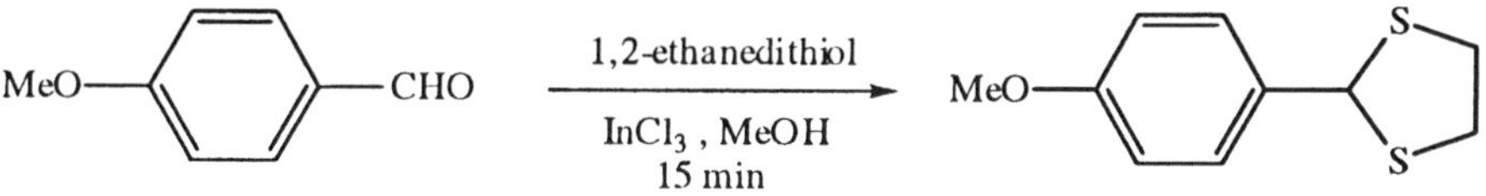

MeO–C6H4–CHO —[1,2-ethanedithiol / $InCl_3$, MeOH, 15 min]→ MeO–C6H4–(1,3-dithiolan-2-yl) 92%

Muthusamy, S.; Babu, S.A.; Gunanathan, C. *Tetrahedron Lett.*, ***2001***, *42*, 359.

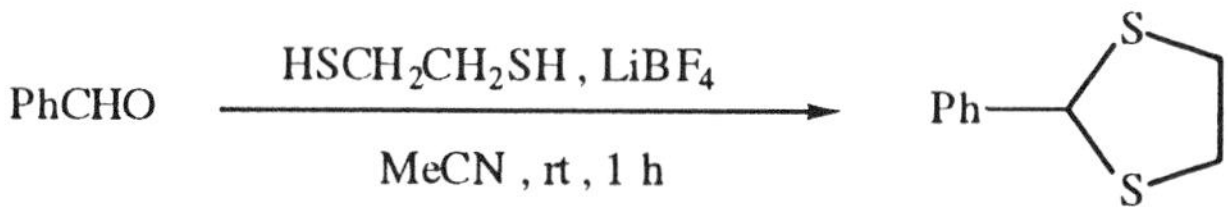

PhCHO —[$HSCH_2CH_2SH$, $LiBF_4$ / MeCN , rt , 1 h]→ 2-Ph-1,3-dithiolane 90%

Yadav, J.S.; Reddy, B.V.S.; Pandey, S.K. *Synlett*, ***2001***, 238.

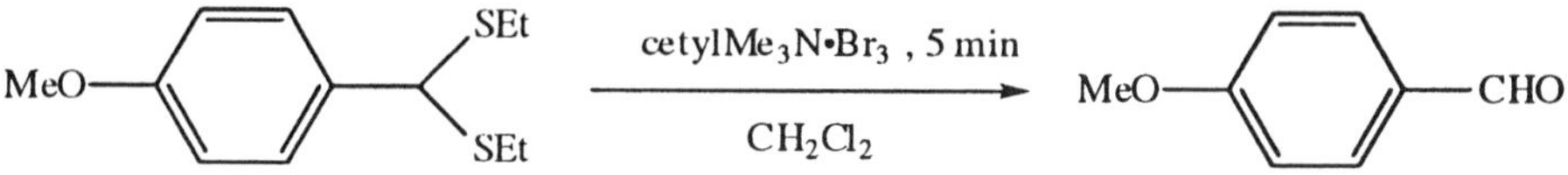

MeO–C6H4–CH(SEt)(SEt) —[cetylMe_3N•Br_3 , 5 min / CH_2Cl_2]→ MeO–C6H4–CHO 93%

Mondal, E.; Bose, G.; Khan, A.T. *Synlett*, ***2001***, 785.

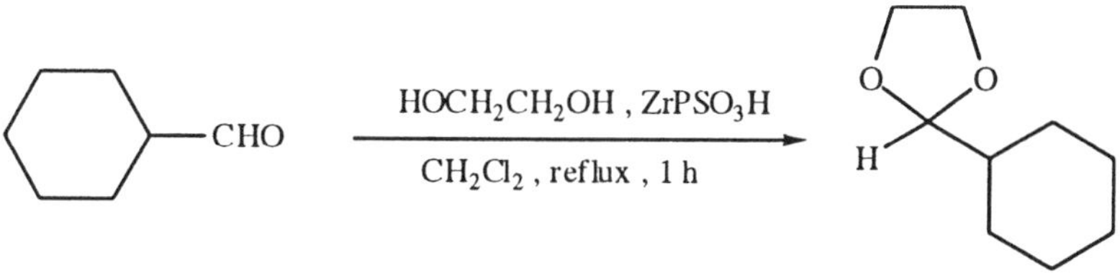

C6H11–CHO —[$HOCH_2CH_2OH$, $ZrPSO_3H$ / CH_2Cl_2 , reflux , 1 h]→ 2-cyclohexyl-1,3-dioxolane 90%

Curini, M.; Epifano, F.; Marcotullio, M.C.; Rosati, O. *Synlett*, ***2001***, 1182.

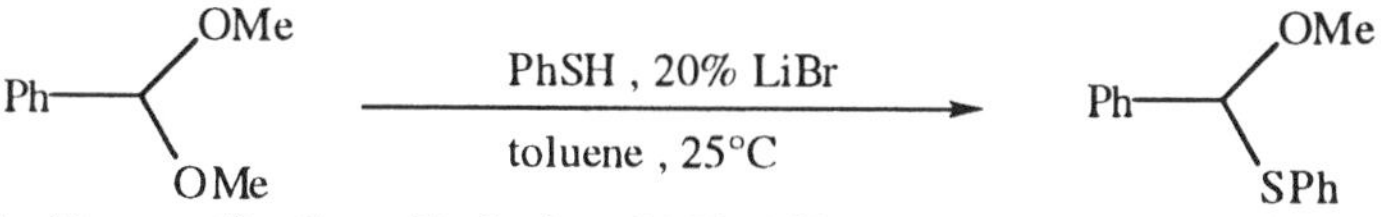

Ph–CH(OMe)(OMe) —[PhSH , 20% LiBr / toluene , 25°C]→ Ph–CH(OMe)(SPh) 90%

Ono, F.; Negoro, R.; Sato, T. *Synlett*, ***2001***, 1581.

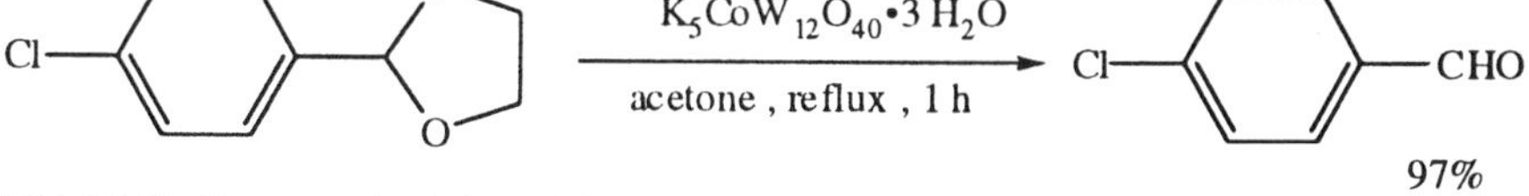

Cl–C6H4–(1,3-dioxolan-2-yl) —[$K_5CoW_{12}O_{40}$•3 H_2O / acetone , reflux , 1 h]→ Cl–C6H4–CHO 97%

Habibi, M.H.; Tangestaninejad, S.; Mohammadpoor-Baltork, I.; Mirkhani, V.; Yadollahi, B. *Tetraheddon Lett.*, ***2001***, *42*, 6771.

Ph–CH(OMe)(OMe) —[$HSCH_2CH_2CH_2SH$, 0.75 h, rt; trichloroisocyanuric acid, 0.75 h, $CHCl_3$]→ 2-phenyl-1,3-dithiane 94%

Firouzabadi, H.; Iranpoor, N.; Hazarkhani, H. *Synlett,* ***2001***, 1641.

PhCHO —[Ac_2O, 10% $LiBF_4$, 40°C; 20 h]→ $PhCH(OAc)_2$ 95%

Sumida, N.; Nishioka, K.; Sato, T. *Synlett,* ***2001***, 1921.

CH_3CH_2CHO —[2 eq EtSH, cat I_2; THF]→ $CH_3CH_2CH(SEt)_2$ 98%

Samajdar, S.; Basu, M.K.; Becker, F.F.; Banik, B.K. *Tetrahedron Lett.*, ***2001***, *42*, 4425.

PhCHO —[0.1% $Bi(OTf)_3$•x H_2O, MeCN; 3 eq Ac_2O]→ $PhCH(OAc)_2$ 91%

Carrigan, M.D.; Eash, K.J.; Oswald, M.C.; Mohan, R.S. *Tetrahedron Lett.*, ***2001***, *42*, 8133.

PhCHO —[Ti^{+4}–montmorillonite; toluene, ethylene glycol, 1 h]→ 2-phenyl-1,3-dioxolane >99%

Kawabata, T.; Mizugaki, T.; Ebitani, K.; Kaneda, K. *Tetrahedron Lett.*, ***2001***, *42*, 8329.

PhCHO —[$HOCH_2CH_2SH$; Amberlyst-15, DCM, 1 h]→ 2-phenyl-1,3-oxathiolane 84%

Ballini, R.; Bosica, G.; Maggi, R.; Mazzacani, A.; Righi, P.; Sartori, G. *Synthesis,* ***2001***, 1826.

PhCHO —[$HSCH_2CH_2SH$, DCM, 2 min; $POCl_3$–montmorillonite, rt]→ 2-phenyl-1,3-dithiolane 96%

Jin, T.-S.; Sun, X.; Ma, Y.-R.; Li, T.-S. *Synth. Commun.*, ***2001***, *31*, 1669.

PhCH=N–OH —[$Zn(NO_3)_2$•6 H_2O, acetone, SiO_2; microwaves, 4 min]→ PhCHO 90%

Tamami, B.; Kiasat, A.R. *Synth. Commun.*, ***2000***, *30*, 4129.

$C_3H_7CHO \xrightarrow[\text{rt , 15 min}]{Ac_2O\text{ , PVC–}FeCl_3} C_3H_7CH(OAc)_2$ 85%

Zolfigol, M.A.; Kiany-Borazjani, M.; Sadeghi, M.M.; Mohammadpoor-Baltork, I.; Memarian, H.R. *Synth. Comm.*, **2000**, *30*, 2919.

PhCHO $\xrightarrow[\text{2 h}]{HS(CH_2)_3SH\text{ , 0.5M }LiClO_4\text{ , ether}}$ 2-phenyl-1,3-dithiane 76%

Tietze, L.F.; Weigand, B.; Wulff, C. *Synthesis*, **2000**, 69.

PhCH(SEt)₂ $\xrightarrow{FeCl_3 \cdot 6\,H_2O\text{ , rt , 15 min}}$ PhCHO 96%

Kamal, A.; Laxman, E.; Reddy, P.S.M.M. *Synlett*, **2000**, 1476.

$PhCH(OMe)_2 \xrightarrow[\text{DCM , rt , 10 min}]{HSCH_2CH_2CH_2SH\text{ , silica chloride}}$ 2-phenyl-1,3-dithiane 94%

Firouzabadi, H.; Iranpoor, N.; Karimi, B.; Hazarkhani, H. *Synlett*, **2000**, 263.

PhCHO $\xrightarrow[\text{DCM , 2 h}]{2.5\%\ Cu(OTf)_2\text{ , }Ac_2O\text{ , rt}}$ $PhCH(OAc)_2$ 96%

Chandra, K.L.; Saravanan, P.; Singh, V.K. *Synlett*, **2000**, 359.

PhCHO $\xrightarrow{Ac_2O\text{ , 10\% NBS , rt , 6 h}}$ $PhCH(OAc)_2$ 95%

Karimi, B.; Seradj, H.; Ebrahimian, G.R. *Synlett*, **2000**, 623.

PhCHO $\xrightarrow[NaSO_4\text{–}SiO_2\text{ , 4 min}]{\text{ethylene glycol , microwaves}}$ 2-phenyl-1,3-dioxolane 85%

Yadav, J.S.; Reddy, B.V.S.; Srinivas, R.; Ramalingam, T. *Synlett*, **2000**, 701.

$Cl\text{–}C_6H_4\text{–}CH{=}N\text{–}OH \xrightarrow[\text{4 min}]{NBS\text{ , }H_2O} Cl\text{–}C_6H_4\text{–}CHO$ 95%

Bandgar, B.P.; Makone, S.S. *Org. Prep. Proceed. Int.*, **2000**, *32*, 391.

PhCHO → ($HOCH_2CH_2SH$, 4% $ZrCl_4$; DCM, rt, 30 min) → 2-phenyl-1,3-oxathiolane 91%

Karimi, B.; Seradj, H. *Synlett*, ***2000***, 805.

2-phenyl-1,3-oxathiolane → (NBS, aq acetone) → PhCHO 93%

Karimi, B.; Seradj, H.; Tabaei, M.H. *Synlett*, ***2000***, 1798.

$PhCH(OMe)_2$ → ($BiCl_3$, MeOH, 30 min) → PhCHO 92%

Sabitha, G.; Babu, R.S.; Reddy, E.V.; Yadav, J.S. *Chem. Lett.*, ***2000***, 1074.

hexanal (CHO) → (Ac_2O, microwaves, 40 sec; Envirocat EPZ-10) → 1,1-diacetoxyhexane (OAc, OAc) 87%

Bandgar, B.P.; Makone, S.S.; Kulkarn, S.R. *Monat. Chem.*, ***2000***, *131*, 417.

4-methylbenzaldehyde oxime (N–OH) → (clay-$(TMS)_2$chromate; DCM, 40 min) → 4-methylbenzaldehyde (CHO) 90%

Heravi, M.M.; Ajami, D.; Tajbakhsh, M.; Ghassemzadeh, M. *Monat. Chem.*, ***2000***, *131*, 1109.

PhCHO → ($HS(CH_2)_3SH$; 0.1 $InBr_3$, DCM, 1 h) → 2-phenyl-1,3-dithiane 80%

Ceschi, M.A.; Felix, L.de A.; Peppe, C. *Tetrahedron Lett.*, ***2000***, *41*, 9695.

$PhCH_2OTHP$ → (Montmorillonite K-10, $Fe(NO_3)_3$; microwaves, 60 sec) → PhCHO 90%

Heravi, M.M.; Ajami, D.; Majtahedi, M.M.; Ghassemzadeh, M. *Tetrahedron Lett.*, ***1999***, *40*, 561.

2-(2-phenylethyl)-1,3-dioxolane → (2.5 eq CAN, 5 min, 70°C; MeCN, H_2O) → $PhCH_2CH_2CHO$ 70%

Ates, A.; Gautier, A.; Leroy, B.; Plancher, J.M.; Quesnel, Y.; Markó, I.E. *Tetrahedron Lett.*, ***1999***, *40*, 1799.

PhCHO —($HSCH_2CH_2CH_2SH$, LiOTf / solvent free , 110°C , 5 min)→ Ph–(1,3-dithiane) quant

Firouzabadi, H.; Karimi, B.; Eslami, S. *Tetrahedron Lett.*, ***1999***, *40*, 4055.

$PhCH(OMe)_2$ —(Magrieve™ , $CHCl_3$ / rt , 10 min)→ PhCHO 95%

forms acid with several substrates

Ko, K.-Y.; Park, S.-T. *Tetrahedron Lett.*, ***1999***, *40*, 6025.

PhCHO —($HS(CH_2)_3SH$. LiBr , 80°C / 15 min (neat))→ Ph–(1,3-dithiane) 99%

Firouzabadi, H.; Iranpoor, N.; Karimi, B. *Synthesis*, ***1999***, 58.

$HO(CH_2)_3(CH_2)_8Me$ —(5% $ReOCl_3(PPh_3)_2$, 23 h / 5% 2,4,6-collidine, 1.2 DMSO / 4 eq ethylene glycol)→ 2-[$(CH_2)_2(CH_2)_8Me$]-1,3-dioxolane 85%

Arterburn, J.B.; Perry, M.C. *Org. Lett.*, ***1999***, *1*, 769.

PhCHO —(3 eq propylene glycol , 1% NBS / $(EtO)_3CH$, 2 eq MeOH , DCM / rt , 6 h)→ Ph–(1,3-dioxane) 95%

Karimi, B.; Ebrahimian, G.R.; Seradj, H. *Org. Lett.*, ***1999***, *1*, 1737.

$PhCH(OMe)_2$ —(5% $ZrCl_4$, DCM , rt / $HSCH_2CH_2SH$)→ Ph–(1,3-dithiolane) quant

Firouzabadi, H.; Iranpoor, N.; Karimi, B. *Synlett*, ***1999***, 319.

PhCHO —($(EtO)_3CH$, $ZrCl_4$, neat / rt)→ $PhCH(OEt)_2$ 97%

Firouzabadi, H.; Iranpoor, N.; Karimi, B. *Synlett*, ***1999***, 321.

C_6H_{13}—CHO —($HSCH_2CH_2SH$, $Cu(OTf)_2$–SiO_2 / neat , 2 h)→ C_6H_{13}–(1,3-dithiolane) 93%

Anand, R.V.; Saravanan, P.; Singh, V.K. *Synlett*, ***1999***, 415.

PhCHO → (1. 3 eq $(EtO)_3CH$, EtOH, rt, DCM, NBS; 2. 10% aq NaOH) → $PhCH(OEt)_2$ 95%

Karimi, B.; Seradj, H.; Ebrahimian, G.-R. *Synlett*, ***1999***, 1456.

PhCHO → ($(EtO)_3CH$, cat DDQ, rt; EtOH, 10 h) → $PhCH(OEt)_2$ 95%

Karimi, B.; Ashtiani, A.M. *Chem. Lett.*, ***1999***, 1199.

PhCHO → (ethylene glycol, CdI_2, 1.5 min; microwaves) → 2-phenyl-1,3-dioxolane 90%

Laskar, D.D.; Prajapati, D.; Sandhu, J.S. *Chem. Lett.*, ***1999***, 1283.

PhCHO → ($HSCH_2CH_2CH_2SH$, HCl_g; AcCl, MeOH, 2 min) → 2-phenyl-1,3-dithiane 95%

Graham, A.E. *Synth. Commun.*, ***1999***, *29*, 697.

$PhCH(SEt)_2$ → (Clayan, microwaves; 110 sec) → PhCHO 89%

Meshram, H.M.; Reddy, G.S.; Sumitra, G.; Yadav, J.S. *Synth. Commun.*, ***1999***, *29*, 1113.

PhCHO → ($(EtO)_3CH$, cat WCl_6; neat, rt, 10 min) → $PhCH(OEt)_2$ 98%

Firouzabadi, H.; Iranpoor, N.; Karimi, B. *Synth. Commun.*, ***1999***, *29*, 2255.

$PhCH{=}CHNO_2$ → (Al–$NiCl_2$–CH_2O, THF) → $PhCH_2CHO$ 77%

Bezbarua, M.S.; Bez, G.; Barua, N.C. *Chem. Lett.*, ***1999***, 325.

CHAPTER 5

PREPARATION OF ALKYLS, METHYLENES AND ARYLS

This chapter lists the conversion of functional groups into methyl, ethyl, propyl, etc. as well as methylene (CH_2), phenyl, etc.

SECTION 61: ALKYLS, METHYLENES AND ARYLS FROM ALKYNES

10% $Ni(cod)_2$, 40% PPh_3

toluene, rt, 0.25 h

C_6F_{13} C_6F_{13} C_6F_{13} C_6F_{13}

65%

Saito, S.; Kawasaki, T.; Tsuboya, N.; Yamamoto, Y. *J. Org. Chem.*, ***2001***, *66*, 796.

Ph

$Pd(PPh_3)_4$, THF, 100°C, 12 h

Ph Ph Bu

Bu Bu

Bu 89%

Gevorgyan, V.; Radhakrishnan, U.; Yakeda, A.; Rubin, M.; Rubin, M.; Yamamoto, Y. *J. Org. Chem.*, ***2001***, *66*, 2885.

N–*t*-Bu

2 eq I_2, DCM

N

rt, 30 min

Ph

Ph

68% I

Huang, Q.; Hunter, J.A.; Larock, R.C. *Org. Lett.*, ***2001***, *3*, 2973.

I N H N tBr

Ph—≡—Ph , DMF

5% $Pd(OAc)_2$, 10 h

5% PPh_3 , NBu_3

100°C

Ph Ph N N H

54%

Zhang, H.; Larock, R.C. *Org. Lett.*, ***2001***, *3*, 3083.

Bu—≡—Bu

1. ≡—SO_2Tol $(C_3H_6)Ti(O\textit{i}\text{-}Pr)_2$, −50°C

2. ≡—SO_2Tol

3. H^+

Bu Bu SO_2Tol

65%

Suzuki, D.; Urabe, H.; Sato, F. *J. Am. Chem. Soc.*, ***2001***, *123*, 7925.

MeO_2C MeO_2C

Bu—≡ , PhH , rt

$[Ir(cod)Cl]_2dppe$, 20min

MeO_2C MeO_2C Bu

84%

Takeuchi, R.; Tanaka, S.; Nakaya, Y. *Tetrahedron Lett.*, ***2001***, *42*, 2991.

HO

5% $RhCl(PPh_3)_3$, 10% $AgBF_4$

$CHCl_3$, 50°C , 2 h

HO

83%

Oh, C.H.; Sung, H.R.; Jung, S.H.; Lim, Y.M. *Tetrahedron Lett.*, ***2001***, *42*, 5493.

OTBS

10% $[Rh(CO)_2Cl]_2$

toluene , 110°C

OTBS

51%

Dankwardt, J.W. *Tetrahedron Lett.*, ***2001***, *42*, 5809.

MeO–CH₂–C≡CH → [≡–CO_2Me; cat Ni(acac)$_2$–PPh$_3$, Et_2Zn, PhOH, THF, rt]

CO_2Me OMe; CO_2Me (95 + MeO CO_2Me; CO_2Me 5) 78%

Mori, N.; Ikeda, S.-i.; Odashima, K. *Chem. Commun.*, ***2001***, 181.

MeO_2C, MeO_2C → [EtO_2C-CN, DCE, 60°C; 2% Ru complex, 30 min] → MeO_2C, MeO_2C, CO_2Et

83%

Yamamoto, Y.; Okuda, S.; Itoh, K. *Chem. Commun.*, ***2001***, 1102.

→ [15% CpCo(CO)$_2$; *o*-xylene, 140°C; hv, *p*-TolCN]

p-Tol, N (1 + *p*-Tol, N 1) 42%

Moretto, A.F.; Zhang, H.-C.; Maryanoff, B.E. *J. Am. Chem. Soc.*, ***2001***, *123*, 3157.

CO_2Me, CO_2Me → [Bu–CH=C=CH$_2$; 5% Ni(dppe)Br$_2$, Zn; MeCN, 80°C, 8 h]

CO_2Me, MeO_2C, Bu, MeO_2C, CO_2Me 80%

Shanmugasundaram, M.; Wu, M.-S.; Cheng, C.-H. *Org. Lett.*, ***2001***, *3*, 4233.

1. $CO_2(CO)_8$, hexanes
2. NMO, DCM, –35°C

Me, OH

98%

Marchueta, I.; Olivella, S.; Solá, L.; Moyano, A.; Pericàs, M.A.; Riera, A. *Org. Lett.*, ***2001***, *3*, 3197.

N-*t*-Bu, Ph

PhI, DMF
5% $Pd_2(dba)_3$

10% PPh_3, $3Na_2Cl_3$
100°

N, Ph, Ph

49%

Daum G.; Larock, R.C. *Org. Lett.*, ***2001***, *3*, 4035.

Bu

1. $Cr(CO)_5$, H_3C, OMe
2. I_2

H, CH_3, O, Bu

71%

Herndon, J.W.; Zhang, Y.; Wang, H.; Wang, K. *Tetrahedron Lett.*, ***2000***, *41*, 8687.

Ph, OH

1. PhMgCl, C_6H_{12}, 80°C
2. DMF, 0°C → rt
3. TsOH, PhH

Ph, Ph, O

92%

Forgione, P.; Wilson, P.D.; Fallis, A.G. *Tetrahedron Lett.*, ***2000***, *41*, 17.

Ph—≡—Ph

decaborane, Pd/C
MeOH, rt

Ph, Ph

97%

Lee, S.H.; Park, Y.T.; Yoon, C.M. *Tetrahedron Lett.*, ***2000***, *41*, 887.

OPh

Bu—≡—≡—Bu
5% $Pd(PPh_3)_4$, THF, 65°C, 8 h

Bu, OPh, Bu

56%

Gevorgyan, V.; Quan, L.G.; Yamamoto, Y. *J. Org. Chem.*, ***2000***, *65*, 568.

Et—≡—Et → 1. Cp_2ZrEt_2 2. MeCN 3. $NiCl_2(PPh_3)_2$, C_3H_7—≡—C_3H_7 → pyridine product (Et, Et, C_3H_7, C_3H_7, Me) 66%

Takahashi, T.; Tsai, F.Y.; Kotora, M. *J. Am. Chem. Soc.*, ***2000***, *122*, 4994.

MeO_2C, MeO_2C diyne → BuC≡CH, DCE, rt, 15 min; 1% Cp*Ru(cod)Cl → MeO_2C, MeO_2C indane, Bu 89%

Yamamoto, Y.; Ogawa, R.; Itoh, K. *Chem.Commun.*, ***2000***, 549.

$(OC)_5Cr$= carbene (N-pyrrolidinyl, CO_2Et) → 1. BuC≡CH, THF, 60°C 2. SiO_2 → Bu, N, HO, CO_2Et 88%

Barluenga, J.; López, L.A.; Martínez, S.; Tomás, M. *Tetrahedron*, ***2000***, *56*, 4967.

diyne → C_6H_{13}—≡, $MoCl_5$, Al; DME, 50°C, 1 h → C_6H_{13} tetrahydronaphthalene 41%

Hara, R.; Guo, Q.; Takahashi, T. *Chem. Lett.*, ***2000***, 140.

O-linked diyne (CO_2Me, CO_2Me) → 2.5% $Pd_2(dba)_3$, DMAD; PPh_3, 0.5 M, toluene, 1h → O, CO_2Me, CO_2Me, CO_2Me, CO_2Me 61%

Yamamoto, Y.; Nagata, A.; Itoh, K. *Tetrahedron Lett.*, ***1999***, *40*, 5035.

MeO_2C CO_2Me MeO_2C CO_2Me → PhMe$_2$SiH, Rh catalyst, CO, toluene → X = SiMe$_2$Ph (86%) + 10% X = H

Ojima, I.; Vu, A.T.; McCullagh, J.V.; Kinoshita, A. *J. Am. Chem. Soc.*, ***1999***, *121*, 3230.

$(OC)_5Cr$ OPh Ph → PhC≡CH, heat → OPh Ph OH 88%

Pulley, S.R.; Sen, S.; Vorogushin, A.; Swanson, E. *Org. Lett.*, ***1999***, *1*, 1721.

SECTION 62: ALKYLS, METHYLENES AND ARYLS FROM ACID DERIVATIVES

Ph Ph CO_2H → hv, HgF_2, Ar, MeCN → Ph Ph Ph Ph 83%

Habibi, M.H.; Farhadi, S. *Tetrahedron Lett.*, ***1999***, *40*, 2821.

SECTION 63: ALKYLS, METHYLENES AND ARYLS FROM ALCOHOLS AND THIOLS

OH C_8H_{17} → 4 eq EtMgBr, Ti(O*i*-Pr)$_4$, ether, rt → Et C_8H_{17} 87%

Kulinkovich, O.G.; Epstein, O.L.; Isakov, V.E.; Khmel'nitskaya, E.A. *Synlett*, ***2001***, 49.

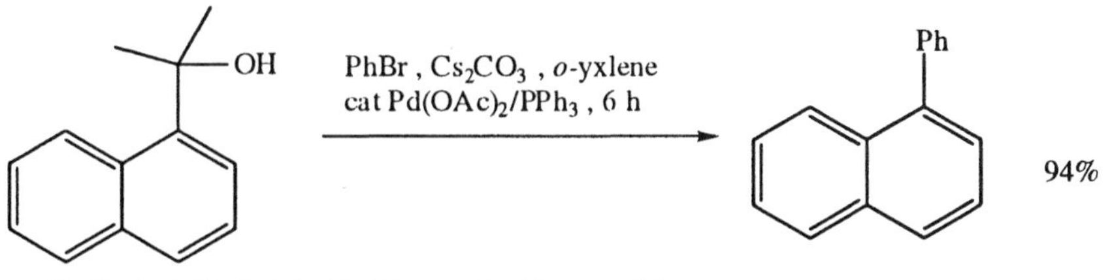

OH

PhBr , Cs_2CO_3 , *o*-yxlene
cat $Pd(OAc)_2/PPh_3$, 6 h

Ph

94%

Terao, Y.; Wakui, H.; Satoh, T.; Miura, M.; Nomura, M.
J. Am. Chem. Soc., ***2001***, *123*, 10407.

SECTION 64: ALKYLS, METHYLENES AND ARYLS FROM ALDEHYDES

n-$C_{11}H_{23}CHO$ —— 3 eq Et_3SiH , 5% $B(C_6F_5)_3$ / DCM , rt , 20 h ——> *n*-$C_{11}H_{23}CH_3$ 96%

Gevorgyan, V.; Rubin, M.; Liu, J.-X.; Yamamoto, Y. *J. Org. Chem.*, ***2001***, *66*, 1672.

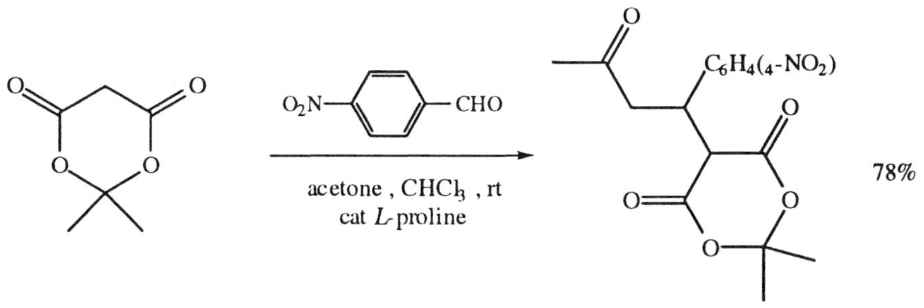

O O O O

O_2N—C6H4—CHO

acetone , $CHCl_3$, rt
cat *L*-proline

O C_6H_4(4-NO_2) O O O O

78%

List, B.; Castello, C. *Synlett*, ***2001***, 1687.

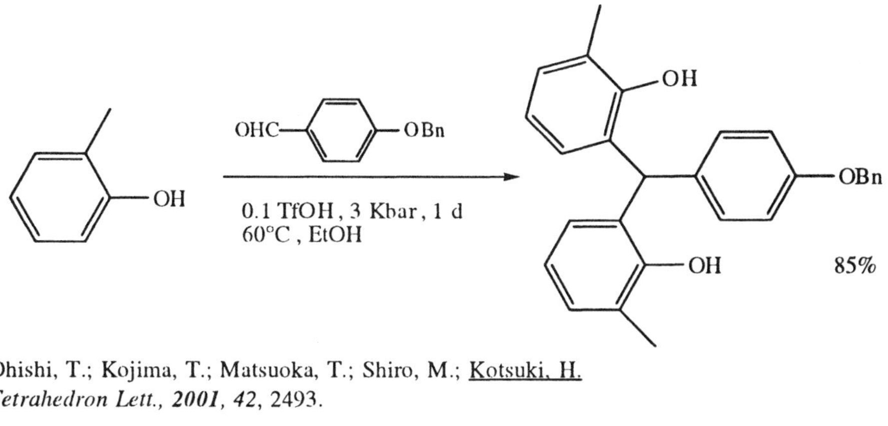

OH

OHC—C6H4—OBn

0.1 TfOH , 3 Kbar , 1 d
60°C , EtOH

OH OBn OH

85%

Ohishi, T.; Kojima, T.; Matsuoka, T.; Shiro, M.; Kotsuki, H.
Tetrahedron Lett., ***2001***, *42*, 2493.

Related Methods: Alkyls, Methylenes and Aryls from Ketones (Section 72)

SECTION 65: ALKYLS, METHYLENES AND ARYLS FROM ALKYLS, METHYLENES AND ARYLS

1. BuLi, TMEDA, ether
2. $C_8H_{17}Br$, ether

OH OH C_8H_{17} 94%

Yong, K.H.; Lotoski, J.A.; Chong, J.M. *J. Org.Chem.*, ***2001***, *66*, 8248.

PhI, hv

Ph

41% conversion (85% yield)

Ho, T.-L.; Ku, C.-K.; Liu, R.S.H. *Tetrahedron Lett.*, ***2001***, *42*, 715.

1. BuLi/LiDMAE (3 eq), 0°C
2. 3 eq $ClSnBu_3$, THF −78°C → 0°C
3. 2 eq 3-bromopyridine, xylene 5% $PdCl_2(PPh_3)_2$, 10% PPh_3 reflux, 12 h

Me N Me N N 70%

Mathieu, J.; Gros, P.; Fort, Y. *Tetrahedron Lett.*, ***2001***, *42*, 1879.

C_6H_{13} Me$_2$ClSiH, hexane, 20% $AlCl_3$ rt, 2 h C_6H_{13} $SiMe_2Cl$ 92%

Nagahara, S.; Yamakawa, T.; Yamamoto, H. *Tetrahedron Lett.*, ***2001***, *42*, 5057.

1. $ClCO_2Ph$, AgOTf rt, MeCN, 30 min
2. $SiMe_3$

N N CO_2Ph quant

Yamaguchi, R.; Nakayasu, T.; Hatano, B.; Nagura, T.; Kozima, S.; Fujita, K.-i. *Tetrahedron*, ***2001***, *57*, 109.

2 [methyl 3-hydroxy-2-naphthoate] —CAN, MeOH / rt, 15 min→ [dimethyl 2,2'-dihydroxy-1,1'-binaphthyl-3,3'-dicarboxylate: CO_2Me, OH, OH, CO_2Me] 94%

Jiang, P.; Lu, S. *Synth. Commun.*, ***2001***, *31*, 131.

[2-naphthol: OH] —2% chiral oxavanadium complex, 2% TMSCl, O_2, $CHCl_3$ / rt, 1 d→ [1,1'-bi-2-naphthol: OH, OH]

82% (51% ee)

Chu, C.-Y.; Huang, D.-R.; Wang, S.-K.; Uang, B.-J. *Chem. Commun.*, ***2001***, 980.

Ph–H —$Pd(OAc)_2$, $TiO(acac)_2$, O_2 / AcOH, 130°C, autoclave, 4 h→ Ph–Ph + Ph–OH + Ph–OAc

(0.76 0.11 0.13) 76%

Okamoto, M.; Yamaji, T. *Chem. Lett.*, ***2001***, 212.

[2,6-dimethylpyridine: N] —e^-, cat $NiBr_2$•dmbp, Zn anode, DMF / Bu_4NBF_4→ [6,6'-dimethyl-2,2'-bipyridine: N, N] 67%

Cassol, T.M.; Demnitz, F.W.J.; Navarro, M.; de. Neves, E.A. *Tetrahedron Lett.*, ***2000***, *41*, 8203.

[anisole: OMe] —ClCH(SEt(Et)CO_2Et / 5% $Yb(OTf)_3$→ [OMe, CH(SEt)CO_2Et]

2.5 : 1 *p:o*

92%

Sinha, S.; Mandal, B.; Chandrasekaran, S. *Tetrahedron Lett.*, ***2000***, *41*, 9109.

Ph—NEt_2 → ($TiCl_4$, DCM, 8 h; 0°C → rt) → Et_2N—C_6H_4—C_6H_4—NEt_2 92%

Periasamy, M.; Jayakumar, K.N.; Bharathi, P. *J. Org. Chem.*, ***2000***, *65*, 3548.

$C_8H_{17}CH(OMe)SiMe_3$ → (allyl-$SiMe_3$; e^-) → $C_8H_{17}CH(OMe)CH_2CH=CH_2$ 80%

Suga, S.; Suzuki, S.; Yamamoto, A.; Yoshida, J.-i. *J. Am. Chem. Soc.*, ***2000***, *122*, 10244.

OH, MeO, C_3H_7 → (methylenecyclohexane; $LiClO_4$, $MeNO_2$) → OH, MeO, Et 71%

Kim, S.; Kitano, Y.; Tada, M.; Chiba, K. *Tetrahedron Lett.*, ***2000***, *41*, 7079.

OH, Me, Me → (*i*-Pr, $Pb(OAc)_3$; 3 eq brucine, toluene, rt, 3h) → OH, *i*-Pr, *i*-Pr, Me, Me 92% (40% ee)

Sato, S.; Kano, T.; Muto, H.; Nakadai, M.; Yamamoto, H. *J. Am. Chem. Soc.*, ***1999***, *121*, 8943.

Ph–CH_3 → (allyl-Br, K_2CO_3, 120°C; *t*-BuOO*t*-Bu) → $PhCH_2CH_2CH=CH_2$ quant

Tanko, J.M.; Sadeghipour, M. *Angew. Chem. Int. Ed.*, ***1999***, *38*, 159.

Si, F, OMe → (PhI, 3 eq TBAF, 50°C, 22 h; 2.5% $(allylPdCl)_2$, 10% P(*o*-Tol)$_3$) → Ph—C_6H_4—OMe 88% + 12% Ph-Ph

Denmark, S.E.; Wu, Z. *Org. Lett.* ***1999***, *1*, 1495.

REVIEWS:

"*cine*- and *tele*-Substitution Reactions," Suwiński, J.; Świerczek, K. *Tetrahedron*, ***2001**, 57*, 1639.

SECTION 66: ALKYLS, METHYLENES AND ARYLS FROM AMIDES

NO ADDITIONAL EXAMPLES

SECTION 67: ALKYLS, METHYLENES AND ARYLS FROM AMINES

Ph–NHNH$_2$•HCl —(PhH, Mn(OAc)$_3$, reflux)→ Ph—Ph 75%

Demir, A.S.; Reis, Ö.; Özgül-Karaaslan, E. *J. Chem. Soc., Perkin Trans. 1*, ***2001***, 3042.

SECTION 68: ALKYLS, METHYLENES AND ARYLS FROM ESTERS

ASYMMETRIC CONVERSIONS

OAc; Ph, Ph —($CH_2(CO_2Me)_2$, LiOAc, BSA, THF, [Pd(η^3-C_3H_5)Cl]$_2$ / chiral imine-phosphine, rt)→ $CH(CO_2Me)_2$; Ph, Ph

96% (92% ee)

Mino, T.; Shiotsuki, M.; Yamamoto, N.; Suenag, T.; Sakamoto, M.; Fujita, T.; Yamashita, M. *J. Org. Chem.*, ***2001**, 66*, 1795.

OCO_2Me —(2% Pd(dba)$_2$, 4% *R*-9-PBN / 3 eq $CH_2(CO_2)_2$, 3 eq BSA, LiOAc, DCE)→ MeO_2C, CO_2Me 95% (51% ee)

Hamada, Y.; Sakaguchi, K.-e.; Hatano, K.; Hara, O. *Tetrahedron Lett.*, ***2001**, 42*, 1297.

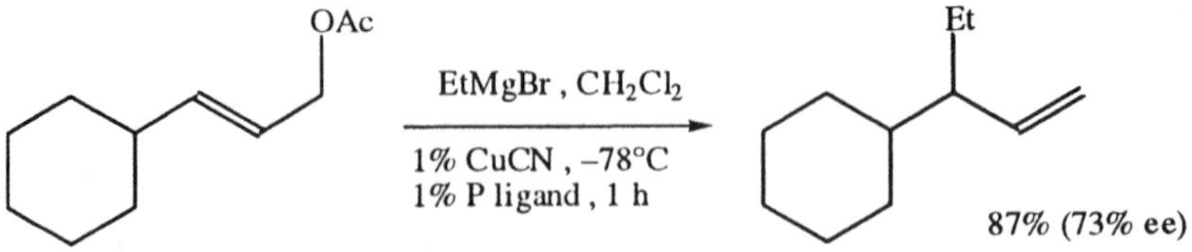

OAc — $EtMgBr$, CH_2Cl_2 / 1% CuCN , –78°C / 1% P ligand , 1 h → Et

87% (73% ee)

Alexakis, A.; Malan, C.; Lea, L.; Benhaim, C.; Fournioux, X. *Synlett,* ***2001***, 927.

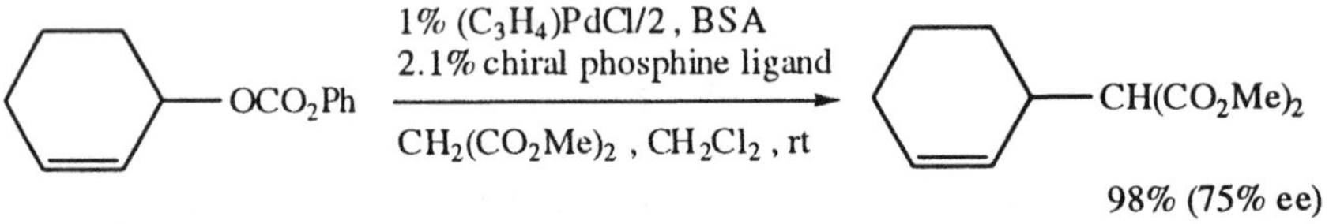

OCO_2Ph — 1% $(C_3H_4)PdCl/2$, BSA / 2.1% chiral phosphine ligand / $CH_2(CO_2Me)_2$, CH_2Cl_2 , rt → $CH(CO_2Me)_2$

98% (75% ee)

Ito, K.; Kashiwagi, R.; Hayashi, S.; Uchida, T.; Katsuki, T. *Synlett,* ***2001***, 284.

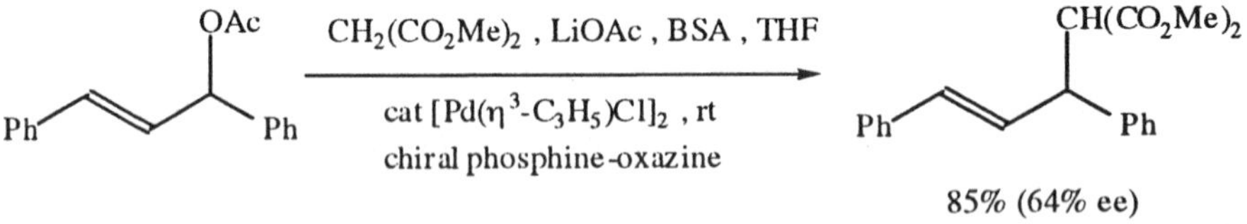

OAc, Ph, Ph — $CH_2(CO_2Me)_2$, LiOAc , BSA , THF / cat $[Pd(\eta^3\text{-}C_3H_5)Cl]_2$, rt / chiral phosphine-oxazine → $CH(CO_2Me)_2$, Ph, Ph

85% (64% ee)

Mino, T.; Hata, S.; Ohtaka, K.; Sakamoto, M.; Fujita, T. *Tetrahedron Lett.,* ***2001***, *42*, 4837.

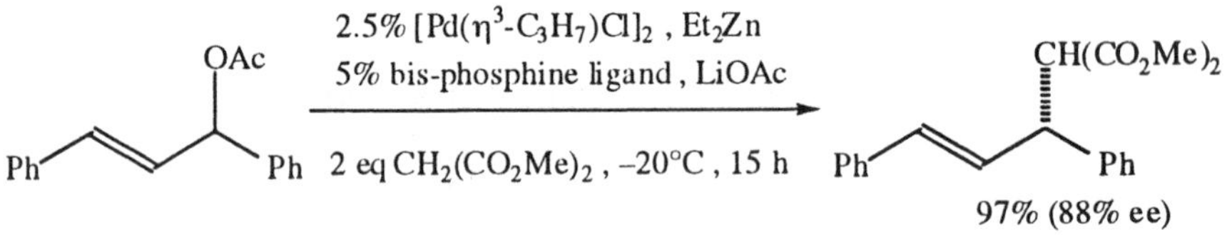

OAc, Ph, Ph — 2.5% $[Pd(\eta^3\text{-}C_3H_7)Cl]_2$, Et_2Zn / 5% bis-phosphine ligand , LiOAc / 2 eq $CH_2(CO_2Me)_2$, –20°C , 15 h → $CH(CO_2Me)_2$, Ph, Ph

97% (88% ee)

Naik, S.; Gopinath, R.; Patel, B.K. *Tetrahedron Lett.,* ***2001***, *42*, 7679.

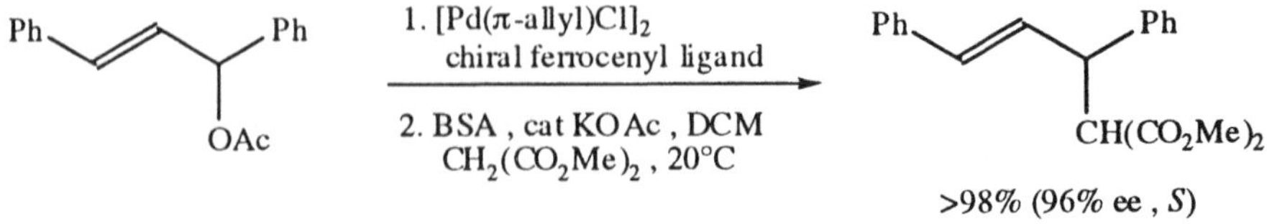

Ph, Ph, OAc — 1. $[Pd(\pi\text{-allyl})Cl]_2$ / chiral ferrocenyl ligand / 2. BSA , cat KOAc , DCM / $CH_2(CO_2Me)_2$, 20°C → Ph, Ph, $CH(CO_2Me)_2$

>98% (96% ee , *S*)

Kang, J.; Lee, J.H.; Choi, J.S. *Tetrahedron Asymm.,* ***2001***, *12*, 33.

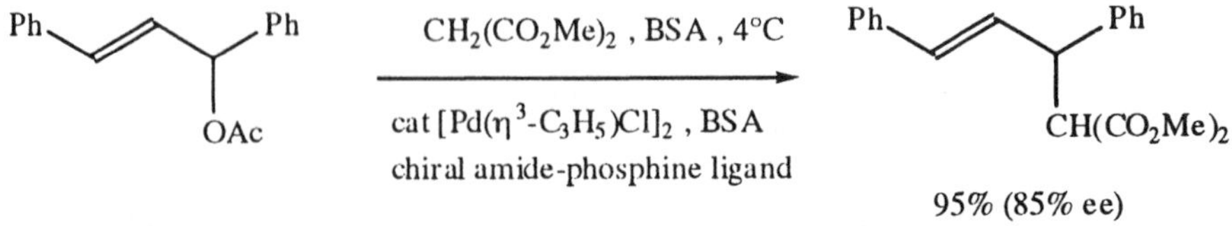

Ph, Ph, OAc — $CH_2(CO_2Me)_2$, BSA , 4°C / cat $[Pd(\eta^3\text{-}C_3H_5)Cl]_2$, BSA / chiral amide-phosphine ligand → Ph, Ph, $CH(CO_2Me)_2$

95% (85% ee)

Mino, T.; Kashihara, K.; Yamashita, M. *Tetrahedron Asymm.,* ***2001***, *12*, 287.

Ph–CH=CH–CH(OAc)–Ph → Ph–CH=CH–CH(CH(CO_2Me)$_2$)–Ph

$CH_2(CO_2Me)_2$, BSA , DCM
chiral pyrrolidine bis-sulfide
7 d , rt , KOAc

quant (81% ee)

Jansat, S.; Gómez, M.; Muller, G.; Diéguez, M.; Aghmiz, A.; Claver, C.; Masdeu-Bultó, A.M.; Flores-Santos, L.; Martin, E.; Maestro, M.A.; Mahía, J. *Tetrahedron Asymm.*, ***2001***, *12*, 1469.

Ph–CH=CH–CH(OAc)–Ph → Ph–CH=CH–CH(CH(CO_2Me)$_2$)–Ph

$CH_2(CO_2Me)_2$, BSA , LiOAc
cat [Pd(η^3-C_3H_7)Cl]$_2$, toluene
chiral amino-phosphine ligand
–20°C , 7 d

94% (94% ee)

Mino, T.; Tanaka, Y.; Sakamoto, M.; Fujita, T. *Tetrahedron Asymm.*, ***2001***, *12*, 2435.

Ph–CH=CH–CH(OAc)–Ph → Ph–CH=CH–CH(CH(CO_2Me)$_2$)–Ph

1% [Pd(allyl)Cl]$_2$, 3 BSA , 2 h
2.5% imino-phosphine ligand
ferrocenyl-allyl amine , ether
3 eq $CH_2(CO_2Me)_2$, LiOAc

98% (97% ee , *S*)

Fukuda, T.; Takehara, A.; Iwao, M. *Tetrahedron Asymm.*, ***2001***, *12*, 2793.

Ph–CH=CH–CH(OAc)–Ph → Ph–CH=CH–CH(CH(CO_2Me)$_2$)–Ph

2.5% chiral thiazoline ligand
2.5% [Pd(η^3-C_3H_7)Cl]$_2$, rt
2 eq BSA , $CH_2(CO_2Me)_2$
10% KOAc , toluene , 3 d

90% (87% ee)

Abrunhosa, I.; Gulea, M.; Levillain, J.; Masson, S. *Tetrahedron Asymm.*, ***2001***, *12*, 2851.

Ph–CH=CH–CH(OAc)–Ph → Ph–CH=CH–CH(CH(CO_2Me)$_2$)–Ph

[(η^3-C_3H_5)PdCl]$_2$, $CH_2(CO_2Me)_2$
chiral pyrrolidinopyridine ligand
BSA , KOAc

quant (84% ee, *R*)

Stranne, R.; Moberg, C. *Eur. J. Org. Chem.*, ***2001***, 2191.

Ph–CH=CH–CH(OAc)–Ph → Ph–CH=CH–CH(CH(CO_2Me)$_2$)–Ph

$CH_2(CO_2Me)_2$, KOAc , BSA
1% Pd[η^3-C_3H_5)Cl$_2$, DCM
2.5% chiral diphosphite , rt

quant (95% ee)

Diéguez, M.; Jansat, S.; Gomez, M.; Ruiz, A.; Muller, G.; Claver, C.
Chem. Commun., ***2001***, 1132.

OAc

Ph Ph

2 eq $CH_2(CO_2Me)_2$, 0°C, 2 d
2 eq BSA, 2 eq KOAc

cat $Pd[\eta^3-(C_3H_5)Cl]_2$
8% light fluorous phosphine ligand

$CH(CO_2Me)_2$

Ph Ph

95% (99% ee)

Cavazzini, M.; Pozzi, G.; Quici, S.; Maillard, D.; Sinou, D. *Chem. Commun.*, ***2001***, 1220.

OAc

Ph Ph

$CH_2(CO_2Me)_2$, $[Pd(\eta^3-C_3H_5)Cl]_2$

chiral phosphine-hydrazone, BSA
LiOAc, DCM, rt, 20 h

$CH(CO_2Me)_2$

Ph Ph

86% (92% ee)

Mino, T.; Ogawa, T.; Yamashita, M. *Heterocycles*, ***2001***, *55*, 453.

Ph Ph

OAc

$CH_2(CO_2Me)_2$, KOAc
N,O-bis-TMS-acetamide

$[Pd(\eta^3-C_3H_5)Cl]_2$
chiral P,N-ligand

Ph Ph

$CH(CO_2Me)_2$

93% (37% ee, *S*)

Arena, C.G.; Drommi, D.; Faraone, F. *Tetrahedron Asymm.*, ***2000***, *11*, 4753.

OAc

$PhB(OH)_2$, $Ni(acac)_2$, THF

Dibal-H, KOH, reflux
ferrocenyl-phosphine ligand

Ph

55% (18% ee)

Chung, K.-G.; Miyake, Y.; Yemura, S. *J. Chem. Soc., Perkin Trans. 1*, ***2000***, 15.

OAc

Ph Ph

$CH_2(CO_2Me)_2$, LiOAc, BSA,
DCM, $[Pd(\eta^3-C_3H_5)Cl]_2$

phosphino-hydroxybinaphthyl ligand

$CH(CO_2Me)_2$

Ph Ph

89% (99% ee)

Kodama, H.; Taiji, T.; Ohta, T.; Furukawa, I. *Tetrahedron Asymm.*, ***2000***, *11*, 4009.

OAc

Ph Ph

$CH_2(CO_2Me)_2$, toluene, LiOAc
1% $Pd(dba)_2$, chiral amino-phosphine

$CH(CO_2Me)_2$

Ph Ph

95% (58% ee)

Gong, L.; Chen, G.; Mi, A.; Jiang, Y.; Fu, F.; Cui, X.; Char, A.S.C. *Tetrahedron Asymm.*, ***2000***, *11*, 4297.

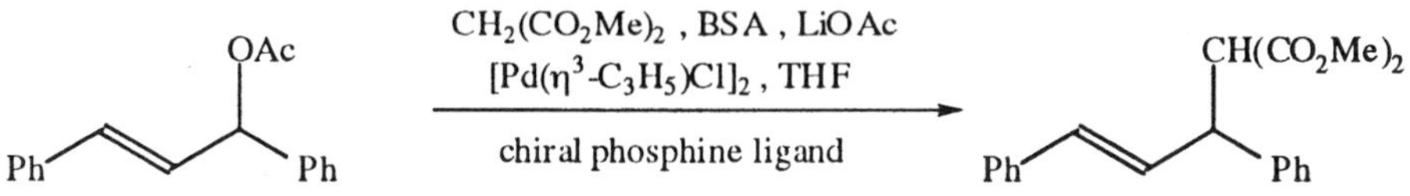

Reagents: $CH_2(CO_2Me)_2$, BSA , LiOAc, $[Pd(\eta^3\text{-}C_3H_5)Cl]_2$, THF, chiral phosphine ligand

Product: $CH(CO_2Me)_2$ — 79% (74% ee)

Mino, T.; Tanaka, Y.; Sakamoto, M.; Fujita, T. *Heterocycles*, ***2000***, *53*, 1485.

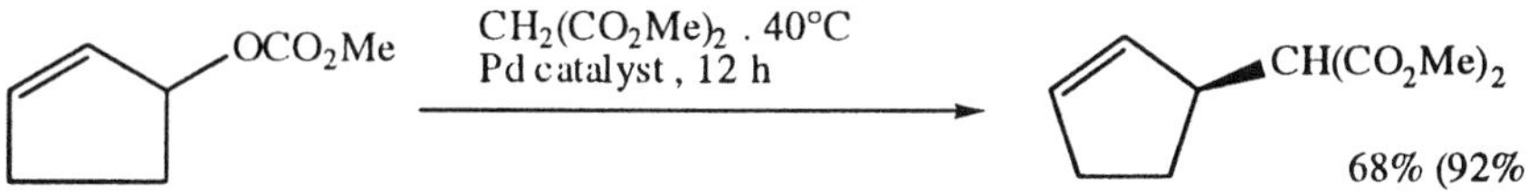

Reagents: $CH_2(CO_2Me)_2$. 40°C, Pd catalyst , 12 h

Substrate: OCO_2Me; product: $CH(CO_2Me)_2$ — 68% (92% ee S)

Uozumi, Y.; Shibatomi, K. *J. Am. Chem. Soc.*, ***2001***, *123*, 2919.

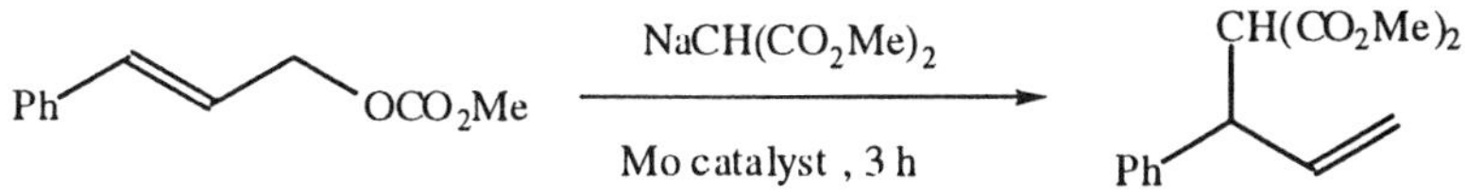

Reagents: $NaCH(CO_2Me)_2$, Mo catalyst , 3 h

Substrate: Ph, OCO_2Me; product: $CH(CO_2Me)_2$, Ph — 88% (99% ee, *S*)

Malkov, A.V.; Spoor, P.; Vinader, V.; Kočvský, P. *Tetrahedron Lett.*, ***2001***, *42*, 509.

NON-ASYMMETRIC CONVERSIONS

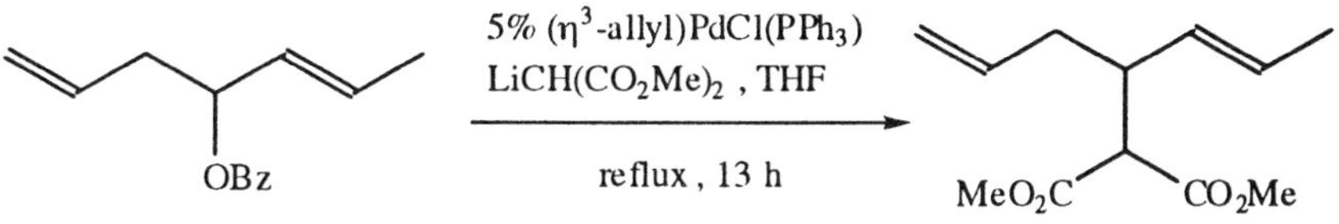

Reagents: 5% $(\eta^3\text{-allyl})PdCl(PPh_3)$, $LiCH(CO_2Me)_2$, THF, reflux , 13 h

Substrate: OBz; product: MeO_2C, CO_2Me — 61% (95:5)

Krafft, M.E.; Sugiura, M.; Abboud, K.A. *J. Am. Chem. Soc.*, ***2001***, *123*, 9174.

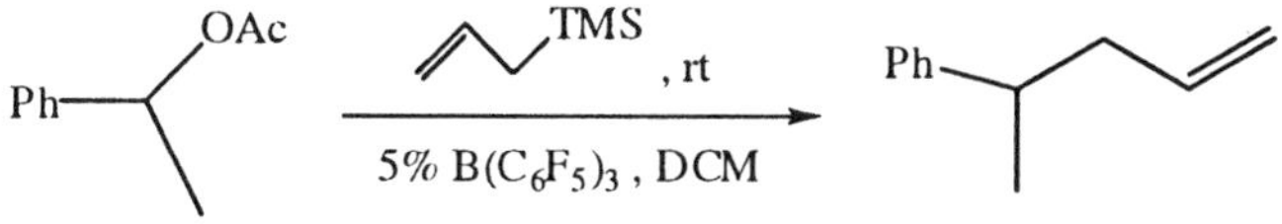

Reagents: allyl-TMS , rt, 5% $B(C_6F_5)_3$, DCM

Substrate: Ph, OAc; product: Ph — 91%

Rubin, M.; Gevorgyan, V. *Org. Lett.* ***2001***, *3*, 2705.

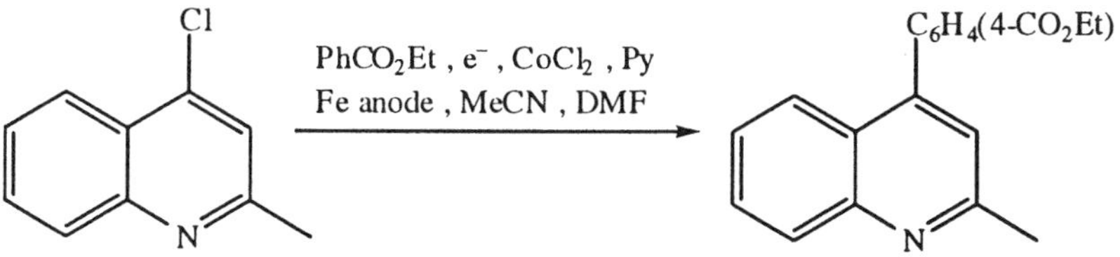

Reagents: $PhCO_2Et$, e^- , $CoCl_2$, Py, Fe anode , MeCN , DMF

Substrate: Cl, N; product: $C_6H_4(4\text{-}CO_2Et)$, N — 56%

Le Gall, E.; Gosmini, C.; Nédélec, J.-Y.; Périchon, J. *Tetrahedron Lett.*, ***2001***, *42*, 267.

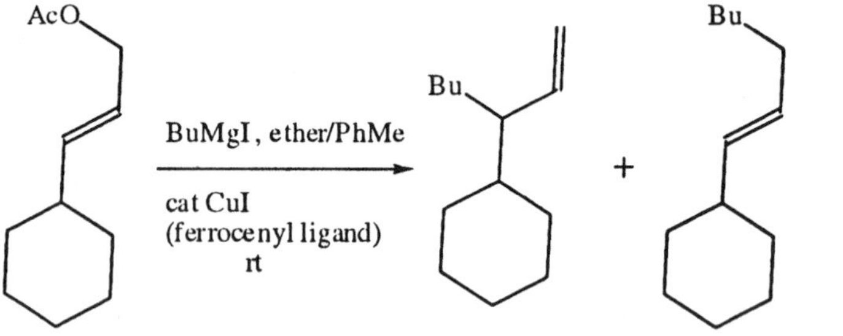

AcO — BuMgI, ether/PhMe; cat CuI (ferrocenyl ligand); rt — Bu + Bu

(98 2) 88%
64% ee

Karlström, A.S.E.; Huerta, F.F.; Muezelaar, G.J.; Bäckvall, J.-E. *Synlett*, ***2001***, 923.

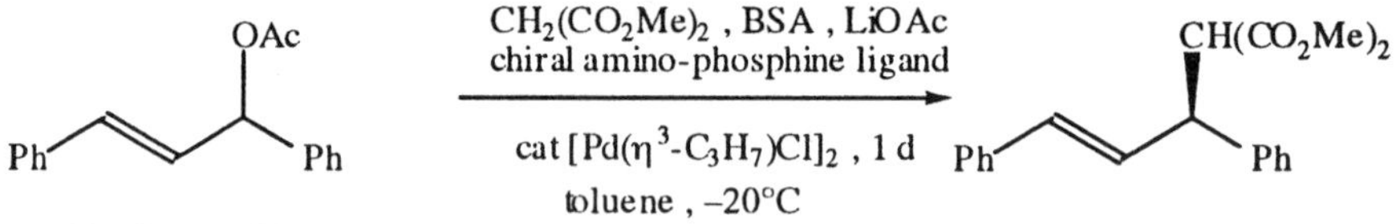

OAc, Ph, Ph — $CH_2(CO_2Me)_2$, BSA, LiOAc; chiral amino-phosphine ligand; cat $[Pd(\eta^3\text{-}C_3H_7)Cl]_2$, 1 d; toluene, –20°C — $CH(CO_2Me)_2$, Ph, Ph

Mino, T.; Tanaka, Y.-i.; Akita, K.; Anada, K.; Sakamoto, M.; Fujita, T. *Tetrahedron Asymm.*, ***2001***, *12*, 1677.

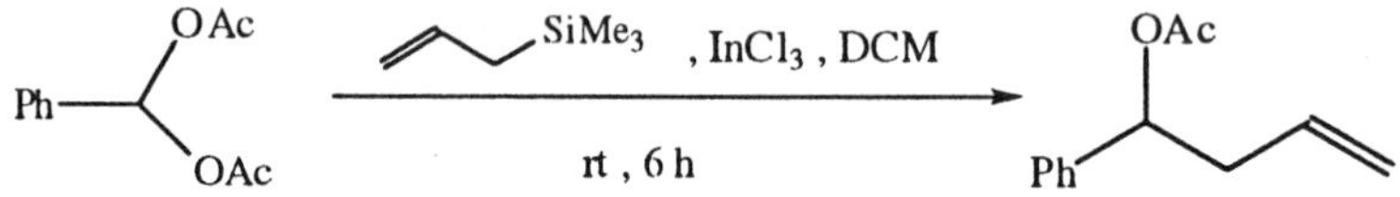

OAc, Ph, OAc — allyl-$SiMe_3$, $InCl_3$, DCM; rt, 6 h — OAc, Ph 90%

Yadav, J.S.; Reddy, B.V.S.; Madhur, Ch.; Sabitha, G. *Chem. Lett.*, ***2001***, 18.

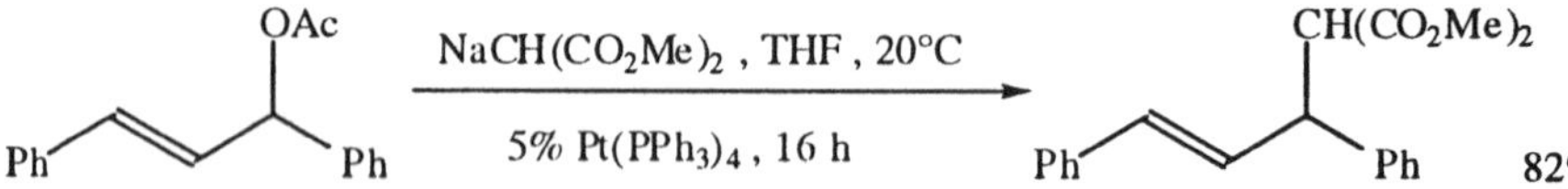

OAc, Ph, Ph — $NaCH(CO_2Me)_2$, THF, 20°C; 5% $Pt(PPh_3)_4$, 16 h — $CH(CO_2Me)_2$, Ph, Ph 82%

Blacker, A.J.; Clarke, M.L.; Loft, M.S.; Mahon, M.F.; Humphries, M.E.; Williams, J.M.J. *Chem. Eur. J.*, ***2000***, *6*, 353.

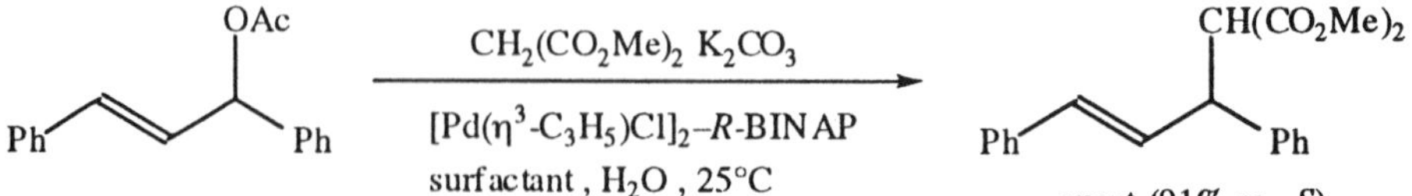

OAc, Ph, Ph — $CH_2(CO_2Me)_2$ K_2CO_3; $[Pd(\eta^3\text{-}C_3H_5)Cl]_2$–*R*-BINAP; surfactant, H_2O, 25°C — $CH(CO_2Me)_2$, Ph, Ph

quant (91% ee, *S*)

Rabeyrin, C.; Nguefack, C.; Sinou, D. *Tetrahedron Lett.*, ***2000***, *41*, 7461.

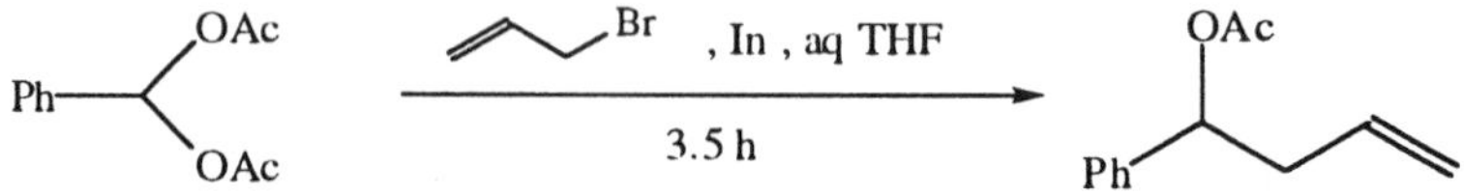

OAc, Ph, OAc — allyl-Br, In, aq THF; 3.5 h — OAc, Ph

82%

Yadav, J.S.; Reddy, B.V.S.; Reddy, G.S.K.K. *Tetrahedron Lett.*, ***2000***, *41*, 2695.

OAc

O

H

$[Ph(Bu)B(OCH(Me)CH(Me)O)]^- Li^+$

cat $NiCl_2(dppp)$, THF
60°C

Ph

O

H

75% (96:4 β:α)

Kobayashi, Y.; Ito, M. *Eur. J. Org. Chem.*, ***2000***, 3393.

Ph OAc

$EtO_2C\frown CO_2Et$
THF , 70°C , 4.5 h

5% $Pd_2(dba)_3$, 0.5 PPh_3
BSA , AcOK

Ph CO_2Et CO_2Et

83%

Poli, G.; Giambastiani, G.; Mordini, A. *J. Org. Chem.*, ***1999***, ***64***, 2962.

OAc

Ph Ph

$CH_2(CO_2Me)_2$, $Pd(OAc)_2$
K_2CO_3 , 5 h

1-butyl-3-methylimidazolium
tetrafluoroborate (ionic liquid)

$CH(CO_2Me)_2$

Ph Ph

91%

Chen, W.; Xu, L.; Chatterton, C.; Xiao, J. *Chem. Commun.*, ***1999***, 1247.

SECTION 69: ALKYLS, METHYLENES AND ARYLS FROM ETHERS, EPOXIDES AND THIOETHERS

The conversion ROR → RR' (R' = alkyl, aryl) is included in this section.

CO_2CEt_3

MeO

MeO OMe

3.8 eq cyclohexylMgBr

toluene , rt , 4 h

CO_2CEt_3

MeO

OMe

90%

Kojima, T.; Ohishi, T.; Yamamoto, I.; Matsuoka, T.; Kotsuki, H.
Tetrahedron Lett., ***2001***, ***42***, 1709.

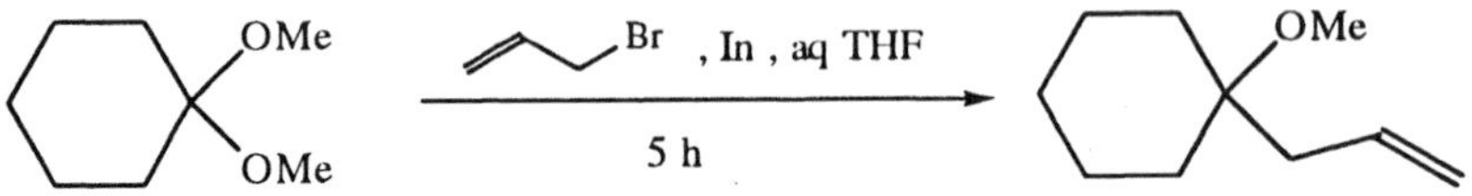

OMe OMe → (allyl bromide), In, aq THF; 5 h → OMe 89%

Kwon, J.S.; Pae, A.N.; Choi, K.I.; Koh, H.Y.; Kim, Y.; Cho, Y.S. *Tetrahedron Lett.*, ***2001***, *42*, 1957.

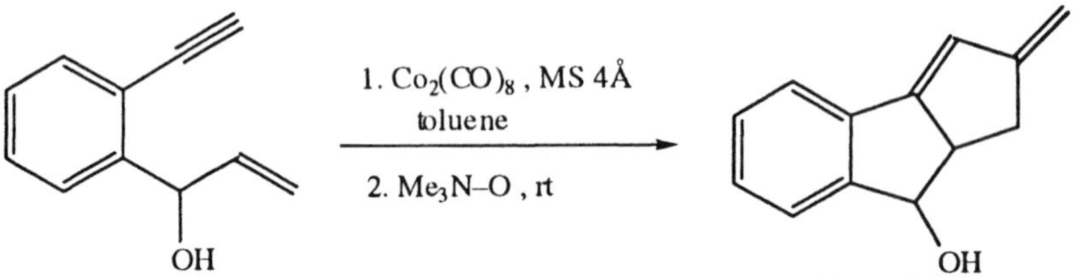

1. $Co_2(CO)_8$, MS 4Å, toluene
2. $Me_3N{-}O$, rt

OH → OH

60% (1:1 *cis:trans*)

Kajikawa, S.; Nishino, H.; Kurosawa, K. *Tetrahedron Lett.*, ***2001***, *42*, 3351.

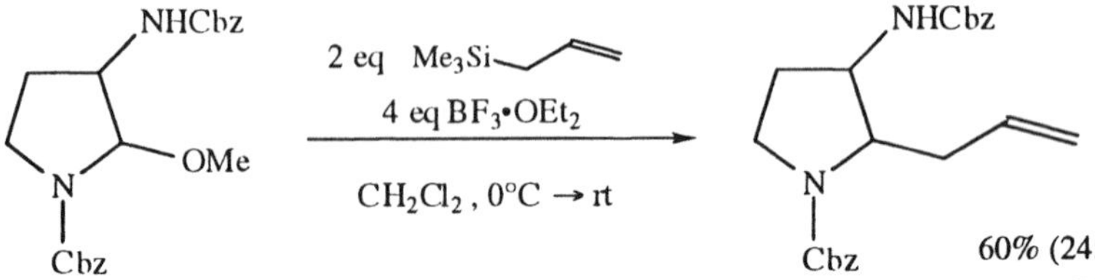

NHCbz, OMe, N, Cbz → 2 eq Me_3Si(allyl), 4 eq $BF_3{\bullet}OEt_2$; CH_2Cl_2, 0°C → rt → NHCbz, N, Cbz

60% (24:76 *cis:trans*)

Matos, M.R.P.N.; Afonso, C.A.M.; Batey, R.A. *Tetrahedron Lett.*, ***2001***, *42*, 7007.

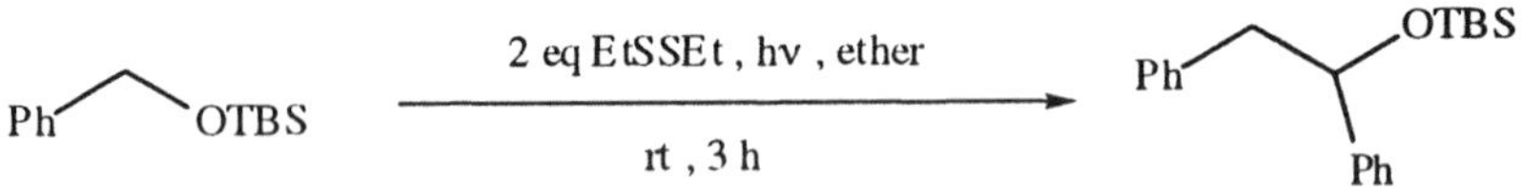

Ph OTBS → 2 eq EtSSEt, hv, ether; rt, 3 h → Ph OTBS Ph

67% (46:54 *dl:meso*)

Fujisawa, H.; Hayakawa, Y.; Sasaki, Y.; Mukaiyama, T. *Chem. Lett.*, ***2001***, 632.

SECTION 70: ALKYLS, METHYLENES AND ARYLS FROM HALIDES AND SULFONATES

The replacement of halogen by alkyl or aryl groups is included in this section. For the conversion of RX → RH (X = halogen) see Section 160 (Hydrides from Halides and Sulfonates).

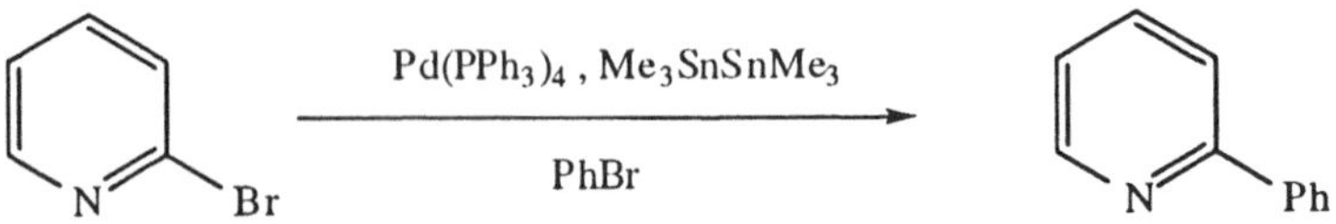

N Br → $Pd(PPh_3)_4$, $Me_3SnSnMe_3$; PhBr → N Ph 45%

Zhang, N.; Thomas, L.; Wu, B. *J. Org. Chem.*, ***2001***, *66*, 1500.

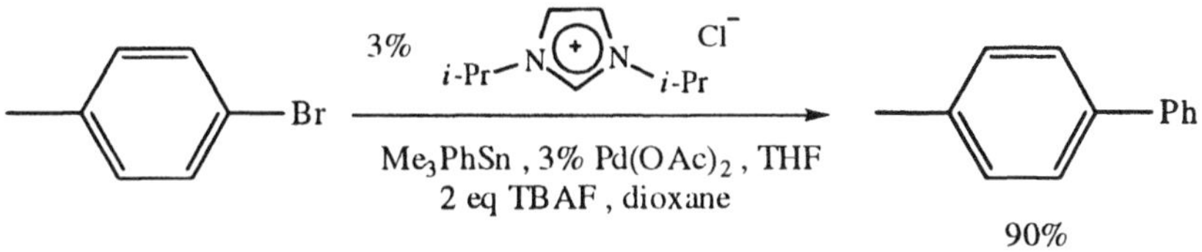

3% i-Pr–N(+)N–i-Pr Cl^-

Me_3PhSn , 3% $Pd(OAc)_2$, THF
2 eq TBAF , dioxane

90%

Grasa, G.A.; Nolan, S.P. *Org. Lett.*, ***2001***, *3*, 119.

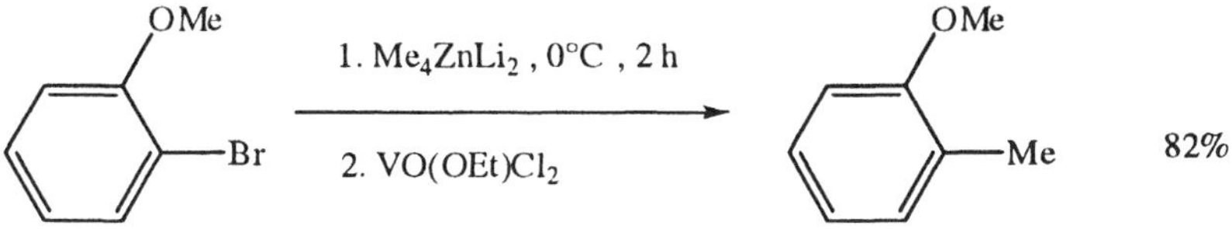

OMe Br

1. Me_4ZnLi_2 , 0°C , 2 h
2. $VO(OEt)Cl_2$

OMe Me 82%

Takada, T.; Sakurai, H.; Hirao, T. *J. Org. Chem.*, ***2001***, *66*, 300.

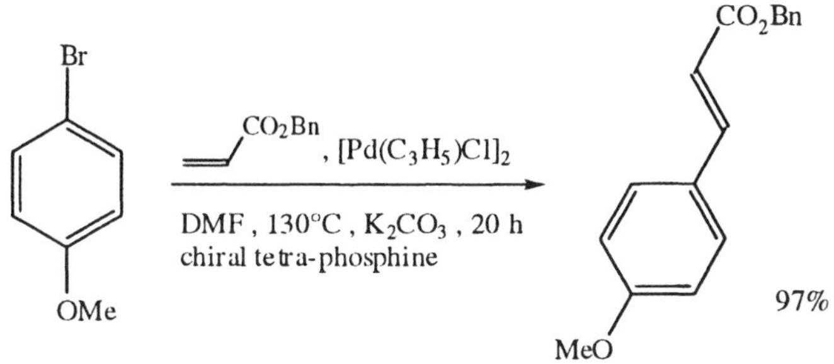

Br OMe

CO_2Bn , $[Pd(C_3H_5)Cl]_2$

DMF , 130°C , K_2CO_3 , 20 h
chiral tetra-phosphine

CO_2Bn MeO 97%

Feuerstein, M.; Doucet, H.; Santelli, M. *J. Org. Chem.*, ***2001***, *66*, 5923.

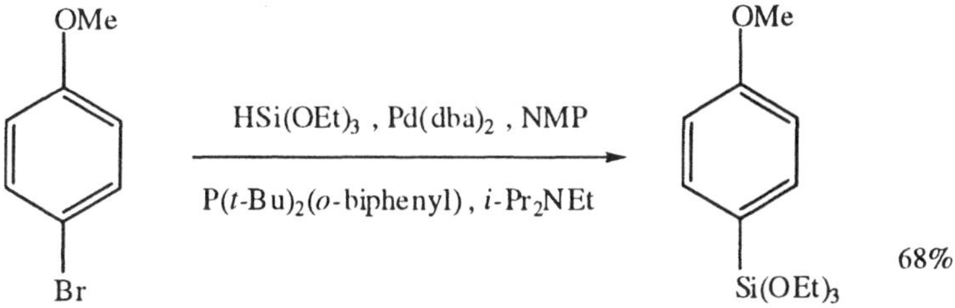

OMe Br

$HSi(OEt)_3$, $Pd(dba)_2$, NMP
$P(t\text{-}Bu)_2(o\text{-biphenyl})$, $i\text{-}Pr_2NEt$

OMe $Si(OEt)_3$ 68%

Manoso, A.S.; DeShong, P. *J. Org. Chem.*, ***2001***, *66*, 7449.

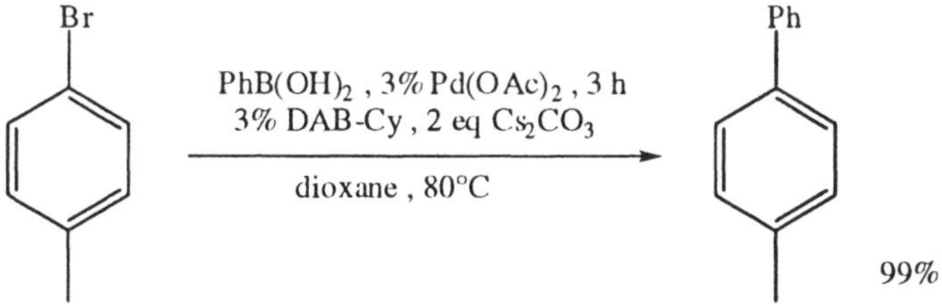

Br

$PhB(OH)_2$, 3% $Pd(OAc)_2$, 3 h
3% DAB-Cy , 2 eq Cs_2CO_3
dioxane , 80°C

Ph 99%

Grasa, G.A.; Hillier, A.C.; Nolan, S.P. *Org. Lett.* ***2001***, *2*, 1077.

$PhCH_2BF_3K$ → $PhCH_2Ph$ 91%
(PhOTf, 3 eq Cs_2CO_3, aq THF, 18h; 9% $PdCl_2(dppf)\bullet CH_2Cl_2$, reflux)

Molander, G.A.; Ito, T. *Org. Lett.* ***2001**, 3,* 393.

$F_3C-C_6H_4-Cl$ → $F_3C-C_6H_4-Ph$ 95%
($PhB(OH)_2$, Pd/C, 80°C; K_2CO_3, DMA/H_2O)

Le Blond, C.R.; Andrews, A.T.; Sun, Y.; Sowa Jr., J.R. *Org. Lett.* ***2001**, 3,* 1555.

1-chloro-4-iodobenzene → 5-(4-chlorophenyl)furan-2-carbaldehyde (CHO) 87%
(furan-2-carbaldehyde, DMF, 110°C; 5% $PdCl_2$, 4 eq KOAc, 10 h; 2 eq Bu_4NBr, 10% PCy_3)

McClure, M.S.; Glover, B.; McSorley, E.; Millar, A.; Osterhout, M.H.; Roschangar, F. *Org. Lett.*, ***2001**, 3,* 1677.

4-iodotoluene (I) → 4-phenyltoluene (Ph) 90%
($PhB(OH)_2$; cat $Pd(PPh_3)_4$, rt, THF; Cu(I) thiophene-2-carboxylate)

Savarin, C.; Liebeskind, L.S. *Org. Lett.*, ***2001**, 3,* 2149.

$MeO-C_6H_4-Br$ → $MeO-C_6H_4-Ph$ 94%
($PhB(OH)_2$, 2% $Pd(OAc)_2$; Na_2CO_3, aq MeCN, rt; 2% water soluble alkyl phosphines)

Shaughnessy, K.H.; Booth, R.S. *Org. Lett.*, ***2001**, 3,* 2757.

4-acetylphenyl OTs → 4-acetylbiphenyl (Ph) 94%
($PhB(OH)_2$, PCy_3; 2 eq K_3PO_4, dioxane; 1.5% $NiCl_2(PCy_3)_2$; 130°C)

Zim, D.; Lando, V.R.; Dupont, J.; Monteiro, A.L. *Org. Lett.*, ***2001**, 3I,* 3049.

$$\xrightarrow[\text{cat } PPh_3 \text{ , NMP , 120°C}]{\text{4-MeC}_6\text{H}_4\text{Br, cat } RuCl_2(\eta^6\text{-}C_6H_6)]_2 \text{ , 20 h}}$$

60% 68%

Oi, S.; Fukita, S.; Hirata, N.; Watanuki, N.; Miyano, S.; Inoue, Y. *Org. Lett.*, ***2001***, *3*, 2579.

1. PPh_3 , TBSOTf , THF
2. In, allyl bromide (Br)

86%

Lee, P.H.; Sung, S.-y.; Lee, K. *Org. Lett.*, ***2001***, *3*, 3201.

2 (2-iodotoluene, I)

1. 5 eq $Pd(OAc)_2$, K_2CO_3 , Bu_4NBr, norbornene
DMF , 105°C
2. PhC≡CPh

Me, Ph, Ph, Me

82%

Catellani, M.; Motti, E.; Baratta, S. *Org. Lett.*, ***2001***, *3*, 3611.

OTf → Ph

$BiPh_3$, K_2CO_3
5% $Pd(PPh_3)_4$
NMP , 100° , 5h

92%

Rao, M.L.N.; Tamazaki, I.; Shimada, S.; Tanaka, T.; Suzuki, Y.; Tanaka, M. *Org. Lett.*, ***2001***, *3*, 4103.

$$PhCH_2Cl \xrightarrow{HSiCl_3 \text{ , } PEt_3 \text{ , 150°C , 2 h}} PhCH_2SiCl_3$$

88%

Cho, Y.S.; Kang, S.-H.; Han, J.-S.; Yoo, B.R.; Jung, I.N. *J. Am. Chem. Soc.*, ***2001***, *123*, 5584.

toluene , reflux

62%

Pyun, D.K.; Lee, C.H.; Ha, H.-J.; Park, C.S.; Chang, J.-W.; Lee, W.K.
Org. Lett., ***2001***, *3*, 4205.

Me_3Si—≡

4.5% $PdCl_2(PPh_3)_2$, 16 h
23% PPh_3 , $CuSO_4$, Al_2O_3
NEt_3 , deoxygenated PhH

66%

Dang, H.; Garcia-Garibay, M.A. *J. Am. Chem. Soc.*, ***2001***, *123*, 355.

34% Ph_3In , THF , reflux

1% $Pd(PPh_3)_2Cl_2$, 3h

95%

Pérez, I.; Sestelo, J.P.; Sarandeses, L.A. *J. Am. Chem. Soc.*, ***2001***, *123*, 4155.

$PhCH_2Cl$ —10 atm CO , PhH , 130°C , 10 h→ $PhCH_2Ph$

95%

Ogoshi, S.; Nakashima, H.; Shimonaka, K.; Kurosawa, H.
J. Am. Chem. Soc., ***2001***, *123*, 8626.

PhCl —9-BBN(B-OMe)(B-$C_{14}H_{29}$)$^-K^+$→ Ph—$(CH_2)_{13}Me$ 86%

2% $Pd(OAc)_2$, 4% IPr-HCl , THF
reflux , 4 h

Fürstner, A.; Leitner, A. *Synlett*, ***2001***, 290.

PhBNa , $Pd(OAc)_2$, MeNHCHO
microwaves , 12 min

86%

Villemin, D.; Gómez-Escalonilla, M.J.; Saint-Clair, J.-F. *Tetrahedron Lett.*, ***2001***, *42*, 635.

PhI → (4-$CH_3C_6H_4B(OH)_2$, KF–Al_2O_3; microwaves, 15 min) → 4-$CH_3C_6H_4$–Ph 98%

Villemin, D.; Caillot, F. *Tetrahedron Lett.*, **2001**, *42*, 639.

6-Chloro-7-azaindole → ($PhB(OH)_2$, dioxane, KF; cat $Pd_2(dba)_3$•$CHCl_3$, 100°C, $P(t\text{-}Bu)_3$) → 6-phenyl-7-azaindole 76%

Allegretti, M.; Arcadi, A.; Marinelli, F.; Nicolini, L. *Synlett*, **2001**, 609.

$PhCHI_2$ → (1. Et_2Zn, THF/NMP; 2. H_3O^+) → $PhCH_2CH_2Et$ 75%

Shibli, A.; Varghese, J.P.; Knochel, P.; Marek, I. *Synlett*, **2001**, 818.

4-Iodoacetophenone → ($(t\text{-}BuSiMePh)_n$, THF; 60°C, 20 h) → 4-phenylacetophenone 67%

Mori, A.; Suguro, M. *Synlett*, **2001**, 845.

MeO–C_6H_4–Cl → ($PhB(OH)_2$, 2 eq K_2CO_3; cat $[PdCl(C_3H_5)]_2$, 130°C; tetraphosphine ligand, 20 h) → MeO–C_6H_4–Ph 95%

Feuerstein, M.; Doucet, H.; Santelli, M. *Synlett*, **2001**, 1458.

2-Bromo-3,4-dimethoxybenzaldehyde → (0.2 $NiCl_2(PPh_3)_2$/Zn, 6 eq NaH, 0.4 PPh_3; toluene, 90°C, 12 h) → biaryl dialdehyde (CHO, OMe, OMe / MeO, OMe, OHC) + 3,4-dimethoxybenzaldehyde

(36 : 64) 95%

+ 2 eq Bu_4NI (80 : 20) 94%

Hong, R.; Hoen, R.; Zhang, J.; Lin, G.-q. *Synlett*, **2001**, 1527.

1. electro-generated Zn/NaPh, DMA, –20°C, 1 h
2. PhI, DMF, 70°C, 3 h, 5% Pd(Po-Tol$_3$)$_2$Cl$_2$

Br–C(=CH$_2$)–CO_2Et → Ph–C(=CH$_2$)–CO_2Et 98%

Jalil, A.A.; Kurono, N.; Tokuda, M. *Synlett*, ***2001***, 1944.

N(pyridin-4-yl)–B(OH)$_2$; cat Pd(dppf)Cl$_2$, 5 h; Na$_2$CO$_3$, reflux, aq DME

Br, O, O → N, O, O 74%

Morris, G.A.; Nguyen, S.T. *Tetrahedron Lett.*, ***2001***, *42*, 2093.

2% perfluoro tosylate, Bu$_4$NCl, phosphine Pd catalyst; OMe, SnBu$_3$, supercritical CO$_2$

Br, MeO$_2$C → MeO, MeO$_2$C 93%

Osswald, T.; Schneider, S.; Wang, S.; Bannwarth, W. *Tetrahedron Lett.*, ***2001***, *42*, 2965.

1. C$_8$H$_{17}$B(OH)$_2$, THF, 3 eq Ag$_2$O, 5% PdCl$_2$(MeCN)$_2$, 20% AsPh$_3$, reflux, 22 h
2. SnMe$_4$, cat PdCl$_2$(P(*o*-Tol$_3$)$_2$, CuI, NMP, 85°C, 18 h

Br, Br, O, O → C$_8$H$_{17}$, C$_8$H$_{17}$, O, O 71x90%

Bellina, F.; Anselmi, C.; Rossi, R. *Tetrahedron Lett.*, ***2001***, *42*, 3851.

GaLn, aq THF; BEt$_3$/O$_2$, 2 h

Br–CH$_2$–CO$_2$Bn → CH$_2$=CH–CH$_2$–CH$_2$–CO$_2$Bn 78%

Usugi, S.-i.; Yoromitsu, H.; Oshima, K. *Tetrahedron Lett.*, ***2001***, *42*, 4535.

2-Bromopyridine → 2-phenylpyridine: $PhB(OH)_2$, K_2CO_3 , 130°C , xylene / $[Pd(C_3H_5)Cl]_2$/tetraphosphine ligand , 20h — quant

Feuerstein, M.; Doucet, H.; Santelli, M. *Tetrahedron Lett.*, ***2001***, *42*, 5659.

Ethyl 6-bromonicotinate → ethyl 6-phenylnicotinate: PhMgCl , 5% $Pd(dba)_2$, 6 h / 10% dppf ,THF , –40°C — 95%

Bonnet, V.; Mongin, F.; Trécourt, F.; Quéguiner, G.; Knochel, P. *Tetrahedron Lett.*, ***2001***, *42*, 5717.

I–C6H4–OMe → Bu–C6H4–OMe: 1. [pinacol(s-Bu)BBu]⁻ Li^+ ; THF , cat $Pd_2(dba)_3$, 80°C, NaOAc; 2. H_2O_2 , NaOH — 69%

Zou, G.; Falck, J.R. *Tetrahedron Lett.*, ***2001***, *42*, 5817.

3-Bromopyridine → 3-phenylpyridine: $PhB(OH)_2$, 5% $Pd(OAc)_2$, 8 h, 15% Na triphenylphosphino-trisulfonate / 2.5 eq i-Pr_2NH , aq MeCN , 80°C — 75%

Dupuis, C.; Adiey, K.; Charruault, L.; Michelet, V.; Savignac, M.; Genêt, J.-P. *Tetrahedron Lett.*, ***2001***, *42*, 6523.

PhBr → Ph—Ph: $PhB(OH)_2$, xylene , K_2CO_3 , 130°C / $[Pd(C_3H_5)Cl]_2$/tetraphosphine ligand — quant

Feuerstein, M.; Doucet, H.; Santelli, M. *Tetrahedron Lett.*, ***2001***, *42*, 6667.

Br–C6H4–OMe → Bu–C6H4–OMe: $BuB(OH)_2$, cat $Pd(dppf)Cl_2$ / 3 eq K_2CO_3 , 2.5 eq Ag_2O, THF , 80°C , sealed tube — 80%

Zou, G.; Reddy, Y.K.; Falck, J.R. *Tetrahedron Lett.*, ***2001***, *42*, 7217.

2 PhBr → Ph—Ph: 10% $Ni(acac)_2$/$Al(acac)_3$, NaH / 2,2'-bpy , THF , reflux — 76%

Massicot, F.; Schneider, R.; Fort, Y.; Illy-Cherrey, S.; Tillement, O. *Tetrahedron*, ***2001***, *57*, 531.

$PhB(OH)_2$, cat $Pd(dba)_2$
cat PPh_3, K_3PO_4, 105°C
overnight

83%

Baillie, C.; Chen, W.; Xiao, J. *Tetrahedron Lett.*, ***2001***, *42*, 9085.

OMs

PhH, 10% $Cu(OTf)_2$
80°C, 4 h

Ph

81%

Singh, R.P.; Kamble, R.M.; Chandra, K.L.; Saravanan, P.; Singh, V.K. *Tetrahedron*, ***2001***, *57*, 241.

NH_2 Cl N N NH_2

$PhB(OH)_2$, cat $Pd(PPh_3)_4$
2M Na_2CO_3, DME

Ph N N NH_2 NH_2

73%

Cooke, G.; de Cremiers, H.A.; Rotello, V.M.; Tarbit, B.; Vanderstraeten, P.E. *Tetrahedron*, ***2001***, *57*, 2787.

O_2N Br

PhI, cat $Pd(OAc)_2$, 130°C
i-$PrNEt_2$, Bu_4NBr, xylene

O_2N Ph

32% + 21% homo-coupling

Hassan, J.; Hathroubi, C.; Gozzi, C.; Lemaire, M. *Tetrahedron*, ***2001***, *57*, 7845.

Cl CHO

OMe $B(OH)_2$, H_2O

2% $Pd(OAc)_2$, 2.5 eq K_2CO_3
Bu_4NBr

MeO CHO

59%

Hesse, S.; Kirsch, G. *Synthesis*, ***2001***, 755.

Br CF_3

$PhSi(OEt)_3$, cat $PdCl_2(MeCN)_2$
aq NaOH, dioxane

Ph CF_3

94%

Murata, M.; Shimazaki, R.; Watanabe, S.; Masuda, Y. *Synthesis*, ***2001***, 2231.

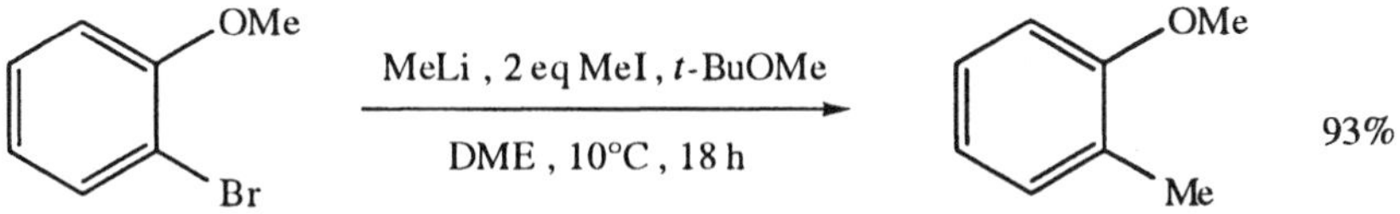
OMe
MeLi , 2 eq MeI , t-BuOMe
DME , 10°C , 18 h
Br
OMe
Me
93%

Andrews, I.P.; Kitteringham, J.; Vogle, M. *Synth. Commun.*, ***2001**, 31,* 2323.

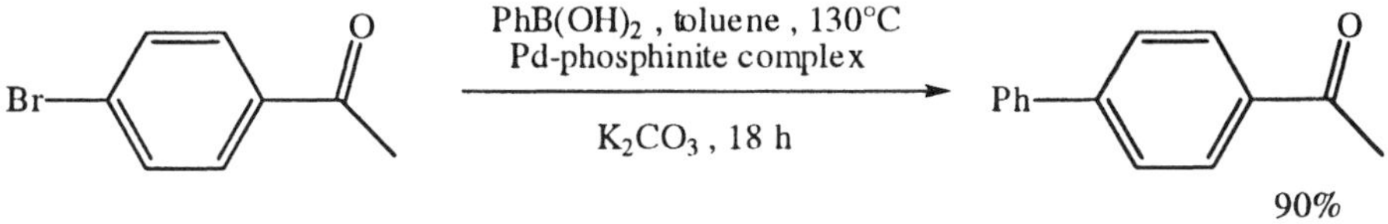
Br
O
PhB(OH)2 , toluene , 130°C
Pd-phosphinite complex
K2CO3 , 18 h
Ph
O
90%

Bedford, R.B.; Welch, S.L. *Chem. Commun.*, ***2001**,* 129.

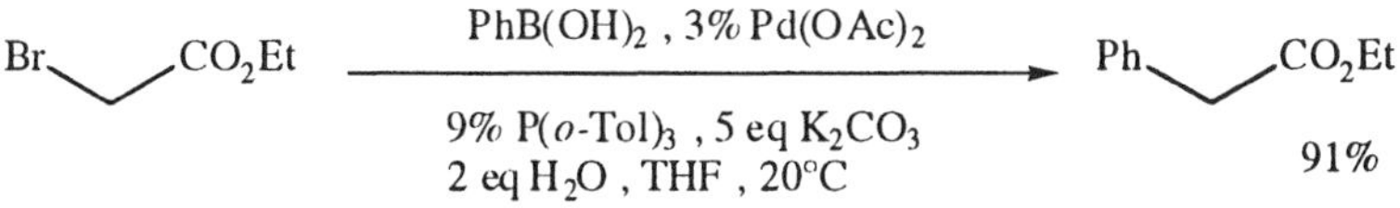
Br
CO2Et
PhB(OH)2 , 3% Pd(OAc)2
9% P(o-Tol)3 , 5 eq K2CO3
2 eq H2O , THF , 20°C
Ph
CO2Et
91%

Gooben, L.J. *Chem. Commun.*, ***2001**,* 669.

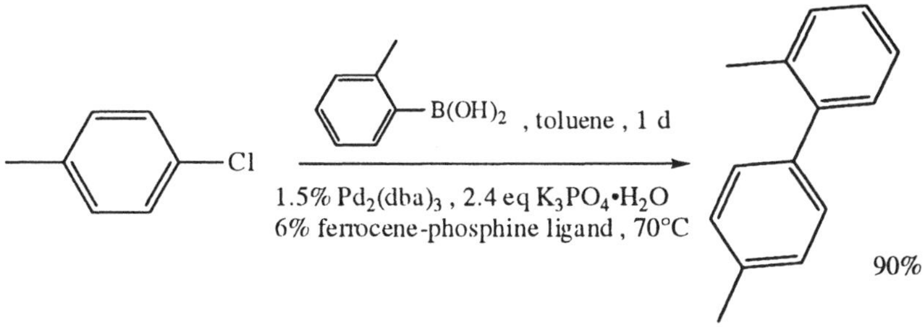
Cl
B(OH)2
, toluene , 1 d
1.5% Pd2(dba)3 , 2.4 eq K3PO4•H2O
6% ferrocene-phosphine ligand , 70°C
90%

Liu, S.-Y.; Choi, M.J.; Fu, G.C. *Chem. Commun.*, ***2001**,* 2408.

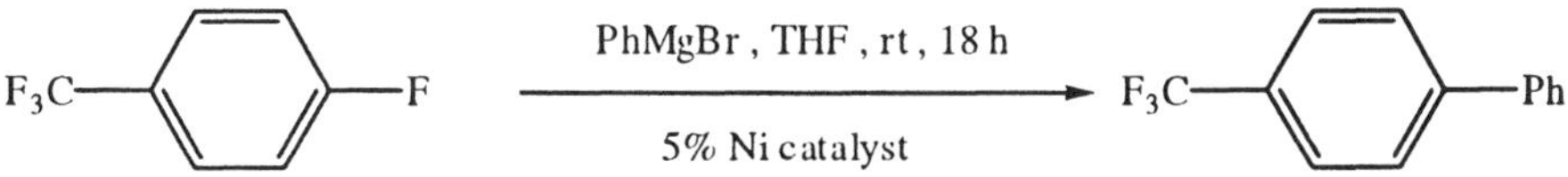
F3C
F
PhMgBr , THF , rt , 18 h
5% Ni catalyst
F3C
Ph

Böhm, V.P.W.; Gstöttmayr, C.W.K.; Weskamp, T.; Hermann, W.A. *Angew. Chem. Int. Ed.*, ***2001**, 40,* 3387.

$$2\ PhBr \xrightarrow[\text{2,2'-bpy , THF , reflux}]{\text{10\% Ni(acac)}_2\text{/Al(acac)}_3\text{ , NaH}} Ph{-}Ph \quad 76\%$$

Massicot, F.; Schneider, R.; Fort, Y.; Illy-Cherrey, S.; Tillement, O. *Tetrahedron*, ***2001**, 57,* 531.

$$PhCH_2Cl \xrightarrow[\text{10\% LiB(C}_6\text{F}_5)_4]{\text{PhH , reflux , 8 h}} PhCH_2Ph \quad 96\%$$

Mukaiyama, T.; Nakano, M.; Kikuchi, W.; Matsuo, J.-i. *Chem. Lett.*, ***2000**,* 1010.

SnBu$_3$ SiMe$_2$ N

I– O

5% $PdCl_2(MeCN)_2$, THF , 1 d
10% $P(C_6F_5)_3$, 50°C

O SiMe$_2$ N

87%

Itami, K.; Kamei, T.; Yoshida, J.-i. *J. Am. Chem. Soc.*, ***2001***, *123*, 8773.

–Cl

$PhB(OH)_2$, toluene , 20 h
cat $Pd(OAc)_2$, 2 eq $BuPAd_2$

K_3PO_4 , 100°C

–Ph

87%

Zapf, A.; Ehrentraut, A.; Beller, M. *Angew. Chem. Int. Ed.*, ***2000***, *39*, 4153.

Br

$PhB(OH)_2$, cat $Pd(PPh_3)_4$
K_2CO_3 , NaCl

solvent free

Ph

96%

Nielsen, S.F.; Peters, D.; Axelsson, O. *Synth. Commun.*, ***2000***, *30*, 3501.

F– –Cl

$PhB(OH)_2$, new Pd catalyst

F– –Ph

67%

Andreu, M.G.; Zapf, A.; Beller, M. *Chem. Commun.*, ***2000***, 2475.

Br

10% $Pd(OAc)_2$, 20% PPh_3
2 eq $PhSi(OMe)_3$, DMF

2 eq TBAF , 85°C

Ph

85%

DeShong, P.; Handy, C.J.; Mowery, M.E. *Pure Appl. Chem.*, ***2000***, *72*, 1655.

F_3C– –Cl

$PhB(OH)_2$, 1% $Pd(OAc)_2$, 18 h
toluene , 120°C , 10% CsF_2

10% P(O–2,4-di-*t*-Bu-phenyl)$_3$

F_3C– –Ph

77%

Zapf, A.; Beller, M. *Chem. Eur. J.*, ***2000***, *6*, 1830.

$PhCH_2Cl$ → PhOH, reflux, 1 h / Envirocat EPZ-10 → Ph, OH

88% (1:4 *o:p*)

Bandgar, B.P.; Kasture, S.P. *Monat. Chem.*, **2000**, *131*, 913.

Br → $PhB(OH)_2$, K_3PO_4, 2% $Pd(OAc)_2$, DMF / 130°C, 1 h → Ph 76%

Zim, D.; Monteiro, A.L.; Dupont, J. *Tetrahedron Lett.*, **2000**, *41*, 8199.

I, I, N, SO_2Ph → $PhB(OH)_2$, 10% $Pd(OAc)_2$ / $P(o\text{-Tol})_3$, K_2CO_3, aq acetone, 70°C, 5 h → Ph, Ph, N, SO_2Ph 83%

Liu, Y.; Gribble, G.W. *Tetrahedron Lett.*, **2000**, *41*, 8717.

Ac, Br → PhH, *p*-xylene, 130°C / $Pd(OAc)_2$, Bu_4NBr, *i*-Pr_2NEt → Ac, Ph 95%

Hassan, J.; Hathroubi, C.; Gozzi, C.; Lemaire, M. *Tetrahedron Lett.*, **2000**, *41*, 8791.

Ph, Cl → 10% ferrocenyl amine ligand, $(t\text{-BuCH}_2)_2Zn$, –30°C, 3 h / 1% $CuBr\text{-}SMe_2$ → Me, Ph 82% (96% ee)

Okano, K.; Murata, K.; Ikariya, T. *Tetrahedron Lett.*, **2000**, *41*, 9277.

Br, O, P—OEt, OEt → $B(OH)_2$, Me / cat $Pd_2(dba)_3$, 70°C, 2 eq K_3PO_4, 80 h → Me, O, P—OEt, OEt

98% (87% ee)

Yin, J.; Buchwald, S.L. *J. Am. Chem. Soc.*, **2000**, *122*, 12051.

Br (bromobenzene) → [bmim]PF_6, 80°C; cat Ni$(PPh_3)_4$ → Ph—Ph 87%

Howarth, J.; James, P.; Dai, J. *Tetrahedron Lett.*, ***2000**, 41*, 10319.

NC—C_6H_4—Cl → $CH_3C_6H_4B(OH)_2$, toluene; 2 eq K_3PO_7•n H_2O, 6% PPh_3; 3% $NiCl_2(PPH_3)_2$, 2 h → NC 87%

Inada, K.; Miyaura, N. *Tetrahedron*, ***2000**, 56*, 8657, 8661.

$CH_3C_6H_4$—Br → 1. 0.5 BuMgLi, THF, 0°C; 2. 1.5 eq $TiCl_4$, –78°C → 0°C → 82%

Inoue, A.; Kitagawa, K.; Shinokubo, H.; Oshima, K. *Tetrahedron*, ***2000**, 56*, 9601.

OMe, Br, CO_2Me → B, O, O; 1.2 eq Tl_2CO_3, 6% $Pd(PPh_3)_4$; PhH, 100 h → MeO, CO_2Me 96%

Chaumeil, H.; Signorella, S.; Le Drian, C. *Tetrahedron*, ***2000**, 56*, 9655.

Ph–I → cat Pd/C, Zn, H_2O, air; 18-crown-6, rt → Ph–Ph 54%

Venkatraman, S.; Li, C.-J. *Tetrahedron Lett.*, ***2000**, 41*, 4831.

$\xrightarrow[\text{0.5\% Pd}_2\text{(bda)}_3\text{, 1\% P(}t\text{-Bu)}_3\text{, 3.3 eq KF}]{(o\text{-Tol)B(OH)}_2\text{, THF, rt}}$ 99%

Cl · O

Littke, A.F.; Dai, C.; Fu, G.C. *J. Am. Chem. Soc.*, ***2000***, *122*, 4020.

PhOTf $\xrightarrow[\text{Pd(PPh}_3)_4\text{, 150°C, 1 min}]{\text{Bu}_4\text{N}^+\text{ (Bn}_3\text{SnF}_2)^-\text{, DMF}}$ $PhCH_2Ph$ 68%

Martínez, A.G.; Barcinia, J.O.; Heras, Md.R.C.; Cerezo, A.d.F. *Org. Lett.*, ***2000***, *2*, 1377.

$PhSi(OMe)_3$ $\xrightarrow[\text{2 eq TBAF, dioxane, THF, 3 h}]{\text{PhBr, 3\% Pd(OAc)}_2\text{, 3\% IPr•HCl}}$ Ph—Ph quant

IPr = 1,3-bis(2,6-diisopropylphenyl)imidazolyl-2-ylidene

Lee, H.M.; Nolan, S.P. *Org. Lett.*, ***2000***, *2*, 2053.

Bu, $B(OH)_2$ $\xrightarrow[\text{2 eq Ag}_2\text{O, 2 eq KOH, 80°C; 3\% PdCl}_2\text{(dppf), dioxane, 2 h}]{\text{Br, Ph}}$ Bu, Ph 82%

Chen, H.; Deng, M.-Z. *J. Org. Chem.*, ***2000***, *65*, 4444.

Br, Br, MeO_2C $\xrightarrow[\text{2.5\% Pd}_2\text{(dba)}_2\text{, 15\% TFP, Na}_2\text{CO}_3]{\text{PhB(OH)}_2\text{, dioxane, 65°C}}$ Br, Ph, MeO_2C 89%

Shen, W. *Synlett*, ***2000***, 737.

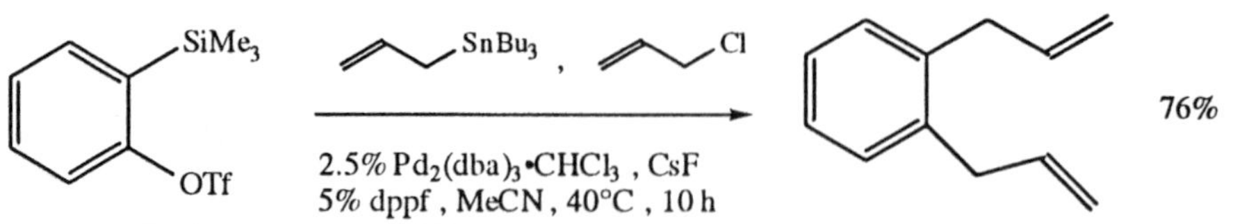

$SiMe_3$ / OTf → (allyl-$SnBu_3$, allyl-Cl; 2.5% $Pd_2(dba)_3 \cdot CHCl_3$, CsF; 5% dppf , MeCN , 40°C , 10 h) 76%

Yoshikawa, E.; Radhakrishnan, K.V.; Yamamoto, Y. *Tetrahedron Lett.*, ***2000***, *41*, 729.

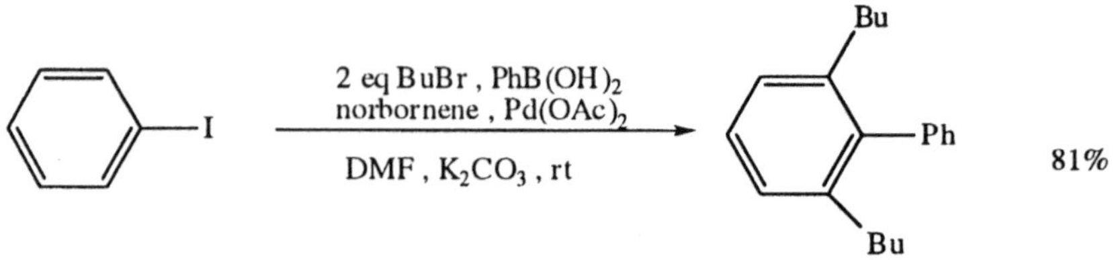

I → (2 eq BuBr , $PhB(OH)_2$; norbornene , $Pd(OAc)_2$; DMF , K_2CO_3 , rt) Bu / Ph / Bu 81%

Catellani, M.; Motti, E.; Minari, M. *Chem. Commun.*, ***2000***, 157.

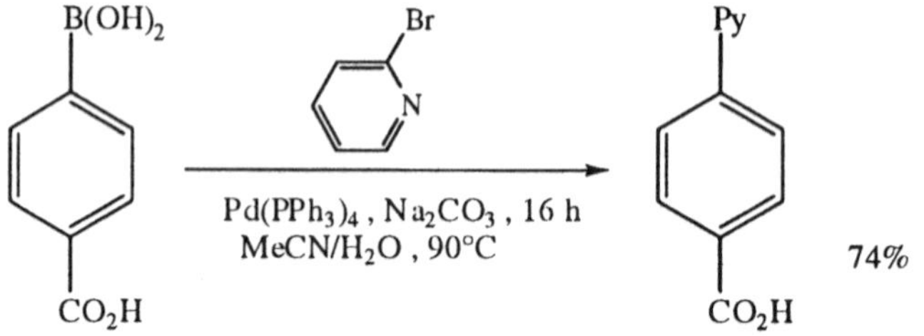

$B(OH)_2$ / CO_2H → (Br / N; $Pd(PPh_3)_4$, Na_2CO_3 , 16 h; $MeCN/H_2O$, 90°C) Py / CO_2H 74%

Gong, Y.; Pauls, H.W. *Synlett*, ***2000***, 829.

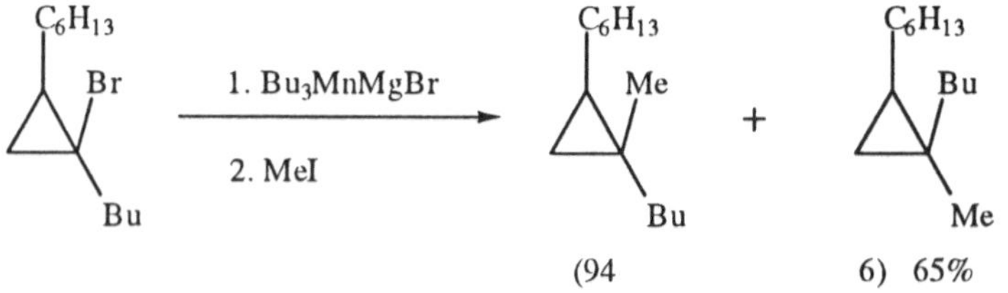

C_6H_{13} / Br / Bu → (1. $Bu_3MnMgBr$; 2. MeI) C_6H_{13} / Me / Bu + C_6H_{13} / Bu / Me

(94 6) 65%

Kakiya, H.; Inoue, R.; Shinokubo, H.; Oshima, K. *Tetrahedron*, ***2000***, *56*, 2131.

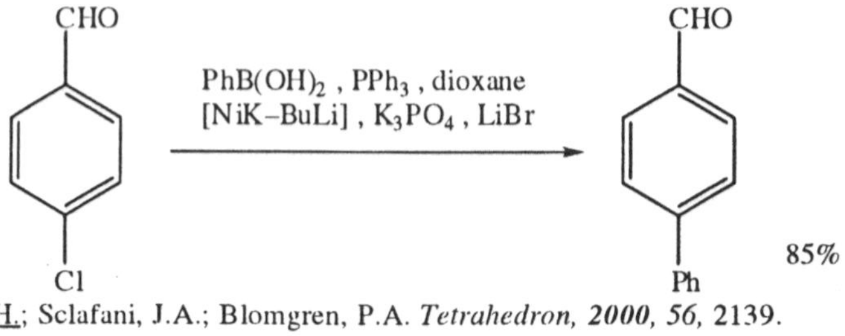

CHO / Cl → ($PhB(OH)_2$, PPh_3 , dioxane; [NiK–BuLi] , K_3PO_4 , LiBr) CHO / Ph 85%

Lipshutz, B.H.; Sclafani, J.A.; Blomgren, P.A. *Tetrahedron*, ***2000***, *56*, 2139.

2 PhI —[$NiCl_2$, $CrCl_2$, Mn , THF , rt / bipyridyl ligand , 70 h]→ Ph—Ph 98%

Chen, C. *Synlett*, ***2000***, 1491.

O_2N–C_6H_3(CF$_3$)–Br —[Me–B–O / O B–Me / Me–B–O; 10% $Pd(PPh_3)_4$, dioxane, K_2CO_3 , reflux]→ O_2N–C_6H_3(CF$_3$)–Me 75%

Gray, M.; Andrews, I.P.; Hook, D.F.; Kitteringham, J.; Vogle, M. *Tetrahedron Lett.*, ***2000***, *41*, 6237.

ZnBr, OMe —[I–C_6H_4–CO_2Et; ionic phosphine liquid , toluene, cat $Pd(dba)_2$, 0.3 h]→ MeO, CO_2Et 91%

Sirieix, J.; Oβberger, M.; Betzemeier, B.; Knochel, P. *Synlett*, ***2000***, 1613.

Ph–Br —[$Me_3Si–SiMe_3$, 3% $Pd(PPh_3)_4$, DMPU / 9% PPh_2Py , 5 eq K_2CO_3 , 2 eq H_2O , 100°C]→ Ph–$SiMe_3$

Gooβen, L.J.; Ferwanah, A.-R.S. *Synlett*, ***2000***, 1801.

Cl, Me —[t-Bu$CO_2(CH_2)_4ZnI$, 5% Ni(0) , LiCl , THF]→ $(CH_2)_4$, O, O, Me 92%

Ni(0) = [$NiCl_2$/2-4 PPh_3/2 BuLi/THF/rt]

Lipshutz, B.H.; Blomgren, P.A.; Kim, S.-K. *Tetrahedron Lett.*, ***1999***, *40*, 197.

PhBr —[0.03 $Ni(CO)_2(PPh_3)_2$, DMSO / 70°C , 6 h]→ Ph—Ph 75%

Leadbeater, N.E.; Resouly, S.M. *Tetrahedron Lett.*, ***1999***, *40*, 4243.

I

, 225°C , 100 bar

aqueous Bu_4NCl

40%

Heck reaction – in aqueous media

67:13:20 1-Ph:2-Ph:3-Ph-cyclohexene

Cron, L.U.; Tinsley, A.S. *Tetrahedron Lett.*, ***1999***, *40*, 227.

N
H

ICH_2CO_2Et , $Na_2S_2O_3$
Bu_4NBr , *t*-BuOMe , hν

methyloxirane

CO_2Et
N
H

90%

Byers, J.H.; Campbell, J.E.; Knapp, F.H.; Thissell, J.G. *Tetrahedron Lett.*, ***1999***, *40*, 2677.

Br
N
H

$SnBu_3$

0.05 $Pd(PPh_3)_4$, 0.2 CuI
60°C , DMF , 30 min

N
H

92%

Benhida, R.; Lecubin, F.; Fourrey, J.-L.; Castellanos, L.R.; Quintero, L.
Tetrahedron Lett., ***1999***, *40*, 5701.

O
N
Br
O
O

$Yb(OTf)_3$, 0.5 Et_3B/O_2

O
C_4H_9
N
Br
O
O

79%

Mero, C.L.; Porter, N.A. *J. Am. Chem. Soc.*, ***1999***, *121*, 5155.

CHO
Cl

Cl ZnI

5% Ni/C , PPh_3 , THF

CHO
CN

80%

Lipshutz, B.H.; Blomgren, P.A. *J. Am. Chem. Soc.*, ***1999***, *121*, 5819.

Me, Br → 1. BuLi, THF, -50°C; 2. Cp_2ZrCMe/Cl, bis-phosphine; -50°C → rt; 3. cat $Pd_2(dba)_3$, Br–C6H4–Cl, 60°C; 4. I_2, CH_2Cl_2 → Me, I, Cl

64% ($m{:}o$ = 80:1)

Frid, M.; Pérez, D.; Peat, A.J.; Buchwald, S.L. *J. Am. Chem. Soc.*, ***1999***, *121*, 9469.

OTMS → $Sc(OTf)_3$, PhH; 80°C, 4 h → Ph 92%

several other catalysts used as well

Kotsuki, H.; Ohishi, T.; Inoue, M.; Kojima, T. *Synthesis*, ***1999***, 603.

O, I → 10% $Pd(dba)_2$, 5 eq Ph_3SiF_2; DMF, 5 h → O, Ph

86%

Mowery, M.E.; DeShong, P. *J. Org. Chem.*, ***1999***, *64*, 3266.

I, Ph → $(C_5H_{11})_2Zn$, THF/NMP; 7% $Ni(acac)_2$, –35°C → C_5H_{11}, Ph

80%

Giovannini, R.; Stüdemann, T.; Devasagayaraj, A.; Dussin, G.; Knochel, P. *J. Org. Chem.*, ***1999***, *64*, 3544.

PEG, O, O, OTf → $PhB(OH)_2$, H_2O; 2.5 eq K_2CO_3; cat $Pd(OAc)_2$; microwaves → PEG, O, O, Ph

>95%

Lettner, C.G.; König, W.A.; Stenzel, W.; Schotten, T. *J. Org. Chem.*, ***1999***, *64*, 3885.

Br → Ph–$B(OH)_2$, H_2O, rt; Bu_4NBr, 2 h → Ph 45%

Bussolari, J.C.; Rehborn, D.C. *Org. Lett.*, ***1999***, *1*, 965.

Br

Br

$(2\text{-furyl})_3P$, $Pd_2(dba)_3$, toluene

2 eq (2-furyl)$SnBu_3$, 100°C

CO_2Me

Br

80%

CO_2Me

Shen, W.; Wang, L. *J. Org. Chem.*, ***1999***, *64*, 8873.

MeO—C_6H_4—$SiMe_2OH$ → PhI, cat Ph$(PPh_3)_4$ / Ag_2O , THF , 60°C , 36 h → MeO—C_6H_4—Ph

80%

Hirabayashi, K.; Kawashima, J.; Nishihara, Y.; Mori, A.; Hiyama, T. *Org. Lett.*, ***1999***, *1*, 299.

Ph—I → 8% Pc/C , air , Zn , overnight / H_2O/acetone , rt → Ph—Ph 94%

Venkatraman, S.; Li, C.-J. *Org. Lett.* ***1999***, *1*, 1133.

Ph—I → 2% $Pd(OAc)_2$, 2% As$(o\text{-Tol})_3$, Cs_2CO_3 , 0.4 M , 75°C , 3 h / 50% hydroquinone → Ph—Ph 96%

Hennings, D.D.; Iwama, T.; Rawal, V.H. *Org. Lett.* ***1999***, *1*, 1205.

3 Me—C_6H_4—I → Ph_3In , THF , reflux, cat $PdCl_2(PPh_3)_2$, 1 h → 3 Me—C_6H_4—Ph

96%

Pérez, I.; Sestelo, J.P.; Sarandeses, L.A. *Org. Lett.* ***1999***, *1*, 1267.

Cl—C_6H_4—OTf → BnZnBr , $Pd(dba)_2$, 2 h / dppf , Bu_4NI , THF , 60°C → Cl—C_6H_4—Bn

94%

Piber, M.; Jensen, A.E.; Rottländer, M.; Knochel, P. *Org. Lett.*, ***1999***, *1*, 13223.

OTf → Ph (naphthalene); 10% $Pd(PPh_3)_4$, NMP, 80°C, 3 h; Et, Et, O, N → Bi—Ph, O, Et, Et; 80%

Rao, M.L.N.; Shimada, S.; Tanaka, M. *Org. Lett.* ***1999***, *1*, 1271.

PhI → (4-methylphenyl)-Ph; $B(OH)_2$, 5% Pd°, KF; neat, Al_2O_3, 80°C; 86%

Kabalka, G.W.; Pagni, R.M.; Hair, C.M. *Org. Lett.*, ***1999***, *1*, 1423.

Br → Ph; 10% $Pd(OAc)_2$, 20% PPh_3; 2 eq $PhSi(OMe)_3$, 2 eq TBAF; DMF, 85°C; 82%

Mowery, M.E.; DeShong, P. *Org. Lett.*, ***1999***, *1*, 2137.

N Cl → N Ph; $PhB(OH)_2$, K_2CO_3, reflux; 5% $Ph(PPh_3)_4$, 8 h; toluene/H_2O/EtOH; 80%

Lohse, O.; Thevenin, P.; Waldvogel, E. *Synlett*, ***1999***, 45.

Ph–I → Ph—Ph; 2.5% $PdCl(\pi\text{-}C_3H_5)_2$, DMSO; TBAF, 120°C, 4.5 h; 82%

Albanese, D.; Landini, D.; Penso, M.; Petricci, S. *Synlett*, ***1999***, 199.

O → O Ph; PhOTf, $Pd_2(dba)_3$•dba, DMF; 12% chiral biaryl bis-phosphine; 3 eq $i\text{-}Pr_2NEt$, 90°C, 20 h; 90% (90% ee)

Tietze, L.F.; Thede, K.; Sannicolò, F. *Chem. Commun.*, ***1999***, 1811.

$Me(CH_2)_8CH_2Cl$ → $Me(CH_2)_8CH_2H$; $NiCl_2$•2 H_2O, Li, 5% dTbb; 55%

dTbb = d,4'-di-*t*-butylbiphenyl

Alonso, F.; Radivoy, G.; Yus, M. *Tetrahedron*, ***1999***, *55*, 4441.

Br / OMe → Bu / OMe

BuI, e^-, 70°C, THF, 3 h

$Pd[P(o\text{-}Tol)_3]_2Cl_2$
electrogenerated reactive Zn

79%

Jurono, N.; Sugita, K.; Takasugi, S.; Tokuda, M. *Tetrahedron*, ***1999***, *55*, 6097.

Br → Ph

$PhB(OH)_2$, dioxane, 8 h

$NiCl_2(bipy)$, Na_3PO_4

70%

Leadbeater, N.E.; Resouly, S.M. *Tetrahedron*, ***1999***, *55*, 11889.

Br, N, 3% $Pd(PPh_3)_4$

Bu, $B(OH)_2$ → Bu, N

3.3 eq $K_3PO_4 \cdot 3\ H_2O$, 32 h
toluene, 85°C

59%

Ma, H.-r.; Wang, X.-h.; Deng, M.-z. *Synth. Commun.*, ***1999***, *29*, 2477.

OH, Br → OH, OH

$Mn_2(CO)_{10}$, DCM

hν

88%

Gilbert, B.C.; Lindsay, C.I.; McGrail, P.T.; Parsons, A.F.; Whittaker, D.T.E. *Synth. Commun.*, ***1999***, *29*, 2711.

N, Me → N, Me, Ph

1. BuLi, THF
2. BF_3

3. PhBr, $PdCl_2(PPh_3)_2$

80%

Ishikura, M.; Agata, I.; Katagiri, N. *J. Heterocyclic Chem.*, ***1999***, *36*, 873.

SECTION 71: ALKYLS, METHYLENES AND ARYLS FROM HYDRIDES

This section lists examples of the reaction of RH → RR' (R,R' = alkyl or aryl). For the reaction C=CH → C=C-R (R = alkyl or aryl), see Section 209 (Alkenes from Alkenes). For alkylations of ketones and esters, see Section 177 (Ketones from Ketones) and Section 113 (Esters from Esters).

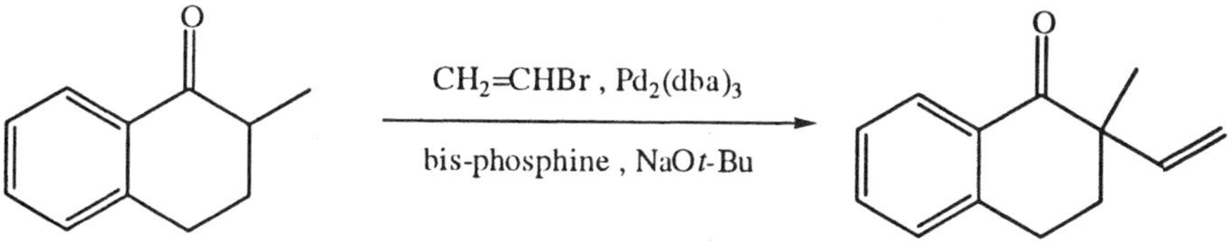

CH_2=CHBr , $Pd_2(dba)_3$

bis-phosphine , NaO*t*-Bu

96% (80% ee)

Chieffi, A.; Kamikawa, K.; Åhman, J.; Fox, J.M.; Buchwald, S.L. *Org. Lett.* ***2001***, *3*, 1897.

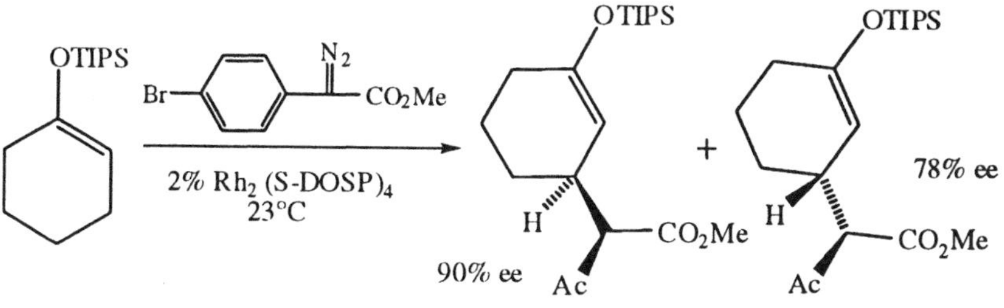

OTIPS

Br, N_2, CO_2Me

2% Rh_2 (S-DOSP)$_4$
23°C

OTIPS, H, CO_2Me, Ac — 90% ee

+

OTIPS, H, CO_2Me, Ac — 78% ee

(65 : 35) 80%

Davies, H.M.L.; Ren, P. *J. Am. Chem. Soc.*, ***2001***, *123*, 2070.

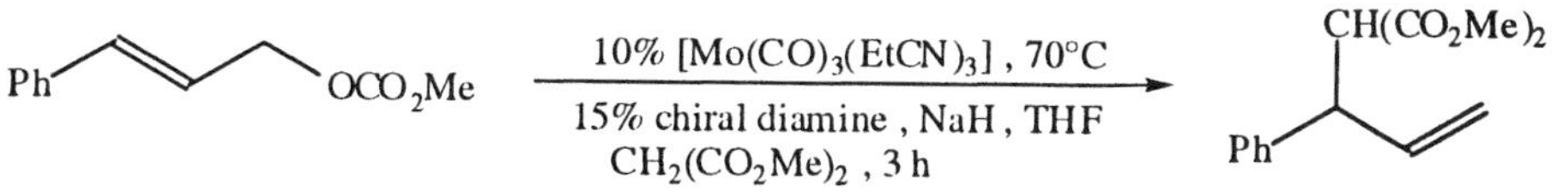

Ph, OCO_2Me

10% [$Mo(CO)_3(EtCN)_3$] , 70°C

15% chiral diamine , NaH , THF
$CH_2(CO_2Me)_2$, 3 h

$CH(CO_2Me)_2$, Ph

70% (99% ee)

Glorius, F.; Neuburger, M.; Pfaltz, A. *Helv. Chim. Acta*, ***2001***, *84*, 3178.

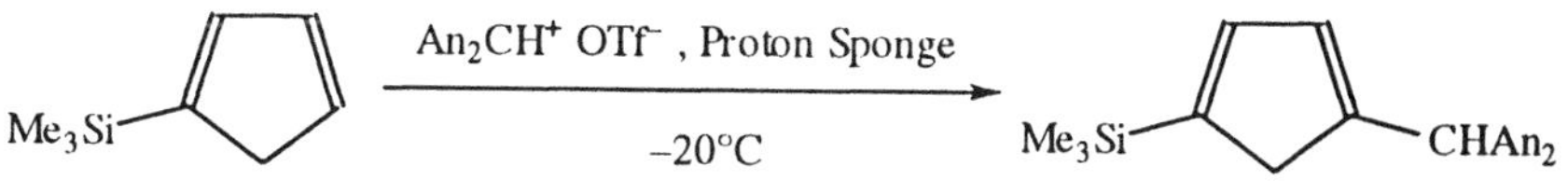

Me_3Si

An_2CH^+ OTf^- , Proton Sponge

−20°C

Me_3Si, $CHAn_2$

An = *p*-(MeO)-C_6H_4–

53%

Herrlich, M.; Hampel, N.; Mayr, H. *Org. Lett.*, ***2001***, *3*, 1629.

1. 1.3 eq *sec*-BuLi, 1.3 eq diamine
−78°C, 5 h
N-Me
2. Me_3SiCl

$SiMe_3$
Boc
41% (53% ee)

Harrison, J.R.; O'Brien, P.; Porter, D.W.; Smith, N.M. *Chem. Commun.*, ***2001***, 1202.

CO_2Et
1. PhOCOCl, EtMgCl
CuBr•DMS, THF
2. DDQ, *i*-PrOAc

Et
CO_2Et
44%

Wallace, D.J.; Gibb, A.D.; Cottrell, I.F.; Kennedy, D.J.; Brands, K.M.; Dolling, U.H. *Synthesis*, ***2001***, 1784.

EtO
cat $ReBr(CO)_5$, *t*-BuCl
DCE, 84°C, 30 min
EtO *t*-Bu
63%
+ 35% 2,6-di-*t*-Bu)

Nishiyama, Y.; Kakushou, F.; Sonoda, N. *Bull. Chem. Soc. Jpn.*, ***2000***, ***73***, 2779.

N_2
Ph CO_2Me
1% Rh catalyst with
DOSP chiral ligands

CO_2Me
Ph
80% (95% ee)

Davies, H.M.L.; Hansen, T.; Churchill, M.R. *J. Am. Chem. Soc.*, ***2000***, ***122***, 3063.

O
CO_2Et
OAc, aq K_2CO_3
1% $Pd(PPh_3)_4$, 30°C
5% TritOOX-100

O
CO_2Et
92%

Kobayashi, S.; Lam, W.W.-L.; Manabe, K. *Tetrahedron Lett.*, ***2000***, ***41***, 6115.

N Cl
1. 3 eq $BuLi–Me_2NCH_2CH_2OLi$
hexane, −78°C
2. 4 eq MeI, THF, 1 h
−78°C → 0°C

Me N Cl
70%

Choppin, S.; Gros, P.; Fort, Y. *Org. Lett.*, ***2000***, ***2***, 803.

Py , $CHCl_3$

98% (7:1 dr)

Elliott, G.I.; Konopelski, J.P.; Olmstead, M.M. *Org. Lett.*, ***1999***, *1*, 1867.

5% Rh catalyst

88%

Lenges, C.P.; Brookhart, M. *J. Am. Chem. Soc.*, ***1999***, *121*, 6616.

, Na , Py_2CuCl_2

ether

66% (40/60 *o/p*)

Tan, C.-Q.; Zheng, X.; Ma, Z.; Gu, Y. *Synth. Commun.*, ***1999***, *29*, 123.

SECTION 72: ALKYLS, METHYLENES AND ARYLS FROM KETONES

The conversions $R_2C=O \rightarrow R\text{-}R$, R_2CH_2, R_2CHR', etc. are listed in this section.

$TiCl_4$, toluene , reflux

65%

Li, Z.; Sun, W.-H.; Jin, C.; Shao, C. *Synlett*, ***2001***, 1947.

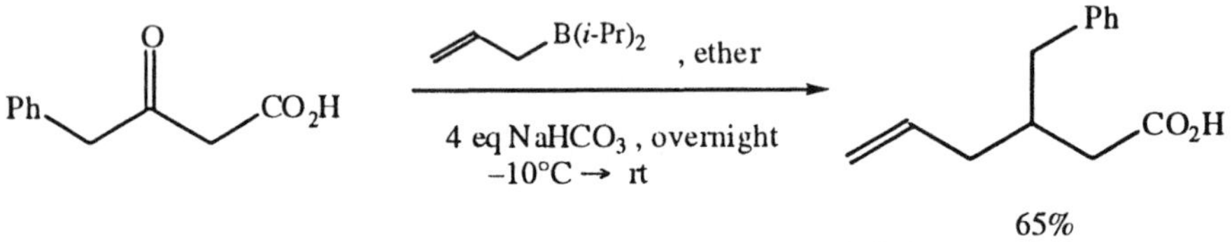

Ph–CH$_2$–CO–CH$_2$–CO$_2$H → (allyl-B(*i*-Pr)$_2$, ether; 4 eq NaHCO$_3$, overnight; –10°C → rt) → 65%

Kabalka, G.W.; Yang, K.; Wang, Z. *Synth. Commun.*, ***2001***, *31*, 511.

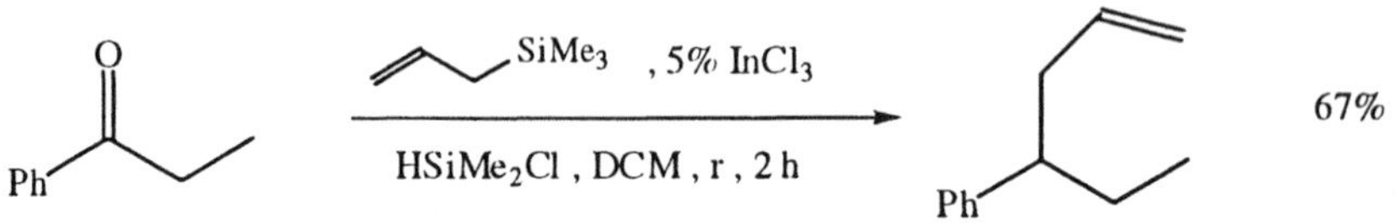

PhCOEt → (allyl-SiMe$_3$, 5% InCl$_3$; HSiMe$_2$Cl, DCM, r, 2 h) → 67%

Yasuda, M.; Ohishi, Y.; Ito, T.; Baba, A. *Tetrahedron Lett.*, ***2000***, *41*, 2425.

SECTION 73 ALKYLS, METHYLENES AND ARYLS FROM NITRILES

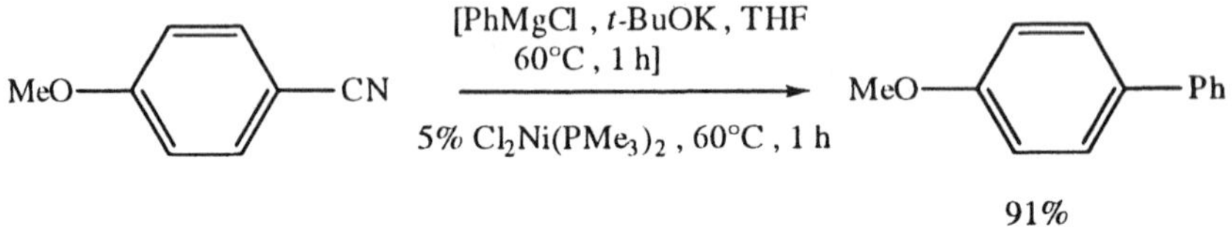

MeO–C$_6$H$_4$–CN → ([PhMgCl, *t*-BuOK, THF 60°C, 1 h]; 5% Cl$_2$Ni(PMe$_3$)$_2$, 60°C, 1 h) → MeO–C$_6$H$_4$–Ph 91%

Miller, J.A. *Tetrahedron Lett.*, ***2001***, *42*, 6991.

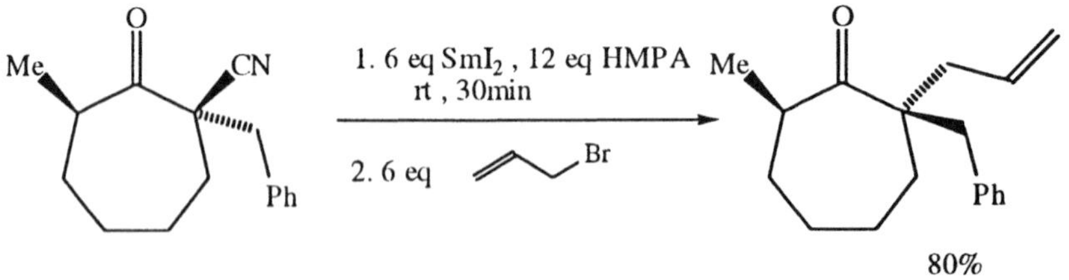

→ (1. 6 eq SmI$_2$, 12 eq HMPA rt, 30min; 2. 6 eq allyl bromide) → 80%

Zhu, J.-L.; Shia, K.-S.; Liu, H.-J. *Tetrahedron Lett.*, ***1999***, *40*, 7055.

SECTION 74: ALKYLS, METHYLENES AND ARYLS FROM ALKENES

The following reaction types are included in this section:

A. Hydrogenation of Alkenes (and Aryls)
B. Formation of Aryls
C. Alkylations and Arylations of Alkenes
D. Conjugate Reduction of Conjugated Aldehydes, Ketones, Acids, Esters and Nitriles
E. Conjugate Alkylations
F. Cyclopropanations, including halocyclopropanations

SECTION 74A: HYDROGENATION OF ALKENES (AND ARYLS)

Reduction of aryls to dienes are listed in Section 377 (Alkene-Alkene).

ASYMMETRIC HYDROGENATIONS

Ph NO_2 enzyme YNAR-I NADPH Ph NO_2 + Ph NO_2

(2 : 3) 31%

>98% ee) >97% ee

Kawai, Y.; Inaba, Y.; Hayashi, M.; Tokitoh, N. *Tetrahedron Lett.*, ***2001***, *42*, 3367.

Ph NO_2 Bakers yeast, glucose, 7 h EtOH, H_2O Ph NO_2

72% (17% ee, *S*)

Kawai, Y.; Inaba, Y.; Tokitoh, N. *Tetrahedron Asymm.*, ***2001***, *12*, 309.

O C_3H_7 Bakers yeast, 2 h O C_3H_7

97% (92% ee)

Filho, E.P.S.; Rodrigues, J.A.R.; Moyam, P.J.S. *Tetrahedron Asymm.*, ***2001***, *12*, 847.

CO_2Me Ph NHAc H_2, cat [Rh(cod)(*S*-BINAPO)] BF_4 DCM, 10 min CO_2Me Ph NHAc

quant (84% ee)

Zhang, F.-Y.; Kwok, W.H.; Chan, A.S.C. *Tetrahedron Asymm.*, ***2001***, *12*, 2337.

MeO_2C CO_2Me H_2, DCM, rt [Rh(cod)(POP)] BF_4 POP = phosphine-phosphite ligand MeO_2C CO_2Me

99.8% ee

Suárez, A.; Pizzano, A. *Tetrahedron Asymm.*, ***2001***, *12*, 2501.

CO_2H CO_2H

polymer-supported BINAP
Ru(BINAP), MeOH

H_2, NEt_3

MeO MeO quant (96% ee)

Fan, Q.H.; Deng, G.-J.; Lin, C.-C.; Chan, A.S.C. *Tetrahedron Asymm.*, **2001**, *12*, 1241.

O O

Bakers yeast

14 h

N N

72% (>99% ee, *S*)

Kawai, Y.; Hayashi, M.; Tokitoh, N. *Tetrahedron Asymm.*, **2001**, *12*, 3007.

50 bar H_2, cat Ir(cod)-chiral oxazoline

Ph Ph

DCM

Ph Ph

quant (97% ee)

Blankenstein, J.; Pfaltz, A. *Angew. Chem. Int. Ed.*, **2001**, *40*, 4445.

[Rh(P–P*)(cod)] BF_4, MeOH
50 bar H_2, rt

N N

P—P* = chiral bis phosphine

Ph Ph

>99% (up to 50% ee)

Tararov, V.I.; Kadyrov, R.; Riermeier, T.H.; Holz, J.; Börner, A. *Tetrahedron Lett.*, **2000**, *41*, 2351.

CO_2Me [Rh(cod)$_2$] PF_6, DCM, rt
phosphine ligand, 1.3 bar H_2 CO_2Me

rt, 20 h

CO_2Me CO_2Me

quant (97.6% ee)

Reetz, M.T.; Mehler, G. *Angew. Chem. Int. Ed.*, **2000**, *39*, 3889.

CO_2H / NHAc → 1% $[Rh(cod)_2]PF_6$, 1.1% chiral bis phosphine; 3 atm H_2 , MeOH , rt , 12 h → CO_2H / NHAc

quant (>99% ee)

Li, W.; Zhang, Z.; Xiao, D.; Zhang, X. *Tetrahedron Lett.*, ***1999***, *40*, 6701.

RR-(EBTHI)-$ZrMe_2$, 1700 psi H_2
$[PhMe_2NH]^+$ $[BC_6F_5]_4^-$
rt , 13-21 h

EBTHI = ethylenebistetrahydroindenyl

87% (93% ee; >99:1 *cis:trans*)

Troutman, M.V.; Appella, D.H.; Buchwald, S.L. *J. Am. Chem. Soc.*, ***1999***, *1221*, 4916.

CO_2Me / Ph / NHAc → 1% adamantyl-bis-P* , MeOH; H_2 , rt , 2 h → CO_2Me / Ph / NHAc

99.9% ee , *R*

Ohashi, A.; Imamoto, T. *Org. Lett.*, ***2001***, *3*, 373.

NON-ASYMMETRIC HYDROGENATIONS

MeO / OBn / MeO → H_2 , Pd black , PhH; 4 h → MeO / OBn / MeO

quant

Maki, S.; Okawa, M.; Matsui, R.; Hirano, T.; Niwa, H. *Synlett*, ***2001***, 1590.

Ph / Ph → *t*-$BuNH_2$•BH_3 , 10% Pd-C; MeOH , sealed vessel; rt , 4 h → Ph / Ph 93%

Couturier, M.; Andresen, B.M.; Tucker, J.L.; Dubé, P.; Brenek, S.J.; Negri, J.T. *Tetrahedron Lett.*, ***2001***, *42*, 2763.

Ph / CO_2H → (polymer)–$CH_2NMe_3^+$ HCO_2^-; cat $RhCl(PPh_3)_3$, DMSO; microwaves → Ph / CO_2H 95%

Desai, B.; Danks, T.N. *Tetrahedron Lett.*, ***2001***, *42*, 5963.

Ph NHCO$_2$Et → Et$_3$SiH, TFA, –10°C; 30 min → Ph NHCO$_2$Et 94%

Masuno, M.N.; Molinski, T.F. *Tetrahedron Lett.*, ***2001***, *42*, 8263.

Me$_3$Si, SiMe$_3$, C$_7$H$_{15}$ → 20% Pd(OH)$_2$/C, H$_2$, 25°C → Me$_3$Si, SiMe$_3$, C$_7$H$_{15}$ + Me$_3$Si, SiMe$_3$, C$_7$H$_{15}$

in hexane (24 : 76)
in DCM (88 : 0)

Hodgson, D.M.; Barker, S.F.; Mace, L.H.; Moran, J.R. *Chem. Commun.*, ***2001***, 153.

Rh-bis-phosphine catalyst on mesoporous silica; H$_2$

Crudden, C.M.; Allen, D.; Mikoluk, M.D.; Sun, J. *Chem. Commun.*, ***2001***, 1154.

OH → cat [RhCl(1,5-cod)]$_2$, 10 bar H$_2$, H$_2$O, supercritical ethane, 62 h; 0.1M Na$_3$PO$_4$, cat Bu$_4$NHSO$_4$, 36°C → OH quant

Bonilla, R.J.; James, B.R.; Jessop, P.G. *Chem. Commun.*, ***2000***, 941.

Raney Ni–Al alloy, 90°C; 1% aq KOH, 5 h → 94%

Tsukinoki, T.; Kanda, T.; Liiu, G.-B.; Ysuzuki, H.; Tashiro, M. *Tetrahedron Lett.*, ***2000***, *41*, 5865.

Ph → NaBH$_4$, NiCl$_2$•6 H$_2$O, hexane; moist Al$_2$O$_3$ → Ph 90%

Yakabe, S.; Hirano, M.; Morimoto, T. *Tetrahedron Lett.*, ***2000***, *41*, 6795.

Ph → H$_2$, *R,R*-Me-DuPhos-Ru, *t*-BuOK, *i*-PrOH → Ph 92% (86% ee, *R*)

Forman, G.S.; Ohkuma, T.; Hems, W.P.; Noyori, R. *Tetrahedron Lett.*, ***2000***, *41*, 9471.

5% Pd/Cl (ethylenediamine)

H_2, THF

N—Cbz → N—Cbz 92%

Hattori, K.; Sajiki, H.; Hirota, K. *Tetrahedron,* ***2000***, *56*, 8433.

PhO Ph

HCO_2NH_4, Ni(R)
ethylene glycol

microwaves, 3 min

75%

Banik, B.K.; Barakat, K.J.; Wagle, D.R.; Manhas, M.S.; Bose, A.K.
J. Org. Chem., ***1999***, *64*, 5746.

H_2 (60 atm), 90°C, 3.5 h
[bmim] (BF_4) ionic liquid

$[H_4Ru_4(\eta^6\text{-}C_6H_6)_4][BF_4]_2$

91%

Dyson, P.J.; Ellis, D.J.; Parker, D.G.; Welton, T. *Chem. Commun.*, ***1999***, 25.

EtO CO_2Et CO_2Et

1. 2 eq —⟨⟩—MgBr

ether, reflux, 3 h
2. aq NH_4Cl

CO_2Et CO_2Et 49%

Kim, Y.M.; Kwon, T.W.; Chung, S.K.; Smith, M.B. *Synth. Commun.*, ***1999***, *29*, 343.

OH Ph CO_2Me

cat $RuCl_2(PPh_3)_3$, 12 h

K_2CO_3, toluene, reflux

O Ph CO_2Me 53%

Basavaiah, D.; Muthukumaran, K. *Synth. Commun.*, ***1999***, *29*, 713.

SECTION 74B: FORMATION OF ARYLS AND HETEROARYLS

Cl—N, O, Cl, O, Cl, , DBU , 2 h

MeO*t*-Bu , –20°C

89%

Tilstam, U.; Harre, M.; Heckrodt, T.; Weinmann, H. *Tetrahedron Lett.*, ***2001***, *42*, 5385.

$2\%\ PdCl_2(MeCN)_2$, 2 h

5 eq NEt_3 , MeCN

89%

Ishikawa, T.; Uedo, E.; Tan, R.; Saito, S. *J. Org. Chem.*, ***2001***, *66*, 186.

$BH_3{\bullet}SMe_2$, THF

Br

Br

Ph OH Ph

98% (94.2% ee , *R*)

Hu, J.-b.; Zhao, G.; Yang, G.-s.; Ding, Z.-d. *J. Org. Chem.*, ***2001***, *66*, 303.

MeO

MeO

1. MgBr, THF , rt , 3 h

2. $BF_3{\bullet}OEt_2$, $MeNO_2$
0°C → rt , 12-15 h

MeS SMe

MeS

79%

SMe

Mohanta, P.K.; Perunceralathan, S.; Ila, H.; Junjappa, H. *J. Org. Chem.*, ***2001***, *66*, 1503.

1. $(CH_2O)_n$, KOH , MeOH
2. $Cu(OAc)_2$, NH_4OAc AcOH

N

64x91%

Pilato, M.L.; Catalano, V.J.; Bell, T.W. *J. Org. Chem.*, ***2001***, *66*, 1525.

C_5H_{11}

$PdCl_2$, $CuCl_2$, BuOH
PhH , 40°C , 12 h

C_5H_{11} C_5H_{11} C_5H_{11}

78%

Li, J.; Jiang, H.; Chen, M. *J. Org. Chem.*, ***2001***, *66*, 3627.

HO

1. BuLi , THF , rt
1 h , $PdCl_2(PPh_3)_2$ EtO—CH(CO_2Et)$_2$
DMSO , PhI
2. KO*t*-Bu , THF

Ph CO_2Et O

53%

Garçon, S.; Vassiliou, S.; Cavicchiioli, M.; Hartmann, B.; Monteiro, N.; Blame, G. *J. Org. Chem.*, ***2001***, *66*, 4069.

Cl
Me_2N^+ PF_6^- NMe_2

$CH_2(CO_2Me)_2$, THF
20% *t*-BuOK

MeO_2C Cl Me N

84%

Marcoux, J.-F.; Marcotte, F.-A.; Wu, J.; Dormer, P.G.; Davies, I.W.; Hughes, D.; Reider, P.J. *J. Org. Chem.*, ***2001***, *66*, 4194.

Ph Ph O

$Pd(OAc)_2$, NEt_3
THF

Ph O Ph

54%

Jeevanandam, A.; Narkunan, K.; Ling, Y.-C. *J. Org. Chem.*, ***2001***, *66*, 6014.

Br Br CHO

1. $Ph_3P{=}C(CO_2Et)CH_2CO_2H$, 50°C , 12 h
2. $ClCO_2Et$, NEt_3
3. NaOH , EtOH

OH Ph Br CO_2Et

70x78%

Serra, S.; Fuganti, C.; More, A. *J. Org. Chem.*, ***2001***, *66*, 7883.

Ph—≡—Ph
5% $Pd(OAc)_2$, 10% PPh_3
5 eq Na_2CO_3 , DMF , 1 d
100°C
96%

Roesch, K.R.; Zhang, H.; Larock, R.C. *J. Org. Chem.*, ***2001***, *66*, 8042.

180°C , 2 h
o-dichlorobenzene
63% 17%

Atienza, C.; Mateo, C.; de Frutos, Ó.; Schavarren, A.M. *Org. Lett.*, ***2001***, *3*, 153.

, THF
20% *t*-BuOK , DABCO
78%

Davies, I.W.; Marcoux, J.-F.; Reider, P.J. *Org. Lett.*, ***2001***, *3*, 209.

C_3H_7—≡—C_3H_7
2.5% $[RhCl(cod)]_2$, toluene
phosphine ligand , NEt_3
110°C , 1 d
73%

Amii, H.; Kishikawa, Y.; Uneyama, K. *Org. Lett.*, ***2001***, *3*, 1109.

NH_4^+ HCO_2^- , 10% Pd/C
PEG-200 , microwaves
30 sec
92%

Rao, H.S.P.; Jothilingam, S. *Tetrahedron Lett.*, ***2001***, *42*, 6595.

HN O Ph OMe I cat $Pd_2(dba)_3$/P(2-furyl)$_3$ NaO*t*-Bu, THF, 5°C MeCN, 40°C Me O N Ph 83%

Arcadi, A.; Cacchi, S.; Cascia, L.; Labrizi, G.; Marinelli, F. *Org. Lett.*, ***2001***, *3*, 2501.

N N N N 150°C, microwaves 20 min N 40%

Díaz-Ortìz, A.; de la Hoz, A.; Prieto, P.; Carrillo, J.R.; Moreno, A.; Neunhooeffer, H. *Synlett*, ***2001***, 236.

Cl OAc CO_2Et Cl 2 eq $EtNO_2$, DMF, 17h 3 eq K_2CO_3, rt → 50°C Cl CO_2Et 75%

Kim, J.N.; Im, Y.J.; Gong, J.H.; Lee, K.Y. *Tetrahedron Lett.*, ***2001***, *42*, 4195.

CO_2Me O CO_2Me O Ph OMe O NaOMe, MeOH, 70°C MeO_2C O OMe Ph OH 72%

Tse, B.; Jones, A.B. *Tetrahedron Lett.*, ***2001***, *42*, 6429.

Bu $Cr(CO)_5$ Ph OMe dioxane, 100°C OH Bu OMe 57%

Jackson, T.J.; Herndon, J.W. *Tetrahedron*, ***2001***, *57*, 3859.

1. acetone, 2 eq SmI_2, THF, 20°C, 6 h
2. 1M HCl

84%

Farcas, S.; Many, J.-L. *Tetrahedron*, **2001**, *57*, 4881.

NHMe

$Mn(OAc)_3$, AcOH

Me

78%

Wu, Y.-L.; Chuang, C.-P.; Lin, P.-Y. *Tetrahedron*, **2001**, *57*, 5543.

NO^+/18-crown-6/$H(NO_3)_2^-$

DCM, rt

quant

Zolfigol, M.A.; Zebarjadian, M.H.; Sadeghi, M.M.; Mohammadpoor-Baltork, I.; Memarian, H.R.; Shamsipur, M. *Synth. Commun.*, **2001**, *31*, 929.

$BaMnO_4$, PhH, reflux

3 min

95%

Memarian, H.R.; Sadeghi, M.M.; Momeni, A.R. *Synth. Commun.*, **2001**, *31*, 2241.

$FeCl_3 \cdot 6\ H_2O$, reflux

aq AcOH, 1 h

60%

Lu, J.; Bai, Y.; Wang, Z.; Yang, B.Q.; Li, W. *Synth. Commun.*, **2001**, *31*, 2625.

$NaHSO_4$, H_2O, wet SiO_2

$Na_2Cr_2O_7 \cdot 2\ H_2O$, DCM
rt, 10 min

97%

Damavandi, J.A.; Zolfigol, M.A.; Karami, B. *Synth. Commun.*, **2001**, *31*, 3183.

MeHN / CO_2Et — $PhI(OAc)_2$, DCM / rt , 2 h → EtO_2C, CO_2Et, N, Me 87%

Zhang, P.-F.; Chen, Z.-C. *Synth. Commun.*, ***2001***, *31*, 1619.

Cp_2 Zr, Ph, Ph — 2 eq Br , 2 eq $ZnBr_2$ / 5% $Pd(PPh_3)_4$, 50°C , 12 h → Ph, Ph 59%

Duan, Z.; Nakajima, K.; Takahashi, T. *Chem. Commun.*, ***2001***, 1672.

OH, NH_2 — acetophenone , KOH / cat Grubbs' catalyst , 80°C → N, Ph 99%

Cho, C.S.; Kim, B.T.; Kim, T.-J.; Shim, S.C. *Chem. Commun.*, ***2001***, 2576.

N, O, C_6F_5, O — cat $Pd(PPh_3)_4$, DMF / Bu_4NCl , 80°C → N 77%

Tsutsui, H.; Narasaka, K. *Chem. Lett.*, ***2001***, 526.

O, OHC, NH — 1. piperidine , Ac_2O , 85°C EtOAc , toluene , 1 h / 2. 150°C , 3 h → O, N 80%

Wei, L.-L.; Hsung, R.P.; Sklenicka, H.M.; Gerasyuto, A.I. *Angew. Chem. Int. Ed.*, ***2001***, *40*, 1516.

Ph Ph N Ph Ph → hv, hexane → Ph N Ph 69%

Campos, P.J.; Caro, M.; Rodríguez, M.A. *Tetrahedron Lett.*, ***2001***, *42*, 3575.

NH_2 → $N(CH_2CH=CH_2)_3$, dioxane, 180°C; $RuCl_3 \cdot n\ H_2O$, dppm; $SnCl_2 \cdot 2\ H_2O$ → N 82%

Cho, C.S.; Oh, B.H.; Shim, S.C.; Oh, D.H. *J. Heterocyclic Chem.* ***2000***, *37*, 1315.

Et Et Et Et $ZrCp_2$ → C_3H_7—≡—Br, THF; cat $NiCl_2(PPh_3)_2$ → Et Et Et Et C_3H_7 C_3H_7 43%

Wang, H.; Tsai, F.-Y.; Takahashi, T. *Chem. Lett.*, ***2000***, 1410.

CN CN → 1. BuNHLi, THF; 2. aq H^+ → NH_2 N Bu 93%

Wang, A.; Zhang, H.; Biehl, E.R. *Heterocycles*, ***2000***, *53*, 291.

CO_2Et N H → $NaNO_2$, $C_2H_2O_4 \cdot 2\ H_2O$; DCM, wet SiO_2, rt → CO_2Et N 98%

Zolfigol, M.A.; Kiany-Borazjani, M.; Sadeghi, M.M.; Mohammadpoor-Baltork, I.; Memarian, H.R. *Synth. Commun.*, ***2000***, *30*, 551.

N-OH Ph → DMF, $POCl_3$, 90°C → Ph CHO N Cl 19%

Amaresh, R.R.; Perumal, P.T. *Synth. Commun.*, ***2000***, *30*, 2269.

O, CO_2Me, O — DMF, $POCl_3$, 22 sec / microwaves → Cl, CO_2Me, Cl, CHO 54%

Selvi, S.; Perumal, P.T. *Synth. Commun.*, ***2000***, *30*, 3925.

Ph, SH — 2% PdI_2–KI, 100°C / dimethylacetamide → Ph, S 58%

Gabriele, B.; Salerno, G.; Fazio, A. *Org. Lett.*, ***2000***, *2*, 351.

Me_2NCO_2t-Bu — 1. *sec*-BuLi, (–)-sparteine, THF, –78°C 2. CuCl•2 LiCl, –55°C 3. Bu, O, Me → Bu, Me, N, Me 64x79%

Dieter, R.K.; Yu, H. *Org. Lett.*, ***2000***, *2*, 2283.

$SiMe_3$, OTf — Cl, CsF, MeCN/THF / 10% $Pd_2(dba)_3$•$CHCl_3$, 20% dppf 60°C → 69%

Yoshikawa, E.; Radhakrishnan, K.V.; Yamamoto, Y. *J. Am. Chem. Soc.*, ***2000***, *122*, 7280.

AcO, Ph, $SiMe_3$, $SiMe_3$ — MeLi, toluene / –20°C → Me_3Si, O, $SiMe_3$ 44%

Brunette, S.R.; Lipton, M.A. *J. Org. Chem.*, ***2000***, *65*, 5114.

NHAc CHO

$CH_2(CN)_2$, microwaves

Al_2O_3 , 8 min

N NH_2 CN

86%

Sharma, U.; Ahmed, S.; Boruah, R.C. *Tetrahedron Lett.*, ***2000***, *41*, 3493.

O

1. $EtNO_2$, K_2CO_3 , H_2O
–60°C , 1 d

2. toluene , reflux , air

O

90x53%

Ballini, R.; Barboni, L.; Bosica, G. *J. Org. Chem.*, ***2000***, *65*, 6261.

Ph NHPh Bt

$Pd(OAc)_2$, PPh_3 , $CuCl_2$

K_2CO_3 , 1 d

Bt = benzotriazyl

N Ph Ph

46%

Katritzky, A.R.; Zhang, L.; Yao, J.; Denisko, O.V. *J. Org. Chem.*, ***2000***, *65*, 8074.

O

2% $AlCl_3$, MeCN , 20°C

OH

65%

Hashmi, A.S.K.; Frost, T.M.; Bats, J.W. *J. Am. Chem. Soc.*, ***2000***, *122*, 11553.

NH_2

$(CH_2{=}CHCH_2)_2\overset{+}{N}(C_3H_7)_2\ Cl^-$
5% $RuCl_2(PPh_3)_3$, 180°C

$SnCl_2$•2 H_2O , H_2O-dioxane
20 h

N

57%

Cho, C.S.; Kim, J.S.; Oh, B.H.; Kim, T.-J.; Shim, S.C.; Yoon, N.S. *Tetrahedron*, ***2000***, *56*, 7747.

O O

SPh , Ag_2CO_3/Celite

MeCN , reflux

O O

51%

Lee, Y.R.; Suk, J.Y.; Kim, B.S. *Org. Lett.*, ***2000***, *2*, 1387.

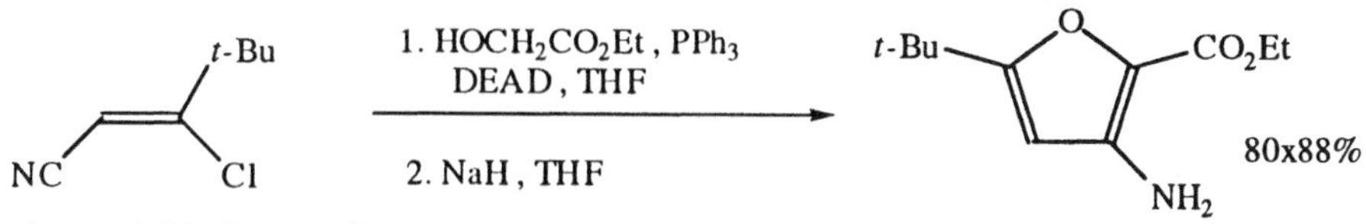

t-Bu, NC, Cl → 1. $HOCH_2CO_2Et$, PPh_3 DEAD, THF 2. NaH, THF → t-Bu, O, CO_2Et, NH_2 80x88%

Redman, A.M.; Dumas, J.; Scott, W.J. *Org. Lett.*, ***2000***, 2, 2061.

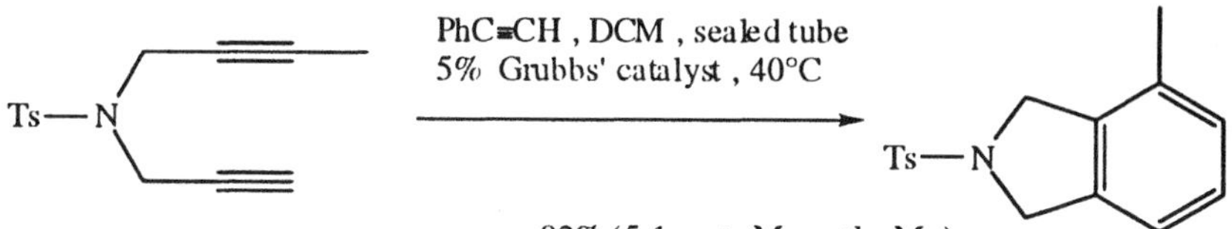

Ts—N → PhC≡CH, DCM, sealed tube 5% Grubbs' catalyst, 40°C → Ts—N

82% (5:1 *meta* Me:*ortho* Me)

Witulski, B.; Stengel, T.; Fernández-Hernandez, J.M. *Chem. Commun.*, ***2000***, 1965.

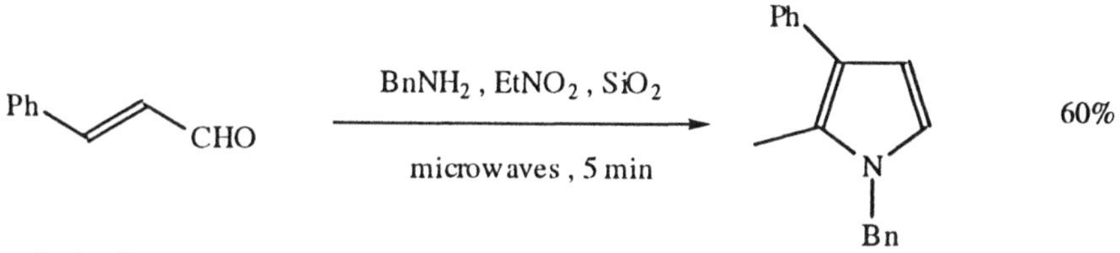

Ph, CHO → $BnNH_2$, $EtNO_2$, SiO_2 microwaves, 5 min → Ph, N, Bn 60%

Ranu, B.C.; Hajra, A.; Jana, U. *Synlett*, ***2000***, 75.

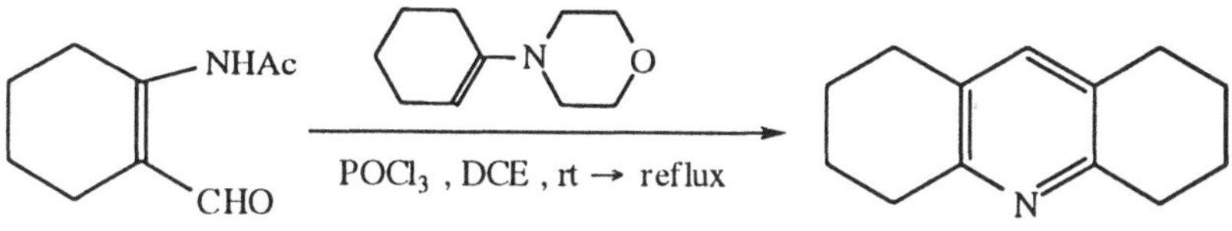

NHAc, CHO → N, O; $POCl_3$, DCE, rt → reflux → N

51%

Bruah, R.C.; Ahmed, S.; Sharma, U.; Sandhu, J.S. *J. Org. Chem.*, ***2000***, *65*, 922.

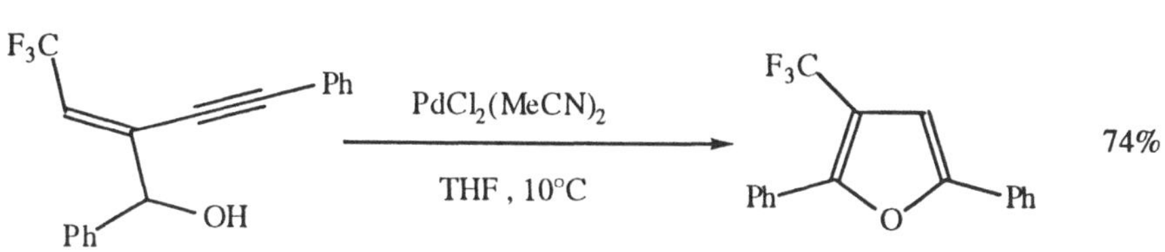

F_3C, Ph, Ph, OH → $PdCl_2(MeCN)_2$ THF, 10°C → F_3C, Ph, O, Ph 74%

Qing, F.L.; Gao, W.-Z.; Ying, J. *J. Org. Chem.*, ***2000***, *65*, 2003.

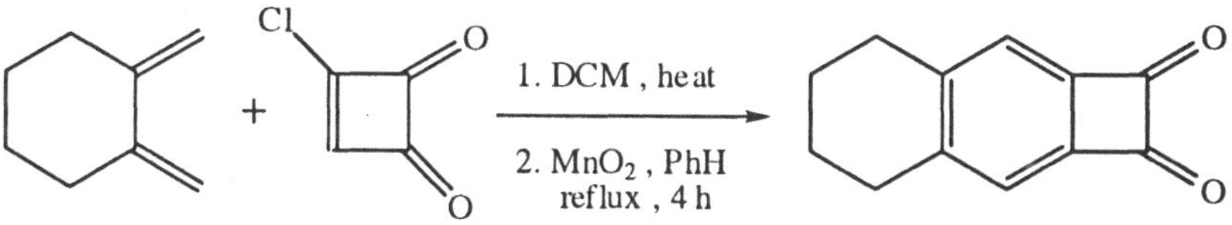

+ Cl, O, O → 1. DCM, heat 2. MnO_2, PhH reflux, 4 h → O, O

53x61%

Schmidt, A.H.; Kircher, G.; Willems, M. *J. Org. Chem.*, ***2000***, *65*, 2379.

1. (pyrrolidine enamine), PhH, reflux, cat p-TsOH, 3 h
2. AcO–CH$_2$C(O)CH$_2$–OAc, PhH, rt, 18 h

34%

Wang, C.; Kohn, H. *Org. Lett.*, ***2000***, *2*, 1773.

OMe

$KMnO_3–Al_2O_3$, acetone

0°C, 1 h

OMe

90%

McBride, C.M.; Chrisman, W.; Harris, C.E.; Singaram, B. *Tetrahedron Lett.*, ***1999***, *40*, 45.

C_5H_{11} I O Ph

palladacycle catalyst

NEt_3, THF, rt

C_5H_{11} O Ph

83%

Luo, F.-T.; Jeevanandam, A.; Bajji, A.C. *Tetrahedron Lett.*, ***1999***, *40*, 121.

C_3H_7–≡–C_3H_7, MeCN
5% $Pd(OAc)_2$, 5% $P(o\text{-Tol})_3$

CsF, 60°C, 4 h

C_3H_7 C_3H_7

63%

Radhakrishnan, K.V.; Yoshikawa, E.; Yamamoto, Y. *Tetrahedron Lett.*, ***1999***, *40*, 7533.

O

1. Et–≡–Et, PhOH
cat $Ni(acac)_2/PPh_3/AlMe_3$

2. DBU, air

O Et Et Et Et

60%

Mori, N.; Ikeda, S.-i.; Sato, Y. *J. Am. Chem. Soc.*, ***1999***, *121*, 2722.

Bu—≡—≡—Bu

5% $Pd(PPh_3)_4$, THF

Bu Bu 89%

Gevorgyan, V.; Takeda, A.a.; Homma, M.; Sadayori, N.; Radhakrishnan, U.; Yamamoto, Y. *J. Am. Chem. Soc.*, ***1999***, *121*, 6391.

Cl OAc

O CO_2Et, THF, reflux

10% $Pt(C_2H_4)(PPh_3)_2$, 2 h

CO_2Et O 62%

Kadota, J.; Komori, S.; Fukumoto, Y.; Murai, S. *J. Org. Chem.*, ***1999***, *64*, 7523.

N—*t*-Bu I

1. PhC≡CH, 2% $PdCl_2(PPh_3)_2$ 1% CuI, NEt_3, 55°C, 2 h

2. 100°C, 10 CuI, DMF, 1 h

N Ph

91%

Roesch, K.R.; Larock, R.C. *Org. Lett.*, ***1999***, *1*, 553.

NC Ph

MeOH, 50°C

N OMe Ph 86%

Suginome, M.; Fukuda, T.; Ito, Y. *Org. Lett.*, ***1999***, *1*, 1977.

S Ph NH_2

0.5 CHO

hν, PhH, 5 h

Ph N 32%

Oda, K.; Nkagami, R.; Nishizono, N.; Machida, M. *Chem. Commun.*, ***1999***, 2371.

O $SiMe_3$

1. THF, *t*-BuOK, 0°C

2. 0.1M HCl, 0°C

O $SiMe_3$ 48%

MaGee, D.I.; Leach, J.D.; Setiadji, S. *Tetrahedron*, ***1999***, *55*, 2847.

Ph–CH=CH–NO_2, THF, 60°C; 5% $Sm(Oi\text{-}Pr)_3$, 3 h; Bu; N; Bu; N; Ph; Et; 63%

Shiraishi, H.; Nishitani, T.; Nishihara, T.; Sakaguchi, S.; Ishii, Y. *Tetrahedron*, ***1999***, *55*, 13957.

O; Ph; Ph; 1. $C_6H_{13}C{\equiv}CB(Oi\text{-}Pr)_3$, 90°C toluene, $BF_3{\bullet}OEt_2$; 2. Li^+, H^+; C_6H_{13}; O; Ph; Ph; 97%

Brown, C.D.; Chong, J.M.; Shen, L. *Tetrahedron*, ***1999***, *55*, 14235.

O; $PhSO_2(CH_2)_5Me$, THF; BuLi, $PhCOCl$, *t*-BuOK; –78°C → rt → reflux; $(CH_2)_5Me$; 75%

Orita, A.; Yaruva, J.; Otera, J. *Angew. Chem. Int. Ed.*, ***1999***, *38*, 2267.

C_6F_5; O; N; O; Ph; cat $Pd(PPh_3)_4$, NEt_3; DMF, 80°C, 1 h; Ph; H; N; 65%

Tsutsui, H.; Narasaka, K. *Chem. Lett.*, ***1999***, 45.

CO_2Me; O; $Ti(Oi\text{-}Pr)_4/N_2/Li$; TMSCl, CsF, rt; THF, 0.8 h; CO_2Me; N; H; 82%

Akashi, M.; Nishida, M.; Mori, M. *Chem. Lett.*, ***1999***, 465.

SECTION 74C: ALKYLATIONS AND ARYLATIONS OF ALKENES

Br; N; $(HO)_2B$; CO_2Me; NHBoc; CsF, DME, cat $Pd(PPh_3)_4$; 80°C, 3.5 h; N; CO_2Me; NHBoc; 54%

Collet, S.; Danion-Bougot, R.; Danion, D. *Synth. Commun.*, ***2001***, *31*, 249.

1. $PhMe_2SiCu(CN)Li$, THF , –30°C
2. allyl–O–P(O)(OPh)$_2$

Ph / SiMe$_2$Ph 73%

Liepins, V.; Bäckvall, J.-E. *Chem. Commun.*, ***2001***, 265.

O / Me_2N ; $(CH_2)_7$ OTBS , DCE
5% Grubbs' catalyst II
40°C , 15 h

O / Me_2N / $(CH_2)_7$ / OTBS
83% (25:1 *E*:*Z*)

Choi, T.-L.; Chatterjee, A.K.; Grubbs, R.H. *Angew. Chem. Int. Ed.*, ***2001***, ***40***, 1277.

PhO_2S ; PhI , $Pd(OAc)_2$, DMF
Ag_2CO_3 , 120°C , 3 d

i-Pr, Ph, PhO_2S + Ph, SO_2Ph
(95 : 5) 86%

Mauleón, P.; Alonso, I.; Carretero, J.C. *Angew. Chem. Int. Ed.*, ***2001***, ***40***, 1291.

(*o*-DPPB)O ; 1.2 eq (acetylacetone) , toluene
1% $[Rh(CO)_2(acac)]$, 50°C
4% $P(OPh)_3$, 02. pyridine , AcOH
1:1 H_2/CO , 20 bar , 5 d

O, O, (*o*-DPPB)O
51% (91:1 *syn*:*anti*)

Breit, B.; Zahn, S.K. *Angew. Chem. Int. Ed.*, ***2001***, ***40***, 1910.

OH ; 10% $In(OTf)_3$, DCM , rt
(80:20 *syn*:*anti*)

OH
79% (68:32 *E*:*Z*)

Loh, T.-P.; Tan, K.-T.; Hu, Q.-Y. *Angew. Chem. Int. Ed.*, ***2001***, ***40***, 2921.

OH

$MeCH(CO_2Me)_2$, DCM , 1 d
cat $Al(O\textit{i}\text{-}Pr)_3$, reflux

OH

CO_2Me

90% CO_2Me

Black, P.J.; Harris, W.; Wiliams, J.M.J. *Angew. Chem. Int. Ed.*, ***2001**, 40*, 4475.

O

Ph

1. 5% CuCl , 5% NaO*t*-Bu , 0°C
5% *S*-*p*-Tol-BINAP , Ph_2SiH_2

toluene
2. 2 eq BnBr , 1.2 eq TBAF , rt
DCM , toluene , 1 d

O

Bn

Ph

67% (94:6 dr)

Yun, J.; Buchwald, S.L. *Org. Lett.*, ***2001**, 3*, 1129.

O O

O N

Me

$Sc(OTf)_3$, O_2
BEt_3 , DCM/THF

–78°C , $SnBr_3$

I

O O

O N

Me

63% (20:1 *anti*:*syn*)

Sibi, M.P.; Chen, J.; Rheault, T.R. *Org. Lett.*, ***2001**, 3*, 3679.

F

F

CF_3

BuLi , THF , 80°C

Bu

F

CF_3

53% (33:1 *Z*:*E*)

Cooper, J.A.; Olivares, C.M.; Sandford, G. *J. Org. Chem.*, ***2001**, 66*, 4887.

SMe

CO_2Et

4 eq $PhMe_2SiH$, 1% H_2PtCl_6

DCE , 23°C , 13 h

SMe

CO_2Et

$SiMe_2Ph$ 94%

Perales, J.B.; van Vranken, D.L. *J. Org. Chem.*, ***2001**, 66*, 7270.

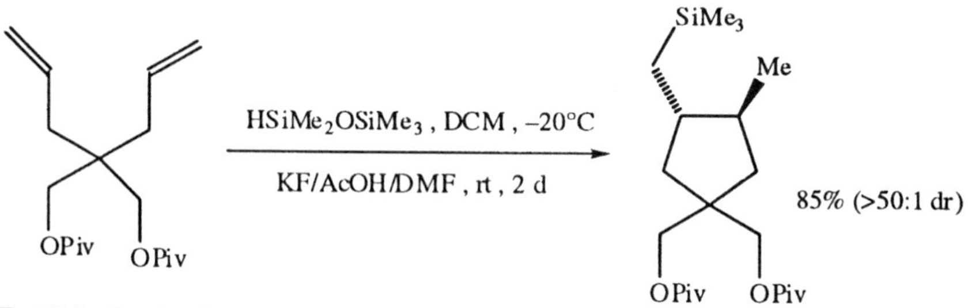

$HSiMe_2OSiMe_3$, DCM , –20°C
KF/AcOH/DMF , rt , 2 d

OPiv OPiv

$SiMe_3$ Me

85% (>50:1 dr)

OPiv OPiv

Pei, T.; Widenhoefer, R.A. *J. Org. Chem.*, ***2001***, *66*, 7639.

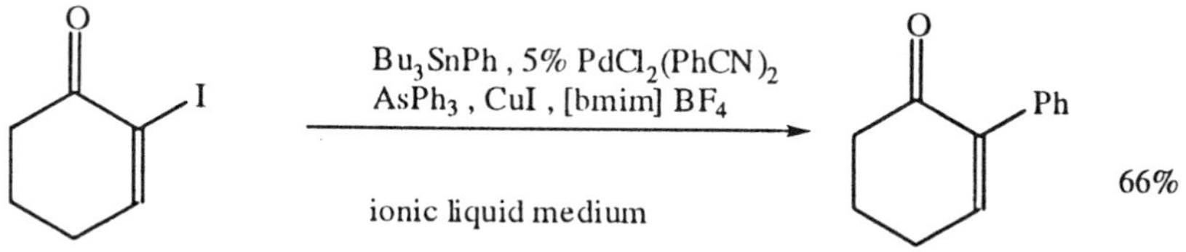

O I

Bu_3SnPh , 5% $PdCl_2(PhCN)_2$
$AsPh_3$, CuI , [bmim] BF_4

ionic liquid medium

O Ph

66%

Handy, S.T.; Zhang, X. *Org. Lett.*, ***2001***, *3*, 233.

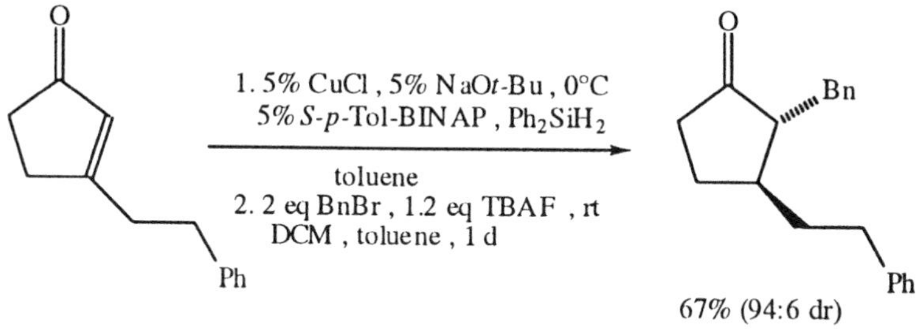

O

1. 5% CuCl , 5% NaO*t*-Bu , 0°C
5% *S*-*p*-Tol-BINAP , Ph_2SiH_2
toluene
2. 2 eq BnBr , 1.2 eq TBAF , rt
DCM , toluene , 1 d

Ph

O Bn

Ph

67% (94:6 dr)

Yun, J.; Buchwald, S.L. *Org. Lett.*, ***2001***, *3*, 1129.

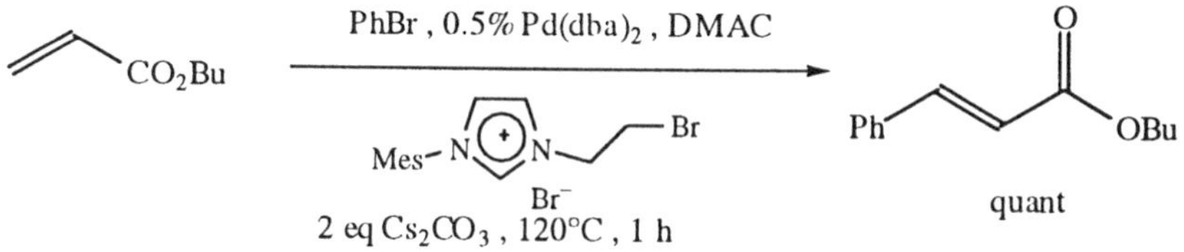

CO_2Bu

PhBr , 0.5% $Pd(dba)_2$, DMAC

Mes–N N Br
Br^-
2 eq Cs_2CO_3 , 120°C , 1 h

O
Ph OBu

quant

Yang, C.; Lee, H.M.; Nolan, S.P. *Org. Lett.*, ***2001***, *3*, 1511.

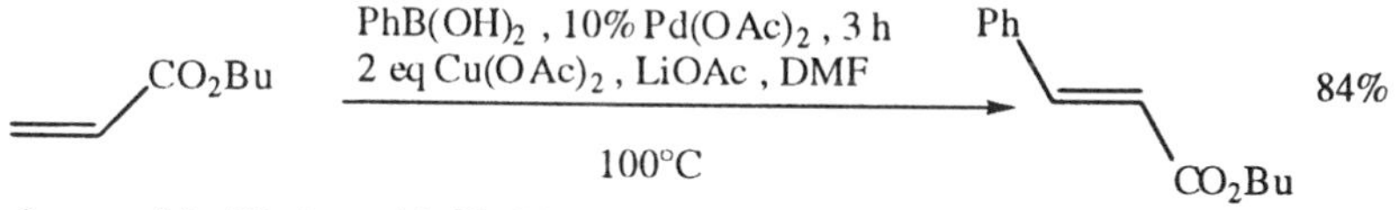

CO_2Bu

$PhB(OH)_2$, 10% $Pd(OAc)_2$, 3 h
2 eq $Cu(OAc)_2$, LiOAc , DMF

100°C

Ph

CO_2Bu

84%

Du, X.; Suguro, M.; Hirabayashi, K.; Mori, A.; Nishikata, T.; Hagiwara, N.; Kawata, K.; Okeda, T.; Wang, H.F.; Fugami, K.; Kosugi, M. *Org. Lett.*, ***2001***, *3*, 3313.

1.5 eq [OMe, $SiMe_2t$-Bu, OMe cyclohexadiene]

0.6 AIBN , hexane (0.25M)

EtO_2C CO_2Et → t-$BuMe_2Si$ CO_2Et CO_2Et

61% (dr=1:1)

Amrein, S.; Timmermann, A.; Studer, A. *Org. Lett.*, ***2001***, *3*, 2357.

$TiCl_4$, DCM
−78°C → rt

Me

55%

Browder, C.C.; Marmsäter, F.P.; West, F.G. *Org. Lett.*, ***2001***, *3*, 3033.

Bu_3SnH , AIBN
PhH , heat

Cl N SO_2Ph O

SO_2Ph 48% + SO_2Ph 15%

Bommerijn, S.; Martin, C.G.; Kennedy, A.R.; Lizos, D.; Murphy, J.A. *Org. Lett.*, ***2001***, *3*, 3405.

Bu SePh O

Bu_3SnH , AIBN
hν , 17°C

Bu O

60% 1/4.5 *cis/trans*

+ 3 eq $AlEt_3$ 72% 7.4.1 *cis/trans*

Ericsson, C.; Engman, L. *Org. Lett.*, ***2001***, *3*, 3459.

$PhN_2 + BF_4^-$ → Ph $B(OH)_2$; imidazolium salt; 2% $Pd(OAc)_2$; THF , rt , 3.5h → Ph Ph 83%

Andrus, M.B.; Song, C. *Org. Lett.*, ***2001***, *3*, 3761.

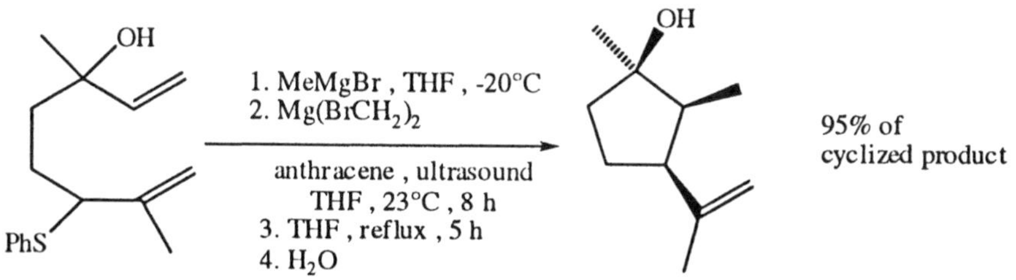

1. MeMgBr, THF, -20°C
2. $Mg(BrCH_2)_2$
anthracene, ultrasound
THF, 23°C, 8 h
3. THF, reflux, 5 h
4. H_2O

OH, PhS

95% of cyclized product

Cheng, D.; Zhu, S.; Yu, Z.; Cohen, T. *J. Am. Chem. Soc.*, ***2001***, *123*, 30.

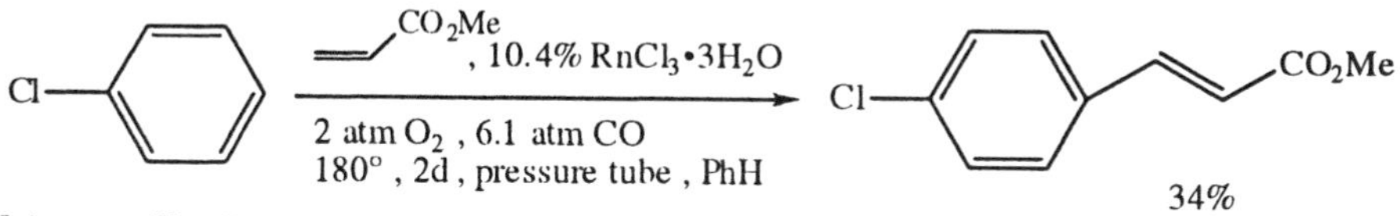

Cl — CO_2Me, 10.4% $RnCl_3 \cdot 3H_2O$
2 atm O_2, 6.1 atm CO
180°, 2d, pressure tube, PhH

CO_2Me 34%

Weissman, H.; Song, X.; Milstein, D. *J. Am. Chem. Soc.*, ***2001***, *123*, 337.

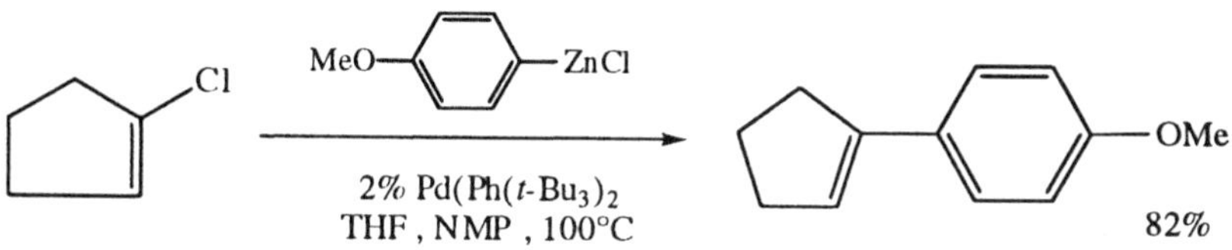

Cl — MeO—C6H4—ZnCl
2% $Pd(Ph(t\text{-}Bu_3)_2$
THF, NMP, 100°C

OMe 82%

Dai, C.; Fu, G.C. *J. Am. Chem. Soc.*, ***2001***, *123*, 2719.

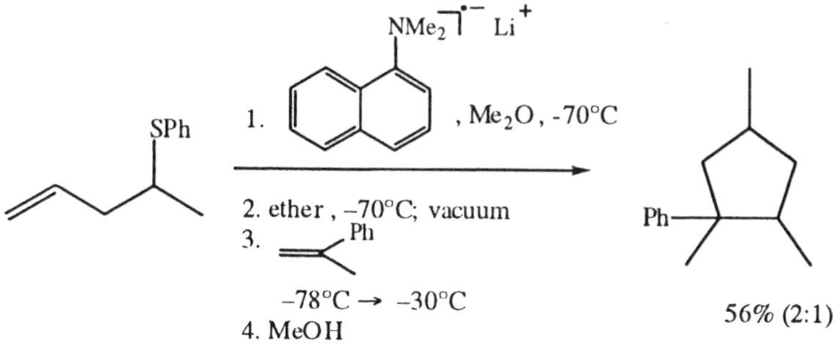

SPh

1. NMe_2 (naphthalene radical anion) Li^+, Me_2O, -70°C
2. ether, −70°C; vacuum
3. Ph (α-methylstyrene)
−78°C → −30°C
4. MeOH

Ph
56% (2:1)

Cohen, T.; Kreethadumrongdat, T.; Liu, X.; Kulkarni, V. *J. Am. Chem. Soc.*, ***2001***, *123*, 3478.

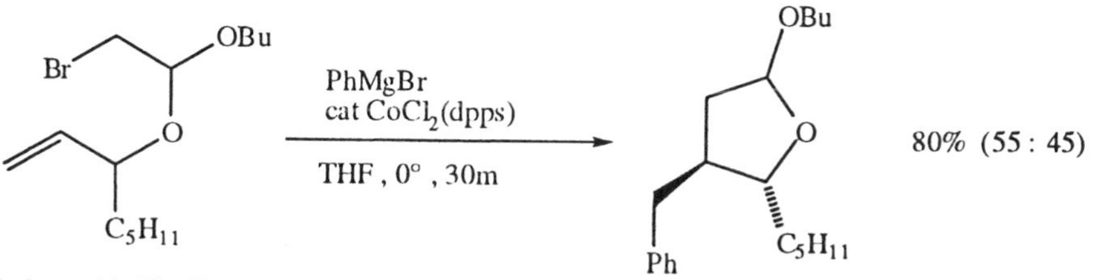

Br, OBu, O, C_5H_{11}

PhMgBr
cat $CoCl_2(dpps)$
THF, 0°, 30m

OBu, O, Ph, C_5H_{11}

80% (55 : 45)

Wakabayashi, K.; Yorimitsu, H.; Oshima, K. *J. Am. Chem. Soc.*, ***2001***, *123*, 5374.

1. *sec*-BuLi, (-)-sparteine, ether
2. CuCN / 2LiCl / THF
3. Ph–CH=CH–Ph, –60°C → rt

N-Boc; Ph; Boc; 90% (>90% ee)

Dieter, R.K.; Topping, C.M.; Chandupatla, K.R.; Lu, K. *J. Am. Chem. Soc.*, ***2001***, *123*, 5132.

CO_2Me; Ph; Br

DABCO, CH_2=CHCN

rt, 7d

Ph; NC; CO_2Me; 67%

Basavaiah, D.; Kumaraguruharan, N.; Sharada, D.S. *Tetrahedron Lett.*, ***2001***, *42*, 85.

OH; I

1. $Ph_2(CH_2$=CH)SiCl, DMAP
2. $Pd(PPh_3)_4$, NEt_3, MeCN

OH; Si–Ph; O; Ph; 89%

Mayasundari, A.; Young, D.G.J. *Tetrahedron Lett.*, ***2001***, *42*, 203.

O + OTf

3% $Pd_2(dba)_3$, rt

NEt_3, PhH, 36 h

O; 93% (86% ee)

Gilbertson, S.R.; Fu, Z.; Xie, D. *Tetrahedron Lett.*, ***2001***, *42*, 365.

Bu; $SiMe_2(2\text{-}Py)_2$

PhI, THF, TBAF, 2 h

$PdCl_2(PhCN)_2$, 50°C

Bu; Ph; 99%

Itami, K.; Nokami, T.; Yoshida, J.-i. *J. Am. Chem. Soc.*, ***2001***, *123*, 5600.

BnN

1. 5% $RhCl(PPh_3)_3$
toluene, 125°C
2. 1N HCl

O; 71%

Thalji, R.K.; Ahrendt, K.A.; Bergman, R.G.; Ellman, J.A.
J. Am. Chem. Soc., ***2001***, *123*, 9692.

PhSiH$_3$, Sm complex
toluene, rt, 3d

H_2PhSi C_8H_{17} + SiPhH$_2$ C_8H_{17}

(95 : 5) 95%

Hou, Z.; Zhang, Y.; Tardif, O.; Wakatsuki, Y. *J. Am. Chem. Soc.*, ***2001***, *123*, 9216.

chiral diamine, THF
2.1 eq EtMgCl, –78°C

76% (79% ee)

Sibi, M.P.; Asano, Y. *J. Am. Chem. Soc.*, ***2001***, *123*, 9708.

cat Ti(TADDOL) complex
$Zn(CH_2I)_2$ MS 4Å
CH_2Cl_2, 0°C

85% (96:4)

Charette, A.B.; Molinaro, C.; Brochu, C. *J. Am. Chem. Soc.*, ***2001***, *123*, 12168.

excess MeO-C$_6$H$_4$-B(OH)$_2$
cat $PdCl_2(PPh_3)_2$, THF, 65°C

97%

Bauer, A.; Miller, M.W.; Vice, S.F.; McCombie, S.W. *Synlett*, ***2001***, 254.

1. 2 eq chiral lithium amide
2. 4 eq BuLi, ether
3. H_3O^+

78% (94% ee, *S*)

Brémond, N.; Mangeney, P.; Normant, J.F. *Tetrahedron Lett.*, ***2001***, *42*, 1883.

$CH_2=C(F)SiPh_2Me$; 5% $Pd(PPh_3)_4$, DMF; 1.5 eq KF, 25°C; 20% CuI, 80 h

60% + 9%

Hanamoto, T.; Kobayashi, T.; Kondo, M. *Synlett*, ***2001***, 281.

In, THF, −78°C; 4 eq MeMgCl, LiCl; 10% $CuCl_2$, 0.5 h

MeO_2C TfO Me $TrO(H_2C)_3$ → MeO_2C Me Me $TrO(H_2C)_3$ 86%

Ide, M.; Nakata, M. *Synlett*, ***2001***, 1511.

MgBr EtO_2C → Bu EtO_2C 60%

I–CH=CH–Bu; cat $Fe(acac)_3$, THF

Dohle, W.; Kopp, F.; Cahiez, G.; Knochel, P. *Synlett*, ***2001***, 1901.

MeO_2C CO_2Me I → MeO_2C MeO_2C I Bu

CH_2=CHBu, BEt_3; $Yb(OTf)_3$, O_2; CH_2Cl_2

Kitagawa, O.; Fujiwara, H.; Tajuchi, T. *Tetrahedron Lett.*, ***2001***, *42*, 2165.

CO_2Et Ph CO_2Et → O $CH(CO_2Et)_2$ Ph

acetone, THF; 20% *S*-1-(2-pyrrolidinylmethyl)-pyrrolidinone

47% (59% ee)

Betancort, J.M.; Sakthivel, K.; Thayumanavan, R.; Barbas III, C.F. *Tetrahedron Lett.*, ***2001***, *42*, 4441.

CHO

O

, MeCN

pyrrolidine , 80°C , 50 h

CHO

O

66%

Hagiwara, H.; Okabe, T.; Hakoda, K.; Hoshi, T.; Ono, H.; Kamat, V.P.; Suzuki, T.; Ando, M. *Tetrahedron Lett.*, **2001**, *42*, 2705.

MeO_2C

MeO_2C

Bu_3SnH , BEt_3 , PhH
2 mM , rt

MeO_2C

CO_2Me

83% (mix of isomers)

Takasu, K.; Maiti, S.; Katsumata, A.; Ihara, M. *Tetrahedron Lett.*, **2001**, *42*, 2157.

NH

PhH , CF_3SO_3H , rt

4 h

Ph NH 86%

Klumpp, D.A.; Beauchamp, P.S.; Sanchez Jr., G.V.; Aguirre, S.; de Leon, S. *Tetrahedron Lett.*, **2001**, *42*, 5821.

AcO OAc

OAc

2 eq PhCH=CH_2 , DCE , 2 d

Grubbs' catalyst , rt , 0.15 M

AcO OAc

Ph

OAc 75%

Forget-Champagne, D.; Mondon, M.; Fonteneau, N.; Gesson, J.-P. *Tetrahedron Lett.*, **2001**, *42*, 7229.

N
Ts

1. 9-BBN , THF , rt , 4 h
2. 3M aq K_3PO_4 , DMF , rt

Br

5% $PdCl_2$(dppf)
DCM , 18 h

47%

N
Ts

Lapinsky, D.J.; Bergmeier, S.C. *Tetrahedron Lett.*, **2001**, *42*, 8583.

1. LiH
2. MeLi/$CeCl_3$/72 h
3. dilute HCl

94%

Bartoli, G.; Dalpozzo, R.; De Nino, A.; Procopio, A.; Sambri, L.; Tagarelli, A. *Tetrahedron Lett.*, ***2001***, *42*, 8833.

10% [Ln(η^2(PhNCPh$_2$)(hmpa)$_6$]
$PhSiH_3$, THF, rt, 20 h

SiH_2Ph 95%

Takaki, K.; Sonoda, K.; Kousaka, T.; Koshoji, G.; Shishido, T.; Takehira, K. *Tetrahedron Lett.*, ***2001***, *42*, 9211.

1. 9-BBN, THF
2. cat $Pd(PPh_3)_4$, K_2CO_3
Br, aq DMF, 65°C

$(CH_2)_6OH$ 93%

Iglesias, B.; Alvarez, R.; de Lera, A.R. *Tetrahedron*, ***2001***, *57*, 3125.

Cp_2ZrCl_2, DMF, 130°C, 6 h
PhBr, cat $Pd(OAc)_2$, PPh_3
Cs_2CO_3

70%

Dyker, G.; Heiermann, J.; Miura, M.; Inoh, J.-I.; Pivsa-Art, S.; Satoh, T.; Nomura, M. *Chem. Eur. J.*, ***2000***, *6*, 3426.

MgCl, $NiCl_2$, ether, 12 h
chiral phosphino-quinocorine

65% (75% ee)

Pellet-Rostaing, S.; Saluzzo, C.; Halle, R.T.; Breuzard, J.; Vial, L.; Le Guyader, F.; Lemaire, M. *Tetrahedron Asymm.*, ***2001***, *12*, 1983.

1. $SiMe_3$, Li, 92 h
3 eq $ZnBr_2$, 20°C
2. H_2O

78% (95:5 *E:Z*)
(73:27 *syn:anti*)

Ferreira, F.; Normant, J.F. *Eur. J. Org. Chem.*, ***2000***, 3581.

Ph–CH=CH$_2$ → 1. Li, 5% DTBB, 3 eq Me_3SiCl, THF, –78°C → 0°C; 2. H_2O → PhCH(SiMe$_3$)CH$_2$SiMe$_3$ 70%

Yus, M.; Martínez, P.; Guijarro, D. *Tetrahedron*, ***2001***, *57*, 10119.

PhCH=CHC(O)CH$_3$ → $Ph_2I^+ BF_4^-$, 5% $Pd(OAc)_2$; 2.5 eq Na_2CO_3, TEBA, DMF, 60°C, 6 h → Ph$_2$C=CHC(O)CH$_3$ 81%

Xia, M.; Chen, Z.C. *Synth. Commun.*, ***2000***, *30*, 1281.

CH$_2$=CHCO$_2$Bu → $PhSiMe_2OH$, cat $Pd(OAc)_2$, 1 d; $Cu(OAc)_2$, LiOAc, DMF, 100°C → PhCH=CHCO$_2$Bu 78%

Hirabayashi, K.; Ando, J.-i.; Kawashima, J.; Nishihara, Y.; Mori, A.; Hiyama, T. *Bull. Chem. Soc. Jpn*, ***2000***, *73*, 1409.

2,5-dihydrofuran → PhOTF, chiral Pd catalyst, *i*-Pr_2NEt, THF, 50°C, 2 d → 2-phenyl-2,3-dihydrofuran 87% (85% ee)

Ogasawara, M.; Yoshida, K.; Hayashi, T. *Heterocyles*, ***2000***, *52*, 195.

CH$_2$=CH(CH$_2$)$_3$SCH$_2$SePh → Bu_3SnH, PhH; AIBN, 80°C → CH$_2$=CH(CH$_2$)$_3$SMe + 3-methylthiolane + thiane

17% 70% 13%

Della, E.W.; Graney, S.D. *Tetrahedron Lett.*, ***2000***, *41*, 7987.

CH$_2$=CHCO$_2$Bn → PhBr, NaOAc, TBAB; 1% Pd-benzothiazole-carbene complex, 130°C, 4 h → PhCH=CHCO$_2$Bn 90%

Calò, V.; Nacci, A.; Lopez, L.; Mannarini, N. *Tetrahedron Lett.*, ***2000***, *41*, 8973.

[2-(Me$_2$NCH$_2$)C$_6$H$_4$PdCl]$_2$ (Pd, Cl/$_2$), NMP, MeO–C$_6$H$_4$–I, CH$_2$=CHCO$_2$Et, K_2CO_3, 150°C, 8 h → MeO–C$_6$H$_4$–CH=CHCO$_2$Et 88%

Iyer, S.; Ramesh, C. *Tetrahedron Lett.*, ***2000***, *41*, 8981.

OTf; CO_2Et — 3.3 $K_3PO_4 \cdot 3H_2O$, 3% $Pd(PPh_3)_4$ / toluene, 100°C, C_5H_{11}–cyclopropyl–$B(OH)_2$ → C_5H_{11}, CO_2Et — 73%

Yao, M.-L.; Deng, M.-Z. *Tetrahedron Lett.*, **2000**, *41*, 9083.

BocHN, C_3H_7 — $Si(OEt)_3$ / $Rh_2(OAc)_4$, toluene, reflux → BocHN, $Si(OEt)_3$, C_3H_7 — 61%

Hewitt, G.W.; Somers, J.J.; Sieburth, S.Mc.N. *Tetrahedron Lett.*, **2000**, *41*, 10175.

Br — 5% $Pd(dba)_2$, 5% PPh_3 / , 2 eq K_2CO_3, DMA, 120°C, 12 h → 82%

Chang, H.-M.; Cheng, C.-H. *J. Org. Chem.*, **2000**, *65*, 1767.

Br, N, Ts — $PhB(OH)_2$, Na_2CO_3, 80°C / cat $Pd(PPh_3)_4$, THF–H_2O → Ph, N, Ts — 75%

Lee, C.-W.; Oh, K.S.; Kim, K.S.; Ahn, K.H. *Org. Lett.*, **2000**, *2*, 1213.

CO_2Et, I — Me, Si, , TBAF, rt / 1% $Pd(dba)_2$, THF → CO_2Et, $H_2C{=}HC$ — 93%

Denmark, S.E.; Wang, Z. *Synthesis*, **2000**, 999.

TfO, O, O — C_5H_{11}–cyclopropyl–$B(OH)_2$, THF / $Pd(MeCN)_2Cl_2$, $AsPh_3$, AgO, 70°C → C_5H_{11}, O, O — 77%

Yao, M.-L.; Deng, M.-Z. *J. Org. Chem.*, **2000**, *65*, 5034.

Bu
B(OH)$_2$
Br–CH=CH–Ph, dioxane, 80°C
3% Pd(PPh$_3$)$_4$, 1.5 eq Ag$_2$O
3.3 eq NaOH (3N)
Bu
Ph
80%

Zhou, S.-m.; Deng, M.-z. *Tetrahedron Lett.*, ***2000***, *41*, 3951.

Et
1. 2.2 eq C$_6$H$_{13}$Li, TMEDA
hexane, 0°C
2. HCl
C$_6$H$_{13}$ C$_6$H$_{13}$
Et + Et
98%

Norsikian, S.; Baudry, M.; Normant, J.F. *Tetrahedron Lett.*, ***2000***, *41*, 6575.

Cl
PhCH=CH$_2$, K$_3$PO$_4$, cat Pd(dba)$_2$
di(1-adamantyl) *n*-butylphosphine
Ph
98%

Ehrentraut, A.; Zapf, A.; Beller, M. *Synlett*, ***2000***, 1577.

O
1. 1.1 eq BF$_3$•OEt$_2$
2 eq Et$_3$SiH
2. 1N HCl
H H
O + O
H H
71% 14%

Giese, S.; West, F.G. *Tetrahedron*, ***2000***, *56*, 10221.

0.2 $Sc(OTf)_3$, [emim] SbF_6
PhH , 20°C , 12 h

C_3H_7 C_3H_7 Ph C_3H_7 + C_3H_7 Ph

[emim] = 1-ethyl-3-methylimidazolium (an ionic liquid) (1.5 : 1) 96%

Song, C.E.; Shim, W.H.; Roh, E.J.; Choi, J.H. *Chem. Commun.*, ***2000***, 1695.

I O

aq H_3PO_2 , $NaHCO_3$, AIBN
EtOH , reflux , 5 h

O 87%

Yorimitsu, H.; Shinokubo, H.; Oshima, K. *Chem. Lett.*, ***2000***, 104.

N *t*-Bu I Ph

[Pd(OAc)$_2$/LiCl]

N Ph 80%

Diederen, J.J.H.; Sinkeldam, R.W.; Frühauf, H.-W.; Hiemstra, H.; Vrieze, K. *Tetrahedron Lett.*, ***1999***, *40*, 4255.

C_5H_{11} C_5H_{11}

1. *i*-PrO_2CCl , $Et_3Al_2Cl_3$
DCM , –15°C → rt
2. H_2O

i-Pr C_5H_{11} C_5H_{11} 67%

Biermann, U.; Metzger, J.O. *Angew. Chem. Int. Ed.*, ***1999***, *38*, 3675.

Br Br

PhZnBr , 5% $Pd(PPh_3)_4$
THF , rt

Br Ph 66%

Villiers, P.; Vicart, N.; Ramondenc, Y.; Plé, G. *Tetrahedron Lett.*, ***1999***, *40*, 8781.

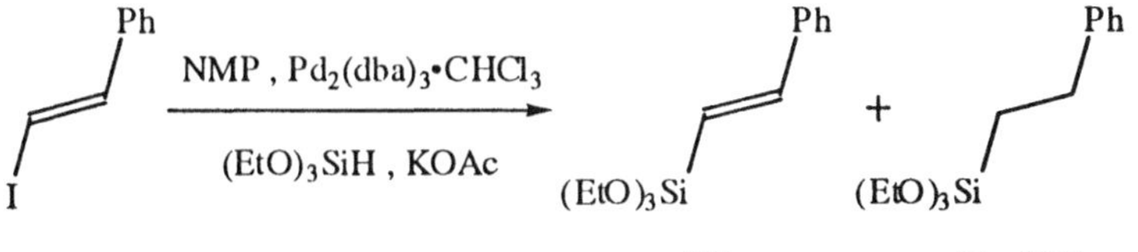
Ph
I
NMP , $Pd_2(dba)_3 \cdot CHCl_3$
$(EtO)_3SiH$, KOAc
Ph
$(EtO)_3Si$
+
Ph
$(EtO)_3Si$

(97 : 3) 81%

Murata, M.; Watanabe, S.; Masuda, Y. *Tetrahedron Lett.*, ***1999***, *40*, 9255.

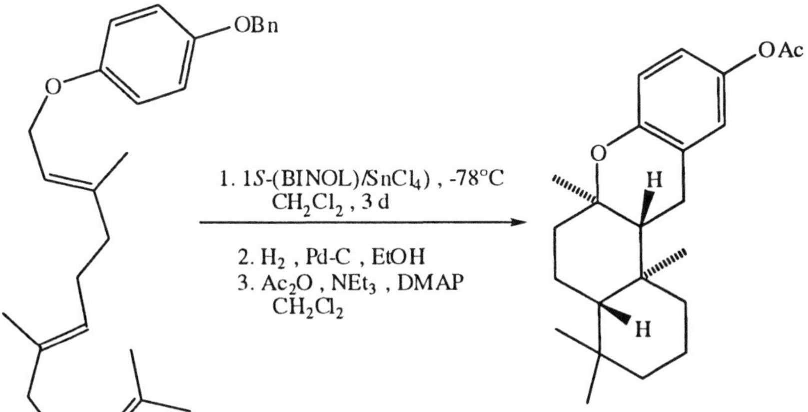
OBn
O
1. 1*S*-(BINOL)/$SnCl_4$) , -78°C
CH_2Cl_2 , 3 d
2. H_2 , Pd-C , EtOH
3. Ac_2O , NEt_3 , DMAP
CH_2Cl_2
OAc
O
H
H

40% overall (44% ee)

Ishihara, K.; Nakamura, S.; Yamamoto, H. *J. Am. Chem. Soc.*, ***1999***, *121*, 4906.

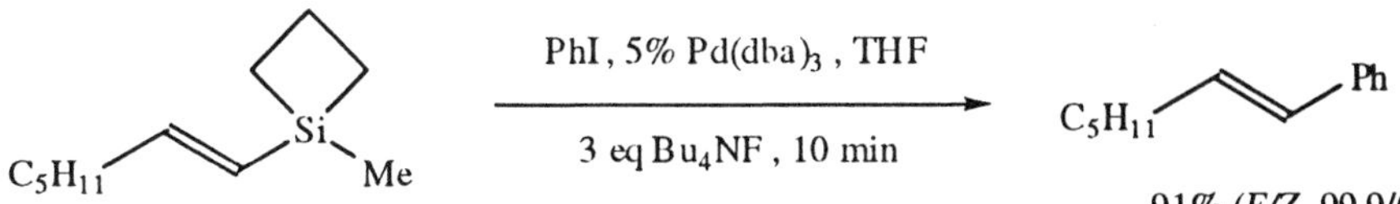
C_5H_{11}
Si
Me
PhI , 5% $Pd(dba)_3$, THF
3 eq Bu_4NF , 10 min
C_5H_{11}
Ph

91% (*E/Z*, 99.9/0.1)

Denmark, S.E.; Choi, J.Y. *J. Am. Chem. Soc.*, ***1999***, *121*, 5821.

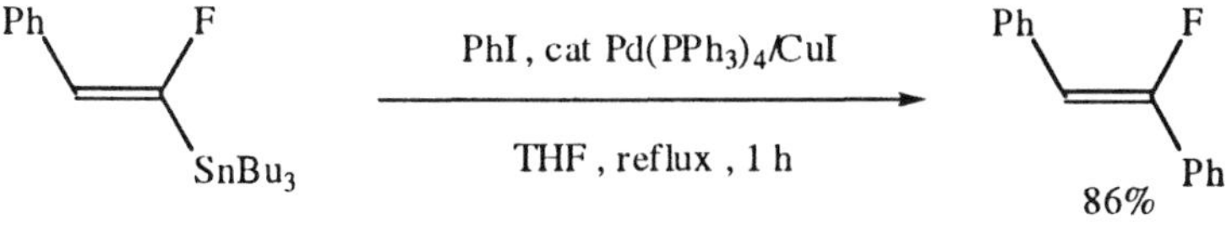
Ph
F
$SnBu_3$
PhI , cat $Pd(PPh_3)_4$/CuI
THF , reflux , 1 h
Ph
F
Ph
86%

Chen, C.; Wilcoxen, K.; Zhu, Y.-F.; Kim, K.-i.; McCarthy, J.R. *J. Org. Chem.*, ***1999***, *64*, 3476.

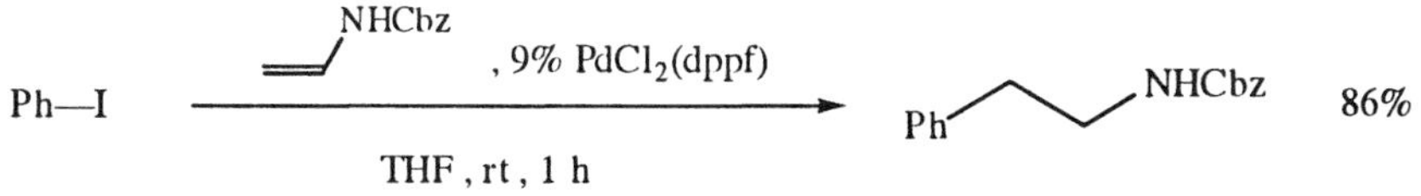
Ph—I
NHCbz
, 9% $PdCl_2(dppf)$
THF , rt , 1 h
Ph
NHCbz
86%

Kamatani, A.; Overman, L.E. *J. Org. Chem.*, ***1999***, *64*, 8743.

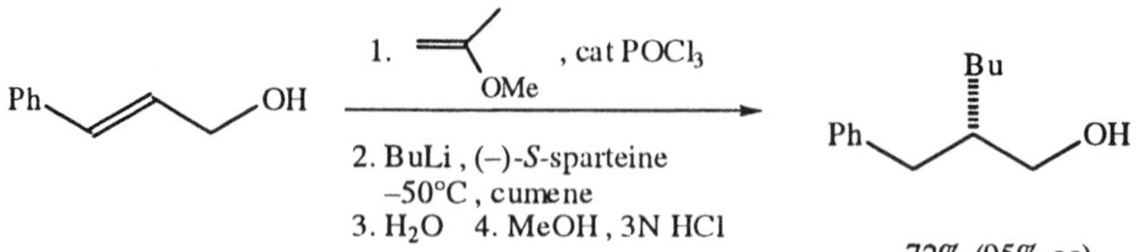

1. (2-methoxypropene), cat $POCl_3$
2. BuLi, (–)-*S*-sparteine –50°C, cumene
3. H_2O 4. MeOH, 3N HCl

72% (95% ee)

Norsikian, S.; Marek, I.; Poisson, J.-F.; Normant, J.F. *J. Org. Chem.*, ***1999***, *64*, 4898.

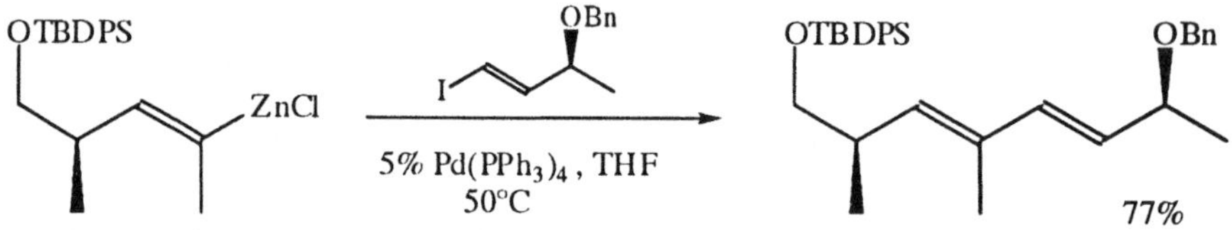

5% $Pd(PPh_3)_4$, THF 50°C

77%

Panek, J.S.; Hu, T. *J. Org. Chem.*, ***1999***, *64*, 4912.

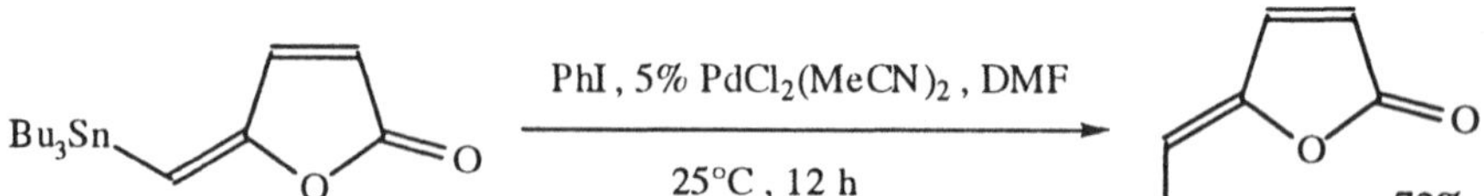

PhI, 5% $PdCl_2(MeCN)_2$, DMF

25°C, 12 h

72%

Rousset, S.; Abarbri, M.; Thibonnet, J.; Duchêne, A.; Parrain, J.-L. *Org. Lett.*, ***1999***, *1*, 701.

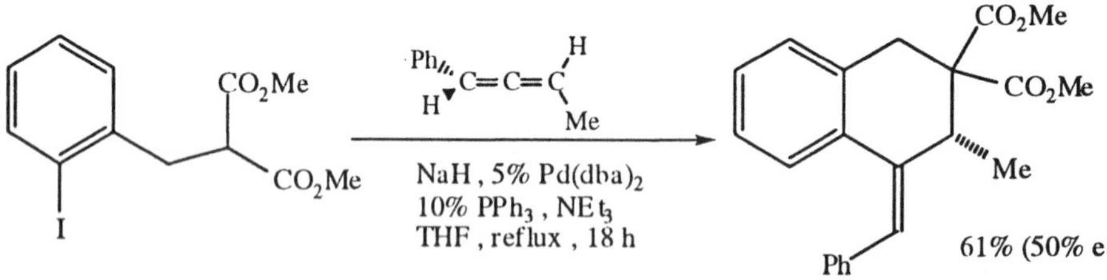

NaH, 5% $Pd(dba)_2$
10% PPh_3, NEt_3
THF, reflux, 18 h

61% (50% ee)

Kato, F.; Hiratsuka, Y.; Mitsui, T.; Watanabe, T.; Hiroi, K. *Heterocycles*, ***1999***, *50*, 83.

SECTION 74D: CONJUGATE REDUCTION OF α,β-UNSATURATED CARBONYL COMPOUNDS AND NITRILES

ASYMMETRIC REDUCTIONS

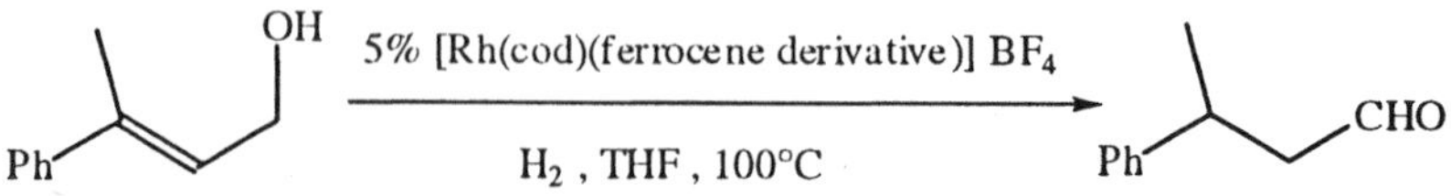

5% [Rh(cod)(ferrocene derivative)] BF_4

H_2, THF, 100°C

91% (75% ee)

Tanaka, K.; Fu, G.C. *J. Org. Chem.*, ***2001***, *66*, 8177.

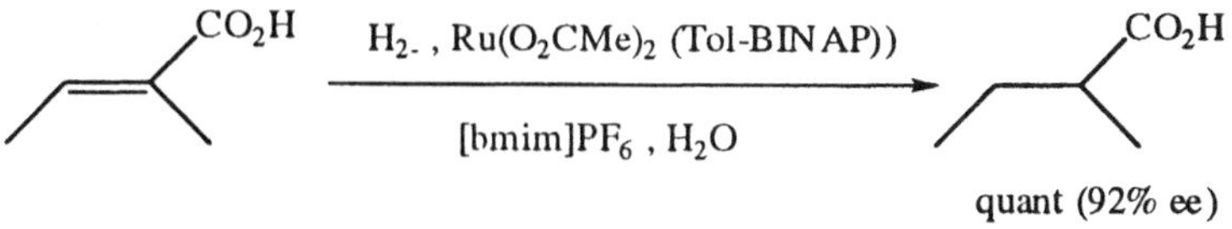

CO_2H — H_2, $Ru(O_2CMe)_2$ (Tol-BINAP)) / [bmim]PF_6, H_2O → CO_2H

quant (92% ee)

Brown, R.A.; Pollet, P.; McKoon, E.; Eckert, C.A.; Liotta, C.L.; Jessop, P.G. *J. Am. Chem. Soc.*, ***2001***, *123*, 1254.

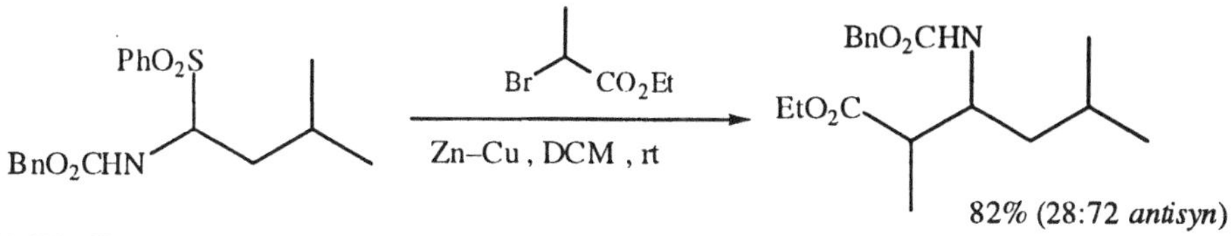

PhO_2S, BnO_2CHN — Br, CO_2Et / Zn–Cu, DCM, rt → BnO_2CHN, EtO_2C

82% (28:72 *anti:syn*)

Li, W.; Zhang, Z.; Xiao, D.; Zhang, C. *J. Org. Chem.*, ***2000***, *65*, 3489.

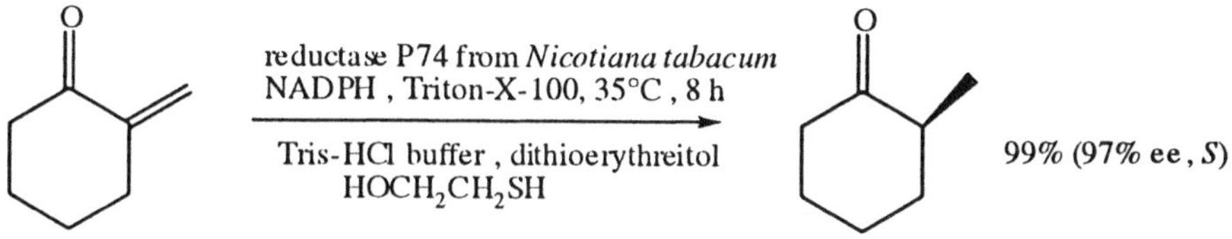

O — reductase P74 from *Nicotiana tabacum*, NADPH, Triton-X-100, 35°C, 8 h / Tris-HCl buffer, dithioerythreitol, $HOCH_2CH_2SH$ → O 99% (97% ee, *S*)

Hirata, T.; Shimoda, K.; Gondai, T. *Chem. Lett.*, ***2000***, 850.

NON-ASYMMETRIC REDUCTIONS

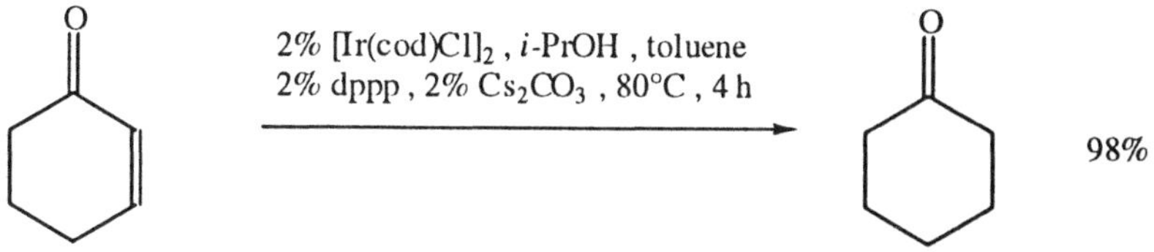

O — 2% $[Ir(cod)Cl]_2$, *i*-PrOH, toluene; 2% dppp, 2% Cs_2CO_3, 80°C, 4 h → O 98%

Sakaguchi, S.; Yamaga, T.; Ishii, Y. *J. Org. Chem.*, ***2001***, *66*, 4710.

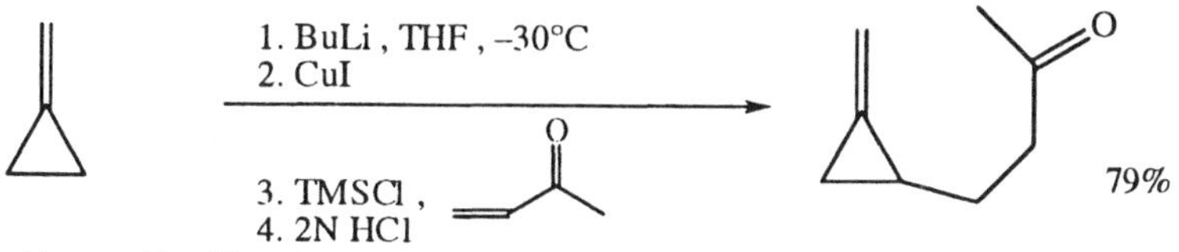

1. BuLi, THF, −30°C
2. CuI
3. TMSCl, O
4. 2N HCl

→ O 79%

Peron, G.; Norton, D.; Kitteringham, J.; Kilburn, J.D. *Tetrahedron Lett.*, ***2001***, *42*, 347.

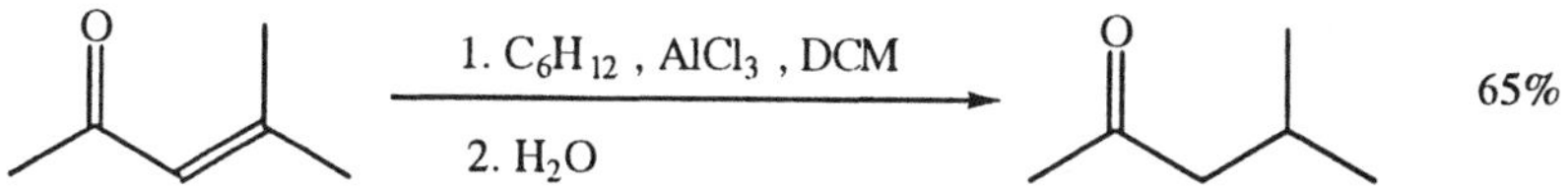

O — 1. C_6H_{12}, $AlCl_3$, DCM / 2. H_2O → O 65%

Shirini, F.; Zolfigol, M.A.; Azadbar, M.R. *Russ. J. Org. Chem.*, ***2001***, *37*, 1600.

Zn , AcOH

94%

Comins, D.L.; Brooks, C.A.; Ingalls, C.L. *J. Org. Chem.*, ***2001***, *66*, 2181.

$[MgBr]^+ [Bu_2SnBr\text{-}I\text{-}H]^-$

THF , rt

90%

Shibata, I.; Suwa, T.; Ryu, K.; Baba, A. *J. Org. Chem.*, ***2001***, *66*, 8690.

$[Rh(ligand)(nbd)] BF_4$, H_2 , THF

2 h

98.7% ee

Yasutake, M.; Gridnev, I.D.; Higashi, N.; Imamoto, T. *Org. Lett.*, ***2001***, *3*, 1701.

In , EtOH , NH_4Cl , H_2O

90°C , 9 h

85%

Ranu, B.C.; Dutta, J.; Guchhait, S.K. *Org. Lett.*, ***2001***, *3*, 2603.

1. 2.5 M aq NaOH , 16 h
cat $PdCl_2$, HCOOH , 65°C

2. 2M HCl

98%

Arterburn, J.B.; Pannala, M.; Gonzalez, A.M.; Chamberlin, R.M. *Tetrahedron Lett.*, ***2000***, *41*, 7847.

Br_3SnH , AcBr

PhH , Δ

50%

Bebbington, D.; Bentley, J.; Nilsson, P.A.; Parsons, A.F. *Tetrahedron Lett.*, ***2000***, *41*, 8941.

3% Mn(dpm)
2 eq $PhSiH_3$
iPrOH

50%

Magnus, P.; Waring, M.J.; Scott, D.A. *Tetahedron Lett.*, ***2000***, *41*, 9731.

CO_2Me Ph $NHCO_2Me$

3% Rh catalyst
15 psi H_2, THF, 6 h

CO_2Me Ph $NHCO_2Me$

93%

Taylor, R.A.; Santora, B.P.; Gagné, M.R. *Org. Lett.*, ***2000***, *2*, 1781.

Ni(R), EtOAc, rt, 25 min
ultrasound

95%

Wang, H.; Lian, H.; Chen, J.; Pan, Y.; Shi, Y. *Synth. Commun.*, ***1999***, *29*, 129.

Dibal–$Co(acac)_2$, THF
–78°C → 0°C, 2 h

72%

Ikeno, T.; Kimura, T.; Ohtsuka, Y.; Yamada, T. *Synlett*, ***1999***, 96.

Ni(R), THF, rt
20 min

90%

Barrero, A.F.; Alvarez-Manzaneda, E.J.; Chahboun, R.; Meneses, R. *Synlett*, ***1999***, 1663.

[$PhMe_2SiH$/0.5 $CuF(PPh_3)_3$•2 EtOH
DMA, 0°C → rt, 2 h

92%

Mori, A.; Fujita, A.; Kajiro, H.; Nishihara, Y.; Hiyama, T. *Tetrahedron*, ***1999***, *55*, 4573.

SECTION 74E: CONJUGATE ALKYLATIONS

ASYMMETRIC ALKYLATIONS

$(F-C_6H_4-BO)_3$, H_2O

dioxane , *R*-BINAP , 40°C
cat $Rh(acac)(C_2H_4)_2$, 12 h

74% (96% ee)

Senda, T.; Ogasawara, M.; Hayashi, T. *J. Org. Chem.*, ***2001***, *66*, 6852.

2 eq $PhB(OH)_2$, aq dioxane
0.5 K_2CO_3 , 100°C , 16 h

3% $Rh^I(acac)(C_2H_4)_2$

62% (89% ee)

Sakuma, S.; Miyaura, N. *J. Org. Chem.*, ***2001***, *66*, 8944.

Ph-$B(OH)_2$

3% $Rh(acac)(C_2H_4)_2$ L*
dioxane , 100° , H_2O , 5 h

quant (97.5:2.5 *S:R*)

Reetz, M.T.; Moulin, D.; Gosberg, A. *Org. Lett.*, ***2001***, *3*, 4083.

CHO

cat chiral imidazolium salt
THF–H_2O , –60°C , 27 h

83% (91% ee)

Paras, N.A.; MacMillan, D.W.C. *J. Am. Chem. Soc.*, ***2001***, *123*, 4370.

Me_2CHNO_2 , NaO*t*-Bu
azacrown ether-phosphine oxide

toluene

43% (95% ee , *R*)

Novák, T.; Tatai, J.; Bakó, P.; Czugler, M.; Feglevich, G.; Tõke, L. *Synlett*, ***2001***, 424.

$PhB(OH)_2$, aq dioxane
1% $Rh(acac)(C_2H_4)_2$
chiral phosphine , 100°C , 1 h

Ph
O
O

99% (96% ee)

Kuriyama, M.; Tomioka, K. *Tetrahedron Lett.*, ***2001***, *42*, 921.

O
Et_2Zn , 2% $Cu(OTf)_2$, toluene
4% phosphoramidite ligand
–20°C , 1.5 h

O
(98% ee , *R*)
Et

Alexakis, A.; Rosset, S.; Allamand, J.; March, S.; Guillen, F.; Benhaim, C. *Synlett*, ***2001***, 1375.

CO_2Me , 20% *t*-BuOK
Ph CO_2Me
10% TIPODAL oxazolines
toluene , –78°C

CO_2Me
Ph CO_2Me
80% (56% ee)

Kim, S.-G.; Ahn, K.H. *Tetrahedron Lett.*, ***2001***, *42*, 4175.

O
$CH_2(CO_2Et)_2$, *t*-BuOK , 20°C
phase transfer catalyst , toluene
Ph Ph

O $CH(CO_2Et)_2$
Ph Ph

74% (35% ee)

Kim, D.Y.; Huh, S.C.; Kim, S.M. *Tetrahedron Lett.*, ***2001***, *42*, 6299.

Ph
10% TRT-N OH
Ph , –30°C
O
Ph Ph
9% $Ni(acac)_2$, 2.2 eq Et_2Zn
MeCN

O Et
Ph Ph

75% (78% ee , *R*)

Shadakshari, U.; Nayak, S.K. *Tetrahedron*, ***2001***, *57*, 8185.

CO_2Et
CO_2Et
1. 0.5% $Cu(OTf)_2$, Et_2Zn
1% TADDOL ligand
toluene , –5°C , 3 h
2. 2N HCl

Et
CO_2Et
CO_2Et
quant (20% ee)

Alexakis, A.; Benhaim, C. *Tetrahedron Asymm.*, ***2001***, *12*, 1151.

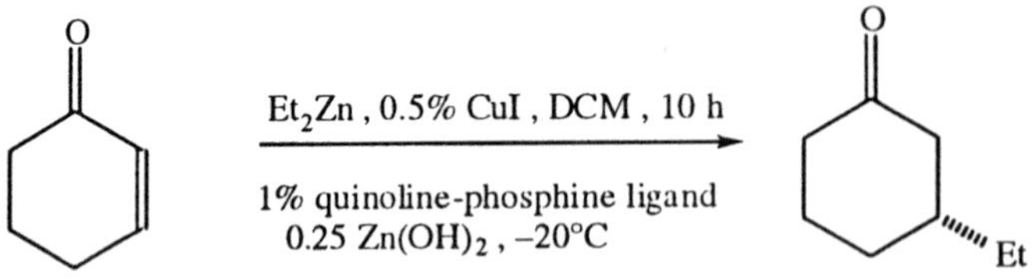

Et_2Zn , 0.5% CuI , DCM , 10 h

1% quinoline-phosphine ligand
0.25 $Zn(OH)_2$, –20°C

Et

quant (53% ee)

Delapierre, G.; Brunel, J.M.; Constantieux, T.; Buono, G.
Tetrahedron Asymm., ***2001***, *12*, 1345.

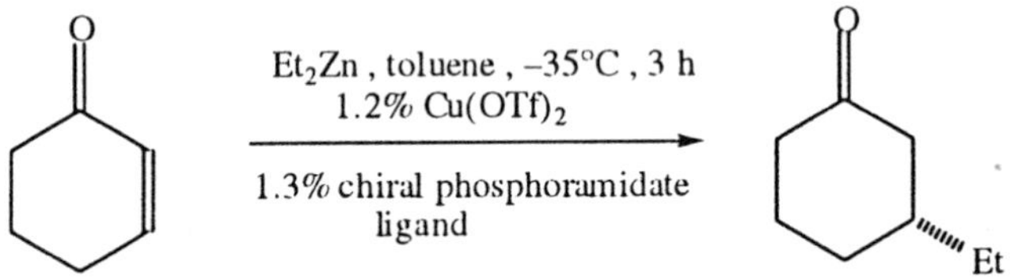

Et_2Zn , toluene , –35°C , 3 h
1.2% $Cu(OTf)_2$

1.3% chiral phosphoramidate ligand

Et

89% ee , *R*

Mandoli, A.; Arnold, L.A.; de Vries, A.H.M.; Salvadori, P.; Feringa, B.L.
Tetrahedron Asymm., ***2001***, *12*, 1929.

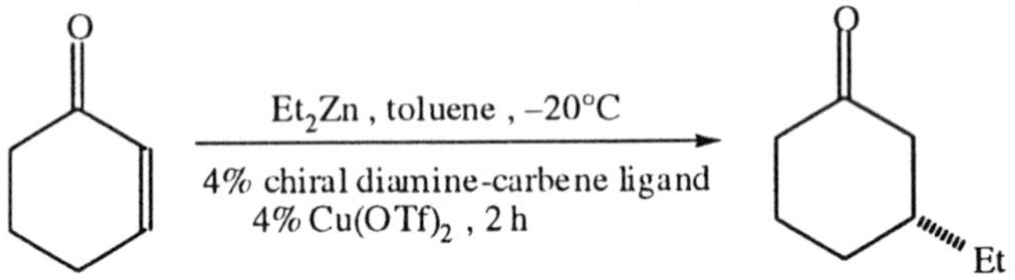

Et_2Zn , toluene , –20°C

4% chiral diamine-carbene ligand
4% $Cu(OTf)_2$, 2 h

Et

92% (22% ee, *S*)

Guillen, F.; Winn, C.L.; Alexakis, A. *Tetrahedron Asymm.*, ***2001***, *12*, 2083.

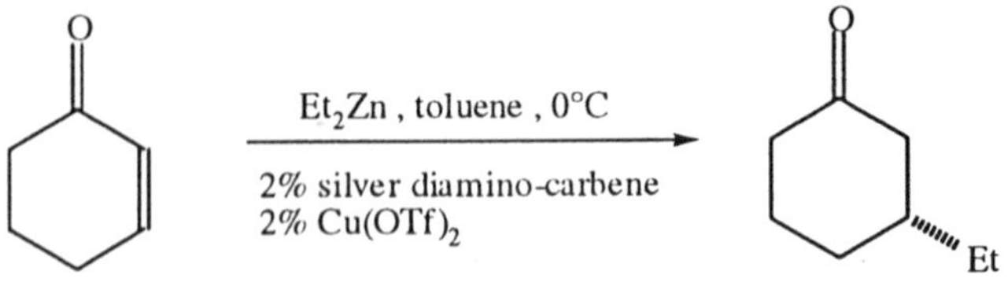

Et_2Zn , toluene , 0°C

2% silver diamino-carbene
2% $Cu(OTf)_2$

Et

98% (23% ee)

Pytkowicz, H.; Roland, S.; Mangeney, P. *Tetrahedron Asymm.*, ***2001***, *12*, 2087.

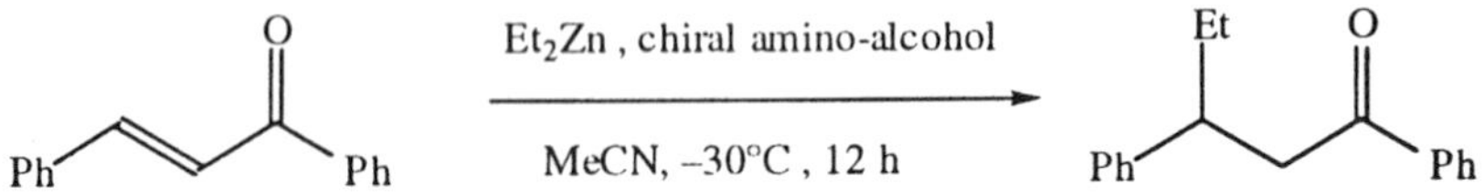

Ph Ph

Et_2Zn , chiral amino-alcohol

MeCN, –30°C , 12 h

Et
Ph Ph

89% (82% ee, *S*)

Tong, P.-E.; Li, P.; Chan, A.S.C. *Tetrahedron Asymm.*, ***2001***, *12*, 2301.

$CH_2(CO_2Me)_2$, Al (salen) complex

THF , 10 min

86% (58% ee)

$CH(CO_2Me)_2$

Jha, S.C.; Joshi, N.N. *Tetrahedron Asymm.*, ***2001***, *12*, 2463.

Et_2Zn , 2% $Cu(OTf)_2$, 16 h
4% phosphonite ligand

toluene , –30°C

Et

quant (41% ee , *R*)

Martorell, A.; Naasz, R.; Feringa, B.L.; Pringle, P.G. *Tetrahedron Asymm.*, ***2001***, *12*, 2497.

Et_2Zn , DCM , 0°C , $Cu(OTf)_2$

chiral amino-phosphite ligand

Et

85% (48% ee , *S*)

Diéguez, M.; Ruiz, A.; Claver, C. *Tetrahedron Asymm.*, ***2001***, *12*, 2861.

Ph

NO_2

15% *L*-proline

acetone,
DMSO , rt

Ph

NO_2

97% (7% ee)

List, B.; Pojarliev, P.; Martin, H.J. *Org. Lett.*, ***2001***, *3*, 2423.

N
H

bis-oxazoline ligand/$Cu(OTf)_2$

CO_2Et

Ph CO_2Et

, THF , 0°C
1 d

Ph

CO_2Et

CO_2Et

N
H

73% (60% ee)

Zhuang, W.; Hansen, T.; Jørgensen, K.A. *Chem. Commun.*, ***2001***, 347.

MeO_2C HN (2-naphth) O

1.3 eq $Mg(ClO_4)_2$, EtI
10 eq Bu_3SnH , DCE

BEt_3 , O_2 , bis-oxazoline
−78°C , 3 h

MeO_2C Et HN (2-naphth) O

79% (84% ee)

Nakamura, I.; Oh, B.H.; Saito, S.; Yamamoto, Y. *Angew. Chem. Int. Ed.*, ***2001***, ***40***, 1298.

O CO_2Bn

O , 0.15 AgOTf, H_2O

0.1 *R*-Tol-BINAP , 0°C

O CO_2Bn O

94% (56% ee)

Kobayashi, S.; Kakumoto, K.; Mori, Y.; Manabe, K. *Isr. J. Chem.*, ***2001***, ***41***, 247.

CO_2Me O

O , DCM , 10h

chiral Ru complex , KOH

O O CO_2Me

>99% (75% ee , *S*)

Suzuki, T.; Torii, T. *Tetrahedron Asymm.*, ***2001***, ***12***, 1077.

O

Et_2Zn , $Cu(OTf)_2$, DCM
chiral xylofuranose ligand

rt , 1 h

O Et

92% (60% ee)

Pàmies, O.; Net, G.; Ruiz, A.; Claver, C.; Woodward, S. *Tetrahedron Asymm.*, ***2000***, ***11***, 871.

O

$Cu(OTf)_2$–chiral phosphite complex

Et_2Zn , toluene , 0°C , 3 h

O Et

quant (73% ee)

Yan, M.; Zhou, Z.-Y.; Chan, A.S.C. *Chem. Commun.*, ***2000***, 115.

Et_2Zn, 5% $Cu(OTf)_2$, toluene, rt

10% chiral phosphine-lactam 4 h

Et

94% (64% ee)

Tomioka, K.; Nakagawa, Y. *Heterocyles*, ***2000***, *52*, 95.

Et_2Zn, $Cu(OTf)_2$, toluene
chiral quinoline binaphthol ligand

−15°C, 5 h

Et

96% (70% ee)

Arena, C.G.; Calabrò, G.P.; Franciò, G.; Faraone, F. *Tetrahedron Asymm.*, ***2000***, *11*, 2387.

Et_2Zn, 2.5% $Cu(OTf)_2$, −30°C

5% chiral phosphoramidite
toluene

Ph Ph Et Ph Ph

85%)89% ee)

Arnold, L.A.; Imbos, R.; Mandoli, A.; de Vries, A.H.M.; Naasz, R.; Feringa, B.L. *Tetrahedron*, ***2000***, *56*, 2865.

Et_2Zn, 3% $Cu(OTf)_2$, toluene, 15 h

cat chiral oxazoline-phosphite, −20°C

Et

99% (up to 86% ee)

Escher, I.H.; Pfaltz, A. *Tetrahedron*, ***2000***, *56*, 2879.

2 eq Et_2Zn, chiral QUIPHOS catalyst

0.5% CuI, DCM, H_2O, −20°C

Et

76% (61% ee)

Delapierre, G.; Constantieux, T.; Brunel, J.M.; Buono, G. *Eur. J. Org. Chem.*, ***2000***, 2507.

$CH_2(CO_2Bn)_2$, 0°C , 45 h

La–linked-BINOL complex

$CH(CO_2Bn)_2$

53% (85% ee)

Kim, Y.S.; Matsunaga, S.; Das, J.; Sekine, A.; Ohshima, T.; Shibasaki, M. *J. Am. Chem. Soc.*, ***2000***, *122*, 6506.

Et_2Zn , MeCN , –30°C
$Ni(acac)_2$

chiral thioether alcohol catalyst

94% (60% ee)

Yin, Y.; Li, X.; Lee, D.-S.; Yang, T.-K. *Tetrahedron Asymm.*, ***2000***, *11*, 3329.

Et_2CuMgI , THF , –5°C

99% (91:1 *S*:*R*)

Schneider, C.; Reese, O. *Synthesis*, ***2000***, 1689.

PhBr

1. BuLi
2. $B(OMe)_3$
3. cyclohexenone , H_2O
3% Rh(I)/*S*-BINAP

71% (99% ee , *S*)

Takaya, Y.; Ogasawara, M.; Hayashi, T. *Tetrahedron Lett.*, ***1999***, *40*, 6957.

1. cat. $Ni(acac)_2$(chiral oxazolidine)
Me_2Zn, TMSCl , triglyme
EtC≡CEt

2. H_3O^+

78% (81% ee)

Ikeda, S.-i.; Cui, D.-M.; Sato, Y. *J. Am. Chem. Soc.*, ***1999***, *121*, 4712.

Et_2Zn , 1% $Cu(OTf)_2$
2% chiral aryl diphosphite
toluene , 0°C , 1 h
quant (90% ee , *S*)

Yan, M.; Chan, A.S.C. *Tetrahedron Lett.*, ***1999***, *40*, 6645.

S (*p*-Tol)BINAP , 5% CuCl
10% NaO*t*-Bu , rt , 1 d
polymethyl hydrosiloxane
Ph CO_2Et
84% (90% ee)

Appella, D.H.; Moritani, Y.; Shitani, R.; Ferreira, E.M.; Buchwald, S.L. *J. Am. Chem. Soc.*, ***1999***, *121*, 9473.

$PhB(OH)_2$, 3% Rh/*S*-BINAP
water , dioxane , 100°C
OEt OEt Ph
94% (96% ee, *S*)

Hayashi, T.; Senda, T.; Takaya, Y.; Ogasawara, M. *J. Am. Chem. Soc.*, ***1999***, *121*, 11591.

Et_2Zn , 1% $Cu(OTf)_2$, 0°C
2% chiral aryl diphosphite
toluene , 3 h
Et
quant (90% ee)

Yan, M.; Yang, L.-W.; Wong, K.-Y.; Chan, A.S.C. *Chem. Commun.*, ***1999***, 11.

$Li[PhB(OMe)_3]^-$, 100°C
aq dioxane , 3 h
3% $Rh(acac)(C_2H_4)_2$/*S*-BINAP
CO_2Me Ph
>99% (89% ee)

Takaya, Y.; Senda, T.; Kurushima, H.; Ogasawara, M.; Hayashi, T. *Tetrahedron Asymm.*, ***1999***, *10*, 4047.

Et_2Zn , $Cu(MeCN)_4BF_4$, −20°C
chiral P-ligand , toluene/DCE
Et
98% (92% ee)

Hu, X.; Chen, H.; Zhang, X. *Angew. Chem. Int. Ed.*, ***1999***, *38*, 3518.

CO_2Me ; O ; O ; , DCM , 10h ; chiral Ru complex , KOH ; O ; O ; CO_2Me ; >99% (75% ee , *S*)

Suzuki, T.; Torii, T. *Tetrahedron Asymm.*, ***2001***, *12*, 1077.

NON-ASYMMETRIC ALKYLATIONS

O ; CO_2Et ; O ; , 140°C , 6 h ; 10% [(O, O, Si–Ph)$_2$]K ; O ; CO_2Et ; O ; 96%

Tateiwa, J.-i.; Hosomi, A. *Eur. J. Org. Chem.*, ***2001***, 1445.

O ; Ph_2SiCl_2 , cat $Rh(cod)_2BF_4$; NaF , H_2O ; O ; Ph ; 97%

Huang, T.-S.; Li, C.-J. *Chem. Commun.*, ***2001***, 2348.

O ; CO_2Me ; Ph ; cat Cu(OTf)(bis-oxazoline) , ; N H ; DCM , –78°C , 2 h ; Ph H CO_2Me ; O ; NH ; quant (88%)

Jensen, K.BV.; Thorhauge, J.; Hazell, R.G.; Jørgensen, K.A. *Angew. Chem. Int. Ed.*, ***2001***, *40*, 160.

0.1% $[RhCl(cod)]_2$, 90°C, H_2O; $TolB(OH)_2$, 6 h — 95%

Itooka, R.; Iguchi, Y.; Miyaura, N. *Chem. Lett.*, ***2001***, 722.

$PhSnMe_3$, H_2O, 50°C, 12 h; 5% $[Rh_2(cod)_2Cl_2]$ — 65%

Venkatraman, S.; Meng, Y.; Li, C.-J. *Tetrahedron Lett.*, ***2001***, *42*, 4459.

allyl-$SiMe_3$, 0.1 $InCl_3$; 5 eq TMSCl, DCM — 73%

Lee, P.H.; Lee, K.; Sung, S.-y.; Chang, S. *J. Org. Chem.*, ***2001***, *66*, 8646.

$PhSnMe_3$, 2h; cat $[Cl(cod)Rh]_2$, H_2O, ultrasound — 82%

Huang, T.-S.; Li, C.-J. *Org. Lett.*, ***2001***, *3*, 2037.

5% $Rh(cod)_2BF_4$, aq dioxane, 100°C; $PhB(OH)_2$, 4 h — 76%

Ramnauth, J.; Poulin, O.; Bratovanov, S.S.; Rakhit, S.; Maddaford, S.P. *Org. Lett.*, ***2001***, *3*, 2571.

Ph-CH=CH-NO_2; 20% chiral diamine, THF, rt, 3 h — 87% (85:15 *syn:anti*) 69% ee

Betancort, J.M.; Barbas III, C.F. *Org. Lett.*, ***2001***, *3*, 3737.

1. 2.2 eq *t*-BuLi, 2.5 eq HMPA; 4 eq TMSCl, –78°C → rt
2. H_2O

91%

Piers, E.; Harrison, C.L.; Zetina-Rocha, C. *Org. Lett.*, ***2001***, *3*, 3245.

1. 2 eq *i*-PrMgCl, ether; $Ti(Oi\text{-}Pr)_4$, –50°C → -40°C
2. $MeCH{=}C(CO_2Et)_2$, ether, 1 h; -40°C → rt

92% (97:3 *anti*:*syn*)

Song, Y.; Okamoto, S.; Sato, F. *Org. Lett.*, ***2001***, *3*, 3543.

$CH_2{=}CHCO_2Me$; AcBr, PhH, 80°C

40%

Miura, K.; Saito, H.; Fujisawa, N.; Wang, D.; Nishikori, H.; Hosomi, A. *Org. Lett.*, ***2001***, *3*, 4055.

$CH_2{=}C(SiPh_2t\text{-}Bu)Cu$, $BF_3{\bullet}OEt_2$; THF, –78°C → 0°C

88%

Barbro, A.; Castreño, P.; García, C.; Pulido, F.J. *J. Org. Chem.*, ***2001***, *66*, 7723.

Ph_3Bi, air, H_2O–THF, 12 h; 5% $Rh_2(cod)_2Cl_2$, 50°C

80%

Venkatraman, S.; Li, C.-J. *Tetrahedron Lett.*, ***2001***, *42*, 781.

$PhPbMe_3$, air, 70°C; cat $Rh_2(cod)_2Cl_2$, H_2O; 15 h

85%

Ding, R.; Chen, Y.-J.; Wang, D.; Li, C.-J. *Synlett*, ***2001***, 1470.

1. 1.2 qe Et_2Zn, $Cu(OTf)_2$
1% phosphoramidite ligand
CH_2Cl_2, -20°C, 15 min
2. $PhCH(OMe)_2$

Ph OMe Et 62%

Alexakis, A.; Trevitt, G.P.; Bernardinelli, G. *J. Am. Chem. Soc.*, ***2001***, *123*, 4358.

I
1. MeMgBr, CuI, ether, 0°C
2. H_2O
Me
58% (1:1.6 *cis:trans*)

Sha, C.-K.; Tseng, C.-T.; Chang, W.-S. *Tetrahedron Lett.*, ***2001***, *42*, 683.

I
SO_2Ph, BEt_3, dioxane
Et N+ H hypophosphite, O_2, rt, 1 h
SO_2Ph
98%

Jang, D.O.; Cho, D.H.; Chung, C.-M. *Synlett*, ***2001***, 1923.

$(BuSe)Cu(CN)BuLi_2$
THF, –78°C
Bu
96%

Zinn, F.K.; Ramos, E.C.; Comasseto, J.V. *Tetrahedron Lett.*, ***2001***, *42*, 2415.

CO_2Et, hv
benzophenone, 2.5 h
CO_2Et
70% (1.33:1 *E:Z*)

Geraghty, N.W.A.; Hannan, J.J. *Tetrahedron Lett.*, ***2001***, *42*, 3211.

CO_2Et
$Bu_4N\ Ph_3SiF_2$, THF
MS, 75°C
CO_2Et
CO_2Et 74% (7:1 *E:Z*)

Xuan, J.X.; Fry, A.J. *Tetrahedron Lett.*, ***2001***, *42*, 3275.

OH; Ph; CO_2Me; TBAB, HCO_2Na, PhBr, 130°C; cat Pd-bithiazole complex, $NaHCO_3$; O; Ph; Ph; 77%

Calò, V.; Nacci, A.; Lopez, L.; Napola, A. *Tetrahedron Lett.*, ***2001***, *42*, 4701.

OLi; Li; 1. (cyclopentenone) O, THF; 2 eq CuI, HMPA; −78°C; 2. aq NH_4Cl; OH; O; 70%

Pastor, I.M.; Yus, M. *Tetrahedron*, ***2001***, *57*, 2371.

O; OTMS; OMe; , 36 h; 10% $SmI_2(thf)_2$, DCM; PhCHO, −60°C; O; OTMS; Ph; CO_2Me; +; O; OTMS; Ph; CO_2Me; (95 : 5) quant

Giuseppone, N.; Collin, J. *Tetrahedron*, ***2001***, *57*, 8989.

N H; O; , $InCl_3$; DCM, rt, 3 h; O; N H; 90%

Yadav, J.S.; Abraham, S.; Reddy, B.V.S.; Sabitha, G. *Synthesis*, ***2001***, 2165.

O; Ph; Ph; NH, H_2O, 2 min; microwaves; N; O; Ph; Ph; 95%

Moghaddam, F.M.; Mohammadi, M.; Hosseinnia, A. *Synth. Commun.*, ***2000***, *30*, 643.

O

$CH_2(CO_2Bn)_2$, THF , 0°

45 h , polymer-bound La-BINOL catalyst

O

CO_2Br

CO_2Br

53%

Matsunaga, S.; Ohshima, T.; Shibasaki, M. *Tetrahedron Lett.*, ***2000***, ***41***, 8473.

1. EtO Li /ether/CuI PBu_3 ,−78°C

2. aq NH_4Cl , −78°C → rt

EtO O O 92%

Bennabi, S.; Narkunan, K.; Rousset, L.; Bouchu, D.; Ciufolini, M.A. *Tetrahedron Lett.*, ***2000***, ***41***, 8873.

O

Et_2Zn , 1% $Ca(OTf)_2$

2.5% aminoethyl[hosphines DCM , 0°C , 5 h

O

91% ee S

Morimoto, T.; Yamaguchi, Y.; Suzuki, M.; Saitoh, A. *Tetrahedron Lett.*, ***2000***, ***41***, 10025.

O

$ZnEt_2$, 2.5 eq CuCN , 0°C

$NH(Bn)SO_2Ph$, toluene , 1 h

O

Et

99%

Kitamura, M.; Miki, T.; Nakano, K.; Noyori, R. *Bull. Chem. Soc. Jpn.*, ***2000***, ***73***, 999.

Br

EtO_2C

O

, e^- , DMF/Py

0.13 $CoBr_2$ + 0.26 bpy , 70°C

O

EtO_2C

64%

Gomes, P.; Gosmini, C.; Nédélec, J.-Y.; Périchon, J. *Tetrahedron Lett.*, ***2000***, ***41***, 3385.

Ph Ph_4Sn , cat $PdCl_2$, 2 eq LiCl

AcOH , air , 25°C , 20 h

Ph Ph O

81%

Ohe, T.; Wakita,T.; Motofusa, S.-i.; Cho, C.S.; Ohe, K.; Uemura, S.
Bull. Chem. Soc. Jpn., ***2000***, *73*, 2149.

1. I—C6H4—Ac , DMF
$Pd(OAc)_2$, KOAc

2. H_2 , 5% Pd/C , EtOAc

O O Ac 64x95%

Arcadi, A.; Chiarini, M.; Marinelli, F.; Berente, Z.; Kollàr, L. *Org. Lett.*, ***2000***, *2*, 69.

PhI, $Pd(OAc)_2$, NEt_3 , 80°C
100 bar CO_2 (supercritical)

Ph Ph O 88%

Cacchi, S.; Fabrizi, G.; Gasparrini, F.; Pace, P.; Villani, C. *Synlett*, ***2000***, 650.

1. catecholborane , cat Me_2NCOMe

2. CH_2=$CHCO_2Me$
150 W lamp , 10°C
3 S N—OCO_2Me

SPy CO_2Me 70%

Ollivier, C.; Renaud, P. *Angew. Chem. Int. Ed.*, ***2000***, *39*, 925.

Ph NO_2

Et_2Zn , 0.5 $Cu(OTf)_2$, –30°C

1% chiral P ligand , toluene

Ph NO_2

quant (64% ee)

Alexakis, A.; Benhaim, C. *Org. Lett.*, ***2000***, *2*, 2579.

O Me_2Zn , $VO(OEt)Cl_2$

THF , –78°C , 20 h

O 78% (62:39 *trans:cis*)

Hirao, T.; Takada, T.; Sakurai, H. *Org. Lett.*, ***2000***, *2*, 3659.

MeO_2C CO_2Me → Ph, MeO_2C, CO_2Me

$PhB(OH)_2$, aq MeOH, 25°C
4% $[Rh(cod)(MeCN)_2]BF_4$, 6 h

84%

Sakuma, S.; Sakai, M.; Itooka, R.; Miyaura, N. *J. Org. Chem.*, ***2000***, *65*, 5951.

Br, O, N, CO_2Et

Bu_3SnH, AIBN, PhH
0.02 M, reflux, 5 h
syringe pump (14 h)

O, N, EtO_2C + O, N, EtO_2C

49% 3%

Lee, E.; Kim, S.K.; Kim, J.Y.; Lim, J. *Tetrahedron Lett.*, ***2000***, *41*, 5915.

MeO_2C CO_2Me

t-BuBr, Bu_3SnH, AIBN
2 eq $Sc(OTf)_3$

t-Bu, MeO_2C, CO_2Me

58% (1:99 *anti:syn*)

Hayen, A.; Koch, R.; Metzger, J.O. *Angew. Chem. Int. Ed.*, ***2000***, *39*, 2758.

CN

$PhCH_2Br$, 2.1 eq $CrCl_3$
Mn, THF, 4.2 eq 4-*t*-Bu-piperidine

Ph CN

85%

Augé, J.; Gil, R.; Kalsey, S. *Tetrahedron Lett.*, ***1999***, *40*, 67.

Ph NO_2

ZnBr, DMF
rt, 1.5 h

Ph NO_2

83%

Kumar, H.M.S.; Reddy, B.V.S.; Reddy, P.T.; Yadav, J.S. *Tetrahedron Lett.*, ***1999***, *40*, 5387.

1. O, BH, O, 10% Me_2NCOMe
DCM, heat, 3 h
2. 3 eq H_2O, DMPU,
DCM, O_2, O

O

86%

Ollivier, C.; Renaud, P. *Chem. Eur. J.*, ***1999***, *5*, 1468.

, PhH

Na_2CO_3 , H_2O

35%

da Silva, F.M.; Gomes, A.K.; Jones Jr. J. *Can. J. Chem.*, ***1999***, *77*, 624.

Ph Ph O

Bu_4GaLi , THF , 0°C → rt

Ph Ph Bu O

78%

Han, Y.; Huang, Y.-Z.; Fang, L.; Tao, W.-T. *Synth. Commun.*, ***1999***, *29*, 867.

Ph CHO

Et_2NTMS , 50°C , 16h

Ph CHO

51%

Hagiwara, H.; Ono, H.; Komatsubara, N.; Hoshi, T.; Suzuki, T.; Ando, M. *Tetrahedron Lett.*, ***1999***, *40*, 6627.

CO_2Et

5.5% bis-oxazoline , 5% $Mg(OTf)_2$
6% NMM , $CHCl_3$, MS , rt

Ph NO_2

Ph NO_2 CO_2Et

95%

Ji, J.; Barnes, D.M.; Zhang, J.; King, S.A.; Wittenberger, S.J.; Morton, H.E. *J. Am. Chem. Soc.*, ***1999***, *121*, 10215.

Ph

, rt , 6 h

$CeCl_3$•7 H_2O , NaI

Ph

98%

Bartoli, G.; Bosco, M.; Bellucci, M.C.; Marcantoni, E.; Sambri, L.; Torregiani, E. *Eur. J. Org. Chem.*, ***1999***, 617.

REVIEWS:

"Catalysis of the Michael Reaction and the Vinylogous Michael Reaction by Ferric Chloride Hexahydrate," Christoffers, J. *Synlett*, ***2001***, 723.

"Rhodium Catalyzed Asymmetric 1,4-Addition of Organoboronic Acids and Their Derivatives to Electron Deficient Olefins," Hayashi, T. *Synlett*, ***2001***, 879.

"Vinylogous Mannich Reactions. Selectivity and Synthetic Utility," Bur, S.K.; Martin, S.F. *Tetrahedron*, ***2001***, *57*, 3221.

"Intramolecular Free Radical Conjugate Additions," Zhang, W. *Tetrahedron*, ***2001***, *57*, 7237.

"Recent Advances in Catalytic Enantioselective Michael Additions," Krause, N.; Hoffmann-Röder, A. *Synthesis*, ***2001***, 171.

"Reactions of Conjugated Haloenoates with Nucleophilic Reagents," Caine, D. *Tetrahedron*, ***2001***, *57*, 2643.

SECTION 74F: Cyclopropanations, including Halocyclopropanations

$(i\text{-Bu})_2Te$, $SiMe_3$; 1. LiTMP, THF, –78°C; 2. Ph–CH=CH–CO_2(-)menthyl → (-)menthylCO_2 ... Ph ... Me_3Si + CO_2(-)menthyl ... Ph ... $SiMe_3$ (92 : 8) 83%

Ye, S.; Tang, Y.; Dai, L.-X. *J. Org. Chem.*, ***2001***, *66*, 5717.

CO_2Me, N_2; CH_2=CHPh, hexane, rt; Rh_2(*S*-DOSP)$_4$ → CO_2Me, Ph

87% (96% de, 89% ee)

Davies, H.M.L.; Townsend, R.J. *J. Org. Chem.*, ***2001***, *66*, 6595.

Ph–CH=N—NHTs; 1. LiHMDS, THF; 2. cat ClFeTPP; 5 eq CH_2=CH-NPhTh → Ph, N, O, O

71%
(33:67 *trans:cis*)

Aggarwal, V.K.; de Vincente, J.; Connert, R.V. *Org. Lett.*, ***2001***, *3*, 2785.

10

1. $EtO_2CCH_2NH_2$•HCl, TPPRh(III), DCM, H_2O, –10°C
2. $NaNO_2$, aq H_2SO_4, rt, 4 d

Ph / CO_2Et

62%

Barrett, A.G.M.; Braddock, D.C.; Lenoir, I.; Tone, H. *J. Org. Chem.*, ***2001***, *66*, 8260.

HO / Et / Et / C

3 eq $Zn(CH_2I)_2$, DME
DCM, –10°C → rt

Me_2NOC / B—Bu / Me_2NOC

Et / Et / OH / H

69% (86% ee)

Charette, A.B.; Jolicoeur, E.; Bydlinski, G.A.S. *Org. Lett.*, ***2001***, *3*, 3293.

CHN_2 / O

Ru catalyst
pentane

O

90% (74% ee)

Barberis, M.; Pérez-Prieto, J.; Stiriba, S.-E.; Lahuerta, P. *Org. Lett.*, ***2001***, *3*, 3317.

NC / $SiMe_3$

1) 2.3 eq BuLi
2) epibromohydrin (O, Br) / $CiClO_4$, THF
3) H_2O

$SiMe_3$ / CN / OH

65% (>98:2 *E:Z*)

Langer, P.; Freifeld, I. *Org. Lett.*, ***2001***, *3*, 3903.

Me_2 Si / Ph / EtO_2C

1. 5 eq Dibal
2. Tf_2O, –78°C
2,6-lutidine

OH / H / Ph / H

86x73%

Taylor, R.E.; Engelhardt, F.C.; Schmitt, M.J.; Yuan, H. *J. Am. Chem. Soc.*, ***2001***, *123*, 2964.

Ph

$EtO_2CCH=N_2$

tris(pyrazolyl)borate-copper(I) catalyst

Ph / CO_2Et

97% >98:2

Díaz-Requejo, M.M.; Belderraín, T.R.; Trofimenko, S.; Pérez, P.J.
J. Am. Chem. Soc., ***2001***, *123*, 3167.

$C_4F_9C(O)OCH_2I$

2 eq Et_2Zn , CD_2Cl_2 , hν

>99%

Charette, A.B.; Beauchemin, A.; Francoeur, S. *J. Am. Chem. Soc.*, ***2001***, *123*, 8139.

10% $PtCl_2$, toluene

80°C

Ts—N Ts N

59%

Fürstner, A.; Stelzer, F.; Szillat, H. *J. Am. Chem. Soc.*, ***2001***, *123*, 11863.

N_2 O O Cl

5% Ru salen complex

hν , THF , 37°C

O H O H Cl

62% (87% ee)

Saha, B.; Uchida, T.; Katsuki, T. *Synlett*, ***2001***, 114.

Ph

N_2CHCO_2t-Bu , Co(II) complex

MeOH , 5% H_2O , 50°C

Ph CO_2t-Bu

89% (83:17 *trans:cis*), 92% ee

Ikeno, T.; Nishizuka, A.; Sato, M.; Yamada, T. *Synlett.*, ***2001***, 406.

$NHSO_2Ph$

1. Me—≡—$IPh^+ OTf^-$

BuLi , THF

2. 1N HCl

O $NHSO_2Ph$

55%

Lee, H.-Y.; Lee, Y.-H. *Synlett*, ***2001***, 1656.

Ph

N_2CHCO_2Et . rt

Ru catalyst , 2,4,6-collidine

Ph CO_2Et + Ph CO_2Et

(30 : 70) 81%

79% ee 82% ee

Zheng, Z.; Yao, X.; Li, C.; Chen, H.; Hu, C. *Tetrahedron Lett.*, ***2001***, *42*, 2847.

Metal salen catalyst, 1 d

NMI, THF, rt

72% (98% ee)

Uchida, T.; Saha, B.; Katsuki, T. *Tetrahedron Lett.*, ***2001***, *42*, 2521.

$Cp_2Ti[P(OEt)_3]_2$

THF, 0°C, 1.5 h

73% (89:11)

Fujiwara, T.; Odaira, M.; Takeda, T. *Tetrahedron Lett.*, ***2001***, *42*, 3369.

Me_3SiCHN_2, DCM, –20°C

cat $[Ru_2(CO)_4(\mu\text{-}OAc)_2]_n$

(32 : 68) 64%

Maas, G.; Seitz, J. *Tetrahedron Lett.*, ***2001***, *42*, 6137.

Cl_2CHCO_2Me, LiHMBS

THF, DBU

62%

Escribano, A.; Pedregal, C.; González, R.; Fernández, A.; Burton, K.; Stephenson, G.A. *Tetrahedron*, ***2001***, *57*, 9423.

N_2CHCO_2Et, DCM

chiral Ru salen catalyst

(26 : 74) 82%

43% ee 44 % ee

Yao, X.; Qiu, M.; Lü, W.; Chen, H.; Zheng, Z. *Tetrahedron Asymm.*, ***2001***, *12*, 197.

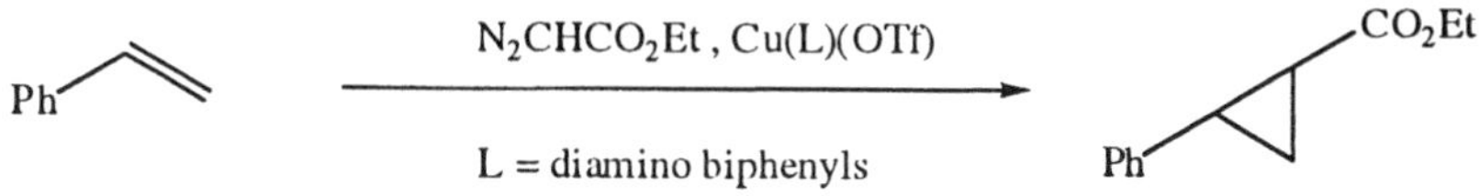

N_2CHCO_2Et, Cu(L)(OTf)

L = diamino biphenyls

Ph — CO_2Et

94% (84:18 *trans:cis*)
(46% ee 57% ee)

Sanders, C.J.; Gillespie, K.M.; Scott, P. *Tetrahedron Asymm.*, **2001**, *12*, 1055.

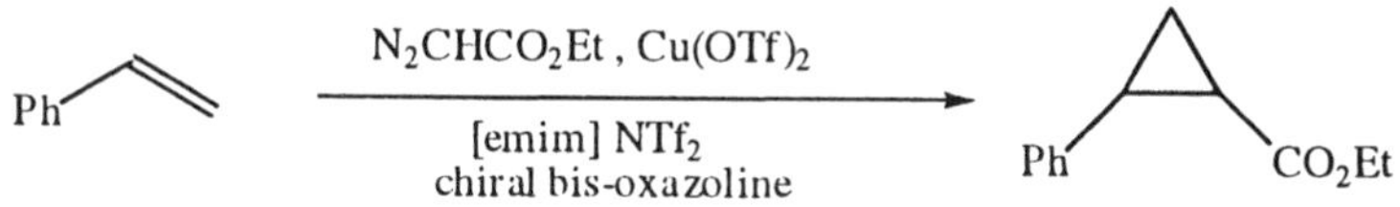

N_2CHCO_2Et, $Cu(OTf)_2$

[emim] NTf_2
chiral bis-oxazoline

in ionic liqluid

Ph — CO_2Et

38% (64 : 36 *trans:cis*)
60% ee 64% ee

Fraile, J.M.; García, J.I.; Herrerías, C.I.; Mayoral, J.A.; Carrié, D.; Vaultier, M. *Tetrahedron Asymm.*, **2001**, *12*, 1891.

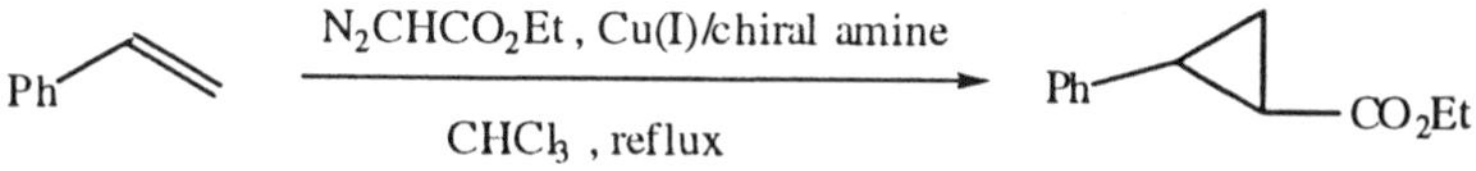

N_2CHCO_2Et, Cu(I)/chiral amine

$CHCl_3$, reflux

Ph — CO_2Et

61% (37:63 *cis:trans*)
36% ee 40% ee

Ma, J.-A.; Wang, L.-X.; Zhang, W.; Zhou, Q.-L. *Tetrahedron Asymm.*, **2001**, *12*, 2801.

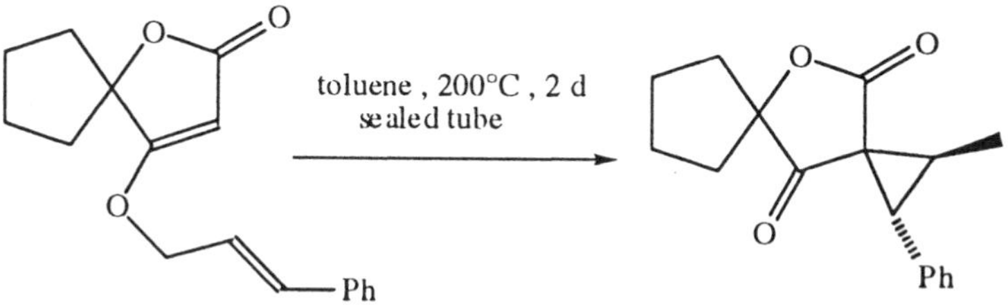

O O O Ph

toluene , 200°C , 2 d
sealed tube

O O O Ph

74%

Schobert, R.; Siegfried, S.; Gordon, G.; Nieuwenhuyzen, M.; Allenmark, S. *Eur. J. Org. Chem.*, **2001**, 1951.

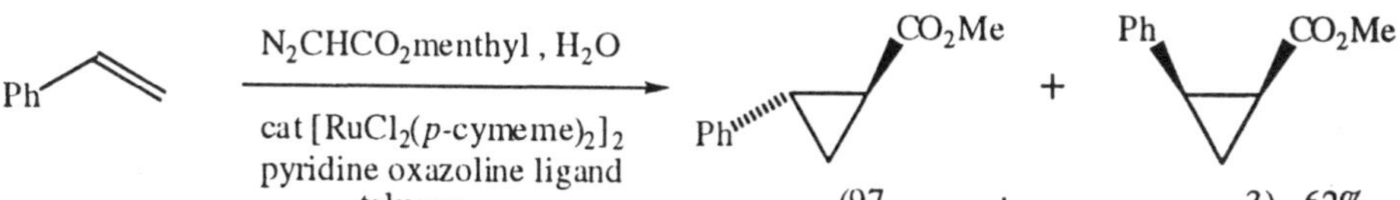

N_2CHCO_2menthyl , H_2O

cat $[RuCl_2(p\text{-cymeme})_2]_2$
pyridine oxazoline ligand
toluene

Ph CO_2Me + Ph CO_2Me

(97 : 3) 62%

Iwasa, S.; Takezawa, F.; Tuchiya, Y.; Nishiyama, H. *Chem. Commun.*, **2001**, 59.

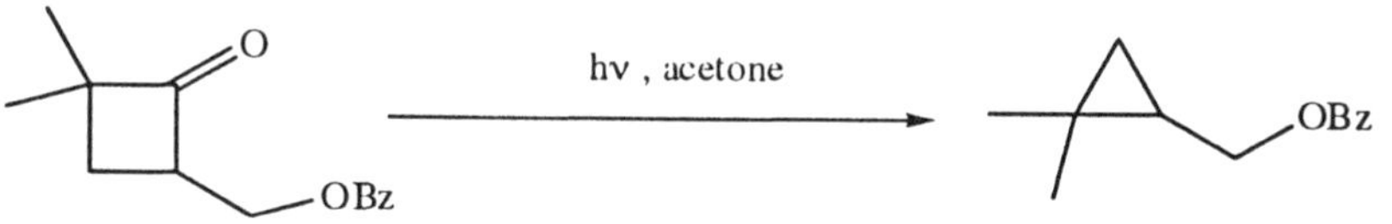

O OBz

hv , acetone

OBz

85%

Ramnauth, J.; Lee-Ruff, E. *Can. J. Chem.*, **2001**, *79*, 114.

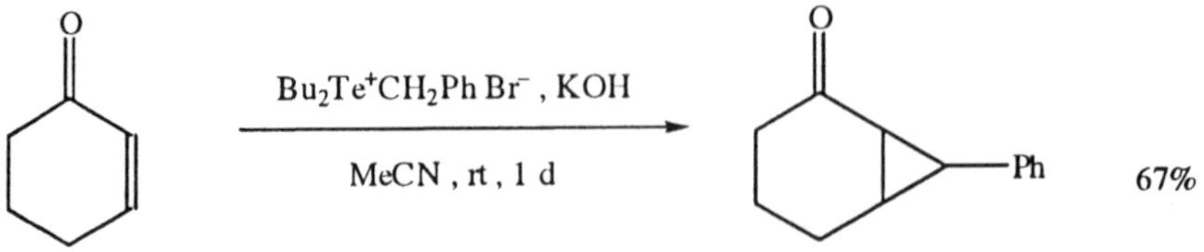

$Bu_2Te^+CH_2Ph\ Br^-$, KOH

MeCN , rt , 1 d

Ph

67%

Guo, X.; Shen, W.; Shao, J.; Zhong, Q. *Synth. Commun.*, ***2000***, *30*, 3275.

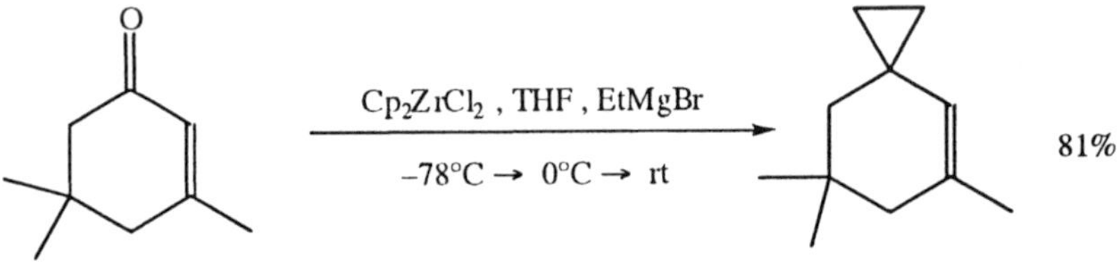

Cp_2ZrCl_2 , THF , EtMgBr

−78°C → 0°C → rt

81%

Gandon, V.; Bertus, P.; Szymoniak, J. *Eur. J. Org. Chem.*, ***2000***, 3713.

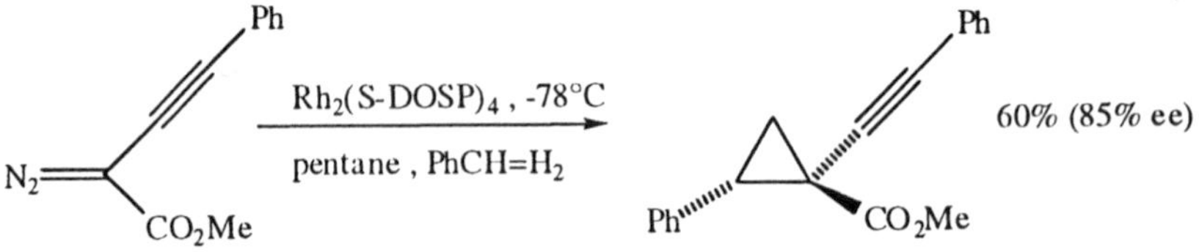

Ph

N_2

CO_2Me

$Rh_2(S\text{-}DOSP)_4$, -78°C

pentane , $PhCH{=}H_2$

Ph

Ph

CO_2Me

60% (85% ee)

Davies, H.M.L.; Boebel, T.A. *Tetrahedron Lett.*, ***2000***, *41*, 8189.

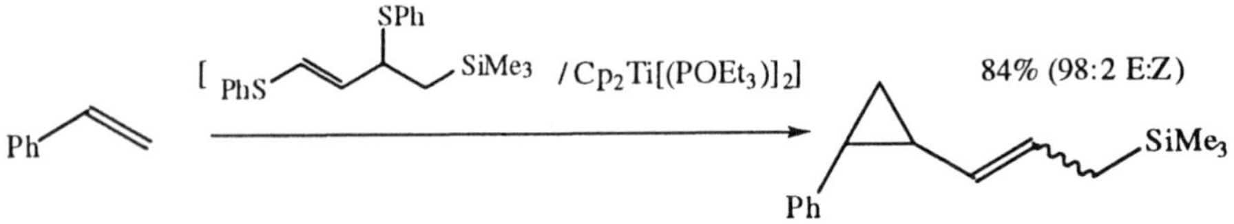

Ph

[PhS ... SPh ... $SiMe_3$ / $Cp_2Ti[(POEt_3)]_2$]

84% (98:2 E:Z)

$SiMe_3$

Ph

Takeda, T.; Takagi, Y.; Saeki, N.; Fujiwara, T. *Tetrahedron Lett.*, ***2000***, *41*, 8377.

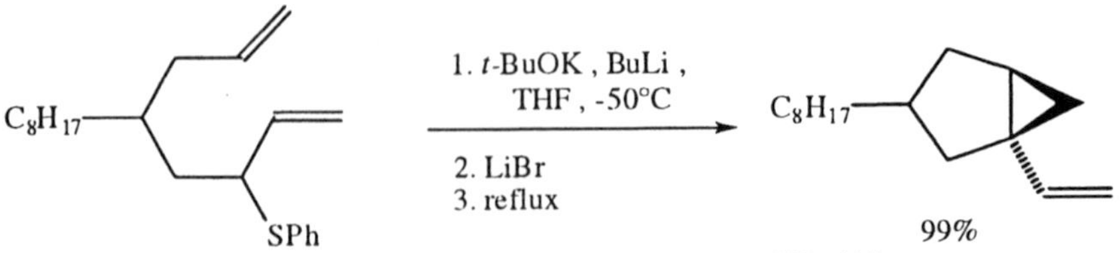

C_8H_{17}

SPh

1. *t*-BuOK , BuLi , THF , -50°C

2. LiBr

3. reflux

C_8H_{17}

99%

Cheng, D.; Know, K.R.; Cohen, T. *J. Am. Chem. Soc.*, ***2000***, *122*, 412.

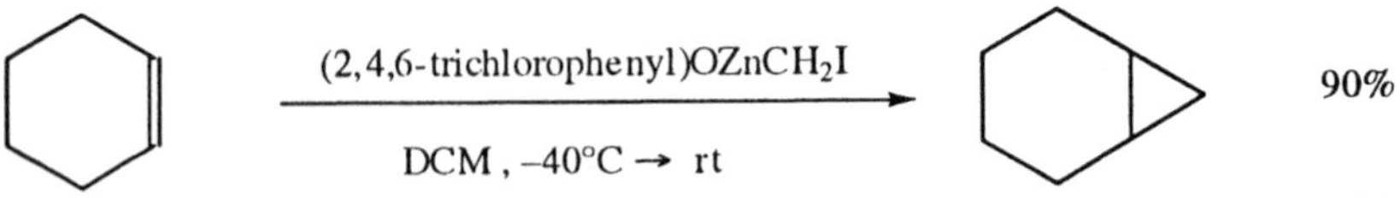

(2,4,6-trichlorophenyl)$OZnCH_2I$

DCM , −40°C → rt

90%

Charette, A.B.; Francoeur, S.; Martel, J.; Wilb, N. *Angew. Chem. Int. Ed.*, ***2000***, *39*, 4539.

$CH_2=CHCO_2Bu$

DMF, NaOAc
TBAB, 110°C, 0.6 h

60%

Calò, V.; Nacci, A.; Lopez, L.; Lerario, V.L. *Tetrahedron Lett.*, ***2000***, ***41***, 8977.

Ph_2S^+–$C^-(CH_3)_2$

$Yb(OTf)_3$, –78°C
DCM, hexane
THF

(10 : 1) 84%

Mamai, A.; Madalengoitia, J.S. *Tetrahedron Lett.*, ***2000***, ***41***, 9009.

N_2CHCO_2t-Bu, THF, 1 d

Co-salen catalyst
catalytic N-Methylimidazole

(98 : 2) 89%

Niimi, T.; Uchida, T.; Irie, R.; Katsuki, T. *Tetrahedron Lett.*, ***2000***, ***41***, 3647.

1. $Cp_2Zr(CH_2=CH_2)$, PMe_3
2. aq H_2SO_4

53%

Bertus, P.; Gandon, V.; Szymoniak, J. *Chem. Commun.*, ***2000***, 171.

$N_2=CHCO_2Et$, $PhNO_2$, 40°C

1% Cu-Schiff base catalyst

65% (91% ee)

Li, Z.; Zheng, Z.; Chen, H. *Tetrahedron Asymm.*, ***2000***, ***11***, 1157.

Ph OH → Et_2Zn , DCM , chiral sulfonamide / Schiff base ligand → Ph OH

88% (98% ee)

Balsells, J.; Walsh, P.J. *J. Org. Chem.*, ***2000***, *65*, 5005.

Bu , DMF , 152°C , 4% BHT / OMe $(OC)_5Cr$ ferrocene → H OMe ferrocene H

40% (89% de)

Barluenga, J.; López, S.; Trabanco, A.A.; Fernández-Acebes, A.; Flórez, J. *J. Am. Chem. Soc.*, ***2000***, *122*, 8145.

O S , $Rh_2(OAc)_4$ / PhMe , $PhCHN_2$, rt

O Ph Ph → O Ph Ph Ph

60% (4:1)

Aggarwal, V.K.; Smith, H.W.; Hynd, G.; Jones, R.V.H.; Fieldhouse, R.; Spey, S.E. *J. Chem. Soc., Perkin Trans. 1*, ***2000***, 3267.

$C_8H_{11}CHO$ → O Ph , THF / $CrCl_2$, 2 h → HO OH Ph C_8H_{17}

93% 58 OH : 42 OH)

Toratsu, C.; Fujii, T.; Suzuki, T.; Takai, K. *Angew. Chem. Int. Ed.*, ***2000***, *39*, 2725.

Ph → N_2CHCO_2t-Bu , hv , 2 d / 5% Ru-salen catalyst , THF → t-BuO_2C Ph + t-BuO_2C Ph

(93 : 7) 45%
97% ee

Uchida, T.; Irie, R.; Katsuki, T. *Synlett*, ***1999***, 1793.

N_2CHCO_2t-Bu; 5% Ru-salen catalyst

Ph, CO_2t-Bu + Ph, CO_2t-Bu

(80 : 20) 53%

81% ee 51% ee

Uchida, T.; Irie, R.; Katsuki, T. *Synlett*, ***1999***, 1163.

Ph, OBn

1. Et_2Zn, hν, CD_2Cl_2, rt
2. CH_2I_2, 0°C
3. Ph OBn

Me_3Si I

\>95%

Charette, A.B.; Beauchemin, A.; Marcoux, J.-F. *Tetrahedron Lett.*, ***1999***, *40*, 33.

$SiMe_3$

1. [Ti(O*i*-Pr)$_3$Cl/*i*-PrMgCl] ether/hexane
2. $PhCO_2Et$

OH, Me_3Si, Ph

88% (93:7 *trans:cis*)

Matsuda, I.; Takeuchi, K.; Itoh, K. *Tetrahedron Lett.*, ***1999***, *40*, 2553.

Ph

N_2CHCO_2Et, 80°C; bis imide Cr(VI) complex

Ph, CO_2Et

90% (*cis/trans* = 0.65)

Jan, D.; Simal, F.; Demonceau, A.; Noels, A.F.; Rufanov, K.A.; Ustynyuk, N.A.; Gourevitch, D.N. *Tetrahedron Lett.*, ***1999***, *40*, 5695.

O, N_2

5% chiral ferrocenyl ligand/Cu(OTf)$_2$; DCE, rt, 8 h

O

51% (63% ee)

Kim, S.-G.; Cho, C.-W.; Ahn, K.H. *Tetrahedron*, ***1999***, *55*, 10029.

Ph–CH(OMe)$_2$

$SnBu_3$; $TiCl_4$, DCM, –78°C, 3 h

OMe, Ph

87%

Sugawara, M.; Yoshida, J.-i. *Chem. Commun.*, ***1999***, 505.

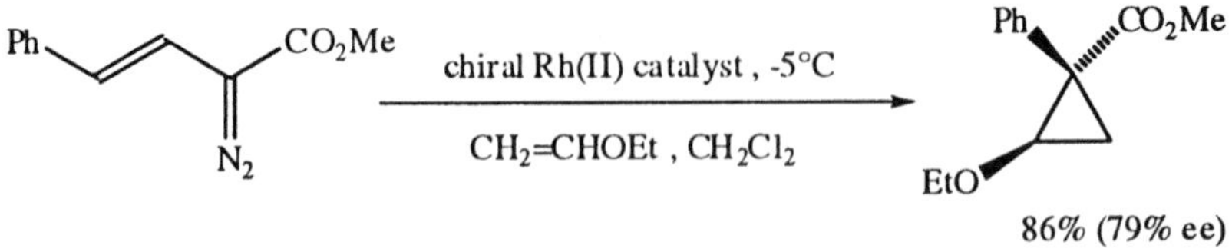

chiral Rh(II) catalyst, -5°C; CH_2=CHOEt, CH_2Cl_2

86% (79% ee)

Davies, H.M.L.; Panaro, S.A. *Tetrahedron Lett.*, ***1999***, *40*, 5287.

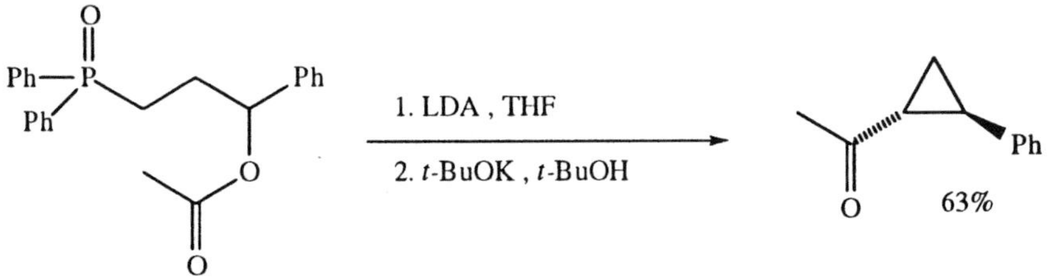

1. LDA, THF
2. *t*-BuOK, *t*-BuOH

63%

Nelson, A.; Warren, S. *J. Chem. Soc., Perkin Trans. 1*, ***1999***, 3425.

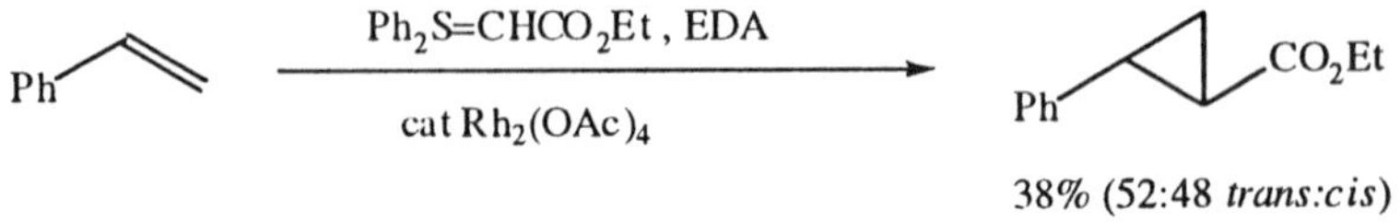

Ph_2S=CHCO$_2$Et, EDA; cat $Rh_2(OAc)_4$

38% (52:48 *trans:cis*)

Müller, P.; Fernandez, D.; Nury, P.; Rossier, J.-C. *Helv. Chim. Acta*, ***1999***, *82*, 935.

SECTION 74G: CYCLOBUTANATIONS, INCLUDING HALOCYCLOBUTANATIONS

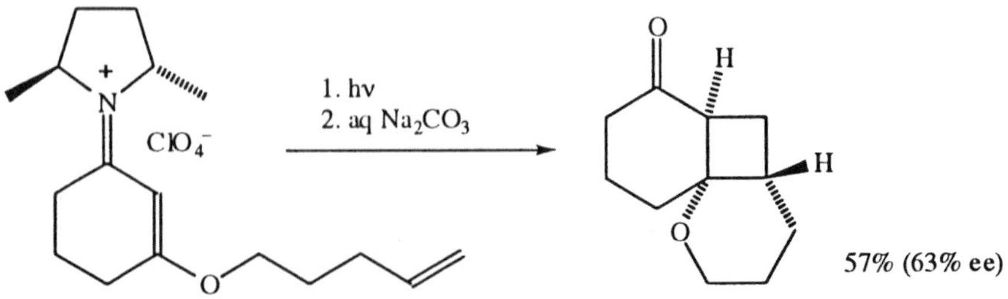

1. hν
2. aq Na_2CO_3

57% (63% ee)

Chen, C.; Chang, V.; Cai, X.; Duesler, E.; Mariano, P.S. *J. Am. Chem. Soc.*, ***2001***, *123*, 6433.

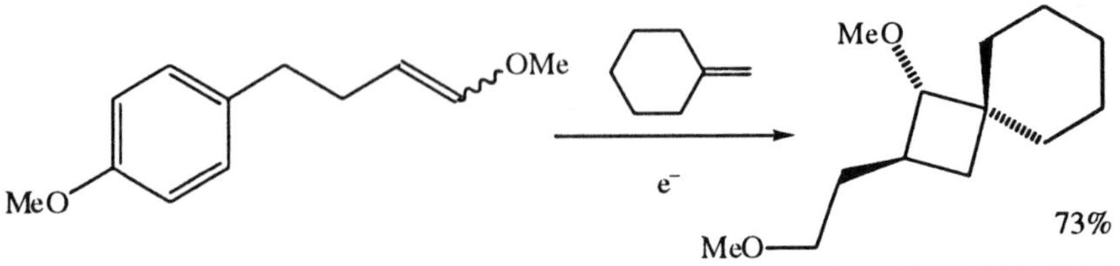

e^-

73%

Chiba, K.; Miura, T.; Kim, S.; Kitano, Y.; Tada, M. *J. Am. Chem. Soc.*, ***2001***, *123*, 11314.

CH_2Cl_2, rt, 30 min

10% $Co(dpm)_2$
4 eq $PhMeSiH_2$

72%

Baik, T.-G.; Luis, A.L.; Wang, L.-C.; Krische, M.J. *J. Am. Chem. Soc.*, ***2001***, *123*, 6716.

ClO_4^- CO_2Me

1. hv. MeCN
2. aq Na_2CO_3

84%

Cai, X.; Chang, V.; Chen, C.; Kim, H.-J.; Mariano, P.S. *Tetrahedron Lett.*, ***2000***, *41*, 9445.

, cat $Pd(OAc)_2$, $P(t\text{-}Bu)_3$

$CsCO_3$, DMF, 80°C, 2.5 h

77%

Terao, Y.; Satoh, T.; Miura, M.; Nomura, M. *Bull. Chem. Soc. Jpn*, ***1999***, *72*, 2345.

REVIEWS:

"Photochemical Dimerization In Solution Of Heterocyclic Substituted Alkenes Bearing An Electron Withdrawing Group," D'auria, M. *Heterocycles*, ***2001***, *54*, 475.

SECTION 75: ALKYLS, METHYLENES AND ARYLS FROM MISCELLANEOUS COMPOUNDS

6 eq PhI, BEt_3, THF

air, rt, 10 min

85%

Liu, J.-T.; Jang, Y.-J.; Shih, Y.-K.; Hu, S.-R.; Chu, C.-M.; Yao, C.-F. *J. Org. Chem.*, ***2001***, *66*, 6021.

$Bu_2CuCNLi_2$, THF

74% (67:33 S_N2':S_N2)

Belelie, J.L.; Chong, J.M. *J. Org. Chem.*, ***2001***, *66*, 5552.

20 eq *i*-PrI, 6 eq Et_3Al
2 eq $(BzO)_2$, ether, rt

96%

Liu, J.-Y.; Liu, J.-T.; Yao, C.-F. *Tetrahedron Lett.*, ***2001***, *42*, 3613.

$PhB(OH)_2$ —cat $Pd(OAc)_2$–2 PPh_3, DMF, 90°C, 2h→ Ph—Ph 83%

Wong, M.S.; Zhang, X.L. *Tetrahedron Lett.*, ***2001***, *42*, 4087.

$Pd(OAc)_2$, $CuCl_2$, 7 min
THF, 23°C

95%

Parrish, J.P.; Flanders, V.L.; Floyd, R.J.; Jung, K.W. *Tetrahedron Lett.*, ***2001***,*4* 2, 7729.

$PhBF_3^-\ NBu_4^+$, aq DME
5% $Pd(OAc)_2$, 5% dppb
Cs_2CO_3, rt, 12 h

90%

Batey, R.A.; Quach, T.D. *Tetrahedron Lett.*, ***2001***, *42*, 9099.

Ph_3BiCl_2, MeCN
5% $PdCl_2$, rt, 3 h

82%

Kang, S.-K.; Ryu, H.-C.; Kim, J.-W. *Synth. Commun.*, ***2001***, *31*, 1021.

Ph_3BiCl_2, MeCN
5% $PdCl_2$, rt, 3 h

79%

Kang, S.-K.; Ryu, H.-C.; Lee, S.-W. *Synth. Commun.*, ***2001***, *31*, 1027.

O S Me$_2$N

1. PhI, cat Pd(0) 76%
2. mcpba, DCM, 0°C
3. cat $Pd(acac)_2$, *i*-PrMgBr
THF

Ph

90% ee

de la Rosa, J.C.; Díaz, N.; Carretero, J.C. *Tetrahedron Lett.*, ***2000***, *41*, 4107.

MeO $\overset{+}{I}$—Ph $\overset{-}{OTf}$

Ph_2TeCl_2, 1% $PdCl_2$
3 eq NaOMe, MeOH/MeCN
rt, 7 h

MeO Ph

88%

Kang, S.-K.; Lee, S.-W.; Kim, M.-S.; Kwon, H.-S. *Synth. Commun.*, ***2001***, *31*, 1721.

$NaBPh_4$ Ph_2SiCl_2, MeCN, rt Ph—Ph 76%

Sakurai, H.; Morimoto, C.; Hirao, T. *Chem. Lett.*, ***2001***, 1084.

ZnI CO_2Me

NCS, TMU, cat $PdCl_2(PPh_3)_2$

CO_2Me CO_2Me

94%

Hossain, K.M.; Kameyama, T.; Shibata, T.; Takagi, K. *Bull. Chem. Soc. Jpn.*, ***2001***, *74*, 2415.

Cl $B(OH)_2$

$PhTeCl_2$, 10% $PdCl_2(PPh_3)_2$
2 eq NaOMe, DME-H_2O
50°C, 5 h

Cl Ph

68%

Kang, S.-K.; Hong, Y.-T.; Kim, D.-H.; Lee, S.-H. *J. Chem. Res. (S)*, ***2001***, 283.

2 CO_2Me ZnI

cat $PdCl_2(PPh_3)_2$, NCS
45°C, 10 h

MeO_2C CO_2Me

84%

Hossain, K.M.; Shibata, T.; Takagi, K. *Synlett*, ***2000***, 1137.

MeO–C_6H_4–N_2^+ BF_4^- ; 2% $pd_2(dba)_3$, MeCN/AcOH, rt

TrO, N, CO_2Me → TrO, N, CO_2Me, OMe

95% (9:1 *anti:syn*)

Servino, E.A.; Correia, C.R.D. *Org. Lett.*, ***2000***, *2*, 3039.

MeO–C_6H_4–$)_2I^+$ BF_4^- ; PhMgBr , cat $Pd(PPh_3)_4$; $ZnCl_2$, THF , rt → MeO–C_6H_4–Ph 86%

Wang, L.; Chen, Z.-C. *Synth. Commun.*, ***2000***, *30*, 3607.

Ph–diazirine ; hv , PhH , $AlCl_3$ → $PhCH_2Ph$ 96%

Moss, R.A.; Fedé, J.-M.; Yan, S. *J. Am. Chem. Soc.*, ***2000***, *122*, 9878.

2 $PhSnBu_3$; 10% $CuCl_2$, 0.5 I_2 , DMF , 100°C , 4h → Ph—Ph 93%

Kang, S.-K.; Baik, T.-G.; Jiao, X.H.; Lee, Y.-T. *Tetrahedron Lett.*, ***1999***, *40*, 2383.

CH_3–C_6H_4–N_2^+ BF_4^- ; Ph–BF_3K , dioxane ; 5% $Pd(OAc)_2$, 3.7 h → CH_3–C_6H_4–Ph 88%

Darses, S.; Michaud, G.; Genêt, J.-P. *Eur. J. Org. Chem.*, ***1999***, 1875.

2-thienyl–$SnBu_3$; PhI , 5% silica-supported Pd ; aq MeCN , 80°C → 2-thienyl–Ph 80%

Kang, S.-K.; Baik, T.-G.; Song, S.-Y. *Synlett*, ***1999***, 327.

2-thienyl–$SnBu_3$; Ph_2I^+ BF_4^- , NMP , 70°C ; 10% $Ni(acac)_2$, 8 h → 2-thienyl–Ph 79%

Kang, S.-K.; Ryu, H.-C.; Lee, S.-W. *J. Chem. Soc., Perkin Trans. 1*, ***1999***, 2661.

PhCH=CH–NO_2 ; $C_5H_{11}ZnI$, THF ; microwaves → PhCH=CH–C_5H_{11} 80%

Hu, Y.; Yu, J.; Yang, S.; Wang, J.-X.; Yin, Y. *Synth. Commun.*, ***1999***, *29*, 1157.

$$Ph_2I^+ BF_4^- \xrightarrow[60°C,\ 1\ h]{PhBF_3K,\ 5\%\ Pd(OAc)_2,\ DME} Ph{-}Ph \quad 99\%$$

Xia, M.; Chen, Z.-C. *Synth. Commun.*, ***1999***, *29*, 2457.

$$MeO{-}C_6H_4{-}I^+Ph\ BF_4^- \xrightarrow[2\ eq\ NaOMe,\ rt,\ 3\ h]{PhPb(OAc)_3,\ CHCl_3,\ 5\%\ Pd_2(dba)_3 \cdot CHCl_3} MeO{-}C_6H_4{-}Ph \quad 75\%$$

Kang, S.-K.; Choi, S.-C.; Baik, T.-G. *Synth. Commun.*, ***1999***, *29*, 2493.

REVIEWS:

"Recent Advances In The Transition Metal Catalyzed Regioselective Approaches To Polysubstituted Benzene Derivatives," Saito, S.; Yamamoto, Y. *Chem. Rev.*, ***2000***, *100*, 294.

"New Mitsunobu Reagents For Carbon–Carbon Bond Formation," Itô, S.; Tsudoda, T. *Pure Appl. Chem.*, ***1999***, *71*, 1053.

CHAPTER 6
PREPARATION OF AMIDES

SECTION 76: AMIDES FROM ALKYNES

Ph

O N Et

toluene
heat

O N Et + O N Et

(1 : 4.3) 56%

Rodríguez, D.; Navarro-Vázquez, A.; Castedo, L.; Domínguez, D.; Saá, C.
J. Am. Chem. Soc., **2001**, *123*, 9168.

$Fe(CO)_5$

1. $NaBH_4$, AcOH
2. PhC≡CH
3. $BuNH_2$
4. $CuCl_2$•2 H_2O

Ph O N O Bu

55%

Remeshkumar, C.; Periasamy, M. *Synlett*, **2000**, 1619.

SECTION 77: AMIDES FROM ACID DERIVATIVES

CO_2H

1. 2 eq LDA
2. PhCN

O NH Ph

70%

Chen, Y.; Li, T.; Sieburth, S.Mc.N. *J. Org. Chem.*, **2001**, *66*, 6826.

Ph~~~CO_2H , toluene; 5% 3,5-$(F_3C)_2C_6H_3B(OH)_2$, 1 h reflux — 59%

Ishihara, K.; Kondo, S.; Yamamoto, H. *Synlett*, ***2001***, 1371.

NMe$_2$ PF_6^- NMe$_2$; NH_3OMe Cl , MeCN , NEt_3 , 3 h — Ph(C=O)NHOMe 86%

Bailén, M.A.; Chinchilla, R.; Dodsworth, D.J.; Nájera, C. *Tetrahedron Lett.*, ***2001***, *42*, 5013.

NH_2; O_2N–C_6H_4–CO_2H; 2 eq polymer-supported carbodiimide; 3 eq polymer-supported DMAP — O_2N ... 96%

Lannuzel, M.; Lamothe, M.; Perez, M. *Tetrahedron Lett.*, ***2001***, *42*, 6703.

Ph~~CO_2H; piperidine; $Me_3N{\bullet}HCl$, NEt_3 , 0°C MeCN , 3 h — 92%

Wakasugi, K.; Nakamura, A.; Tanabe, Y. *Tetrahedron Lett.*, ***2001***, *42*, 7427.

OBn CO_2H; NCO; DCM , DMAP , 3 h 0°C → rt — OBn 75%

Schuemacher, A.C.; Hoffmann, R.W. *Synthesis*, ***2001***, 243.

PhCOOH → (1. polymer-bound PPh_3, DCM, 5 eq CCl_3CN, rt, 3 h; 2. $PhNH_2$, polymer-bound morpholine) → PhC(O)NHPh 75%

Buchstaller, H.P.; Ebert, H.M.; Anlauf, U. *Synth. Commun.*, ***2001***, *31*, 1001.

Phthalic anhydride → (H_2NCHO, microwaves, 2 min) → phthalimide (NH) 97%

Peng, Y.; Song, G.; Qian, X. *Synth. Commun.*, ***2001***, *31*, 1927.

$PhCO_2H$ → (2,2'-dithiobis(pyridine N-oxide); Bu_3P, NEt_3, DCM, 16 h; (MeO)MeNH•HCl) → PhC(O)N(OMe)Me 80%

Banwell, M.; Smith, J. *Synth. Commun.*, ***2001***, *31*, 2011.

$PhOCH_2CO_2H$ → (1. Cl_3CCN, PPh_3 0°C; 2. PMPCH=NPh, 0°C → rt) → β-lactam (PhO, PMP; N-Ph) 61%

Govande, V.V.; Arun, M.; Deshmukh, A.R.A.S.; Bhawal, B.M. *Synth. Commun.*, ***2000***, *30*, 4177.

MeO–C₆H₄–C(O)Cl → (DMF, reflux, 12 h) → MeO–C₆H₄–C(O)NMe_2 92%

Lee, W.S.; Park, K.H.; Yoon, Y.-J. *Synth. Commun.*, ***2000***, *30*, 4241.

$PhCH_2CO_2H$ → (PhCH=N-PMP, CH_2Cl_2, 12 h; NEt_3, triphosgene, -40°C → rt) → β-lactam (Ph, Ph; N-PMP) 95%

Krishnaswamy, D.; Bhawal, B.M.; Desmukh, A.R.A.S. *Tetrahedron Lett.*, ***2000***, *41*, 417.

CO_2H — 1. 2 eq BuLi , 2 eq $NHEt_2$; 2. EtCN — 65%

Brun, E.M.; Gil, S.; Mestres, R.; Parva, M. *Synthesis*, ***2000***, 273.

$PhCO_2H$ — NMe_2 PF_6^-, N+, S, NMe_2, O; NH_4Cl , DIEA , DMF rt , 30 min — Ph NH_2 O 83%

Bailen, M.A.; Chinchilla, R.; Dodsworth, D.J.; Nájera, C. *Tetrahedron Lett.*, ***2000***, *41*, 9809.

Ph CO_2H — di-(2-pyridyl)carbonate , DMAP , ether; $BnNH_2$, rt — Ph NHBn O 91%

Shiina, I.; Suenaga, Y.; Nakano, M.; Mukaiyama, T. *Bull. Chem. Soc. Jpn.*, ***2000***, *73*, 2811.

O OH O — HN NH; HMDS , 110°C (neat) — O N NH O 93%

Chou, W.-C.; Chou, M.-C.; Lu, Y.-Y.; Chen, S.-F. *Tetrahedron Lett.*, ***1999***, *40*, 3419.

Br CO_2H — 1. 2 eq PPh_3/2 eq CCl_3CN rt , 1 h; 2. NH_2•HCl NEt_3 , rt — O NH Br 90%

Jang, D.O.; Park, D.J.; Kim, J. *Tetrahedron Lett.*, ***1999***, *40*, 5323.

PhO–C(=S)–Cl → (NEt_3, DCE, 20°C, 1 h) → PhO–C(=S)–NEt_2 93%

Milan, D.S.; Prager, R.H. *Aust. J. Chem.*, ***1999***, *52*, 841.

SECTION 78: AMIDES FROM ALCOHOLS AND THIOLS

$PhCH_2OH$ → ($Mg(HSO_4)_2$, PhCN; 90°C, 1.5 h) → $PhCH_2NHBz$ 91%

Salehi, P.; Khodaie, M.M.; Zolfigol, M.A.; Keyvan, A. *Synth. Commun.*, ***2001***, *31*, 1947.

Ph_2CHOH → ($Fe(ClO_4)_3$, SiO_2; MeCN, reflux, 30 min) → $Ph_2CHNHAc$ 94%

Salehi, P. Motlagh, A.R. *Synth. Commun.*, ***2000***, *30*, 665.

$PhCH_2OH$ → (Fe^{+3}–Montmorillonite K10; MeN, 70°C, 20 h) → $PhCH_2NHAc$ 90%

Lakouraj, M.M.; Movassagh, B.; Fasihi, J. *Synth. Commun.*, ***2000***, *30*, 821.

SECTION 79: AMIDES FROM ALDEHYDES

$CH_3C(=O)CH_2CO_2Et$ → (PhCHO, urea, MeCN; $BiCl_3$, heat, 5 h) → 5-EtO_2C-6-methyl-4-phenyl-3,4-dihydropyrimidin-2(1H)-one 95%

Ramalinga, K.; Vijayalakshmi, P.; Kaimal, T.N.B. *Synlett*, ***2001***, 863.

PhCHO → (1. $BocNH_2$, TsNa, aq MeOH, HCO_2H, rt, 1 d; 2. $NaBH_4$, THF, rt) → $PhCH_2NHBoc$ 65x73%

Bernacka, E.; Klepacz, A.; Zeierzak, A. *Tetrahedron Lett.*, ***2001***, *42*, 5093.

PhCHO → (Al_2O_3, $MeSO_3H$, $NH_4OH•HCl$; 140°C, 3 h) → $PhC(=O)NH_2$ 95%

Sharghi, H.; Sarvari, M.H. *J. Chem.Res. (S)*, ***2001***, 446.

PhCHO → [ethyl acetoacetate, MeCN, heat; urea (H_2NCONH_2), $Mn(OAc)_3•2H_2O$, 2 h] → 5-ethoxycarbonyl-6-methyl-4-phenyl-3,4-dihydropyrimidin-2(1H)-one (EtO_2C, Ph, NH, N-H, O) 96%

Kumar, K.A.; Kasthuraiah, M.; Reddy, C.S.; Reddy, C.D. *Tetrahedron Lett.*, ***2001***, *42*, 7873.

PhCHO → [piperidine, NMO, K_2CO_3; cat $[Rh(cod)_2]BF_4$, toluene, 140°C, 20 h] → Ph(C=O)N-piperidine 96%

Tillack, A.; Rudloff, I.; Beller, M. *Eur. J. Org. Chem.*, ***2001***, 523.

PhCHO → [ethyl acetoacetate (CO_2Et), CAN; urea, MeOH, ultrasound] → 5-ethoxycarbonyl-6-methyl-4-phenyl-3,4-dihydropyrimidin-2(1H)-one (EtO_2C, Ph, NH, N-H, O) 92%

Yadav, J.S.; Reddy, B.V.S.; Reddy, K.B.; Raj, K.S.; Prasad, A.R. *J. Chem. Soc., Perkin Trans. 1*, ***2001***, 1939.

3,4-methylenedioxybenzaldehyde (CHO) → [$Ph_3P=C=C=O$, THF, 2 d; $Me_2CHCH_2NH_2$] → N-isobutyl-3-(3,4-methylenedioxyphenyl)acrylamide (NH, O) 50%

Schobert, R.; Siegfried, S.; Gordon, G.J. *J. Chem. Soc., Perkin Trans. 1*, ***2001***, 2393.

BuCHO → [1. *p*-$TolSO_2Na$, HCOOH, H_2O, $PhSO_2NH_2$; 2. $NaHCO_3$, H_2O/CH_2Cl_2] → $BuCH=NSO_2Ph$ 66%

Chemla, F.; Hebbe, V.; Normant, J.-F. *Synthesis*, ***2000***, 75.

PhCHO → ($CH_2=CHCH_2SiMe_3$, $TrClO_4$, 0°C / $Me_3SiNHCO_2Et$) → PhCH($NHCO_2Et$)$CH_2CH=CH_2$ 96%

Niimi, L.; Serita, K.-i.; Hiraoka, S.; Yokozawa, T. *Tetrahedron Lett.*, ***2000***, *41*, 7075.

SECTION 80: AMIDES FROM ALKYLS, METHYLENES AND ARYLS

NO ADDITIONAL EXAMPLES

SECTION 81: AMIDES FROM AMIDES

Conjugate reductions of unsaturated amides are listed in Section 74D (Alkyls from Alkenes).

1. Li, NH_3, THF, *t*-BuOH, −78°C; 2. MeI — 70%

Guo, Z.; Schultz, A.G. *J. Org. Chem.*, ***2001***, *66*, 2154.

5% $Pd(OAc)_2$, 5% PCy_3, 1.5 eq NaO*t*-Bu, dioxane, 70°C, 5 h — 68%

Lee, S.; Hartwig, J.F. *J. Org. Chem.*, ***2001***, *66*, 3402.

SO_2NHEt — 1. BuLi, THF, 0°C, 15 min; 2. MeI, 0°C — SO_2NHEt 92%

MacNeil, S.L.; Familoni, OB.; Snieckus, V. *J. Org. Chem.*, ***2001***, *66*, 3662.

SO_2NH_2 — 1. *t*-BuOCl, NaOH, H_2O, rt, 1 h; 2. PTAB, MeCN, rt, 1 d — N, SO_2 70%

Dauban, P.; Dodd, R.H. *Tetrahedron Lett.*, ***2001***, *42*, 1037.

Br—C6H4—CHO

cat $Pd_2(dba)_3$, Xantphos, 16 h
toluene, 120°C

OHC

87%

Cacchi, S.; Fabrizi, G.; Goggiamani, A.; Zappia, G. *Org. Lett.*, ***2001***, *3*, 2539.

1. BuLi, (-)-sparteine
toluene, –78°C

2. O_2N–CH=CH–Ph

MeO — N — Boc

O_2N — Ph — Ph — N(Ac)Boc

90%r(94:6 dr)

Johnson, T.A.; Curtis, M.D.; Beak, P. *J. Am. Chem. Soc.*, ***2001***, *123*, 1230.

NHBoc

$SiMe_2$*i*-Pr

1. LHMDS
2. Me_3SiCl, NEt_3
3. 1N HCl

i-$PrMe_2Si$ — MeO_2C — NHBoc + *i*-$PrMe_2Si$ — MeO_2C — NHBoc

(22 : 1) 60%

Mohamed, M.; Brook, M.A. *Tetrahedron Lett.*, ***2001***, *42*, 191.

H_2N—$S(O_2)$—O

2% $Rh_2(OAc)_4$, $PhI(OAc)_2$

2.3 MgO, CH_2Cl_2, 40°C

HN—$S(O_2)$—O

90%

Espino, C.G.; Wehn, P.M.; Chow, J.; DuBois, J. *J. Am. Chem. Soc.*, ***2001***, *123*, 6935.

H—C(O)—N(Me)—CH_2—OMe

1. LDA
2. Me_3SiCl
3. CsF, 145°C, Et_3SiCl

Et_3Si—C(O)—N(Me)—CH_2—OMe

61x84%

Cunico, R.F. *Tetrahedron Lett.*, ***2001***, *42*, 1423.

1. 2 eq $TiCl_4$, CH_2Cl_2, 20°C
2. H_2/Pd-C, EtOH
Bu$_3$SnH, AIBN
3. Na/NH_3, EtOH, –78°C

85x98%

Ito, T.; Yamazaki, N.; Kibayashi, C. *Synlett*, ***2001***, 1506.

$PhI(O_2CCF_3)_2$, 1.5 h
aq MeCN

96%

Yu, C.; Jiang, Y.; Liu, B.; Hu, L. *Tetrahedron Lett.*, ***2001***, *42*, 1449.

Et_2Zn, 2 eq $ClCH_2I$, toluene, rt
2 eq tetrahydrothiophene, 16 h

68%

Aggarwal, V.K.; Stenson, R.A.; Jones, R.V.H.; Fieldhouse, R.; Blacker, J. *Tetrahedron Lett.*, ***2001***, *42*, 1587.

$PhCH_2Cl$, Cs_2CO_3, TBAI
DMF, 23°C, 5 h

98%

Salvatore, R.N.; Shin, S.I.; Flanders, V.L.; Jung, K.W. *TetrahedronLett.*, ***2001***, *42*, 1799.

toluene, reflux
5 h

86%

Alcaide, B.; Rodríguez-Ranera, C.; Rodríguez-Vicente, A. *Tetrahedron Lett.*, ***2001***, *42*, 3081.

e⁻ , EtMgCl , ether

−72°C , 20 min

N, CO_2Me → N, Et, CO_2Me 65%

Suga, S.; Okajima, M.; Yoshida, J.-i. *Tetrahedron Lett.*, ***2001***, *42*, 2173.

$In(OTf)_3$, PhH , DCM

rt , 4.5 h

Ts, N, Ph → Ph, Ph, NHTs 75%

Yadav, J.S.; Reddy, B.V.S.; Rao, R.S.; Veerendhar, G.; Nagaiah, K. *Tetrahedron Lett.*, ***2001***, *42*, 8067.

N–Ts

$Sii\text{-}Pr_3$

$BF_3{\cdot}OEt_2$, toluene
rt , 30 min

Ts, N, $Sii\text{-}Pr_3$

72% (56:46)

Akiyama, T.; Sugano, M.; Kagoshima, H. *Tetrahedron Lett.*, ***2001***, *42*, 3889.

30% BuBr , CH_2Cl_2 , rt
$N(CH_2\text{-}2\text{-pyridyl})_3$, 1 h

Br, O, N, Bn → Br, O, N, Bn 96%

Clark, A.J.; Battle, G.M.; Bridge, A. *Tetrahedron Lett.*, ***2001***, *42*, 4409.

$PhI(OAc)_2$, I_2 , 3 h

DCE , ultrasound , 40°C

SO_2NHEt → SO_2NH_2 69%

Katohgi, M.; Togo, H. *Tetrahedron*, ***2001***, *57*, 7481.

1. 2 eq NaH , DMF
0°C → rt

2. 1,3-diiodopropane

O, N, H, Ts → O, N, H, Ts 60%

Costa, A.; Nájera, C.; Sansano, J.M. *Tetrahedron Asymm.*, ***2001***, *12*, 2205.

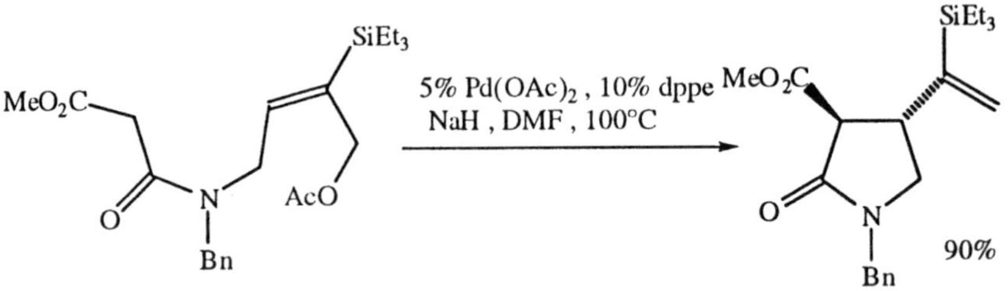
SiEt3
MeO2C
O
N
Bn
AcO
5% Pd(OAc)2 , 10% dppe
NaH , DMF , 100°C
SiEt3
MeO2C
O
N
Bn
90%

Poli, G.; Giambastiani, G.; Malacria, M.; Thorimbert, S. *Tetrahedron Lett.*, ***2001***, *42*, 6287.

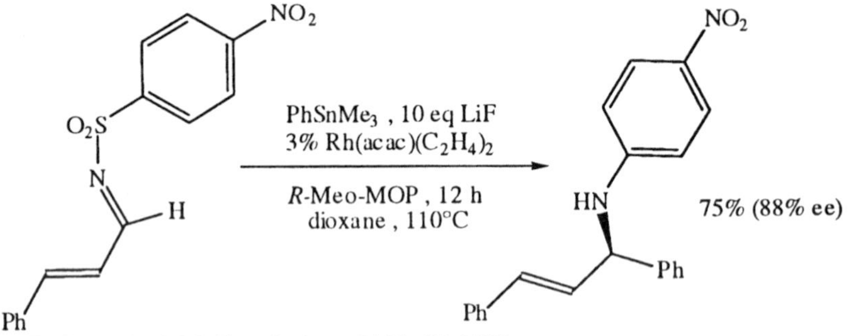
NO2
O2S
N
H
Ph
PhSnMe3 , 10 eq LiF
3% Rh(acac)(C2H4)2
R-Meo-MOP , 12 h
dioxane , 110°C
NO2
HN
Ph
Ph
75% (88% ee)

Hayashi, T.; Ishigedani, M. *Tetrahedron*, ***2001***, *57*, 2589.

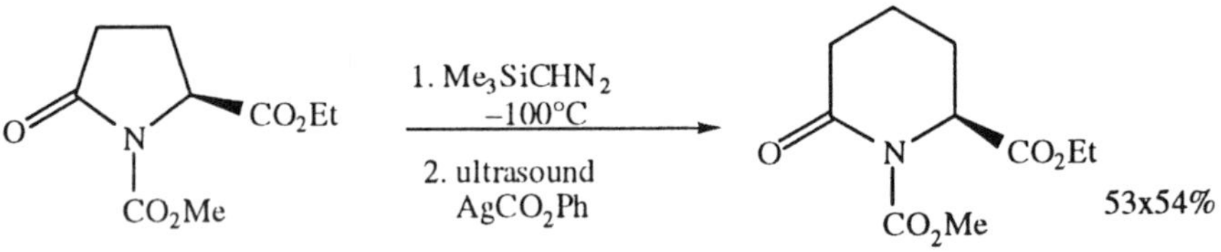
O
N
CO2Et
CO2Me
1. Me3SiCHN2
−100°C
2. ultrasound
AgCO2Ph
O
N
CO2Et
CO2Me
53x54%

Coutts, I.G.C.; Saint, R.E.; Saint, S.L.; Chambers-Asman, D.M *Synthesis*, ***2001***, 247.

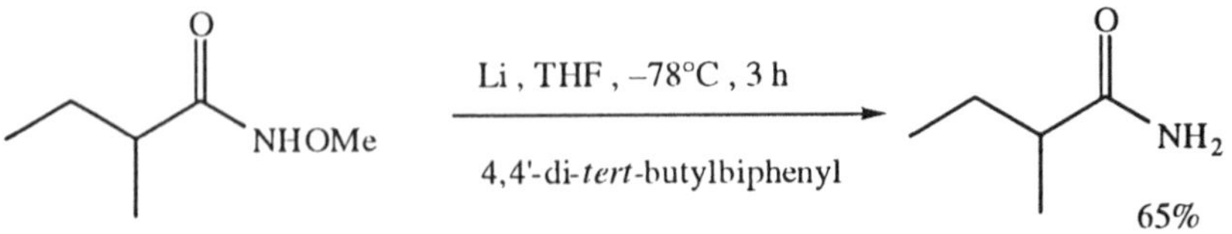
O
NHOMe
Li , THF , −78°C , 3 h
4,4'-di-tert-butylbiphenyl
O
NH2
65%

Yus, M.; Radivoy, G.; Alonso, F. *Synthesis*, ***2001***, 914.

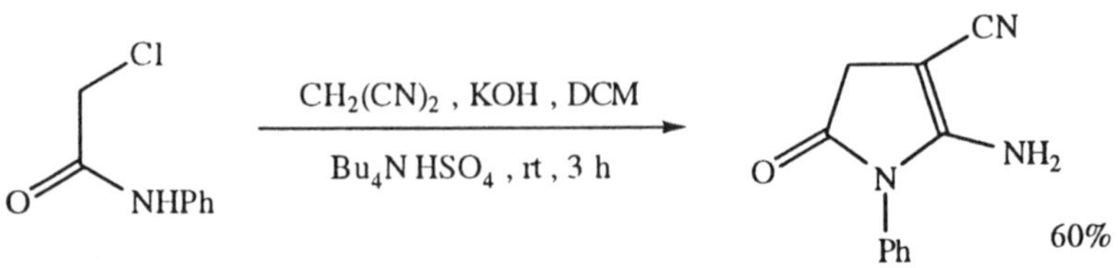
Cl
O
NHPh
CH2(CN)2 , KOH , DCM
Bu4N HSO4 , rt , 3 h
CN
O
N
NH2
Ph
60%

Dave, C.G.; Parikh, V.A. *Synth. Commun.*, ***2001***, *31*, 1301.

3 eq [polymer-supported thiophosphonamide reagent]; 200°C, toluene, 15 min microwaves — PhC(O)NMe$_2$ → PhC(S)NMe$_2$ 95%

Ley, S.V.; Leach, A.G.; Storer, R.I. *J. Chem. Soc., Perkin Trans, 1*, *2001*, 358.

4 eq mcpba, 8 eq NaH; *i*-PrOH, rt — SePh, NHBz → N–Bz 85%

Ward, V.R.; Cooper, M.A.; Ward, A.D. *J. Chem. Soc., Perkin Trans. 1*, *2001*, 944.

Bu_3SnH, AIBN; toluene, reflux — I, O, N–Bn, O → H, O, N–Bn, H, O 54% + H, O, N–Bn, OH, O 27%

El Bialy, S.A.A.; Ohtan, S.; Sato, T.; Ikeda, M. *Heterocycles*, *2001*, *54*, 1021.

5 eq LDA, ether; –78°C — Ts–N → TsHN 60% + O 15%

Arjona, O.; Menchaca, R.; Plumet, J. *Heterocycles*, *2001*, *55*, 5.

7% $[PdCl(C_3H_5)]_2$, DCM; $AgSbF_6$, PPh_3, rt, 18 h — Ns, N → N, Ns 60%

Bothe, U.; Rudbeck, H.C.; Tanner, D.; Johannsen, M.
J. Chem. Soc., Perkin Trans. 1, *2001*, 3305.

MeMgBr, THF-toluene; chiral Cu catalyst, 0°C — N–Ts → NHTs, Me 52% (91% ee)

Müller, P.; Nury, P. *Helv. Chim. Acta*, *2001*, *84*, 662.

PhCH=N–Ts → ($PhCH=N-NTs^- Na^+$, 1% $Rh_2(OAc)_4$; 20% chiral sulfide, dioxane; 10% $BnNEt_3Cl$, 40°C) → N-Ts 2,3-diphenylaziridine

68% (2.5:1 *trans:cis*)

Aggarwal, V.K.; Alonso, E.; Fang, G.; Ferrara, M.; Hynd, G.; Procelloni, M. *Angew. Chem. Int. Ed.*, ***2001***, *40*, 1433.

2 eq LiHMDS, DCM; –78°C → rt

85%

Lindström, U.M.; Somfai, P. *Chem. Eur. J.*, ***2001***, *7*, 94.

cyclopentene, BF_3•OEt_2; DCM, –78°C

(1 : 1) 45%

Ungureanu, I.; Klotz, P.; Mann, A. *Angew. Chem. Int. Ed.*, ***2000***, *39*, 4615.

1. BuLi, THF, –78°C; 2. –78°C → rt, allyl bromide

98% (9.8:1 de)

He, S.; Kozmin, S.A.; Rawal, V.H. *J. Am. Chem. Soc.*, ***2000***, *122*, 190.

$TiCl_4$, DCM; rt

75%

Frank, K.E.; Aubé, J. *J. Org. Chem.*, ***2000***, *65*, 655.

NO_2 O_2S N F_3C

PhSnMe$_3$, LiF , 110°C

3% Rh(acac)(C$_3$H$_4$)$_2$
6% R-MOP , dioxane , 10 h

NO_2 O_2S HN Ph F_3C

82% (92% ee, *S*)

Hayashi, T.; Ishigedani, M. *J. Am. Chem. Soc.*, ***2000***, *122*, 976.

N—Ts

5% TBAF , THF , 4 h

Me$_3$SiN$_3$, rt

NHTs N_3

>99%

Wu, J.; Hou, X.-L.; Dai, L.-X. *J. Org. Chem.*, ***2000***, *65*, 1344.

NHTs SiMe$_2$Bn

5% TiCl$_4$, CHCl$_3$

rt , 3.3 h

SiMe$_2$Bn N—Ts

93%

Miura, K.; Hondo, T.; Nakagawa, T.; Takahashi, T.; Hosomi, A. *Org. Lett.*, ***2000***, *2*, 385.

MeO$_2$C Ph Ph O N H Ph Me

Mn(OAc)$_3$

Ph OAc MeO$_2$C Ph O N Ph Me H

78% (54:46 *anti:syn*)

D'Annibale, A.; Nanni, D.; Trogolo, C.; Umani, F. *Org. Lett.*, ***2000***, *2*, 401.

PhSO$_2$NHEt

3 eq PhI(OAc)$_2$, I$_2$, DCE , 3 h

ultrasound , 40°C

PhSO$_2$NH$_2$ 71%

Katohgi, M.; Yokoyama, M.; Togo, H. *Synlett*, ***2000***, 1055.

t-BuSOCl , DCM
i-Pr_2NEt , $(PhSe)_2$
−50°C → rt

SePh

59%

Lin, X.; Stein, D.; Weinreb, S.M. *Tetrahedron Lett.*, **2000**, *41*, 2333.

Bt
N
Bn

Me_3Si
BF_3•OEt_2 , DCM

Bt = benzotriazol-1-yl

90%

Katritzky, A.R.; Mehta, S.; He, H.-Y.; Cui, X. *J. Org. Chem.*, **2000**, *65*, 4364.

N–SO_2Ph
Ph H

$PhMe_2SiH$, 10% $B(C_6F_5)_3$
toluene , rt , 30 min

$NHSO_2Ph$
Ph H

93%

Blackwell, J.M.; Sonmor, E.R.; Scoccitti, T.; Piers, W.E. *Org. Lett.*, **2000**, *2*, 3921.

Ph N–SO_2Ph

Br , In
THF

$NHSO_2Ph$
Ph

99%

Lu, W.; Chan, T.H. *J. Org. Chem.*, **2000**, *65*, 8589.

Ph N–Ts

2.5 eq $Me_3SiCH=N_2$
dioxane , 40°C

Ph N–Ts
Me_3Si

72% (95:5 *cis:trans*)

Aggarwal, V.K.; Ferrara, M. *Org. Lett.*, **2000**, *2*, 4107.

$NHSO_2Ph$
N
MeO

1. MeO–C6H4–$B(OH)_2$, 90°C
cat $[Rh(cod)MeCN)_2]BF_4$, dioxane
2. H_2O

$NHSO_2Ph$
MeO

79%

Ueda, M.; Saito, A.; Miyayra, N. *Synlett*, **2000**, 1637.

$BF_3 \bullet OEt_2$, toluene

microwaves 54-56°C

(>99 : 1) >99%

Cardillo, G.; Gentilucci, L.; Gianotti, M.; Tolomelli, A. *Synlett*, **2000**, 1309.

$Mn(OAc)_3$

95%

Wu, Y.-L.; Chuang, C.-P.; Lin, P.-Y. *Tetrahedron*, **2000**, *56*, 6209.

1. LHMDS, –78°C

2. MeI

(93 : 7) 65%

Bull, S.D.; Davies, S.G.; Nicholson, R.L.; Sanganee, H.J.; Smith, A.D. *Tetrahedron Asymm.*, **2000**, *11*, 3475.

3 eq AcOH

$Na(OAc)_3BH$, MS 4Å

DCE, 15 h, 0°C → rt

64%

Beshore, D.C.; Dinsmore, C.J. *Tetrahedron Lett.*, **2000**, *41*, 8735.

Ph–CH=N–Ts → (Et_2Zn, $Cu(OTf)_2$, toluene; chiral phosphine ligand, O_2) → PhCH(Et)NHTs 98% (93% ee)

Fujihara, H.; Nagai, K.; Tomioka, K. *J. Am. Chem. Soc.*, ***2000***, *122*, 12055.

Pyridine-2-carboxamide (NH_2) → ($PhSi(OMe)_3$, DMF; $Cu(OAc)_2$, rt) → N-phenylpyridine-2-carboxamide (NHPh) 61%

Lam, P.Y.S.; Deudon, S.; Hauptman, E.; Clark, C.G. *Tetrahedron Lett.*, ***2000***, *42*, 2427.

$TolO_2S$–N=CH–Ph → (Me_3SiCHN_2; 3h) → 2-Me_3Si-3-Ph-1-(SO_2Tol)aziridine 75%

Hori, R.; Aoyama, T.; Shioiri, T. *Tetrahedron Lett.*, ***2000***, *41*, 9455.

PhC(=S)NHBn → (Caro's acid/SiO_2, MeCN) → PhC(=O)NHBn 84%

Movassagh, B.; Lakouraj, M.M.; Ghodrati, K. *Synth. Commun.*, ***2000***, *30*, 2353.

PhCH=NTs → (vinyl oxirane, 3% $Pd(PPh_3)_4$, THF; rt, 1 h) → 2-Ph-3-Ts-4-vinyl-oxazolidine 97%

Shim, J.-G.; Yamamoto, Y. *Heterocycles*, ***2000***, *52*, 885.

N-Bn-N-(cyclohex-3-enyl)-CCl_3-acetamide → ($(TMS)_3SiH$, AIBN, PhH; reflux, 3 h) → N-Bn bicyclic lactam 66%

Quirante, J.; Escolano, C.; Diaba, F.; Bonjoch, J. *Heterocycles*, ***1999***, *50*, 731.

PhBr , $Pd(OAc)_2$, dppf , 2 d

NaO*t*-Bu , toluene , 120°C

N—H → N—Ph 82%

Shakespeaare, W.C. *Tetrahedron Lett., 1999, 40*, 2035.

PhCHO , TFA , Et_3SiH

toluene , 120°C , 36 h

Ph NH_2 → Ph N H Ph 92%

Dubé, D.; Scholte, A.A. *Tetrahedron Lett., 1999, 40*, 2295.

CO_2Et Ph Cl O N H OTBDPS

Bu_3SnH , AIBN , toluene , heat

Ph CO_2Et O N H OTBDPS 79%

dr (8.9:3.5:1)

Bryans, J.S.; Large, J.M.; Parsons, A.F. *Tetrahedron Lett., 1999, 40*, 3487.

t-Bu N O Ph

1. 2 eq BuLi, THF , -78°C → rt
2. BnBr

3. 10% oxalic acid , reflux

O Bn Ph Bu

72% (73% ee)

Hanson, P.R.; Probst, D.A.; Robinson, R.E.; Yau, M. *Tetrahedron Lett., 1999, 40*, 4761.

O O N—H Ph

KF/Al_2O_3 , BuBr

O O N—Bu Ph

83%

Blass, B.; Drowns, M.; Harris, C.L.; Liu, S.; Portlock, D.E. *Tetrahedron Lett., 1999, 40*, 6545.

1. PhCOMe, $Ti(OEt)_4$, THF, 75°C, 10 h
2. $NaBH_4$, -48°C

78% (96:4 dr)

Borg, G.; Cogan, D.A.; Ellman, J.A. *Tetrahedron Lett.*, ***1999***, *40*, 6709.

Me_2NH, toluene, autoclave
160°C

75%

Yang, Y.; Lu, S. *Org. Prep. Proceed. Int.*, ***1999***, *31*, 559.

Bu_3SnH, AIBN, PhH
reflux, 65°C

65% (1:1)

Ikeda, M.; Shikaura, J.; Maekawa, N.; Daibuzono, K.; Teranishi, H.; Teraoka, Y.; Oda, N.; Ishibashi, H. *Heterocycles*, ***1999***, *50*, 31.

$Rh_2(OAc)_4$, MeCN
reflux, 10 h

58%

Lee, Y.R.; Suk, J.Y.; Kim, B.S. *Tetrahedron Lett.*, ***1999***, *40*, 8219.

1. *sec*-BuLi
2. EtI

55%

McCombie, S.W.; Lin, S.-I.; Vice, S.F. *Tetrahedron Lett.*, ***1999***, *40*, 8767.

AIBN , Bu_3SnH
toluene , 60°C

(6 : 1)

Jones, K.; Brunton, S.A.; Gosain, R. *Tetrahedron Lett.*, ***1999***, *40*, 8935.

$PhSnMe_3$, THF , 60°C , 19 h
2% $[Rh(cod)(MeCN)_2]BF_4$

98%

Oi, S.; Moro, M.; Fukuhara, H.; Kawanishi, T.; Inoue, Y. *Tetrahedron Lett.*, ***1999***, *40*, 9259.

polymethylhydrosiloxane
Pd–C , EtOH

96%

Chandrasekhar, S.; Ahmed, M. *Tetrahedron Lett.*, ***1999***, *40*, 9325.

MeMgBr , 30% CuI , 1.5 h
ether

52% (91% ee)

(with 10% cat — 91% & 25% ee)

Muller, P.; Nury, P. *Org. Lett.*, ***1999***, *1*, 439.

1. PhLi , $AlMe_3$
2. HCl/MeOH
3. BzCl , NEt_3

65x80%

Cogan, D.A.; Ellman, J.A. *J. Am. Chem. Soc.*, ***1999***, *121*, 268.

3.3% $Pd(OAc)_2$, 5% MOP
K_2CO_3 , toluene , 36 h
100°C

82%

MOP = 2-methoxy-2'-diphenylphosphino-bis-1,1'-biphenyl

Yang, B.H.; Buchwald, S.L. *Org. Lett.*, ***1999***, *1*, 35.

Me O Ph N •H_2O OH Me NH_2

1. 3.2 eq LHMDS, 3.2 eq LiCl, THF, 0°C
2. MeI, 0°C, 1.5 h

Me O Ph N NH_2 OH Me Me

91% (93% de)

Myers, A.G.; Schnider, P.; Kwon, S.; Kung, D.W. *J. Org. Chem.*, ***1999***, *64*, 3322.

Ph Me_3Sn N Boc Cl

1. BuLi, (-)-sparteine, ether, -78°C, 2 h
2. -25°C, 1 h

N Boc Ph

41% (89:11 er)

Serino, C.; Stehle, N.; Park, Y.S.; Florio, S.; Beak, P. *J. Org. Chem.*, ***1999***, *64*, 1160.

REVIEWS:

"Synthetic Catalysis of Amide Isomerization," Cox, C.; Lectka, T. *Accts. Chem. Res.*, ***2000***, *33*, 849.

SECTION 82: AMIDES FROM AMINES

Ph NH_2 → BnCl, CO_2, Cs_2CO_3 / TBAI, DMF, 23°C, 5 h → Ph H N OBn O

96%

Salvatore, R.N.; Shin, S.I.; Nagle, A.S.; Jung, K.W. *J. Org. Chem.*, ***2001***, *66*, 1035.

Tr N OTBS → TFA, Et_3SiH, DCM / 1 h → H N OTBS

85%

Vedejs, E.; Klapars, A.; Warner, D.L.; Weiss, A.H. *J. Org. Chem.*, ***2001***, *66*, 7542.

$Ph_2C=N—OH$ → t-$BuSO_2Cl$, ether, NEt_3 / −35°C → rt → $Ph_2C=N—SO_2t$-Bu

Artmann III, G.D.; Bartolozzi, A.; Franck, R.W.; Weinreb, S.M. *Synlett*, ***2001***, 232.

$PhNH_2$ → ($PhCH(CO_2Et)_2$, microwaves; 15 min) → 4-hydroxy-3-phenyl-2-quinolinone (OH, Ph, N H, O) 83%

Lange, J.H.M.; Verveer, P.C.; Osnabrug, S.J.M.; Visser, G.M. *Tetrahedron Lett.*, ***2001***, *42*, 1367.

I, F_3C, NH_2 → (1. NaHMDS, THF, 0°C → rt, 2 h; 2. $PhCO_2Me$) → I, F_3C, H N, Ph, O 94%

Wang, J.; Rosingana, M.; Discordia, R.P.; Soundararajan, N.; Polniaszek, R. *Synlett.*, ***2001***, 1485.

cyclohexyl-NH_2 → (1. O_2N-C_6H_4-O_2CNHBn, NEt_3, CH_2Cl_2, 20 min; 2. H_2/Pd, AcOH) → cyclohexyl-N(H)-C(=O)-NH_2 92%

Liu, Q.; Luedtke, N.W.; Tor, Y. *Tetrahedron Lett.*, ***2001***, *42*, 1445.

Ph$CH_2CH_2NH_2$ → (1. Cs_2CO_3, CO_2, MeI, 23°C, TBAI, DMF; 2. BnBr, Cs_2CO_3) → PhCH_2CH_2N(Me)C(=O)OBn 87%

Salvatore, R.N.; Ledger, J.A.; Jung, K.W. *Tetrahedron Lett.*, ***2001***, *42*, 6023.

N-Bn aziridine (OTBDMS) → (DME, acetone, CO; cat $Co_2(CO)_8$, 100°C, 4 h) → β-lactam (OTBDMS, O, N, Bn) quant

Davoli, P.; Forni, A.; Moretti, I.; Prati, F.; Torre, G. *Tetrahedron*, ***2001***, *57*, 1801.

$PhCH_2NH_2$ → (BnBr, CO_2, DBU, H_2O) → $BnNHCO_2Bn$ (6%) + $PhCH_2NBn_2$ (60%)

Shi, M.; Shen, Y.-M. *Helv. Chim. Acta*, ***2001***, *84*, 3357.

$C_3H_7CO_2Me$, hexane
amino-cyclase I , rt

HN C3H7

53%

Youshko, M.I.; van Rantwijk, F.; Sheldon, R.A. *Tetrahedron Asymm.*, ***2001***, *12*, 3267.

NO_2 NO_2

98% HCO_2H , 100°C , 3 h

72%

NH_2 NHCHO

Chakrabarty, M.; Khasnobis, S.; Harigaya, Y.; Konda,Y. *Synth. Commun.*, ***2000***, *30*, 187.

NH_2 EtO_2C

NH

microwaves , 10 min

N N

84%

Suri, O.P.; Satti, N.K.; Suri, K.A. *Synth. Commun.*, ***2000***, *30*, 3709.

1. N O NEt_4^+ (via electrolysis)

$C_6H_{11}NH_2$ $C_6H_{11}NHCO_2Et$

2. CO_2
3. EtI

80%

Feroci, M.; Inesi, A.; Rossi, L. *Tetrahedron Lett.*, ***2000***, *41*, 963.

Ac_2O , 120°C , BF_3•OEt_2

N—i-Pr N—Ac 76%

14 h

Dave, P.R.; Kumar, K.A.; Duddu, R.; Axenrod, T.; Dai, R.; Das, K.K.; Guan, X.-P.; Sun, J.; Trivedi, N.J.; Gilardi, R.D. *J. Org. Chem.*, ***2000***, *65*, 1207.

Br Br Ph

Ph

heat , toluene

N

overnight NH 40%

TMSO O

Bandini, E.; Favi, G.; Martelli, G.; Panunzio, M.; Piersanti, G. *Org. Lett.*, ***2000***, *2*, 1077.

1. PhOC(O)Cl; 2. allyl bromide (crotyl bromide), In, DMF, 16h — 75% (30:70)

Loh, T.-P.; Lye, P.-L.; Wang, R.-B.; Sim, K.-Y. *Tetrahedron Lett.*, **2000**, *41*, 7779.

$PhNH_2$ → $PhNHCHO$ 96%

1.5g HCO_2NH_2; MeCN, reflux, 11h

Reddy, P.G.; Kumar, G.D.K.; Baskaran, S. *Tetrahedron Lett.*, **2000**, *41*, 9149.

1. BuLi, THF, –45°C
2. 1 atm CO, –45°C → –30°C
3. satd aq NH_4Cl
–30°C → 20°C

MeO — MeO — 84%

Iwamoto, K.; Chatani, N.; Murai, S. *J. Org. Chem.*, **2000**, *65*, 7944.

Cy — N—Cy — O

2% $Ru_3(CO)_{12}$, 2 atm CO, 20 h; toluene, 160°C — 71%

Kamitani, A.; Chatani, N.; Morimoto, T.; Murai, S. *J. Org. Chem.*, **2000**, *65*, 930.

Ph NH2 → Ph $CHCO_2Me$ 85%

$(MeO)_2C{=}O$, γ-Al_2O_3; reflux, 2 d

Vauthey, I.; Valot, F.; Gozzi, C.; Fache, F.; Lemaire, M. *Tetrahedron Lett.*, **2000**, *41*, 6347.

Ph NH_2 → Ph NHAc 95%

$C_6F_5(Me)NCO_2Ac$, THF; rt, 50 min

Kondo, K.; Sekimoto, E.; Nakao, J.; Murakami, Y. *Tetrahedron*, **2000**, *56*, 5843.

MeO_2CO—(cis-CH=CH)—OCO_2Me

$[Pd(\eta^3\text{-}C_3H_7)Cl]_2$, dppf, DCM, rt

NH_2 N–C_6H_{13} 67%

Tanimori, S.; Kirihata, M. *Tetrahedron Lett.*, ***2000***, *41*, 6785.

N—Ph

$KMnO_4$, DCE, reflux, 6 h

$BnNEt_3Cl$

N—Ph 70%

Markgraf, J.H.; Stickney, C.A. *J. Heterocyclic Chem.*, ***2000***, *37*, 109.

OTBDMS Et N Bn

$Co_2(CO)_8$, DMF

500 psi CO, 105°C

Et OTBDMS O N Bn + OTBDMS Et O N Bn

(83 : 17) 93%

Davoli, P.; Prati, F. *Heterocyles*, ***2000***, *53*, 2379.

Me_2NH

BnCl, K_2CO_3, 5% Bu_4NBr, 2 h

onium salt, supercritical CO_2
80 atm, 100°C

Me_2NCO_2Bn 82%

Yoshida, M.; Hara, N.; Okuyama, S. *Chem. Commun.*, ***2000***, 151.

MeO N H

1. DCI, cat DMAP
AcCN, heat

2. BnOH

MeO N O BnHN 70%

also generates esters from alcohols

Macor, J.E.; Cuff, A.; Cornelius, L. *Tetrahedron Lett.*, ***1999***, *40*, 2733.

Et_2NH

$PhNO_2$, 3 eq CO, Se

O Et_2N NHPh 76%

Yang, Y.; Lu, S. *Tetrahedron Lett.*, ***1999***, *40*, 4845.

PhSNEt$_2$ → [5% Pd(PPh$_3$)$_4$, Py, CO; 80°C, 10 h] → PhS–C(=O)–NEt$_2$ 90%

Kuniyasu, H.; Hiraike, H.; Morita, M.; Tanaka, A.; Sugoh, K.; Kurosawa, H. *J. Org. Chem.*, ***1999***, *64*, 7305.

PhNH$_2$ → [1. BuLi, THF, 0°C → rt; 2. EtCO$_2$Et, –78°C] → PhNHCOEt 95%

Ooi, T.; Tayama, E.; Yamada, M.; Maruoka, K. *Synlett*, ***1999***, 729.

SmI$_2$, THF, 10 min
pivalic acid

73%

Honda, T.; Ishikawa, F. *Chem. Commun.*, ***1999***, 1065.

BnCl, cat PdCl$_2$(Ph$_3$)$_2$
CO, MeCN

48%

Cho, C.S.; Jiang, L.H.; Shim, S.C. *Synth. Commun.*, ***1999***, *29*, 2695.

, NEt$_3$, CO
cat PdCl$_2$(PPh$_3$)$_2$, 100°C
MeCN, 1 d

80%

Cho, C.S.; Wu, X.; Jiang, L.H.; Shim, S.C.; Kim, H.R. *J. Heterocyclic Chem.*, ***1999***, *36*, 297.

SECTION 83: AMIDES FROM ESTERS

2.5% [Pd(C$_3$H$_5$)Cl]$_2$, 5% dppf
NaN(CHO)$_2$, MeCN, 60°C

86%

Wang, Y.; Dink, K. *J. Org. Chem.*, ***2001***, *66*, 3238.

cat $RhCl(PPh_3)_3$, 30°C
$P(OMe)_3$, toluene
$LiN(SO_2C_6H_4\text{-}4\text{-}Me)$

OCO_2Me → Tol, N, O_2 S, OMe 96%

Evans, P.A.; Robinson, J.E.; Moffett, K.K. *Org. Lett.*, ***2001***, *3*, 3269.

MeO_2C, N, OMe, O → MeO_2C, N, NHBn, O 94%

$BnNH_2$, THF, rt
0.5 $MgBr_2$, 1 h

Guo, Z.; Dowdy, E.D.; Li, W.-S.; Polniaszek, R.; Delaney, E. *Tetrahedron Lett.*, ***2001***, *42*, 1843.

Ph, O, NH_2, O → O, HN, O, Ph 74%

5% $Rh_2(tpa)_4$, 23 eq MgO, DCE
1.4 eq $PhI(OAc)_2$, 40°C, 12 h

Espino, C.G.; DuBois, J. *Angew. Chem. Int. Ed.*, ***2001***, *40*, 598.

O, Ph, O → H, N, Ph, OEt, O 81%

cat $Pd(PPh_3)_4$, EtOH
1.2 eq $(PhO)_2P(O)N_3$
2 eq NaN_3, 80°C

Okumoto, H.; Nishihara, S.; Yamamoto, S.; Hino, H.; Nozawa, A.; Suzuki, A. *Synlett*, ***2000***, 991.

PhCOOH → PhCONH*t*-Bu 89x75%

1. $Bt\text{-}SO_2Me$, NEt_3
2. *t*-$BuNH_2$

Bt = benzotriazyl

Katritzky, A.R.; He, H.-Y.; Suzuki, K. *J. Org. Chem.*, ***2000***, *65*, 8210.

O, O, OEt → Ph, EtO, NH, N, H, O 95%

PhCHO, H_2N, O, NH_2
$LaCl_3 \cdot 7H_2O$
$EtOH_2$ reflux, 5h

Lu, J.; Bai, Y.; Wang, Z.; Yang, B.; Ma, H. *Tetrahedron Lett.*, ***2000***, *41*, 9075.

[$TolO_2S$–CH_2–C(=O)–NH–Bn / 2NaH, THF]

CH_2=CH–CO_2Me → 3-($TolO_2S$)-N-Bn-piperidine-2,6-dione 90%

Chang, M.-Y.; Chang, B.-R.; Tai, H.-M.; Chang, N.-C. *Tetrahedron Lett.*, ***2000***, *41*, 10273.

$PhCO_2Me$ —($PhNH_2$, microwaves, *t*-BuOH; 1 min)→ PhCONHPh 83%

Varma, R.S.; Naicker, K.P. *Tetrahedron Lett.*, ***1999***, *40*, 6177.

MeO_2CO–CH(C_3H_7)–CH=CH_2 —(1. BnNHTs, LiHMDS; 2. $Rh(PPh_3)_3Cl$, THF, $P(OMe)_3$, 30°C)→ Bn(Ts)N–CH(C_3H_7)–CH=CH_2 + Bn(Ts)N–CH_2–CH=CH–C_3H_7

(3 : 1) 84%

Evans, P.A.; Robinson, J.E.; Nelson, J.D. *J. Am. Chem. Soc.*, ***1999***, *121*, 6761.

SECTION 84: AMIDES FROM ETHERS, EPOXIDES AND THIOETHERS

2-methyl-2-vinyloxetane —(Ph—N=C=O, 3% $Pd(PPh_3)_4$; THF, rt)→ 4-methyl-3-phenyl-4-vinyl-1,3-oxazinan-2-one 57%

Larksarp, C.; Alper, H. *J. Org. Chem.*, ***1999***, *64*, 4152.

C_6H_{13}-oxirane —(1. *p*-$TolSO_2NH_2$, dioxane, K_2CO_3, cat TEBA; 2. TsCl, Py, DCM; 3. K_2CO_3, MeCN)→ C_6H_{13}-aziridine N–Ts 91%

Albanese, D.; Landini, D.; Penso, M.; Petricci, S. *Tetrahedron*, ***1999***, *55*, 6387.

SECTION 85: AMIDES FROM HALIDES AND SULFONATES

F_3C–C_6H_4–Br → (H_2NCHO, DMAP, 120°C; 10% $PdCl_2(PPh_3)_2$, 18 h; dioxane, autoclave, 5 bar CO) → F_3C–C_6H_4–$C(O)NH_2$ 71%

Schnyder, A.; Beller, M.; Mehltretter, G.; Nsenda, T.; Studer, M.; Indolese, A.F. *J. Org. Chem.*, ***2001***, *66*, 4311.

Ph-I → (NH_2NHBoc, 1% CuI; 10% 1,10-phenanthroline; Cs_2CO_3, DMF, 80°C) → Ph–N(Boc)–NH_2 97%

Wolter, M.; Klapars, A.; Buchwald, S.L. *Org. Lett.*, ***2001***, *3*, 3803.

NC–C_6H_4–Br → (TBSO-CH2-pyrrolidin-2-one (N-H); $Pd_2(bda)_3$, xantphos, 3.5 h; Cs_2CO_3, dioxane, 105°C) → NC–C_6H_4–N(pyrrolidinone, TBSO–CH2) 95%

Browning, R.G.; Mahmud, H.; Badarinarayana, V.; Lovely, C.J. *Tetrahedron Lett.*, ***2001***, *42*, 7155.

t-Bu–C_6H_4–OTf → ($BzNH_2$, 1.4 Cs_2CO_3, 16 h; xantphos/cat $Pd_2(dba)_3$; dioxane, 100°C) → *t*-Bu–C_6H_4–NHBz 94%

Yin, J.; Buchwald, S.L. *Org. Lett.*, ***2000***, *2*, 1101.

Ph-CH=NPh → (BnBr, BNAH, hv, 4 h) → $PhCH_2$–N(Ph)–CH_2Ph 87%

BNAH = 1-benzyl-1,4-dihydronicotinamide

Jin, M.; Zhang, D.; Yang, L.; Lium. Y.; Liu, Z. *Tetrahedron Lett.*, ***2000***, *41*, 7357.

I–$CH_2CH_2CH_2$–CH=CH–CO_2Et → (*p*-$TsNH_2$, K_2CO_3; DMF) → N-Ts-pyrrolidin-2-yl–CH_2CO_2Et 62%

Bunce, R.A.; Allison, J.C. *Synth. Commun.*, ***1999***, *29*, 2175.

SECTION 86: AMIDES FROM HYDRIDES

MeO MeO

$EtO_2C—N=N—CO_2Et$
$InCl_3$–SiO_2, microwaves
3 min

MeO MeO CO_2Et N $NHCO_2Et$ 88%

Yadav, J.S.; Subba Reddy, B.V.; Kumar, G.M.; Madan, C. *Synlett*, ***2001***, 1781.

TBDMSO N Boc CO_2Me

cat RuO_2, $NaIO_4$
aq EtOAc

TBDMSO O N Boc CO_2Me >95%

Zhang, X.; Schmitt, A.C.; Jiang, W. *Tetrahedron Lett.*, ***2001***, *42*, 5335.

Ph

$PhI(OAc)_2$, $TsNH_2$
fluorinated Mn porphyrin

NHTs Ph
81% (58% conversion)

Yu, X.-Q.; Huang, J.-S.; Zhou, X.-G.; Che, C.-M. *Org. Lett.*, ***2000***, *2*, 2233.

2% Ru-amine complex
PhI=NTs

NHTs
90% (57% conversion)

Au, S.-M.; Huang, J.-S.; Che, C.-M.; Yu, W.-Y. *J. Org. Chem.*, ***2000***, *65*, 7858.

SECTION 87: AMIDES FROM KETONES

O Ph Ph

$TsNH_2$, 0.5 $TiCl_4$, 2 eq NEt_3
DCE, reflux, 5 h

Ts N Ph Ph 83%

Ram, R.N.; Khan, A.A. *Synth. Commun.*, ***2001***, *31*, 841.

$H_2N(C{=}S)NH_2$, $PhSO_2Na{\cdot}H_2O$

HCOOH , rt , 6 d

70%

Liepa, A.J.; Wright, D.M.J. *Aust. J. Chem.*, ***2000***, *53*, 73.

BnN_3 , $TiCl_4$, DCM

0°C → rt

88%

Desai, P.; Schildknegt, K.; Agrios, K.A.; Mossman, C.; Milligan, G.L.; Aubé, J. *J. Am. Chem. Soc.*, ***2000***, *122*, 7226.

1. $LiCH_2NMeBoc$, –78°C
2. MeOTf, –78°C
3. satd $NaHCO_3$, –78°C → rt

4. TFA, CH_2Cl_2
5. 155-160°C

53x75%

Zhang, S.; Liebeskind, L.S. *J. Org. Chem.*, ***1999***, *64*, 4042.

1. Ph(OH)CH$CH_2CH_2N_3$

–82°C → rt
2. KOH

98%

Furness, K.; Aubé, J. *Org. Lett.*, ***1999***, *1*, 495.

Ts–N=CH–CO_2Et , 5 min

$Cp_2Co^+ Co(CO)_4^-$
benzotrifluoride

85%

Wack, H.; Drury III, W.J.; Taggi, A.E.; Ferraris, D.; Lectka, T. *Org. Lett.*, ***1999***, *1*, 1985.

O

Me

1. Ph–CH(OH)–CH$_2$CH$_2$–N$_3$ (OH, Ph, N_3)

−82°C → rt

2. KOH

O, N, OH, Me, Ph

98%

Furness, K.; Aubé, J. *Org. Lett.*, ***1999***, *1*, 495.

SECTION 88: AMIDES FROM NITRILES

Ph—CN → (1. $H_2PO_3Se^-$, MeOH, H_2O, rt, overnight; 2. aq K_2CO_3) → Ph–C(=Se)–NH$_2$ 50%

Kamiński, R.; Glass, R.S.; Skowrońska, A. *Synthesis*, ***2001***, 1308.

PhCN → ($NaBO_3$, 4 H_2O, aq EtOH; microwaves, 2 min) → Ph–C(=O)–NH$_2$ 85%

Sharifi, A.; Mohsenzadeh, F.; Mojtahedi, M.M.; Saidi, M.R.; Balalaie, S. *Synth. Commun.*, ***2001***, *31*, 431.

MeCN → ($C_3H_7NH_2$, H_2O, DME, 1 d; 0.1% Pt(II)-phosphine oxide complex; 160°C) → $AcNHC_3H_7$ 57%

Cobley, C.J.; van den Heuvel, M.; Abbadi, A.; de Vries, J.G. *Tetrahedron Lett.*, ***2000***, *41*, 2467.

$PhCH_2CN$ → (Ac_2O, $NaBH_4/NiCl_2$) → $PhCH_2NHAc$ 94%

Caddick, S.; deK. Haynes, A.K.; Judd, D.B.; Williams, M.R.V. *Tetrahedron Lett.*, ***2000***, *41*, 3513.

Ph-CN → (aq NaOH, PEG-400; microwaves, 45 sec) → Ph–C(=O)–NH$_2$ 83%

Bendale, P.M.; Khadilkar, B.M. *Synth. Commun.*, ***2000***, *30*, 1713.

SECTION 89: AMIDES FROM ALKENES

p-Tol-$SO_2N(Br)Na$, 5% $CuCl_2$

microwaves

81%

Chanda, B.M.; Vyas, R.; Bedekar, A.V. *J. Org. Chem.*, ***2001***, *66*, 30.

5

PhI=NTs , 10% TP*Na , 10% CuCl

MeCN , rt , 12 h

84%

TP = tris-(3,5-dimethylpyrazoyl)borate

Handy, S.T.; Czopp, M. *Org. Lett.*, ***2001***, *3*, 1423.

H_2NO_2S–C_6H_4–OMe

PhI=O , 10% $Cu(MeCN)_4PF_6$

MS 3Å , MeCN

78%

Dauben, P.; Sanière, L.; Tarrade, A.; Dodd, R.H. *J. Am. Chem. Soc.*, ***2001***, *123*, 7707.

PhI=NTs , $Cu(MeCN)_4ClO_4$

chiral binaphthyl-imine ligand

DCM , 20°C , 4 h

92% (22% ee, *S*)

Shi, M.; Wang, C.-J.; Chan, A.S.C. *Tetrahedron Asymm.*, ***2001***, *12*, 3105.

$NsONHCO_2Et$, CaO , DCM

rt , 30 min

91%

Fioravanti, S.; Morreale, A.; Pellacani, L.; Tardella, P.A. *Synthesis*, ***2001***, 1975.

PhI=NTs , $MeReO_3$, MeCN , 5 h

28%

Jeon, H.-J.; Nguyen, S.B.T. *Chem. Commun.*, ***2001***, 235.

1. NaOH , urethane , PrOH
$K_2[OsO_2(OH)_4]$, $(DHQD)_2PHAL$
1,3-dichloro-5,5-dimethyl-hydantoin , H_2O
2. NaOH

Ph HN O O + Ph HN O O

(2 : 1) 65%
88% ee 73% ee

Barta, N.S.; Sidler, D.R.; Somerville, K.B.; Weissman, S.A.; Larsen, R.D.; Reider, P.J. *Org. Lett.*, ***2000***, *2*, 2821.

nitrido Mn-salen complex
0.5 Py , 1.2 Ts_2O , DCM
Py-N-oxide , 0°C , 30 min

N–Ts 82%

Nishimura, M.; Minakata, S.; Thongchant, S.; Ryu, I.; Komatsu, M. *Tetrahedron Lett.*, ***2000***, *41*, 7089.

Ph chloramine-T , MeCN
10% Py–HBr_3 , 25°C

Ts N Ph 60%

Ali, S.I.; Nikalje, M.D.; Sudalai, A. *Org. Lett.*, ***1999***, *1*, 705.

Ph 1. $PhI{=}NSO_2CH_2CH_2TMS$, CuOTf
MeCN , MS 4Å
2. 4 eq TASF , rt

NH Ph 68x62%

Dauben, P.; Dodd, R.H. *J. Org. Chem.*, ***1999***, *64*, 5304.

SECTION 90: AMIDES FROM MISCELLANEOUS COMPOUNDS

N_2 O O N–Bn , microwaves
o-dichlorobenzene , 180°C

C_3H_7 O O N Bn 54%

Linder, M.R.; Podlech, J. *Org. Lett.*, ***2001***, *3*, 1849.

hv , MeCN , 18 h

O 48%

Bourguet, E.; Baneres, J.-L.; Girard, J.-P.; Parello, J.; Vidal, J.-P.; Lusinchi, X.; Declercq, J.-P. *Org. Lett.*, ***2001***, *3*, 3067.

OH N Ph

H_2 , AcOEt , $NiBr_2Et$

Pd/C , toluene , 60°C , 5d
Candida antactica lipase B

NHAc Ph 80% (98% ee)

Choi, Y.K.; Kim, M.J.; Ahn, Y.; Kim, M.-J. *Org. Lett.*, ***2001***, *3*, 4099.

N—OH

$BuPyBF_4$, PCl_5, 80°C

Py = pyridinium

O NH 99%

Peng, J.; Deng, Y. *Tetrahedron Lett.*, ***2001***, *42*, 403.

N—OH Ph NH_2

NaOMe , MeOH , Rose bengal

hv , O_2

O Ph NH_2 74%

Öcal, N.; Erden, I. *Tetrahedron Lett.*, ***2001***, *42*, 4765.

N—OH•HCl Ph Ph

neat , 105°C , 8 h

O Ph NHPh 96%

Chandrasekhar, S.; Gopalaiah, K. *Tetrahedron Lett.*, ***2001***, *42*, 8123.

=N-OH

$FeCl_3$, solvent free

80°C , 3 h

NH =O 82%

Khodaei, M.M.; Meybodi, F.A.; Rezai, N.; Salehi, P. *Synth. Commun.*, ***2001***, *31*, 2047.

$SnBu_3$

PhI=NTs , $Cu(OTf)_2$

MeCN , rt

NHTs 72%

Kim, D.Y.; Kim, H.S.; Choi, Y.J.; Mang, J.Y.; Lee, K. *Synth. Commun.*, ***2001***, *31*, 2463.

$PhNO_2$ —[Ac_2O, In, $InCl_3$ / $CHCl_3$, 20 h]→ $PhN(OAc)_2$ 81% (+ 13% PhNHAc)

Kim, B.H.; Cheong, J.W.; Han, R.; Jun, Y.M.; Baik, W.; Lee, B.M. *Synth. Commun.*, ***2001***, *31*, 3577.

N–OH, Ph, Et —[2.5% $[RhCl(cod)]_2$, 25% TFA, 15% $P(o\text{-Tol})_3$, DCM, reflux / 3 h]→ PhHN, O, Et 78% + Ph, O, NHEt 29%

Arisawa, M.; Yamaguchi, M. *Org. Lett.*, ***2001***, *3*, 311.

=N-OH —[[bmim] PF_6, P_2O_5 / 75°C, 16 h]→ NH, O 95%

ionic liquid medium

Ren, R.X.; Zueva, L.D.; Ou, X. *Tetrahedron Lett.*, ***2001***, *42*, 8441.

O, N_2, NHBoc —[TFA, DCE, –78°C]→ HN, O 69%

Yang, H.; Jurkauskas, V.; Mackintosh, N.; Mogrenn, T.; Stephenson, C.R.J.; Foster, K.; Brown, W.; Roberts, E. *Can.J. Chem.*, ***2000***, *78*, 800.

Ph, N_3 —[Fe/NH_4Cl, MeOH, 15 min, ultrasound, $ClCO_2Et$]→ Ph, $NHCO_2Et$ 90% (65% in 12 h without sonication

Chandrasekhar, S.; Narsihmulu, Ch. *Tetrahedron Lett.*, ***2000***, *41*, 7969.

Et_2N, O, Cl —[BuMgCl, THF / 5% $NiCl_2(PPh_3)_2$, 2h]→ Et_2N, O, Bu 72%

Lemoucheux, L.; Rouden, J.; Lasne, M.-C. *Tetrahedron Lett.*, ***2000***, *41*, 9997.

$N\text{–}OSO_2Et$, Ph, Ph —[$BF_3\cdot OEt_2$, DCM, 25°C, 6 h]→ Ph, O, NHPh 98%

Anilkumar, R.; Chandrasekhar, S. *Tetrahedron Lett.*, ***2000***, *41*, 5427.

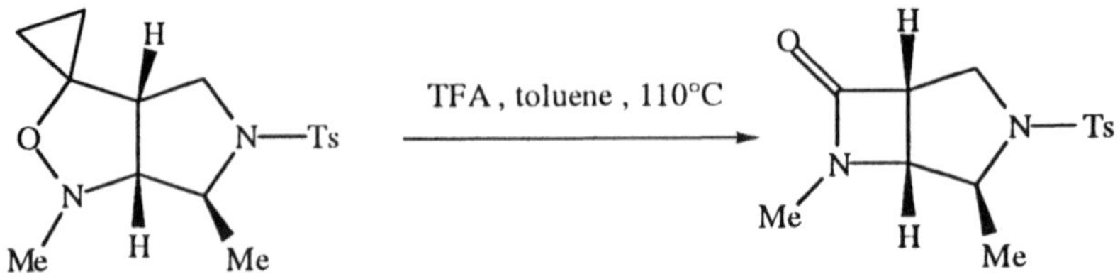

TFA, toluene, 110°C

Cordero, F.M.; Pisaneschi, F.; Goti, A.; Ollivier, J.; Salaün, J.; Brandi, A. *J. Am. Chem. Soc.*, ***2000***, *122*, 8077.

BocN(Li)–SO_2Ph → (BuLi, –78°C → 0°C) BuNHBoc 71%

Dembech, P.; Seconi, G.; Ricci, A. *Chem. Eur. J.*, ***2000***, *6*, 1281.

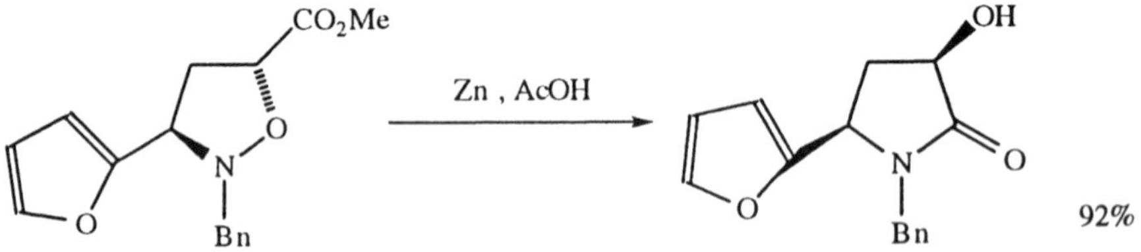

Zn, AcOH 92%

Merino, P.; Anoro, S.; Merchan, F.; Téjero, T. *Heterocycles*, ***2000***, *53*, 861.

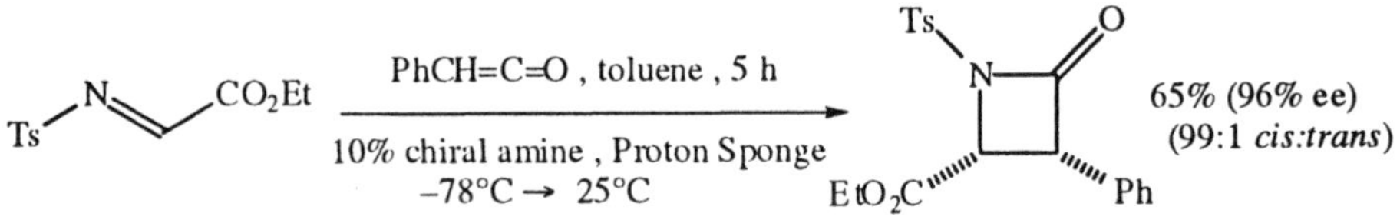

PhCH=C=O, toluene, 5 h; 10% chiral amine, Proton Sponge, –78°C → 25°C 65% (96% ee) (99:1 *cis:trans*)

Taggi, A.E.; Hafez, A.M.; Wack, H.; Young, B.; Drury III, W.J.; Lectka, T. *J. Am. Chem. Soc.*, ***2000***, *122*, 7831.

PhCH=N–OH → (Boc_2O, polymethylhydrosiloxane; Pd-C, EtOH, heat) $PhCH_2NHBoc$ 85%

Chandrasekhar, S.; Reddy, M.V.; Chandraiah, L. *Synlett*, ***2000***, 1351.

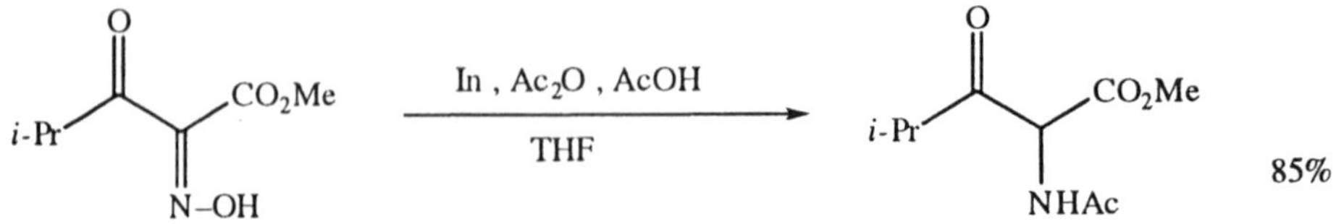

In, Ac_2O, AcOH; THF 85%

Harrison, J.R.; Moody, C.J.; Pitts, M.R. *Synlett*, ***2000***, 1601.

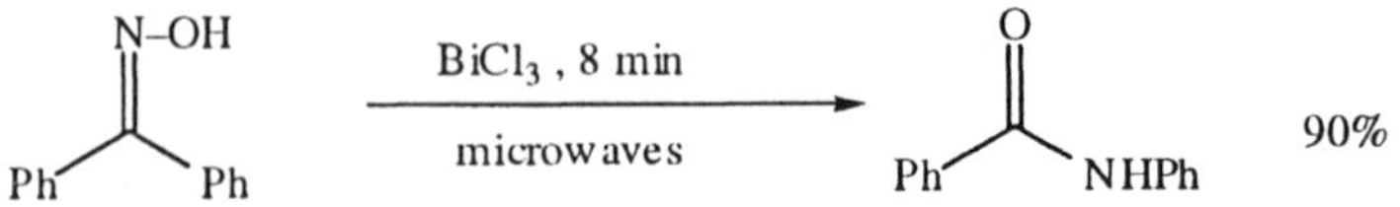

$BiCl_3$, 8 min; microwaves 90%

Thakur, A.J.; Boruah, A.; Prajapati, D.; Sandhu, J.S. *Synth. Commun.*, ***2000***, *30*, 2105.

1. TEAPC, MeCN-TEAP, rt, 1 h
2. 5 eq TsCl, 25°C, 10 h

TEAPC = $(Et_4N)_2$ peroxydicarbonate

86%

Feroci, M.; Inesi, A.; Mucciante, V.; Rossi, L. *Tetrahedron Lett.*, ***1999***, *40*, 6059.

PhCH=NOH → PhCONH$_2$ quant
SiO_2, microwaves, 140°C, 1 min

Loupy, A.; Régnier, S. *Tetrahedron Lett.*, ***1999***, *40*, 6221.

1. PMe_3, THF, rt
2. BnO_2CCl, rt
3. H_2O, pH 7

Ph–CH_2–N_3 → Ph–CH_2–$NHCO_2Me$ 97%

Ariza, X.; Urpí, F.; Vilarrasa, J. *Tetrahedron Lett.*, ***1999***, *40*, 7515.

PhN_3 → PhNHAc 90%
Me_3SiCl, Ac_2O, heat, 10 min

Barua, A.; Bez, G.; Barua, N.C. *Synlett*, ***1999***, 553.

1. BuMgCl, cat CuCN•2 LiCl
HMPA-THF, 30 min
2. H_3O^+
3. PhCOCl, NEt_3

96%

Tsutsui, H.; Ichikawa, T.; Narasaka, K. *Bull. Chem. Soc. Jpn.*, ***1999***, *72*, 1869.

SECTION 90A: PROTECTION OF AMIDES

BnBr, Bu_3P
1,1'-(azodicarbonyl)dipiperidine

86%

Winum, J.-Y.; Barragan, V.; Montero, J.-L. *Tetrahedron Lett.*, ***2001***, *42*, 601.

CHAPTER 7
PREPARATION OF AMINES

SECTION 91: AMINES FROM ALKYNES

C_6H_{13}—≡ → ($PhNH_2$, PCy_3, toluene; 1.5% $[Rh(cod)_2]BF_4$) → N–Ph imine of C_6H_{13} methyl ketone 79%

Hartung, C.G.; Tillack, A.; Trauthwein, H.; Beller, M. *J. Org. Chem.*, ***2001***, *66*, 6339.

OMe, Et, O → (100 psi NH_3; 100°C, 14 h) → NH_2, N, Et 82%

Hegde, V.B.; Renga, J.M.; Owen, J.M. *Tetrahedron Lett.*, ***2001***, *42*, 1847.

Et—≡—Et → (1. $PhNH_2$, 3% Cp_2TiMe_2, toluene, microwaves, 3 h; 2. H_2, 5% Pd/C, THF, 25°C, 3 d) → NHPh, Et, Et 54%

Bytschkov, I.; Doye, S. *Eur. J. Org. Chem.*, ***2001***, 4411.

Ph—≡—Ph → (1. Ph_2CHN_2, 3% Cp_2TiMe_2, 110°C; 2. H_2, Pd/C, THF, 25°C, 3 d) → NH_2, Ph, Ph 67%

Haak, E.; Siebeneicher, H.; Doye, S. *Org. Lett.*, ***2000***, *2*, 1935.

Ph—≡—Ph → ($PhNH_2$, 1% Cp_2TiMe_2; C_6D_6, 90°C) → Ph, N–Ph, Ph 52%

Haak, E.; Bytschkov, I.; Doye, S. *Angew. Chem. Int. Ed.*, ***1999***, *38*, 3389.

SECTION 92: AMINES FROM ACID DERIVATIVES

NO ADDITIONAL EXAMPLES

SECTION 93: AMINES FROM ALCOHOLS AND THIOLS

PhNCO

N OH → N NHPh 69%

Katritzky, A.R.; Huang, T.-B.; Voronkov, M.V. *J. Org. Chem.*, ***2001***, *66*, 1043.

BuOH → 1. $(EtO)_2P(O)NHCO_2t$-Bu , DIAD, TPP , THF , rt , 2 h; 2. TsOH , H_2O , EtOH , reflux → $BuNH_3^+$ OTs^- 76%

Klepacz, A.; Zwiwezak, A. *Synth. Commun.*, ***2001***, *31*, 1683.

Ph, Ph, OH → 1. NH_2CO_2Ph , TFA; 2. LiOH , MeCN , H_2O, 50°C , 3h → Ph, Ph, NH_2 80x95%

Laurent, M.; Marchand-Brynaert, J. *Synthesis*, ***2000***, 667.

$PhCH_2OH$ → 1. NaN_3 , CCl_4–DMF; 2. > 2 eq PPh_3 → Ph—NH_2 95%

Reddy, G.V.S.; Rao, G.V.; Subramanyam, R.V.K.; Iyengar, D.S. *Synth. Commun.*, ***2000***, *30*, 2233.

SECTION 94: AMINES FROM ALDEHYDES

CHO, Bn, NBn_2 → 1. *p*-anisidine , DCM , MS 4Å; 2. TMSI, BuLi , THF; −78°C → rt → PMP, N, Bn, NBn_2 + PMP, N, Bn, NBn_2

PMP = *p*-methoxyphenyl

(4 : 1) 88%

Reetz, M.T.; Lee, W.K. *Org. Lett.*, ***2001***, *3*, 3119.

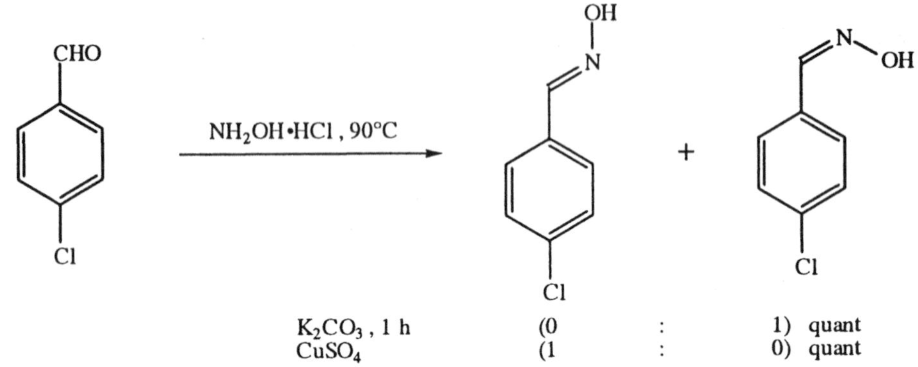

CHO, Cl; $NH_2OH{\bullet}HCl$, 90°C; OH, N; N, OH; Cl; Cl

K_2CO_3, 1 h	(0	:	1) quant
$CuSO_4$	(1	:	0) quant

Sharghi, H.; Sarvari, M.H. *Synlett*, ***2001***, 99.

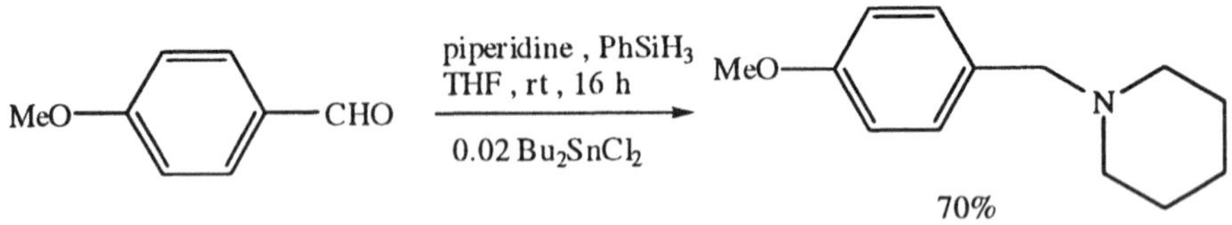

MeO, CHO; piperidine, $PhSiH_3$, THF, rt, 16 h; 0.02 Bu_2SnCl_2; MeO, N; 70%

Apodaca, R.; Xiao, W. *Org. Lett.*, ***2001***, *3*, 1745.

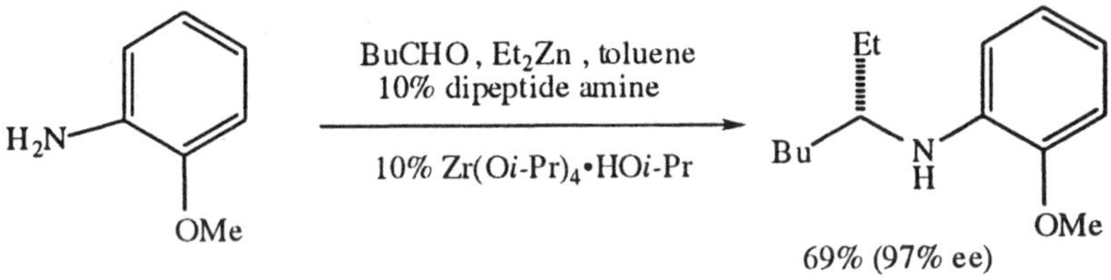

H_2N, OMe; BuCHO, Et_2Zn, toluene, 10% dipeptide amine; 10% $Zr(Oi\text{-}Pr)_4{\bullet}HOi\text{-}Pr$; Et, Bu, N, H, OMe; 69% (97% ee)

Porter, J.R.; Traverse, J.F.; Hoveyda, A.H.; Snapper, M.L. *J. Am. Chem. Soc.*, ***2001***, *123*, 10409.

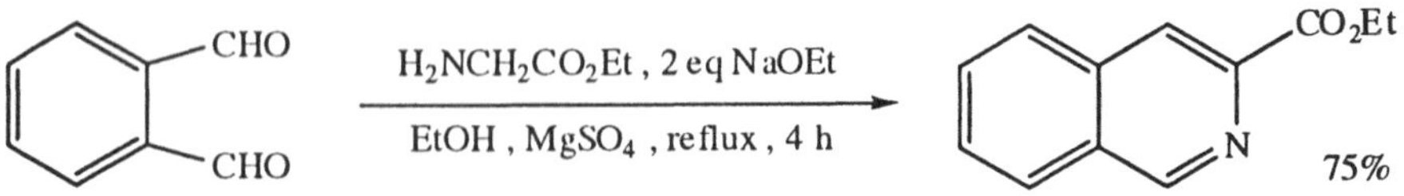

CHO, CHO; $H_2NCH_2CO_2Et$, 2 eq NaOEt; EtOH, $MgSO_4$, reflux, 4 h; CO_2Et, N; 75%

Meziane, M.A.A.; Royer, S.; Bazureau, J.P. *Tetrahedron Lett.*, ***2001***, *42*, 1017.

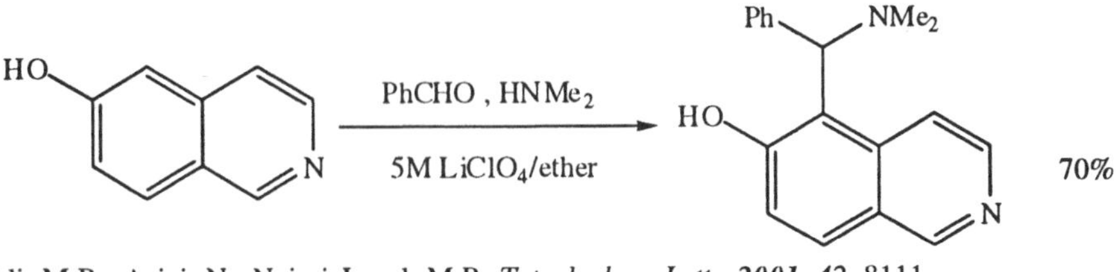

HO, N; PhCHO, $HNMe_2$; 5M $LiClO_4$/ether; Ph, NMe_2, HO, N; 70%

Saidi, M.R.; Azizi, N.; Naimi-Jamal, M.R. *Tetrahedron Lett.*, ***2001***, *42*, 8111.

PhCHO $\xrightarrow[CH_2Cl_2\text{ , }0°C \rightarrow rt]{0.2\text{ eq }TiCl_4\text{ , }Et_2NSiHMe_2\text{ , }36\text{ h}}$ $PhCH_2NEt_2$ 90%

Miura, K.; Ootsuka, K.; Suda, S.; Nishikori, H.; Hosomi, A. *Synlett, **2001***, 1617.

PhCHO $\xrightarrow[2.\ NaBH_4\text{ , rt}]{1.\ Ti(Oi\text{-}Pr)_4\text{ , }Me_2NH\text{ , MeOH , rt}}$ $PhCH_2NMe_2$ 90%

Bhattacharyya, S. *Synth. Commun., **2000**, 30*, 2001.

ClH•H_2N … CO_2H (leucine) → (cyclohexanecarboxaldehyde; H_2 , 20% $Pd(OH)_2/C$, EtOH , rt) → N-cyclohexylmethyl leucine (NH, CO_2H) 93%

Song, Y.; Sercel, A.D.; Johnson, D.R.; Colbry, N.L.; Sun, K.-L.; Roth, B.D. *Tetrahedron Lett., **2000**, 41*, 8225.

PhCHO $\xrightarrow[BnNH_2\text{ , THF , 5 h}]{\text{polymethylhydrosiloxane-}Ti(Oi\text{-}Pr)_4}$ Bn_2NH 90%

Chandrasekhar, S.; Reddy, Ch.R.; Ahmed, M. *Synlett, **2000***, 1655.

MeO-C6H4-NH_2 $\xrightarrow[\text{toluene , reflux; 6M HCl}]{\text{CH=CH-CHO}}$ 6-MeO-2-Me-quinoline (N, Me) 61%

Matsugi, M.; Tabussa, F.; Minamikawa, J.-i. *Tetrahedron Lett., **2000**, 41*, 8523.

PhCHO $\xrightarrow[Bu_2SnClH\text{–HMPA , 1 h}]{H_2N\text{–}C_6H_4\text{–Cl , THF , 0°C}}$ Ph–CH(H_2N)–C6H4–Cl 99%

Suwa, T.; Sugiyama, E.; Shibata, I.; Baba, A. *Synthesis, **2000***, 789.

PhCHO $\xrightarrow[NaBH_4\text{ , rt}]{BuNH_2\text{ , }NiCl_2•6\ H_2O/MeOH}$ $PhCH_2NHBu$ 80%

Saxena, I.; Borah, R.; Sarma, J.C. *J. Chem. Soc., Perkin Trans. 1, **2000***, 503.

PhCHO → ($BnONH_2$, BEt_3, DCM, 20°C / BF_3•OEt_2, $MgSO_4$, 1 d) → Ph–CH(NHOBn)–Et 83%

Miyabe, H.; Yamakawa, K.; Yoshioka, N.; Naito, T. *Tetrahedron*, ***1999***, *55*, 11209.

2-methylbenzaldehyde (CHO) → (1. Me_2N–$SiMe_3$, $LiClO_4$/ether; 2. $Me_2PhSiCl$, rt) → 2-MeC$_6$H$_4$–CH(NMe_2)–$SiMe_2Ph$ 74%

Naimi-Jamal, M.R.; Mojtahedi, M.M.; Ipaktschi, J.; Saidi, M.R. *J. Chem. Soc., Perkin Trans. 1*, ***1999***, 3709.

Related Methods: Section 102 (Amines from Ketones)

SECTION 95: AMINES FROM ALKYLS, METHYLENES AND ARYLS

NO ADDITIONAL EXAMPLES

SECTION 96: AMINES FROM AMIDES

N-(diphenylamino)phthalimide (N—NPh_2) → (5 eq $AlCl_3$, PhH / rt, 4 h) → carbazole 81%

Kikugawa, Y.; Aoki, Y.; Sakamoto, T. *J. Org. Chem.*, ***2001***, *66*, 8612.

3,3-dimethylpiperidin-2-one → (1. MeLi, DME, –78°C; 2. TMSOTf, –78°C; 3. *t*-BuLi, ether, –20°C → 25°C) → 2-methyl-3,3-dimethyl-3,4,5,6-tetrahydropyridine 58%

Ahn, Y.; Cardenas, G.I.; Yang, J.; Romo, D. *Org. Lett.*, ***2001***, *3*, 7511.

O N — 1% $Re_2(CO)_{10}$, 5% Et_2NH; toluene , 100°C , 16 h; 3.5 eq $HSiEt_3$ → N — 96%

Igarashi, M.; Fuchikami, T. *Tetrahedron Lett.*, ***2001***, *42*, 1945.

O N H — 1. Me_3SiCl; 2. PhMgBr → Ph N — 90%

Coindet, C.; Comel, A.; Kirsch, G. *Tetrahedron Lett.*, ***2001***, *42*, 6101.

$PhNHSO_2$(2-pyridyl) — 10 eq Mg , MeOH , 0°C; 3 h → $PhNH_2$ — 86%

Pak, C.S.; Lim, D.S. *Synth. Commun.*, ***2001***, *31*, 2209.

N H O — 2 eq SmBr , THF; rt , 5 min → N H — 95%

Li, Z.; Zhang, Y. *Tetrahedron Lett.*, ***2001***, *42*, 8507.

N H O — 1. $(CH_2{=}CHCH_2)_3B$, THF , 65°C; 2. MeOH; 3. aq NaOH → N H — 90%

Bubnov, Y.N.; Pastukhov, F.V.; Yampolsky, I.V.; Ignatenko, A.V. *Eur. J. Org. Chem.*, ***2000***, 1503.

O N Br — 1. *t*-BuLi , ether , –78°C; 2. $LiAlH_4$, ether , reflux → N — 47%

Jones, K.; Storey, J.M.D. *J. Chem. Soc., Perkin Trans. 1*, ***2000***, 769.

O H_2N NH_2 — Cl , aq NaOH , 80°C; 40 h → $(CH_2{=}CHCH_2)_3N$ — 56%

Sachinvala, N.; Winsor, D.L.; Maskos, K.; Grimm, C.; Hamed, O.; Vigo, T.L.; Bertoniere, N.R. *J. Org. Chem.*, ***2000***, *65*, 9234.

1. $B(CH_2CH=CH_2)_3$, THF, 65°C, 2 h
2. MeOH
3. aq NaOH

90%

Bubnov, Yu.N.; Klimkina, E.V.; Zhun, I.V.; Pastukhov, F.V.; Yampolsky, I.V. *Pure Appl. Chem.*, ***2000***, *72*, 1641.

TMSCl, NaI, MeCN
heat, 3-4 h

88%

Sabitha, G.; Reddy, B.V.S.; Abraham, S.; Yadav, J.S. *Tetrahedron Lett.*, ***1999***, *40*, 1569.

2.1 eq 9-BBN, THF
reflux, 1 h

96%

Collins, C.J.; Lanz, M.; Singaram, B. *Tetrahedron Lett.*, ***1999***, *40*, 3673.

1. Na naphthalenide, DME, -78°C
2. H_2O

89%

Bergmeier, S.C.; Seth, P.P. *Tetrahedron Lett.*, ***1999***, *40*, 6181.

$LiH_3B \cdot NMe_2$, THF, 65°C, 2 h

86%

Flaniken, J.M.; Collins, C.J.; Lanz, MM.; Singaram, B. *Org. Lett.*, ***1999***, *1*, 799.

$BH_3 \cdot THF$, THF, heat

81%

Akula, M.R.; Kabalka, G.W. *Org. Prep. Proceed. Int.*, ***1999***, *31*, 214.

Related Methods: Section 105A (Protection of Amines)

SECTION 97: AMINES FROM AMINES

Piperidine (N–H) → PhCl, 3 eq t-BuOK → N-Ph piperidine 92%

Beller, M.; Breindl, C.; Riermeier, T.H.; Tillack, A. *J. Org. Chem.*, ***2001***, *66*, 1403.

Imidazole (NH) → $PhB(OH)_2$, H_2O, rt, O_2; cat $[Cu(OH)(tmeda)]_2Cl_2$, overnight → N-Ph imidazole 43%

Collman, J.P.; Zhong, M.; Zeng, L.; Costanzo, S. *J. Org. Chem.*, ***2001***, *66*, 528.

2-($SiMe_3$)pyridine → 1. t-BuLi, ether, –78°C, 30 min; 2. $PhCH_2Br$ → 2-pyridyl-Si(Me_2)-CH_2CH_2Ph 99%

Itami, K.; Kame, T.; Mitsudo, K.; Nokami, T.; Yoshida, J.-i. *J. Org. Chem.*, ***2001***, *66*, 3970.

PhCH=N–$(CH_2)_3$–Cl → 1. Li, DTBB; 2. H_2O → 2-Ph pyrrolidine (N–H) >99%

Yus, M.; Soler, T.; Foubelo, F. *J. Org. Chem.*, ***2001***, *66*, 6207.

F_3C–C(=N-PMP)–CO_2Et → $Pd(O_2CCF_3)_2$, R-BINAP, rt; 100 atm H_2, CF_3CH_2OH, 1 d → F_3C–CH(NHPMP)–CO_2Et >99% (88% ee, R)

Abe, H.; Amii, H.; Uneyama, K. *Org. Lett.*, ***2001***, *3*, 313.

2-Naphthyl–C(=$NHBn^+$ BF_4^-)–CH_3 → 56 psi H_2 DCM, rt; 2% CpRu(SS-Chiraphos)H, 2 d → 2-Naphthyl–CH(NHBn)–CH_3 90% (60% ee)

Magee, M.P.; Norton, J.R. *J. Am. Chem. Soc.*, ***2001***, *123*, 1778.

Ph N OMe

10% peptidic ligand
10% $Zr(Oi\text{-}Pr)_4$

Et_2Zn , i-PrOH
toluene

Et Ph N H OMe

84% (93% ee)

Porter, J.R.; Traverse, J.F.; Hoveyda, A.H.; Snapper, M.L.
J. Am. Chem. Soc., ***2001***, *123*, 984.

N N

5%$[RhCl(coe)_2]_2$

2.5% PCy_3 , 60°
THF , 20h

N N 70%

Tan, K.L.; Bergman, R.G.; Ellman, J.A. *J. Am. Chem. Soc.* ***2001***, *123*, 2685.

O N Ph

1. Li / THF , –20°C , 4 h
2. H_2O

HO N Ph

>95%

Azzena, U.; Pilo, L.; Piras, E. *Tetrahedron Lett.*, ***2001***, *42*, 129.

I

H N

1% CuCl , 10% chiral diamine
2.1 eq K_3PO_4 , 110°C , 1d
dioxane

N

99%

Klapars, A.; Antilla, J.C.; Huang, X.; Buchwald, S.L. *J. Am. Chem. Soc.*, ***2001***, *123*, 7727.

MeO_2CO Ph

pyrrolidine , EtOH , reflux
2% $[Ir(cod)Cl]_2$, 8% $P(OPh)_3$

3 h

Ph N + Ph N

(96 : 4) 95%

Takeuchi, R.; Ue, N.; Tanabe, K.; Yamashita, K.; Shiga, N.
J. Am. Chem. Soc., ***2001***, *123*, 9525.

PhOTf, $Pd(OAc)_2$, dppf, DABCO, DMSO, 80°C — $N(Bn)_2$ — 74%

Wu, J.; Marcoux, J.-F.; Davies, I.W.; Reider, P.J. *Tetrahedron Lett.*, ***2001***, *42*, 159.

$SiCl_3$, DMF, 0°C, 18 h — NHBz, NHNHBz — 79% (1:99 *syn:anti*)

Hirabayashi, R.; Ogawa, C.; Sugiura, M.; Kobayashi, S. *J. Am. Chem. Soc.*, ***2001***, *123*, 9493.

EtI, $InCl_3$, $Mn_2(CO)_{10}$, hv — 85%

Friestad, G.K.; Qin, J. *J. Am. Chem. Soc.*, ***2001***, *123*, 9922.

2 eq MeMgBr, toluene, rt, cat $Sc(OTf)_3$, 3 h — 90%

Saito, S.; Hatanaka, K.; Yamamoto, H. *Synlett*, ***2001***, 1859.

Cl_3SiH, CH_2Cl_2, N-CHO — 79%

Iwasaki, F.; Onomura, O.; Mishima, K.; Kanematsu, T.; Maki, T.; Matsumura, Y. *Tetrahedron Lett.*, ***2001***, *42*, 2525.

$PhCHBr_2$, Mg, ether — C_3H_7 — 60% (61:39 *cis:trans*)

Biscoe, M.R.; Fry, A.J. *Tetrahedron Lett.*, ***2001***, *42*, 2759.

cat $Ru_3(CO)_{12}$, CO , 21 h

$CH_2{=}CH_2$, *i*-PrOH , 140°C

Et

Et

92% (54:46)

Chatani, N.; Asaumi, T.; Yorimitsu, S.; Ikeda, T.; Kakiuchi, F.; Murai, S. *J. Am. Chem. Soc.*, ***2001***, *123*, 10935.

NH_2

$Y[N(TMS)_2]_3$, C_6D_6 , 24°C

6 h

NH

>95%

Kim, Y.K.; Livinghouse, T.; Bercaw, J.W. *Tetrahedron Lett.*, ***2001***, *42*, 2933.

1. hv (254 nm)

2. NH_2

NH*i*-Pr

Likhotvorik, I.T.; Tippmann, E.; Platz, M.S. *Tetrahedron Lett.*, ***2001***, *42*, 3049.

EtO_2C

EtO_2C

=N—*p*-An

1. EtMgBr , THF , –78°C

2. air , aq KOH , EtOH

p-An—NHEt

91x84%

Niwa, Y.; Takayama, K.; Shimizu, M. *Tetrahedron Lett.*, ***2001***, *42*, 5473.

$PhNH_2$

1. $ZnEt_2$, $Cu(acac)_2$, DCM-hexane

2. 10% aq HCl

3. 10% aq NaOH

PhNHEt + $PhNEt_2$

(9 : 1) 95%

Brielles, C.; Harnett, J.J.; Doris, E. *Tetrahedron Lett.*, ***2001***, *42*, 8301.

MeO

MeO

N

(–)-sparteine , MeLi , 20 h

toluene , –76°C → –15°C

MeO

MeO

NH

Me

77% (32% ee)

Chrzanowska, M.; Sokolowska, J. *Tetrahedron Asymm.*, ***2001***, *12*, 1435.

H_2N–cyclopropyl → PhBr, cat $Pd_2(dba)_3$, DINAP / NaO*t*-Bu, toluene, 80°C, 1 d → PhHN–cyclopropyl 53%

Cai, W.; Loeppky, R.N. *Tetrahedron, 2001, 57*, 2953.

aniline (NH_2) → $(HOCH_2CH_2)_3N\cdot HCl$, 20 h, cat $RuH_2(PPh_3)_4$, 180°C / $SnCl_2\cdot 2\ H_2O$, aq dioxane → indole (N–H) 85%

Cho, C.S.; Kim, J.H.; Kim, T.-J.; Shim, S.C. *Tetrahedron, 2001, 57*, 3321.

Ph_2NH → BnCl, CsF–Celite, MeCN → PH_2NBn 70%

Hayat, S.; Rahman, A.-u.; Choudhary, M.I.; Khan, K.M.; Schumann, W.; Bayer, E. *Tetrahedron, 2001, 57*, 9951.

N–OBn, C_3H_7 → 2.5 (Ph, Ph, O, *i*-Pr, HN–BH) • BH_3 / THF → NH_2, C_3H_7 98% (68% ee)

Fontaine, E.; Namane, C.; Meneyrol, J.; Geslin, M.; Serva, L.; Roussey, E.; Tissandié, S.; Maftouh, M.; Roger, P. *Tetrahedron Asymm., 2001, 12*, 2185.

Ph–CH=N–Ph → (dihydropyran, O), LiBF₄, MeCN → (O, HN, Ph) 88% (85:15 *trarnscis*)

Yadav, J.S.; Reddy, B.V.S.; Madhuri, C.R.; Sabitha, G. *Synthesis, 2001*, 1065.

H–N (7-azabicyclo[2.2.1]heptane) → (2-bromopyridine: Br, N), $Pd_2(dba)_3\cdot$dppf / NaO*t*-Bu, dioxane, 110°C, 6 h → N-(2-pyridyl) (N, N) 66%

Cheng, J.; Trudell, M.L. *Org. Lett., 2001, 3*, 1371.

PhNH$_2$ —[PhB(OH)$_2$, 2,6-lutidine, air / 5% Cu(OAc)$_2$, 10% myristic acid]→ PhNHPh 79%

Antilla, J.C.; Buchwald, S.L. *Org. Lett.*, ***2001***, *3*, 2077.

Br N Ph —[SmBr$_2$–HMPA, rt / 15 min]→ Br NH Ph Me >99%

Knettle, B.W.; Flowers II, R.A. *Org. Lett.*, ***2001***, *3*, 2321.

H$_2$N —[cat L$_2$LnEL(TMS)$_2$ / C$_6$D$_6$, 125°C, 9 h]→ N H >95%

Ryu, J.-S.; Marks, T.J.; McDonald, F.E. *Org. Lett.*, ***2001***, *3*, 3091.

Ph N CONH$_2$ Ph H —[CH$_2$=CHCH$_2$ZnBr / THF, 0°C → rt]→ Ph HN CONH$_2$ Ph H 93% (>99:1 *RR:SS*)

van der Sluis, M.; Dalmolen, J.; de Lange, B.; Kaptein, B.; Kellogg, R.M.; Broxterman, Q.B. *Org. Lett.*, ***2001***, *3*, 3943.

H H N Ph —[hv / 6h]→ N Ph 70%

Campos, P.J.; Soldevilla, A.; Sampedro, D.; Rodríguez, M.A. *Org. Lett.*, ***2001***, *3*, 4087.

Ph$_2$NH + PhI —[10% Cu(neocup)(PPh$_3$)Br / *t*-BuOK, toluene, 110°C, 6 h]→ Ph$_2$N-Ph 78%

Gujadhur, R.K.; Bates, C.G.; Venkataraman, D. *Org. Lett.*, ***2001***, *3*, 4315.

CO$_2$*i*-Pr *i*-PrO$_2$C N —[1. Na, NH$_3$ 2. EtI / 3. aq NH$_4$Cl]→ CO$_2$*i*-Pr *i*-PrO$_2$C Et N H 98%

Donohoe, T.J.; McRiner, A.J.; Helliwell, M.; Sheldrake, P. *J. Chem. Soc., Perkin Trans. 1*, ***2001***, 1435.

Ph—N(NH$_2$)(Me) → (PhH, 5 eq $AlCl_3$, rt, 1 h) → Ph–C$_6$H$_4$–NHMe 92%

Ohwada, A.; Nara, S.; Sakamoto, T.; Kikugawa, Y.
J. Chem. Soc., Perkin Trans. 1, ***2001***, 3064.

2-(PhCH=N)phenol → (chiral Zr catalyst, toluene; CH$_2$=C(Me)CH$_2$SnBu$_3$, 0°C) → 2-(NH–CH(Ph)CH$_2$C(Me)=CH$_2$)phenol 74% (55% ee)

Gastner, T.; Ishitani, H.; Akiyama, R.; Kobayashi, S. *Angew. Chem. Int. Ed.*, ***2001***, *40*, 1896.

PhCH=N–Ph → (CH$_2$=CHCH$_2$Br, Yb, TMSCl; THF) → PhCH(NHPh)CH$_2$CH=CH$_2$ 75%

Su, W.; Li, J.; Zhang, Y. *Synth. Commun.*, ***2001***, *31*, 273.

PhCH=N–Ph → (2.5 eq In, NH_4Cl, EtOH; reflux, 10 h) → $PhCH_2NHPh$ 80%

Banik, B.K.; Hackfeld, L.; Becker, F.F. *Synth. Commun.*, ***2001***, *31*, 1581.

PhCH=N–Ph → (CH$_2$=CHCH$_2$Br, SmI_2; THF, 6 h, 0°C → rt) → PhCH(NHPh)CH$_2$CH=CH$_2$ 95%

Kim, B.H.; Han, R.; Park, R.J.; Bai, K.H.; Jun, Y.M.; Baik, W.
Synth. Commun., ***2001***, *31*, 2297.

1,2,3,4-Tetrahydroquinoline (N–H) → (decaborane, rt, 30 min; 37% HCHO in MeOH) → N-Me-1,2,3,4-tetrahydroquinoline 91%

Jung, Y.J.; Bae, J.W.; Yoon, C.-O.M.; Yoo, B.W.; Yoon, C.M.
Synth. Commun., ***2001***, *31*, 3417.

$PhNH_2$ → (Bu_4NBr, cat $Ru_3(CO)_{12}$, dioxane; $SnCl_2$•2 H_2O, 180°C, 20 h) → PhNHBu (+ 5% $PhNBu_2$) 68%

Cho, C.S.; Kim, J.S.; Kim, H.S.; Kim, T.-J.; Shim, S.C. *Synth. Commun.*, ***2001***, *31*, 3791.

Ph(C=N–Ph)CH3 → (2 eq EtMgCl, THF, 20°C; 10% Cp_2ZrCl_2, 1.5 h) → Ph–C(CH3)(Et)–NHPh 78%

Gandon, V.; Bertus, P.; Szymoniak, J. *Eur. J. Org. Chem.*, ***2001***, 3677.

Ph(C=N–Ph)CH3 → (H_2, ferrocenyl-binaphthyl complex of Ir; 40 h) → PhCH(NHPh)CH3 quant (84% ee)

Xiao, D.; Zhang, X. *Angew. Chem. Int. Ed.*, ***2001***, *40*, 3425.

PhCH=NPh → ($Et_3NH^+ HCO_2^-$, HCOOH; microwaves, 4 min) → $PhCH_2NHPh$ 90%

Moghaddam, F.M.; Khakshoor, O.; Ghaffarzadeh, M. *J. Chem. Res. (S)*, ***2001***, 525.

Imine (PhCH=N– of valinol) → (1. allyl bromide, DMF, rt, 1 h; 2. AcOEt, 1N HCl) → homoallylic amine (Ph, NH, OH) 95% (>99% de)

Yanada, R.; Kaieda, A.; Takemoto, Y. *J. Org. Chem.*, ***2001***, *66*, 7516.

$BnNH_2$ → (2.5 eq PhCHO, 4 eq HCOOH; microwaves, 5 min) → BnBHBn 59% (+ 40% Bn_3N)

Torchy, S.; Barbry, D. *J. Chem. Res. (S)*, ***2001***, 292.

N-benzylidene aniline (O_2N-aryl) → (allyl–Si(*i*-Pr)$_3$; $SnCl_4$, toluene, 0°C, MS 4Å) → tetrahydroquinoline (HN, Si(*i*-Pr)$_3$, O_2N) 91% (84:16 *cis:trans*)

Akiyama, T.; Suzuki, M.; Kagoshima, H. *Heterocycles*, ***2000***, *52*, 529.

$\text{cyclohexylidene}=N-C_3H_7$ —(1-benzotriazolyl-CH=$\overset{+}{N}Me_2$ $\bar{C}l$; THF, rt, 4 h)→ 4,5,6,7-tetrahydro-1H-indazole 66%

Katritzky, A.R.; Denisenko, A.; Denisenko, S.N.; Arend, M.
J. Heterocyclic Chem., ***2000***, *37*, 1309.

$Bn_2N-C_3H_7$ —(CAN, aq MeCN, rt)→ $BnNH-C_3H_7$ 69%

Bull, S.D.; Davies, S.G.; Fenton, G.; Mulvaney, A.W.; Prasad, R.S.; Smith, A.D.
J. Chem. Soc., Perkin Trans. 1, ***2000***, 3765.

$BuNH_2$ —(Ph-Bi(OAc)$_2$ dibenzobismole; $Cu(OAc)_2$, THF, 50°C, 14 h)→ BuNHPh 64%

Fedorov, A.Yu.; Finet, J.-P. *J. Chem. Soc., Perkin Trans. 1*, ***2000***, 3775.

1,2-diaminopropane —($C_5OH \cdot H_2O$; Ph(CH$_2$)$_3$Br; MS 4Å, DMF, 23°C, 28 h)→ $CH_3CH(NH_2)CH_2CH_2CH_2NH(CH_2)_3Ph$ 77%

Salvatore, R.N.; Schmidt, S.E.; Shin, S.I.; Nagle, A.S.; Worrell, J.H.; Jung, K.W.
Tetrahedron Lett., ***2000***, *41*, 9705.

N-Bn cyclohexene aziridine —($TMSN_3$, MeCN; rt, 2.5h)→ trans-2-azido-N-benzylcyclohexylamine ($NHBn$, N_3) 99%

Chandrasekhar, M.; Sekar, G.; Singh, V.K. *Tetrahedron Lett.*, ***2000***, *41*, 10079.

PhN=C(Ph)–C(Ph)=N–OH —(Ac_2O; reflux)→ 2,3-diphenylquinoxaline 57%

Xekou Koulotakis, N.P.; Hadjiantoniou-Maroulis, C.P.; Maroulis, A.J.
Tetrahedron Lett., ***2000***, *41*, 10299.

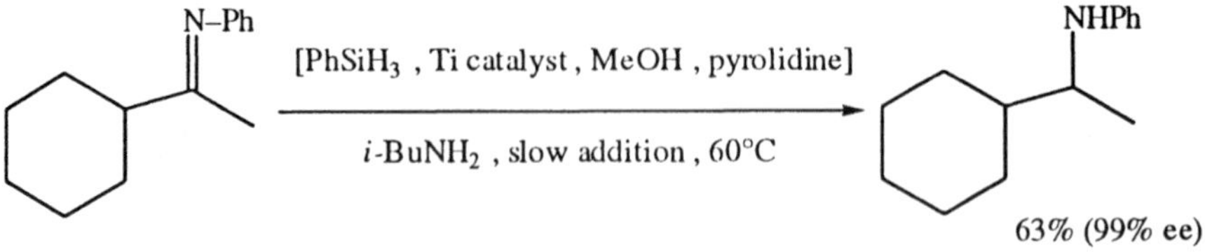

N–Ph

[$PhSiH_3$, Ti catalyst , MeOH , pyrolidine]

i-$BuNH_2$, slow addition , 60°C

NHPh

63% (99% ee)

Hansen, M.C.; Buchwald, S.L. *Org. Lett.*, ***2000***, *2*, 713.

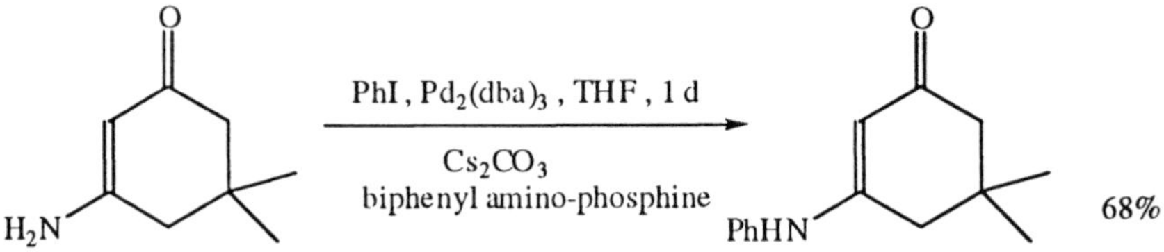

O

H_2N

PhI , $Pd_2(dba)_3$, THF , 1 d

Cs_2CO_3
biphenyl amino-phosphine

O

PhHN

68%

Edmondson, S.D.; Mastracchio, A.; Parmee, E.R. *Org. Lett.*, ***2000***, *2*, 1109.

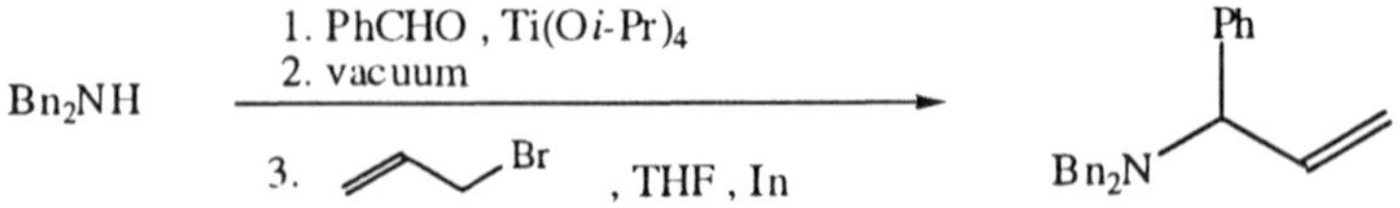

Bn_2NH

1. PhCHO , $Ti(Oi\text{-}Pr)_4$
2. vacuum

3. Br , THF , In

Ph

Bn_2N

Choucair, B.; Léon, H.; Miré, M.-A.; Lebreton, C.; Mosset, P. *Org. Lett.*, ***2000***, *2*, 1851.

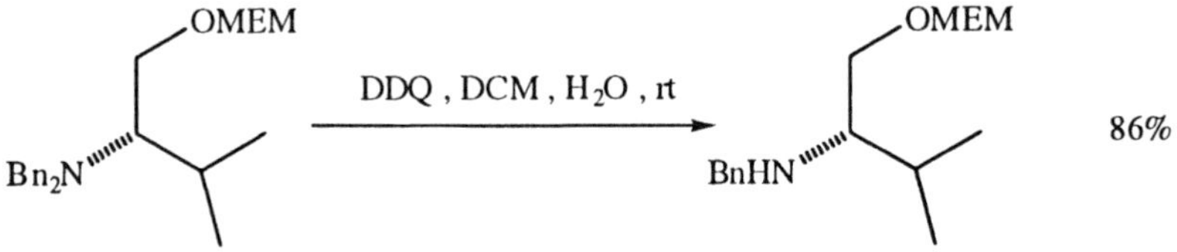

OMEM

Bn_2N

DDQ , DCM , H_2O , rt

OMEM

BnHN

86%

Hungerhoff, B.; Samanta, S.S.; Roels, J.; Metz, P. *Synlett*, ***2000***, 77.

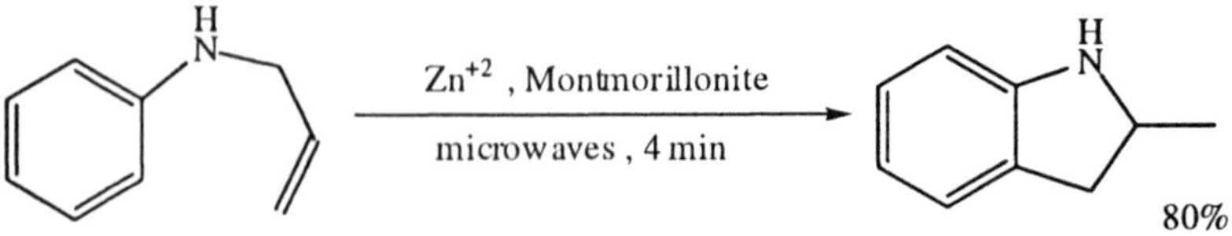

H
N

Zn^{+2} , Montmorillonite

microwaves , 4 min

H
N

80%

Yadav, J.S.; Reddy, B.V.S.; Rasheed, M.A.; Kumar, H.M.S. *Synlett*, ***2000***, 487.

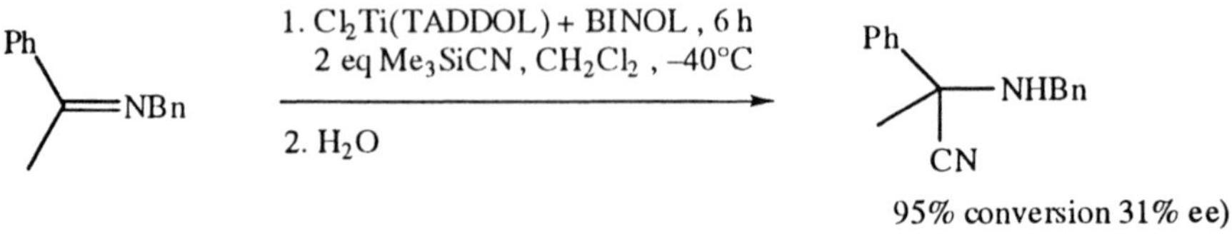

Ph

NBn

1. $Cl_2Ti(TADDOL)$ + BINOL , 6 h
2 eq Me_3SiCN , CH_2Cl_2 , –40°C

2. H_2O

Ph

NHBn

CN

95% conversion 31% ee)

Byrne, J.J.; Chavarot, M.; Chavant, P.-Y.; Vallée, Y. *Tetrahedron Lett.*, ***2000***, *41*, 873.

$Ph_2I^+ BF_4^-$, 10% CuI

2 eq Na_2CO_3, DCM, rt, 6 h

NH → N—Ph 65%

Kang, S.-K.; Lee, S.-H.; Lee, D. *Synlett, **2000***, 1022.

30% aq H_2O_2, acetone

0.1% $Mn_2O(OAc)_2$(TPTN)

Ph, Ph → O, Ph, Ph

Bull, S.D.; Davies, S.G.; Fenton, G.; Mulvaney, A.W.; Prasad, R.S.; Smith, A.D. *Chem. Commun.*, ***2000***, 337.

5% Cp_2ZrCl_2, THF, 6 h
2 eq BuMgCl, 22°C

OTs

92%

de Armas, J.; Kolis, S.P.; Hoveyda, A.H. *J. Am. Chem. Soc.*, ***2000**, 122*, 5977.

Br, N

1. 2.2 eq *t*-BuLi, –78°C
pentane–ether
2. (–)-sparteine

3. –40°C, 1.5 h
4. MeOH

Me, N

69% (93:7 er)

Bailey, W.F.; Mealy, M.J. *J. Am. Chem. Soc.*, ***2000**, 122*, 6787.

Br, N, Bn

1. *t*-BuLi, (–)-sparteine
toluene, –90°C

2. H_3O^+

Me, N, Bn

85% (87% ee)

Gil, G.S.; Groth, U.M. *J. Am. Chem. Soc.*, ***2000**, 122*, 6789.

EtO_2C, CO_2Et, N, H

nicotinium dichromate

MeCN, reflux, 15 min

EtO_2C, CO_2Et, N

97%

Sadeghi, M.M.; Mohammadpoor-Baltork, I.; Memarian, H.R.; Subhani, S. *Synth. Commun.*, ***2000**, 30*, 1661.

NH_2 , Br , toluene , Cl

2.1 NaO*t*-Bu
cat $Pd_2(dba)_3$, 80°C
cat aryl phosphine

N 84%

Harris, M.C.; Buchwald, S.L. *J. Org. Chem.*, ***2000***, *65*, 5327.

EtO_2C CO_2Et N H

$Mg(HSO_4)_2$, $NaNO_2$
DCM , rt , 10 min

EtO_2C CO_2Et N

75%

Zolfigol, M.A.; Kiany-Borazjani, M.; Sadeghi, M.M.; Mohammadpoor-Baltork, I.; Memarian, H.R. *Synth. Commun.*, ***2000***, *30*, 3919.

C_3H_7 N H X

NaH_2PO_3 , HCHO

C_3H_7 N Me X

X = CN 90%
X = CO_2Me 83%

Davis, B.A.; Durden, D.A. *Synth. Commun.*, ***2000***, *30*, 3353.

NH_2 NO_2

$Bu_4N HSO_4$, $(MeO)_2SO_2$
toluene , aq NaOH

NHMe NO_2 98%

Voskresensky, S.; Makosza, M. *Synth. Commun.*, ***2000***, *30*, 3523.

N–*p*-Tol Ph H

2 eq MeLi , toluene , –75°C
chiral bis-oxazoline ligand

NH–*p*-Tol Ph H

97% (73% ee)

Denmark, S.E.; Stiff, C.M. *J. Org. Chem.*, ***2000***, *65*, 5875.

Bu_2NH —Oxone , 80°C , SiO_2 , 8 h→ Bu_2N–OH 98%

Fields, J.D.; Kropp, P.J. *J. Org. Chem.*, ***2000***, *65*, 5937.

1. (allyl)$_3$B , 36°C , 3 h
2. aq NaOH

80%

Bubnov, Y.N.; Zhun, I.V.; Klimkina, E.V.; Igantenko, A.V.; Starikova, Z.A. *Eur. J. Org. Chem.*, ***2000***, 3323.

$PhNH_2$ → PhHN(allyl) 98%

allyl-OH , $Pd(OAc)_2$, 80°C
PPh_3 , $Ti(O\textit{i}\text{-}Pr)_4$, MS 4Å

Yang, S.-C.; Yu, C.-L.; Tsai, Y.-C. *Tetrahedron Lett.*, ***2000***, ***41***, 7097.

i-Pr
$Pb(OAc)_3$
NH_2
NH_2
i-Pr

88%

Saito, S.; Kano,T.; Ohyabu, Y.; Yamamoto, H. *Synlett.*, ***2000***, 1676.

Bu_3Sn
N
Bn

1. BuLi , hexane/ether/THF
-78°C → rt
2. allyl-Br

Bn—N

60%

Coldham,I.; Fernàndez, J.-C.; Snowden, D.J. *Tetrahedron Lett.*, ***1999***, ***40***, 1819.

$PhCH_2NH_2$ → $PhCH_2NHC_3H_7$

1-propanol , 300°C
γ-Al_2O_3

80% conversion
77% monoalkylation
5% dialkylation

Valot, F.; Fache, F.; Jacquot, R.; Spagnol, M.; Lemaire, M. *Tetrahedron Lett.*, ***1999***, ***40***, 3689.

$BuNH_2$ → Bu_2NH

$[Fe(CN)_5NO]^{-2}$, MeCN
N_2 , rt

Doctorovich, F.; Trápani, C. *Tetrahedron Lett.*, ***1999***, ***40***, 4635.

PhCH=N—Ph → $PhCH_2$NHPh 58%

Sm , I_2 , rt , 30 min

some imines gave dimers

Banik, B.K.; Zegrocka, O.; Banik, I.; Hackfeld, L.; Becker, F.F. *Tetrahedron Lett.*, ***1999***, ***40***, 6731.

Bn, N, Ph; allyl-$SiMe_3$; 5 eq $(\pi C_3H_5PdCl)_2$, TBAF, 31 h; hexane/THF, 8 h; NHBn, Ph; 69%

Nakamura, K.; Nakamura, H.; Yamamoto, Y. *J. Org. Chem.*, ***1999***, ***64***, 2614.

H, NH_2, Et, C_3H_7, H; 1. 2% $(Me_5Cp)_2SmCH(TMS)_2$ silica; 2. $Pd(OH)_2/C$, H_2; H, NH, Et, C_5H_{11}; 88%

Arredondo, V.M.; Tian, S.; McDonald, F.E.; Marks, T.J. *J. Am. Chem. Soc.*, ***1999***, ***121***, 3633.

Ph, N, Ph, Me; 0.15% chiral Ir catalyst, H_2; CH_2Cl_2, 40°C, 20 h; HN, Ph, Ph, Me; 99.7% (83% ee, *R*)

Kainz, S.; Brinkmann, A.; Leitner, W.; Pfaltz, A. *J. Am. Chem. Soc.*, ***1999***, ***121***, 6421.

NHBz, N, Ph, H; allyl-$SiCl_3$; DMF, rt, 1 h; NHBz, HN, Ph; 96%

Kobayashi, S.; Hirabayashi, R. *J. Am. Chem. Soc.*, ***1999***, ***121***, 6940.

$C_5H_{11}NH_2$; Ph_3SiH; $(Ph_2CH)PhN{-}Yb(hmpa)_4$; $C_5H_{11}NHSiPh_3$; 61%

Takaki, K.; Kamata, T.; Miura, Y.; Shishido, T.; Takehira, K. *J. Org. Chem.*, ***1999***, ***64***, 3891.

Ph, N, Ph; $TiCl_4$, NEt_3; Ph, N, Ph, Ph; 90%

Periasamy, M.; Srinivas, G.; Bharathi, P. *J. Org. Chem.*, ***1999***, ***64***, 4204.

Ph, N, Ph, H; allyl-$SiMe_3$, 5 h; 1% TBAF, THF, MS 4Å; NHPh, Ph; 93%

Wang, D.-K.; Zhou, Y.-G.; Tang, Y.; Hou, X.-L.; Dai, L.-X. *J. Org. Chem.*, ***1999***, ***64***, 4233.

1. $Pd(OAc)_2$, rac BINAP , 80°C
NaO*t*-Bu
toluene Br—C6H4—CF_3

2. $Pd(OAc)_2$,NaO*t*-Bu , 80°C
chiral phosphine
toluene 2-bromopyridine

46%

Harris, M.C.; Geis, O.; Buchwald, S.L. *J. Org. Chem.*, ***1999***, *64*, 6019.

TsCl , Py , DMAP

DCM , rt , 12 h

77%

Chandrasekhar, S.; Mohanty, P.K.; Harikishan, K.; Sasmal, P.K. *Org. Lett.*, ***1999***, *1*, 877.

$W(CO)_6$, $CO/I_2/K_2CO_3$

DCM

48%

McCusker, J.E.; Grasso, C.A.; Main, A.D.; McElwee-White, L. *Org. Lett.*, ***1999***, *1*, 961.

K_2CO_3 , aq EtOH , 100°C

99%

Tortolani, D.R.; Poss, M.A. *Org. Lett.*, ***1999***, *1*, 1261.

PhC≡CPh , Na_2CO_3 , 8 h

5% $Pd(OAc)_2$, LiCl
DMF , 100°C

85%

Roesch, K.R.; Larock, R.C. *Org. Lett.*, ***1999***, *1*, 1551.

$Ph_2CHNHC_3H_7$ → (1. DDQ, PhH, MS 4Å, 60°C, 1 h; 2. 0.1N HCl) → $C_3H_7NH_3^+\ Cl^-$ 71%

Sampson, P.B.; Honek, J.F. *Org. Lett.*, ***1999***, *1*, 1395.

(TMSN$_3$, acetone; 5% Cr complex) → NHCH$_2$Ar, N$_3$ quant (70% ee)

Ar = 2,4-dinitrophenyl

Li, Z.; Fernández, M.; Jacobsen, E.N. *Org. Lett.*, ***1999***, *1*, 1611.

Ph NH$_2$ → (BuBr, CsOH•H$_2$O, 21 h; DMF, MS 4Å, 23°C) → Ph NHBu 89%

Salvatore, R.N.; Nagle, A.S.; Schmidt, S.E.; Jung, K.W. *Org. Lett.*, ***1999***, *1*, 1893.

Ph N–Ph → (1.5 GeEt$_3$, AcOH; 0.5 BF$_3$•OEt$_2$, MeCN, 0°C) → NHPh, Ph 90%

Akiyama, T.; Iwai, J.; Onuma, Y.; Kagoshima, H. *Chem. Commun.*, ***1999***, 2191.

O, N, Bn → (N–Et; cat MgI$_2$, THF) → N, Et, O, N, Bn + N, Et, O, N, Bn (91 : 1) 55%

Alper, P.B.; Meyers, C.; Lerchner, A.; Siegel, D.R.; Carreira, E.M. *Angew. Chem. Int. Ed.*, ***1999***, *38*, 3186.

NH$_2$ → (1. N, N, N, OH, EtOH, 60°C; 2. AlCl$_3$, CHCl$_3$, 25°C; 3. NaBH$_4$, MeOH, 25°C) → N, Me 92x87x70%

Locher, C.; Peerzada, N. *J. Chem. Soc., Perkin Trans. 1*, ***1999***, 179.

PhHN–N=C(CH$_3$)(2-pyridyl) → ($ZnCl_2$, microwaves, Montmorillonite K10; 143°C, 3.5 min) → 2-(2-pyridyl)indole 60%

Lipińska, T.; Guibé-Jampel, E.; Petit, A.; Loupy, A. *Synth. Commun.*, ***1999***, *29*, 1349.

$PhNH_2$ → (2 eq EtBr, DMSO, K_2CO_3; rt, 6 h) → $PhNEt_2$ 95%

Srivastava, S.K.; Chauhan, P.M.S.; Bhaduri, A.P. *Synth. Commun.*, ***1999***, *29*, 2085.

Pyrrolidine (NH) → (BnBr, K_2CO_2–Al_2O_3; microwaves, 5 min) → N-Bn pyrrolidine 98%

Jasiinghani, H.G.; Khadilkar, B.M. *Synth. Commun.*, ***1999***, *29*, 3693.

Ph–N=CH–Ph → (—(Si(Me)H—O—)$_n$—; $ZnCl_2$, ether) → Ph–NH–CH$_2$–Ph 67%

Chandrasekhar, S.; Reddy, M.V.; Chandraiah, L. *Synth. Commun*, ***1999***, *29*, 3981.

2-methyl cyclic imine (azepine) → (1. LDA, THF, –30°C; 2. MeI) → 2-ethyl cyclic imine 44%

Brimble, M.A.; Gorsuch, S. *Aust. J. Chem.*, ***1999***, *52*, 965.

$H_2N(CH_2)_3$–NH–$(CH_2)_9$–CH=CH$_2$ → (cat [Rh]/BiPhePhos, PhH; H_2/CO, 80°C, 20 h) → bicyclic NH/N macrocycle 80%

Bergmann, D.J.; Campi, E.M.; Jackson, W.R.; Patti, A.F. *Aust. J. Chem.*, ***1999***, *52*, 1131.

C_3H_7–CH=CH–CH$_2$CH$_2$–CH(NHMe)–C_7H_{15} → (1. NCS, toluene, rt, 30 min; 2. Bu_3SnH, AIBN, toluene, reflux, 1 h; 3. 10% aq KF) → N-Me 2-Bu-5-C_7H_{15} pyrrolidine 59%

Senboku, H.; Hasegawa, H.; Orito, K.; Tokuda, M. *Heterocycles*, ***1999***, *50*, 333.

REVIEWS:

"Amines Via Nucleophilic 1, 2-Addition To Ketimines. Construction Of Nitrogen-Substituted Quaternary Carbon Atoms," Steinig, A.G.; Spero, D.M. *Org. Prep. Proceed.. Int.*, ***2000***, *32*, 205.

"Catalytic Enantioselective Addition To Imines," Kobayashi, Sh.; Ishitani, H. *Chem. Rev.*, ***1999***, *99*, 1069.

SECTION 98: AMINES FROM ESTERS

Ph OMs → 1. LiH_3BNMe_2, THF, 25°C; 2. 3M HCl, MeOH → Ph NMe_2 92%

Thomas, S.; Huynh, T.; Enriquez-Rios, V.; Singaram, B. *Org. Lett.*, ***2001***, *3*, 3915.

Ph Ph OCO_2Et → $(C_3H_7)_2NH$, 5% Ru catalyst; CH_2Cl_2, 20°C, 10 h → Ph Ph $N(C_3H_7)_2$ 89% (74% ee)

Matsushima, Y.; Onitsuka, K.; Kondo, T.; Mitsudo, T.-a.; Takahashi, S. *J. Am. Chem. Soc.*, ***2001***, *12*, 10405.

OAc Ph Ph → cat $Pd_2(dba)_3$, $BnNH_2$; *S*-BINAP, THF, 2 d → NHBn Ph Ph 79% (39% ee, *Sx*)

Kodama, H.; Taiji, T.; Ohta, T.; Furukawa, I. *Synlett*, ***2001***, 385.

AcO Ph Ph → $C_6H_{13}NHTs$, rt; cat $TiCl_4$, 5 h → $C_6H_{13}HN$ Ph Ph 41%

Mahrwald, R.; Quint, S. *Tetrahedron Lett.*, ***2001***, *42*, 1655.

OAc → $(C_3H_7)_2NH$, H_2O, 25°C; cat $[Pd(C_3H_5)Cl]_2$, 20 h; tetraphosphine ligand → $N(C_3H_7)_2$

Feuerstein, M.; Laurenti, D.; Doucet, H.; Santelli, M. *Tetrahedron Lett.*, ***2001***, *42*, 2313.

SECTION 99: AMINES FROM ETHERS, EPOXIDES AND THIOETHERS

NO ADDITIONAL EXAMPLES

SECTION 100: AMINES FROM HALIDES AND SULFONATES

Ph—I → (Et_2NH, $Pd(OAc)_2$, phosphine ligand; NaO*t*-Bu, toluene, 80°C) → Ph—NEt_2 85%

Ali, M.H.; Buchwald, S.L. *J. Org. Chem.*, ***2001***, *66*, 2560.

4-*t*-Bu-bromobenzene → (PhNHMe, 1.5 eq NaO*t*-Bu; 1% $Pd(OAc)_2$, toluene, 80°C; polymer-bound phosphine ligand) → 4-*t*-Bu-C6H4-N(Ph)Me 93%

Parrish, C.A.; Buchwald, S.L. *J. Org. Chem.*, ***2001***, *66*, 3820.

4-chlorotoluene → (4% 1,3-diaryl-imidazolium, PhNHMe; 1% $Pd_2(dba)_3$, *t*-BuOK; dioxane, 100°C, 3 h) → 4-Me-C6H4-N(Me)Ph 98%

Ar = 2,6-diisopropylphenyl

Grasa. G.A.; Viciu, M.S.; Huang, J.; Nolan, S.P. *J. Org. Chem.*, ***2001***, *66*, 7729.

3-bromothiophene → ($C_6H_{13}NH_2$, $Pd(OAc)_2$; $P(t\text{-}Bu)_3$, NaO*t*-Bu, xylene) → 3-(NHC_6H_{13})thiophene 37% + N,N-di(3-thienyl)-NHC_6H_{13} 20%

Ogawa. L.; Radke, K.R.; Rothstein, S.D.; Rasmussen, S.C. *J. Org. Chem.*, ***2001***, *66*, 9067.

Ph—Br → (1. TMS_2NLi, 2% $Pd(dba)_2$, P*t*-Bu_3; rt, 16 h; 2. HCl) → Ph—NH_2 92%

Lee, S.; Jørgensen, M.; Hartwig, J.F. *Org. Lett.*, ***2001***, *3*, 2729.

1. Ph_3SiNH_2, toluene, .5% $Pd_2(dba)_3$, 1.3 LiHMDS, 1.2% (2-Ph-C$_6$H$_4$-PCy$_3$)
2. H_3O^+

OMe, Cl → NH_2, OMe 98%

Huang, X.; Buchwald, S.L. *Org. Lett.*, ***2001***, *3*, 3417.

piperidine, THF, NaH/*t*-AmONa, 10% Ni(0)/bipy, reflux, 15 min

Cl, Cl → N, Cl + N, N (75 : 25) 68%

Desmarets, C.; Schneider, R.; Fort, Y. *Tetrahedron Lett.*, ***2001***, *42*, 247.

Br, N—OMe, BuO, O

e^-, cobaloxime, MeOH, Et_4NOTs, NaOH, rt

NHOMe, BuO, O 61% (5.6:1)

Inokuchi, T.; Kawafuchi, H. *Synlett*, ***2001***, 421.

Br, N, N, AcHN

Cu, NH_3, ethylene glycol

NH_2, N, N, AcHN 86%

Lang, F.; Zewge, D.; Houpis, I.N.; Volante, R.P. *Tetrahedron Lett.*, ***2001***, *42*, 3251.

PhCl

pyrrolidine, NaH, *t*-BuOH, 3 h, 2% $Ni(acac)_2$, 8% dihydroimidazoline carbene ligand

Ph—N 96%

Gradel, B.; Brenner, E.; Schneider, R.; Fort, Y. *Tetrahedron Lett.*, ***2001***, *42*, 5689.

F–C$_6$H$_4$–CHO → (pyrrolidine , K_2CO_3 / DMSO , ultrasound, 15 min) → 4-pyrrolidinyl–C$_6$H$_4$–CHO 92%

Magdolen, P.; Mečiarová, M.; Toma, Š. *Tetrahedron*, ***2001***, *57*, 4781.

Cl, Cl, N; H_2N–C$_6$H$_4$–NO_2 → (Pd(OAc)$_2$, BINAP , 18 h / K_2CO_3 , toluene , reflux) → Cl, N, N–H, NO_2 80%

Junckers, T.H.M.; Maes, B.U.W.; Lemière, G.L.F.; Dommisse, R. *Tetrahedron*, ***2001***, *57*, 7027.

NHBoc, ZnI → (allyl Br , cat CuBr-SMe_2 / DMF , –10°C → rt , 14 h) → NHBoc 51%

Hunter, C.; Jackson, R.F.W.; Rami, H.K. *J. Chem. Soc., Perkin Trans. 1*, ***2001***, 1349.

4-Me–C$_6$H$_4$–NH_2 → (2-OMe–C$_6$H$_4$–Br , toluene , 3 h / cat $PdCl_2$, bisphosphine ligand, NaO*t*-Bu , 80°C) → H N, OMe 97%

Zhang, X.-X.; Harris, M.C.; Sadighi, KJ.P.; Buchwald, S.L. *Can. J. Chem.*, ***2001***, *79*, 1799.

OMe, Br → (PhNHMe , 0.5% $Pd_2(dba)_3$ / BINAP , NaO*t*-Bu , 80°C, toluene , 29 h) → OMe, NMePh 75%

Wolfe, J.P.; Buchwald, S.L. *J. Org. Chem.*, ***2000***, *65*, 1144.

I → (2.5% $Pd_2(dba)_3$, indole, toluene , *t*-BuONa , 100°C / biphenyl amino-phosphine) → N 90%

Old, D.W.; Harris, M.C.; Buchwald, S.L. *Org. Lett.*, ***2000***, *2*, 1403.

HN O , cat Ni(II)/C

NC–C6H4–Cl → NC–C6H4–N(morpholine) 88%

dppf , LiO*t*-Bu , toluene
BuLi , 2.5 h

Lipshutz, B.H.; Ueda, H. *Angew. Chem. Int. Ed.*, ***2000***, *39*, 4492.

1. NaN_3 , MeCN
2. PPh_3
3. PhCHO

$PhCH_2Br$ → $PhCH_2N{=}CHPh$ 95%

Vaněk, P.; Klán, P. *Synth. Commun.*, ***2000***, *30*, 1503.

NH

Cl–C6H4–CHO → N–C6H4–CHO 60%

basic Al_2O_3 , microwaves

Kidwai, M.; Sapra, P.; Dave, B. *Synth. Commun.*, ***2000***, *30*, 4479.

OTf → PPh_2 51%

PPh_3 , 10% $Pd(OAc)_2$
DMF , 110°C , 2h

Kwong, F.Y.; Lai, C.W.; Tian, Y.; Chan, K.S. *Tetrahedron Lett.*, ***2000***, *41*, 10285.

Cl → NBu_2 86%

Bu_2NH , $Pd(dba)_2$, DME
t-BuONa , rt , 20 h

Stauffer, S.R.; Lee, S.; Stambuli, J.P.; Hauck, S.I.; Hartwig, J.F. *Org. Lett.*, ***2000***, *2*, 1423.

Ph O → Ph N 99%

$\overset{+}{N}H$ $\overset{-}{C}lO_4$, Bu_3SnH
DMF , rt , 4 h

Suwa, T.; Sugiyama, E.; Shibata, I.; Baba, A. *Synlett*, ***2000***, 556.

Cl → NMePh 99%

PhNHMe , *t*-BuOK
dioxane , $Pd_2(dba)_3$

Ar–N⊕N–Ar Cl^- Ar = 2,6-di-*t*-Bu-C_6H_3

Huang, J.; Grasa, G.; Nolan, S.P. *Org. Lett.*, ***1999***, *1*, 1307.

HN O , NaO*t*-Bu , 105°C

6% , 2% Pd(dba)$_2$

P(*c*-C_6H_{11})$_2$

Cl N O

92%

Bei, X.; Guram, A.S.; Turner, H.W.; Weinberg, W.H. *Tetrahedron Lett.*, ***1999***, *40*, 1237.

OMe

NH_2 , toluene , 80°C , 30 min

Br NH OMe

Pd(OAc)$_2$,

O

PPh_2 PPh_2

92%

Guari, Y.; van Es, D.S.; Reek, J.N.H.; Kamer, P.C.J.; van Leeuven, P.W.N.M. *Tetrahedron Lett.*, ***1999***, *40*, 3789.

Br

P(*o*-Tol)$_3$, 2 eq NaO*t*-Bu
2% Pd(dba)$_2$, dioxane

3 HN O

reflux

Br

(10 : 2) 76%

Beletskaya, I.P.; Bessmertnykh, A.G.; Guilard, R. *Tetrahedron Lett.*, ***1999***, *40*, 6393.

N—Li

2 I

THF , –40°C

N

81%

Tripathy, S.; Le Blanc, R.; Durst, T. *Org. Lett.*, ***1999***, *1*, 1973.

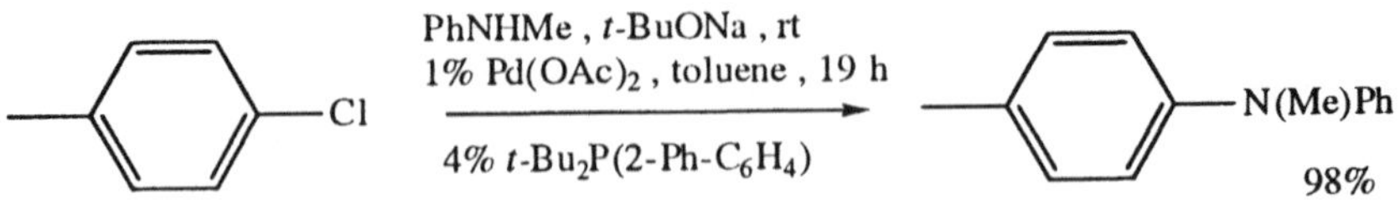

PhNHMe , *t*-BuONa , rt
1% $Pd(OAc)_2$, toluene , 19 h
4% *t*-$Bu_2P(2\text{-}Ph\text{-}C_6H_4)$

Cl → N(Me)Ph

98%

Wolfe, J.P.; Buchwald, S.L. *Angew. Chem. Int. Ed.*, ***1999***, *38*, 2413.

SECTION 101: AMINES FROM HYDRIDES

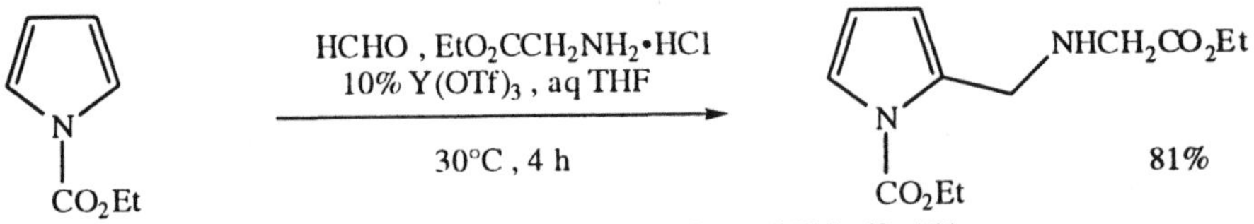

HCHO , $EtO_2CCH_2NH_2$•HCl
10% $Y(OTf)_3$, aq THF
30°C , 4 h

N, CO_2Et → $NHCH_2CO_2Et$, N, CO_2Et

81%

Zhang, C.; Dong, J.; Cheng, T.; Li, R. *Tetrahedron Lett.*, ***2001***, *42*, 461.

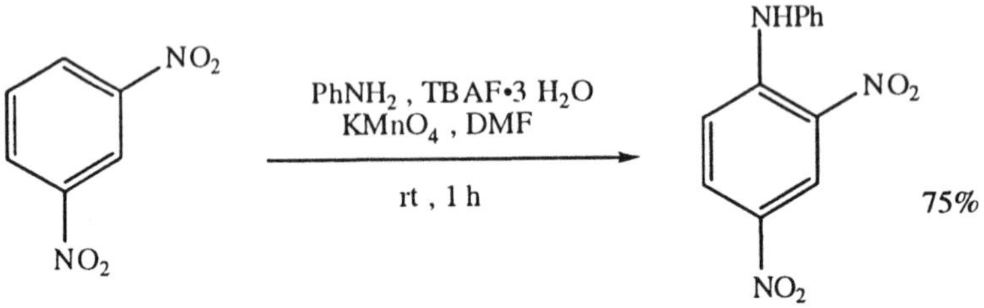

NO_2, NO_2

$PhNH_2$, TBAF•3 H_2O
$KMnO_4$, DMF
rt , 1 h

NHPh, NO_2, NO_2

75%

Huertas, I.; Gallardo, I.; Marguet, J. *Tetrahedron Lett.*, ***2001***, *42*, 3439.

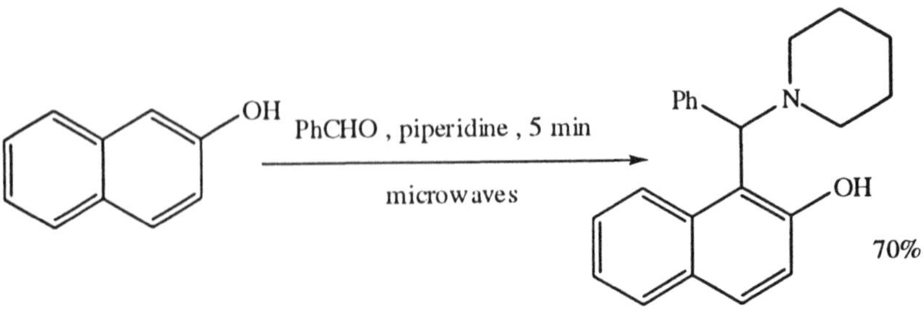

OH

PhCHO , piperidine , 5 min
microwaves

Ph, N, OH

70%

Sharifi, A.; Mirzaei, M.; Naimi-Jamal, M. *Monat. Chem.*, ***2001***, *132*, 875.

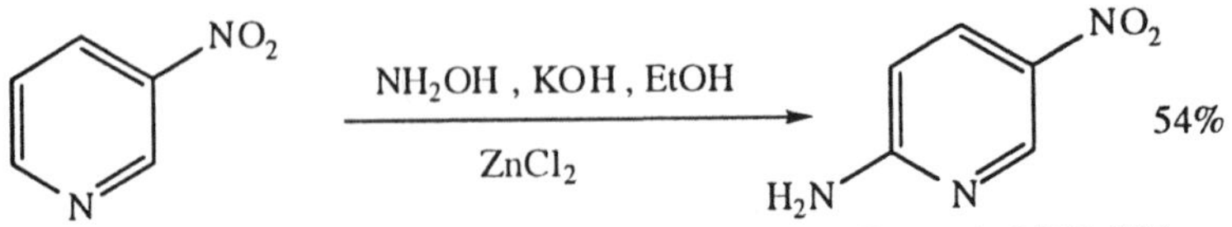

NO_2, N

NH_2OH , KOH , EtOH
$ZnCl_2$

NO_2, H_2N, N

54%

Bakke, J.M.; Svensen, H.; Trevisan, R. *J. Chem. Soc., Perkin Trans. 1*, ***2001***, 376.

O_2N — TBAF•3 H_2O, $PhNH_2$ / hν, DMF, rt, 3 h — NO_2 → O_2N, NHPh, NO_2 54%

Juertas, I.; Gallardo, I.; Marquet, J. *Tetrahedron Lett.*, ***2000***, *41*, 279.

O — NH_2 / 1% $Pd_2(dba)_3$, 150°C / 1% BINAP, 12 h → O, H N 90%

Hicks, F.A.; Brookhart, M. *Org. Lett.*, ***2000***, *2*, 219.

O, Ph, Ph, O — $MeONH_2$, NaOMe / MeOH, 4 h → O, Ph, NH_2, Ph, O 83%

Seko, S.; Miyake, K. *Synth. Commun.*, ***1999***, *29*, 2487.

SECTION 102: AMINES FROM KETONES

O, Ph, N — 1. *t*-$BuSONH_2$, $Ti(OEt)_4$ / 2. MeMgBr / 3. HCl/MeOH → Me, NH_2, Ph, N quant (60:40 -:+)

Shaw, A.W.; de Solms, S.J. *Tetraahedron Lett.*, ***2001***, *42*, 7173.

O — 1. Ph_2NNH_2, PhH, MS 4Å, 6 h / 2. TFAA, 0°C, 3 h / 3. heat, 3 h → N, Ph 81%

Miyata, O.; Kimura, Y.; Naito, T. *Synthesis*, ***2001***, 1635.

1. $Me_2N-C_6H_4-NO_2$ DMSO, KF/Al_2O_3 2. MeI

Me_2N MeO 60%

Moskalev, N.; Mąkosza, M. *Chem. Commun.*, ***2001***, 1248.

$PhNH_2$, Montmorillonite KSF

10 h

N Ph 96%

Samajdar, S.; Becker, F.F.; Banik, B.K. *Heterocycles*, ***2001***, *55*, 1019.

aq NH_4Cl, 250°C

2 d

N C_5H_{11} 74%

Kotsuki, H.; Mehta, B.K.; Yanigisawa, K. *Synlett*, ***2001***, 1323.

Tol-NH_2, $B_{10}H_{14}$, MeOH, 1 h

NHTol 97%

Bae, J.W.; Lee, S.H.; Cho, Y.J.; Yoon, C.M. *J. Chem. Soc., Perkin Trans. 1*, ***2000***, 145.

$BnNH_2$, microwaves

30 sec

N Bn 90%

Danks, T.N. *Tetrahedron Lett.*, ***1999***, *40*, 3957.

NNH_2 Ph Ph

1. 0.1% $Pd(OAc)_2$, 1.4 NaO*t*-Bu 0.1% Xantphos, toluene, 80°C

2. cyclohexanone, EtOH $TsOH•OH_2$, reflux

Ph N H 92%

Wagaw, S.; Yang, B.H.; Buchwald, S.L. *J. Am. Chem. Soc.*, ***1999***, *121*, 10251.

1. NH_4Cl, $Ti(Oi\text{-}Pr)_4$, EtOH, NEt_3, rt, 10 h
2. $NaBH_4$, rt, 7 h

83%

Bhattacharyya, S.; Neidigh, K.A.; Avery, M.A.; Williamson, J.S. *Synlett*, ***1999***, 1781.

$NH_2NHMe{\bullet}HCl$, EtOH, reflux

(3 : 1) 70%

Martins, M.A.P.; Freitag, R.A.; da Rosa, A.; Flores, A.F.C.; Zanatta, N.; Benacorse, H.G. *J. Heterocylic Chem.*, ***1999***, *36*, 217.

Related Methods: Section 94 (Amines from Aldehydes)

SECTION 103: AMINES FROM NITRILES

PhCN → $PhCH_2NH_2$ 75%

$LiBH_3{\bullet}NMe_2$, THF, 65°C, 12 h

Thomas, S.; Collins, C.J.; Cuzens, J.R.; Spiciarich, D.; Goralski, C.T.; Singaram, B. *J. Org. Chem.*, ***2001***, *66*, 1999.

5% Ni/SO_2, 10 atm H_2, 2M NH_3/MeOH, 120°C, 6 h

97%

Takamizawa, S.; Wakasa, N.; Fuchikami, T. *Synlett*, ***2001***, 1623.

PhCN → PhCH-NNHTs 96%

Ni(R), NaH_2PO_2, $TsNHNH_2$, AcOH, H_2O, Py, rt

Tóth, M.; Sumsák, L. *Tetrahedron Lett.*, ***2001***, *42*, 2723.

sponge Ni, H_2, 28% aq NH_3, EtOH, rt, 5 h

86%

Tanaka, K.; Nagasawa, M.; Kasuga, Y.; Sakamura, H.; Takuma, Y.; Iwatani, K. *Tetrahedron Lett.*, ***1999***, *40*, 5885.

Ph–CH$_2$–CN → Ph–CH$_2$–C(cyclopropyl)–NH$_2$ 70%

1. EtMgBr , Ti(O*i*-Pr)$_4$, ether
2. BF_3•OEt_2

Bertus, P.; Szymoniak, J. *Chem. Commun.*, ***2001***, 1792.

SECTION 104: AMINES FROM ALKENES

cyclohexyl-NH_2 , CO/H_2
cat [Rh$^+$(cod)(η^6-PhBPh$_3$)$^-$]

Ph 78% HN Ph + HN Ph 22%

Lin, Y.-S.; El Ali, B.; Alper, H. *Tetrahedron Lett.*. ***2001***, *42*, 2423.

cat diphosphino-cyclobutene Pd complex
toluene , $PhNH_2$, 5 h

Ph NHPh 92%

Minami, T.; Okamoto, H.; Ikeda, S.; Tanaka, R.; Ozawa, F.; Yoshifuji, M. *Angew. Chem. Int. Ed.*, ***2001***, *40*, 4501.

piperidine , cat BuLi , THF
–78°C → rt

Ph N 91%

Hartung, C.G.; Breindl, C.; Tillack, A.; Beller, M. *Tetrahedron*, ***2000***, *56*, 5157.

OEt → Et_2N OEt 75%

Et_2NH , CO/H_2
[Rh(cod)Cl]$_2$

Rische, T.; Bärfacker, L.; Eilbracht, P. *Eur. J. Org. Chem.*, ***1999***, 653.

SECTION 105: AMINES FROM MISCELLANEOUS COMPOUNDS

$Ph_3P{=}O$ → Ph_3P 97%

1. MeOTf , DME , rt , 2 h
2. $LiAlH_4$, 0°C , 3 h

Imamoto, T.; Kikuchi, S.-i.; Miura, T.; Wada, Y. *Org. Lett.*, ***2001***, *3*, 87.

HCO_2NH_4 , 10% Pd-C

MeOH , rt , 14 h

NH•HCl 95%

Zacharie, B.; Moreau, N.; Dockendorff, C. *J. Org. Chem.*, ***2001***, *66*, 5264.

NH_2 Br

1. acetophenone , PhH
2. Bu_3SnH , AIBN , PhH , 80°C

N Ph 86%

Johnston, J.N.; Plotkin, M.A.; Viswanathan, R.; Prabhakaran, E.N. *Org. Lett.*, ***2001***, *3*, 1009.

CHO N NH_2

MeO–P(O)(OMe) O Bu

NaOH , MeOH

N N Bu

95%

Hsiao, Y.; Rivera, R.; Yasuda, N.; Hughes, D.L.; Reider, P.J. *Org. Lett.*, ***2001***, *3*, 1101.

O OPh P O OPh Ph NO_2 N H

Ph_3SnH , AIBN , PhH

heat

Ph N

85%

Crich, D.; Ranganathan, K.; Huang, X. *Org. Lett.*, ***2001***, *3*, 1917.

Ph PhHC═N+ O−

1. $AlCl_3$•6 H_2O , KI , aq MeCN

2. $Na_2S_2O_3$, H_2O , rt

Ph PhHC═N 90%

Boruah, M.; Konwar, D. *Synlett*, ***2001***, 795.

N_3

2 eq PMe_3 , toluene , 10 eq MeI

CH_2Cl_2 , 25°C , 1.5 h

NHMe 76%

Kato, H.; Ohmori, K.; Suzuki, K. *Synlett*, ***2001***, 1003.

1. MeO–B(pinacol)
2. NaOH , H_2O_2
3. MeLi
4. NH_2OSO_3H

high % ee

Hoffmann, R.W.; Hölzer, B.; Knopff, O. *Org. Lett.*, ***2001**, 3,* 1945.

PhNHNH$_2$ → PhNHBoc 85%

$(t\text{-BuO})_2C{=}O$, 10% Pd-C , 3 h

EtOH , 3 PHMS

Chandrasekhar, S.; Reddy, Ch.R.; Rao, R.J. *Synlett*, ***2001***, 1561.

BuLi , THF , –78°C

46% (98% ee) + 13% of 2 other isomers

Pearson, W.H.; Stevens, E.P.; Aponick, A. *Tetrahedron Lett.*, ***2001**, 42*, 7361.

Ph_2SiH_2 , 1% $RuCl_2(PPh_3)$

AgOTf, DME , rt , 25 h

62% (83% ee)

Takei, I.; Nishibayashi, Y.; Ishii, Y.; Mizobe, Y.; Uemura, S.; Hidai, M. *Chem. Commun.*, ***2001***, 2360.

2.5 eq $BF_3{\cdot}OEt_2$, O_2
MS 4Å , CD_3CN , 1 h

0°C → 70°C

73% (+ 23 PhCHO)

Shimizu, M.; Makino, H. *Tetrahedron Lett.*, ***2001**, 42*, 8865.

Ph, +N–O⁻ =CH(Ph) ; allyl bromide, Sm, TBAB; DMF, H_2O, 1.5 h → Ph–N(OH)–CH(Ph)CH$_2$CH=CH$_2$ 85%

Laskar, D.D.; Prajapati, D.; Sandhu, J.S. *Tetrahedron Lett.*, ***2001***, *42*, 7883.

EtO_2C–CO–C(=PPh$_3$)–CO_2Et ; 1. NCS; 2. heat → EtO_2C-azirine(Cl)(CO_2Et) 90x98%

Pinhoe Melo, T.M.V.S.; Lopes, C.S.J.; Cardoso, A.L.; Gonsalves, A.M.d'A.R. *Tetrahedron*, ***2001***, *57*, 6203.

$PhNO_2$; 4 eq In, HCl, aq THF, 30 min → $PhNH_2$ 84%

Lee, J.G.; Choi, K.I.; Koh, H.Y.; Kim, Y.; Kang, Y.; Cho, Y.S. *Synthesis*, ***2001***, 81.

Pyridine N-oxide ; LiCl, $NaBH_4$, THF, 0°C; rt → pyridine 98%

Ram, S.R.; Chary, K.P.; Iyengar, D.S. *Synth.Commun.*, ***2000***, *30*, 3511.

$PhCH_2N_3$; $ZrCl_4$, $NaBH_4$, THF → $PhCH_2NH_2$ 95%

Chary, K.P.; Ram, S.R.; Salhuddin, S.; Iyengar, D.S. *Synth. Commun.*, ***2000***, *30*, 3559.

$PhCH_2N_3$; $NaBH_4$, LiCl, THF; 0°C, rt → $PhCH_2NH_2$ 95%

Ram, S.R.; Chary, K.P.; Iyengar, D.S. *Synth. Commun.*, ***2000***, *30*, 4495.

Cl–C_6H_4–N_3 ; $FeSO_4$•7 H_2O, NH_3, rt; MeOH, 2.5 h → Cl–C_6H_4–NH_2 98%

Kamal, A.; Laxman, E.; Arifuddin, M. *Tetrahedron Lett.*, ***2000***, *41*, 7743.

Ph, +N–O⁻ =CH(Ph) ; allyl–InBr; DMF, H_2O, rt, 3 h → Ph–N(OH)–CH(Ph)CH$_2$CH=CH$_2$ 90%

Kumar, H.M.S.; Anjaneyulu, S.; Reddy, E.J.; Yadav, J.S. *Tetrahedron Lett.*, ***2000***, *41*, 9311.

N-OH ... OH; H_2 / cat 48% HBr → NH_2 ... OH + NH_2 ... OH

(23 : 1) 94%

Davies, I.W.; Taylor, M.; Marcoux, J.-F.; Matty, L.; Wu, J.; Hughes, D.; Reider, P.J. *Tetrahedron Lett.*, ***2000***, *41*, 8021.

$PhCH_2N_3$ → 5% [bis(benzenedithiolato)Sn], $NaBH_4$, aq THF, 15°C, 30 min → $PhCH_2NH_2$ 97%

Bosch, I.; Costa, A.M.; Martín, M.; Urpí, F.; Vilarrasa, J. *Org. Lett.*, ***2000***, *2*, 397.

Pyridine N-oxide → In, aq NH_4Cl, MeOH / heat, 4 h → pyridine 93%

Yadav, J.S.; Reddy, B.V.S.; Reddy, M.M. *Tetrahedron Lett.*, ***2000***, *41*, 2663.

4-Methylpyridine N-oxide → cat [Re(O)Me$_2$ dithiolate], PPh_3, H_2O/PhH, 5% TBAB → 4-methylpyridine quant

Wang, Y.; Espenson, J.H. *Org. Lett.*, ***2000***, *2*, 3525.

Ph-CH(N_3)-CH_2OH → $NaBH_4$, $CoCl_2$•6 H_2O / H_2O, 10 min → Ph-CH(NH_2)-CH_2OH 97%

Fringuelli, F.; Pizzo, F.; Vaccaro, L. *Synthesis*, ***2000***, 646.

PhN_3 → $FeCl_3$–Zn, EtOH, rt, 4 h → $PhNH_2$ 98%

Pathak, D.; Laskar, D.D.; Prajapati, D.; Sandhu, J.S. *Chem. Lett.*, ***2000***, 816.

$PhCH_2N_3$ → NH_2NMe_2, $FeCl_3$•6 H_2O / MeCN, reflux, 8 h → $PhCH$=N–NMe_2 quant

Barrett, I.C.; Langille, J.D.; Kerr, M.A. *J. Org. Chem.*, ***2000***, *65*, 6268.

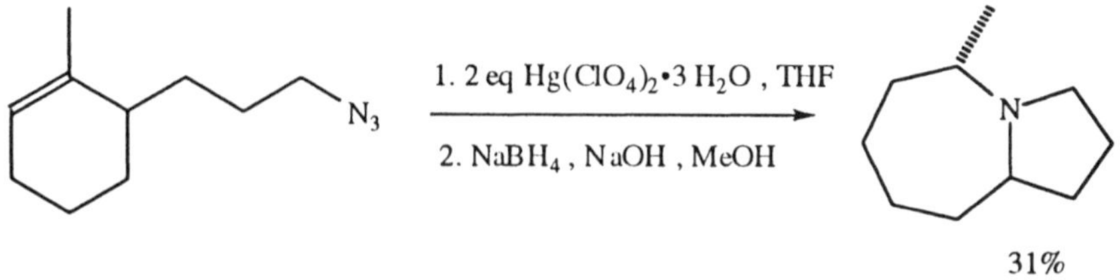

1. 2 eq $Hg(ClO_4)_2 \cdot 3\ H_2O$, THF
2. $NaBH_4$, NaOH, MeOH

N_3

31%

Pearson, W.H.; Hutta, D.A.; Fang, W.-K. *J. Org. Chem.*, ***2000***, *65*, 8326.

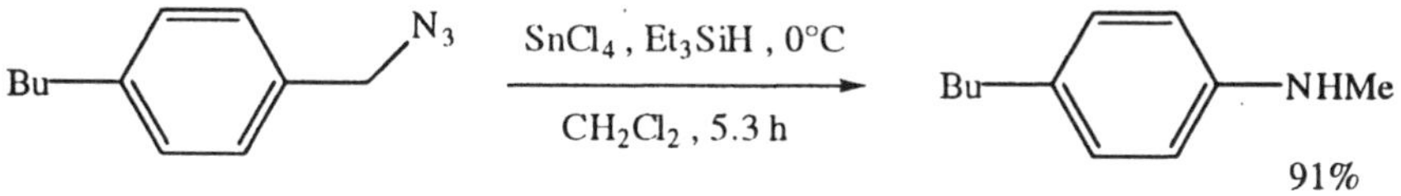

Bu ... N_3 — $SnCl_4$, Et_3SiH, 0°C / CH_2Cl_2, 5.3 h → Bu ... NHMe

91%

Lopez, F.J.; Nitzan, D. *Tetrahedron Lett.*, ***1999***, *40*, 2071.

$PhCH_2N_3$ — In, NH_4Cl, EtOH, reflux, 1 h → $PhCH_2NH_2$ 98%

Reddy, G.V.; Rao, G.V.; Iyengar, D.S. *Tetrahedron Lett.*, ***1999***, *40*, 3937.

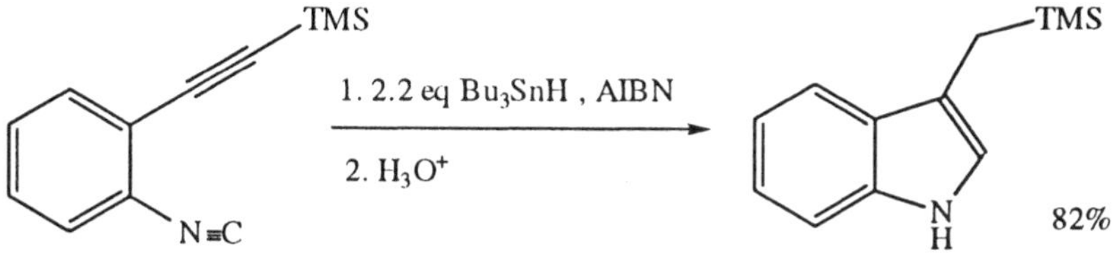

TMS, N≡C — 1. 2.2 eq Bu_3SnH, AIBN; 2. H_3O^+ → TMS, N H 82%

Rainer, J.D.; Kennedy, A.R.; Chase, E. *Tetrahedron Lett.*, ***1999***, *40*, 6325.

PhN_3 — EtMgBr, ether, 30 min → PhNHEt 90%

Kumar, H.M.S.; Reddy, B.V.S.; Anjaneyulu, S.; Yadav, J.S. *Tetrahedron Lett.*, ***1999***, *40*, 8305.

PhN_3 — Br, Zn, DMF / rt, 40 min → PhHN 82%

Kumar, H.M.S.; Anjaneyulu, S.; Reddy, B.V.S.; Yadav, J.S. *Synlett*, ***1999***, 551.

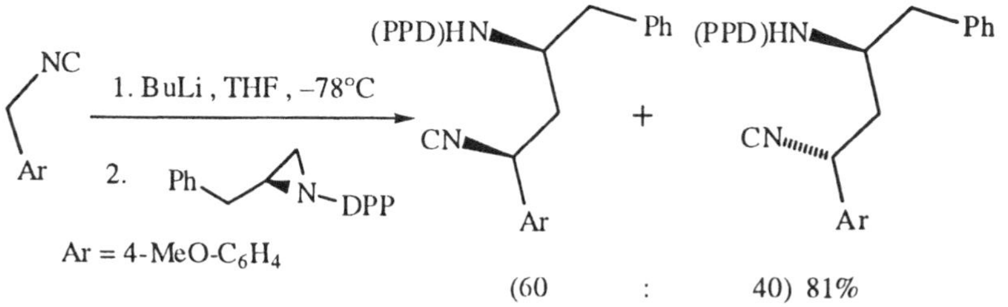

NC, Ar — 1. BuLi, THF, –78°C; 2. Ph, N–DPP → (PPD)HN, Ph, CN, Ar + (PPD)HN, Ph, CN, Ar

$Ar = 4\text{-}MeO\text{-}C_6H_4$

(60 : 40) 81%

Kaiser, A.; Balbi, M. *Tetrahedron Asymm.*, ***1999***, *10*, 1001.

MCM–silylamine–Pd(II) complex, MeOH, rt: cyclohexyl–N_3 → cyclohexyl–NH_2 92%

Kantam, M.L.; Chowdari, N.S.; Rahman, S.; Choudary, B.M. *Synlett*, ***1999***, 1413.

AMINES FROM NITRO COMPOUNDS

$PhNO_2$ → $PhNH_2$ (porphyrinato-iron / $NaBH_4$, $HOCH_2CH_2Br$, rt) 95%

Wilkinson, H.S.; Tanoury, G.J.; Wald, S.A.; Senanayake, C.H. *Tetrahedron Lett.*, ***2001***, *42*, 167.

$PhNO_2$ → $PhNH_2$ ($Me_3N–BH_3$, $Pd(OH)_2/C$, MeOH, rt, 6 h) 90%

Couturier, M.; Tucker, J.L.; Andresen, B.M.; Dubé, P.; Brenek, S.J.; Negri, J.T. *Tetrahedron Lett.*, ***2001***, *42*, 2285.

$PhNO_2$ → $PhNH_2$ (N_2H_4•H_2•H_2O, Al_2O_3, $FeCl_3$•6 H_2O, microwaves, 5+2 min, 108°C) 89%

Vass, A.; Dudás, J.; Tóth, J.; Varma, R.S. *Tetrahedron Lett.*, ***2001***, *42*, 5347.

O_2N–C_6H_4–CN → H_2N–C_6H_4–CN (57% HI, 90,C, 4 h) 85%

Kumar, J.S.D.; Ho, M.M.; Toyokuni, T. *Tetrahedron Lett.*, ***2001***, *42*, 5601.

$PhNO_2$ → $PhNH_2$ (FeS–NH_4Cl, aq MeOH, reflux, 4 h) 81%

Desai, D.G.; Swami, S.S.; Dabhade, S.K.; Ghagare, M.G. *Synth. Commun.*, ***2001***, *31*, 1249.

PhCHO → PhN(allyl)$_2$ + PhN(allyl)O-allyl (allyl bromide, In, aq MeOH, 12 h) (16 : 1) 51%

Kang, K.H.; Choi, K.I.; Koh, H.Y.; Kim, Y.; Chung, B.Y.; Choi, Y.S. *Synth. Commun.*, ***2001***, *31*, 2277.

$$PhNO_2 \xrightarrow[\text{MeOH, rt, 3 h}]{\text{2 eq Sm, 5\% 1,5-dioctyl-4,4'-bipyridinium dibromide}} PhNH_2 \quad 83\%$$

Yu, C.; Liu, B.; Hu, L. *J. Org. Chem.*, ***2001***, *66*, 919.

$$O_2N\text{-}C_6H_4\text{-}CO_2H \xrightarrow[\text{MeOH, reflux, <30 min}]{B_{10}H_{14}\text{, Pd-C, cat AcOH}} H_2N\text{-}C_6H_4\text{-}CO_2H \quad 97\%$$

Bae, J.W.; Cho, Y.J.; Lee, S.H.; Yoon, C.M. *Tetrahedron Lett.*, ***2000***, *41*, 175.

$$Ph\text{-}NO_2 \xrightarrow[\text{MeOH, 20 h}]{\text{allyl bromide, 4 eq Sn}} Ph\text{-}NBn_2 \quad 72\%$$

Bieber, L.W.; da Costa, R.C.; da Silva, M.F. *Tetrahedron Lett.*, ***2000***, *41*, 4827.

$$PhNO_2 \xrightarrow{ZrCl_4\text{, NaBH}_4\text{, THF, 6 h}} PhNH_2 \quad 92\%$$

Chary, K.P.; Ram, S.R.; Iyengar, D.S. *Synlett*, ***2000***, 683.

$$PhNO_2 \xrightarrow[\text{microwaves}]{NaH_2PO_2/FeSO_4\bullet 7\ H_2O\text{, 50 sec}} PhNH_2 \quad 78\%$$

Meshram, H.M.; Ganesh, Y.S.S.; Sekhar, K.C.; Yadav, J.S. *Synlett*, ***2000***, 993.

$$\text{1-nitronaphthalene} \xrightarrow[\text{MeOH, ultrasound, 10 min}]{\text{4 eq Sm, 20 eq NH}_4\text{Cl}} \text{1-naphthylamine} \quad 88\%$$

Basu, M.K.; Becker, F.F.; Banik, B.K. *Tetrahedron Lett.*, ***2000***, *41*, 5603.

$$PhNO_2 \xrightarrow[\text{2.5 h}]{Cu(BH_2S_3)_2\text{, THF, reflux}} PhNH_2 \quad 90\%$$

also from ArN_3

Firouzabadi, H.; Tamami, B.; Kiasat, A.R. *Synth. Commun.*, ***2000***, *30*, 587.

$$PhNO_2 \xrightarrow[\text{aq NaOH, Na}_2\text{CO}_3\text{, 313°K, 13 h}]{Ni(NO_3)\bullet 6\ H_2O\text{, Al(NO}_3)\bullet 9\ H_2O} PhN_2 \quad 95\%$$

Jyothi, T.M.; Raja, T.; Talawar, M.B.; Sreekumar, K.; Sugunan, S.; Rao, B.S. *Synth. Commun.*, ***2000***, *30*, 1573.

$$EtNO_2 \xrightarrow[\text{10 min}]{\text{HCOOH, Ni(R), MeOH}} EtNH_2 \quad 50\%$$

Gowda, D.C.; Gowda, A.S.P.; Baba, A.R.; Gowda, S. *Synth. Commun.*, ***2000***, *30*, 2889.

$$PhNO_2 \xrightarrow{HCO_2NH_4\text{ , MeOH , rt , 1 h}} PhNH_2 \quad 90\%$$

Gowda, D.C.; Mahesh, B. *Synth. Commun.*, ***2000***, *30*, 3639.

$$PhNO_2 \xrightarrow[\text{5 h}]{\text{In , }NH_4Cl\text{ , aq EtOH}} PhNH_2 \quad 90\%$$

Banik, B.K.; Suhendra, M.; Banik, I.; Becker, F.F. *Synth. Commun.*, ***2000***, *30*, 3745.

5 atm CO, DMF; 0.1 Se, 10 NEt_3; 100°C, 5 h

NO_2 → NH 65%

Nishiyama, Y.; Maema, R.; Ohno, K.; Hirose, M.; Sonoda, N. *Tetrahedron Lett.*, ***1999***, *40*, 5717.

$$PhNO_2 \xrightarrow[\text{4 h}]{\text{Sm , trace }I_2\text{ , aq THF-}NH_4Cl} PhNH_2 \;(56\%) + PhNHNHPh \;(20\%)$$

Wang, L.; Zhou, L.; Zhang, Y. *Synlett*, ***1999***, 1065.

$$O_2N\text{–}C_6H_4\text{–}CO_2H \xrightarrow[\text{THF , 8 eq TMSCl}]{0.2\ CrCl_2\text{ , 16 eq Mn powder}} H_2N\text{–}C_6H_4\text{–}CO_2H \quad 83\%$$

Hari, A.; Miller, B.L. *Angew. Chem. Int. Ed.*, ***1999***, *38*, 2777.

$$PhNO_2 \xrightarrow[\text{100°C , 30 min}]{FeCl_3\text{–Zn–DMF–}H_2O} PhNH_2 \quad 92\%$$

Desai, D.G.; Swami, S.S.; Hapase, S.B. *Synth. Commun.*, ***1999***, *29*, 1033.

REVIEWS:

"[3+2] Synthesis of Pyrrolizidine and Indolizidine Alkaloids," Broggini, G.; Zecchi, G. *Synthesis*, ***1999***, 905.

"Synthesis of Secondary Amines," Salvatore, R.N.; Yoon, C.H.; Jung, K.W. *Tetrahedron*, ***2001***, *57*, 7785.

"A Comparison of Imine-Forming Methodologies," Love, B.E.; Boston, T.S.; Nguyen, B.T.; Rorer, J.R. *Org. Prep. Proceed. Int.*, ***1999***, *31*, 399.

SECTION 105A: PROTECTION OF AMINES

Grubbs' catalyst
toluene , reflux
5 h
77%

Alcaide, B.; Almendros, P.; Alonso, J.M.; Aly, M.F. *Org. Lett.*, ***2001***, *3*, 3781.

CO_2H
$NHCOCF_3$
1. BnOH , DCC , DMAP , MeCN
2. $NaBH_4$, EtOH
CO_2Bn
$NHCOCF_3$
85%

Barker, D.; McLeod, M.D.; Brimble, M.A.; Savage, G.P. *Tetrahedron Lett.*, ***2001***, *42*, 1785.

MeO NHCbz
i-PrOH , 4 eq HCO_2NH_4
10% Pd/C
microwaves (1 min x 3 cycles)
MeO NH_2
95%

Daga, M.C.; Taddei, M.; Varchi, G. *Tetrahedron Lett.*, ***2001***, *42*, 5191.

O
PhHN O
10%$[Ni(bipy)_3](BF_4)_2$
e^- , Zn anode , DMF
$PhNH_2$ 78%

Franco, D.; Duñach, E. *Tetrahedron Lett.*, ***2000***, *41*, 7333.

N
SO_2NMe_2
e^- , DMF
N
H
76% + HSO_2NMe_2

Largeron, M.; Farrell, B.; Rousseau, J.-F.; Fleury, M.-B.; Potier, P.; Dodd, R.H. *Tetrahedron Lett.*, ***2000***, *41*, 9403.

NH_2 — CF_3CO_2Et, 0.1 DMAP, 20 h, 85°C, THF → N(H)C(O)CF_3 78%

Prashad, M.; Hu, B.; Har, D.; Repič, O.; Blacklock, T.J. *Tetrahedron Lett.*, ***2000***, *41*, 9957.

AcO, NBn_2 — O, MeO, OMe; hν, aq MeCN → Bn_2NH 60% + OH, MeO, OMe

Cossy, J.; Rakotoarisoa, H. *Tetrahedron lett.*, ***2000***, *41*, 2097.

BnO, N–Bn — , 10% Pd/C; EtOH, 22°C → BnO, N–H 72%

Bajwa, J.S.; Slade, J.; Repič, O.; Blacklock, T. *Tetrahedron Lett.*, ***2000***, *41*, 6021.

Ph_2N — $TiCl_3$•Li–THF, THF, 25°C → Ph_2NH 75%

Rele, S.; Talukdar, S.; Banerji, A. *Tetrahedron Lett.*, ***1999***, *40*, 767.

NH_2 — $Ph_2Si(CH_2CH_2OTs)_2$, NEt_3, DMF → N, $SiPh_2$ 90%; TBAF, THF

Kim, B.M.; Cho, J.H. *Tetrahedron Lett.*, ***1999***, *40*, 5333.

Ph, NHBoc — $AlCl_3$, CH_2Cl_2, 2 h → Ph, NH_2 88%

Bose, D.S.; Lakshminarayana, V. *Synthesis*, ***1999***, 66.

Ph, NH_2 — $BnO_2CN(SO_2CF_3)(4\text{-}CF_3\text{-}C_6H_4)$; THF, rt, 0.1 h → Ph, NHCbz 95%

Yasuhara, T.; Nagaoka, Y.; Tomioka, K. *J. Chem. Soc., Perkin Trans. 1*, ***1999***, 2233.

NH_2

1. CO_2 , NEt_3
CH_2Cl_2 , –78°C

2. TIPS-OTf , –78°C

OH

NH O(Si*i*-Pr)$_3$

O

87%

OH

TBAF , THF , 0°C (88%)

Lipshutz, B.H.; Papa, P.; Keith, J.M. *J. Org. Chem.*, ***1999***, *64*, 3792.

CHAPTER 8
PREPARATION OF ESTERS

SECTION 106: ESTERS FROM ALKYNES

Ph—≡—SMe → (TsOH, silica, DCM; 40°C, 2 h) → $PhCH_2C(O)SMe$ 86%

Braga, A.L.; Martins, T.L.C.; Silveira, C.C.; Rodrigues, O.E.D. *Tetrahedron*, ***2001***, *57*, 3297.

Bu—≡ → (3 eq CO, H_2O, PdI_2/KI, 80°C; dioxane, 15 h) → 3-butyl-2(5H)-furanone 65%

Gabriele, B.; Salerno, G.; Costa, M.; Chiusoli, G.P. *Tetrahedron Lett.*, ***1999***, *40*, 989.

Bu—≡ → (10 atm CO, CO_2, 200 eq H_2O; PdI_2, 10 eq KI, dioxane, 15 h) → 3-butyl-2(5H)-furanone 77%

Gabriele, B.; Salerno, G.; Costa, M.; Chiusoli, G.P. *Chem. Commun.*, ***1999***, 1381.

SECTION 107: ESTERS FROM ACID DERIVATIVES

The following types of reactions are found in this section:

1. Esters from the reaction of alcohols with carboxylic acids, acid halides and anhydrides.
2. Lactones from hydroxy acids
3. Esters from carboxylic acids and halides, sulfoxides and miscellaneous compounds

$PhCH_2CO_2H$ → (1. BCl_3; 2. MeOH) → $PhCH_2CO_2Me$ 94%

Dyke, C.A.; Bryson, T.A. *Tetrahedron Lett.*, ***2001***, *42*, 3959.

$Br(CH_2)_5CO_2H$ → (CBr_4, MeOH, hν, rt, 1 d; selective for non-conjugated acids) → $Br(CH_2)_5CO_2Me$ 98%

Lee, A.S.-Y.; Yang, H.-C.; Su, F.-Y. *Tetrahedron Lett.*, ***2001***, *42*, 301.

CHO…CO_2H → (3 eq 2-chloro-N-methylpyridinium iodide (I^-), MeCN; 4 eq NEt_3, 25°C) → bicyclic β-lactone 55%

Coretz, G.S.; Tennyson, R.L.; Romo, D. *J. Am. Chem. Soc.*, ***2001***, *123*, 7945.

p-Tolyl–SO_3H → (CH_2Cl_2; polymer-bound triazene, Cl, N=N–NHBn) → p-Tolyl–SO_2Bn 84%

Vignola, N.; Dahmen, S.; Enders, D.; Bräse, S. *Tetrahedron Lett.*, ***2001***, *42*, 7833.

$Ph(CH_2)_2CO_2H$ → (BnBr, 5 eq $KF{\bullet}2H_2O$, 3 h; 5% Bu_4NHSO_4, THF, rt) → $Ph(CH_2)_2CO_2Bn$ 99%

Ooi, T.; Sugimoto, H.; Doda, K.; Maruoka, K. *Tetrahedron Lett.*, ***2001***, *42*, 9245.

$PhCO_2H$ → (*t*-BuOH, Al_2O_3, Zn dust, DCM; sulfur sublimate, rt) → $PhCO_2t$-Bu 90%

Karmakar, D.; Das, P.J. *Synth. Commun.*, ***2001***, *31*, 535.

$PhCO_2H$ → (EtOH, 2% $Fe(SO_4)_3{\bullet}x\ H_2O$; cat H_2SO_4) → $PhCO_2Et$ 97%

Xu, Q.-h.; Liu, W.-y.; Chen, B.-h.; Ma, Y.-x. *Synth. Commun.*, ***2001***, *31*, 2113.

$PhCO_2H$ → (1. MeOH, 65°C, H_2O, 1 h, microwaves; 2. H_2SO_4) → $PhCO_2Me$ 84%

Zhang, Z.; Zhou, L.; Zhang, M.; Wu, H.; Chen, Z. *Synth. Commun.*, ***2001***, *31*, 2435.

2-Pyridyl–CO_2H → (MsCl, Py, 0°C, 40 min) → 2-Pyridyl–CO_2Me 71%

Siddiqui, B.S.; Begum, F.; Begum, S. *Tetrahedron Lett.*, ***2001***, *42*, 9059.

CO_2H —— toluene, 100°C / cat [Ru{PhCH=C(Ph)C≡CPh)(PMe*i*-Pr_2}] ——→ 97%

Jiménez-Tenorio, M.; Puerta, M.C.; Valerga, P.; Moreno-Dorado, F.J.; Guerra, F.M.; Massanet, G.M. *Chem. Commun.*, ***2001***, 2324.

CO_2H / Br —— KOH, DMSO, MeI / rt, 2 h ——→ CO_2Me / Br 94%

Avila-Zárraga, J.G.; Mariínez, R. *Synth. Commun.*, ***2001***, ***31***, 2177.

$EtCO_2H$ —— BuOH, $NaHSO_4•H_2O$ / reflux, 30 min ——→ $EtCO_2Bu$ 96%

Li, Y.-Q. *Synth. Commun.*, ***1999***, *29*, 3901.

PhSSPh —— Ac_2O, Zn/$AlCl_3$, DMF / 65°C, 4 h ——→ PhSAc 80%

Movassagh, B.; La Kouraj, M.M.; Fadaei, Z. *J. Chem. Res. (S)*, ***2001***, 22.

PhCO_2H —— $C_8H_{17}OH$ / $Me_3N•HCl$, NEt_3, 0°C / MeCN, 3 h ——→ Ph$CO_2C_8H_{17}$ 93%

Wakasugi, K.; Nakamura, A.; Tanabe, Y. *Tetrahedron Lett.*, ***2001***, ***42***, 7427.

PhCO_2H —— $C_8H_{17}OH$, toluene, 80°C, 4 h / diphenylammonium triflate ——→ Ph$CO_2C_8H_{17}$ 93%

Wakasugi, K.; Misaki, T.; Yamada, K.; Tanabe, Y. *Tetrahedron Lett.* ***2000***, ***41***, 5249.

$MeCO_2H$ —— BnOH, act. C, microwaves / 12 sec ——→ $MeCO_2Bn$ 83%

Sagar, A.D.; Shinde, N.A.; Bandgar, B.P. *Org. Prep. Proceed. Int.*, ***2000***, ***32***, 287.

Ph(C=O)OH —— BnSH, 4 eq P_4H_{10}, toluene / 110°C, 4 h ——→ Ph(C=S)SBn 96%

Sudalai, A.; Kanagasabapathy, S.; Benicewicz, B.C. *Org. Lett.* ***2000***, ***2***, 3213.

Reaction: 2-(carboxymethyl)benzoic acid (CO_2H, CO_2H) → $NaHSO_4$–SiO_2, MeOH, 45 h → 2-(methoxycarbonylmethyl)benzoic acid (CO_2H, CO_2Me) 96%

Das, B.; Venkataiah, B.; Madhusudhan, P. *Synlett*, ***2000***, 59.

Reaction: long-chain CO_2H → 1. $SOCl_2$, Amberlyst A21, DCM; 2. MeOH → long-chain CO_2Me 89%

Girard, C.; Tranchant, I.; Nioré, P.-A.; Herscovici, J. *Synlett*, ***2000***, 1577.

$$PhCH_2CO_2H \xrightarrow[\text{rt, 6 h}]{NaHCO_4\text{–}SiO_2\text{, EtOAc}} PhCH_2CO_2Et \quad 95\%$$

Das, B.; Venkataiah, B. *Synthesis*, ***2000***, 1671.

Reaction: $Bu_2C{=}CHCO_2H$ → $Br^+(coll)_2PF_6$, CH_2Cl_2 → 3-bromo-4,4-dibutyloxetan-2-one (Bu, Bu, Br, O) 60%

Homsi, F.; Rousseau, G. *J. Org. Chem.*, ***1999***, *64*, 81.

$$PhCO_2H \xrightarrow{\text{EtOH, 2 eq Py, DCM, rt}} PhCO_2Et \quad 95\%$$

Reagent: 2-chloro-1,3-dimethylimidazolinium chloride (Cl, Me–N, N-Me, Cl^-)

Isobe, T.; Ishikawa, T *J. Org. Chem.*, ***1999***, *64*, 6984.

$$PhCO_2H \xrightarrow[\text{2. } Me_2SO_4]{\text{1. LiOH, } H_2O} PhCO_2Me \quad 96\%$$

Chakraborti, A.K.; Basak, A.; Grover, V. *J. Org. Chem.*, ***1999***, *64*, 8014.

Reaction: PhCH(OH)CO_2H → MeOH, $Fe_2(SO_4)_3{\cdot}xH_2O$, reflux, 3.5 h → PhCH(OH)CO_2Me 83%

Zhang, G.-S.; Gong, H. *Synth. Commun.*, ***1999***, *29*, 1547.

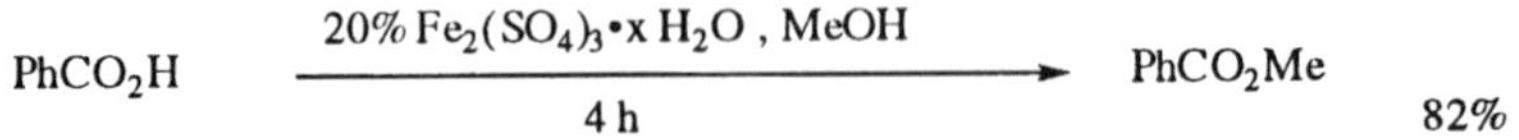

$PhCO_2H$ —— 20% $Fe_2(SO_4)_3 \cdot x\, H_2O$, MeOH; 4 h ——> $PhCO_2Me$ 82%

Zhang, G.-S. *Synth. Commun.*, ***1999***, *29*, 607.

Further examples of the reaction $RCO_2H + R'OH \rightarrow RCO_2R'$ are included in Section 108 (Esters from Alcohols and Phenols) and in Section 30A (Protection of Carboxylic Acids).

SECTION 108: ESTERS FROM ALCOHOLS AND THIOLS

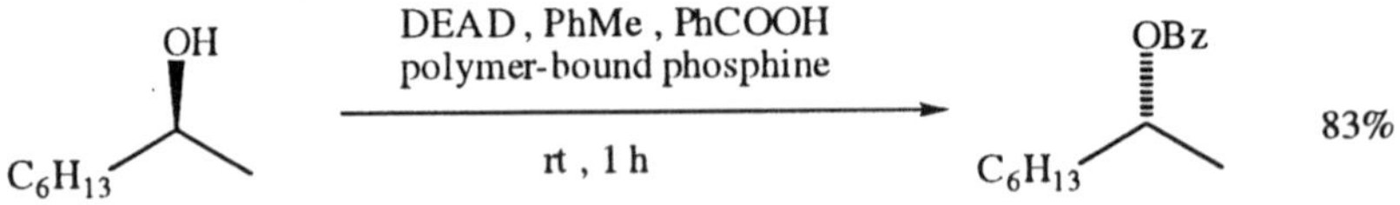

OH, C_6H_{13} —— DEAD, PhMe, PhCOOH, polymer-bound phosphine; rt, 1 h ——> OBz, C_6H_{13} 83%

Charette, A.B.; Janes, M.K.; Boezio, A.A. *J. Org. Chem.*, ***2001***, *66*, 2178.

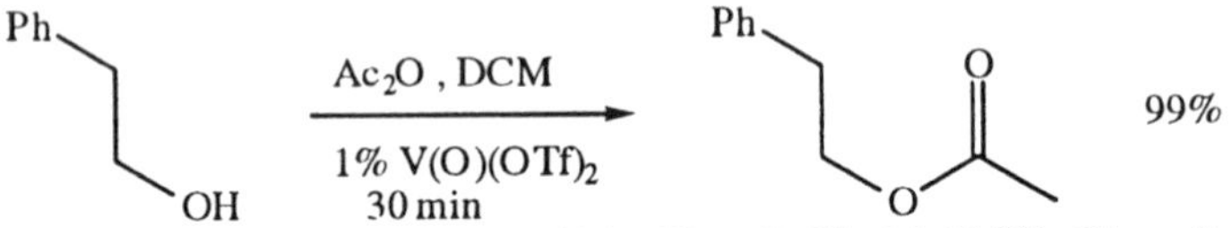

Ph, OH —— Ac_2O, DCM; 1% $V(O)(OTf)_2$, 30 min ——> Ph, O, O 99%

Chen, C.-T.; Kuo, J.-H.; Li, C.-H.; Barhate, N.B.; Hon, S.-W.; Li, T.-W.; Chao, S.-D.; Liu, C.-C.; Li, Y.-C.; Chang, I.-H.; Lin. J.-S.; Liu, C.-J.; Chou, Y.-C. *Org. Lett.*, ***2001***, *3*, 3729.

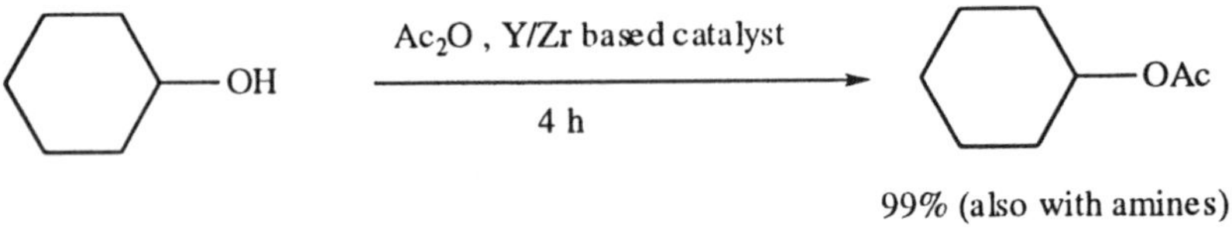

OH —— Ac_2O, Y/Zr based catalyst; 4 h ——> OAc

99% (also with amines)

Kumar, P.; Pandey, R.K.; Bodas, M.S.; Dongare, M.K. *Synlett*, ***2001***, 206.

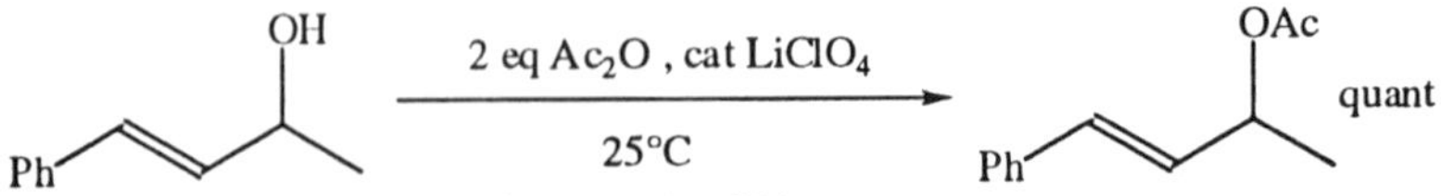

OH, Ph —— 2 eq Ac_2O, cat $LiClO_4$; 25°C ——> OAc, Ph quant

Nakae, Y.; Kusaki, I.; Sato, T. *Synlett*, ***2001***, 1584.

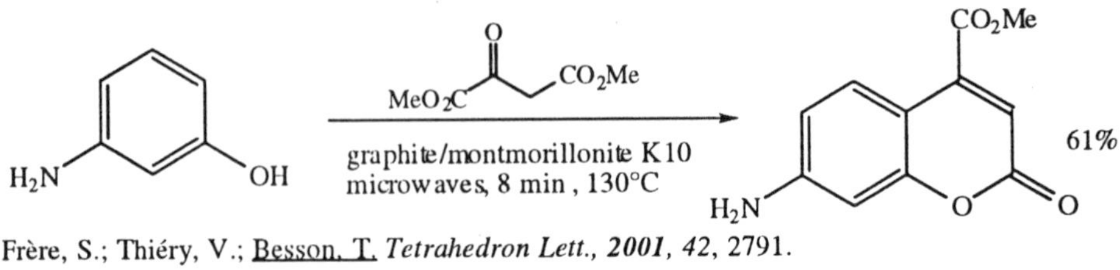

H_2N, OH —— MeO_2C, O, CO_2Me; graphite/montmorillonite K10, microwaves, 8 min, 130°C ——> CO_2Me, H_2N, O, O 61%

Frère, S.; Thiéry, V.; Besson, T. *Tetrahedron Lett.*, ***2001***, *42*, 2791.

$PhCH_2OH$ —[Ac_2O, cat NBS, CH_2Cl_2; rt, 4 h]→ $PhCH_2OAc$ 94%

Karimi, B.; Seradj, H. *Synlett*, ***2001***, 519.

MeO–CH₂CH₂–OH —[Ag_2O, TsCl, KI, 20°C]→ MeO–CH₂CH₂–OTs 60%

Bouzide, A.; Le Berre, N.; Sauvé, G. *Tetrahedron Lett.*, ***2001***, *42*, 8781.

Resorcinol —[[bmim]Cl•2 $AlCl_3$, 30°C; EtOAc, 10 min]→ 7-hydroxy-4-methylcoumarin 95%

Potdar, M.K.; Mohile, S.S.; Salunkhe, M.M. *Tetrahedron Lett.*, ***2001***, *42*, 9285.

α-Terpineol —[HO_2C–CH₂CH₂–CO_2Me; DCC, DMAP, DCM; 1.5 GPa, rt, 1 d]→ terpinyl ester (O–C(=O)–CH₂CH₂–CO_2Me) 92%

Shimizu, T.; Hiramoto, K.; Nakata, T. *Synthesis*, ***2001***, 1027.

Ph–CH(OH)–CO_2H —[TiO_4/SO_4^{-2}, MeOH; reflux, 2 h]→ Ph–CH(OH)–CO_2Me 96%

Jin, T.-S.; Ma, Y.-R.; Li, Y.; Sun, X.; Li, T.-S. *Synth. Commun.*, ***2001***, *31*, 2051.

1,2-Benzenedimethanol —[Pd(II)-hydrotalcite, Py, O_2; toluene, 80°C, 28 h]→ phthalide 56%

Kakiuchi, N.; Nishimura, T.; Inoue, M.; Uemura, S. *Bull. Chem. Soc. Jpn.*, ***2001***, *74*, 165.

$PhCH_2OH$ —[AcOH, $BiCl_3$, 1 h]→ $PhCH_2OAc$ 95%

Mohammadpoor-Baltork, I.; Khosropour, A.R.; Aliyan, H. *J. Chem. Res. (S)*, ***2001***, 280.

1. TBAF / THF
2. TPAP / NMO

80% (45% ee)

Leroy, B.; Dumeunier, R.; Markó, I.E. *Tetrahedron Lett.*, **2000**, *41*, 10215.

$ClSO_2NH_2$, $AcNMe_2$

98%

Okada, M.; Iwashita, S.; Koizumi, N. *Tetrahedron Lett.*, **2000**, *41*, 7047.

$Et_3C\text{-}OH$ → 1. $ClCH_2CN$, AcOH, H_2SO_4; 2. thiourea, AcOH/EtOH → $Et_3C\text{-}NH_3Cl$

Jirgensons, A.; Kauss, V.; Kalvinsh, I.; Gold, M.R. *Synthesis*, **2000**, 1709.

BuBr, Cs_2CO_3
TBAI, DMF, 93°C, 1 d

91%

Chu, F.; Dueno, E.E.; Jung, K.W. *Tetrahedron Lett.*, **1999**, *40*, 1847.

, PhH, 3 NEt_3
ultrasound, 3 h

89%

Caldwell, J.J.; Harrity, J.P.A.; Heron, N.M.; Kerr, W.J.; McKendry, S.; Middlemiss, D. *Tetrahedron Lett.*, **1999**, *40*, 3481.

BnCl, CO_2, TBAI, 4 h
Cs_2CO_3, DMF, 23°C

94%

Kim, S.-I.; Chu, F.; Dueno, E.E.; Jung, K.W. *J. Org. Chem.*, **1999**, *64*, 4578.

$C_{18}H_{37}OH$ —[N-trifluoroacetylsuccinimide; THF, reflux, 5 h]→ $C_{18}H_{37}O_2CCF_3$ 95%

Katritzky, A.R.; Yang, B.; Qiu, G.; Zhang, Z. *Synthesis*, ***1999***, 55.

$PhCH_2OH$ —[1. CO_2, Bu_3P, CyTMG, DMF; 2. CBr_4, 2 h]→ $PhCH_2OCO_2CH_2Ph$ 67%

Kadokawa, J.-i.; Habu, H.; Fukumachi, S.; Karasu, M.; Tagaya, H.; Chiba, K. *J. Chem. Soc., Perkin Trans. 1*, ***1999***, 2205.

Further examples of the reaction ROH → RCO_2R' are included in Section 107 (Esters from Acid Derivatives) and in Section 45A (Protection of Alcohols and Phenols).

SECTION 109: ESTERS FROM ALDEHYDES

—[20% SmI_2, 10% *i*-PrSH]→ quant

Hsu J.-L.; Fang, J.-M. *J. Org. Chem.*, ***2001***, *66*, 8673.

CHO, CO_2allyl —[1. e^-; 2. H_3O^+]→ O, O 95%

Franco, D.; Duñach, E. *Synlett*, ***2001***, 806.

CHO —[1% OAl(O*i*-Pr)$_2$, OAl(O*i*-Pr)$_2$; DCM, rt, 2 h]→ O, O*i*-Pr 99%

Simpura, I.; Nevalainen, V. *Tetrahedron*, ***2001***, *57*, 9867.

CHO, CHO —[CaO, PhH, 313°K, 15 min]→ O, O quant

Seki, T.; Hattori, H. *Chem. Commun.* ***2001***, 2510.

PhCHO $\xrightarrow{MeOH, H_2O_2, cat V_2O_5}$ $PhCO_2Me$ quant

Gopinath, R.; Patel, B.K. *Org. Lett.*, ***2000***, *2*, 577.

, $AlCl_3/i-Pr_2NEt$

$AgSbF_6$, CH_2Cl_2

O O Ph CHO Cl Ph 93%

Nelson, S.G.; Wan, Z.; Peelen, T.J.; Spencer, K.L. *Tetrahedron Lett.*, ***1999***, *40*, 6535.

CHO a biphenylene Al catalyst

toluene, 21°C, 15 min

O O 99%

Ooi, T.; Miura, T.; Takaya, K.; Maruoka, K. *Tetrahedron Lett.*, ***1999***, *40*, 7695.

$OSiEt_3$ Me SPy, DCM

$ZnCl_2$, 25°C

CHO Ph O O Ph Me

<30% (>19:1 *trans:cis*)

with $SnCl_4$, −78°C 62% (>19:1 *cis:trans*)

Wang, Y.; Zhao, C.; Romo, D. *Org. Lett.*, ***1999***, *1*, 1197.

O OHC cat (SmI_2 + *i*-PrSH) O O 88%

Hsu, J.-L.; Chen, C.-T.; Fang, J.-M. *Org. Lett.*, ***1999***, *1*, 1989.

CHO Br $C_3H_7CO_2Na$, CO $PdCl_2(PPh_3)_2$, 80°C

PhH, 18 h

C_3H_7 O O O O 62%

Cho, C.S.; Bark, D.Y.; Shim, S.C. *J. Heterocyclic Chem.*, ***1999***, *36*, 289.

PhCHO → (1. 2% chiral bicylic amino-alcohol Et_2Zn, hexane/toluene, 0°C, 3 h; 2. Ac_2O) → OAc, Ph 98% (98% ee)

Nugent, W.A. *Chem. Commun.*, ***1999***, 1369.

CHO, CO_2Me → (1. $(pinyl)_2Ballyl$, ether −100°C, 1 h; 2. *p*-TsOH, heat) → O, O 82x90% (>99% ee)

Ramachandran, P.V.; Krzeminski, M.P.; Reddy, M.V.R.; Brown, H.C. *Tetrahedron Asymm.*, ***1999***, *10*, 11.

PhCHO → (1. Br, CO_2H, In, 20°C THF/H_2O; 2. 6M HCl) → Ph, O, O 75%

Choudhury, P.K.; Foubelo, F.; Yus, M. *Tetrahedron*, ***1999***, *55*, 10779.

PhCHO → (1. $SiMe_3$, 0.5 $TaCl_5$, 0°C DCM, 1 h; 2. Ac_2O, rt, 12 h) → OAc, Ph 70%

Chandrasekhar, S.; Mohanty, P.K.; Raza, A. *Synth. Commun.*, ***1999***, *29*, 257.

Related Methods: Section 117 (Esters from Ketones)

SECTION 110: ESTERS FROM ALKYLS, METHYLENES AND ARYLS

No examples of the reaction R-R → RCO_2R' or $R'CO_2R$ (R,R' = alkyl, aryl, etc.) occur in the literature. For the reaction R-H → RCO_2R' or $R'CO_2R$, see Section 116 (Esters from Hydrides).

(1. 1 atm CO, CBr_4/2 $AlBr_3$ CH_2Br_2, −40°C, 2 h; 2. *i*-PrOH) → CO_2*i*-Pr 82%

Akhrem, I.; Afanas'eva, L.; Petrovskii, P.; Vitt, S.; Orlinkov, A. *Tetrahedron Lett.*, ***2000***, *41*, 9903.

SECTION 111: ESTERS FROM AMIDES

10% $MgBr_2$, MeOH; 8 h — 97%

Orita, A.; Nagano, Y.; Hirano, J.; Otera, J. *Synlett*, ***2001***, 637.

$CuBr_2$/LiO*t*-Bu, THF; 20 min — 88%

Yamaguchi, J.-i.; Aoyagi, T.; Fujikura, R.; Suyama, T. *Chem. Lett.*, ***2001***, 466.

OH; $(CH_2)_4OH$; HO_2C; N=N; 2; CN; H_2O, 75°C, 16 h — 95%

Yorimitsu, H.; Wakabayashi, K.; Shinokubo, H.; Oshima, K. *Bull. Chem. Soc. Jpn.*, ***2001***, *74*, 1963.

OH; H_2O, 75°C, 16 h; HO_2C; N=N; CO_2H; CN; CN — 95%

water soluble radical initiator

Yorimitsu, H.; Wakabayashi, K.; Shinokubo, H.; Oshima, K. *Tetrahedron Lett.*, ***1999***, *40*, 519.

1. CAN; 2. MeOH, rt — 97%

Štefane, B.; Kočevar, M.; Polanc, S. *Tetrahedron Lett.*, ***1999***, *40*, 4429.

SECTION 112: ESTERS FROM AMINES

NO ADDITIONAL EXAMPLES

SECTION 113: ESTERS FROM ESTERS

Conjugate reductions and conjugate alkylations of unsaturated esters are found in Section 74 (Alkyls from Alkenes).

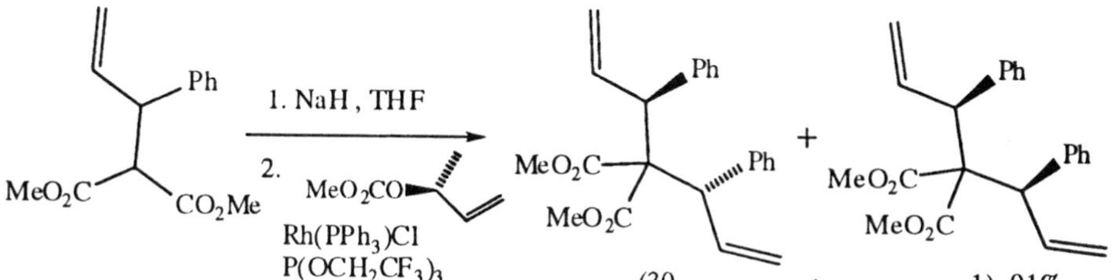

1. NaH, THF; 2. MeO_2CO (allylic carbonate), $Rh(PPh_3)Cl$, $P(OCH_2CF_3)_3$

(30 : 1) 91%

Evans, P.A.; Kennedy, L.J. *J. Am. Chem. Soc.*, ***2001***, *123*, 1234.

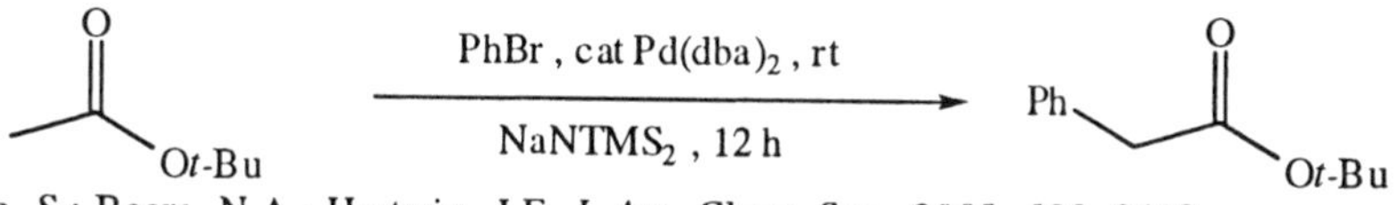

PhBr, cat $Pd(dba)_2$, rt; $NaNTMS_2$, 12 h

87%

Lee, S.; Beare, N.A.; Hartwig, J.F. *J. Am. Chem. Soc.*, ***2001***, *123*, 8410.

$EtCO_2Me$ → (t-BuOK, ether, 0°C) → $EtCO_2t$-Bu 71%

Vasin, V.A.; Razin, V.V. *Synlett*, ***2001***, 658.

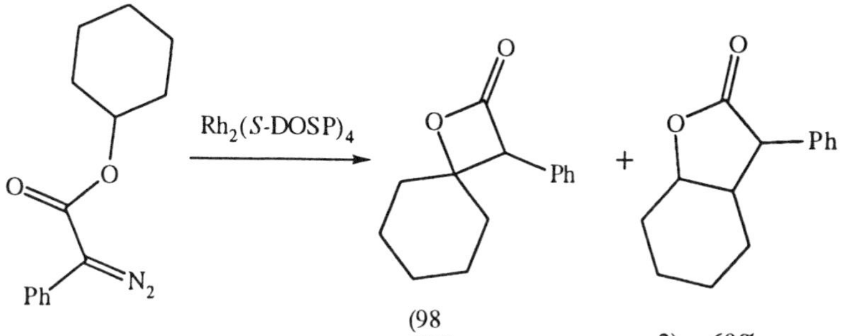

$Rh_2(S\text{-DOSP})_4$

(98 : 2) 69%
63% ee

Doyle, M.P.; May, E.J. *Synlett*, ***2001***, 967.

$CH_3COCH_2CO_2Et$ —(BuOH , $LiClO_4$ / toluene , heat , 2 h)→ $CH_3COCH_2CO_2Bu$ 88%

Bandgar, B.P.; Sadavarte, V.S.; Uppalla, L.S. *Synlett, 2001*, 1338.

$CH_3COCH_2CO_2Me$ —(C_3H_7OH , cat NBS , 100°C / toluene , 3 h)→ $CH_3COCH_2CO_2C_3H_7$ 94%

Bandgar, B.P.; Uppalla, L.S.; Sadavarte, V.S. *Synlett, 2001*, 1715.

CH_3COCH_2COOEt —($FeSO_4$, BuOH , toluene / reflux , 2 h)→ CH_3COCH_2COOBu 85%

also with $CuSO_4$ (88%)

Bandgar, B.P.; Sadavarte, V.S.; Uppalla, L.S. *Synth. Commun., 2001, 31*, 2063.

$CH_3COCH_2CO_2Et$ —(C_3H_7OH , $NaBPh_4$, toluene / reflux , 2 h)→ $CH_3COCH_2CO_2C_3H_7$ 90%

Bandgar, B.P.; Sadavarte, V.S.; Uppalla, L.S. *Chem. Lett., 2001*, 894.

$CH_3COCH_2CO_2Et$ —(BuOH , Zn , toluene / heat , 2 h)→ $CH_3COCH_2CO_2Bn$ 98%

Bandgar, B.P.; Sadavarte, V.S.; Uppalla, L.S. *J. Chem. Res. (S), 2001*, 16.

2-acetyl methyl benzoate (CO_2Me) —(Ru catalyst , *i*-PrOH / 20°C , 23 h)→ 3-methylphthalide 99% (94% ee , *S*)

Evaraere, K.; Scheffler, J.-L.; Mortreux, A.; Carpentier, J.-F. *Tetrahedron Lett., 2001, 42*, 1899.

$PhSO_2CF_3$ —(MeMgBr , THF , 1 h / 0°C → rt)→ Ph—SO_2Me 77%

Steensma, R.W.; Galabi, S.; Tagat, J.R.; McCombie, S.W. *Tetrahedron Lett., 2001, 42*, 2281.

cat $[Ru(aceneCl_2)]_2$; chiral ligand, t-BuOH, t-BuOK

>99% (67% ee)

Everaere, K.; Mortreux, A.; Bulliard, M.; Brusse, J.; van der Gen, A.; Nowogrocki, G.; Carpentier, J.-F. *Eur. J. Org. Chem.*, ***2001***, 275.

OBu → Ot-Bu

$TBAF \cdot 3\ H_2O$, BnBr; THF, rt, 1.5 h

Ph OSiMe$_3$ → Ph OBn

96%

Ooi, T.; Sugimoto, H.; Maruoka, K. *Heterocycles, **2001**, 51*, 593.

SmI_2, THF; 5 eq EtOH

Ph OEt → Ph lactone

70%

Williams, D.B.G.; Blann, K.; Holzapfel, C.W. *Synth. Commun.*, ***2001**, 31*, 203.

BuOH, Amberlyst-15; toluene, reflux, 2 h

CO_2Me → CO_2Bu

85%

Chavan, S.P.; Subbarao, Y.T.; Dantale, S.W.; Sivappa, R. *Synth. Commun.*, ***2001**, 31*, 289.

Me_4Sn, 2.5% $Pd_2(dba)_3$, 80°C; 10% $AsPh_3$, 10% CuI, NMP, 87 h

Ph → Ph, Me

58%

Rossi, R.; Bellina, F.; Raugei, E. *Synlett*, ***2000***, 1749.

1. dilauryl peroxide, PhH, reflux
2. 5 eq $EtSO_2N_3$, PhCl

dilauryl peroxide, 100°C

I → N_3

80% (74:26 *exo:endo*)

Ollivier, C.; Renaud, P. *J. Am. Chem. Soc.*, ***2000**, 122*, 6496.

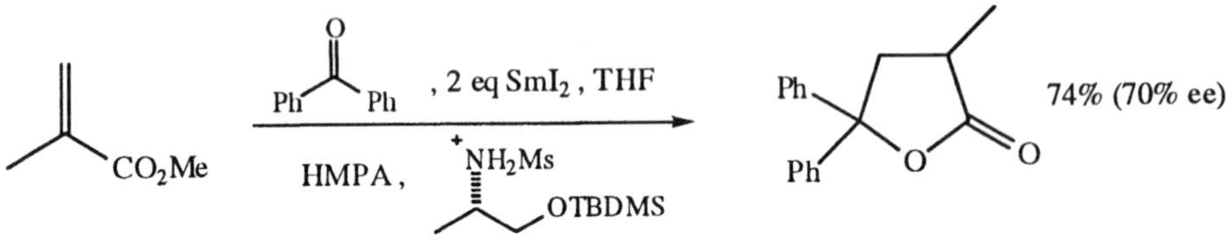

CO_2Me — PhC(O)Ph, 2 eq SmI_2, THF / HMPA, $\overset{+}{N}H_2Ms$ OTBDMS → Ph Ph O O 74% (70% ee)

Wang, W.; Xu, M.-H.; Lei, X.-S.; Lin, G.-Q. *Org. Lett.*, ***2000***, *2*, 3773.

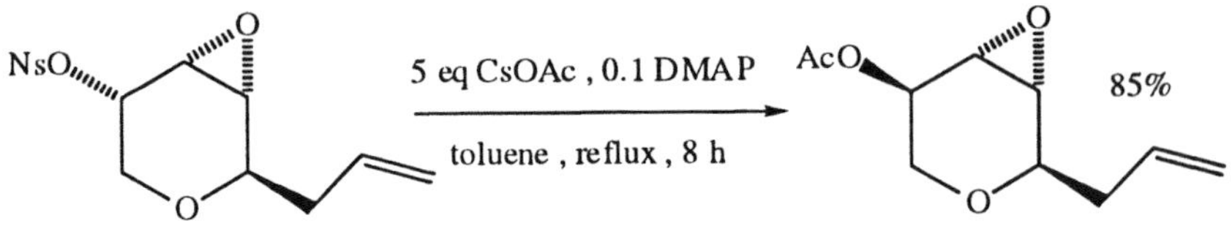

NsO — 5 eq CsOAc, 0.1 DMAP / toluene, reflux, 8 h → AcO 85%

Ns = nitrobenzenesulfonyl

Hawryluk, N.A.; Snider, B.B. *J. Org. Chem.*, ***2000***, *65*, 8379.

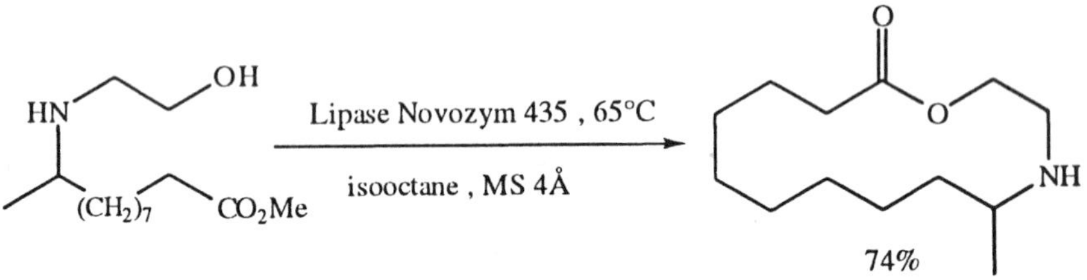

HN OH $(CH_2)_7$ CO_2Me — Lipase Novozym 435, 65°C / isooctane, MS 4Å → O O NH 74%

Lu, W.; Sih, C.J. *Tetrahedron Lett.*, ***1999***, *40*, 4965.

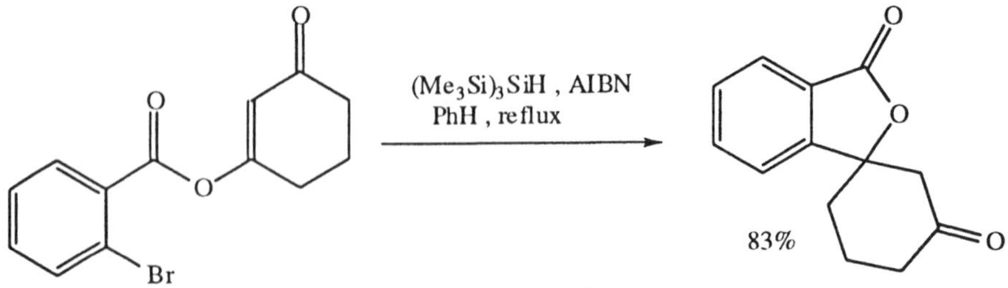

O O O Br — $(Me_3Si)_3SiH$, AIBN PhH, reflux → O O O 83%

Zhang, W.; Pugh, G. *Tetrahedron Lett.*, ***1999***, *40*, 7595.

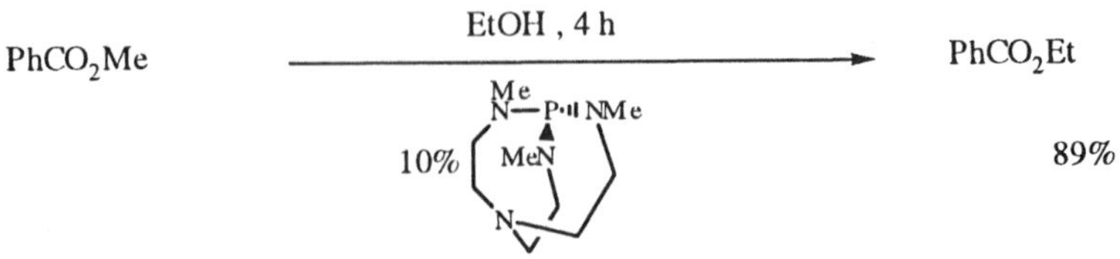

$PhCO_2Me$ — EtOH, 4 h / 10% Me N—P NMe MeN N → $PhCO_2Et$ 89%

Ilankumaran, P.; Verkade, J.G. *J. Org. Chem.*, ***1999***, *64*, 3086.

Br

73%

Enholm, E.J.; Gallagher, M.E.; Moran, K.M.; Lombardi, J.S.; Schulte II, J.P. *Org. Lett.* ***1999***, *1*, 689.

BuOH , Mo–ZrO_2 , toluene

reflux

OMe OBu

98%

Reddy, B.M.; Reddy, V.R.; Manohar, B. *Synth. Commun.*, ***1999***, *29*, 1235.

REVIEWS:

"Stereochemistry of the Alkylation of Lactones," Ibrahim-Ouali, M.; Parrain, J.-L.; Santelli, M. *Org. Prep. Proceed. Int.*, ***1999***, *31*, 467.

SECTION 114: ESTERS FROM ETHERS, EPOXIDES AND THIOETHERS

2% PPNCo$(CO)_4$, 900 psi CO

BF_3•OEt_2 , DME , 80°C , 1 d

Bu

66%

PPN = bis(triphenylphosphine) iminium

Lee, J.T.; Thomas, P.J.; Alper, H. *J. Org. Chem.*, ***2001***, *66*, 5424.

BuOBu —— 1. $MoCl_5$, DCE , 80°C; 2. BzCl , 80°C ——> $PhCO_2Bu$ 75%

Guo, Q.; Miyaji, T.; Gao, G.; Hara, R.; Takahashi, T. *Chem. Commun.*, ***2001***, 1018.

MCPBA

CH_2Cl_2 , 25° , 1h

Me OMe OMe EtO_2C

65%

Piccialli, V.; Graziano, M.L. *Tetrahedron Lett.*, ***2001***, *42*, 93.

1. cat $Pd(OAc)_2$, cat $P(OPh)_3$ THF

2. MeMgBr, –78°C, 1 h

TBDMS, MeO_2C, O, Me, TBDMS

78% (74% de)

Marion, F.; Le Fol, R.; Courillon, C.; Malacria, M. *Synlett*, ***2001***, 138.

$(OC)_3(Cp)W$

25% BF_3•OEt_2, CH_2Cl_2 –40°C

76%

Madhushaw, R.J.; Li, C.-L.; Shen, K.-H.; Hu, C.-C.; Liu, R.-S. *J. Am. Chem. Soc.*, ***2001***, *123*, 7427.

Bu, O, OH

$Pb(OAc)_4$/$Cu(OAc)_2$

Py, PhH, reflux

Bu, O, O

78%

Rigby, J.H.; Payen, A.; Warshakoon, N. *Tetrahedron Lett.*, ***2001***, *42*, 2047.

BnOBn → BnOAc 94%

$FeCl_3$-Montmorillonite K10

Ac_2O, 70°C, 22 h

Lakouraj, M.M.; Movassagh, B.; Fasihi, J. *J. Chem. Soc. (S)*, ***2001***, 378.

Ph, $OCPh_3$

1. 2 eq AcCl, DCM, 0°C, 18 h

2. aq $NaHCO_3$

Ph, OAc

98%

Bergmeier, S.C.; Arason, K.M. *Tetrahedron Lett.*, ***2000***, *41*, 5799.

O_2N, O

CO_2, 5% NaBr, NMP 100°C, 30 h

[18% with Ph-oxirane]

O_2N, O, O, O

57%

Iwasaki, T.; Kihara, N.; Endo, T. *Bull. Chem. Soc. Jpn.*, ***2000***, *73*, 713.

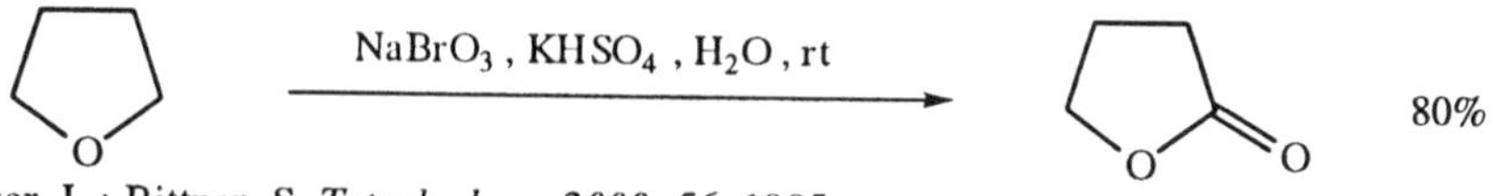

$NaBrO_3$, $KHSO_4$, H_2O, rt

80%

Metsger, L.; Bittner, S. *Tetrahedron, **2000**, 56*, 1905.

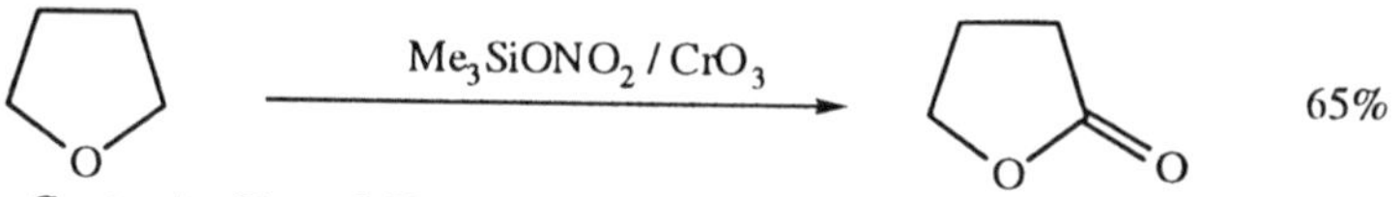

Me_3SiONO_2 / CrO_3

65%

Shahi, S.P.; Gupta, A.; Pitre, S.V.; Reddy, M.V.R.; Kumareswaran, R.; Vankar, Y.D. *J. Org. Chem.* ***1999**, 64*, 4509.

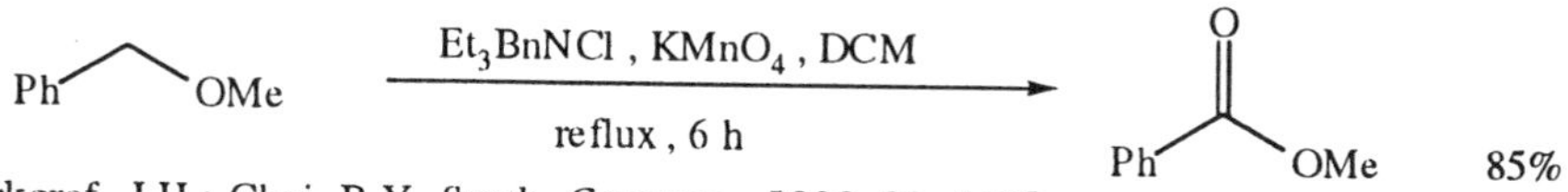

Ph OMe → Et_3BnNCl, $KMnO_4$, DCM; reflux, 6 h → Ph C(O) OMe 85%

Markgraf, J.H.; Choi, B.Y. *Synth. Commun., **1999**, 29*, 2405.

SECTION 115: ESTERS FROM HALIDES AND SULFONATES

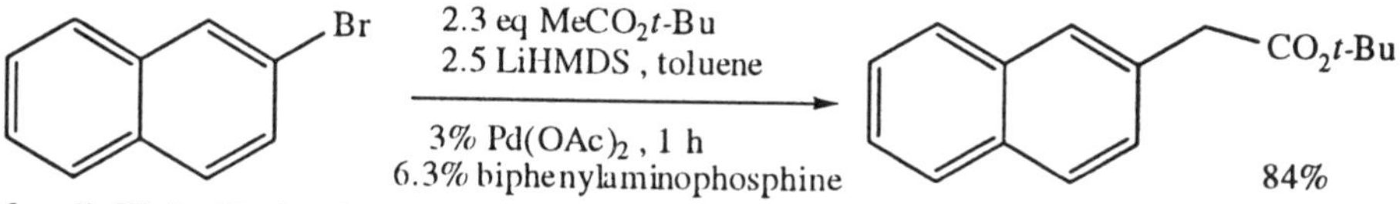

Br → 2.3 eq $MeCO_2t$-Bu, 2.5 LiHMDS, toluene; 3% $Pd(OAc)_2$, 1 h, 6.3% biphenylaminophosphine → CO_2t-Bu 84%

Moradi, W.A.; Buchwald, S.L. *J. Am. Chem. Soc., **2001**, 123*, 7996.

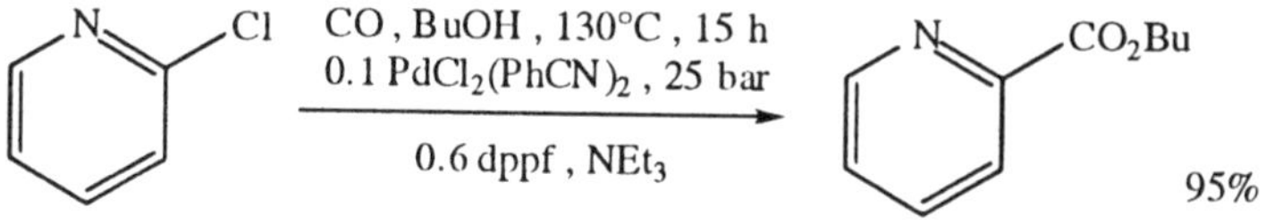

N Cl → CO, BuOH, 130°C, 15 h, 0.1 $PdCl_2(PhCN)_2$, 25 bar; 0.6 dppf, NEt_3 → N CO_2Bu 95%

Beller, M.; Mägerlein, W.; Indolese, A.F.; Fischer, C. *Synthesis, **2001***, 1098.

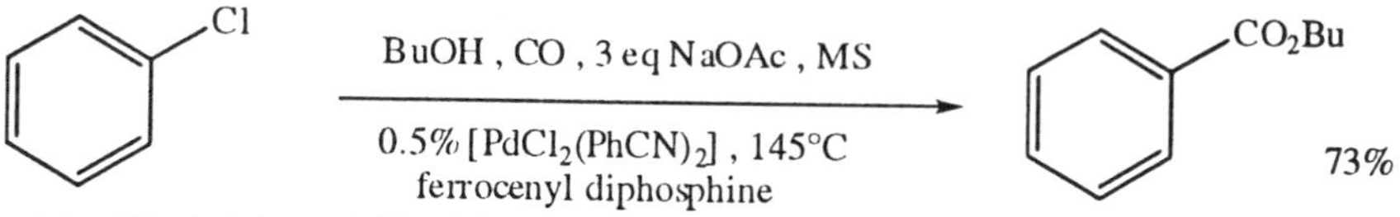

Cl → BuOH, CO, 3 eq NaOAc, MS; 0.5% $[PdCl_2(PhCN)_2]$, 145°C, ferrocenyl diphosphine → CO_2Bu 73%

Mägerlein, W.; Indolese, A.F.; Beller, M. *Angew. Chem. Int. Ed., **2001**, 40*, 2856.

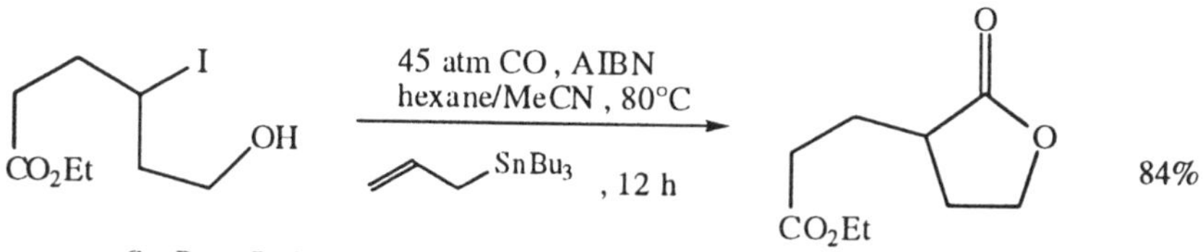

I, OH, CO_2Et → 45 atm CO, AIBN, hexane/MeCN, 80°C; $SnBu_3$, 12 h → O, O, CO_2Et 84%

Kreimerman, S.; Ryu, I.; Minakata, S.; Komatsu, M. *Org. Lett., **2000**, 2*, 389.

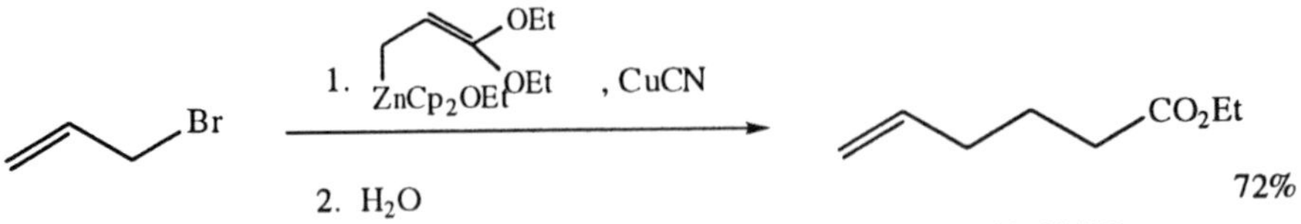

1. ZnCp$_2$OEt (OEt)$_2$ reagent, CuCN; 2. H_2O — 72%

Sato, A.; Ito, H.; Yamaguchi, Y.; Taguchi, T. *Tetrahedron Lett.*, **2000**, *41*, 10239.

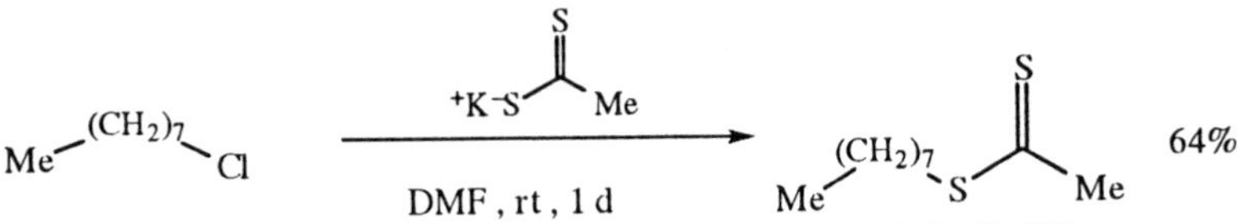

Me(CH$_2$)$_7$Cl → Me(CH$_2$)$_7$SC(S)Me, $^+K^-S$C(S)Me, DMF, rt, 1 d — 64%

Zheng, T.-C.; Burkart, M.; Richardson, D.E. *Tetrahedron Lett.*, **1999**, *40*, 603.

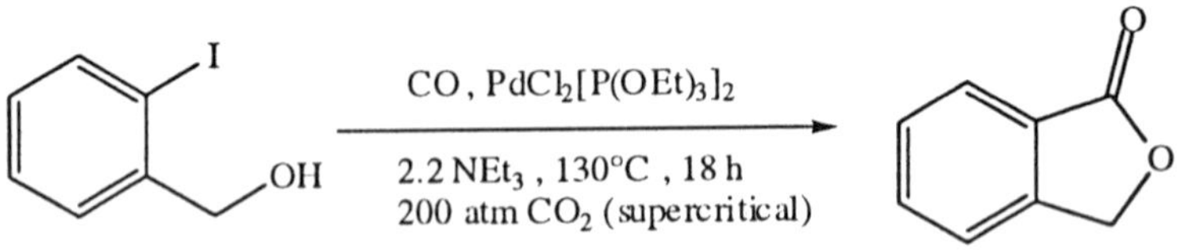

CO, $PdCl_2[P(OEt)_3]_2$; 2.2 NEt_3, 130°C, 18 h; 200 atm CO_2 (supercritical)

Kayaki, Y.; Noguchi, Y.; Iwasa, S.; Ikariya, T.; Noyori, R. *Chem. Commun.*, **1999**, 1235.

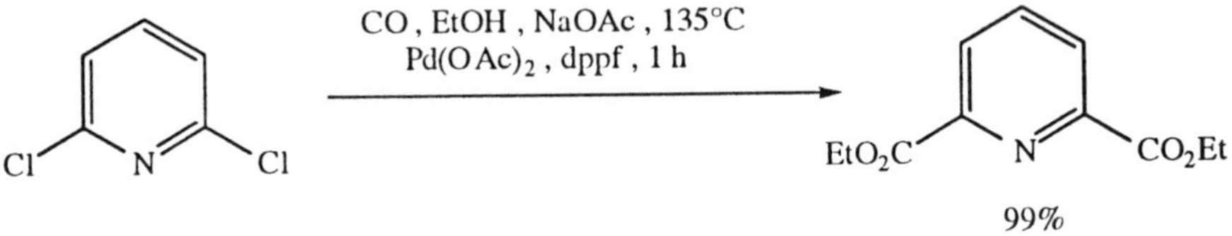

CO, EtOH, NaOAc, 135°C; $Pd(OAc)_2$, dppf, 1 h — 99%

Bessard, Y.; Crettaz, R. *Heterocycles*, **1999**, *51*, 2589.

Related Methods: Section 25 (Acid Derivatives from Halides).

SECTION 116: ESTERS FROM HYDRIDES

This section contains examples of the reaction R-H → RCO_2R' or $R'CO_2R$ (R = alkyl, aryl, etc.).

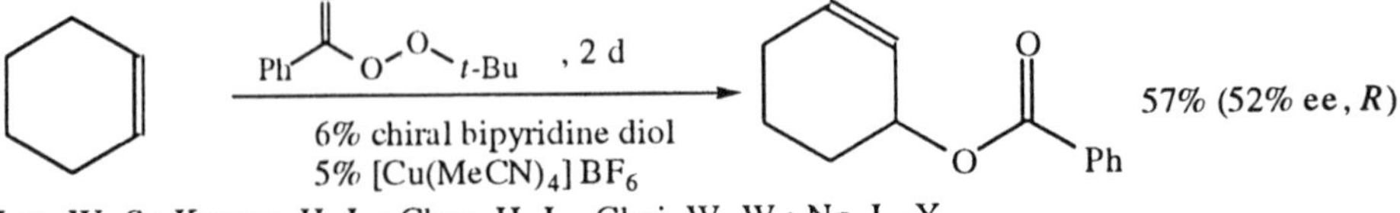

PhC(O)OO*t*-Bu, 2 d; 6% chiral bipyridine diol; 5% $[Cu(MeCN)_4]BF_6$ — 57% (52% ee, *R*)

Lee, W.-S.; Kwong, H.-L.; Chan, H.-L.; Choi, W.-W.; Ng, L.-Y. *Tetrahedron Asymm.*, **2001**, *12*, 1007.

1% $RuCl_3 \cdot n\ H_2O$, $MeCO_3H$

TFA, DCM, rt, 4 h

O_2CCF_3 + O

77% 13%

Komiya, N.; Noji, S.; Murahashi, S.-I. *Chem. Commun.*, ***2001***, 65.

1. $Tl(OTf)_3$, DMF, 60°C

2. H_2O

Ph O O H O 94%

Lee, J.C.; Jin, Y.S.; Choi, J.-H. *Chem. Commun.*, ***2001***, 956.

O CO_2Bn

, 30°C, 8 h

0.1 $Sc(O_3SOC_{12}H_{25})_3$

CO_2Bn 81%

Mori, Y.; Kakumoto, K.; Manabe, K.; Kobayashi, S. *Tetrahedron Lett.*, ***2000***, *41*, 3107.

$PhCO_3t$-Bu, 5% $Cu(OTf)_2$

7.5% tris(oxazoline) ligand
acetone, 0°C, 2 d

O Ph

80% (83% ee)

Kohmura, Y.; Katsuki, T. *Tetrahedron Lett.*, ***2000***, *41*, 3941.

SeO_2, Ac_2O, O_2

reflux, 12 h

AcO + OAc

(19 : 1) 36%

Koltun, E.S.; Kass, S.R. *Synthesis*, ***2000***, 1366.

CBr_4–2 $AlBr_3$, CO

i-PrOH, -20°C, 30 min

t-Bu O *i*-Pr + O O*i*-Pr

(84 : 16) <7%

Akhrem, I.; Orlinkov, A.; Afanas'eva, L.; Petrovskii, P.; Vitt, S. *Tetrahedron Lett.*, ***1999***, *40*, 5897.

Also via: Section 26 (Acid Derivatives) and Section 41 (Alcohols).

SECTION 117: ESTERS FROM KETONES

engineered *E. coli* cells

61% (>98% ee)

Mihovilovic, M.D.; Chen, G.; Wang, S.; Kyte, B.; Rochon, F.; Kayser, M.M.; Stewart, J.D. *J. Org. Chem.*, ***2001***, *66*, 733.

1% (3,5-di-CF_3-C_6H_3–Se)$_2$

2 eq 60% H_2O_2, CF_3CH_2OH
20°C, 1 h

90%

ten Brink, G.-J.; Vis, J.-M.; Arends, I.W.C.E.; Sheldon, R.A. *J. Org. Chem.*, ***2001***, *66*, 2429.

CO_2Me, 2 eq SmI_2, THF

chiral proton source
–78°C → –10°C

990% (39% ee)

Xu, M.-H.; Wang, W.; Xia, L.-J.; Lin, G.-Q. *J. Org. Chem.*, ***2001***, *66*, 3953.

50% MgI_2, 50% *R*-BINOL, –25°C
CH_2Cl_2, $PhCMe_2OOH$, 8 h

65%

Bolm, C.; Beckmann, O.; Cosp, A.; Palazzi, C. *Synlett*, ***2001***, 1461.

H_2O_2, Ph-AsO_3H_2

$(CF_3)_2CHOH$, 240 min

85%

Berkessel, A.; Andreae, M.R.M. *Tetrahedron Lett.*, ***2001***, *42*, 2293.

1. *S*-BINOL-$AlMe_2Cl$
hexane/toluene, rt

2. cumene hydroperoxide
–25°C → rt

71%

Bolm, C.; Beckmann, O.; Palazzi, C. *Can. J. Chem.*, ***2001***, *79*, 1593.

Ph O → Et_2Zn , chiral amino-alcohol; toluene , –78°C → –26°C → Ph O O 75% (39% ee)

Shinohara, T.; Fujioka, S.; Kotsuki, H. *Heterocycles*, ***2001***, *55*, 237.

O Ph SiMe → HNO_3 , MeOH; 2 h → O Ph OMe quant

Patrocínio, A.F.; Moran, P.J.S. *Synth. Commun.*, ***2000***, *30*, 1419.

O

1) $LiSnMe_3$
2) O , BF_3• OEt_2
–78°C , 1 h
3. $Pb(OAc)_4$, $CaCO_3$
4. H_2 $ThCl(PPh_3)_3$

O O 45%

Posner, G.H.; Wang, Q.; Halford, B.A.; Elias, J.S.; Maxwell, J.P. *Tetrahedron Lett.*, ***2000***, *41*, 9655.

O → $PdCl_2$, CO; $CuCl_2$, MeOH → CO_2Me CO_2Me + CO_2Me Cl

(85 : 15) 73%

Hamed, O.; El-Qisairi, A.; Henry, P.M. *Tetrahedron Lett.*, ***2000***, *41*, 3021.

O → polymer-supported peroxyacid → O O 68%

Lambert, A.; Elings, J.A.; Macquarrie, D.J.; Carr, G.; Clark, J.H. *Synlett*, ***2000***, 1052.

O Ph CO_2Me → cat $Ru_3(CO)_{12}$, ethylene; 5 atm CO , toluene , 160°C; $P(4\text{-}CF_3\text{-}C_6H_4)_3$, 20 h → Ph MeO_2C O O 94%

Chatani, N.; Tobisu, M.; Asaumi, T.; Fukumoto, Y.; Murai, S. *J. Am. Chem. Soc.*, ***1999***, *121*, 7160.

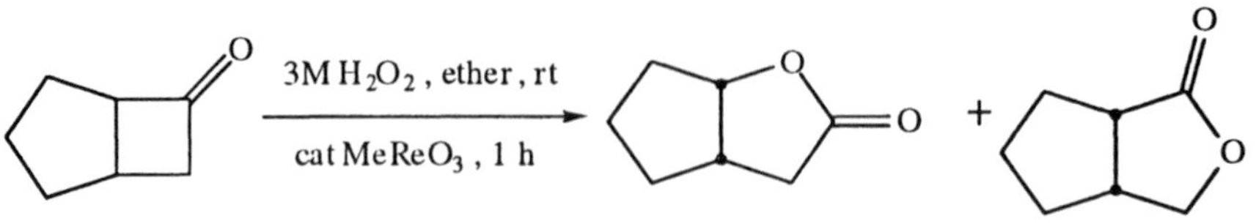

3M H_2O_2, ether, rt

cat $MeReO_3$, 1 h

(94 : 6) 80%

Phillips, A.M.F.; Romão, C. *Eur. J. Org. Chem.*, ***1999***, 1767.

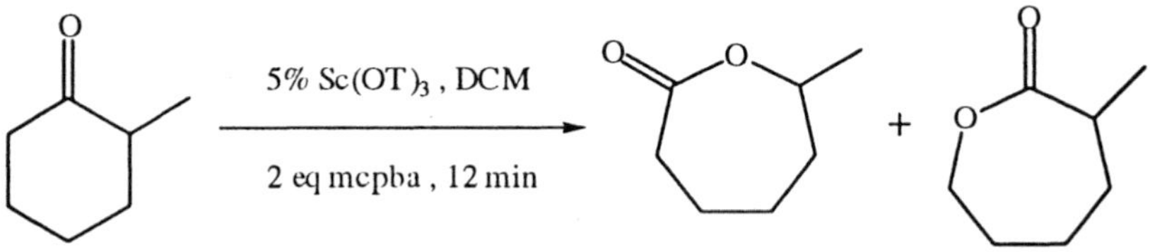

5% $Sc(OT)_3$, DCM

2 eq mcpba, 12 min

(96 : 4) 90%

Kotsuki, H.; Arimura, K.; Araki, T.; Shinohara, T. *Synlett*, ***1999***, 462.

REVIEWS:

"100 Years of Baeyer-Villiger Oxidations," Renz, M.; Meunier, B. *Eur. J. Org. Chem.*, ***1999***, 737.

Also via: Section 27 (Acid Derivatives).

SECTION 118: ESTERS FROM NITRILES

NO ADDITIONAL EXAMPLES

SECTION 119: ESTERS FROM ALKENES

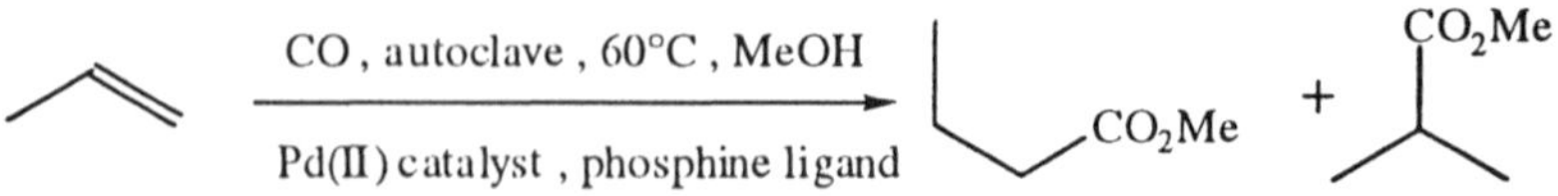

CO, autoclave, 60°C, MeOH

Pd(II) catalyst, phosphine ligand

CO_2Me 81% + CO_2Me 19%

Pugh, R.I.; Drent, E.; Pringle, P.G. *Chem. Commun.*, ***2001***, 1476.

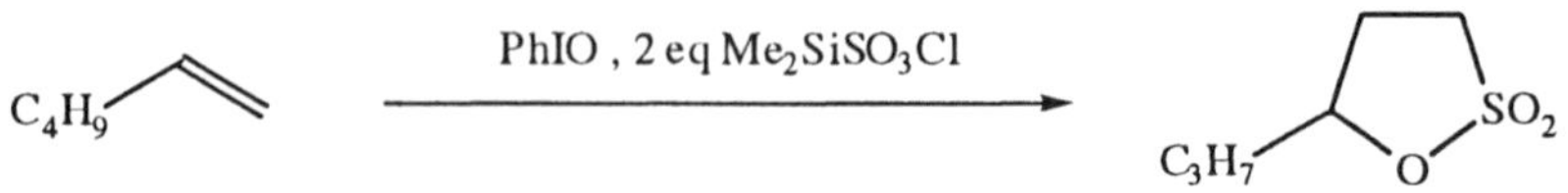

C_4H_9

PhIO, 2 eq Me_2SiSO_3Cl

C_3H_7 O SO_2 68%

Bassindale, A.R.; Katampe, I.; Maesano, M.G.; Patel, P.; Taylor, P.G. *Tetrahedron Lett.*, ***1999***, *40*, 7417.

1. 1 atm CO, THF, 1% $Ni(cod)_2$, 0°C, 2 eq DBU, 4 h; 2. aq HCl

Ph ... → Ph—CH(CO_2Me)—CH=CH—CH$_2$—COMe 96%

Takimoto, M.; Mori, M. *J. Am. Chem. Soc.*, ***2001***, *123*, 2895.

O_3, hexane, reflux; –50°C → 0°C, 5 min

32% + 35% (CO_2H)

Barrero, A.F.; Alvarez-Manzaneda, E.J.; Chahboun, R.; Cuerva, J.M.; Segovia, A. *Synlett*, ***2000***, 1269.

Ph, OH, Me, Me — $Pd_2(dba)_3$/chiral phosphine, 800 psi CO/H_2, 80°C; CH_2Cl_2, 1 d → Me, Me, Ph, O, O 87% (95% ee, *R*)

Cao, P.; Zhang, X. *J. Am. Chem. Soc.*, ***1999***, *121*, 7708.

Also via: Section 44 (Alcohols).

SECTION 120: ESTERS FROM MISCELLANEOUS COMPOUNDS

Me_3Si, Ph, C=O — PhSH, PhH, –78°C, Chinchona alkaloid catalyst → Me_3Si, O, SPh, Ph 97% (91% ee)

Blake, A.J.; Friend, C.L.; Outram, R.J.; Simpkins, N.S.; Whitehead, A.J. *Tetrahedron Lett.*, ***2001***, *42*, 2877.

PhSeSePh —AcCl, Sm/cat $CoCl_2$, THF→ PhSeAc 79%

Chen, R.; Zhang, Y. *Synth. Commun.*, ***2000***, *30*, 1331.

mcpba, 4 h
DCM, rt

(4 : 1) 70%

Hunt, K.W.; Grieco, P.A. *Org. Lett.*, ***2000***, *2*, 1717.

Merrifield resin—//—O

1. 5 eq Bu_3SnH, 0.5 AIBN
PhH, reflux
2. Jones reagent

36%

Watanabe, Y.; Ishikawa, S.; Takao, G.; Toru, T. *Tetrahedron Lett.*, ***1999***, *40*, 3411.

PhSSPh

1. PBu_3, H_2O, DMF, 1 h
2. Ac_2O, 1 h

2 PhS(C=O)Me 95x97%

Ayers, J.T.; Anderson, S.R. *Synth. Commun.*, ***1999***, *29*, 351.

BF_3•OEt_2, H_2S, reflux
dioxane, 4 h

67%

Nair, S.K.; Askokan, C.V. *Synth. Commun.*, ***1999***, *29*, 791.

PhSSPh —(Ac_2O, Sm/cat $CoCl_2$; THF, 50°C, 4 h)→ PhSAc 75%

Chen, R.; Zhang, Y. *Synth. Commun.*, ***1999***, *29*, 3699.

PhSeSePh —(BzCl, Sm/$CrCl_3$, THF, rt)→ PhSeBz 70%

Liu, Y.; Zhang, Y. *Synth. Commun.*, ***1999***, *29*, 4043.

REVIEWS:

"Carboxylic Acids and Esters," Franklin, A.S. *J. Chem. Soc., Perkin Trans. 1*, ***1999***, 3537

CHAPTER 9

PREPARATION OF ETHERS, EPOXIDES AND THIOETHERS

SECTION 121: ETHERS, EPOXIDES AND THIOETHERS FROM ALKYNES

t-Bu—≡—t-Bu → (1. S_2Cl_2, CH_2Cl_2; 2. distil) → 56%

Nakayama, J.; Takahashi, K.; Watanabe, T.; Sugihara, Y.; Ishii, A.
Tetrahedron Lett., ***2000***, *41*, 8349.

SECTION 122: ETHERS, EPOXIDES AND THIOETHERS FROM ACID DERIVATIVES

NO ADDITIONAL EXAMPLES

SECTION 123: ETHERS, EPOXIDES AND THIOETHERS FROM ALCOHOLS AND THIOLS

→ ($BF_3{\bullet}OEt_2$) → 97%

Hardouin, C.; Taran, F.; Doris, E. *J. Org. Chem.*, ***2001***, *66*, 4450.

O_2N–C6H4–OH → (MeI, 5M aq NaOH; cat $Me(C_8H_{17})_3NCl$) → O_2N–C6H4–OMe 90%

Eynde, J.J.V.; Mailleux, I. *Synth. Commun.*, ***2001***, *31*, 1.

HO — Ph; .1 $In(OTf)_3$; .1 Ph⁀CHO; DCM, 25°C, 240 h → O + O (81 : 19) 54%; Ph, Ph, Ph

Loh, T.-P.; Hu, Q.-Y.; Ma, L.-T. *J. Am. Chem. Soc.*, ***2001***, *123*, 2450.

OH, Br; 2% $Pd(OAc)_2$, 1.5 eq Cs_2CO_3; cat chiral binaphthyl ligand, toluene 50°C, 26 h → O 82%

Kuwabe, S.-i.; Torraca, K.E.; Buchwald, S.L. *J. Am. Chem. Soc.*, ***2001***, *123*, 12202.

HO, HO, OH; acetone, 4°C, 1 d; TsOH → HO, HO, O 63%

Guiso, M.; Marra, C.; Cavarischia, C. *Tetrahedron Lett.*, ***2001***, *42*, 6531.

OH; ⁀OH; DCE, cat *p*-TsOH → O 90%

Ishino, Y.; Mihara, M.; Hayakawa, N.; Miyata, T.; Kaneko, Y.; Miyata, T. *Synth. Commun.*, ***2001***, *31*, 439.

OAc, HO, TBTPS-O; 4% $Pd(dba)_2$, (+)-DPPIO; THF, rt → O, OTBTPS 98% (81:19 *cis:trans*)

Hara, O.; Fujii, K.; Hamada, Y.; Sakagami, Y. *Heterocycles*, ***2001***, *54*, 419.

PhSH → [$(4\text{-Me-}C_6H_4)I^+ BF_4^-$, THF, rt; cat $Pd(PPh_3)_4$, Na_2CO_3] → $PhS(4\text{-MeC}_6H_4)$ 95%

Wang, L.; Chen, Z.-C. *Synth. Commun.*, ***2001***, *31*, 1227.

CHO → [Ph, S^+, O; CH_2Cl_2] → O, Ph

quant (97:3 *trans:cis*)

Solladié-Cavallo, A.; Bouérat, L.; Roje, M. *Tetrahedron Lett.*, ***2000***, *41*, 7309.

OH, Ph, Cl → [1. PhBr, PPh_3, DEAD THF, rt; 2. BuLi, THF −50°C → rt] → O, Ph

78 +83%

Hodgetts, K.J. *Tetrahedron Lett.*, ***2000***, *41*, 8655.

Ph, OH → [10% $MeAl(NTf)_2$, DCM, 21°C; 0.2M, 0.5 h] → Ph, O, Ph >99%

Ooi, T.; Ichikawa, H.; Itagaki, Y.; Maruoka, K. *Heterocycles*, ***2000***, *52*, 575.

OH → [PhCHO, $InCl_3$, $SnCl_3$; DCM, rt, 4 h] → Cl, O, Ph

34% *cis* (8% *trans*)

Li, J.; Li, C.-J. *Heterocycles*, ***2000***, *53*, 1691.

CO_2H, $(CH_2)_{12}$, OH → [O, N=C=N; DMAP, DMAP•HCl, $CHCl_3$] → O, $(CH_2)_9$, O

77% (+ 13% diolide)

Keck, G.E.; Sanchez, C.; Wager, C.A. *Tetrahedron Lett.*, ***2000***, *41*, 8673.

PhI, cat CuBr, toluene
$(Me_2N)_3P{=}N{-}P(NMe_3)_2{=}NEt$
reflux, 1-4 h

Cl-C6H4-SH → Cl-C6H4-SPh 91%

Palomo, C.; Oiarbide, M.; López, R.; Gómez-Bengoa, E. *Tetrahedron Lett.*, ***2000***, ***41***, 1283.

$PhB(OH)_2$, 1.5 eq $Cu(OAc)_2$
DMF, 3 eq Py, MS 4Å
reflux, 3 h

SH → SPh 88%

Herradura, P.S.; Pendola, K.A.; Guy, R.K. *Org. Lett.*, ***2000***, *2*, 2019.

Me_2SO_4, NaOH, 8 h
PEG-400, heat

O_2N-C6H4-OH → O_2N-C6H4-OMe 95%

Cao, Y.-Q.; Pei, B.-G. *Synth. Commun.*, ***2000***, *30*, 1759.

BuBr, CsOH, TBAI
MS 4Å, DMF, 23°C, 10 h

Ph~~~OH → Ph~~~OBu 92%

Dueno, E.E.; Chu, F.; Kim, S.-I.; Jung, K.W. *Tetrahedron Lett.*, ***1999***, *40*, 1843.

Ph~~OH
PMe_3, ADDP, Im-H

Ph~SH → Ph~S~~Ph 79%

ADDP = 1,1'(azodicarbonyl)dipiperidine

Falck, J.R.; Lai, J.-Y.; Cho, S.-D.; Yu, J. *Tetrahedron Lett.*, ***1999***, *40*, 2903.

, $LiClO_4$–$MeNO_2$
Montmorillonite/H_2O
rt, 2 d

OH, OH → O >99%

Chiba, K.; Hirano, T.; Kitano, Y.; Tada, M. *Chem. Commun.*, ***1999***, 691.

i-PrOH, zeolite–$KMnO_4$
DCE, rt, 30 min

Ph~~OH → Ph~~O*i*-Pr 90%

Gadhwal, S.; Boruah, A.; Prajapati, D.; Sandhu, J.S. *Synth. Commun.*, ***1999***, *29*, 1921.

1. Et_3NHCl, MeCN, rt, 15 min
2. EtI, overnight

PhSH → Ph—S—Et 93%

Feroci, M.; Inesi, A.; Rossi, L. *Synth. Commun.*, ***1999***, *29*, 2611.

OH, OMe → , e^-, AcOH / 3M $LiClO_4$, $MeNO_2$ → OMe 97%

Chiba, K.; Fukuda, M.; Kim, S.; Kitano, Y.; Tada, M. *J. Org. Chem.*, ***1999***, *64*, 7654.

MeO, OH, O, Ph → $CH_2(NMe)_2$, EtOH / reflux, 3 h → MeO, O, O, Ph 70%

Valenti, P.; Belluti, F.; Rampa, A.; Bisi, A. *Synth. Commun.*, ***1999***, *29*, 3895.

Ph—⟨⟩—OH → BnBr, Cs_2CO_3, DMF / 23°C, 1 h → Ph—⟨⟩—OBn 92%

Parrilsh, J.P.; Sudaresan, B.; Jung, K.W. *Synth. Commun.*, ***1999***, *29*, 4423.

$PhCH_2OH$ → 1. Mg, I_2, reflux / 2. Tf_2O, 5°C → $PhCH_2OCH_2Ph$ 80%

Nishiyama, T.; Kameyama, H.; Maekawa, H.; Watanuki, K. *Can. J. Chem.*, ***1999***, *77*, 258.

SECTION 124: ETHERS, EPOXIDES AND THIOETHERS FROM ALDEHYDES

PhCHO → (Me, S, Me), BnBr, NaOH / t-BuOH, H_2O, 2 d → (O, Ph, Ph) 92%

Zanardi, J.; Leriverend, C.; Aubert, D.; Julienne, K.; Metzner, P. *J. Org. Chem.*, ***2001***, *66*, 5620.

PhCHO → $PhSeSiMe_3$, PhH / Bu_3SnH, cat AIBN, 80°C, 18 h → $PhCH_2OSiMe_3$ 78%

Nishiyama, Y.; Kajimoto, H.; Kotani, K.; Sonoda, N. *Org. Lett.*, ***2001***, *3*, 3087.

PhCHO → ($Ph(MeO_2C)C{=}N_2$, 1% $Rh_2(OAc)_4$, DCM) → Ph, CO_2Me, Ph epoxide 66%

Doyle, M.P.; Hu, W.; Timmons, D.J. *Org. Lett.*, ***2001***, *3*, 933.

PhCHO → (H, OH, H, S, *p*-Tol; K_2CO_3, rt, 4d, CH_2Cl_2) → Ph, Ph epoxide 54% (94:6 *trans:cis*) 5% ee SS

Saito, T.; Akiba, D.; Sakairi, M.; Kanazawa, S. *Tetrahedron Lett.*, ***2001***, *42*, 57.

OH → (MeCHO, PhH, 0.1 M, hv) → HO, O + HO, O (81 : 19)

Griesbeck, A.G.; Bondock, S. *J. Am. Chem. Soc.*, ***2001***, *123*, 6191.

CHO, CO_2Me → (Rh catalyst, $HSiEt_3$, toluene, 50°C) → $OSiEt_3$, CO_2Me + $OSiEt_3$, CO_2Me

+ $RhCl(PPh_3)_3$ (3 : 1) 81%
+ $RhH(PPh_3)_4$ (1 : 2) 81%

Emiabata-Smith, D.; McKillop, A.; Mills, C.; Motherwell, W.B.; Whitehead, A.J. *Synlett*, ***2001***, 1302.

Bu, O, O → (1. $Ti(Oi\text{-}Pr)_4$/2 *i*-PrMgBr; 2. O, CHO; 3. aq HCl) → Bu, O, O 74%

Teng, X.; Wada, T.; Okamoto, S.; Sato, F. *Tetrahedron Lett.*, ***2001***, *42*, 5501.

PhSH , p-TsOH

SPh

54%

Ishino, Y.; Mihara, M.; Kawai, H. *Synlett*, ***2001***, 1317.

OH

Ph CHO

In(OTf)$_3$, DCM , 0°C

Ph

O

95% (65:35 *trans:cis*)

In(OTf)$_3$, DCM

Ph CHO

40°C

O

Ph

65%

Loh, T.-P.; Hu, Q.Y.; Tan, K.-T.; Cheng, H.-S. *Org. Lett.*, ***2001***, *3*, 2669.

Ph Ph

OsO_4 , $K_3Fe(CN)_6$, K_2CO_3
polymer-bound DHQD-PHAL

t-BuOH , H_2O , 0°C , 18 h

OH

Ph Ph

OH

92% (99% ee)

Federov, A.Yu.; Carrara, F.; Finet, J.-P. *Tetrahedron Lett.*, ***2001***, *42*, 5875.

CHO

CH_2I_2 , MeLi , THF

0°C → rt

O

72%

Concellón, J.M.; Cuervo, H.; Fernández-Fano, R. *Tetrahedron* , ***2001***, *57*, 8983.

PhCHO

PhCH=N–NTs$^-$ Na$^+$, 1% $Rh_2(OAc)_4$

20% tetrahydrothiophene , MeCN
20% $BnNEt_3Cl$, 40°C , 3 h

O

Ph Ph

95% (>98:2 *trans:cis*)

Aggarwal, V.K.; Alonso, E.; Hyund, G.; Lydon, K.M.; Palmer, M.J.; Porcelloni, M.; Studley, J.R. *Angew. Chem. Int. Ed.*, ***2001***, *40*, 1430.

OH, CHO → 2 eq $Me_2C(OMe)_2$, $Sc(OTf)_3$, DCM, rt, 20 min → O, OMe, OMe, OMe 92%

Yadav, J.S.; Reddy, B.V.S.; Rao, T.P. *Tetrahedron Lett.*, ***2000***, *41*, 7943.

$SiMe_3$ → PhCHO, THF, rt, 3 d; 0.2 (*i*-Pr, N–P, N, *i*-Pr, N, *i*-Pr, N) → OTMS, Ph 74%

Wang, Z.; Kisanga, P.; Verkade, J.G. *J. Org. Chem.*, ***1999***, *64*, 6459.

PhCHO → Me_3Si / $SnBu_3$, $InCl_3$, DCM → Ph, O, Ph 68% (8:1)

Viswanathan, G.S.; Yang, J.; Li, C.-J. *Org. Lett.*, ***1999***, *1*, 993.

CHO → 4 eq BnBr, 0°C, MeCN, 2.8 eq NaOH, 21 h (O, O, S, Ph, Ph) → O, Ph 78% (92% ee)

Hayakawa, R.; Shimizu, M. *Synlett*, ***1999***, 1388.

PhCHO → (S), 2 eq BnBr, 2 eq KOH, MeCN–H_2O, rt, 1 d → O, Ph, Ph 92% (88% de/84% ee)

Julienne, K.; Metzner, P.; Henryon, V. *J. Chem. Soc., Perkin Trans. 1*, ***1999***, 731.

SECTION 125: ETHERS, EPOXIDES AND THIOETHERS FROM ALKYLS, METHYLENES AND ARYLS

NO ADDITIONAL EXAMPLES

SECTION 126: ETHERS, EPOXIDES AND THIOETHERS FROM AMIDES

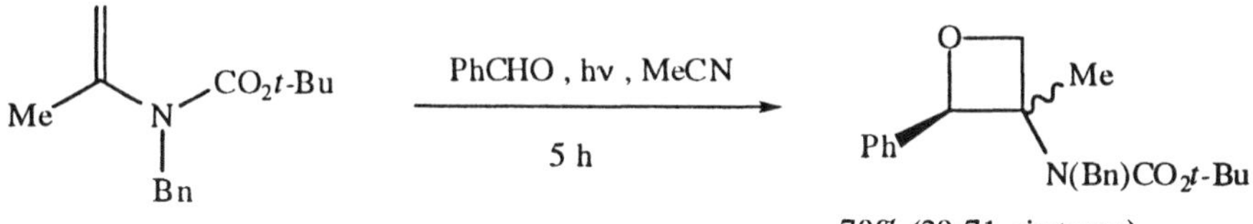

CO2t-Bu
Me
N
Bn
PhCHO, hv, MeCN
5 h
O
Me
Ph
N(Bn)CO2t-Bu

70% (29:71 *cis:trans*)

Bach, T.; Schröder, J. *Synthesis*, ***2001***, 1117.

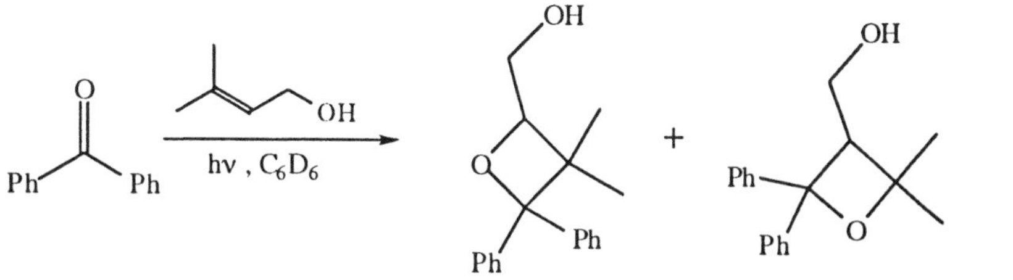

O
Ph
Ph
OH
hv, C6D6
OH
O
Ph
Ph
+
OH
Ph
Ph
O

(>95 : 5) 58% conversion

Adam, W.; Stegmann, V.R. *Synthesis*, ***2001***, 1203.

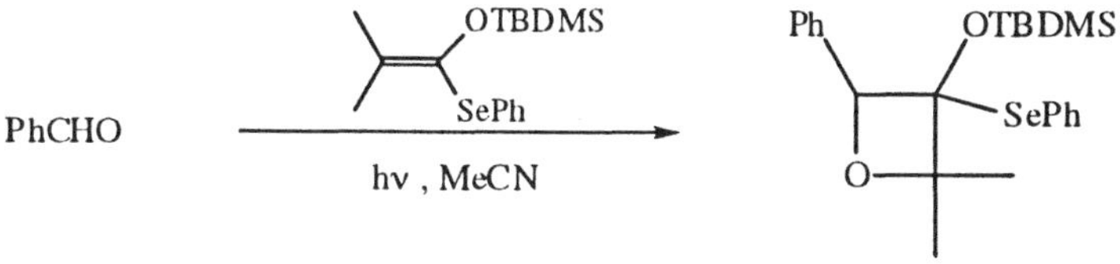

PhCHO
OTBDMS
SePh
hv, MeCN
Ph
OTBDMS
SePh
O

81% (62:38 *trans:cis*) 80% conversion

Abe, M.; Tachibana, K.; Fujimoto, K.; Nojima, M. *Synthesis*, ***2001***, 1243.

SECTION 127: ETHERS, EPOXIDES AND THIOETHERS FROM AMINES

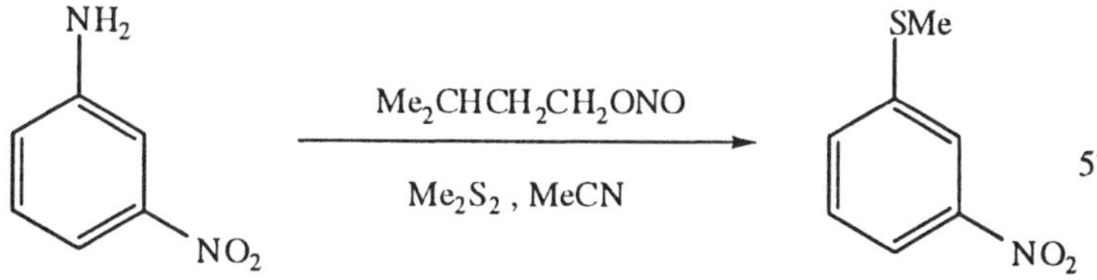

NH2
NO2
Me2CHCH2CH2ONO
Me2S2, MeCN
SMe
NO2

57%

Allaire, F.S.; Lyga, J.W. *Synth. Commun.*, ***2001***, ***31***, 1857.

N—NHTs $\xrightarrow[\text{reflux}]{t\text{-BuOK}, t\text{-BuOH}}$ Ot-Bu 82%

Chandrasekhar, S.; Rajaiah, G.; Chandraiah, L.; Swamy, D.N. *Synlett*, ***2001***, 1779.

SECTION 128: ETHERS, EPOXIDES AND THIOETHERS FROM ESTERS

1% $[RuCl_2(CO)_3]_2$, EtI, 16 h; 5% Et_3NH, toluene, 100°C; Et_3SiH — OMe → $OSiEt_3$, OMe 57%

Igarashi, M.; Mizuno, R.; Fuchikami, T. *Tetrahedron Lett.*, ***2001***, *42*, 2149.

Ph ... OMe, O $\xrightarrow[\text{DCM}]{TiCl_4, \text{ TMSOTf}, Et_3SiH}$ Ph ... OMe 81%

Yato, M.; Homma, K.; Ishida, A. *Tetrahedron*, ***2001***, *57*, 5353.

1. i-Bu_2AlH; 2. PhSeH, BF_3•OEt_2; 3. H_2O — SePh 91% (85:15 *trans:cis*)

Schmitt, A.; Reissig, H.-U. *Synthesis*, ***2001***, 867.

OCO_2Me, C_3H_7 — 1. PhOH, THF, NaHMDS; 2. $Rh(PPh_3)_3Cl$, $P(OMe)_3$ → OPh, C_3H_7 + OPh, C_3H_7 (70 : 1) 96%

Evans, P.A.; Leahy, D.K. *J. Am. Chem. Soc.*, ***2000***, *122*, 5012.

S, MeO, Ph — 1. Ph_3SnH, Ph_2SiH_2, rt, 0.5 BEt_3, 1 h; 2. AIBN, 110°C → MeO Ph 94%

Jang, D.O.; Song, S.H. *Synlett*, ***2000***, 811.

OCO_2Me — cat Cp*RuCl(cod), MeCN, BuSH, rt, 1 h → SBu 97%

Kondo, T.; Morisaki, Y.; Uenoyama, S.-y.; Wada, K.; Mitsudo, T.-a. *J. Am. Chem. Soc.*, ***1999***, *121*, 8657.

2.2 eq Ph_3SnH , rt

BEt_3 , toluene
30 min

97%

MaGee, D.I.; Leach, J.D.; Setiadji, S. *Tetrahedron*, ***1999***, *55*, 2847.

SECTION 129: ETHERS, EPOXIDES AND THIOETHERS FROM ETHERS, EPOXIDES AND THIOETHERS

$C_{10}H_{21}$ (epoxide) → Me_3SiCl, hexane, 2 h; *sec*-BuLi/chiral diamine, −90°C → −50°C → $C_{10}H_{21}$ (epoxide) $SiMe_3$

73%

Hodgson, D.M.; Norsikian, S.L.M. *Org. Lett.*, ***2001***, *3*, 461.

OCMeCy$_2$ (acrylate)

1% CuOTf, EtOAc, rt
1.1% *S,S*-bis-ferrocene heterocycles

CO_2CMeCy_2, Ph

74% (84:16 *cis:trans*)

Lo, M.M.-C.; Fu, G.C. *Tetrahedron*, ***2001***, *57*, 2621.

$Ph_2CH—OMe$ → C_3H_7OH, $Fe(ClO_4)_3$; reflux, 150 min → $Ph_2CHOC_3H_7$ 97%

Salehi, P.; Irandoost, M.; Seddighi, B.; Behbahani, F.K.; Tahmasebi, D.P. *Synth. Commun.*, ***2000***, *30*, 1743.

PhS (allyl) → I–$(CH_2)_5$–I, $NaNH_2$–DME; 0°C, 2 h → SPh (1-vinylcyclohexyl) 75%

Choppin, S.; Gros, P.; Fort, Y. *Synth. Commun.*, ***2000***, *30*, 795.

Ph, $SnBu_3$, O

1. 1.5 eq BuLi, THF, −78°C
2. 3 eq LiCl, −78°C → 0°C

Ph, O

63% (86:14 *trans:cis*)
>95% ee for both

Komine, N.; Tomooka, K.; Nakai, T. *Heterocycles*, ***2000***, *52*, 1071.

$CH_2=CHCH_2SiMe_3$, -50°C
$SnCl_4$, CH_2Cl_2

OBn SiMe$_3$

79% (1:1 *cis:trans*)

Sugita, Y.; Kimura, Y.; Yokie, I. *Tetrahedron Lett.*, ***1999***, *40*, 5877.

MeO—C$_6$H$_4$—MgBr

THF , 2 eq EtMgBr
Me_3CCH_2I , rt , 5 h

76%

Inoue, A.; Shinokubo, H.; Oshima, K. *Synlett*, ***1999***, 1582.

$TMSN_3$, PhH
Mn salen-immobilized mesoporous meterial

OTMS
N_3

90%

Kantam, M.L.; Choudary, B.M.; Barathi, B. *Synth. Commun.*, ***1999***, *29*, 1121.

0.2 $Sn(PPh_3)(ClO_4)_2$, MeCN
$(NH_2)_2C=S$, reflux , 45 min

93%

Tangestaninejad, S.; Mirkhani, V. *Synth. Commun.*, ***1999***, *29*, 2079.

Ph OCHO

2.5 eq SmI_2 , THF–HMPA
2.5 eq TES–Cl

Ph $OSiEt_3$

97%

Honda, T.; Ishikawa, F. *Synth. Commun.*, ***1999***, *29*, 3323.

Ph Se Ph

$AlBr_3$, DCM , rt , 2 h

Se
Ph

75%

Abe, H.; Koshiba, N.; Yamasaki, A.; Harayama, T. *Heterocycles*, ***1999***, *51*, 2301.

NH_4SCN , P4VP–Ce(OTf)$_4$
MeCN , rt , 15 min

90%

Iranpoor, N.; Tamami, B.; Shekarriz, M. *Synth. Commun.*, ***1999***, *29*, 3313.

REVIEWS:

"The Brook Rearrangement In Tandem Bond Formation Strategies," Moser, W.H. *Tetrahedron,* ***2001**, 57,* 2065.

SECTION 130: ETHERS, EPOXIDES AND THIOETHERS FROM HALIDES AND SULFONATES

NaO*t*-Bu , cat $Pd(OAc)_2$, 100°C
toluene , phosphine ligand

Bu—C6H4—Cl → Bu—C6H4—O*t*-Bu

92%

Parrish, C.A.; Buchwald, S.L. *J. Org. Chem.*, ***2001**, 66*, 2498.

Br, Et

BuOH , 2% $Pd(OAc)_2$
2.5% phosphine ligand

toluene , 2.5 eq Cs_2CO_3

BuO, Et

63%

Torraca, K.E.; Huang, X.; Parrish, C.A.; Buchwald, S.L.
J. Am. Chem. Soc., ***2001**, 123*, 10770.

Br

OH

NMP , 70°C , 6 h

O

O_2N

88%

NO_2

Gujadhur, R.; Venkataraman, D. *Synth. Commun.*, ***2001**, 31*, 2865.

Cl

1. *i*-PrMgBr , THF , –10°C

2. PhCHO , –10°C → reflux

I

O

82%

Ph

Delacroix, T.; Bérillon, L.; Cahiez, G.; Knochel, P. *J. Org. Chem.*, ***2000**, 65*, 8108.

Br

CS_2 , $NaBH_4$, THF

Br

S

S

83%

Wan, Y.; Kurchan, A.N.; Barnhurst, L.A.; Kutateladze, A.G. *Org. Lett.*, ***2000**, 2*, 1133.

$PhSeSnBu_3$, toluene, 80°C

5% $Pd(Ph_3)_4$

—SePh

Nishiyama, Y.; Tokunaga, K.; Sonoda, N. *Org. Lett.*, ***1999***, *1*, 1725.

Related Methods: Section 123 (Ethers from Alcohols).

SECTION 131: ETHERS, EPOXIDES AND THIOETHERS FROM HYDRIDES

NO ADDITIONAL EXAMPLES

SECTION 132: ETHERS, EPOXIDES AND THIOETHERS FROM KETONES

O Ph Ph

hν

solid state, 10h

O Ph H Ph

59%

Kang, T.; Scheffer, J.R. *Org. Lett.*, ***2001***, *3*, 3361.

O Ph Ph

O Ph Ph

SmI_2, Sm, rt, 20 min

Ph Ph Ph O Ph

85%

Ma, Y.; Zhang, Y.; Chen, J. *Synthesis*, ***2001***, 1004.

O

$C_8H_{17}OH$, Pd/C, H_2 stream

105°C

$O-C_8H_{17}$

97%

Fujii, Y.; Furugaki, H.; Yano, S.; Kita, K. *Chem. Lett.*, ***2000***, 926.

O Ph

[t-$BuSiMe_2CFBr_2$/BuLi]

THF

t-$BuMe_2Si$ O Ph F

94%

Shimizu, M.; Hata, T.; Hiyama, T. *Heterocycles*, ***2000***, *52*, 707.

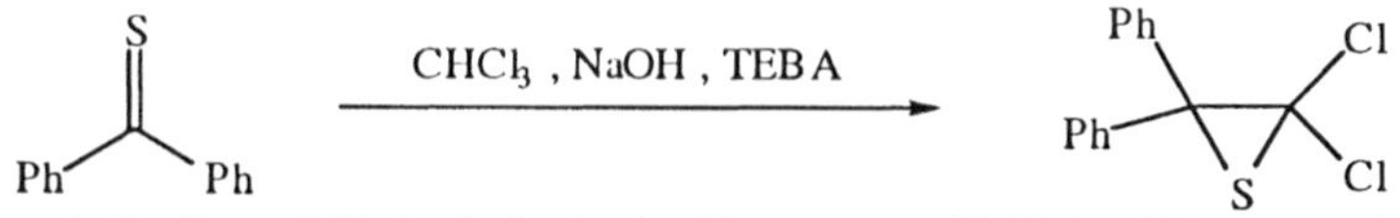
S
Ph
Ph
$CHCl_3$, NaOH , TEBA
Ph
Cl
Ph
S
Cl

80%

Mlossteń, G.; Romański, J.; Swiatek, A.; Hemgartner, H. *Helv. Chim. Acta*, ***1999***, *82*, 946.

Related Methods: Section 124 (Epoxides from Aldehydes).

SECTION 133: ETHERS, EPOXIDES AND THIOETHERS FROM NITRILES

NO ADDITIONAL EXAMPLES

SECTION 134: ETHERS, EPOXIDES AND THIOETHERS FROM ALKENES

ASYMMETRIC COMPOUNDS

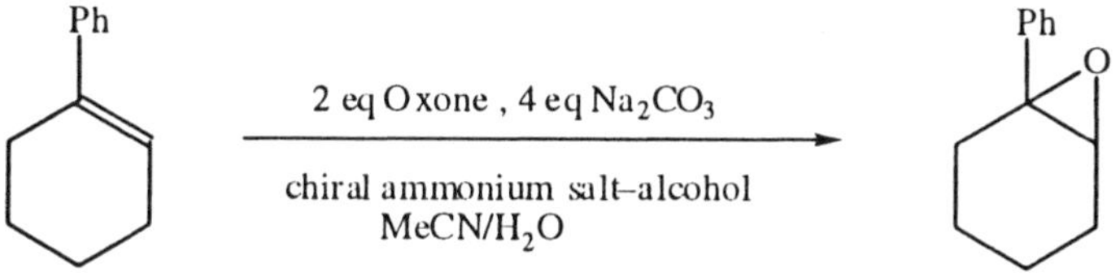
Ph
2 eq Oxone , 4 eq Na_2CO_3
chiral ammonium salt–alcohol
$MeCN/H_2O$
Ph
O

64% (30% ee)

Page, P.C.B.; Rassias, G.A.; Barros, D.; Ardakani, A.; Buckley, B.; Bethell, D.; Smith, T.A.D.; Slawin, A.M.Z. *J. Org. Chem.*, ***2001***, *66*, 6926.

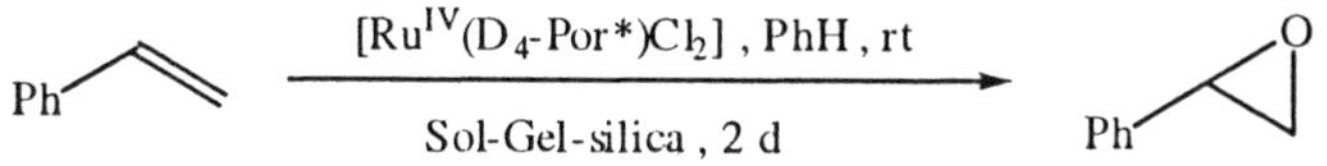
Ph
$[Ru^{IV}(D_4\text{-Por*})Cl_2]$, PhH , rt
Sol-Gel-silica , 2 d
O
Ph

75% (70% ee)

Zhang, R.; Yu, W.-Y.; Wong, K.-y.; Che, C.-M. *J. Org. Chem.*, ***2001***, *66*, 8145.

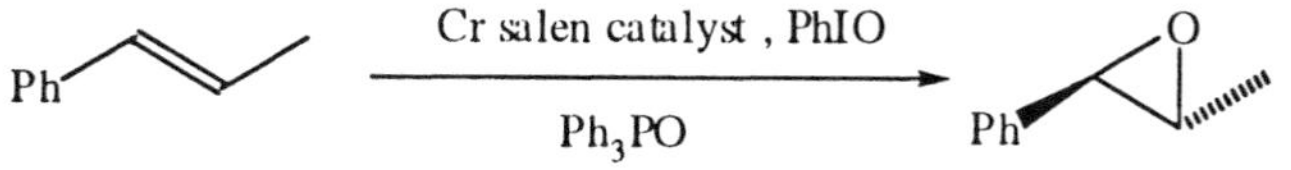
Ph
Cr salen catalyst , PhIO
Ph_3PO
O
Ph

45% (92% ee)

Daly, A.M.; Renehan, M.F.; Gilheany, D.G. *Org. Lett.*, ***2001***, *3*, 663.

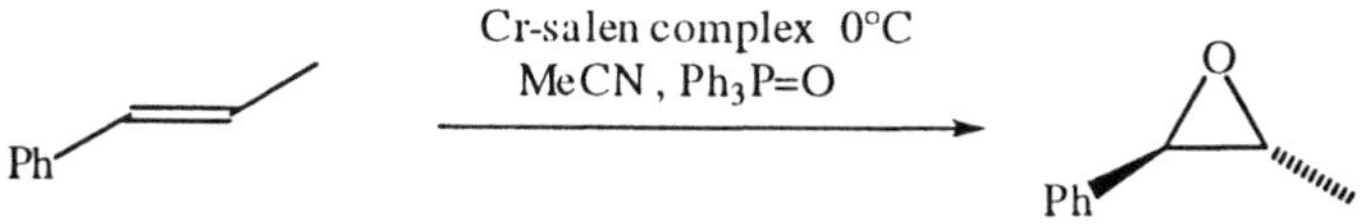
Ph
Cr-salen complex 0°C
MeCN , $Ph_3P=O$
O
Ph

72% ee

O'Mahony, C.P.; McGarrigle, E.M.; Renehan, M.F.; Ryan, K.M.; Kerrigan, N.J.; Bousquet, C.; Gilheany, D.G. *Org. Lett.*, ***2001***, *3*, 3435.

0.1 t-$BuCH_2CHO$, 3 h , rt
0.1 bis-amidopyrrolidine
MeCN , $NaHCO_3$, H_2O

93% (65% ee)
80% conversion

Wong, M.-K.; Ho, L.-M.; Zheng, Y.-S.; Ho, C.-Y.; Yang, D. *Org. Lett.*, ***2001***, *3*, 2587.

5% (R) • Ln catalyst
TBHP - decane
MS 4Å , THF , 25 min

99% (96% ee)

Nemoto, T.; Ohshima, T.; Yamaguchi, K.; Shibasaki, M.
J. Am. Chem. Soc., ***2001***, *123*, 2725.

1. 20% La-*S*-BINOL-O=$AlPh_3$ complex
2.4 eq TBHP , toluene , THF
MS 4Å , rt
2. MeOH

86% (91% ee)

Nemoto, T.; Ohshima, T.; Shibasaki, M. *J. Am. Chem. Soc.*, ***2001***, *123*, 9474.

Mn salen catalyst , NaOCl
CH_2Cl_2 , Na_2HPO_4 (pH 11.3)
4-phenyl pyridine N-oxide

53% (65% ee , *R*)

Ahn, K.-H.; Park, S.W.; Choi, S.; Kim, H.-J.; Moon, C.J. *Tetrahedron Lett.*, ***2001***, *42*, 2485.

Mn salen complex , PyNO
CH_2Cl_2 , 5.5 h

quant (60% ee , *RS*)

Kureshy, R.I.; Khan, N.H.; Abdi, S.H.R.; Patel, S.T.; Jasra, R.V.
Tetrahedron Lett., ***2001***, *42*, 2915.

urea–H_2O_2 , DBU
190 h
immobilized poly-L-leucine on polystyrene

76% (84% ee)

Bentley, P.A.; Bickley, J.F.; Roberts, S.M.; Steiner, A. *Tetrahedron Lett.*, ***2001***, *42*, 3741.

H_2O_2 , MeCN , K_2CO_3
EDTA , 0°C , 12 h

93% (92% ee)

Shu, L.; Shi, Y. *Tetrahedron, **2001**, 57*, 5213.

cumene hydroperoxide , THF

5% BINOL catalyst

90% (82% ee , *R,S*)

Chen, R.; Qian, C.; de Vries, J.G. *Tetrahedron, **2001**, 57*, 9837.

MePhCH–OOH , toluene , 0°C

10% chiral phase transfer catalyst

62% ee , *S,R*)

Adam, W.; Rao, P.B.; Degen, H.-G.; Saha-Möller, C.R. *Tetrahedron Asymm.*, ***2001**, 12*, 121.

urea–H_2O_2 , DCM , MeOH
dimeric Mn-salen catalyst

NH_4OAc , 6 h

68% (23% ee , *R*)

Kureshy, R.I.; Khan, N.H.; Abdi, S.H.R.; Patel, S.T.; Jasra, R.V.
Tetrahedron Asymm., ***2001**, 12*, 433.

mesoporous silica , OsO_4
Cinchona alkaloid , 0°C , 1 d

quant (99% ee)

Bortolini, O.; Fogagnolo, M.; Fantin, G.; Maietti, S.; Medici, A.
Tetrahedron Asymm., ***2001**, 12*, 1113.

mesoporous silica , OsO_4
Cinchona alkaloid , 0°C , 1 d

quant (99% ee)

Lee, H.M.; Kim, S.-W.; Hyeon, T.; Kim, B.M. *Tetrahedron Asymm.*, ***2001**, 12*, 1537.

Oxone , $NaHCO_3$, aq Na_2(edta)

MeCN , chiral ketone, rt , 1 d

quant (48% ee , *R*)

Armstrong, A.; Moss, W.O.; Reeves, J.R. *Tetrahedron Asymm.*, ***2001**, 12*, 2779.

β^3-Leu-α-Leu-R , aq NaOH , DBU

toluene , THF , urea-H_2O_2 , 1 d

92% (70% ee)

Coffey, P.E.; Drauz, K.-H.; Roberts, S.M.; Skidmore, J.; Smith, J.A.
Chem. Commun., ***2001***, 2330.

urea–H_2O_2 , DBU
190 h

immobilized poly-L-leucine on polystyrene

76% (84% ee)

Bentley, P.A.; Bickley, J.F.; Roberts, S.M.; Steiner, A. *Tetrahedron Lett.*, ***2001***, *42*, 3741.

, ether , rt , 6 h

1.2% Ru-salen catalyst

57% (84% ee)

Nakata, K.; Takeda, T.; Mihara, J.; Hamada, T.; Irie, R.; Katsuki, T.
Chem. Eur. J., ***2001***, *7*, 3776.

Oxone , dioxane , $NaHCO_3$

chiral cyclic diamino-ketone

71% (70% ee)

Matsumoto, K.; Tomioka, K. *Heterocycles*, ***2001***, *54*, 615.

Oxone , chiral ketone

K_2CO_3 , DME-DMM

quant (81% ee)

Tian, H.; She, X.; Xu, J.; Shi, Y. *Org. Lett.*, ***2001***, *3*, 1929.

1. 20% La-*S*-BINOL-O=$AlPh_3$ complex
2.4 eq TBHP , toluene , THF

MS 4Å , rt

2. MeOH

86% (91% ee)

Nemoto, T.; Ohsima, T.; Shibasaki, M. *J. Am. Chem. Soc.*, ***2001***, *123*, 9474.

cat (Mg tartrate complex: EtO_2C, CO_2Et, O, O, Mg), t-BuOOH; MS 4Å, 2 h

Bu, O → Bu, O, O 54% (75%)

Jacques, O.; Richards, S.J.; Jackson, R.F.W. *Chem. Commun.*, ***2001***, 2712.

Ph, OH → $VO(Oi\text{-}Pr)_3$, toluene, –20°C; chiral hydroxamic acid ligand → Ph, O, OH

96% (91% ee)

Hoshino, Y.; Murase, N.; Oishi, M.; Yamamoto, H. *Bull. Chem. Soc. Jpn.*, ***2000***, ***73***, 1653.

N → *E. coli* JM109 (pTAB19) → N, O 95% >99% ee

Bernasconi, S.; Orsini, F.; Sello, G.; Colmegna, A.; Galli, E.; Bestetti, G. *Tetrahedron Lett.*, ***2000***, ***41***, 9157.

Ph, O, Ph → OOH, Ph, KOH, MeCN; –40°C → Ph, O, O, Ph

99% (51% ee)

Adam, W.; Rao, P.B.; Degen, H.-G.; Saha-Möller, C.R. *J. Am. Chem. Soc.*, ***2000***, ***122***, 5654.

O → polymer-bound chiral Mn(III)-salen complex; NaOCl, DCM, 0°C; 4-Ph-pyridine N-oxide → O, O

90% (95% ee)

Song, C.E.; Roh, E.J.; Yu, B.M.; Chi, D.Y.; Kim, S.C.; Lee, K.J. *Chem. Commun.*, ***2000***, 615.

Ph → Oxone, K_2CO_3, DME/DMM; pH 8 buffer, –10°C; chiral dioxirane → O, Ph 87% (91% ee)

Tian, H.; She, X.; Shu, L.; Yu, H.; Shi, Y. *J. Am. Chem. Soc.*, ***2000***, ***112***, 11551.

Ph–CH=CH–Ph → 2% Mn-salen complex, PhIO / PhCl, 0°C → trans-stilbene oxide

82% (66% ee)

Nishikori, H.; Ohta, C.; Katsuki, T. *Synlett*, ***2000***, 1557.

1-phenylcyclohexene → cat dihydroquinolinium tetraphenylborate salts / chiral amine, 4 eq Oxone, 8 eq Na_2CO_3, 0°C, aq MeCN → 1-phenylcyclohexene oxide

68% (77% ee)

Page, P.C.B.; Rassias, G.A.; Barros, D.; Bethell, D.; Schilling, M.B. *J. Chem. Soc., Perkin Trans.1*, ***2000***, 3325.

(Z)-alkenyl OAc → chloroperoxidase, *t*-BuOOH / sodium citrate buffer, rt → epoxide OAc

64% (95% ee)

Hu, S.; Hager, L.P. *Tetrahedron Lett.*, ***1999***, *40*, 1641.

Ph–CH=CH–Me → Cr (salen) complex, MeCN / -10°C → Ph, Me epoxide

83% ee

Ryan, K.M.; Bousquet, C.; Gilheany, D.C. *Tetrahedron Lett.*, ***1999***, *40*, 3613.

PhCO–CH=CH–Ph → polyleucine urea-H_2O_2, DBU, THF / 150 min → epoxy ketone

73% (58% ee)

Allen, J.V.; Drauz, K.-H.; Flood, R.W.; Roberts, S.M.; Skidmore, J. *Tetrahedron Lett.*, ***1999***, *40*, 5417.

Ph–CH=CH–Me → H_2O_2, MeCN, 0°C, EDTA, K_2CO_3 / cat (fructose-derived ketone) → epoxide

84% (92% ee, *RR*)

Shu, L.; Shi, Y. *Tetrahedron Lett.*, ***1999***, *40*, 8721.

10% N-anthracenylmethyl substituted Cinchona alkaloid

11% aq NaOCl, toluene, 2 d

90% (81% ee)

Lygo, B.; Wainwright, P.G. *Tetrahedron, **1999**, 55*, 6289.

poly-*L*-leucine/silica gel

UHP, DBU, THF, 14 h

78% (93% ee)

Geller, T.; Roberts, S.M. *J. Chem. Soc., Perkin Trans. 1, **1999***, 1397.

NON-ASYMMETRIC COMPOUNDS

CTAB, VA-061, H_2O
10 eq EPHP, 80°C, 2 h

MeO MeO 98% (55:45 *trans:cis*)

Kita, Y.; Nambu, H.; Ramesh, N.G.; Anilkumar, G.; Matsugi, M. *Org. Lett.*, ***2001***, *3*, 1157.

CF_3CH_2OH, 60% H_2O_2

5% Na_2HPO_4, reflux, 20h

67%

van Vliet, M.C.A.; Arends, I.W.C.E.; Sheldon, R.A. *Synlett*, ***2001***, 248.

OH OH

Ti(IV) complex, TBHP

CH_2Cl_2, MS 4Å, rt, 1 h

91%

Massa, A.; D'Ambrosi, A.; Proto, A.; Screttri, A. *Tetrahedron Lett.*, ***2001***, *42*, 1995.

H_2WO_4/fluoroapotite

urea–H_2O_2

O

62%

Ichihara, J. *Tetrahedron Lett.*, **2001**, *42*, 695.

$(CF_3)_2C(OH)OOH$

$(CF_3)_2CHOH$, 1 h

O

80% (>99% after 21 h)

van Vliet, M.C.A.; Arends, I.W.C.E.; Sheldon, R.A. *Synlett*, **2001**, 1305.

Ph

PhSH, PhH, 80°C, 1 d

Ph SPh + Ph SPh

2% 97%

+ montmorillonite K10 80% 6%

Kanagasabapathy, S.; Sudalai, A.; Benicewicz, B.C. *Tetrahedron Lett.*, **2001**, *42*, 3791.

C_5H_{11}

H_2O_2, Ph-AsO_3H_2

$(CF_3)_2CHOH$, 4.5 h

C_5H_{11} O

95%

Berkessel, A.; Andreae, M.R.M. *Tetrahedron Lett.*, **2001**, *42*, 2293.

Ph Ph O

Amberlyst A-26(OH^-), H_2O_2

dioxane, rt, 6.5 h

O Ph Ph O

96%

Lakouraj, M.M.; Movassagh, B.; Bahrami, K. *Synth. Commun.*, **2001**, *31*, 1237.

PhO N N N S

CD_3CN, 25°C, 15 h

S

45%

Adam, W.; Bargon, R.M. *Eur. J. Org. Chem.*, **2001**, 1959.

OH

amphiphilic sugars, H_2O_2, Mo^{+6}

2°C, 1 d

O OH

75%

Denis, C.; Misbahi, K.; Kerbal, A.; Ferrières, V.; Plusquellec, D. *Chem. Commun.*, **2001**, 2460.

cat $(3,5-(CF_3)_2C_6H_3Se)_2$

cat NaOAc , 60% H_2O_2 , 1 h

CF_3CH_2OH , 20°C

Bu → epoxide Bu 74%

ten Brink, G.-J.; Fernandes, B.C.M.; van Vliet, M.C.A.; Arends, I.W.C.E.; Sheldon, R.A. *J. Chem. Soc., Perkin Trans. 1*, ***2001***, 224.

Bu, Cl, N, t-Bu

$TiCl_3$–$AlMe_3$, DCM

–78°C

Bu, N, Cl, t-Bu + Bu, N, Cl, t-Bu

(90 : 10) quant

Sjöholm, Å.; Hemmerling, M.; Pradeille, N.; Somfai, P. *J. Chem. Soc., Perkin Trans. 1*, ***2001***, 891.

NHw, S, CO_2Me

O, Cl

NEt_3 , PhH

N, S, CO_2Me 80%

Pitts, M.R.; Harrison, J.R.; Moody, C.J. *J. Chem. Soc., Perkin Trans. 1*, ***2001***, 955.

0.25 S_8 , cat $Mo(O)(S_2CNEt_2)_2$

acetone , 56°C , 15 h

S 69%

+ 20% cyclooctene

Sinha, J.; Layek, S.; Mandal, G.C.; Bhattacharjee, M. *Chem. Commun.*, ***2001***, 1916.

C_8H_{17}

10 atm N_2O , PhCl , 140°C

5% Ru-porphyrin complex

O, C_8H_{17} 63%

Yamada, T.; Hashimoto, K.; Kitaichi, Y.; Suzuki, K.; Ikeno, T. *Chem. Lett.*, ***2001***, 268.

$CHCl_3$, $FeSO_4$, SiO_2, 31°C

O

quant

Monfared, H.H.; Ghorbani, M. *Monat. Chem.*, ***2001***, *132*, 989.

O

1% F_3C C_6F_{13}

hexafluoro-2-propanol–H_2O
$NaHCO_3$, Oxone

O

81%

Legros, J.; Crousse, B.; Bourdon, J.; Bonnet-Delpon, D.; Bégué, J.-P.
Tetrahedron Lett., ***2001***, *42*, 4463.

Ph OH , O_2, cat $(CF_3)_2C=O$

Me C_5H_{11}

cat N-hydroxyphthalimide, PhCN
80°C, 1 d

O

Me C_5H_{11}

58% (+ 45% acetophenone)

Iwahama, T.; Sakaguchi, S.; Ishii, Y. *Heterocycles*, ***2000***, *52*, 693.

Ph Ph

Oxone, MeCN–aq EDTA, rt

polymer-bound ketone

O

Ph Ph

94%

Song, C.E.; Lim, J.S.; Kim, S.C.; Lee, K.-J.; Chi, D.Y. *Chem. Commun.*, ***2000***, 2415.

Ph

2 eq mcpba, CH_2Cl_2, –78°C

5% $Cu(MeCN)_4PF_6$, 8 h

O

Ph

77%

Andrus, M.B.; Poehlein, B.W. *Tetrahedron Lett.*, ***2000***, *41*, 1013.

O

H_2O_2, $NaHCO_3$, D_2O, 25°C

O

O

70% conversion (99%)

Yao, H.; Richardson, D.E. *J. Am. Chem. Soc.*, ***2000***, *122*, 3220.

Ph Ph

4 eq PhCHO, $BnEt_3NCl$, rt

1.1 eq chloramine M, 1 d

O

Ph Ph

63%

Yang, D.; Zhang, C.; Wang, X.-C. *J. Am. Chem. Soc.*, ***2000***, *122*, 4039.

PhMe$_2$Si, OTBS, Ph → (BnOCH$_2$CHO, −78°C; BF$_3$•OEt$_2$, DCM) → PhMe$_2$Si, O, OBn, OTBS, Ph 78% (>20:1)

Micalizio, G.C.; Roush, W.R. *Org. Lett.*, ***2000***, *2*, 461.

Ph, O, Ph → (30% aq H_2O_2, LiOH•H_2O, rt; ether, polymer-bound quaternary ammonium salts, 20 h) → Ph, O, O, Ph 98%

Anand, R.V.; Singh, V.K. *Synlett*, ***2000***, 807.

Bu, Bu → (1% MeReO$_3$, 2.5 eq TFA; 2.5 eq Na percarbonate, 6 h; 12% pyrazole) → Bu, O, Bu 85%

Vaino, A.R. *J. Org. Chem.*, ***2000***, *65*, 4210.

C_5H_{11} → (O_2, tetralin, PhEt, PhCN, 60°C, 2 h; 10% N-hydroxyphthalimide, 0.1% Co(OAc)$_2$, 5% Mo(CO)$_6$) → C_5H_{11}, O 79% (99:1 *trans:cis*)

Iwahama, T.; Hatta, G.; Sakaguchi, S.; Ishii, Y. *Chem. Commun.*, ***2000***, 163.

Ph, Ph → (30% aq H_2O_2, acetone; 0.1% Mn$_2$O(OAc)$_2$(TPTN)) → Ph, O, Ph

Brinksma, J.; Hage, R.; Kerschner, J.; Feringa, B.L. *Chem. Commun.*, ***2000***, 537.

cyclohexene → (2% MeReO$_3$, rt, 8 h; urea–H_2O_2) → O 95%

Owens, G.S.; Abu-Omar, M.M. *Chem. Commun.*, ***2000***, 1165.

OH, Ph, C_5H_{11} → (Me$_2$C(Cp)$_2$TiCl$_2$, *t*-BuOOH, 28 h; hexane, 60°C) → OH, O, Ph, C_5H_{11} 80% (44:56 *erythro:threo*)

Sala, G.D.; Giordano, L.; Lattanzi, A.; Porto, A.; Scettri, A. *Tetrahedron*, ***2000***, *56*, 3567.

Oxone , acetone , $NaHCO_3$
0°C , 30 min

HO HO 91% (2:3 α:β)

Ferraz, H.M.C.; Muzzi, R.M.; de O. Viera, T.; Viertler, H. *Tetrahedron Lett.*, **2000**, *41*, 5021.

CO_2Et N F H O

Ph Ph 10% , 10% Oxone Ph Ph quant

$KHSO_5$, $NaHCO_3$, MeCN
aq Na_2EDTA , 2 h

Armstrong, A.; Hayter, B,R.; Moss, W.O.; Reeves, J.R.; Wailer, J.S. *Tetrahedron Asymm.*, **2000**, *11*, 2057.

60% H_2O_2 , CF_3CH_2OH
60°C , 20h
O 99%

Neimann, K.; Neumann, R. *Org. Lett.*, **2000**, *2*, 2861.

Me_2CHCH_2CHO , O_2
DCM , Pd/SiO_2 , rt , 6.5 h
O 98%

Gao, H.; Angelici, R.J. *Synth. Commun.*, **2000**, *30*, 1239.

Co catalyst , 60°C
$Me_3CCH(Me)CH=CH_2$
O 96%

Hunter, R.; Turner, P.; Rimmer, S. *Synth. Commun.*, **2000**, *30*, 4461.

30% aq H_2O_2 (2 eq), 0.5% $MeReO_3$
10% 3-fluoropyridine , DCM
Ph Ph O 97%

Adolfsson, H.; Copéret, C.; Chiang, J.P.; Yudin, A.K. *J. Org. Chem.*, **2000**, *65*, 8651.

H_2O_2 , CF_3COMe , MeCN
aq K_2CO_3
Ph Ph O 93%

Shu, L.; Shi, Y. *J. Org. Chem.*, **2000**, *65*, 8807.

15% H_2O_2, Ph_3SiOEt
$Me_2N-CH(OBn)_2$, H_2O
$[(\pi-C_5H_5N)^+(CH_2)_{10}Me]_3PW_{12}O_{40}$
modified SiO_2

quant

Sakamoto, T.; Pac, C. *Tetrahedron Lett.*, ***2000***, *41*, 10009.

2 eq Oxone, pyrrolidine, 0.5 Py
aq MeCN, 10 eq $NaHCO_3$

Ph, O

95%

Adamo, M.F.A.; Aggarwal, V.K.; Sage, M.A. *J. Am. Chem. Soc.*, ***2000***, *122*, 8317.

OH

$Ti(OCH_2CH_2O)_2$
hexane, 16 h

O, OH

47%

Massa, A.; Screttri, A. *Synlett*, ***2000***, 1348.

OH
0.1 N^+ BF_4^-
Ph
Oxone, 4 eq $NaHCO_3$, 25°C
aq MeCN, 16 h

O, Ph

81%

Minakata, S.; Takemiya, A.; Nakamura, K.; Ryu, I.; Komatsu, M. *Synlett*, ***2000***, 1810.

Bu_2O, 60% H_2O_2 in CF_3CH_2OH
2% $PhAs(CH_2CH_2C_8H_{17})_2$, reflux

O

95%

van Vliet, M.C.A.; Arends, I.W.C.E.; Sheldon, R.A. *Tetrahedron Lett.*, ***1999***, *40*, 5239.

Ph, Ph

Oxone, aq MeCN

O, O, 30% Bn–N, O, N, O, Bn

O, Ph, Ph

70%

Carnell, A.J.; Johnstone, R.A.W.; Parsy, C.C.; Sanderson, W.R.
Tetrahedron Lett., ***1999***, *40*, 8029.

C_5H_{11} (alkene) → $Mg_{10}Al_2(OH)_{24}CO_3$, DCE; 30% aq H_2O_2 , *i*-PrCONH$_2$; Na dedecyl sulfate , 70°C , 1 d → C_5H_{11} (epoxide) 80%

Yamaguchi, K.; Ebitani, K.; Kaneda, K. *J. Org. Chem.*, ***1999***, *64*, 2966.

moist MMPP; DCM , 7 h → 93%

MMPP = monoperoxyphthalic acid

Foti, C.J.; Fields, J.D.; Kropp, P.J. *Org. Lett.*, ***1999***, *1*, 903.

C_5H_{11} (alkene) → Novozym-435 , H_2O_2; $(MeO)_2C=O$ → C_5H_{11} (epoxide) 67%

Klaas, M.Rg.; Warwel, S. *Org. Lett.*, ***1999***, *1*, 1025.

60% H_2O_2 , $C_8F_{17}Br$/PhH , 70°C; 5% 3,5-(C_8F_{17})$_2$C$_6$H$_3$SeBu → 92%

Betzemeier, B.; Lhermitte, F.; Knochel, P. *Synlett*, ***1999***, 489.

Ph (styrene) → PhSH , H-Rho-zeolite , 5 h; hexane , 7 h → Ph–CH$_2$CH$_2$–SPh 69%

Kumar, P.; Pandey, R.K.; Hegde, V.R. *Synlett*, ***1999***, 1921.

H_2O_2 , DCM–EtOAc; 5% perfluoroheptadecan-9-one → quant

van Vliet, M.C.A.; Arends, I.W.C.E.; Sheldon, R.A. *Chem. Commun.*, ***1999***, 263.

0.1 $MeReO_3$, CF_3CF_2OH; 0.5 pyrazole , 60% H_2O_2; 21 h → >99%

van Vliet, M.C.A.; Arends, I.W.C.E.; Sheldon, R.A. *Chem. Commun.*, ***1999***, 821.

H_2/CO (400 psi), PhH
$[Rh(OAc)_2]_2$, 20 h

80%

Bergmann, D.J.; Campi, E.M.; Jackson, W.R.; Patti, A.F. *Chem. Commun.*, ***1999***, 1279.

Oxone, $NaHCO_3$, H_2O
MeCN, 10% chiral iminium salt

89%

Armstrong, A.; Ahmed, G.; Garnett, I.; Goacoleu, K.; Wailes, J.S. *Tetrahedron*, ***1999***, ***55***, 2341.

amino acid-Schiff base Mn complex
O_2, 80°C, 11.5 h

65%

Wang, R.-M.; Hao, C.-J.; Wang, Y.-P. *Synth. Commun.*, ***1999***, ***29***, 1409.

SECTION 135: ETHERS, EPOXIDES AND THIOETHERS FROM MISCELLANEOUS COMPOUNDS

1.5 eq $HGaCl_2$, BEt_3
THF, 0°C, 3-5 h

97% (84:16)

Mikami, S.; Fujita, K.; Nakamura, T.; Yorimitsu, H.; Shinokubo, H.; Matsubara, S.; Oshima, K. *Org. Lett.*, ***2001***, ***3***, 1853.

Sm–NH_4Cl, MeOH
ultrasound, 4h

89%

Yadav, J.S.; Subba Reddy, B.V.; Srinivas, C.; Srihari, P. *Synlett*, ***2001***, 854.

PhSeSePh → 2 EtSePh quant
2 eq EtI, Zn/H_2O
K_2HPO_4

Bieber, L.W.; de Sá, A.C.P.F.; Menezes, P.H.; Gonçalves, S.M.C. *Tetrahedron Lett.*, ***2001***, ***42***, 4597.

$PhC(=N_2)CO_2Me$ → (PhCHO, $Rh_2(OAc)_4$, rt, 1 h; hexane/CH_2Cl_2) → 2,3-diphenyl-2-(CO_2Me)oxirane 76%

Davies, H.M.L.; De Meese, J. *Tetrahedron Lett.*, ***2001***, *42*, 6803.

TolSSTol → (1. Cp_2TiH; 2. $Ph_2I^+X^-$, THF-HMPA, 60°C) → TolSPh 78%

Huang, X.; Wu, L.-L.; Xu, X.-H. *Synth. Commun.*, ***2001***, *31*, 1871.

$PhCH_2S(O)CH_2Ph$ → (2,6-dihydroxypyridine, MeCN, reflux, 4 h) → $PhCH_2SCH_2Ph$ 98%

Miller, S.J.; Collier, T.R.; Wu, W. *Tetrahedron Lett.*, ***2000***, *41*, 3781.

$PhS(O)Ph$ → (2 eq $TiCl_4$, MeCN; 0°C, 15 min) → PhSPh 94%

Shimizu, M.; Shibuya, K.; Hayakawa, R. *Synlett*, ***2000***, 1437.

1-nitro-2-(ethylsulfinyl)cyclohexene → (TMS–SePh, MeOH; DCM) → 1-nitro-2-(phenylseleno)cyclohexene 76%

Abe, H.; Fujii, H.; Yamasaki, A.; Kinome, Y.; Takeuchi, Y.; Harayama, T. *Synth. Commun.*, ***2000***, *30*, 543.

4,5-diphenyl-1,3,2-dioxathiolane 2,2-dioxide (O_2) → (NaOH, THF, MeOH; rt, 1 h) → 2,3-diphenyloxirane 86%

Jang, D.O.; Joo, Y.H.; Cho, D.H. *Synth. Commun.*, ***2000***, *30*, 4489.

$PhS(O)Ph$ → (WCl_6/Zn/THF, 10 min) → PhSPh 96%

Firouzabadi, H.; Karimi, B. *Synthesis*, ***1999***, 500.

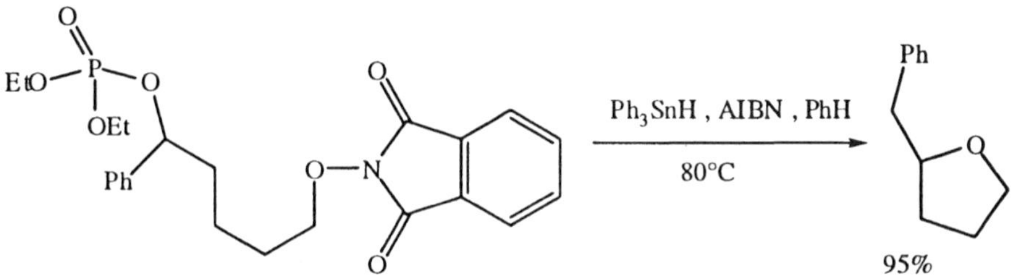

Ph$_3$SnH , AIBN , PhH

80°C

95%

Crich, D.X.; Huang, X.; Newcomb, M. *Org. Lett.*, ***1999***, *1*, 225.

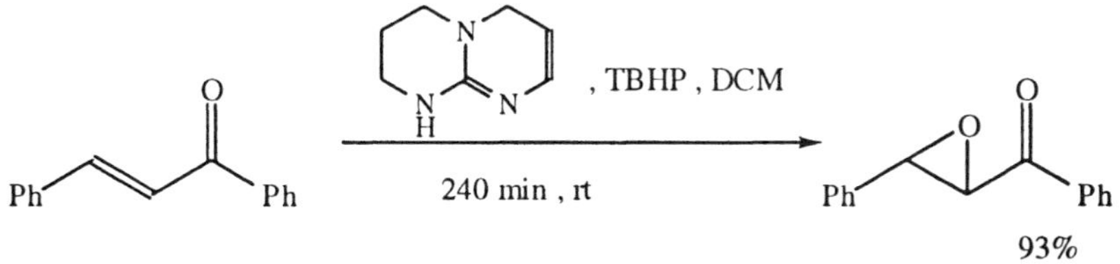

, TBHP , DCM

240 min , rt

93%

Genski,T.; Macdonald, G.; Wei, X.; Lewis, N.; Taylor, R.J.K. *Synlett*, ***1999***, 795.

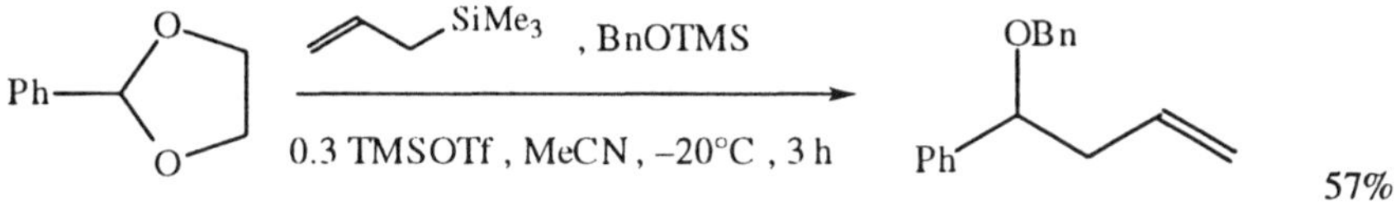

, BnOTMS

0.3 TMSOTf , MeCN , –20°C , 3 h

57%

Suzuki, T.; Oriyama, T. *Synth. Commun.*, ***1999***, *29*, 1263.

CHAPTER 10

PREPARATION OF HALIDES AND SULFONATES

SECTION 136: HALIDES AND SULFONATES FROM ALKYNES

NO ADDITIONAL EXAMPLES

SECTION 137: HALIDES AND SULFONATES FROM ACID DERIVATIVES

Ph–CH=CH–CO_2H → (3 eq Oxone, reflux, NaBr, 3 h; aq MeCN, Na_2CO_3, rt) → Ph–CH=CH–Br 88%

You, H.-W.; Lee, K.-J. *Synlett,* ***2001***, 105.

2-oxo-2H-pyran-3-carboxylic acid (CO_2H) → (NBS, MeCN, LiOAc) → 3-bromo-2H-pyran-2-one (Br) 42%

Cho, C.-G.; Park, J.-S.; Jung, I.-H.; Lee, H. *Tetrahedron Lett.,* ***2001***, *42*, 1065.

Ph–CH=CH–CO_2H → (LiBr, aq MeCN, CAN, rt) → Ph–CH=CH–Br

Roy, S.C.; Guin, C.; Maiti, G. *Tetrahedron Lett.,* ***2001***, *42*, 9253.

Ph–CH=CH–CO_2H → (Na_2MoO_4, KBr, H_2O_2; H_2O) → Ph–CH=CH–Br 65%

Sinha, J.; Layek, S.; Mandal, G.C.; Bhattacharjee, M. *Chem. Commun.,* ***2001***, 1916.

Ph–CH=CH–CO_2H —[NBS, LiOAc, MeCN, H_2O; microwaves, 1 min]→ Ph–CH=CH–Br

84% (95:5 *E*:*Z*)

Kuang, C.; Senboku, H.; Tokuda, M. *Synlett*, ***2000***, 1439.

Ph–≡–CO_2H —[*N*-iodosuccinimide, 10% $Bu_4N^{+-}O_2CCF_3$; CH_2Cl_2, 15 min]→ Ph–≡–I

99% + succinimide

Naskar, D.; Roy, S. *J. Org. Chem.*, ***1999***, *64*, 6896.

SECTION 138: HALIDES AND SULFONATES FROM ALCOHOLS AND THIOLS

1-butanol (OH) —[bmiCl, $MeSO_3H$; rt, 1d]→ 1-chlorobutane (Cl) 98%

bmiCl= ionic liquid

Ren, R.X.; Wu, J.X. *Org. Lett.*, ***2001***, *3*, 3727.

$PhCH_2OH$ —[$BF_3•OEt_2$, KI, dioxane; rt, 5 min]→ $PhCH_2I$ 95%

Bandgar, B.P.; Sadavarte, V.S.; Uppalla, L.S. *Tetrahedron Lett.*, ***2001***, *42*, 951.

Ph–CH₂–SH —[NBS, PPh_3, CH_2Cl_2; rt, 5 h]→ Ph–CH₂–Br 78%

Iranpoor, N.; Firouzabadi, H.; Aghapour, G. *Synlett*, ***2001***, 1176.

cyclopentanol (OH) —[Cl_3CCCl_3, diphos, rt; THF, 3 h]→ cyclopentyl chloride (Cl)

Pollastri, M.P.; Sagal, J.F.; Chang, G. *Tetrahedron Lett.*, ***2001***, *42*, 2459.

2-hydroxy-6-methyl-3-cyanopyridine (CN, N, OH) —[P_2O_5, Bu_4NBr, toluene; 100°C, 1 h]→ 2-bromo-6-methyl-3-cyanopyridine (CN, N, Br) 75%

Kato, Y.; Okada, S.; Tomimoto, K.; Mase, T. *Tetrahedron Lett.*, ***2001***, *42*, 4849.

$PhCH_2OH$ —[$MeSO_3H$, NaI, DCM; reflux, 15 min]→ $PhCH_2I$ 92%

Ram, A.; Ramesh, G.; Laxman, N. *Synth. Commun.*, ***2001***, *31*, 827.

$C_{10}H_{21}$–OH —[IF_5 , Et_3N–3 KF , heptane; 80°C , 2 d]→ $C_{10}H_{21}$–F 32%

Yoneda, N.; Fukuhara. T. *Chem. Lett.*, ***2001***, 222.

2,2-dimethylbutan-1-ol —[$CeCl_3$•7 H_2O , NaI; MeCN , reflux]→ 2-iodo-2-methylbutane (I) 89%

Di Deo, M.; Marcantoni, D.; Torregiani, E.; Bartoli, G.; Bellucci, M.C.; Bosco, M.; Sambri, L. *J. Org. Chem.*, ***2000***, *65*, 2830.

$HO(CH_2)_8OH$ —[48% HBr; cyclohexane]→ $HO(CH_2)_8Br$ 85%

Chong, J.M.; Heuft, M.A.; Rabbat, P. *J. Org. Chem.*, ***2000***, *65*, 5837.

C_4H_9–OH —[$Me_2\overset{+}{N}$=C(Cl)SPh $\overset{-}{Cl}$, DCM]→ C_4H_9–Cl 97%

Gomez, L.; Gellibert, F.; Wagner, A.; Mioskowski, C. *Tetrahedron Lett.*, ***2000***, *41*, 6049.

$C_7H_{15}CH_2OH$ —[TsCl , $Me_2N(CH_2)_3NMe_2$; MeCN , 10°C]→ $C_7H_{15}CH_2OTs$ 94%

Yoshida, Y.; Shimonishi, K.; Sakakura, Y.; Koada, S.; Aso, N.; Tanabe, Y. *Synthesis*, ***1999***, 1633.

$C_7H_{15}CH_2$–OH —[2-chloro-1,3-dimethylimidazolinium chloride (Me–N, N–Me, Cl, Cl^-); NEt_3 , CH_2Cl_2 , rt , 12 h]→ $C_7H_{15}CH_2$–Cl 93%

Isobe, T.; Ishikawa, T. *J. Org. Chem.*, ***1999***, *64*, 5832.

$PhCH_2OH$ —[$SOCl_2$, benzotriazole , rt; dry DCM , 10 min]→ $PhCH_2Cl$ 95%

Chaudhari, S.S.; Akamanchi, K.G. *Synlett*, ***1999***, 1763.

SECTION 139: HALIDES AND SULFONATES FROM ALDEHYDES

PhCHO —[BCl_3 , hexane , reflux]→ Ph—$CHCl_2$ 90%

Kabalka, G.W.; Wu, Z. *Tetrahedron Lett.*, ***2000***, *41*, 579.

$$\text{PhCHO} \xrightarrow[\text{rt , 3 h}]{\text{cyclohexyl–BBr}_2\text{ , hexane}} \text{PhCH}_2\text{Br} \quad 85\%$$

Kabalka, G.W.; Wu, Z.; Ju, Y. *Tetrahedron Lett.*, ***2000***, *41*, 5161.

$$\text{PhCHO} \xrightarrow[\text{2 h , rt}]{\text{BuBCl}_2\text{ , O}_2\text{ , hexane}} \text{PhCH(Cl)Bu} \quad 52\%$$

Kabalka, G.W.; Wu, Z.; Ju, Y. *Tetrahedron Lett.*, ***2001***, *42*, 6239.

SECTION 140: HALIDES AND SULFONATES FROM ALKYLS, METHYLENES AND ARYLS

For the conversion R-H → R-Halogen, see Section 146 (Halides from Hydrides).

NO ADDITIONAL EXAMPLES

SECTION 141: HALIDES AND SULFONATES FROM AMIDES

NO ADDITIONAL EXAMPLES

SECTION 142: HALIDES AND SULFONATES FROM AMINES

NO ADDITIONAL EXAMPLES

SECTION 143: HALIDES AND SULFONATES FROM ESTERS

$$\text{cyclohexyl–O–C(O)Cl} \xrightarrow[\text{115°C} \rightarrow \text{120°C , 4 h}]{\text{0.01\% hexabutylguanidinum chloride}} \text{cyclohexyl–Cl} \quad 93\%$$

Violleau, F.; Thiébaud, S.; Borredon, E.; Le Gars, P. *Synth. Commun.*, ***2001***, *31*, 367.

$$\text{PhCH}_2\text{CH(OCS}_2\text{Me)CH}_2\text{CH}_2\text{CH}_3 \xrightarrow[\text{1 h}]{\text{70\% HF/Py , NIS , DCM}} \text{PhCH}_2\text{CH(F)CH}_2\text{CH}_2\text{CH}_3 \quad 70\%$$

Kanie, K.; Tanaka, Y.; Suzuki, K.; Kuroboshi, M.; Hiyama, T.
Bull. Chem. Soc. Jpn., ***2000***, *73*, 471.

SECTION 144: HALIDES AND SULFONATES FROM ETHERS, EPOXIDES AND THIOETHERS

$CeCl_3$•7 H_2O , NaI , $MeNO_2$

propanedithiol , 30 min

O

I

86%

Bartoli, G.; Cupone, G.; Dalpozzo, R.; De Nino, A.; Maiuolo, L.; Marcantoni, E.; Procopio, A. *Synlett*, ***2001***, 1897.

SECTION 145: HALIDES AND SULFONATES FROM HALIDES AND SULFONATES

H_3C $(CH_2)_6$ Br → Me_2SiF_2Ph NBu_4 , MeCN / reflux , 1 d → H_3C $(CH_2)_6$ F 72%

Kvíčala, J.; Mysík, P.; Paleta, O. *Synlett*, ***2001***, 547.

MeO—C6H4—Br → 1. BuLi , THF , –78°C; 2. I (iodomethyl lactone) O , 90 min → MeO—C6H4—I 68%

Harrowven, D.C.; Nunn, M.I.T.; Fenwick, D.R. *Tetrahedron Lett.*, ***2001***, *42*, 7501.

$Me(CH_2)_{13}$—I → *p*-Tol-IF_2 , NEt_3–THF , DCM / fluororesin reaction vessel , rt → $Me(CH_2)_{13}$—F 72%

Sawaguchi, M.; Hara, S.; Nakamura, Y.; Ayuba, S.; Fukuhara, T.; Yoneda, N. *Tetrahedron*, ***2001***, *57*, 3315.

$Me(CH_2)_{13}$—I → e^- , Et_3N–n HF , DCM → $Me(CH_2)_{13}$—F 80%

Sawaguchi, M.; Ayuba, S.; Nakamura, Y.; Fukuhara, T.; Hara, S.; Yoneda, N. *Synlett*, ***2000***, 999.

SECTION 146: HALIDES AND SULFONATES FROM HYDRIDES

α-Halogenations of aldehydes, ketones and acids are found in Sections 338 (Halide-Aldehyde), 369 (Halide-Ketone), 359 (Halide-Esters) and 319 (Halide-Acids).

Ph, Br — In, EtOH, aq NH_4Cl, 16 h → Ph — 95% (95:5 *E:Z*)

Ranu, B.C.; Samanta, S.; Guchhait, S.K. *J. Org. Chem.*, ***2001***, *66*, 4102.

NH_2 — $BnNEt_3^+ ICl_2^-$, DCM/MeOH → I — NH_2 99%

Kesynkin, D.V.; Tour, J.M. *Org. Lett.*, ***2001***, *3*, 991.

CBr_4 / CH_2Cl_2, 50% NaOH, 25°C, 1 d, 10% Bu_4NBr → Br 75%

also prepared iodide and chloride

Fokin, A.A.; Lauenstein, O.; Gunchenko, P.A.; Schreiner, P.R. *J. Am. Chem. Soc.*, ***2001***, *123*, 1842.

NH_2, MeO — SO_2Cl_2, AcOH, rt → NH_2•HCl, Cl, MeO 63%

Yu, G.; Mason, H.J.; Wu, X.; Endo, M.; Douglas, J.; Macor, J.E. *Tetrahedron Lett.*, ***2001***, *42*, 3247.

OMe — CAN, MeCN, LiBr, rt → OMe 70% (3:7 *o:p*)

Roy, S.C.; Guin, C.; Rana, K.K.; Maiti, G. *Tetrahedron Lett.*, ***2001***, *42*, 6941.

N, NH_2 — NBS, MeCN, rt, 17 h → Br, N, NH_2 93%

Cañibano, V.; Rodríguez, J.F.; Santos, M.; Sanz-Tejefor, A.; Carreño, M.C.; González, G.; García-Ruano, J.L. *Synthesis*, ***2001***, 2175.

OMe — Oxone, KBr, H_2O, MeCN, rt → Br — OMe 89%

Tamhankar, B.V.; Desai, U.V.; Mane, R.B.; Wadgaonkar, P.P.; Bedekar, A.V. *Synth. Commun.*, ***2001***, *31*, 2021.

$POCl_3$, NEt_3 , DCM

40°C , 1 h

90%

Jung, J.-C.; Jung, Y.-J.; Park, O.-S. *Synth. Commun.*, ***2001***, *31*, 2507.

$NaIO_4$, I_2 , cat H_2SO_4

EtOH , 65°C , 1 h

75%

Bonesi, S.M.; Erra-Balsells, R. *J. Heterocyclic Chem.*, ***2001***, *38*, 77.

1. LiHMDS , THF
2. fluorinated Cinchona alkaloid

−78°C

86% (76% ee)

Mohar, B.; Baudoux, J.; Plaquevent, J.-C.; Cahard, D. *Angew. Chem. Int. Ed.*, ***2001***, *40*, 4214.

1. NCI , 90% H_2SO_4 , rt , 20°C

2. ice water

79%

Chaikovskii, V.K.; Shorokhodov, V.I.; Filimonov, V.D. *Russ. J. Org. Chem.*, ***2001***, *37*, 1503.

Br_2 , SiO_2 , CCl_4 , rt

94%

Ghiaci, M.; Asghari, J. *Bull. Chem. Soc. Jpn.*, ***2001***, *74*, 1151.

ICl , ZnO , DCM ,rt

5% $Cp_2FeB[3.5(CF_3)_2C_3H_6]_4$

8 h

–I

68% 47/53 (o/p)

Mukaiyama, T.; Kitagawa, H.; Matsuo, J.-i. *Tetrahedron Lett.*, ***2000***, ***41***, 9383.

NH_2 Br

KBr , $NaBO_3$• 4 H_2O

AcOH , rt , 145 h

NH_2 Br Br

84%

Roche, D.; Prasad, K.; Repic, O.; Blacklock, T.J. *Tetrahedron Lett.*, ***2000***, ***41***, 2083.

t-BuOBr , HNaX-zeolite

ether , CCl_4 , 25°C , 1 h

Br–

49%

Smith, K.; El-Hiti, G.A.; Hammond, M.E.W.; Bahzad, D.; Li, Z.; Siquet, C. *J. Chem. Soc., Perkin Trans. 1*, ***2000***, 2745.

–OMe

NBS , acetone , cat HCl

1 min

Br– –OMe

98%

Andersh, B.; Murphy, D.L.; Olson, R.J. *Synth. Commun.*, ***2000***, ***30***, 2091.

–OMe

HZSM-5 , AcOH , 2 h

KBr , 30% H_2O_2 , rt

Br– –OMe

98%

Narender, N.; Srinivasu, P.; Kulkarni, S.J.; Raghavan, K.V. *Synth. Commun.*, ***2000***, ***30***, 3669.

Ph–H

I_2 , $NaIO_4$, H_2SO_4

AcOH , Ac_2O

Ph–I

65%

Luliński, P.; Skulski, L. *Bull. Chem. Soc. Jpn.*, ***2000***, ***73***, 951.

NC– –NH_2

NCS , *i*-PrOH

Cl

NC– –NH_2

96%

Zanka, A.; Kubota, A. *Synlett*, ***1999***, 1984.

NaOH , CHI_3 , 1 d , rt

92%

Schreiner, P.R.; Lauenstein, O.; Butova, E.D.; Fokin, A.A. *Angew. Chem. Int. Ed.*, ***1999***, *38*, 2786.

OTs

$(PhMe_2Si)_2Cu(CN)Li_2$

THF , –78°C → rt

$SiMe_2Ph$

75%

Chambers, R.D.; Parsons, M.; Sandford, G.; Skinner, C.J.; Atherton, M.J.; Moilliet, J.S. *J. Chem. Soc., Perkin Trans. 1*, ***1999***, 803.

OMe

t-BuOOH , HBr , H_2O_2

Br OMe 76%

Barhate, N.B.; Gajare, A.S.; Wakharkar, R.D.; Bedekar, A.V. *Tetrahedon*, ***1999***, *55*, 11127.

S Bu

NBS , AcOH , 8 min

S Br Bu

91%

Hoffmann, K.J.; Carlsen, P.H.J. *Synth. Commun.*, ***1999***, *29*, 1607.

PhH $\xrightarrow[AcOH,\ Ac_2O,\ 10°C]{H_2SO_4,\ MnO_2,\ I_2}$ PhI 63%

Luliński, P.; Skulski, L. *Bull. Chem. Soc. Jpn.*, ***1999***, *72*, 115.

MeO

–$I(OAc)_2$, I_2 , AcOEt

rt , dark , 16 h

MeO I

56%

Togo, H.; Abe, S.; Nogami, G.; Yokoyama, M. *Bull. Chem. Soc. Jpn.*, ***1999***, *72*, 2351.

SECTION 147: HALIDES AND SULFONATES FROM KETONES

NO ADDITIONAL EXAMPLES

SECTION 148: HALIDES AND SULFONATES FROM NITRILES

NO ADDITIONAL EXAMPLES

SECTION 149: HALIDES AND SULFONATES FROM ALKENES

For halocyclopropanations, see Section 74E (Alkyls from Alkenes).

$Mg(ClO_4)_2$, BEt_3/O_2
chiral bis-oxazoline catalyst
toluene, –78C, 5 h

O, CO_2Et, Br — O, CO_2Et, Br

67% (94% ee)

Yang, D.; Gu, S.; Yan, Y.-L.; Zhu, N.-Y.; Cheung, K.-K.
J. Am. Chem. Soc., ***2001***, *123*, 8612.

chiral Ru catalyst, CCl_4
rt, 1 d

Ph — CCl_3, Cl, Ph

90%

Simal, F.; Wlodarczak, L.; Demonceau, A.; Noels, A.F. *Eur. J. Org. Chem.*, ***2001***, 2689.

Me_2N-C_6H_4-Cl
hv, MeCN
Ar = 4-dimethylaminophenyl

Bu — Ar, Cl, Bu 35% + Cl, Ar, Bu 18%

Mella, M.; Coppo, P.; Guizzardo, B.; Fagnoni, M.; Freccero, M.; Albini, A.
J. Org. Chem., ***2001***, *66*, 6344.

1. 5% Cp_2ZrCl_2, $C_8H_{17}OTs$
THF, 55°C
MgBr
2. NBS, 0°C

Ph — C_8H_{17}, Ph, Br

73%

de Armas, J.; Hoveyda, A.H. *Org. Lett.*, ***2001***, *3*, 2097.

zirconium dihydrate reagent
THF, rt, 4 h

C_5H_{11} — C_5H_{11}, Br, C_5H_{11}, Br

82%

Wipf, P.; Wang, X. *Tetrahedron Lett.*, ***2000***, *41*, 8237.

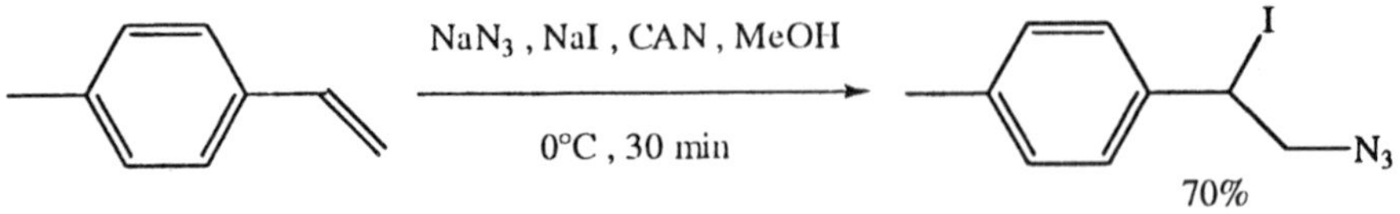

NaN_3, NaI, CAN, MeOH

0°C, 30 min

I

N_3

70%

Nair, V.; George, T.G.; Sheeba, V.; Augustine, A.; Balagopal, L.; Nair, L.G. *Synlett*, ***2000***, 1597.

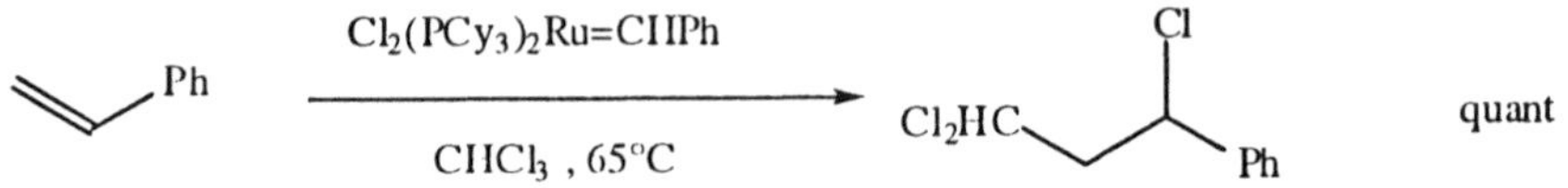

Ph

$Cl_2(PCy_3)_2Ru{=}CHPh$

$CHCl_3$, 65°C

Cl

Cl_2HC

Ph

quant

Tallarico, J.A.; Malnick, L.M.; Snapper, M.L. *J. Org. Chem.*, ***1999***, *64*, 344.

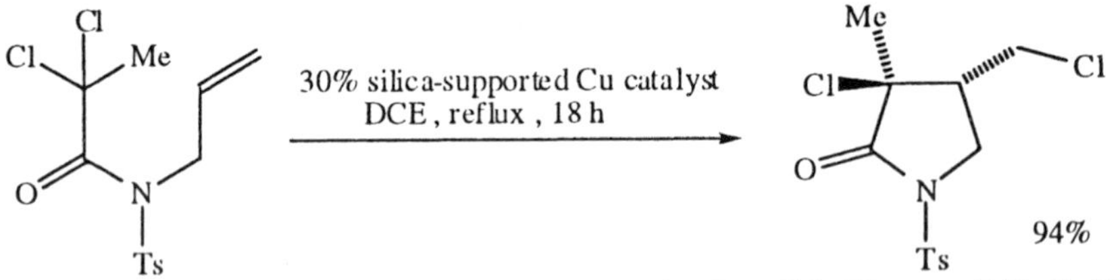

Cl Cl Me O N Ts

30% silica-supported Cu catalyst
DCE, reflux, 18 h

Me Cl Cl O N Ts

94%

Clark, A.J.; Filik, R.P.; Haddleton, D.M.; Radigue, A.; Sanders, C.J.; Thomas, G.H.; Smith, M.E. *J. Org. Chem.*, ***1999***, *64*, 8954.

SECTION 150: HALIDES AND SULFONATES FROM MISCELLANEOUS COMPOUNDS

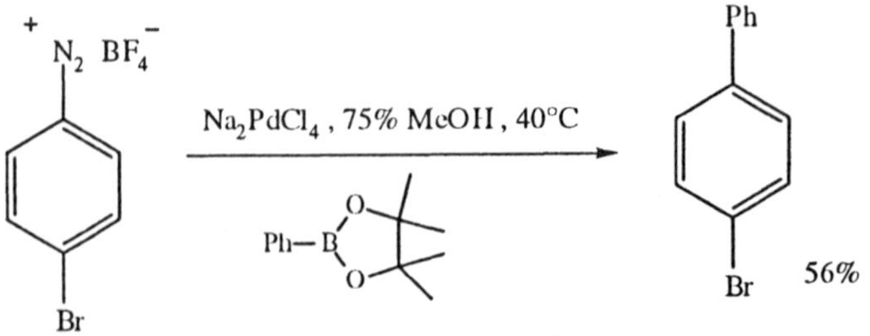

$N_2^+ BF_4^-$

Br

Na_2PdCl_4, 75% MeOH, 40°C

Ph—B(O)(O) (pinacol ester)

Ph

Br

56%

Willis, D.M.; Strongin, R.M. *Tetrahedron Lett.*, ***2000***, *41*, 6271.

CHAPTER 11

PREPARATION OF HYDRIDES

This chapter lists hydrogenolysis and related reactions by which functional groups are replaced by hydrogen: e.g. $RCH_2X \rightarrow RCH_2\text{-}H$ or R-H.

SECTION 151: HYDRIDES FROM ALKYNES

NO ADDITIONAL EXAMPLES

SECTION 152: HYDRIDES FROM ACID DERIVATIVES

This section lists examples of decarboxylations ($RCO_2H \rightarrow R\text{-}H$) and related reactions.

CO_2H microwaves, 15 min CO_2H 98%

Ph CO_2H 190°C Ph

Zara, C.L.; Jin, T.; Giguere, R.J. *Synth. Commun.*, ***2000***, *30*, 2099.

SECTION 153: HYDRIDES FROM ALCOHOLS AND THIOLS

This section lists examples of the hydrogenolysis of alcohols and phenols (ROH → R-H).

OH $InCl_3$, 2 eq Ph_2SiHCl, DCM

Ph Ph rt, 2 h Ph Ph 87%

Yasuda, M.; Onishi, Y.; Ueba, M.; Miyai, T.; Baba, A. *J. Org. Chem.*, ***2001***, *66*, 7741.

OH H_3PO_2, cat I_2, AcOH

Ph Ph 60°C Ph Ph quant

Gordon, P.E.; Fry, A.J. *Tetrahedron Lett.*, ***2001***, *42*, 831.

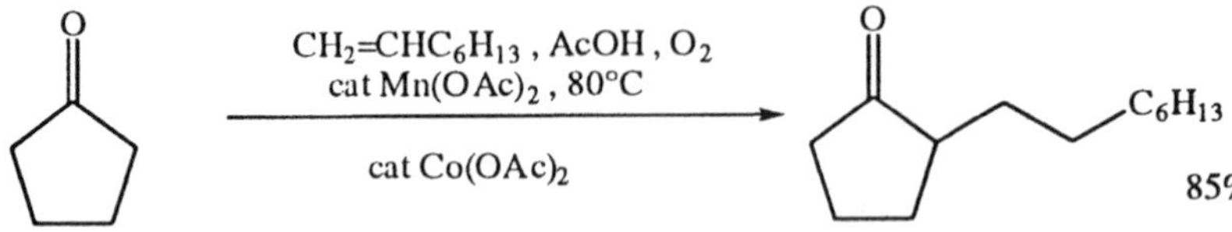

CH_2=CHC_6H_{13} , AcOH , O_2
cat $Mn(OAc)_2$, 80°C
cat $Co(OAc)_2$

C_6H_{13}

85%

Crudden, C.M.; Allen, D.; Mikeluk, M.D.; Sun, J. *Chem. Commun.*, **2001**, 1154.

Also via: Section 160 (Halides and Sulfonates).

SECTION 154: HYDRIDES FROM ALDEHYDES

For the conversion RCHO → R-Me, etc., see Section 64 (Alkyls from Aldehydes).

NO ADDITIONAL EXAMPLES

SECTION 155: HYDRIDES FROM ALKYLS, METHYLENES AND ARYLS

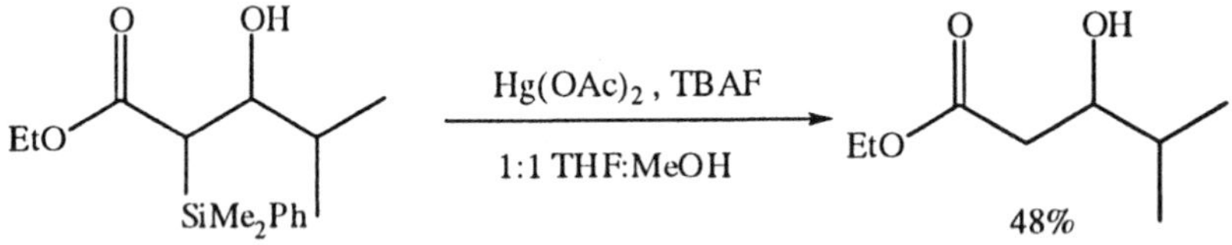

O OH

EtO

$SiMe_2Ph$

$Hg(OAc)_2$, TBAF

1:1 THF:MeOH

O OH

EtO

48%

Poliskie, G.M.; Mader, M.M.; van Well, R. *Tetrahedron Lett.*, ***1999**, 40*, 589.

SECTION 156: HYDRIDES FROM AMIDES

NO ADDITIONAL EXAMPLES

SECTION 157: HYDRIDES FROM AMINES

This section lists examples of the conversion RNH_2 (or R_2NH) → R-H.

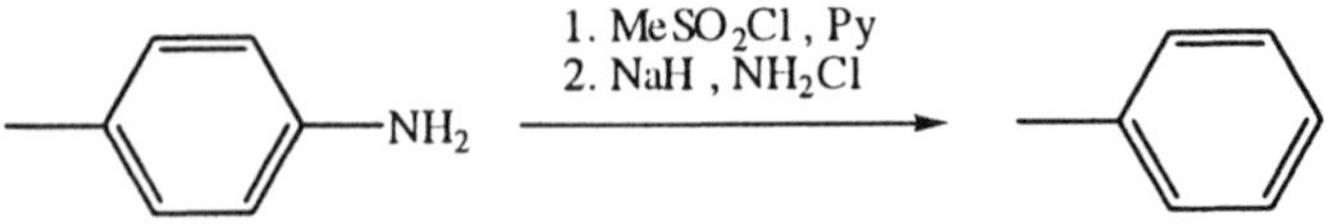

NH_2

1. $MeSO_2Cl$, Py
2. NaH , NH_2Cl

96x87%

Wang, Y.; Guziec Jr., F.S. *J. Org. Chem.*, ***2001**, 66*, 8293.

NaNO$_2$, AcOH , EtOH

aq Na bisulfite , rt , 20 h

80%

Geoffroy, O.J.; Morinelli, T.A.; Meier, G.B. *Tetrahedron Lett.*, ***2001***, *42*, 5367.

2 eq Li , THF , 1 d , rt

quant

Azzena, U.; Dessanti, F.; Melloni, G.; Pisano, L. *Tetrahedron Lett.*, ***1999***, *40*, 8291.

SECTION 158: HYDRIDES FROM ESTERS

This section lists examples of the reactions RCO$_2$R' → R-H and RCO$_2$R' → R'H.

HCO$_2$NH$_2$
cat Rh$_3$(CO)$_{12}$

dioxane , 160°C
20 h

76%

Chatani, N.; Tatamidani, H.; Ie, Y.; Kakiuchi, F.; Murai, S.
J. Am. Chem. Soc., ***2001***, *123*, 4849.

Ph$_2$SiH$_2$, (*t*-BuO)$_2$

140°C , 20 h

68%

Tang, D.Q.; Kim. J.; Cho, D.H.; Chung, C.-M. *Tetrahedron Lett.*, ***2001***, *42*, 1073.

1. 10% Cp$_2$TiCl$_2$, THF
i-PrMgBr , rt , 4 h

2. 2N HCl

71%

Yu, Y.; Zhang, Y. *Synth. Commun.*, ***1999***, *29*, 243.

SECTION 159: HYDRIDES FROM ETHERS, EPOXIDES AND THIOETHERS

This section lists examples of the reaction R-O-R' → R-H.

SPh, Ph, OBu → 5 eq SmI_2, *t*-BuOH, 1 d; PhH–HMPA → Ph, OBu 80%

Nakata, D.; Kusaka, C.; Tani, S.; Kunishima, M. *Tetrahedron Lett.*, ***2001***, *42*, 415.

Ph, $OSiMe_2$ → 1. "Cp_2Zr" 2. D_2O; "Cp_2Zr" = (Cp_2ZrCl_2/2 eq BuLi/THF, –78°C) → Ph, D 80%

Ganchegui, B.; Bertus, P.; Szymoniak, J. *Synlett*, ***2001***, 123.

SECTION 160: HYDRIDES FROM HALIDES AND SULFONATES

This section lists the reductions of halides and sulfonates, R-X → R-H.

Br, HN, O, N → DCO_2K, H_2O, DMSO; cat $Pd(OAc)_2$, 20 sec microwaves → D, HN, O, N >95%

Jones, J.R.; Lockley, W.J.S.; Lu, S.-Y.; Thompson, S.P. *Tetrahedron Lett.*, ***2001***, *42*, 331.

Me, NH_2, Br, Br, NH_2 → AcOH, HBr, aniline; reflux, 4 h → Me, NH_2, NH_2 70%

Choi, H.; Chi, D.Y. *J. Am. Chem .Soc.*, ***2001***, *123*, 9202.

2-chloronaphthalene → naphthalene: Ni/C, PPh_3, K_2CO_3 / $Me_2NH{\bullet}BH_3$, 7.5 h; 99%

Lipshutz, B.H.; Tomioka, T.; Sato, K. *Synlett*, ***2001***, 970.

$PhCH_2Br$ → $PhCH_3$: $InCl_3$, Bu_3SnH, THF / rt, 2 h; 86%

Inoue, K.; Sawada, A.; Shibata, I.; Baba, A. *Tetrahedron Lett.*, ***2001***, *42*, 4661.

Br (mesityl bromide) → ^2H (deuterated mesitylene): 1. BuLi, THF, –78°C; 2. 9-(^{2}H]-9-Ph-fluorene / –78°C → rt; 89%

Cintrat, J.-C.; Pillon, F.; Rousseau, B. *Tetrahedron Lett.*, ***2001***, *42*, 5001.

NC–C$_6$H$_4$–Cl → NC–C$_6$H$_5$: 3% $NiCl_2(PPh_3)_2$, 3% PPh_3, 18 h / $Me_2NH{\bullet}BH_3$, K_2CO_3, MeCN, 40°C; quant

Lipshutz, B.H.; Tomioka, T.; Pfeiffer, S.S. *Tetrahedron Lett.*, ***2001***, *42*, 7737.

PhCOCH$_2$Cl → PhCOCH$_3$: decaborane, 10% Pd/C / MeOH, rt, 5 h; 94%

Lee, S.H.; Jung, Y.J.; Cho, Y.J.; Yoon, C.-O.M.; Hwang, H.-J.; Yoon, C.M. *Synth. Commun.*, ***2001***, *31*, 2251.

$C_{12}H_{25}$–I → $C_{12}H_{25}$–H: (2-furyl)$_3$GeH, BEt_3, THF, 1 h; 99%

Nakamura, T.; Yorimitsu, H.; Shinokubo, H.; Oshima, K. *Bull. Chem. Soc. Jpn.*, ***2001***, *74*, 747.

Ph(CH$_2$)$_3$Br → Ph(CH$_2$)$_2$CH$_3$: Ni(R), THF, reflux, 12 h; 89%

Barrero, A.F.; Alvarez-Manzaneda, E.J.; Chahboun, R.; Meneses, R.; Romera, J.L. *Synlett*, ***2001***, 485.

OH, I, i-Pr, CO_2Bn — Al–Hg → OH, i-Pr, CO_2Bn 93%

Wang, Y.-C.; Yan, T.-H. *Chem. Commun.*, ***2000***, 545.

O, Cl, Ph — In , H_2O , 80°C , 3 h / 0.01M sodium dedecylsulfate → O, Ph >99%

Park, L.; Keum, G.; Kang, S.B.; Kim, K.S.; Kim, Y.
J. Chem. Soc., Perkin Trans. 1, ***2000***, 4462.

C_8H_{17}, Br — cat [$Ni(OAc)_2$/$Al(acac)_3$] , NaH / THF , reflux , 15 min , *t*-BuOH → C_8H_{17} quant

Massicot, F.; Schneider, R.; Fort, Y.; Illy-Cherrey, S.; Tillement, O.
Tetrahedron, ***2000***, *56*, 4765.

N–*i*-Pr, Br, Ph — $SnCl_2$, MeOH , rt , 5 h → N–*i*-Pr, Ph 93%

Aelterman, W.; Eeckhaut, A.; De Kimpe, N. *Synlett*, ***2000***, 1283.

Ph, Br, Br — 4 eq $(MeO)_2P(O)H$, 16 h / 4.5 eq NEt_3 , DMF , 70°C → Ph, Br 67%

Abbas, S.; Hayes, C.J.; Worden, S. *Tetrahedron Lett.*, ***2000***, *41*, 3215.

Ph, Br — (polymer)–O–$SnBu_2Cl$ / DMA , AIBN , 80°C / $NaBH_4$, 6.5 h → Ph 93%

Enholm, E.J.; Schulte II, J.P. *Org. Lett.*, ***1999***, *1*, 1275.

$C_{12}H_{25}$–I — (2-furyl)$_3$GeH , BEt_3 / THF , 1 h → $C_{12}H_{25}$–H 92%

Nakamura, T.; Yorimitsu, H.; Shinokubo, H.; Oshima, K. *Synlett*, ***1999***, 1415.

MeO–C6H4–Cl —[3 eq *i*-PrMgBr , THF , rt / 10% $(C_5H_5)_2TiCl_2$, 2 d]→ MeO–C6H5 93%

Hara, R.; Sato, K.; Sun, W.-H.; Takahashi, T. *Chem. Commun.*, ***1999***, 845.

$PhCH_2Br$ —[Bu_4NBH_4 , THF , rt , 5 min]→ $PhCH_3$ 90%

Narasimhan, S.; Swarnalakshmi, S.; Balakumar, R.; Velmathi, S. *Synth. Commun.*, ***1999***, *29*, 685.

Bromocyclohexane —[Pd/MCM-41 , EtOH , H_2]→ cyclohexane 43%

Kantam, M.L.; Rahman, A.; Bandyopadhyay, T.; Haritha, Y. *Synth. Commun.*, ***1999***, *29*, 691.

SECTION 161: HYDRIDES FROM HYDRIDES

NO ADDITIONAL EXAMPLES

SECTION 162: HYDRIDES FROM KETONES

This section lists examples of the reaction $R_2C\text{-}(C{=}O)R \rightarrow R_2C\text{-}H(R)$.

Cyclopentane —[cat ($C_{12}H_{25}O_2C$-substituted N-hydroxyphthalimide, N-OH) / Co(II) , Mn(II) , 100°C , 14 h]→ cyclopentanone (41%) + cyclopentanol (OH) (13%) + glutaric acid (CO_2H, CO_2H) (46%)

Sawatari, N.; Yokota, T.; Sakaguchi, S.; Ishii, Y. *J. Org. Chem.*, ***2001***, *66*, 7889.

2-Acetylcyclopentane-1,3-dione —[2.5 eq $NaBH_3CN$, THF / 2N aq H^+]→ 2-ethylcyclopentane-1,3-dione 75%

Pashkovsky, F.S.; Lokot, I.P.; Lakhvich, F.A. *Synlett*, ***2001***, 1391.

H_3PO_2, cat I_2 / AcOH, reflux — quant

Hicks, L.D.; Han, J.K.; Fry, A.J. *Tetrahedron Lett.*, ***2000***, *41*, 7817.

Pt, K-10, 50 bar H_2 / 170°C, 15 h — 98%

Török, B.; London, G.; Bartók, M. *Synlett, **2000***, 631.

cat $RuCl_2(PPh_3)_3$, PhH / *t*-BuOOH, 20°C — 87%

Murahashi, S.-I.; Komiya, N.; Oda, Y.; Kuwabara, T.; Naota, T.
J. Org. Chem., ***2000***, *65*, 9186.

$KMnO_4$, $CuSO_4$•5 H_2O / DCM, ultrasound, 3 h — 70%

Mečiarová, M.; Toma, S.; Heribanová, A. *Tetrahedron, **2000***, *56*, 8561.

Me_2SiClH, 5% $InCl_3$ / DCM, 25°C, 2 h — 99%

Miyai, T.; Ueba, M.; Baba, A. *Synlett, **1999***, 182.

1. N_2H_4, toluene, microwaves 16 min / 2. KOH, microwaves, 30 min — 95%

Gadhwal, S.; Baruah, M.; Sandhu, J.S. *Synlett, **1999***, 1573.

SECTION 163: HYDRIDES FROM NITRILES

This section lists examples of the reaction, R-C≡N → R-H (includes reactions of isonitriles (R-N≡C).

NO ADDITIONAL EXAMPLES

SECTION 164: HYDRIDES FROM ALKENES

NO ADDITIONAL EXAMPLES

SECTION 165: HYDRIDES FROM MISCELLANEOUS COMPOUNDS

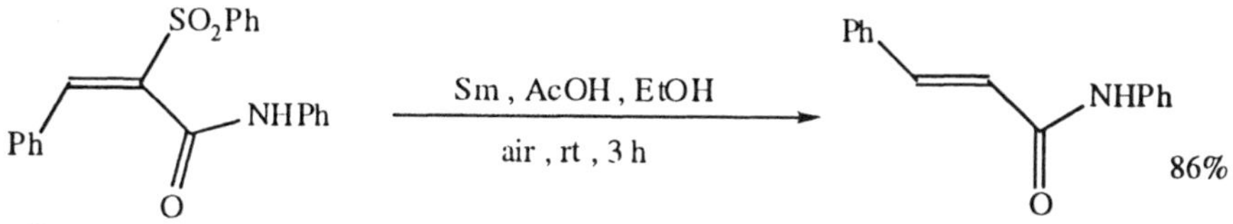

SO_2Ph

Ph NHPh O

Sm , AcOH , EtOH

air , rt , 3 h

Ph NHPh O

86%

Liu, Y.; Zhang, Y. *Org. Prep. Proceed. Int.*, ***2001***, *33*, 376.

CHAPTER 12
PREPARATION OF KETONES

SECTION 166: KETONES FROM ALKYNES

1) $CO_2(CO)_8$, hexanes
2) NMO , DCM
, –38°C

Ph
+
Ph
50%
15%

Marchueta, I.; Verdaguer, X.; Moyano, A.; Pericàs, M.A.; Riera, A. *Org. Lett.*, ***2001***, *3*, 3193.

NO_2
$SnCl_2$, aq EtOH
NH_2
75%

Bosch, E.; Jefferies, L. *Tetrahedron Lett.*, ***2001***, *42*, 8141.

Ph
10% $RuCp(MeCN)_3PF_6$, acetone
10% CSA , 1.3 H_2O
MeO_2C CO_2Me
Ph
MeO_2C
CO_2Me
75%

Trost, B.M.; Brown, R.E.; Toste, F.D. *J. Am. Chem. Soc.*, ***2000***, *122*, 5877.

Ph—≡ → (cat Tf_2NH, H_2O, dioxane; 100°C, 2 d) → PhC(O)Me 97%

Tsuchimoto, T.; Joya, T.; Shirakawa, E.; Kawakami, Y. *Synlett*, ***2000***, 1777.

Ph—≡—Ph → (1. $PhNH_2$, 1% Cp_2TiMe_2, 100°C, toluene; 2. SiO_2, DCM) → $PhCH_2C(O)Ph$ 92%

Haak, E.; Bytschkov, I.; Doye, S. *Angew. Chem. Int. Ed.*, ***1999***, *38*, 3389.

SECTION 167: KETONES FROM ACID DERIVATIVES

PhC(O)OH → (1. CDMT, NMM, THF, rt; 2. EtMgX, CuI, 0°C) → PhC(O)Et 98%

CDMT = 2-chloro-4,6-dimethoxy-[1,3,5]-triazine

De Luca, L.; Giacomelli, G.; Porcheddu, A. *Org. Lett.*, ***2001***, *3*, 1519.

$HO_2C(CH_2)_4CO_2H$ → (graphite, 450°C, 760 mm) → cyclopentanone 85%

microwaves (2x2 min)x6, 450°C 90%

Barrero, A.F.; Alvarez-Manzaneda, E.J.; Chahboun, R.; Meneses, R.; Romera, J.L. *Synlett*, ***2001***, 485.

2-(NHBoc)-3-F-C_6H_3C(O)OH → (MeMgBr, THF, 50°C) → 2-(NHBoc)-3-F-C_6H_3C(O)Me 80%

<5% when F is replaced with H

Zhang, P.; Terefenko, E.A.; Slavin, J. *Tetrahedron Lett.*, ***2001***, *42*, 2097.

4-Br-C_6H_4C(O)Cl → ($NaBPh_4$, KF, acetone; cat $Pd(PPh_3)_2Cl_2$, 9 min, microwaves) → 4-Br-C_6H_4C(O)Ph 98%

Wang, J.-X.; Wei, B.; Hu, Y.; Liu, Z.; Yang, Y. *Synth. Commun.*, ***2001***, *31*, 3885.

Bu_3Sn (methallyl) + PhCOCl, $InCl_3$, MeCN, PPh_3, rt, 2.5 h → PhCOCH$_2$C(=CH$_2$)CH$_3$ 99%

Inoue, K.; Shimizu, Y.; Shibata, I.; Baba, A. *Synlett*, ***2001***, 1659.

PhCOOH → $PhB(OH)_2$, 3% $Pd(F_6\text{-acac})_2$, 60°C; 9% PCy_3, 2 eq Na_2CO_3, 20 h, disuccinoyl anhydride → PhCOPh 90%

Gooßen, L.J.; Ghosh, K. *Chem. Commun.*, ***2001***, 2084.

PhCOBr → e^-, $Bu_4N\ ClO_4$ → PhCOCOPh + PhC($PhCO_2$)=C(O_2CPh)Ph

Pb cathode/MeCN 65% 24%

Sn cathode/MeCN 0% 78%

Kise, N.; Ueda, N. *Bull. Chem. Soc. Jpn.*, ***2001***, *74*, 755.

PhCOCl → 1. $Pd(PPh_3)_4$ 2. KOAc 3. BBu_3 → PhCOBu 74%

Kabalka, G.W.; Malladi, R.R.; Tejedor, D.; Kelley, S. *Tetrahedron Lett.*, ***2000***, *41*, 999.

PhC(OH)(CH$_3$)CO_2H → mesoporous silica FSM-16, hν, hexane, rt, 12 h → PhCOCH$_3$ 78%

Itoh, A.; Kodama, T.; Inagaki, S.; Masaki, Y. *Org. Lett.*, ***2000***, *2*, 331.

Bu-cyclopropyl-$B(OH)_2$ → BzCl, cat $PdCl_2(dppf)$, Ag_2O, K_2CO_3, 80°C → Bu-cyclopropyl-COPh 78%

Chen, H.; Deng, M.-Z. *Org. Lett.*, ***2000***, *2*, 1649.

ClCOPh → $PhB(OH)_2$, Cs_2CO_3, toluene, 5% $Pd(PPh_3)_4$, 100°C → PhCOPh 80%

Haddach, M.; McCarthy, J.R. *Tetrahedron Lett.*, ***1999***, *40*, 3109.

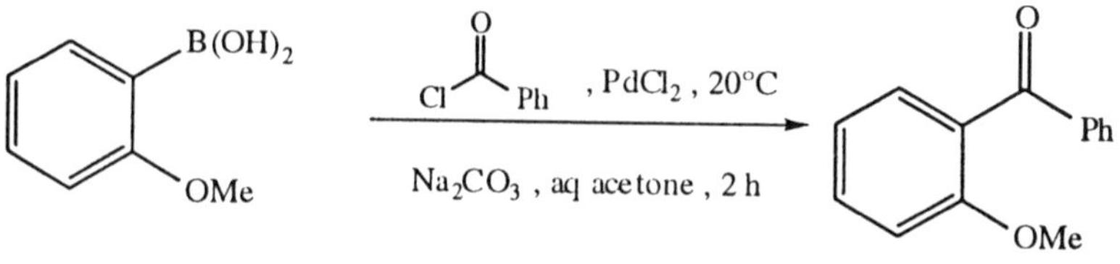

B(OH)$_2$, OMe → Cl(C=O)Ph, $PdCl_2$, 20°C; Na_2CO_3, aq acetone, 2 h → Ph, OMe

80%

Bumagin, N.A.; Korolev, D.N. *Tetraheron Lett.*, ***1999***, *40*, 3057.

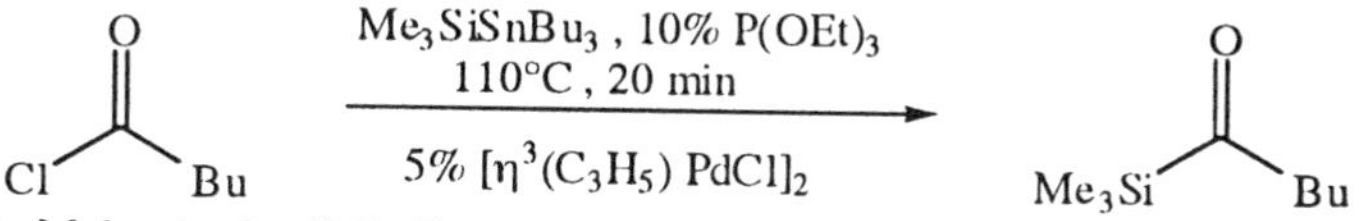

Cl(C=O)Bu → $Me_3SiSnBu_3$, 10% $P(OEt)_3$, 110°C, 20 min; 5% $[\eta^3(C_3H_5)PdCl]_2$ → Me_3Si(C=O)Bu 63%

Geng, F.; Maleczka Jr., R.E. *Tetrahedron Lett.*, ***1999***, *40*, 3113.

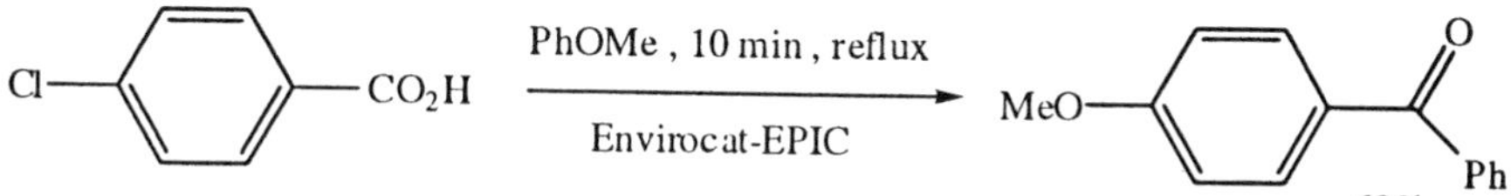

Cl–C_6H_4–CO_2H → PhOMe, 10 min, reflux; Envirocat-EPIC → MeO–C_6H_4–(C=O)Ph

69%

Bandgar, B.P.; Sadvarte, V.S. *Synth. Commun.*, ***1999***, *29*, 2587.

SECTION 168: KETONES FROM ALCOHOLS AND THIOLS

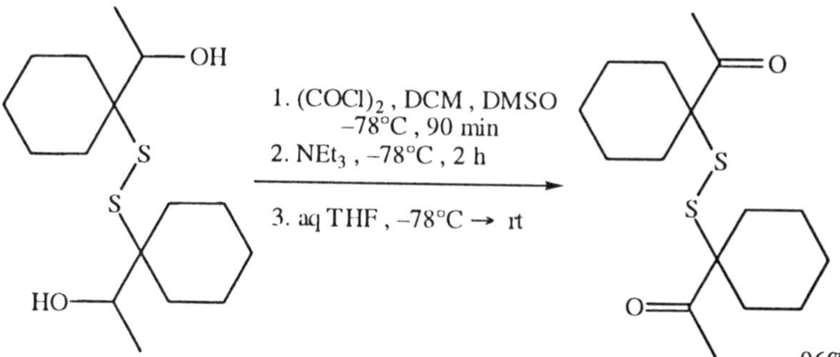

OH, S, S, HO → 1. $(COCl)_2$, DCM, DMSO, –78°C, 90 min; 2. NEt_3, –78°C, 2 h; 3. aq THF, –78°C → rt → O, S, S, O

96%

Fang, X.; Bandarage, U.K.; Wang, T.; Schroeder, J.D.; Garvey, D.S. *J. Org. Chem.*, ***2001***, *66*, 4019.

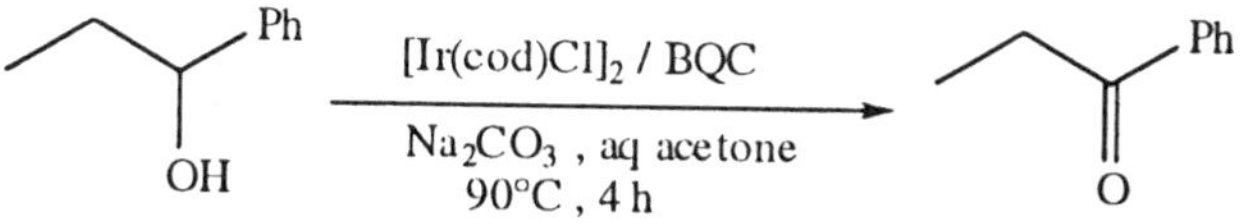

Ph, OH → $[Ir(cod)Cl]_2$ / BQC; Na_2CO_3, aq acetone, 90°C, 4 h → Ph, O 98%

BQC = 2,2'-biquinoline

Ajjou, A.N. *Tetrahedron Lett.*, ***2001***, *42*, 13.

$RuCl_2(PPh_3)_3$, TEMPO, O_2 / 100°C, PhCl, 7 h — 98%

Dijksman, A.; Marino-González, A.; Payerar, A.M.; Arends, I.W.C.E.; Sheldon, R.A.
J. Am. Chem. Soc., ***2001***, *123*, 6826.

$K_2[OsO_2(OH)_4]$, O_2, 12 h / aq *t*-BuOH — 95%

Döbler, C.; Mehltretter, G.M.; Sundermeier, U.; Eckert, M.; Militzer, H.-C.; Beller, M.
Tetrahedron Lett., ***2001***, *42*, 8447.

5% $VO(acac)_2$, MeCN / 80°C, 3 h — 98%

Maeda, Y.; Makiuchi, N.; Matsumura, S.; Nishimura, T.; Uemura, S.
Tetrahedron Lett., ***2001***, *42*, 8877.

5% $Fe(CO)_5$, hν (Pyrex) / pentane, 2 h — 93%

Cherkaoui, H.; Soufiaoui, M.; Grée, R. *Tetrahedron*, ***2001***, *57*, 2379.

$KMnO_4$–Al_2O_3–H_2O / hexane, 4 h — 55%

Patrocínio, A.F.; Moran, P.J.S. *Synth. Commun.*, ***2001***, *31*, 2457.

Montmorillonite / $Bi(NO_3)_3$, 5 min — 81%

Samajdar, S.; Becker, F.F.; Banik, B.K. *Synth. Commun.*, ***2001***, *31*, 2691.

silica gel supported Jones reagent / DCM — 93%

Ali, M.H.; Wiggin, C.J. *Synth. Commun.*, ***2001***, *31*, 3383.

PhI, 2% $Pd(OAc)_2$, $NaHCO_3$ / Bu_4NCl, H_2O, 20 h — 87%

Zhao, H.; Cai, M.-Z.; Hu, R.-H.; Song, C.-S. *Synth. Commun.*, ***2001***, *31*, 3665.

C_5H_{11} OH —[$[RuCl_2(PPh_3)_3/BuLi]$, THF, 20 min]→ C_5H_{11} O quant

Uma, R.; Davies, M.K.; Crévisy, C.; Grée, R. *Eur. J. Org. Chem.*, ***2001***, 3141.

O Ph —[$Ru(tmp)(O)_2$, DCE, 120°C; 10 atm N_2O]→ OH Ph 92%

Hashimoto, K.; Kitaichi, Y.; Tanaka, H.; Ikeno, T.; Yamada, T. *Chem. Lett.*, ***2001***, 922.

OH Ph Ph O —[1. $Fe(NO_3)_3 \cdot 9\ H_2O$, 100°C, 15 min; 2. acetone, H_2O]→ O Ph Ph O 95%

Zhao, Y.-W.; Wang, Y.-L. *J. Chem. Res. (S)*, ***2001***, 70.

OH Ph PhS —[1. 1.1 eq P_4–*t*-Bu, THF/hexane 0°C → rt, overnight; 2. pH 7 buffer]→ Ph O Ph 58%

P_4–*t*-Bu = *t*-BuN=P[—N=P(NMe_2)$_3$]$_3$

Mamdani, H.T.; Hartley, R.C. *Tetrahedron Lett.*, ***2000***, ***41***, 747.

OH Ph —[PhIO, PPNO; Cr(III) salen catalyst]→ O Ph quant

Adam, W.; Gelalcha, F.G.; Saha-Möller, C.R.; Stegmann, V.R. *J. Org. Chem.*, ***2000***, ***65***, 1915.

OH —[$PhI(OAc)_2$, DCM, 20°C, 5 h; Cr(III)-salen Cl]→ O 99%

Adam, W.; Hajra, S.; Herderich, M.; Saha-Möller, C.R. *Org. Lett.*, ***2000***, ***2***, 2773.

OH C_6H_{13} —[10% N-hydroxyphthalmide, EtOAc 0.5% $Co(OAc)_2$, 5% mcpba; 70°C, 3 h]→ O C_6H_{13} 90%

Iwahama, T.; Yoshino, Y.; Keitoku, T.; Sakaguchi, S.; Ishii, Y. *J. Org. Chem.*, ***2000***, ***65***, 6502.

OH / Ph — Al_2O_3 , Br_2–Al_2O_3 , DCM → Ph–C(=O)– 83%

Love, B.E.; Nguyen, B.T. *Synth. Commun.*, ***2000***, *30*, 963.

$Bi(NO_3)_3 \cdot 5\ H_2O$, AcOH / 4% $Cu(OAc)_2$, H_2O → 99%

Tymonko, S.A.; Nattier, B.A.; Mohan, R.S. *Tetrahedron Lett.*, ***1999***, *40*, 7657.

t-Bu, Ph, OH — PhBr , 1% $Pd_2(dba)_3 \cdot CHCl_3$, 100°C / 2% *R*-(+)-BINAP , dioxane , K_2CO_3 → *t*-Bu, O, Ph, Ph 96%

Nishimura, T.; Uemura, S. *J. Am. Chem. Soc.*, ***1999***, *121*, 11010.

OH, C_6H_{13} — 1% mcpba , 1% TEMPO / 1% Bu_4NBr , CH_2Cl_2 , rt → O, C_6H_{13} 92%

Rychnovsky, S.D.; Vaidyanathan, R. *J. Org. Chem.*, ***1999***, *64*, 310.

OH, Ph, Ph, O — 1% $VO(OEt)Cl_2$, rt / O_2 , MeCN , 1.5 h → O, Ph, Ph, O 94%

Kirahara, M.; Ochiai, Y.; Takizawa, S.; Takahata, H.; Nemoto, H. *Chem. Commun.*, ***1999***, 1387.

OH, C_6H_{13} — $RuCl_2(PPh_3)_3$, TEMPO / O_2 , 100°C , PhCl , 9 h → O, C_6H_{13} 90%

Dijksman, A.; Arends, I.W.C.E.; Sheldon, R.A. *Chem. Commun.*, ***1999***, 1591.

O, Ph, Ph, OH — $FeCl_3 \cdot 6\ H_2O$, solid state / 90°C → O, Ph, Ph, O 95%

Zhou, Y.-M.; Ye, X.-R.; Xin, X.-Q. *Synth. Commun.*, ***1999***, *29*, 2229.

OH, Ph — dimethyldioxirane, acetone, 25°C → PhC(O)CH$_3$ 98%

Baumstark, A.L.; Kovac, F.; Vasquez, P.C. *Can. J. Chem.*, ***1999***, *77*, 308.

OH, C$_3$H$_7$ — aq H_2O_2, Na_2WO_4, 90°C, 4 h; $Me(C_8H_{17})_3NHSO_4$ → CH$_3$C(O)C$_3$H$_7$ 77%

Sato, K.; Aoki, M.; Takagi, J.; Zimmerman, K.; Noyori, R. *Bull. Chem. Soc. Jpn.*, ***1999***, *72*, 2287.

REVIEWS:

"Selective Oxidation of Secondary Alcohols," Arterburn, J.B. *Tetrahedron*, ***2001***, *57*, 9765.

Related Methods: Section 48 (Aldehydes from Alcohols and Phenols).

SECTION 169: KETONES FROM ALDEHYDES

Ph–CH=CH–CHO — 1. PhLi, THF 2. TMSCl → Ph–CH(TMS)–CH$_2$–C(O)–Ph 98%

Nudelman, N.S.; García, G.V. *J. Org. Chem.*, ***2001***, *66*, 1387.

Br, Ph (CH$_2$=C(Br)Ph) — PhCHO, BF_3•OEt_2, CH_2Cl_2; rt, 1 d → Ph–CH=CH–C(O)–Ph 73%

Chan, P.W.H.; Kamijo, S.; Yamamoto, Y. *Synlett*, ***2001***, 910.

PhCHO — CH_2=CHBu, 5% $Rh(PPh_3)_3Cl$; 20% 2-amino-3-picoline, microwaves 10 min → Ph–C(O)–CH$_2$CH$_2$–Bu 75%

Jun, C.-H.; Chung, J.-H.; Lee, D.-Y.; Loupy, A.; Chatti, S. *Tetrahedron Lett.*, ***2001***, *42*, 4803.

BuCHO — 1-butene, toluene, 80°C, cat $(BzO)_2$; cat N-hydroxyphthalimide → Bu–C(O)–CH$_2$CH$_2$CH$_2$CH$_3$ 67%

Tsujimoto, S.; Iwahama, T.; Sakaguchi, S.; Ishii, Y. *Chem. Commun.*, ***2001***, 2352.

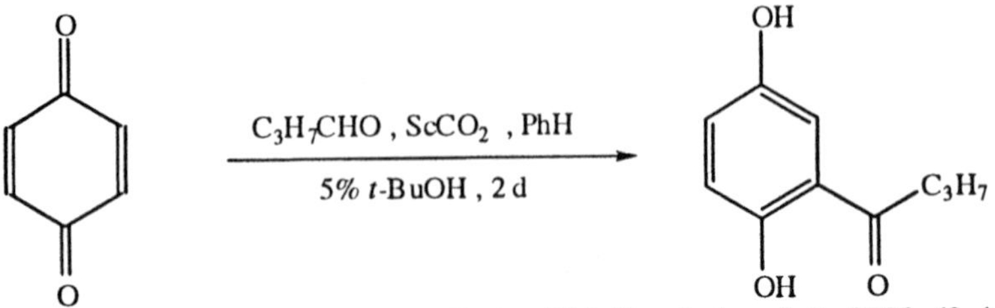

C_3H_7CHO, $ScCO_2$, PhH; 5% t-BuOH, 2 d — 81%

Pacut, R.; Grimm, M.L.; Kraus, G.A.; Tanko, J.M. *Tetrahedron Lett.*, ***2001***, *42*, 1415.

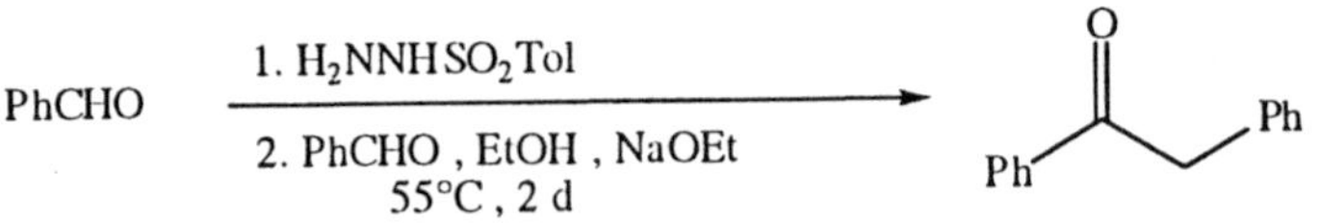

PhCHO — 1. H_2NNHSO_2Tol; 2. PhCHO, EtOH, NaOEt, 55°C, 2 d — 42%

Angle, S.R.; Neitzel, M.L. *J. Org. Chem.*, ***2000***, *65*, 6458.

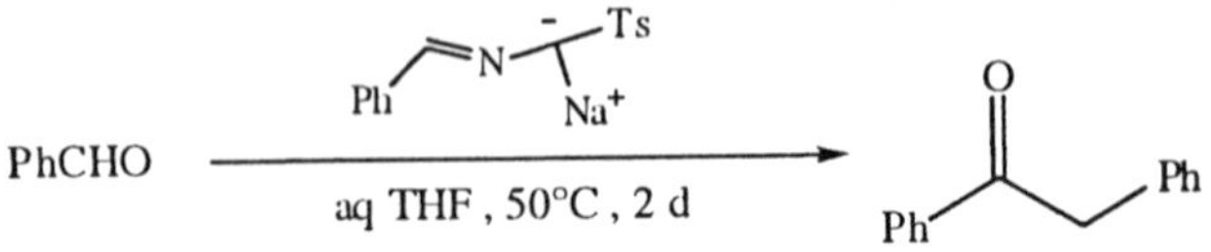

PhCHO — aq THF, 50°C, 2 d — 59%

Aggarwal, V.K.; de Vincente, J.; Pelotier, B.; Holmes, I.P.; Bonnert, R.V. *Tetrahedron Lett.*, ***2000***, *41*, 10327.

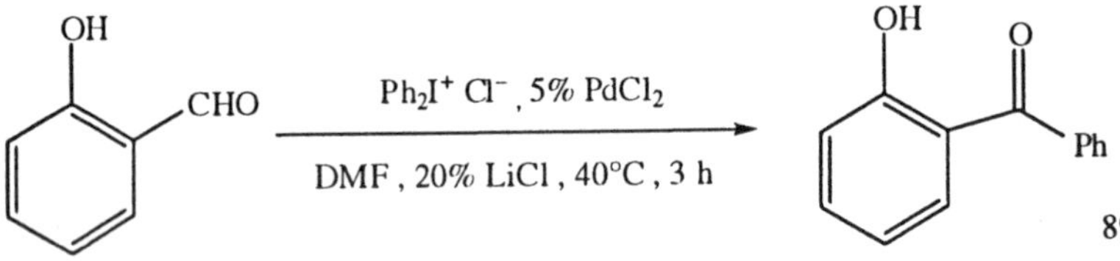

$Ph_2I^+ Cl^-$, 5% $PdCl_2$; DMF, 20% LiCl, 40°C, 3 h — 89%

Xia, M.; Chen, Z.C. *Synth. Commun.*, ***2000***, *30*, 531.

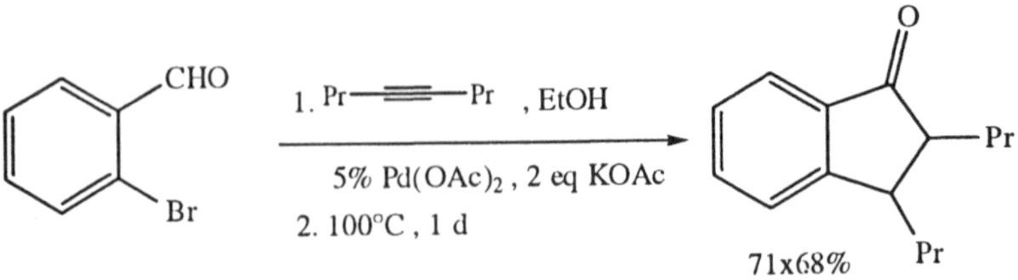

1. Pr—≡—Pr, EtOH; 5% $Pd(OAc)_2$, 2 eq KOAc; 2. 100°C, 1 d — 71x68%

Gevorgyan, V.; Quan, L.G.; Yamamoto, Y. *Tetrahedron Lett.*, ***1999***, *40*, 4089.

SECTION 170: KETONES FROM ALKYLS, METHYLENES AND ARYLS

This section lists examples of the reaction, $R\text{-}CH_2\text{-}R' \rightarrow R(C{=}O)\text{-}R'$.

NO ADDITIONAL EXAMPLES

SECTION 171: KETONES FROM AMIDES

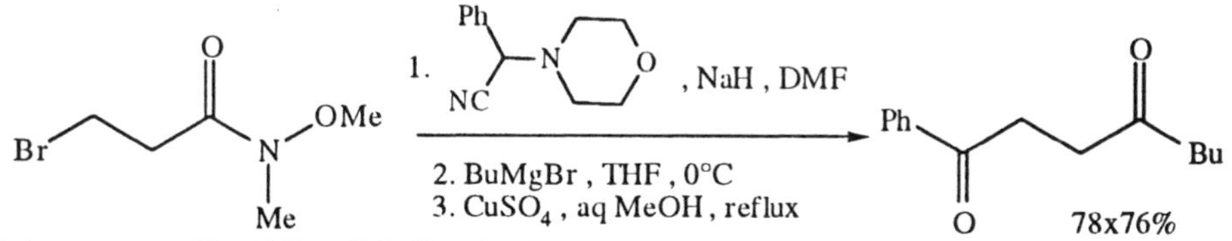

1. (Ph)(NC)CH–N-morpholine, NaH, DMF
2. BuMgBr, THF, 0°C
3. $CuSO_4$, aq MeOH, reflux

78x76%

Selvamurugan, V.; Aidhen, I.S. *Tetrahedron*, ***2001***, *57*, 6065.

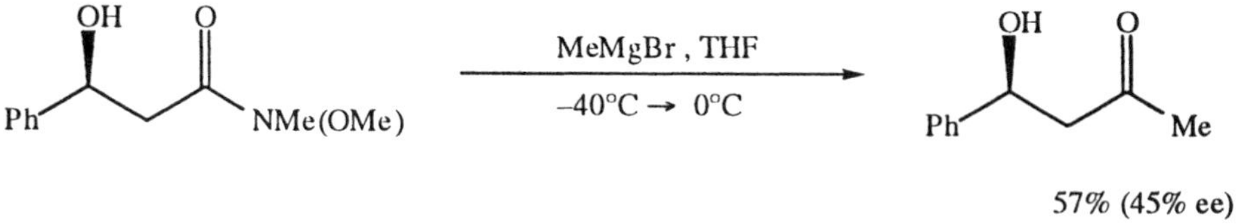

MeMgBr, THF
–40°C → 0°C

57% (45% ee)

Andrés, J.M.; Pedrosa, R.; Pérez-Encabo, A. *Tetrahedron*, ***2000***, *56*, 1217.

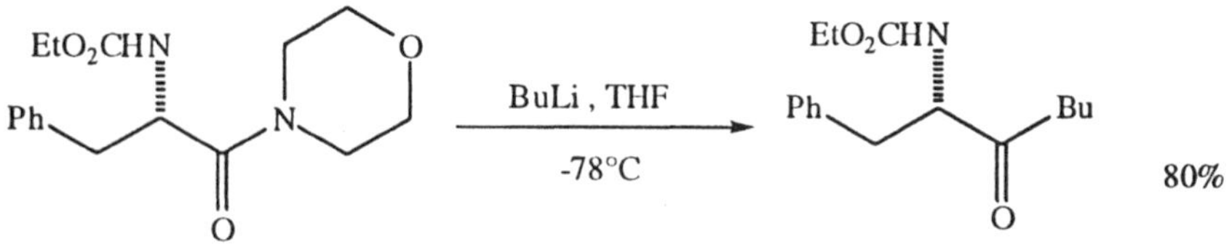

BuLi, THF
-78°C

80%

Sengupta, S.; Mondal, S.; Das, D. *Tetrahedron Lett.*, ***1999***, *40*, 4107.

REVIEWS:

"Base Catalysis in the Willgerodt-Kindler Reaction," Renard, M.; Lambert, D.; Isa, M. *Org. Prep. Proceed. Int.*, ***2001***, *33*, 335.

SECTION 172: KETONES FROM AMINES

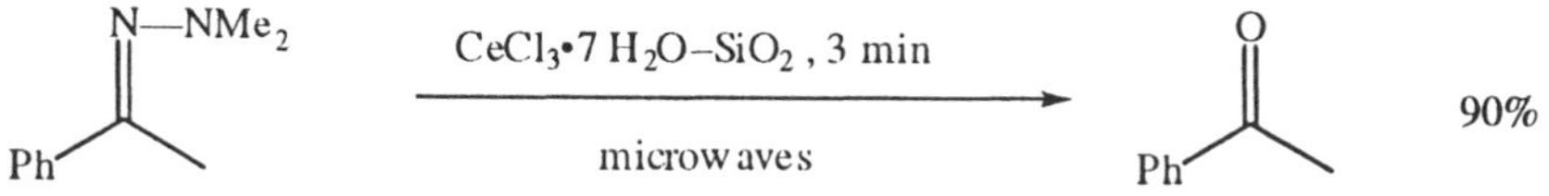

$CeCl_3$•7 H_2O–SiO_2, 3 min
microwaves

90%

Yadav, J.S.; Subba Reddy, B.V.; Reddy, M.S.K.; Sabitha, G. *Synlett*, ***2001***, 1134.

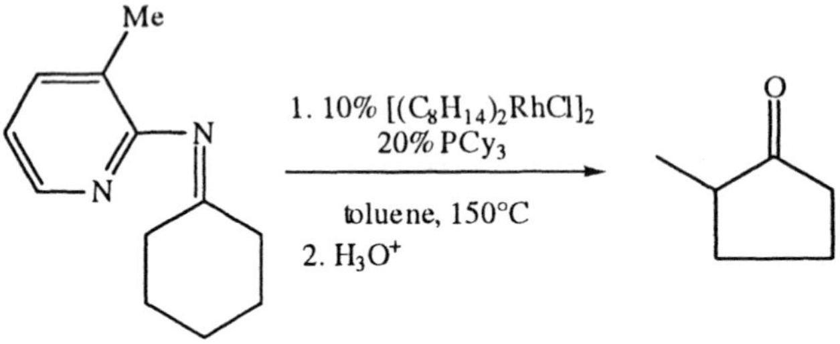

Me

N

N

1. 10% $[(C_8H_{14})_2RhCl]_2$
20% PCy_3

toluene, 150°C
2. H_3O^+

O

21%

Jun, C.-H.; Lee, H.; Lim, S.-G. *J. Am. Chem. Soc.*, ***2001***, *123*, 751.

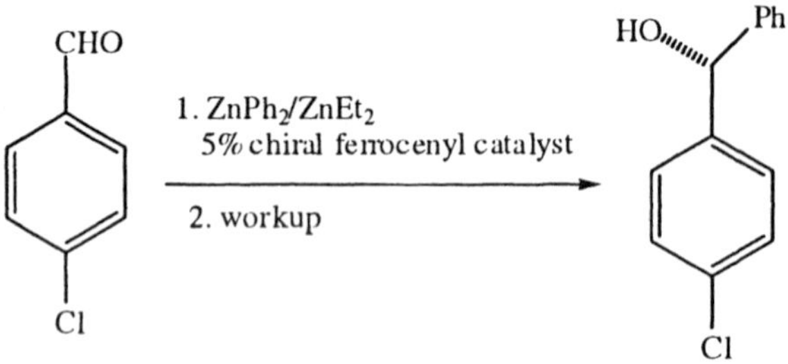

CHO

Cl

1. $ZnPh_2/ZnEt_2$
5% chiral ferrocenyl catalyst

2. workup

HO Ph

Cl

92% (95% ee)

Jun, C.-H.; Hong, J.-B.; Kim, Y.-H.; Chung, K.-Y. *Angew. Chem. Int. Ed.*, ***2000***, *39*, 3440.

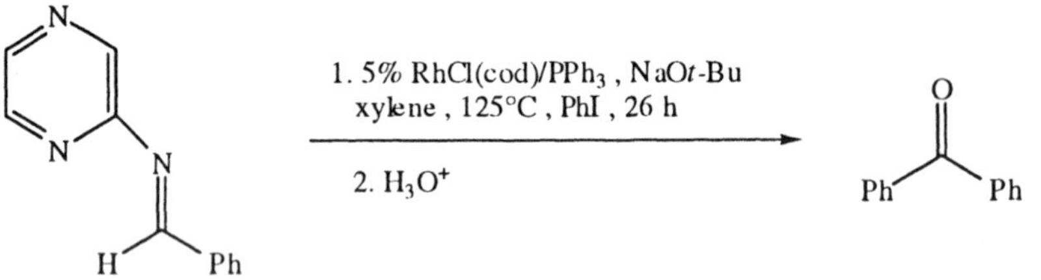

N

N

N

H Ph

1. 5% RhCl(cod)/PPh_3 , NaO*t*-Bu
xylene , 125°C , PhI , 26 h

2. H_3O^+

O

Ph Ph

86%

Ishiyamama, T.; Hartung, J. *J. Am. Chem. Soc.*, ***2000***, *122*, 12043.

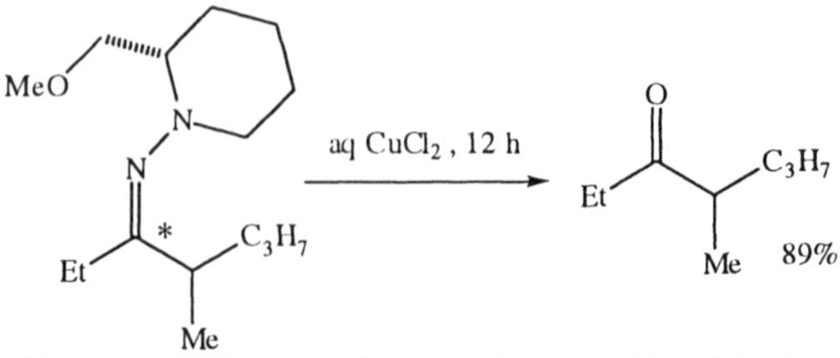

MeO

N

N

Et * C_3H_7

Me

aq $CuCl_2$, 12 h

O

Et C_3H_7

Me 89% (99% ee)

Enders, D.; Hundertmark, T.; Lazny, R. *Synth. Commun.*, ***1999***, *29*, 27.

SECTION 173: KETONES FROM ESTERS

PhSe, O, N—Me

CH_2=$CHCO_2Me$, PhH
Bu_3SnH , reflux

CO_2Me, O, N—Me

62%

Bennasar, M.-L.; Roca, T.; Griera, R.; Bosch, J. *Org. Lett.*, ***2001***, *3*, 1697.

S, O, O, CO_2Me, O_2N

5 eq Mg , MeOH
reflux , 30 min

CO_2Me, OH, O_2N

75%

Rho, H.S.; Ko, B.-S. *Synth. Commun.*, ***2001***, *31*, 283.

EtS, O, OH

$PhB(OH)_2$, 1% Pd_2dba_3 , THF
3% TFP , 50°C , 18 h

1.6 eq S CO_2Cu

Ph, O, OH

81%

Liebeskind, L.S.; Srogl, J. *J. Am. Chem. Soc.*, ***2000***, *122*, 11260.

EtO_2C, OTBDPS, OH

1. $ClTi(Oi\text{-}Pr)_3$, EtMgBr
2. MsCl , Py

OTBDPS, O

62x58%

Cho, S.Y.; Cha, J.-K. *Org. Lett.* ***2000***, *2*, 1337.

esterase from *Marchantia polymorpha*

79% (>99% ee)

Hirata, T.; Shimoda, K.; Kawano, T. *Tetrahedron Asymm.* ***2000***, *11*, 1063.

$PhB(OH)_2$, Pd catalyst
6 eq K_2CO_3 , NaI

DMA , 95°C

88% (+ tetrahydrothiophene)

Savarin, C.; Srogl, J.; Liebeskind, L.S. *Org. Lett.*, ***2000***, 2, 3229.

5 eq CF_3SO_3H , PhH

85°C , 8 h

75%

Hwang, J.P.; Prakash, G.K.S.; Olah, G.A. *Tetrahedron*, ***2000***, *56*, 7199.

1. BuMgBr , THF , 0°C

2. PhLi , THF , 0°C → rt

76%

Lee, N.R.; Lee, K.I. *Synth. Commun.*, ***1999***, *29*, 1249.

SECTION 174: KETONES FROM ETHERS, EPOXIDES AND THIOETHERS

$Al(OC_6F_5)_3$, DCM

0°C

96%

Kita, Y.; Furkawa, A.; Futamura, J.; Ueda, K.; Sawama, Y.; Hamamoto, H.; Fujioka, H. *J. Org. Chem.*, ***2001***, *66*, 8779.

O NMe2 — 1. PhBr, $Pd(OAc)_2$, dppp, TlOAc, 80°C; 2. PhBr, 5% $Pd(OAc)_2$, PPh_3, 100°C; 3. aq HCl, TBME → Ph, Ph, O, Ph 65%

Nilsson, P.; Larhed, M.; Hallberg, A. *J. Am. Chem. Soc.*, ***2001***, *123*, 8217.

OTMS — $NaBrO_3$, NH_4Cl, 80°C; aq MeCN, 45 min → O 75%

Shaabani, A.; Karimi, A.-R. *Synth. Commun.*, ***2001***, *31*, 759.

O — Bi, cat $Cu(OTf)_2$, O_2; DMSO, 100°C → O, O 74%

Antoniotti, S.; Duñach, E. *Chem. Commun.*, ***2001***, 2566.

OTMS — $Ph_3Sb(OAc)_2$, cat $PdCl_2$; DME/MeCN, H_2O, rt, 6 h → O 68%

Kang, S.-K.; Ryu, H.-C.; Hong, Y.-T. *J. Chem. Soc., Perkin Trans. 1*, ***2000***, 3350.

O, Ph — 70% *t*-BuOOH, cat CrO_3; DCM, rt, 1 h → O, Ph 77%

Chandrasekhar, S.; Mohanty, P.K.; Ramachander, T. *Synlett*, ***1999***, 1063.

$OSiMe_3$ — BnBr, $ZnCl_2$/acidic Al_2O_3 → O, Bn 65%

Kad, G.L.; Singh, V.; Khurana, A.; Chaudhary, S.; Singh, J. *Synth. Commun.*, ***1999***, *29*, 3439.

SECTION 175: KETONES FROM HALIDES AND SULFONATES

Ph, Cl — PhI, $Fe(CO)_5$, DMF, e^-; Bu_4NBF_4, cat $NiBr_2(bpy)$ → O, Ph, Ph 82%

Dolhem, E.; Barhdadi, R.; Folest, J.C.; Nédélec, J.Y.; Troupel, M. *Tetrahedron*, ***2001***, *57*, 525.

Cl–C$_6$H$_4$–OTf → (1 atm CO, SnMe$_4$, 3% Pd(OAc)$_2$, dppp; NEt$_3$, DMF, 60°C) → Cl–C$_6$H$_4$–C(O)CH$_3$ 83%

Garrido, F.; Raeppel, S.; Mann, A.; Lautens, M. *Tetrahedron Lett.*, ***2001***, *42*, 265.

$Ph_2MeSi(SiMePh_2)CCl_2$ → (1. EtMgBr; 2. CuCN•2 LiCl, –78°C → 0°C; 3. air, aq NH_4Cl) → Et–C(O)–$SiMePh_2$ 75%

Inoue, A.; Kondo, J.; Shinokubo, H.; Oshima, K. *J. Am. Chem. Soc.*, ***2001***, *123*, 11109.

4-Iodopyridine → (cat $PdCl_2/PPh_3$, 5 bar CO; $PhB(OH)_2$, K_2CO_3, 80°C, 20 h) → Ph–C(O)–(4-pyridyl) 80%

Couve-Bonnaire, S.; Carpentier, J.-F.; Mortreux, A.; Castanet, Y.
Tetrahedron Lett., ***2001***, *42*, 3689.

3-Bromoquinoline → (1. cat Pd(dba)$_2$, PPh$_3$, rt, toluene, 15 min; 2. toluene, 110°C, CH_2=C(OEt)$SnBu_3$; 3. 1M HCl, rt, 1 d) → 3-acetylquinoline 88%

Legros, J.-Y.; Primault, G.; Fiaud, J.-C. *Tetrahedron*, ***2001***, *57*, 2507.

PhC(O)CH$_2$Br → (PhC(O)CN, SmI$_2$; THF, rt, 1 h) → PhC(O)CH$_2$C(O)Ph 82%

Back, H.S.; Yoo, B.W.; Keum, S.R.; Yoon, C.M.; Kim, S.H.; Kim, J.H.
Synth. Commun., ***2000***, *30*, 31.

Prenyl bromide → (Ph(CH$_2$)$_4$C(O)Zr(Cl)Cp$_2$; CuI, DMF, 0°C, 10 h) → Ph(CH$_2$)$_4$C(O)CH$_2$CH=C(CH$_3$)$_2$ + Ph(CH$_2$)$_4$C(O)C(CH$_3$)$_2$CH=CH$_2$ (29 : 62) 91%

Hanzawa, Y.; Narita, K.; Taguchi, T. *Tetrahedron Lett.*, ***2000***, *41*, 109.

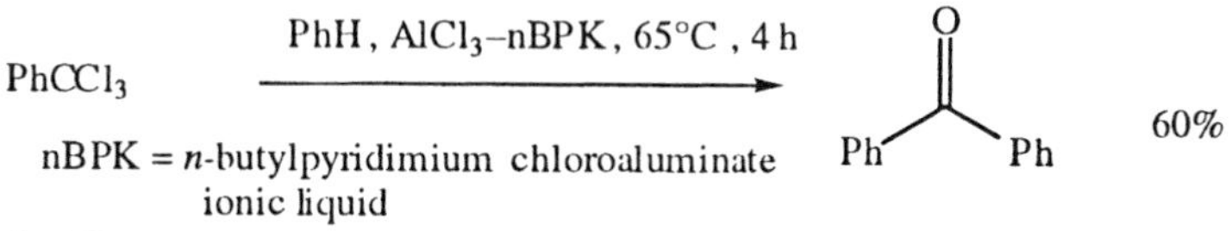

$PhCCl_3$ → (PhH, $AlCl_3$–nBPK, 65°C, 4 h) → PhC(O)Ph 60%

nBPK = *n*-butylpyridimium chloroaluminate ionic liquid

Rebeiro, G.L.; Khadilkar, B.M. *Synth. Commun.*, **2000**, *30*, 1605.

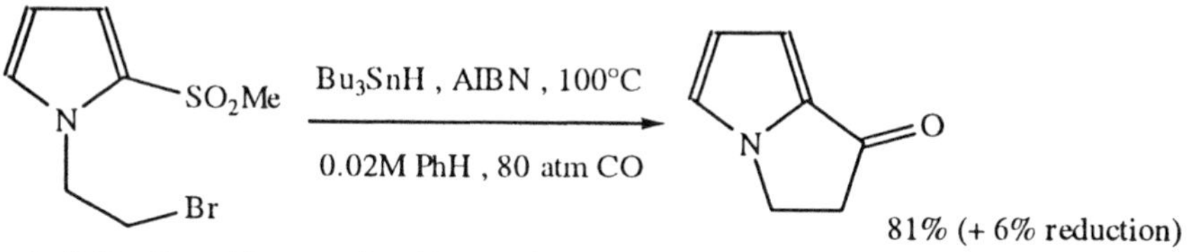

(Bu$_3$SnH, AIBN, 100°C; 0.02M PhH, 80 atm CO) 81% (+ 6% reduction)

Miranda, L.D.; Cruz-Almanza, R.; Alvarez-García, A.; Muchowski, J.M. *Tetrahedron Lett.*, **2000**, *41*, 3035.

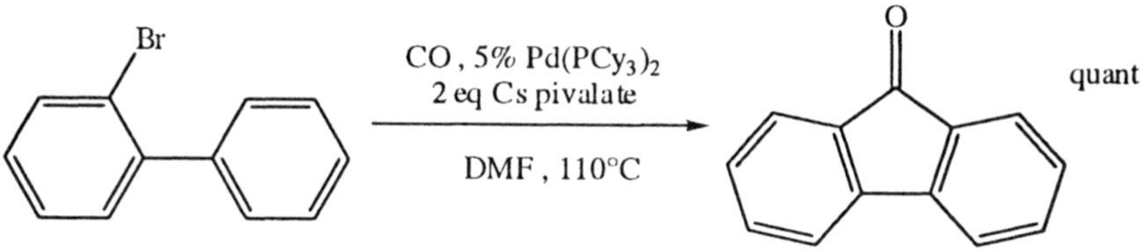

(CO, 5% Pd(PCy_3)$_2$, 2 eq Cs pivalate; DMF, 110°C) quant

Campo, M.A.; Larock, R.C. *Org. Lett.*, **2000**, *2*, 3675.

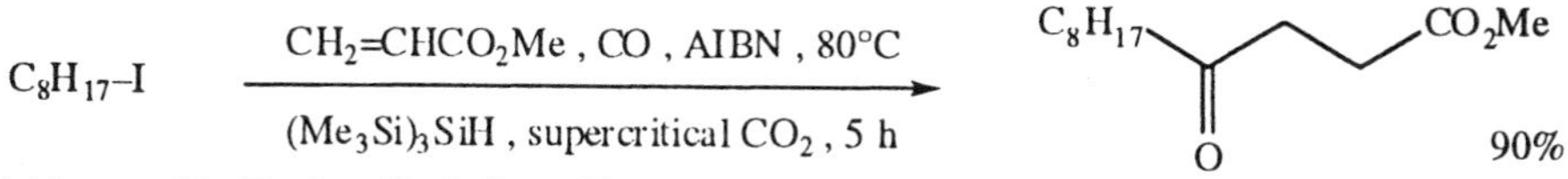

C_8H_{17}–I → (CH_2=$CHCO_2Me$, CO, AIBN, 80°C; ($Me_3Si)_3SiH$, supercritical CO_2, 5 h) → $C_8H_{17}C(O)CH_2CH_2CO_2Me$ 90%

Kishimoto, Y.; Ikariya, T. *J. Org. Chem.*, **2000**, *65*, 7656.

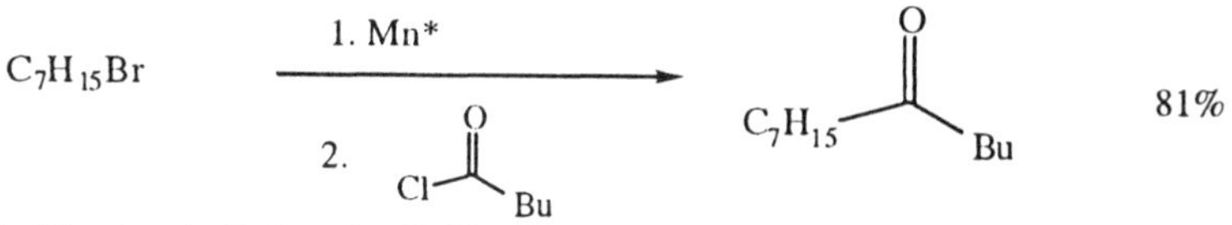

$C_7H_{15}Br$ → (1. Mn*; 2. ClC(O)Bu) → $C_7H_{15}C(O)Bu$ 81%

Cahiez, G.; Martin, A.; Delacroix, T. *Tetrahedron Lett.*, **1999**, *40*, 6407.

Related Methods: Section 177 (Ketones from Ketones).
Section 55 (Aldehydes from Halides).

SECTION 176: KETONES FROM HYDRIDES

This section lists examples of the replacement of hydrogen by ketonic groups, R-H → R(C=O)-R'. For the oxidation of methylenes, R_2CH_2 → $R_2C=O$, see section 170 (Ketones from Alkyls).

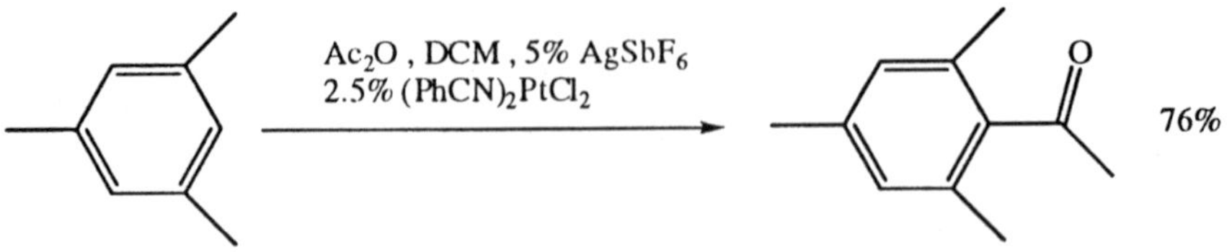

Ac_2O , DCM , 5% $AgSbF_6$
2.5% $(PhCN)_2PtCl_2$

O

76%

Fürstner, A.; Voigtländer, D.; Schrader, W.; Giebel, D.; Reetz, M.T. *Org. Lett.*, ***2001***, *3*, 417.

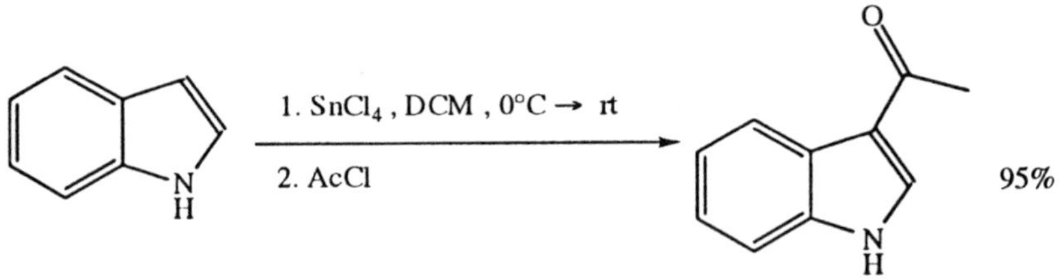

N
H

1. $SnCl_4$, DCM , 0°C → rt
2. AcCl

O
N
H

95%

Ottoni, O.; de V.F. Neder, A.; Dias, A.K.B.; Cruz, R.P.A.; Aquino, L.B. *Org. Lett.*, ***2001***, *3*, 1005.

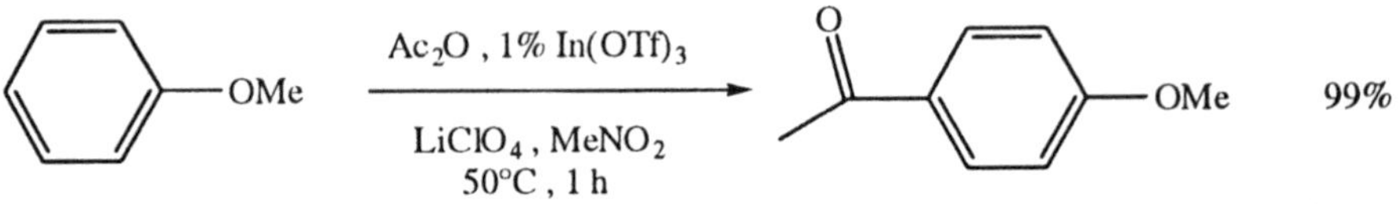

OMe

Ac_2O , 1% $In(OTf)_3$
$LiClO_4$, $MeNO_2$
50°C , 1 h

O
OMe

99%

Chapman, C.J.; Frost, C.G.; Hartley, J.P.; Whittle, A.J. *Tetrahedron Lett.*, ***2001***, *42*, 773.

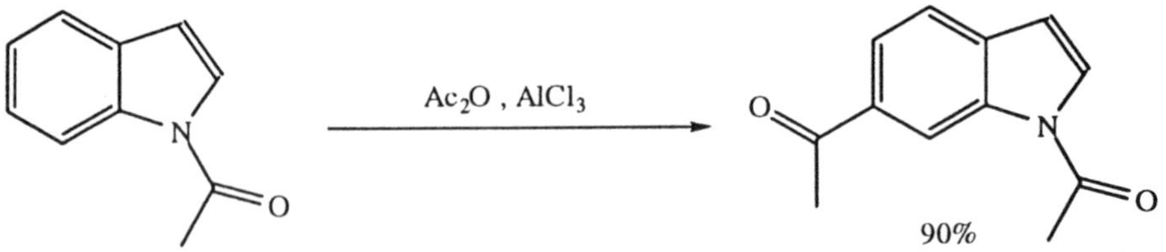

N
O

Ac_2O , $AlCl_3$

O
N
O

90%

Cruz, R.P.A.; Ottoni, O.; Abella, C.A.M.; Aquino, L.B. *Tetrahedron Lett.*, ***2001***, *42*, 1467.

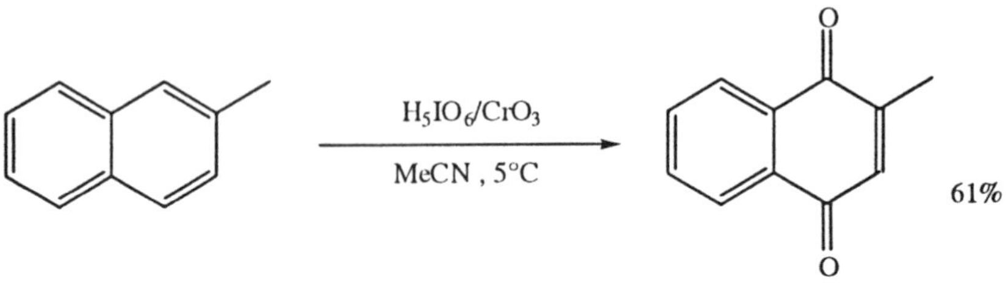

H_5IO_6/CrO_3
MeCN , 5°C

O
O

61%

Yamazaki, S. *Tetrahedron Lett.*, ***2001***, *42*, 3355.

hv (>495 nm), O_2, rt
ε-tetraphenylphorphin
CH_2Cl_2, rt, 2.5 h

OH N O O N 82%

Cossy, J.; Belotti, D. *Tetrahedron Lett.*, ***2001***, *42*, 4329.

$KMnO_4/CuSO_4\bullet5\ H_2O$
neat, 1. 3h

O 90%

Shaabani, A.; Lee, D.G. *Tetrahedron Lett.*, ***2001***, *42*, 5833.

Ce(OTf)$_4$, MeCN, rrt

Ph Ph O Ph OH

+ 1.5% H_2O	9.5%	85.7%
+ 16.4% H_2O	99.7%	–

Laali, K.K.; Herbert, M.; Cushnyr, B.; Bhatt, A.; Terrano, D. *J. Chem. Soc., Perkin Trans. 1*, ***2001***, 578.

Ac_2O, DCM, 5% $AgSbF_6$
2.5% $(PhCN)_2PtCl_2$

O 76%

Fürstner, A.; Voigtländer, D.; Schrader, W.; Giebel, D.; Reetz, M.T. *Org. Lett.*, ***2001***, *3*, 417.

1. Cl, O, O, Cl, $AlCl_3$
2. , rt, 14 h

O 77%

Taber, D.F.; Sethuraman, M.R. *J. Org. Chem.*, ***2000***, *65*, 254.

$Ru_3(CO)_2$, CH_2=CH_2

CO , 20 atm , toluene
160°C , 20 h

O 98%

Ie, Y.; Chatani, N.; Ogo, T.; Marshall, D.R.; Fukuyama, T.; Kakiuchi, F.; Murai, S.
J. Org. Chem., ***2000***, *65*, 1475.

BzCl , $Ga(ONf)_3$

Nf = nonafluorobutanesulfonate

Ph 99%

Matsuo, J.-i.; Odashima, K.; Kobayashi, S. *Synlett*, ***2000***, 403.

cis[Ru^{VI}(6,6'-Cl_2bpy$)_2O_2](ClO_4)_2$

MeCN , rt , N_2 , 30 min

50%

Che, C.-M.; Cheng, K.-W.; Chan, M.C.W.; Lau, T.-C.; Mak, C.-K.
J. Org. Chem., ***2000***, *65*, 7996.

Ph Ph

1. NaOCl , NaBr ,TBAH-HSO_4

2. H_2O

Ph Ph 86%

Clark, J.H.; Grigoropoulou, G.; Scott, K. *Synth. Commun.*, ***2000***, *30*, 3731.

MeO

Ac_2O , cat $Sc(OTf)_3$, 4 h

$MeNO_2$, 50°C

MeO 89%

Kawada, A.; Mitamura, S.; Matsuo, J.-i.; Tsuchiya, T.; Kobayashi, S.
Bull. Chem. Soc. Jpn., ***2000***, *73*, 2325.

mcpba , $NaHCO_3$, air

CH_2Cl_2 , rt

88% O

Ma, D.; Xia, C.; Tian, H. *Tetrahedron Lett.*, ***1999***, *40*, 8915.

Ph⌒Ph → ($NaBrO_3$, CeO_2, 95°C, 1 d; aq dioxane) → PhC(O)Ph 90%

Shi, Q.-Z.; Wang, J.-G.; Cai, K. *Synth. Commun.*, ***1999***, *29*, 1177.

1\. Na, $C_3H_7CO_2Me$ THF, 25°C; 2. aq HCl → 55%

C_3H_7

Periasamy, M.; Reddy, M.R.; Bharathi, P. *Synth. Commun.*, ***1999***, *29*, 677.

Ph⌒Ph → (CrO_3–SiO_2, 140°C; 5 h) → PhC(O)Ph 68%

Bose, D.S.; Sunder, K.S. *Synth. Commun.*, ***1999***, *29*, 4295.

$CH_2(CO_2H)_2$, PPA, 2 h; 60-80°C → 92%

Renault, O.; Dallemagne, P.; Rault, S. *Org. Prep. Proceed. Int.*, ***1999***, *31*, 324.

SECTION 177: KETONES FROM KETONES

This section contains alkylations of ketones and protected ketones, ketone transpositions and annulations, ring expansions and ring openings and dimerizations. Conjugate reductions and Michael alkylations of enone are listed in Section 74 (Alkyls from Alkenes).

For the preparation of enamines or imines from ketones, see Section 356 (Amine-Alkene).

1\. LDA, THF, –78°C
2\. 2.5% $[Pd(\eta^3\text{-}C_3H_5)Cl]_2$, THF
7.5% bis-ferrocene-bis-phosphine ligand, 3 h, allyl-O_2COEt

90% (82% ee)

You, S.-L.; Hou, X.-L.; Dai, L.-X.; Zhu, X.-Z. *Org. Lett.*, ***2001***, *3*, 3149.

OTMS

1. EtOK, –78°C, 2.5 h
2. cat BuLi, LiBr, HMPA
–78C, 12 h, allyl iodide (I)

61%

Yu, W.; Jin, Z. *Tetrahedron Lett.*, **2001**, *42*, 369.

Ph O SiMe$_3$

Ph–CH=CH–CH$_2$O$_2$CCF$_3$
cat Pd(η^3-$C_6H_5CH{=}CHCH_2$)(CF_3CO_2)$_2$
THF, reflux

Ph O Ph

60%

Obora, Y.; Ogawa, Y.; Imai, Y.; Kawamura, T.; Tsuji, Y.
J. Am. Chem. Soc., **2001**, *123*, 10489.

OBn O O

2 eq OTMS Ph
BF_3•OEt_2, CH_2Cl_2, –78°C

OBn O Ph O

85% (88:12 *trans:cis*)

Sugita, Y.; Kimura, C.; Hosoya, H.; Yamadoi, S.; Yokoe, I. *Tetrahedron Lett.*, **2001**, *42*, 1095.

OH O

HO allyl alcohol, 2.4 BEt_3
5% Pd(OAc)$_2$, 10% PPh_3
THF, rt, 5 h

OH O

81%

Horino, Y.; Naito, M.; Kimura, M.; Tanaka, S.; Tamaru, Y. *Tetrahedron Lett.*, **2001**, *42*, 3113.

O O CHN$_2$

$TiCl_4$, DCM
0°C, 2 min

O O

84%

Muthusamy, S.; Babu, S.A.; Gunanathan, C. *Synth. Commun.*, **2001**, *31*, 1205.

O Ph

NBu_3, dioxane, 5% $RuCl_3$•n H_2O
3 eq PPh_3, 180°C, 40 h

O Ph Bu

70%

Cho, C.S.; Kim, B.T.; Lee, M.J.; Kim, T.-J.; Shim, S.C.
Angew. Chem. Int. Ed., **2001**, *40*, 958.

OLi

1. 0.1 chiral imide
2. 2,6-di-*t*-Bu-phenol

THF , –78°C

O

H

>99% (90% ee)

Yanagisawa, A.; Watanabe, T.; Kikuchi, T.; Yamamoto, H. *J. Org. Chem.*, ***2000***, *65*, 2979.

1. Tf_2O , 0°C , DCM
2 (*i*-Pr)C(=O)CH$_2$Ph
3. *t*-BuOK , DMSO

N

Ph

O

N

96x91%

Katritzky, A.R.; Zhang, S.; Kurz, T.; Wang, M. *Org. Lett.*, ***2001***, *3*, 2807.

Cl

P(*t*-Bu)$_3$ (2-biphenyl), 1% $Pd(OAc)_2$

PhC(=O)CH$_2$CH$_3$, 1.3 NaO*t*-Bu ,

toluene , 80°C , 12 h

O

Ph

91%

Fox, J.M.; Huang, X.; Chieffi, A.; Buchwald, S.L. *J. Am. Chem. Soc.*, ***2000***, *122*, 1360.

Ph

O

$C_8H_{17}Br$, 4N NaOH , 100°C

calix(n)arene catalyst , 10 h

C_8H_{17}

Ph

O

86%

Shimizu, S.; Suzuki, T.; Sasaki, Y.; Hirai, C. *Synlett*, ***2000***, 1664.

O

1. LDA
2. 2.5% $Pd_2(dba)_3$•$CHCl_3$

AcO–CH(Ph)–CH=CH–Ph , dppf

Ph

Ph

O

\+

Ph

Ph

O

(87 : 13)

Braun, M.; Laicher, F.; Meier, T. *Angew. Chem. Int. Ed.*, ***2000***, *39*, 3494.

1% $Cu(acac)_2$, 2.5 eq Bu_3SnH

PhH , reflux , 5 h

Ph CHN$_2$ → Ph 85%

Tan, Z.; Qu, Z.; Chen, B.; Wang, J. *Tetrahedron*, ***2000***, *56*, 7457.

CH_2=CHC_6H_{13} , AcOH , O_2
cat $Mn(OAc)_2$, 80°C

cat $Co(OAc)_2$

C_6H_{13} 85%

Iwahama, T.; Sakaguchi, S.; Ishii, Y. *Chem. Commun.*, ***2000***, 2317.

2% $Pd(dba)_2$/ferrocenyl phosphine ligand
PhBr , NaO*t*-Bu , THF

25°C , rt

Ph → Ph Ph 99%

Kawatsura, M.; Hartwig, J.F. *J. Am. Chem. Soc.*, ***1999***, *121*, 1473.

5% chiral ligand , 2 eq LDA
2.5% $[(\eta^3\text{-}C_3H_5)PdCl]_2$

Me_3SnCl , 30 min , DME
OAc

99% (86% ee)

Trost, B.M.; Schroeder, G.M. *J. Am. Chem. Soc.*, ***1999***, *121*, 6759.

LDA , HMPA

Ph_3BiCl_2

Ph 57%

Arnauld, T.; Barton, D.H.R.; Normant, J.-F.; Doris, E. *J. Org. Chem.*, ***1999***, *64*, 6915.

BnBr , Bu_4NF , aq THF

23°C

Bn 89%

Christoffers, J. *Synth. Commun.*, ***1999***, *29*, 117.

Clayfen , microwaves

90 sec

95%

Varma, R.S.; Kumar, D. *Synth. Commun.*, ***1999***, *29*, 1333.

Related Methods: Section 49 (Aldehydes from Aldehydes).

SECTION 178: KETONES FROM NITRILES

PhCN

, cat SmI_2

cat NiI_2 , 30 min , rt

68%

Kang, H.-Y.; Song, S.-E. *Tetrahedron Lett.*, ***2000***, *41*, 937.

CN

1. Br , 4 eq Zn , 0.4 $AlCl_3$

THF , 2 h

2. 2M HCl

61%

Lee, A.S.-Y.; Lin, L.-S. *Tetrahedron Lett.*, ***2000***, *41*, 8803.

SECTION 179: KETONES FROM ALKENES

O_2 , $MeSO_3H$, aq MeCN

10% $Pd(OAc)_2$, 15% NPMoV/Cl

50°C , 6 h

NPMoV = molybdovanadophosphate

85% + 1%

Kishi, A.; Higashino, T.; Sakaguchi, S.; Ishii, Y. *Tetrahedron Lett.*, ***2000***, *41*, 99.

$C_{10}H_{21}$

i-PrOH , 5% $Pd(OAc)_2$, 20% Py

toluene , O_2 , 60°C

$C_{10}H_{21}$

70%

Nishimura, T.; Kakiuchi, N.; Onoue, T.; Ohe, K.; Uemura, S.
J. Chem. Soc., Perkin Trans. 1, ***2000***, 1915.

CO_2Me

Ph

$OSnBu_3$

Bn_4NBr , THF , reflux , 12 h

Ph CO_2Me

>99%

Yasuda, M.; Ohigashi, N.; Shibata, I.; Baba, A. *J. Org. Chem.*, ***1999***, *64*, 2180.

See also: Section 134 (Ethers from Alkenes).
Section 174 (Ketones from Ethers).

SECTION 180: KETONES FROM MISCELLANEOUS COMPOUNDS

Conjugate reductions and reductive alkylations of enones are listed in Section 74 (Alkyls from Alkenes).

Br

1. CH_2=CHOBu , cat dppp , NEt_3
2.5% $Pd(OAc)_2$, 100°C , 18 h
[bmim] BF_4
2. HCl

ionic liquid medium

O

95%

Xu, L.; Chen, W.; Ross, J.; Xiao, J. *Org. Lett.*, ***2001***, *3*, 295.

O
CO_2t-Bu
PPh_3

MMPP , CH_2Cl_2 , rt , 8 h

O
CO_2t-Bu
O

85%

MMPP = magnesium monoperoxy phthalate

Lee, K.; Im, J.-M. *Tetrahedron Lett.*, ***2001***, *42*, 1539.

=N—OH

$KMnO_4$, Al_2O_3 , acetone
rt , 2 h

=O

93%

Chrisman, W.; Blankinship, M.J.; Taylor, B.; Harris, C.E. *Tetrahedron Lett.*, ***2001***, *42*, 4775.

O_2N

1. PhLi , THF , –78°C , 1 d
2. AcOH , 0°C

Ph
O

72x95%

Santos, R.P.; Lopes, R.S.C.; Lopes, C.C. *Synthesis*, ***2001***, 845.

N-OH
Ph

0.5 $Bi(NO_3)_3$•5 H_2O , aq acetone
0.1 $Cu(OAc)_2$•xH_2O , 2 h
Montomorillonite K-10

O
Ph

88%

Nattier, B.A.; Eash, K.J.; Mohan, R.S. *Synthesis*, ***2001***, 1010.

N–OH, Ph, Ph → ($AlCl_3$, 30 sec; benzyl-DABCO Cr_2O_3, Bn) → Ph(C=O)Ph 89%

Hakipour, A.R.; Mallakpour, S.E.; Mohammadpoor-Baltork, I.; Khoee, S. *Synth. Commun.*, ***2001***, *31*, 1187.

$B(OH)_2$ → (Ac_2O, dioxane, sealed tube; cat $[Rh(CH_2{=}CH_2)Cl]_2$, DME, 16 h) → 68%

Frost, C.G.; Wadsworth, K.J. *Chem. Commun.*, ***2001***, 2316.

2 $PhPb(OAc)_3$ → (CO, 5 eq NaOMe, $CHCl_3$; 5% $Pd(dba)_3{\cdot}CHCl_3$, MeCN, rt) → Ph(C=O)Ph 70%

Kang, S.-K.; Ryu, H.-C.; Choi, S.-C. *Synth. Commun.*, ***2001***, *31*, 1035.

N–OH, Ph → ($[(NO_3)_3Ce]_3{\cdot}H_2IO_6$, MeCN; reflux, 5 min) → Ph(C=O) 93%

Shirini, F.; Azadbar, M.R. *Synth. Commun.*, ***2001***, *31*, 3775.

N–OH, Ph → (polymer–$I(O_2CCF_3)_2$; aq THF, rt, 40 min) → Ph(C=O) 83%

Chen, D.-J.; Cheng, D.-P.; Chen, Z.-C. *Synth. Commun.*, ***2001***, *31*, 3847.

$Ph_2C{=}N{-}OH$ → (PEG-NO_2, rt, 4.5 h) → $Ph_2C{=}O$ 80%

Liu, X.; Zhang, Q.; Zhang, S.; Zhang, J. *Org. Prep. Proceed. Int.*, ***2001***, *33*, 87.

C_5H_{11}(C=O)O(C=O)C_5H_{11} → ($B(OH)_2$, PPh_3; cat $Pd(OAc)_2$, THF; 2.5 eq H_2O, 16 h) → (C=O)C_5H_{11} 98%

Gooßen, L.J.; Ghosh, K. *Angew. Chem. Int. Ed.*, ***2001***, 40.

N–OH; Ph; $(NH_4)_2S_2O_8$, $AgNO_3$, hexane; Montmorillonite , 50°C , 2.5 h; O; Ph; 98%

Hirano, M.; Kojima, K.; Yakabe, S.; Morimoto, T. *J. Chem. Res. (S)*, ***2001***, 277.

N–OH; Ph; $Et_4N^+ MnO_4^-$, neat , 40 min; O; Ph; 80%

Bigdeli, M.A.; Nikje, M.M.A.; Heravi, M.M. *J. Chem. Res. (S)*, ***2001***, 496.

$(C_5Me_5)_2Ti$; CO , 45°C , pentane; 12 h; O; >95%

Carter, C.A.G.; Greidanus, G.; Chen, J.-X.; Stryker, J.M.
J. Am. Chem. Soc., ***2001***, *123*, 8872.

I; CO_2Me; TBSO; Ph; 1.2 CsF , 2 eq $BnEt_3NCl$; 2 eq K_2CO_3 , MeCN , 92°C; 10% $Pd(PPh_3)_4$, 20% PPh_3; MeO_2C; Ph; O; 47%

Khan, F.A.; Czerwonka, R.; Reissig, H.-U. *Eur. J. Org. Chem.*, ***2000***, 3607.

$NNMe_2$; Ph; $Pd(OAc)_2/SnCl_2$; aq DMF; O; Ph; 75%

Mino, T.; Hirota, T.; Fujita, N.; Yamashita, M. *Synthesis*, ***2000***, 69.

NOH; Ph; Ni(II) complex , O_2 , Me_3CCHO; PhF , 5 h; O; Ph; 90%

Blay, G.; Benach, E.; Fernándz, I.; Galletero, S.; Pedro, J.R.; Ruiz, R. *Synthesis*, ***2000***, 403.

$N–NMe_2$; $MeReO_3$, H_2O_2; MeCN , AcOH; O; 85%

Stanković, S.; Espenson, J.H. *J. Org. Chem.*, ***2000***, *65*, 2218.

$SbCl_3$, 9 sec

=N-NHCONH$_2$

microwaves

=O

70%

Mitra, A.K.; De, A.; Karchaudhuri, N. *Synth. Commun.*, ***2000***, *30*, 1651.

N-OH

SiO_2-CrO_3 , toluene

75°C , 2 h

Ph Ph

O

Ph Ph

89%

Bendale, P.M.; Khaditkar, B.M. *Synth. Commun.*, ***2000***, *30*, 665.

N-OH

Caro's acid , SiO_2 , AcOH

25°C

Br

O

Br

94%

Movassagh, B.; Lakouraj, M.M.; Ghodrati, K. *Synth. Commun.*, ***2000***, *30*, 4501.

t-Bu

N O

Ph

1. 2 eq BuLi, THF , -78°C → rt
2. BnBr

3. 10% oxalic acid , reflux

O

Bn

Ph

Bu

72% (73% ee)

Dwyer, M.P.; Price, D.A.; Lamar, J.E.; Meyers, A.I. *Tetrahedron Lett.*, ***1999***, *40*, 4765.

Ph =NOH

1. $Co_2(CO)_8$, NEt_3

2. aq MeOH

Ph =O

98%

Mukai, C.; Nomura, I.; Kataoka, O.; Hanaoka, M. *Synthesis*, ***1999***, 1872.

Bt

Bt = benzotriazole

1. $Pd(OAc)_2$, PPh_3 , $ZnBr_2$/PhH

2. N

3. H_2O , reflux

O

80-90% (96:4 *E:Z*)

Katritzky, A.R.; Huang, Z.; Fang, Y. *J. Org. Chem.*, ***1999***, *64*, 7625.

$(OC)_5Cr$=C(OCH$_2$CH=CH$_2$)Tol —[1% $Pd(PPh_3)_4$, 1 atm CO; DCM, 0°C, 17 h]→ Tol-CO-CH$_2$CH=CH$_2$ 71%

Sakurai, H.; Tanabe, K.; Narasaka, K. *Chem. Lett.*, ***1999***, 75.

Cyclohexanone semicarbazone (N—$NHCONH_2$) —[NH_4CrO_3Cl, Al_2O_3; DCM, 40°C, 3 h]→ cyclohexanone 99%

Zhang, G.-S.; Gong, H.; Yang, D.-H.; Chen, M.-F. *Synth. Commun.*, ***1999***, *29*, 1165.

Cyclopentanone oxime (=N-OH) —[zeolite-$KMnO_4$, DCE; rt, 30 min]→ cyclopentanone (=O) 90%

Jadhav, V.K.; Wadgaonkar, P.P.; Joshi, P.L.; Salunkhe, M.M.
Synth. Commun., ***1999***, *29*, 1989.

Cyclohexanone semicarbazone (=N-$NHCONH_2$) —[Me_3NH^+ CrO_3Cl^-; Al_2O_3, ether, 30 min]→ cyclohexanone (=O) 87%

Gong, H.; Zhang, G.-S. *Synth. Commun.*, ***1999***, *29*, 2591.

PhCH(NO_2)CH_3 —[NaH, THF, $Me_3SiOOSiMe_3$; rt, 1 d]→ PhCOCH$_3$ 70%

Shahi, S.P.; Vankar, Y.D. *Synth. Commun.*, ***1999***, *29*, 4321.

Ph_2C=N-NHPh —[$KHCO_3$, Oxone, acetone; 45 min]→ Ph_2C=O 98%

Hajipour, A.R.; Mahboubghah, N. *Org. Prep. Proceed. Int.* ***1999***, *31*, 112.

Ph_3B —[cat $Pd(OAc)_2$, MeOH, O_2; 25°C, 15 h]→ Ph—Ph 93%

Ohe, T.; Tanaka, T.; Kuroda, M.; Cho, C.S.; Ohe, K.; Uemura, S.
Bull. Chem .Soc. Jpn., ***1999***, *72*, 1851.

REVIEWS:

"Recent Advances Into The Enantioselective Protonation Of Prostereogenic Enol Derivatives,"
Eames, J.; Weerasooriya, N. *Tetrahedron Asymm.*, ***2001***, *12*, 1.

"Regeneration Of Carbonyl Compounds From The Corresponding Oximes," Corsaro, A.; Chiacchio, U.; Pistarà, V. *Synthesis*, ***2001***, 1903.

SECTION 180A: PROTECTION OF KETONES

$PhI(O_2CCF_3)_2$, CF_3COOH

MeCN-H_2O

85%

Fleming, F.F.; Funk, L.; Altundas, R.; Tu, Y. *J. Org. Chem.*, ***2001***, *66*, 6502.

HO_2CCHO , Amberlyst 15

microwaves , 1 min

94%

Chavan, S.P.; Soni, P.; Kamat, S.K. *Synlett*, ***2001***, 1251.

—$(C_6H_4SO_2N(Br)CH_2)_2$—

CCl_4 , rt , 2 h

95%

Khazaei, A.; Vaghei, R.G.; Tajbakhsh, M. *Tetrahedron Lett.*, ***2001***, *42*, 5099.

cat $TiCl_4$, NEt_3 , MeOH

0°C , 30 min

83%

Clerici, A.; Pastori, N.; Porta, O. *Tetrahedron*, ***2001***, *57*, 217.

$HOCH_2CH_2OH$, 110°C , 6 h

Pt-Mo/ZrO_2 catalyst

98%

Reddy, B.M.; Reddy, V.R.; Giridhar, D. *Synth. Commun.*, ***2001***, *31*, 1819.

acetone , reflux , cat $Bi(OTf)_3$

97%

Mohammadpoor-Baltork, I.; Khosropour, A.R.; Aliyan, H. *Synth. Commun.*, ***2001***, *31*, 3411.

aq acetone , $PhI(OAc)_2$; rt , 3 min — 92%

Shi, X.-X.; Wu, Q.-Q. *Synth. Commun.*, ***2000***, *30*, 4081.

1. 0.1 $VOCl_3$, CF_3CH_2OH , O_2 reflux , 5 h ; 2. H_2O — quant

Kirahara, M.; Ochiai, Y.; Arai, N.; Takizawa, S. ; Momose, T.; Nemoto, H. *Tetrahedron Lett.*, ***1999***, *40*, 9055.

$HOCH_2CH_2SH$, DCM ; zeolite HS7-360 , 15 h — quant

Ballini, R.; Barboni, L.; Maggi, R.; Sartori, G. *Synth. Commun.*, ***1999***, *29*, 713.

$PhI(OTf)_3$, NaI , DCM ; rt , 15 min — 91%

Chen, L.-C.; Wang, H.-M. *Org. Prep. Proceed. Int.*, ***1999***, *31*, 562.

$FeSO_4{\cdot}7\,H_2O$, $CHCl_3$; rt , 45 min — 90%

Nasreen, A.; Adapa, S.R. *Org. Prep. Proceed. Int.*, ***1999***, *31*, 573.

See Section 362 (Ester-Alkene) for the formation of enol esters and Section 367 (Ether-Alkenes) for the formation of enol ethers. Many of the methods in Section 60A (Protection of Aldehydes) are also applicable to ketones.

CHAPTER 13
PREPARATION OF NITRILES

SECTION 181: NITRILES FROM ALKYNES

NO ADDITIONAL EXAMPLES

SECTION 182: NITRILES FROM ACID DERIVATIVES

NO ADDITIONAL EXAMPLES

SECTION 183: NITRILES FROM ALCOHOLS AND THIOLS

NO ADDITIONAL EXAMPLES

SECTION 184: NITRILES FROM ALDEHYDES

PhCHO —(NH_3 , $MgSO_4$, O_2 , 16 h)→ PhCN 89%

Lai, G.; Bhamare, N.K.; Anderson, W.K. *Synlett*, ***2001***, 230.

PhCHO —(I_2 , NH_3/H_2O , THF , rt , 30min)→ Ph—CN 96%

Talukdar, S.; Hsu, J.-L.; Chou, T.-C.; Fang, J.-M. *Tetrahedron Lett.*, ***2001***, ***42***, 1103.

PhCHO —($CH_2(CN)_2$, urotropine / microwaves , 2 min)→ Ph–CH=C(CN)CN 87%

Yadav, J.S.; Reddy, B.V.S.; Madan, Ch. *J. Chem. Res. (S)*, ***2001***, 190.

Ph–CH=CH∿CHO —[NH_3, CuCl, *i*-PrOH; 50% H_2O_2, 3 h]→ Ph–CH=CH∿CN 76%

Erman, M.B.; Snow, J.W.; Williams, M.J. *Tetrahedron Lett.*, ***2000***, *41*, 6749.

HO–C_6H_4–CHO —[NH_2OH •HCl, NH_4OAc; microwaves, 1 min]→ HO–C_6H_4–CN 92%

Das, B.; Ramesh, C.; Madhusudhan, P. *Synlett*, ***2000***, 1599.

BuCHO —[NH_4HCO_3, aq KOH, MeCN, 1.5 h; $(Bu_4N)_2S_2O_8$, $Cu(HCO_2)Ni(CO_2H)_2$]→ Bu–CN 89%

Chen, F.-E.; Fu, H.; Meng, G.; Cheng, Y.; Lü, Y.-X. *Synthesis*, ***2000***, 1519.

PhCHO —[NH_2OH•HCl, toluene; FeOO280]→ PhCN 72%

Bajpai, A.R.; Deshpande, A.B.; Samant, S.D. *Synth. Commun.*, ***2000***, *30*, 2785.

PhCHO —[NH_2OH•HCl, 220°C, NEt_3; phthalic anhydride, microwaves, 3 min]→ PhCN 84%

Veverková, E.; Toma, Š. *Synth. Commun.*, ***2000***, *30*, 3109.

PhCHO —[NH_2OH•HCl, NMP, 115°C, 5 h]→ PhCN 86%

Kumar, H.M.S.; Reddy, B.V.S.; Reddy, P.T.; Yadav, J.S. *Synthesis*, ***1999***, 586.

PhCHO —[NH_2OH•HCl, microwaves; $NaHSO_4$•SiO_2, 2 min]→ PhCN 84%

Das, B.; Madhusudhan, P.; Venkataiah, B. *Synlett*, ***1999***, 1569.

PhCHO —[$PhNHCO_2NH_2$•TsOH, THF; reflux]→ PhCN 90%

Coşkun, N.; Arikan, N. *Tetrahedron*, ***1999***, *55*, 11943.

MeO–C_6H_4–CHO —[H_2NOH•H_2O, 15 min; N-methylpyrrolidinone]→ MeO–C_6H_4–CN 98%

Chakraborti, A.K.; Kaur, G. *Tetrahedron*, ***1999***, *55*, 13265.

SECTION 185: NITRILES FROM ALKYLS, METHYLENES AND ARYLS

Cl—C6H3(Cl)—Me → (400°C, NH_3, 3/2 O_2; cat DC–108) Cl—C6H3(Cl)—CN 86%

Qiong, Z.; Chi, H.; Guangyong, X.; Chongwen, X.; Yuanyin, C. *Synth. Commun.*, ***1999***, *29*, 2349.

SECTION 186: NITRILES FROM AMIDES

PyBOP, *i*-Pr_2NEt, DCM 95%

PyBOP = benzotriazol-1-yloxytris(pyrrolidino)phosphonium hexafluorophosphate

Bose, D.S.; Narsaiah, A.V. *Synthesis*, ***2001***, 373.

Ph–C(O)–NH_2 → (aq NaOH, TEBA, $CHCl_3$; ultrasound) PhCN 84%

Sivakumar, M.; Senthilkumar, P.; Pandit, A.B. *Synth. Commun.*, ***2001***, *31*, 2583.

Ph–CH_2–C(O)–NH_2 → ($Cl_3C–CH_2OCOCl$; Et_3N) Ph–CH_2–CN 84%

Bose, D.S.; Kumar, K.K. *Synth. Commun.*, ***2000***, *30*, 3047.

2,4-dimethoxybenzamide → (PhO–C(S)–Cl; Py, <5°C → rt, 5 h) 2,4-dimethoxybenzonitrile 91%

Bose, D.S.; Goud, P.R. *Tetrahedron Lett.*, ***1999***, *40*, 747.

Ph–C(=O)–NH$_2$ —[Bu_2SnO , microwaves , toluene]→ PhCN 90%

Bose, D.S.; Jayalakshmi, B. *J. Org. Chem.*, ***1999***, *64*, 1717.

Ph–C(=O)–NH$_2$ —[$(CF_3SO_2)_2O$, NEt_3 , rt / CH_2Cl_2 , 15 min]→ PhCN 90%

Bose, D.S.; Jayalakshmi, B. *Synthesis*, ***1999***, 64.

Ph–C(=O)–NH$_2$ —[Bu_2SnO , toluene / reflux]→ PhCN 86%

Bose, D.S.; Jayalakshmi, B.; Goud, P.R. *Synthesis*, ***1999***, 1724.

4-Br–C$_6$H$_4$–C(=O)NH$_2$ —[EDCI , Py , DCM , rt]→ 4-Br–C$_6$H$_4$–CN 84%

EDCI= 1-(3-dimethylaminopropyl)-3-ethyl-B-carbodiimide hydrochloride

Bose, D.S.; Sunder, K.S. *Synth. Commun.*, ***1999***, *29*, 4235.

SECTION 187: NITRILES FROM AMINES

$PhCH_2NH_2$ —[Me_3NO , OsO_4 , THF/H_2O/Py / rt , 12 h]→ PhCN 32%

Gao, S.; Herzig, D.; Wang, B. *Synthesis*, ***2001***, 544.

SECTION 188: NITRILES FROM ESTERS

NO ADDITIONAL EXAMPLES

SECTION 189: NITRILES FROM ETHERS, EPOXIDES AND THIOETHERS

NO ADDITIONAL EXAMPLES

SECTION 190: NITRILES FROM HALIDES AND SULFONATES

2-chlorobenzonitrile → (2 eq KCN, cat $Pd(OAc)_2$ / cat dppp, toluene, 160°C) → phthalonitrile 91%

Sundermeier, M.; Zapf, A.; Beller, M.; Sans, J. *Tetrahedron Lett.*, ***2001***, *42*, 6707.

PhI → ($[NaBH_3CN/catechol]$, K_3PO_4 / cat $Pd(PPh_3)_4/PPh_3$, THF, reflux, 1 d) → PhCN 78%

Jiang, B.; Kan, Y.; Zhang, A. *Tetrahedron*, ***2001***, *57*, 1581.

C_4H_9—Br → (NaCN, PEG-400 / reflux, 6 h) → C_4H_9—CN 96%

Cao, Y.-Q.; Chen, B.-H.; Pei, B.-G. *Synth. Commun.*, ***2001***, *31*, 2203.

4-chloroanisole (Cl, OMe) → ($Zn(CN)_2$, DMF, 150°C, 4 h / cat $Pd_2(dba)_3$, cat dppf, cat Zn) → 4-methoxybenzonitrile (CN, OMe) 88%

Jin, F.; Confalone, P.N. *Tetrahedron Lett.*, ***2000***, *41*, 3271.

2-(2-bromobenzyl)malononitrile (Br, CN, CN) → (cyclohexyl-S ring, NC–C(Et)(Me)–N=N–C(Et)(Me)–CN / syringe pump, 9 h) → 2-(2-cyanoethyl)benzonitrile (CN, CN) 72%

Bowman, W.R.; Bridge, C.F.; Brookes, P. *Tetrahedron Lett.*, ***2000***, *41*, 8989.

MeO–C_6H_4–I → (CuCN, cat $Pd_2(dba)_3$, dppf / THF, reflux, 1 h) → MeO–C_6H_4–CN 93%

Sakamoto, T.; Ohsawa, K. *J. Chem. Soc., Perkin Trans. 1*, ***1999***, 2323.

SECTION 191: NITRILES FROM HYDRIDES

NO ADDITIONAL EXAMPLES

SECTION 192: NITRILES FROM KETONES

NO ADDITIONAL EXAMPLES

SECTION 193: NITRILES FROM NITRILES

Conjugate reductions and Michael alkylations of alkene nitriles are found in Section 74D (Alkyls from Alkenes).

OMe; F; 4 eq i-PrCN, KHMDS; toluene, 60°C, 18 h; OMe; CN; 95%

Caron, S.; Vazquez, E.; Wojcik, J.M. *J. Am. Chem. Soc.*, ***2000***, *122*, 712.

CN; Ph; CH_2=CHPh, DMSO; 30% t-buOK, 25°C; Ph; Ph; CN; 77%

Rodriguez, A.L.; Bunlaksananusorn, T.; Knochel, P. *Org. Lett.*, ***2000***, *2*, 3285.

Ph-I; $CH_2(CN)_2$, 2.2 eq t-BuOK; 0.2 $NiBr_2(PPh_3)_2$, 0.4 PPh_3; 0.6 Zn, THF, 2 d; Ph; CN; CN; 70%

Cristau, H.J.; Vogel, R.; Taillefer, M.; Gadras, A. *Tetrahedron Lett.*, ***2000***, *41*, 8457.

PhHC; CN; SO_2Ph; 6 eq SmI_2, THF, MeOH; 60°C, 6 h; Ph; CN; 67%

Guo, H.; Zhang, Y. *Synth. Commun.*, ***2000***, *30*, 1879.

SECTION 194: NITRILES FROM ALKENES

Ph-CH=CH$_2$ → [$Me_2C(OH)CN$, $Ni(cod)_2$, toluene / chiral diphosphite complex, 100°C , 1 d] → PhCH(CH$_3$)CN >98%

Yan, M.; Xu, Q.-Y.; Chan, A.S.C. *Tetrahedron Asymm.*, **2000**, *11*, 845.

SECTION 195: NITRILES FROM MISCELLANEOUS COMPOUNDS

Ph—CH=N—OH → [2% $[RuCl_2(p\text{-cymeme})_2]$, 10 min / MeCN, MS 4Å , 80°C] → Ph—CN 94%

Yang, S.H.; Chang, S. *Org. Lett.*, **2001**, *3*, 4209.

MeO—C$_6$H$_4$—CH=N–OH → [TsCl , Py , $CHCl_3$ / Al_2O_3 , microwaves] → MeO—C$_6$H$_4$—CN 64%

Ghiaci, M.; Bakhtiari, K. *Synth. Commun.*, **2001**, *31*, 1803.

PhCH=N–OH → [$MeO_2(CN)SO_2NEt_3$ / 25°C , DCM , 2h] → PhCN 66%

Jose, B.; Saulatha, M.S.; Pillai, P.M.; Prathapan, S. *Synth. Commun.*, **2000**, *30*, 1509.

PhCH=N–OH → [ferric sulfate , PhH , reflux / 3 h] → PhCN 81%

Desai, D.G.; Swami, S.S.; Mahale, G.D. *Synth. Commun.*, **2000**, *30*, 1623.

PhCH=N–OH → [$InCl_3$, MeCN , reflux , 1.5 h] → PhCN 98%

Barman, D.C.; Thakur, A.J.; Prajapati, D.; Sandhu, J.S. *Chem. Lett.*, **2000**, 1196.

PhCH=N–OH → [$SOCl_2$, DCM , rt , 15 min / benzotriazole] → Ph—CN 96%

Chaudhari, S.S.; Akamanchi, K.G. *Synth. Commun.*, **1999**, *29*, 1741.

$PhCH{=}N{-}NMe_2$ $\xrightarrow[\text{microwaves}]{\text{Oxone , wet } Al_2O_3 \text{ , 6 min}}$ PhCN 78%

Ramalingam, T.; Reddy, B.V.S.; Srinivas, R.; Yadav, J.S. *Synth. Commun.*, ***2000***, *30*, 4507.

PhCH=N–OH $\xrightarrow[\text{80°C , 10 h}]{TiCl_3(OTf) \text{ , sealed tube}}$ PhCN 90%

Iranpoor, N.; Zeynizadeh, B. *Synth. Commun.*, ***1999***, *29*, 2747.

N–OH, Cl, Cl $\xrightarrow[\text{6 min}]{\text{EPZ–10 , microwaves}}$ CN, Cl, Cl 68%

Bandgar, B.P.; Sadavarte, V.S.; Sabu, K.R. *Synth. Commun.*, ***1999***, *29*, 3409.

CHAPTER 14
PREPARATION OF ALKENES

SECTION 196: ALKENES FROM ALKYNES

$190°C$, 2.5 h

83%

Herz, H.-G.; Schatz, J.; Maas, G. *J. Org. Chem.*, ***2001***, *66*, 3176.

H_2, Lindlar catalyst, DMF
diethylenetriamine, rt, 18 h

NH_2 Ph

NH_2 97.3% Ph

Campos, K.R.; Cai, D.; Journet, M.; Kowal, J.J.; Larsen, R.D.; Reider, P.J. *J. Org. Chem.* ***2001***, *66*, 3634.

$PhMe_2SiCl$ → (CH_2=CHMgBr, THF; cat Cp_2TiCl_2) → $PhMe_2Si$ … $SiMe_2Ph$

94% (74:26 *E:Z*)

Watanabe, H.; Terao, J.; Kambe, N. *Org. Lett.*, ***2001***, *3*, 1733.

EtO_2C EtO_2C → (4% $[IrCl(CO)_3]_n$, 80°C; toluene, 2 h) → EtO_2C EtO_2C

93%

Chatani, N.; Inoue, H.; Morimoto, T.; Muto, T.; Murai, S. *J. Org. Chem.*, ***2001***, *66*, 4433.

C_5H_{11} (alkyne); 1-iodonaphthalene, THF, rt; $HSiMe_2OSiMe_2H$, t-Bu_3P-Pt(DVDS), 2 eq Bu_4NF, rt, 5% $Pd(dba)_2$; C_5H_{11} + C_5H_{11} (1.5 : 1) 82%; 98% E

Denmark, S.E.; Wang, Z. *Org. Lett.*, ***2001***, *3*, 1073.

S-PMP; 5% $dppeFeCl_2$, DCE; 1.25 eq Me_3SiCHN_2, 83°C, 7 h; $SiMe_3$, PMP-S; 78%

Prabharasuth, R.; Van Vranken, D.L. *J. Org. Chem.*, ***2001***, *66*, 5256.

Ph—≡—Ph; 2 eq PhI, 5% $Pd(OAc)_2$, NaOAc, Bu_4NCl, DMF, 100°C, 1 d; Ph; 62%

Larock, R.C.; Tian, Q. *J. Org. Chem.*, ***2001***, *66*, 7372.

OH; OAc; 1. TBDPSCl, imidazole, DMF, rt; 2. 2% $Pd_2(dba)_3$•$CHCl_3$, AcOH, 23% PPh_3, toluene, reflux; OTBDPS; OAc; OAc; 90 X 85%

El-Sayed, E.; Anand, N.K.; Carreira, E.M. *Org. Lett.*, ***2001***, *3*, 3017.

Et—≡—Et; I, CO_2Et; 5% $Pd(OAc)_2$, Ag_2CO_3, MeCN, 20°C, 3h; Et, Et, Et, Et, CO_2Et; 72%

Kotora, M.; Matsumura, H.; Gao, G.; Takahashi, T. *Org. Lett.*, ***2001***, *3*, 3467.

OTIPS, Ph, Ph, EtO_2C, CO_2Et → 10% $W(CO)_5$•THF, THF; 3 eq H_2O → Ph, Ph, CO_2Et, CO_2Et

99% (>95:5 *exo:endo*)

Iwasawa, N.; Maeyama, K.; Kusama, H. *Org. Lett.*, ***2001***, *3*, 3871.

TMS, O → $Co_2(CO)_8$, TBHP; DME, heat → TMS, H, O, H

65%

Ajamian, A.; Gleason, J.L. *Org. Lett.*, ***2001***, *3*, 4161.

OH → I, $(CH_2)_8$, CO_2Me; 6% Me_3SnCl, aq KF 11 h; cat TBAF, PMHS, ether; 1% $PdCl_2(PPh_3)_2$, 1% $Pd_2(dba)_3$; 4% (2-furyl)$_3$P, 37°C → OH, CO_2Me, $(CH_2)_8$

73%

Maleczka Jr., R.E.; Gallagher, W.P. *Org. Lett.*, ***2001***, *3*, 4173.

OH → Ph, Br; 6% Me_3SnCl; 1% $PdCl_2(PPh_3)_2$; 1% Pd_2dba_3, 4% (2-furyl)$_3$P; aq Na_2CO_3, PMHS, ether; 37°C, 15 h → Ph, Ph

90%

Gallagher, W.P.; Terstiege, I.; Maleczka Jr., R.E. *J. Am. Chem. Soc.*, ***2001***, *123*, 3194.

C_3H_7—≡—C_3H_7 → 1% Ru catalyst, CH_2Cl_2; $(EtO)_3SiH$ → C_3H_7, $Si(OEt)_3$, C_3H_7

99%

Trost, B.M.; Ball, Z.T. *J. Am. Chem. Soc.*, ***2001***, *123*, 12726.

PhS(C=O)Ph → (C_6H_{13}–C≡CH, cat $Pt(PPh_3)_4$, 15 h, toluene, reflux) → PhS(C₆H₁₃)C=CHPh 77%

Sugoh, K.; Kuniyasu, H.; Sugae, T.; Ohtaka, A.; Takai, Y.; Tanaka, A.; Machino, C.; Kambe, N.; Hurosawa, H. *J. Am. Chem. Soc.*, ***2001***, *123*, 5108.

Me_3Si / Me_3Si → (1. allyl bromide, ether, 5 h; 2 eq $Ti(Oi\text{-}Pr)_4$/2 i-PrMgBr, –15°C → –40°C; 2. H_3O^+) → Me_3Si / $SiMe_3$ 88%

Okamoto, S.; Subbaraj, K.; Sato, F. *J. Am. Chem. Soc.*, ***2001***, *123*, 4857.

C_3H_7–C≡C–C_3H_7 → ($PhB(OH)_2$, aq dioxane; 3% $Rh(acac)(C_2H_4)_2$, dppb, 100°C, 3 h) → Ph, C_3H_7, C_3H_7 87%

Hayashi, T.; Inoue, K.; Tankguchi, N.; Ogasawara, M. *J. Am. Chem. Soc.*, ***2001***, *123*, 9918.

Br, Et → (BuOH, 2% $Pd(OAc)_2$, 2.5% phosphine ligand; toluene, 2.5 eq Cs_2CO_3) → BuO, Et 63%

Asao, N.; Shimada, T.; Shimada, T.; Yamamoto, Y. *J. Am. Chem. Soc.*, ***2001***, *12*, 10899.

Ph, OMe, OMe → (Me_3SiI, toluene; –50°C → 10°C) → I, Ph, OMe + Ph, I, OMe (18 : 82) 84%

Takami, K.; Yorimitsu, H.; Shinokubo, H.; Matsubara, S.; Oshima, K. *Synlett*, ***2001***, 293.

$HSiCl_3$, 20°C, 18 h
1% $[PdCl(\pi\text{-}C_3H_5)]_2$ PFFOMe

t-Bu, Cl_3Si, H, Me

81% (72% ee, *S*)

Vo-Tranh, G.; Boucard, V.; Sauriat-Dorizon, H.; Guibé, F. *Synlett*, ***2001***, 37.

0.5% $[RuCl_2(p\text{-cymene})]_2$, 80°C
1% IMeSHCl, Cs_2CO_3, toluene

O, O

86%

Ackermann, L.; Bruneau, S.; Dixneuf, P.H. *Synlett*, ***2001***, 397.

Ph

Ph_2PH, cat $Pd(PPh_3)_4$
MeCN, 130°C, 18 h

Ph, Ph_2P + Ph, PPh_2

(0 : 100) 95%

+ $Ni(acac)_2$, $(EtO)_2PH{=}O$ (95 : 5) 90%

Kazankova, M.A.; Efimova, I.V.; Kochetkov, A.N.; Afanas'ev, V.V.; Beletskaya, I.P.; Dixneuf, P.H. *Synlett*, ***2001***, 497.

Ph

$PhICl_2$, $Pb(SCN)_2$
CH_2Cl_2, 5°C

Ph, SCN, NCS, H

58%

Prakash, O.; Sharma, V.; Batra, H.; Moriarty, R.M. *Tetrahedron Lett.*, ***2001***, *42*, 553.

Me, $SiMe_3$

$PhCH(OMe)_2$
5% $TrClO_4$, CH_2Cl_2
–50°C, 2 h

MeO, Ph, Me

quant

Niimi, L.; Shiino, K.; Hiraoka, S.; Yokozawa, T. *Tetrahedron Lett.*, ***2001***, *42*, 1721.

OBn

1. 10% $Fe(acac)_3$, 3 eq BuLi, Tol
2. 1N HCl

Bu, OBn

97%

Hojo, M.; Murakami, Y.; Aihara, H.; Sakuragi, R.; Baba, Y.; Hosomi, A. *Angew. Chem. Int. Ed.*, ***2001***, *40*, 621.

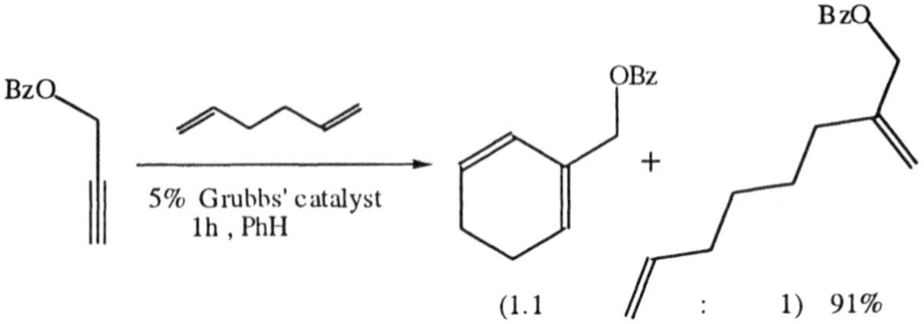

BzO

5% Grubbs' catalyst
1h , PhH

OBz

+

BzO

(1.1 : 1) 91%

Smulik, J.A.; Diver, S.T. *Tetrahedron Lett.*, **2001**, *42*, 171.

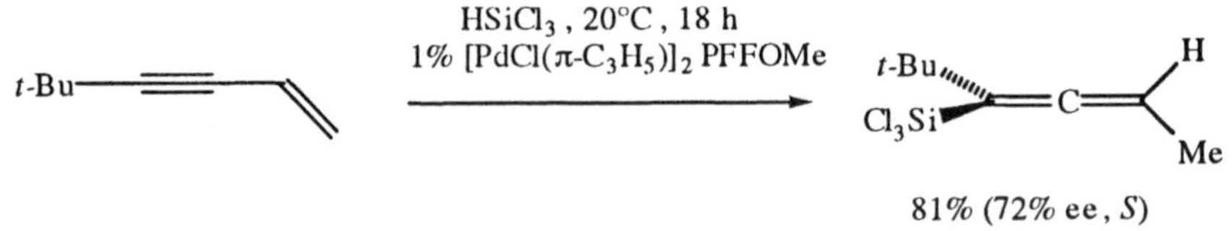

t-Bu

$HSiCl_3$, 20°C , 18 h
1% $[PdCl(\pi\text{-}C_3H_5)]_2$ PFFOMe

t-Bu, Cl_3Si, C, H, Me

81% (72% ee, *S*)

Han, J.W.; Tokunaga, N.; Hayashi, T. *J. Am. Chem. Soc.*, **2001**, *123*, 12915.

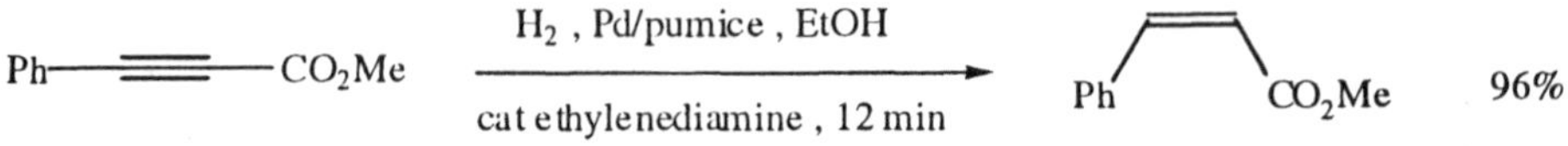

Ph—≡—CO_2Me

H_2 , Pd/pumice , EtOH
cat ethylenediamine , 12 min

Ph CO_2Me 96%

Gruttadauria, M.; Liotta, L.F.; Noto, R.; Deganello, G. *Tetrahedron Lett.*, **2001**, *42*, 2015.

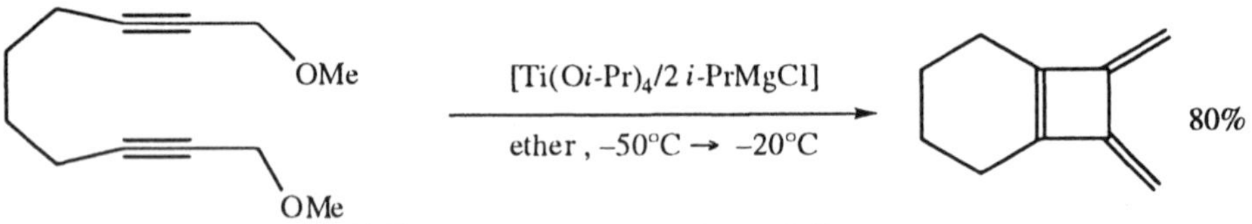

OMe
OMe

$[Ti(Oi\text{-}Pr)_4/2\ i\text{-}PrMgCl]$
ether , –50°C → –20°C

80%

Delas, C.; Urabe, H.; Sato, F. *Tetrahedron Lett.*, **2001**, *42*, 4147.

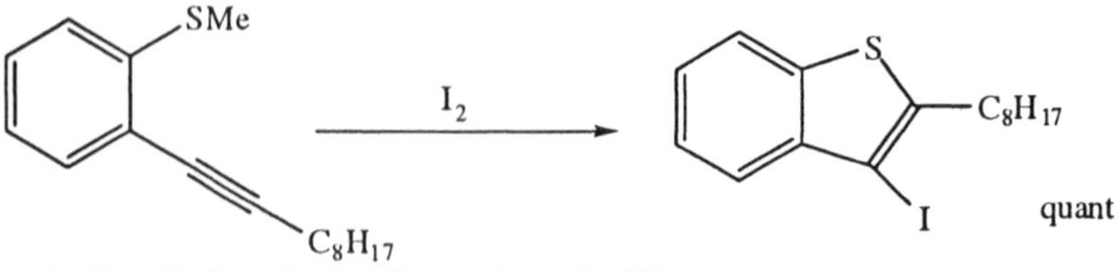

SMe
C_8H_{17}

I_2

S
C_8H_{17}
I

quant

Larock, R.C.; Yue, D. *Tetrahedron Lett.*, **2001**, *42*, 6011.

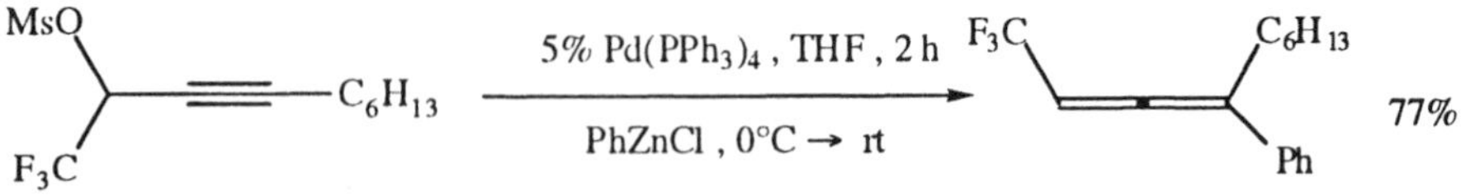

MsO
F_3C
C_6H_{13}

5% $Pd(PPh_3)_4$, THF , 2 h
PhZnCl , 0°C → rt

F_3C C_6H_{13}
Ph

77%

Konno, T.; Tanikawa, M.; Ishihara, T.; Yamanaka, H. *Chem. Lett.*, **2000**, 1360.

OMe

Ph

cat $Pd(OAc)_2/PPh_3$

MeONa , MeOH
25°C , 1 d

Ph

42%

OMe

+

Ph

18%

OMe

Wu, M.-J.; Wei, L.-M.; Lin, C.-F.; Leou, S.-P.; Wei, L.-L. *Tetrahedron*, ***2001***, *57*, 7839.

Ph

O

5% $[Rh(diphos)Cl]_2$, DCE

$AgSbF_6$, rt , 2 h

Ph

O

62% (96% ee)

Cao, P.; Zhang, X. *Angew. Chem. Int. Ed.*, ***2000***, *39*, 4104.

OAc

t-Bu_2CuLi–LiCN , PBu_3

–80°C

t-Bu

92% (25:75 *E*:*Z*)

Krause, N.; Purpura, M. *Angew. Chem. Int. Ed.*, ***2000***, *39*, 4355.

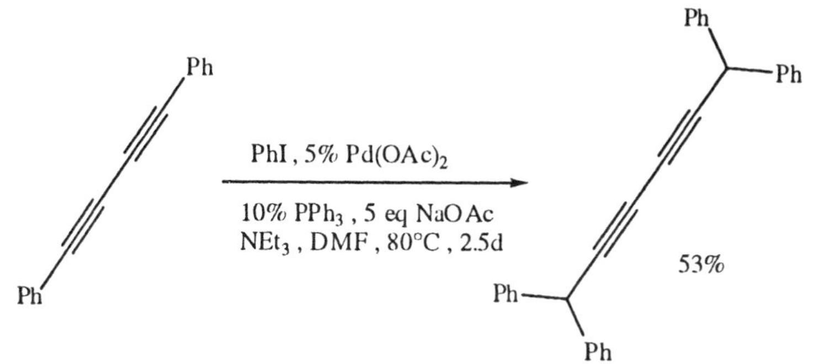

Dyker, G.; Borowski, S.; Henkel, G.; Kellner, A.; Dix, I.; Jones, P.G.
Tetrahedron Lett., ***2000***, *41*, 8259.

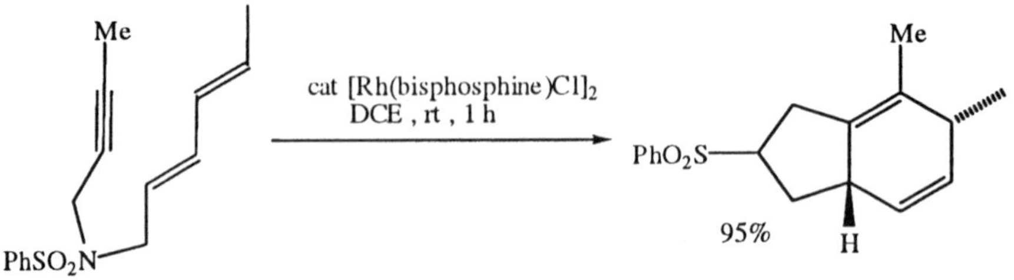

Me

cat $[Rh(bisphosphine)Cl]_2$
DCE, rt, 1 h

$PhSO_2N$

PhO_2S

95%

Wang, B.; Cao, P.; Zhang, X. *Tetrahedron Lett.*, **2000**, *41*, 8041.

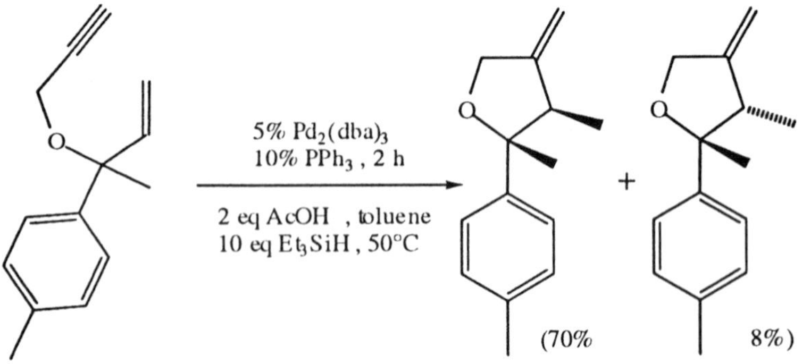

5% $Pd_2(dba)_3$
10% PPh_3, 2 h

2 eq AcOH, toluene
10 eq Et_3SiH, 50°C

(70% 8%)

Oh, C.H.; Han, J.W.; Kim, J.S.; Um, S.Y.; Jung, H.H.; Jang, W.H.; Won, H.S. *Tetrahedron Lett.*, **2000**, *41*, 8365.

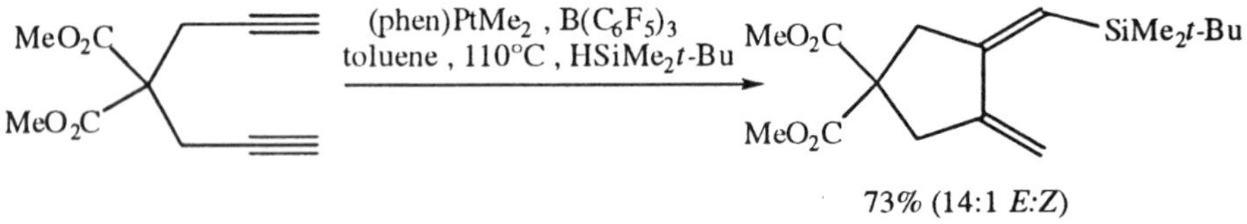

MeO_2C
MeO_2C

(phen)$PtMe_2$, $B(C_6F_5)_3$
toluene, 110°C, $HSiMe_2t$-Bu

$SiMe_2t$-Bu

73% (14:1 *E:Z*)

Madine, J.W.; Wang, X.; Widenhoefer, R.A. *Org. Lett.*, **2001**, *3*, 385.

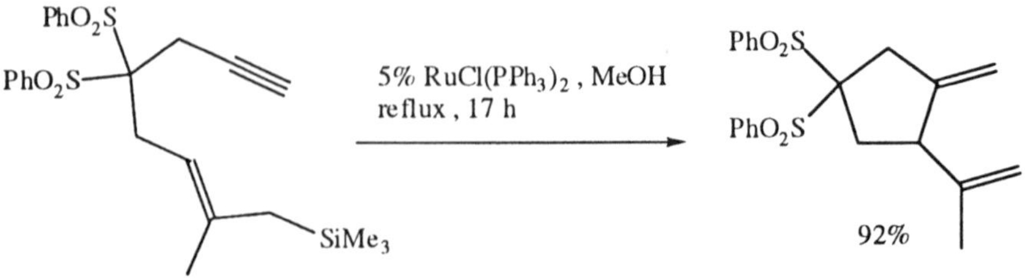

PhO_2S
PhO_2S

5% $RuCl(PPh_3)_2$, MeOH
reflux, 17 h

$SiMe_3$

92%

Fernández-Rivas, C.; Méndez, M.; Echavarren, A.M. *J. Am. Chem. Soc.*, **2000**, *122*, 1221.

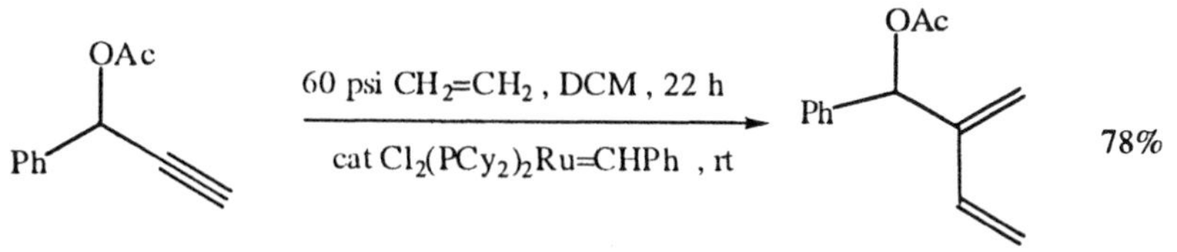
OAc
Ph
60 psi CH_2=CH_2 , DCM , 22 h
cat $Cl_2(PCy_2)_2$Ru=CHPh , rt
OAc
Ph
78%

Smulik, J.A.; Diver, S.T. *J. Org. Chem.*, ***2000***, *65*, 1788.

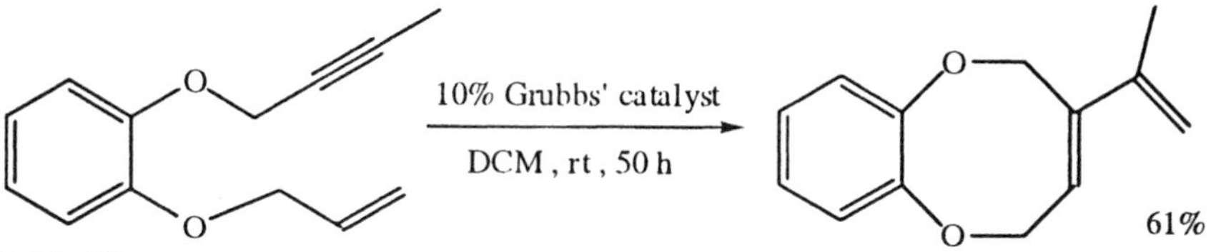
O
O
10% Grubbs' catalyst
DCM , rt , 50 h
O
O
61%

Mori, M.; Kitamura, T.; Sakaibara, N.; Sato, Y. *Org. Lett.*, ***2000***, *2*, 543.

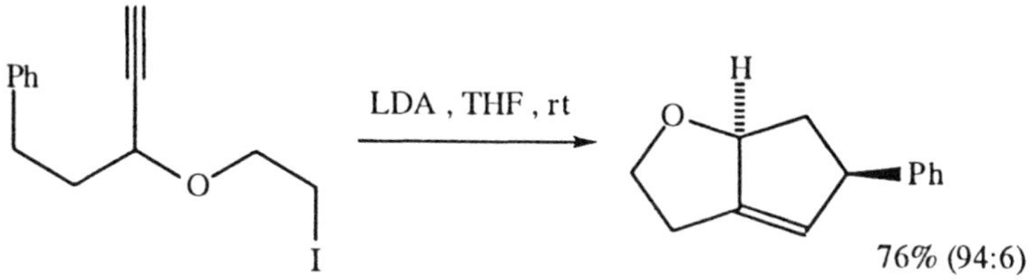
Ph
O
I
LDA , THF , rt
H
O
Ph
76% (94:6)

Harada, T.; Fujiwara, T.; Iwazaki, K.; Oku, A. *Org. Lett.*, ***2000***, *2*, 1855.

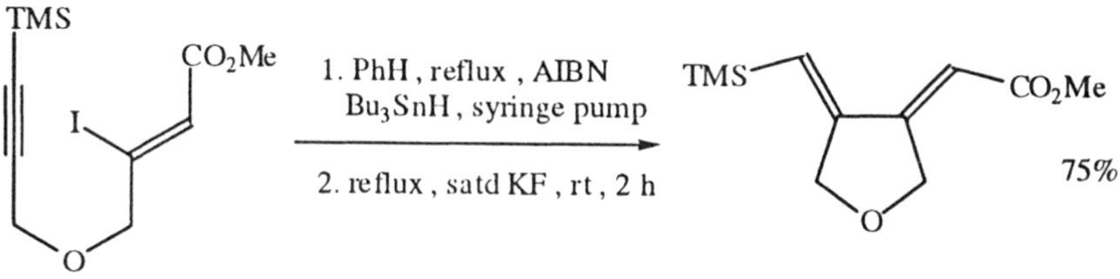
TMS
CO_2Me
I
O
1. PhH , reflux , AIBN
Bu_3SnH , syringe pump
2. reflux , satd KF , rt , 2 h
TMS
CO_2Me
O
75%

Sha, C.-K.; Zhan, Z.-P.; Wang, F.-S. *Org. Lett.*, ***2000***, *2*, 2011.

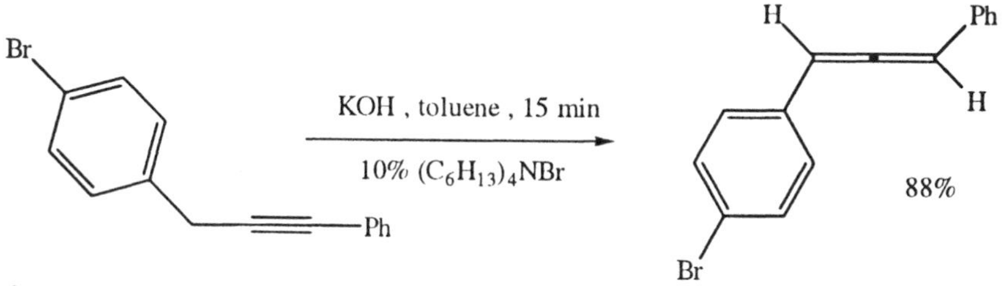
Br
Ph
KOH , toluene , 15 min
10% $(C_6H_{13})_4NBr$
H
Ph
H
Br
88%

Oku, M.; Arai, S.; Katayama, K.; Shioiri, T. *Synlett*, ***2000***, 493.

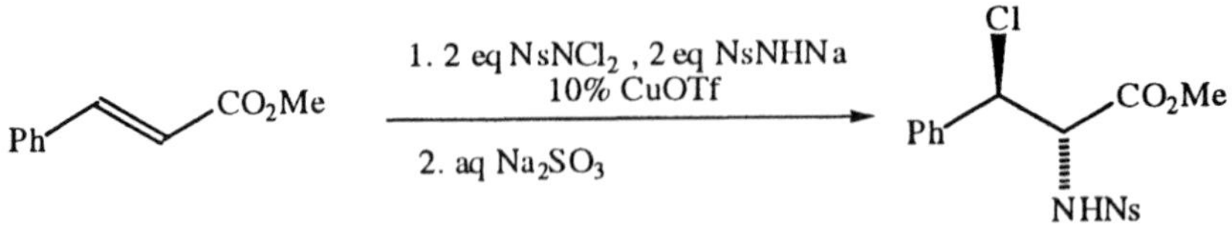
Ph
CO_2Me
1. 2 eq $NsNCl_2$, 2 eq NsNHNa
10% CuOTf
2. aq Na_2SO_3
Cl
Ph
CO_2Me
NHNs

76% (20:1 *anti:syn*)

Li, G.; Wei, H.-X.; Kim, S.H. *Org. Lett.*, ***2000***, *2*, 2249.

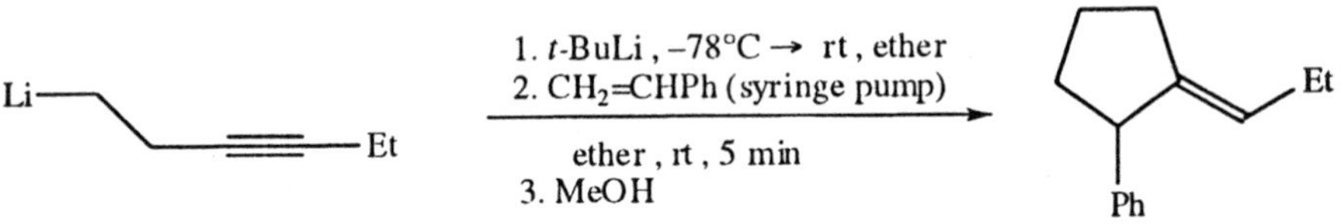
Li
Et
1. *t*-BuLi , –78°C → rt , ether
2. CH_2=CHPh (syringe pump)
ether , rt , 5 min
3. MeOH
Et
Ph

61% (97:3 *E:Z*)

Wei, X.; Taylor, R.J.K. *Angew. Chem. Int. Ed.*, ***2000***, *39*, 409.

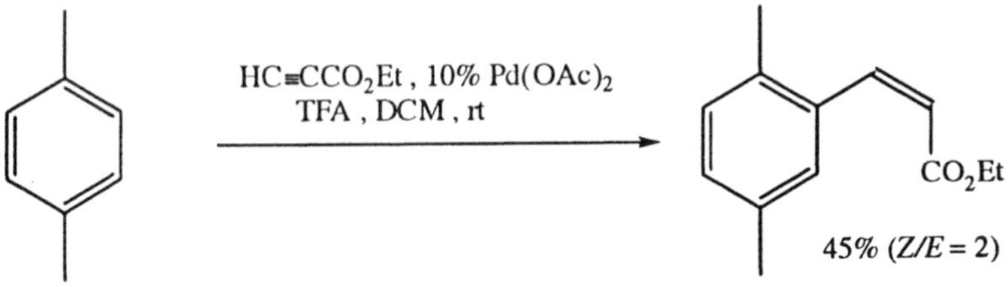
HC≡CCO_2Et , 10% $Pd(OAc)_2$
TFA , DCM , rt
CO_2Et

45% (*Z/E* = 2)

Jia, C.; Lu, W.; Oyamada, J.; Kitamura, T.; Matsuda, K.; Irie, M.; Fujiwara, Y. *J. Am. Chem. Soc.*, ***2000***, *122*, 7252.

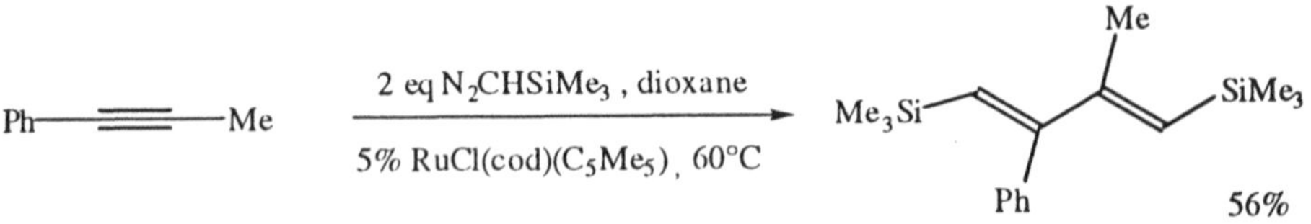
Ph
Me
2 eq $N_2CHSiMe_3$, dioxane
5% RuCl(cod)(C_5Me_5) , 60°C
Me
Me_3Si
$SiMe_3$
Ph

56%

Le Paih, J.; Dérien, S.; Özdemir, I.; Dixneuf, P.H. *J. Am. Chem. Soc.*, ***2000***, *122*, 7400.

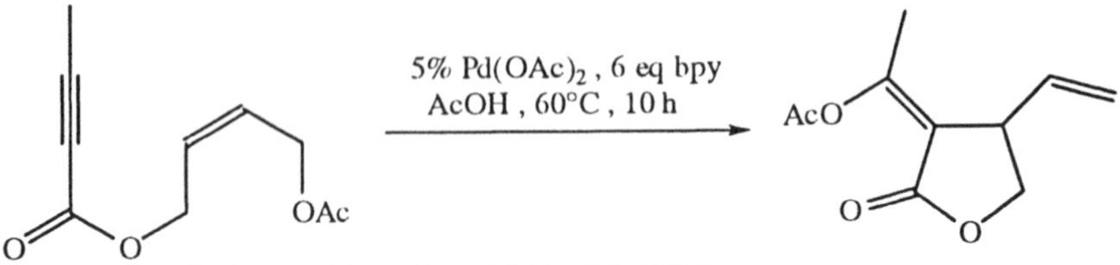
O
O
OAc
5% $Pd(OAc)_2$, 6 eq bpy
AcOH , 60°C , 10 h
AcO
O
O

87%

Zhang, Q.; Lu, X. *J. Am. Chem. Soc.*, ***2000***, *122*, 7604.

$$\text{I-C}_6\text{H}_4\text{-CO}_2\text{Et} \xrightarrow[\text{1\% Pd(dba)}_2\text{, THF}]{\text{1-methyl-1-vinylsiletane, TBAF, rt}} \text{H}_2\text{C=HC-C}_6\text{H}_4\text{-CO}_2\text{Et} \quad 93\%$$

Chatani, N.; Inoue, H.; Ikeda, T.; Murai, S. *J. Org. Chem.*, ***2000***, *65*, 4913.

$$\text{C}_6\text{H}_{13}\text{-C}\equiv\text{C-C(CH}_3)_2\text{-OCO}_2\text{Me} \xrightarrow[\text{THF, reflux, 1 h}]{\text{Me}_3\text{SiCN, 5\% Pd(PPh}_3)_4} (\text{C}_6\text{H}_{13})(\text{NC})\text{C=C=C(CH}_3)_2 \quad 91\%$$

Tsuji, Y.; Tankguchi, M.; Yasuda, T.; Kawamura, T.; Obora, Y. *Org. Lett.*, ***2000***, *2*, 2635.

$$\text{pyrrole} \xrightarrow[\text{Pd(OAc)}_2\text{, AcOH, rt}]{\text{Ph-C}\equiv\text{C-CO}_2\text{Et}} \text{2-pyrrolyl-C(Ph)=CH-CO}_2\text{Et} \quad 78\%$$

Lu, W.; Jia, C.; Kitamura, T.; Fujiwara, Y. *Org. Lett.*, ***2000***, *2*, 2927.

$$\text{AcO-norbornene} \xrightarrow[\text{cat Cp*RuCl(cod), 60h}]{\text{Ph-C}\equiv\text{C-CO}_2\text{Et, THF, 80°C}} \text{cyclobutene (Ph, CO}_2\text{Et)} + \text{cyclobutene (CO}_2\text{Et, Ph)} \quad (2.8 : 1) \quad 68\%$$

Jordan, R.W.; Tam, W. *Org. Lett.*, ***2000***, *2*, 3031.

$$\text{Ph-C}\equiv\text{C-Ph} \xrightarrow[\text{2 eq NaOAc, Bu}_4\text{NCl, 100°C}]{\text{PhI, 5\% Pd(OAc)}_2\text{, 10\% PPh}_3\text{, DMF}} \text{9-benzylidenefluorene} \quad 62\%$$

Tian, Q.; Larock, R.C. *Org. Lett.*, ***2000***, *2*, 3329.

$$t\text{-Bu-C}\equiv\text{CH} \xrightarrow[\text{THF, microwaves; 2. Pd(PPh}_3)_4\text{, PhCH=CHBr, microwaves}]{\text{1. BuSnCl, aq KF, PMHS, cat TBAF, Pd(PPh}_3)_4} \text{PhCH=CH-CH=CH-}t\text{-Bu} \quad 86\%$$

Maleczka Jr., R.E.; Lavis, J.M.; Clark, D.H.; Gallaagher, W.P. *Org. Lett.*, ***2000***, *2*, 3655.

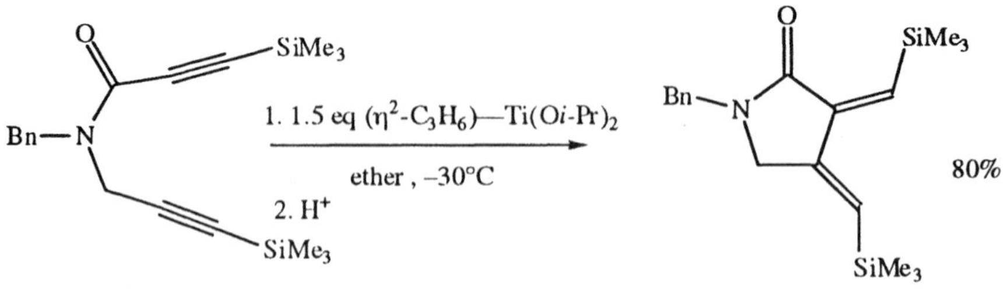

1. 1.5 eq $(\eta^2\text{-}C_3H_6)$—Ti(O*i*-Pr)$_2$ ether, –30°C 2. H^+

80%

Urabe, H.; Nakajima, R.; Sato, F. *Org. Lett.*, ***2000***, *2*, 3481.

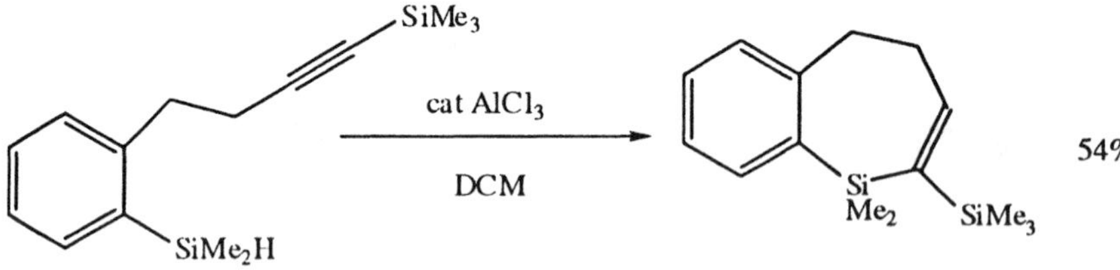

cat $AlCl_3$ DCM

54%

Sudo, T.; Asao, N.; Yamamoto, Y. *J. Org. Chem.*, ***2000***, *65*, 8919.

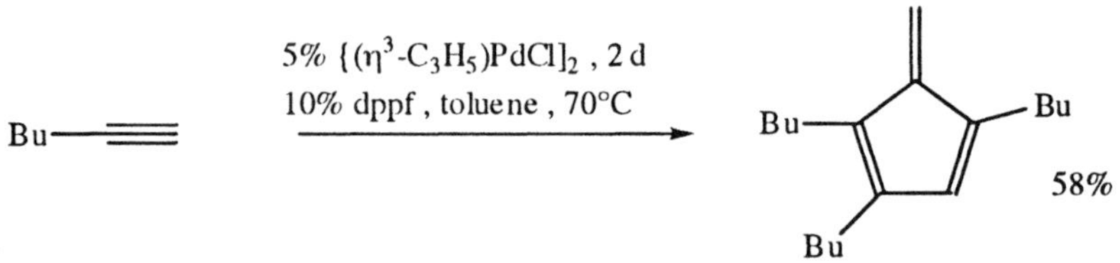

5% $\{(\eta^3\text{-}C_3H_5)PdCl\}_2$, 2 d 10% dppf, toluene, 70°C

58%

Radhakrishnan, U.; Gevorgyan, V.; Yamamoto, Y. *Tetrahedron Lett.*, ***2000***, *41*, 1971.

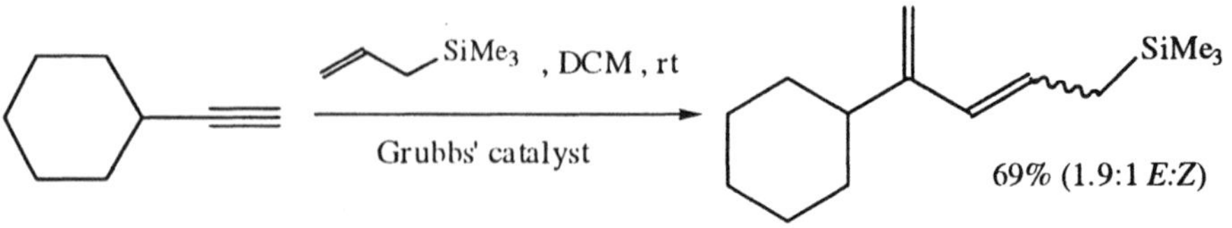

$SiMe_3$, DCM, rt Grubbs' catalyst

69% (1.9:1 *E:Z*)

Stragies, R.; Voigtmann, U.; Blechert, S. *Tetrahedron Lett.*, ***2000***, *41*, 5465.

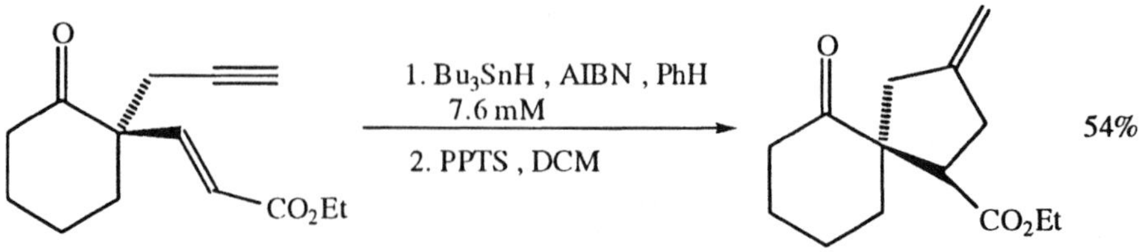

1. Bu_3SnH, AIBN, PhH 7.6 mM 2. PPTS, DCM

54%

Robertson, J.; Lam, H.W.; Abazi, S.; Roseblade, S.; Lusch, R.K. *Tetrahedron*, ***2000***, *56*, 8959.

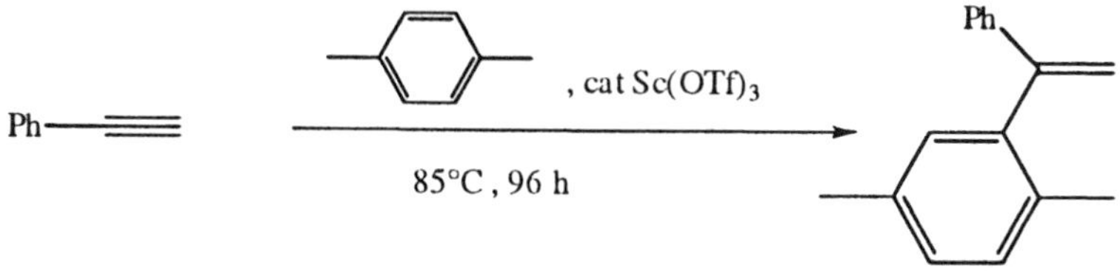

Ph—≡ → (p-xylene, cat $Sc(OTf)_3$; 85°C, 96 h) 92%

Tsuchimoto, T.; Maeda, T.; Shirakawa, E.; Kawakami, Y. *Chem. Commun.*, ***2000***, 1573.

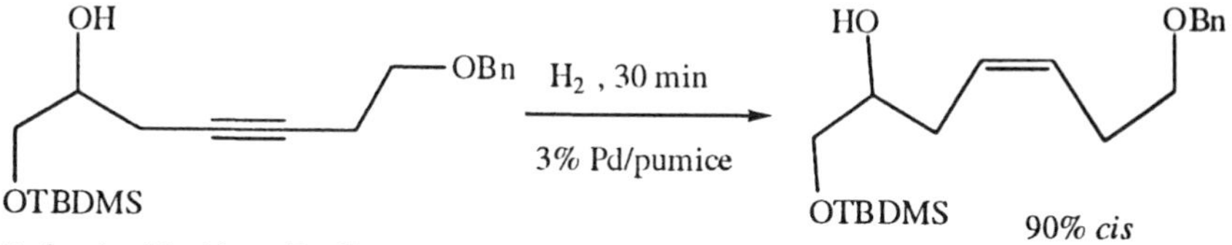

OH, OTBDMS, OBn → (H_2, 30 min; 3% Pd/pumice) HO, OTBDMS, OBn 90% *cis*

Gruttadauria, M.; Noto, R.; Deganello, G.; Liotta, L.F. *Tetrahedron Lett.*, ***1999***, ***40***, 2857.

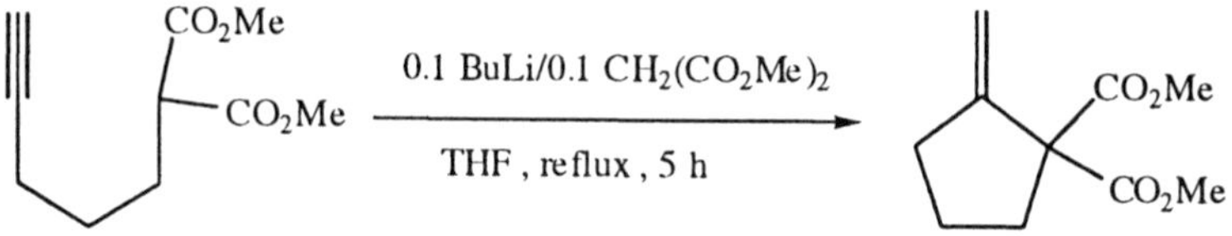

CO_2Me, CO_2Me → (0.1 BuLi/0.1 $CH_2(CO_2Me)_2$; THF, reflux, 5 h) CO_2Me, CO_2Me 86%

Kitagawa, O.; Suzuki, T.; Fujiwara, H.; Fujita, M.; Taguchi, T. *Teterahedron Lett.*, ***1999***, ***40***, 4585.

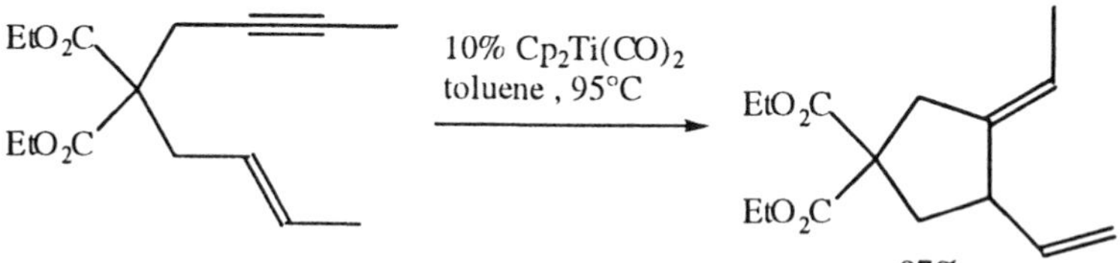

EtO_2C, EtO_2C → (10% $Cp_2Ti(CO)_2$, toluene, 95°C) EtO_2C, EtO_2C 97%

Sturla, S.J.; Kablaoui, N.M.; Buchwald, S.L. *J. Am. Chem. Soc.*, ***1999***, ***121***, 1976.

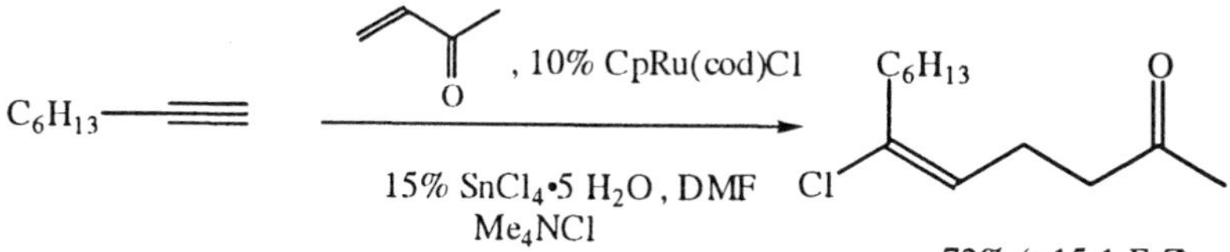

C_6H_{13}—≡ → (methyl vinyl ketone (O), 10% CpRu(cod)Cl; 15% $SnCl_4$•5 H_2O, DMF, Me_4NCl) C_6H_{13}, Cl, O 72% (>15:1 *E*:*Z*)

Trost, B.M.; Pinkerton, A.B. *J. Am. Chem. Soc.*, ***1999***, ***121***, 1988.

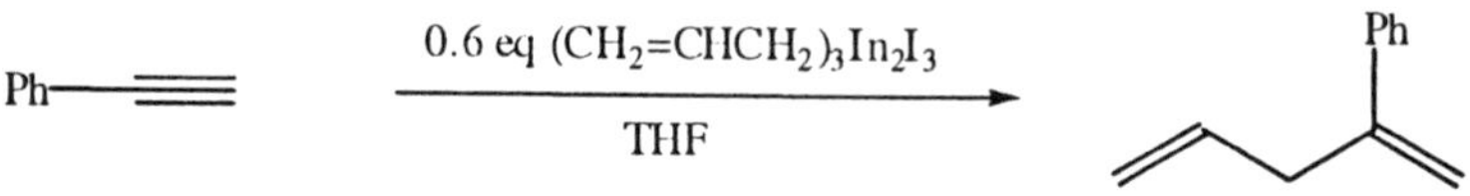

Ph—≡ → (0.6 eq $(CH_2{=}CHCH_2)_3In_2I_3$; THF) Ph 94%

Fujiwara, N.; Yamamoto, Y. *J. Org. Chem.* ***1999***, ***64***, 4095.

$SiMe_3$ — 0.5 $EtAlCl_2$, CH_2Cl_2 / -78°C → $SiMe_3$ 69%

Asao, N.; Shimada, T.; Yamamoto, Y. *J. Am. Chem. Soc.*, ***1999***, ***121***, 3797.

OH — 1. 8 Br, 3 eq In, THF / 2. 32% HCl → OH + OH

(>97 : 3) 97%

Klaps, E.; Schmid, W. *J. Org. Chem.*, ***1999***, ***64***, 7537.

Ph—≡—Ph — 1. Cp_2ZrEt_2 2. I_2 / 3. CuCl → Ph Ph 64%

Takahashi, T.; Shen, B.; Nakajima, K.; Xi, Z. *J. Org. Chem.*, ***1999***, ***64***, 8706.

C_6H_{11} N — $Cp*_2YMe$–THF, rt / $PhSiH_3$, 1 d → N C_6H_{11} H SiH_2Ph 43%

Molander, G.A.; Corrette, C.P. *J. Org. Chem.*, ***1999***, ***64***, 9697.

PhC≡CH, [(naphth)Rh(cod)] BF_4 / DCM, 15°C, 15 min → Ph 83%

Paik, S.-J.; Son, S.U.; Chung, Y.K. *Org. Lett.*, ***1999***, ***1***, 2045.

AcO Me Me — –80°C / Bu_2CuLi•LiI, ether → Me Me Bu 90% (67:33 *E*:*Z*)

Purpura, M.; Krause, N. *Eur. J. Org. Chem.*, ***1999***, 267.

CO_2Et (alkyne) 1. PPh_3, DCM, acetoacetanilide (NHPh); 2. PhH, reflux; 3. toluene, reflux → EtO_2C–CH=C(CO_2Et)–C(CH_3)=CH–C(=O)NHPh

92x53x43%

Yavari, I.; Asghari, S. *Tetrahedron*, ***1999***, *55*, 11853.

cyclopentene oxide → 1.2 eq (NH-azabicyclic pyrrolidinylmethyl amine), 2 eq LDA, 5 eq DBU, THF, rt, 1 d → cyclopent-2-en-1-ol (OH)

78% (95% ee)

Fleming, I.; de Marigorta, E.M. *J. Chem. Soc., Perkin Trans. 1*, ***1999***, 889.

REVIEWS:

"Metal-Mediated Carbometallation Of Alkynes And Alkenes Containing Adjacent Heteroatoms," Fallis, A.G.; Forgione, P. *Tetrahedron*, ***2001***, *57*, 5899.

SECTION 197: ALKENES FROM ACID DERIVATIVES

NO ADDITIONAL EXAMPLES

SECTION 198: ALKENES FROM ALCOHOLS AND THIOLS

C_6H_{13}–CH(OH)–CH_2–CN → 1. MeMgCl, THF, rt, 10 h; 2. aq NH_4Cl → C_6H_{12}–CH=CH–CN

71% (1:1 *E:Z*)

Fleming, F.F.; Shook, B.C. *Tetrahedron Lett.*, ***2000***, *41*, 8847.

Ph–CH(OH)–CH_2–CH=CH–CHO → cat $HfCl_4(thf)_2$, MeCN, reflux, 1 d → Ph–CH=CH–CH=CH–CHO

99%

Saito, S.; Nagahara, T.; Yamamoto, H. *Synlett*, ***2001***, 1690.

$PhCH_2OH$ → (o-iodoxybenzoic acid, 1.5 h; $Ph_3P{=}CHCO_2Et$, DMSO) → Ph–CH=CH–CO_2Et 98%

Maiti, A.; Yadav, J.S. *Synth. Commun.*, ***2001***, *31*, 1499.

Br–CH=CH–CH_2OH → ($Ph_3P{=}CHCO_2Et$; MnO_2, DCM) → Br–CH=CH–CH=CH–CO_2Et 68% + Br–CH=CH–CH=CH–CO_2Et (EtO_2C) 16%

Wei, X.; Taylor, R.J.K. *J. Org. Chem.*, ***2000***, *65*, 616.

(1. $BF_3{\cdot}OEt_2$, –78°C → rt; 2. allyl bromide, Br)

70% (40:60 *cis:trans*)

Schmitt, A.; Reißig, H.-U. *Eur. J Org. Chem.*, ***2000***, 3893.

OH ($Ph_3P{=}CHCO_2Me$, MnO_2, DCM; toluene, rt, 4 h) → CO_2Me

80% (>99% *E*)

Blackburn, L.; Wei, X.; Taylor, R.J.K. *Chem. Commun.*, ***1999***, 1337.

SECTION 199: ALKENES FROM ALDEHYDES

CHPh, Me, Me, P, N, N, N–Me, N → (PhCHO, THF, rt, 15 min; *E*-selective) → Ph–CH=CH–Ph

91%

Wang, Z.; Zhang, G.; Guzei, I.; Verkade, J.G. *J. Org. Chem.*, ***2001***, *66*, 3521.

CHO, OBn → ($Bu_3P{=}$CH–CO_2Me; DCM, toluene) → COMe, OBr 78% (91 : 9 *E:Z*)

Harcken, C.; Martin, S.F. *Org. Lett.*, ***2001***, *3*, 3591.

PhCHO → ($TNSCH=N_2$, DCM; 1-2% Rh catalyst) Ph–CH=CH–OTMS 62% 1:3.7 (*E*:*Z*)

Dias, E.L.; Brookhart, M.; White, P.S. *J. Am. Chem. Soc.*, ***2001***, *123*, 2442.

PhCHO → ($InCl_3$–Zn, MeCN, 8 h; reflux) Ph–CH=CH–Ph + Ph–CH=CH–Ph (98 : 2) 93%

Barman, D.C.; Thakur, A.J.; Prajapati, D.; Sandhu, J.S. *Synlett*, ***2001***, 515.

PhCHO → ($EtO_2CCH_2P(O)(OEt)_2$, NEt_3; 8 Kbar, MeCN, 1 d) Ph–CH=CH–CO_2Et 85%

Has-Becker, S.; Bodmann, K.; Kreuder, R.; Santoni, G.; Rein, T.; Reiser, O. *Synlett*, ***2001***, 1395.

MeO–C_6H_4–CHO → (1. $LiCH_2SiMe_3$; 2. Et_2AlCl, aq THF) MeO–C_6H_4–CH=CH–$SiMe_3$ 86%

Kwan, M.L.; Yeung, C.W.; Breno, K.L.; Doxsee, K.M. *Tetrahedron Lett.*, ***2001***, *42*, 1411.

CH=CH–CHO → ($PyrO_2S$–(CH₂)₄–OTIPS; KHMDS, toluene, rt) diene–OTIPS + diene–OTIPS (91 : 9) 67%

Charette, A.B.; Berthelette, C.; St.-Martin, D. *Tetrahedron Lett.*, ***2001***, *42*, 5149.

PhCHO → (1. $MeNO_2$, AcOH–NH_4OAc, 100°C, 3 h; 2. BEt_3, Et_2O–H_2O) Ph–CH=CH–Et 79%

Liu, J.-T.; Yao, C.-F. *Tetrahedron Lett.*, ***2001***, *42*, 6147.

PhCHO → ($MeCO_2Et$, $P(RNCH_2CH_2)_3N$; *i*-BuCN, 40°C, 2 h) Ph–CH=CH–CO_2Et 96%

Kisanga, P.; D'Sa, B.; Verkade, J. *Tetrahedron*, ***2001***, *57*, 8047.

PhCHO → (EtO_2C–CH(Me)–P(O)(OEt)$_2$; LiO*t*-Bu, hexane, 25°C, 1 h) → Ph–CH=C(Me)–CO_2Et 93% (*E/Z* = 140)

Petroski, R.J.; Weisleder, D. *Synth. Commun.*, ***2001***, *31*, 89.

2-furaldehyde (CHO) → ($CH_2(CO_2H)_2$, EtOAc, 3.5 h; graphite, ultrasound, 43°C) → 2-furyl–CH=CHCO$_2$H 77%

Li, J.-T.; Zang, H.-J.; Fang, J.-Y.; Li, I.-J.; Li, T.-S. *Synth. Commun.*, ***2001***, *31*, 653.

Ph–CH(OH)–CH(Me)–CH=CH$_2$ → (10% PhCHO, 10% Sn(OTf)$_2$, DCM, 0°C, 30 min) → Ph–CH(OH)–CH$_2$–CH=CH–Ph 76%

Nokami, J.; Anthony, L.; Sumida, S.-i. *Chem. Eur. J.*, ***2000***, *6*, 2909.

PhCHO → ([$(PhO)_2P(O)CH_2CO_2Et$, DBU, NaI, 2 HMPA, –78°C, 2 h]; –78°C → 0°C) → Ph–CH=CH–CO_2Et 86% (93:7 *Z:E*)

Ando, K.; Oishi, T.; Hirama, M.; Ohno, H.; Ibuka, T. *J. Org. Chem.*, ***2000***, *65*, 4745.

PhCHO → ($AlCl_3$–Zn, MeCN, reflux; 10 h) → PhCH=CHPh 92% (99:1 *E:Z*)

Dutta, D.K.; Konwar, D. *Tetrahedron Lett.*, ***2000***, *41*, 6227.

$CH_3C(O)CH_2P(O)(OCH_2CF_3)_2$ → (PhCHO, KHMDS, THF, 18-crown-6, -78°C, 1 h) → Ph–CH=CH–C(O)CH$_3$ 95% (100% Z)

Yu, W.; Su, M.; Jin, Z. *Tetrahedron Lett.*, ***1999***, *40*, 6725.

Ph–CH$_2$CH$_2$–CHO → (CHI_3, cat $CrCl_3(thf)_3$; Zn, Me_3SiCl, dioxane, 25°C, 4h) → Ph–CH$_2$CH$_2$–CH=CH–I 84% (95:5 *E:Z*)

Takai, K.; Ichiguchi, T.; Hikasa, S. *Synlett*, ***1999***, 1268.

O_2N—C6H4—CHO

TMG, MeCN, 4H

CO_2Et

OEt

quant

O_2N

TMG = N,N,N',N'-tetramethylguanidium

Barrett, A.G.M.; Cramp, S.M.; Roberts, R.S.; Zecri, F.J. *Org. Lett.*, ***1999***, *1*, 579.

Related Methods: Section 207 (Alkenes from Ketones).

SECTION 200: ALKENES FROM ALKYLS, METHYLENES AND ARYLS

This section contains dehydrogenations to form alkenes and unsaturated ketones, esters and amides. It also includes the conversion of aromatic rings to alkenes. Reduction of aryls to dienes is found in Section 377 (Alkene-Alkene). Hydrogenation of aryls to alkanes and dehydrogenations to form aryls are included in Section 74 (Alkyls from Alkenes).

1. LDA / TMSCl
2. NBS, -78°C

3. TBAF, -78°C

TMS OBn

74%

OBn

Dumeunier, R.; Markó, I.E. *Tetrahedron Lett.*, ***2000***, *41*, 10219.

SECTION 201: ALKENES FROM AMIDES

Related Methods: Section 65 (Alkyls from Alkyls).
Section 74 (Alkyls from Alkenes).

NO ADDITIONAL EXAMPLES

SECTION 202: ALKENES FROM AMINES

BnO OBn CH_2=CHMgBr, THF 60°C, 3 h BnO OBn

Azi—N N—Azi 76%

Azi = 2-phenyl *N*-aziridinyl group

Kim, S.; Ohi, D.H.; Yoon, J.-Y.; Cheong, J.H. *J. Am. Chem. Soc.*, ***1999***, *121*, 5330.

Ph Ph N H 2 eq N_2O_4, NEt_3, THF −43°C, 10 min Ph Ph 89%

Lee, K.; Kim, Y.H. *Synth. Commun.*, ***1999***, *29*, 1241.

SECTION 203: ALKENES FROM ESTERS

HO OAc THF HO Et + HO Et

EtCu(CN)(MgCl), 0°C	4%	83%
EtMgCl/cat CuCN, −18C	86%	5%

Ito, M.; Murugesh, M.G.; Kobayashi, Y. *Tetrahedron Lett.*, ***2001***, *42*, 423.

Ph O S S 10% $CF_3SO_3^-$ $\overset{+}{H}N$ toluene, heat Ph 98%

Fauré-Tromeur, M.; Zard, S.Z. *Tetrahedron Lett.*, ***1999***, *40*, 1305.

OMe

10% $Mo(CO)_3(EtCN)_3$
15% chiral bis-amide

NaH , $CH_2(CO_2Me)_2$
THF , 70°C

MeO_2C CO_2Me + CO_2Me CO_2Me

(5 : 1) 85%
94%ee

Glorius, F.; Pfaltz, A. *Org. Lett.*, ***1999***, *1*, 141.

SECTION 204: ALKENES FROM ETHERS, EPOXIDES AND THIOETHERS

$(Me_2N)_2P(O)Cl$, H_2O
150°C , 30 min
92%

Demir, A.S. *Tetrahedron*, ***2001***, *57*, 227.

NaI , acetone
Amberlyst 15
98%

Righi, G.; Bovicelli, P.; Sperandio, A. *Tetrahedron*, ***2000***, *56*, 1733.

1. NCS , CCl_4 , reflux , 3 h
2. mcpba , rt , 18 h
66%

MacGee, D.I.; Beck, E.J. *J. Org. Chem.*, ***2000***, *65*, 8367.

CO_2Et
LiI , Amberlyst-15 , acetone
rt , 6 h
C_5H_{11} CO_2Et C_5H_{11} 80%

Antonioletti, R.; Bovicelli, P.; Fazzolari, E.; Righi, G. *Tetrahedron Lett.*, ***2000***, *41*, 9315.

Co salen complex
10 eq Na(Hg)
THF, 6 h
quant (NMR exp)

Isobe, H.; Branchaud, B.P. *Tetrahedron Lett.*, ***1999***, *40*, 8747.

$MeReO_3$, H_2S, PPh_3
MeCN, 5 min
quant

Jacob, J.; Espenson, J.H. *Chem. Commun.*, ***1999***, 1003.

SECTION 205: ALKENES FROM HALIDES AND SULFONATES

I⁀⁀⁀CO_2Me, MeCN
$Pd(OAc)_2$, Cs_2CO_3, reflux
tri-2-furylphosphine norbornene
CO_2Me
62%

Lautens, M.; Paquin, J.-F.; Piguel, S.; Dahlmann, M. *J. Org. Chem.*, ***2001***, *66*, 8127.

OAc
Bu
$SiMe_3$
Cl
1. excess SmI_2, THF, reflux
2. 0.1 M aq HCl
Bu $SiMe_3$
96% (>98% Z)

Concellón, J.M.; Bernad, P.L.; Bardales, E. *Org. Lett.*, ***2001***, *3*, 937.

$(CH_2{=}CH)_2InCl$
I Me
.25% $Pd(dba)_3{\bullet}CHCl_3$
1.5% $P(2furyl)_3$
THF/H_2O reflux, 3h
Me
96%

Takami, K.; Yorimitsu, H.; Shinobulo, H.; Matsubara, S.; Oshima, K. *Org. Lett.*, ***2001***, *3*, 1997.

Br
O
O
1. (polymer)–C_6H_4–PPh_2
PhCHO, toluene, 150°C
2. aq K_2CO_2, 125°C, 5min
3. DMF, 180°C, 5min
O
O
Ph
53%

Westman, J. *Org. Lett.*, ***2001***, *3*, 3745.

$$\text{PhCH}_2\text{Cl} \xrightarrow[\substack{5\%\ Pd_2(dba)_3\ ,\ acetone \\ 5\%\ CHCl_3\ ,\ rt\ ,\ 1\ d \\ 40\%\ PPh_3}]{\text{CH}_2\text{=CHCH}_2\text{SnBn}_3}$$ 80%

Bao, M.; Nakamura, H.; Yamamoto, Y. *J. Am. Chem. Soc.*, ***2001***, ***123***, 759.

$$\text{Ph-CH=CH-CH}_2\text{Cl} \xrightarrow[\text{THF , DMF}]{CrCl_3\text{–}LiAlH_4\ ,\ i\text{-PrOH}} \text{PhCH}_2\text{CH=CH}_2$$ 97.5%

Omoto, M.; Kato, N.; Sogon, T.; Mori, A. *Tetrahedron Lett.*, ***2001***, ***42***, 939.

$$\text{Ph(Br)C=N-OBn} \xrightarrow[\substack{5\%\ Pd(PPh_3)_4\ ,\ toluene \\ 110°C\ ,\ 30\ min}]{CH_2\text{=}CHSnBu_3\ ,\ 2\ eq\ KF}$$ 90%

Chang, S.; Lee, M.; Kim, S. *Synlett*, ***2001***, 1557.

$$\text{PhCHBr-CHBr-CO}_2\text{H} \xrightarrow[\text{30 sec}]{NEt_3\ ,\ DMF\ ,\ microwaves} \text{PhCH=CHBr}$$ 95% (>99:1 *Z:E*)

Kuang, C.; Senboku, H.; Tokuda, M. *Tetrahedron Lett.*, ***2001***, ***42***, 3893.

$$\text{Br-(CH}_2)_5\text{-Br} \xrightarrow[\text{100 torr}]{HMPA\ ,\ 100°C} \text{CH}_2\text{=CH(CH}_2)_3\text{Br}$$ 78%

Hoye, T.R.; Van Veidhuizen, J.J.; Vos, T.J.; Zhao, P. *Synth. Commun.*, ***2001***, ***31***, 1367.

$$\text{TolHN-CO-CHBr-CHBr-Ph} \xrightarrow{h\nu\ ,\ MeOH\ ,\ 30\ min} \text{TolHN-CO-CH=CH-Ph}$$ 75%

Aruna, S.; Kalyanakumar, R.; Ramakrishnan, V.T. *Synth. Commun.*, ***2001***, ***31***, 3125.

$$\text{PhCHBr-CHBr-Ph (meso)} \xrightarrow{HMPA\ ,\ 160°C\ ,\ N_2\ ,\ 15\ min} \text{PhCH=CHPh}$$ 93% (97:3 *E:Z*)

Khurana, J.M.; Bansal, G.; Chauhan, S. *Bull. Chem. Soc. Jpn.*, ***2001***, ***74***, 1089.

Br

$Me_3SiCH{=}N_2$, $CHCl_3$
2.5% $Pd_2(dba)_3{\cdot}CHCl_3$
15% $AsPh_3$, *i*-$PrNEt_2$
reflux , 12 h

MeO_2C — MeO_2C — 56%

Greenman, K.L.; Carter, D.S.; Van Vranken, D.L. *Tetrahedron*, ***2001***, *57*, 5219.

Br N Ph O Ph

$CH_2{=}CHSnBu_3$, 2% $Pd_2(dba)_3$
20% tri-2-furylphosphine , NMP
65°C , 16 h

N Ph O Ph

98%

Clapham, B.; Sutherland, A.J. *J. Org. Chem.*, ***2001***, *66*, 9033.

Ph Cl Cl

Cp_2Zr Et Et Et Et
50°C , 12h , THF
2.5 DMPU , 2 eq CuCl

Ph Et Et Et Et

63%

Duan, Z.; Sun, W.-H.; Liiu, Y.; Takahashi, T. *Tetrahedron Lett.*, ***2000***, *41*, 7471.

Br Br

500°C

Br

90%

Werstiuk, N.H.; Roy, C.D. *Tetrahedron Lett.*, ***2001***, *42*, 3255.

Br Ph Ph Br

Fe , MeOH , reflux , 30 min

Ph Ph

90%

Thakur, A.J.; Boruah, A.; Baruah, B.; Sandhu, J.S. *Synth. Commun.*, ***2000***, *30*, 157.

PhI

DABCO , Br_4NCl
cat $Pd(dba)_2$, MeCN
$SiMe_3$, rt , 1d

Ph $SiMe_3$ 75%

Jeffery, T. *Tetrahedron Lett.*, ***2000***, *41*, 8445.

Br, Ph, Ph, Br → Sm/THF, TMSCl (trace H_2O), 5 h → Ph, Ph 82%

Xu, X.; Lu, P.; Zhang, Y. *Synth. Commun.*, ***2000***, *30*, 1917.

C_5H_{11}, Br → Me_2C=CHMgBr, 20°C, 3% Li_2CuCl_4, THF, 4 eq NMP → C_5H_{11} 75%

Cahiez, G.; Chaboche, C.; Jézéquel, M. *Tetrahedron*, ***2000***, *56*, 2733.

Br, N, + → CH_2=$CHSnBu_3$, rt, 5% $Pd(PPh_3)_4$, DMF, 10% CuI, Ar → N, + 55%

Barchín, B.M.; Valenciano, J.; Cuadro, A.M.; Alvarez-Builla, J.; Vaquero, J.J. *Org. Lett.*, ***1999***, *1*, 545.

O, Cl → CH_2=$CHSnBu_3$, 80°C, 1.5% $Pd_2(dba)_3$, 6% P(*t*-Bu)$_3$, 2.2 eq CsF, dioxane, 12 h → O 87%

Littke, A.F.; Fu, G.C. *Angew. Chem. Int. Ed.*, ***1999***, *38*, 2411.

Br, Ph, CO_2H, Br → 1. Br_2, $CHCl_3$, 2. TEA, DMF, rt → Ph, Br 83%

Kim, S.H.; Wei, H.-X.; Willis, S.; Li, G. *Synth. Commun.*, ***1999***, *29*, 4179.

CO_2Et, CO_2Et → 1. Br, Br, NaH, DMF, 2. DBU, 75-80°C → CO_2Et, CO_2Et 70x52%

Bunce, R.A.; Burns, S.E. *Org. Prep. Proceed. Int.*, ***1999***, *31*, 99.

SECTION 206: ALKENES FROM HYDRIDES

For conversions of methylenes to alkenes (RCH_2R' → $RR'C$=CH_2), see Section 200 (Alkenes from Alkyls).

NO ADDITIONAL EXAMPLES

SECTION 207: ALKENES FROM KETONES

O

Ph

$TiCl_3$, Li , naphthalene

THF , 3 h

Ph

Ph

89% (56:44 *E:Z*)

Rele, S.; Talukdar, S.; Banerji, A.; Chattopadhyay, S. *J. Org. Chem.*, ***2001***, *66*, 2990.

EtO_2C

O

1. BuC≡COLi , -78° , THF

2. H^+, PhH, reflux

EtO_2C

Bu

H

87%

Shindo, M.; Matsumoto, K.; Sato, Y.; Shishido, K. *Org. Lett.*, ***2001***, *3*, 2029.

O

1. LDA ,

EtO_2C CHO

$PO(OEt)_2$

2. Ac_2O

3. NaH

EtO_2C

OAc

81%

Kraus, G.A.; Jones, C. *Synlett*, ***2001***, 793.

Bu CO_2Et

Br Br

1. Li , Naphth , –78°C , 0.5 h

2. 0°C , 30 min

3. benzophenone , rt

Bu CO_2H

Ph Ph

89%

Shindo, M.; Koretsune, R.; Yokota, W.; Itoh, K.; Shishido, K. *Tetrahedron Lett.*, ***2001***, *42*, 8357.

O

Bu

1. $EtSO_2Ph$, BuLi

2. Na(Hg)

3. 5% HMPA , SmI_2

THF , 0°C , 1 h

Bu

86x69%

Markó, I.E.; Murphy, F.; Kumps, L.; Ales, A.; Touillaux, R.; Craig, D.; Carballares, S.; Dolan, S. *Tetrahedron*, ***2001***, *57*, 2609.

C_3H_7 C_3H_7 I C_3H_7 I C_3H_7

1. 2 eq BuLi , ether , –78°C , 1 h
2. cyclohexanone , –78°C , 30 min

C_3H_7 C_3H_7 C_3H_7 C_3H_7

75%

Xi, Z.; Song, Q.; Chen, J.; Guan, H.; Li, P. *Angew. Chem. Int. Ed.*, ***2001***, *40*, 1913.

Ph Ph O

1. KH , THF , 0°C , *t*-BuSiMe$_2$Cl
2. 3.3 eq LDA , THF , 0°C
3. 3.5 eq TMSCl

Ph Ph SiMe$_3$ SiMe$_3$

84%

Langer, P.; Döring, M.; Seyferth, D.; Görls, H. *Chem. Eur. J.*, ***2001***, *8*, 573.

Ph O Ph

[PhS SPh SiMe$_3$ / Cp$_2$Ti[(POEt$_3$)]$_2$]

Me$_3$Si Ph Ph

58% (89:11 , *E*:*Z*)

Takeda, T.; Takagi, Y.; Saeki, N.; Fujiwara, T. *Tetrahedron Lett.*, ***2000***, *41*, 8377.

Related Methods: Section 199 (Alkenes from Aldehydes).

SECTION 208: ALKENES FROM NITRILES

Ph O CN

Br X

In , aq THF

Ph O X + Ph O X

X = H 69% –
X = Me – 73%

Yoo, B.-W.; Lee, S.-J.; Choi, K.-H.; Keum, S.-R.; Ko, J.-J.; Choi, K.-I.; Kim, J.-H. *Tetrahedron Lett.*, ***2001***, *42*, 7287.

1.5 eq CH_2=CHMgBr , 1 h

1.6 eq $AgBF_4$, –78°C → 0°C

91%

Agami,C.; Couty, F.; Evano, G. *Org. Lett.*, ***2000***, *2*, 2085.

SECTION 209: ALKENES FROM ALKENES

2%[Rh(cod)Cl]$_2$
8% TPPDS

$PhB(OH)_2$, Na_2CO_3
phase transfer agent , H_2O

80%

TPPDS = Ph(P—C_6H_4—SO_2K)$_2$

Lautens, M.; Roy, A.; Fukuoka, K.; Fagnou, K.; Martín-Matute, B. *J. Am. Chem. Soc.*, ***2001***, *123*, 5358.

, 140°C , NaOAc

Pd-NaY zeolite , DMAC

89% *trans* (<1% *cis*)

Djakovitch, L.; Koehler, K. *J. Am. Chem. Soc.*, ***2001***, *123*, 5990.

2 eq H_2O

2.5% $Pd_2(dba)_3$•dba , 1 d
10% PPh_3 , DMF , 23°C

63%

Castaño, A.M.; Méndez, M.; Ruano, M.; Eschavarren, A.M. *J. Org. Chem.*, ***2001***, *66*, 589.

Et_2Zn , 2% phosphoramide

1% CuBr , Me_2S , ether
–40°C → –10°C , 18 h

(47 : 53) quant

Malda, H.; van Zijl, A.W.; Arnold, L.A.; Feringa, B.L. *Org. Lett.*, ***2001***, *3*, 1169.

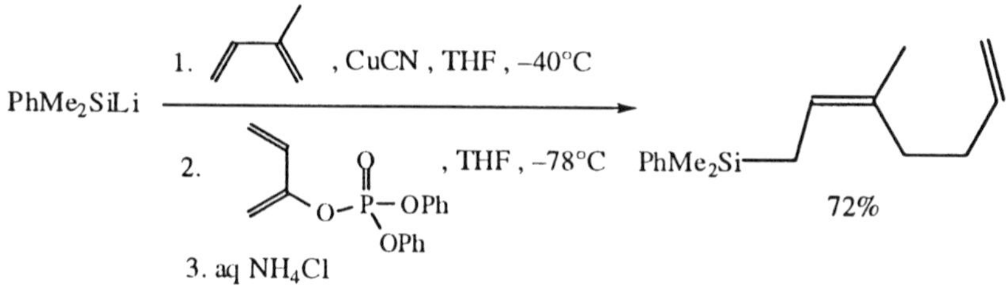

1. , CuCN , THF , –40°C

PhMe$_2$SiLi

2. O–P(=O)–OPh, OPh , THF , –78°C PhMe$_2$Si

72%

3. aq NH_4Cl

Liepins, V.; Bäckvall, J.-E. *Org. Lett.*, ***2001***, *3*, 1861.

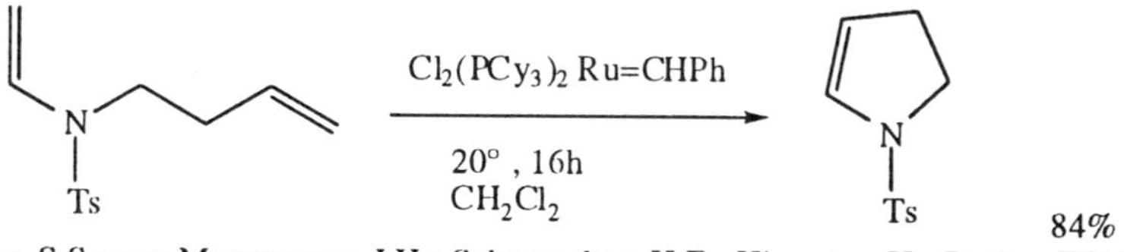

N, Ts

$Cl_2(PCy_3)_2$ Ru=CHPh

20° , 16h

CH_2Cl_2

N, Ts

84%

Kinderman, S.S.; van Maarseveen, J.H.; Schoemaker, H.E.; Hiemstra, H.; Rutjes, F.P.J.T. *Org. Lett.*, ***2001***, *3*, 2045.

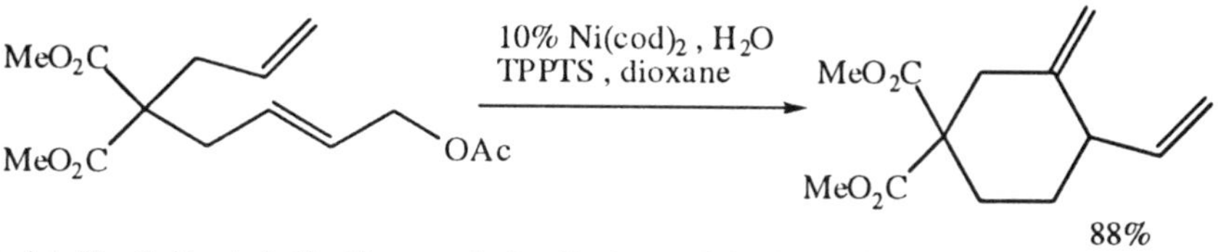

MeO_2C

MeO_2C

OAc

10% Ni(cod)$_2$, H_2O

TPPTS , dioxane

MeO_2C

MeO_2C

88%

Michelet, V.; Galland, J.-C.; Charruault, L.; Savignac, M.; Genêt, J.-P. *Org. Lett.* ***2001***, *3*, 2065.

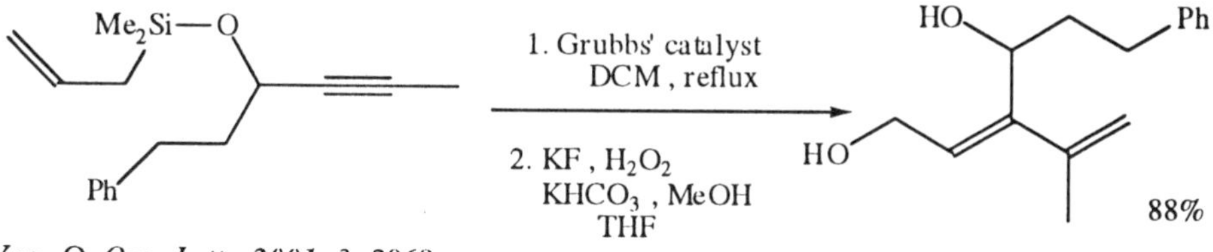

Me$_2$Si—O

Ph

1. Grubbs' catalyst

DCM , reflux

2. KF , H_2O_2

$KHCO_3$, MeOH

THF

HO

Ph

HO

88%

Yao, Q. *Org. Lett.*, ***2001***, *3*, 2069.

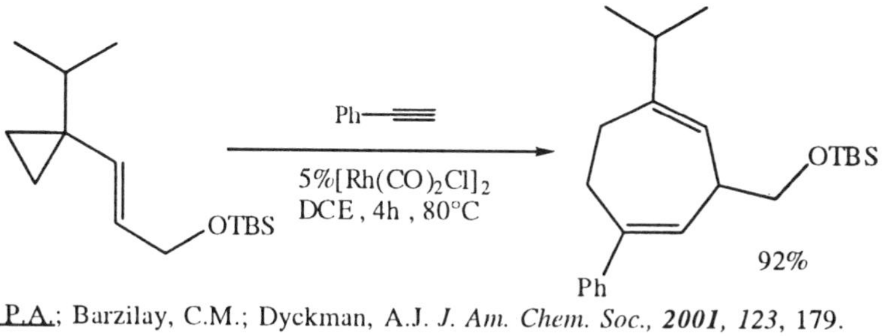

OTBS

Ph—≡

5%[Rh(CO)$_2$Cl]$_2$

DCE , 4h , 80°C

OTBS

92%

Ph

Wender, P.A.; Barzilay, C.M.; Dyckman, A.J. *J. Am. Chem. Soc.*, ***2001***, *123*, 179.

5% Grubbs' catalyst

[bmim] PF_6, 80°C, 1 h

N Tos → N Tos quant

Buijsman, R.C.; van Vuuren, E.; Sterrenburg, J.G. *Org. Lett.*, **2001**, *3*, 3785.

OTBS → OTPS

5% CH_2=CHPh

0.3 M $CHCl_3$
3% Grubbs' catalyst

Ph

60% (>95% E)

Wright, D.L.; Usher, L.C.; Estrella-Jimenez, M. *Org. Lett.*, **2001**, *3*, 4275.

5% Grubbs' catalyst

Ph → Ph Ph

62%

Verbicky, C.A.; Zercher, C.K. *Tetrahedron Lett.*, **2000**, *41*, 8723.

OTBDMS

$[CpRu(N{\equiv}CMe)_3]^+ PF_6^-$

OTBDS

(34 : 1) 76%

OTBDS

Trost, B.M.; Surivet, J.-P.; Toste, F.D. *J. Am. Chem. Soc.*, **2001**, *123*, 2897.

Ph O

5% Mo catalyst

80°C, 3h

O Ph

93% (74% ee)

Cefalo, D.R.; Kiely, A.F.; Wuchrer, M.; Jamieson, J.Y.; Schrock, R.R.; Hoveyda, A.H. *J. Am. Chem. Soc.*, **2001**, *123*, 3139.

toluene , 80°C

2 h

80%

Ishikawa, T.; Senzaki, M.; Kadoya, R.; Morimoto, T.; Miyake, N.; Iawa, M.; Saito, S. *J. Am. Chem. Soc.*, **2001**, *123*, 4607.

20% $RuCl_2(PCy_3)_2$=CHPh

PhH , rt , 1 h

89%

Rainer, J.D.; Cox, J.M.; Allwein, S.P. *Tetrahedron Lett.*, **2001**, *42*, 179.

1% $[Ru(cod)Cl_2]_n$, 1 d
i-PrOH , 90°C

78%

Yamamoto, Y.; Nakagai, Y.-i.; Ohkoshi, N.; Itoh, K. *J. Am. Chem. Soc.*, **2001**, *123*, 6372.

PhBr → CH_2=$CHCO_2Bu$, 2 eq Cs_2CO_3 / 2% $Pd(OAc)_2$, 4% IMesHCl / DMF , 120°C , 1 h → CO_2Bu 97%

Yang, C.; Nolan, S.P. *Synlett*, **2001**, 1539.

Ph → C_6F_5Br , $CaCO_3$, 1d / 1% $(Bu_4N)_2[Pd_2(\mu Br)_2)C_6F_5)Br_2]$ → Ph—C_6F_5 98%

Albéniz, A.C.; Espinet, P.; Martín-Ruiz, B.; Milstein, D. *J. Am. Chem. Soc.*, **2001**, *123*, 11504.

(700 psi) CH_2=CH_2 → 2 eq $SnMe_4$, 45°C , cat $TaCl_5$ / PhCl , autoclave , 4 h → C_4H_9CH=CH_2 94%

Andes, C.; Harkins, S.B.; Murtuza, S.; Oyler, K.; Sen, A. *J. Am. Chem. Soc.*, **2001**, *123*, 7423.

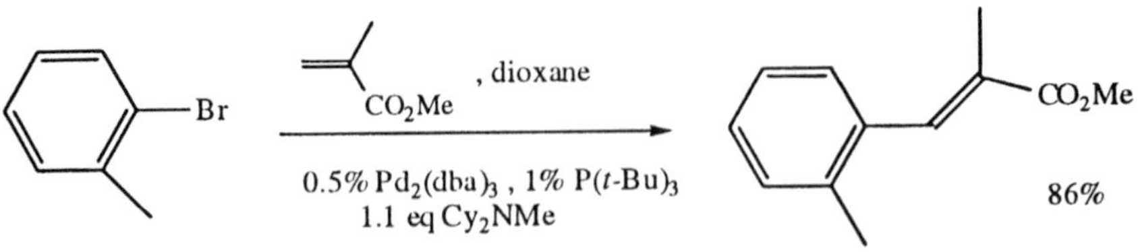

Br

, dioxane

CO_2Me

0.5% $Pd_2(dba)_3$, 1% $P(t\text{-}Bu)_3$
1.1 eq Cy_2NMe

CO_2Me

86%

Littke, A.F.; Fu, G.C. *J. Am. Chem. Soc.*, ***2001***, *123*, 6989.

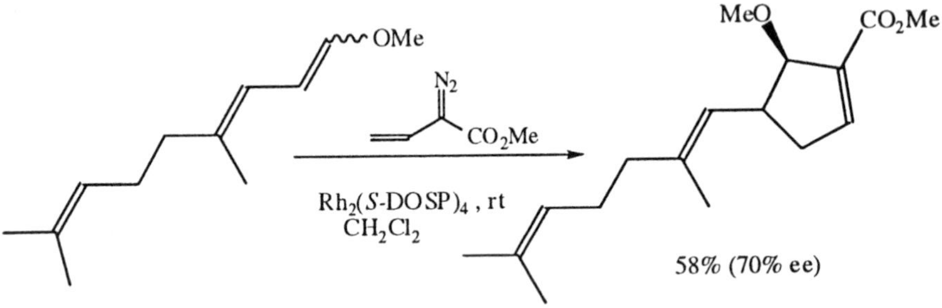

OMe

N_2

CO_2Me

$Rh_2(S\text{-}DOSP)_4$, rt
CH_2Cl_2

MeO CO_2Me

58% (70% ee)

Davies, H.M.L.; Xiang, B.; Kong, N.; Stafford, D.G. *J. Am. Chem. Soc.*, ***2001***, *123*, 7461.

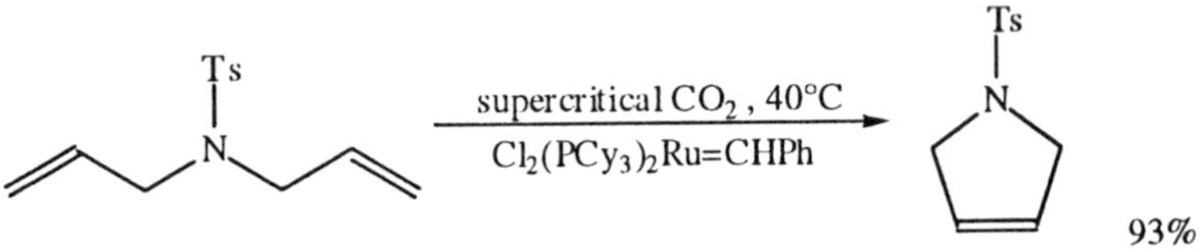

Ts

N

supercritical CO_2, 40°C
$Cl_2(PCy_3)_2Ru{=}CHPh$

Ts

N

93%

Fürstner, A.; Ackermann, L.; Beck, K.; Hori, H.; Koch, D.; Langemann, K.; Liebl, M.; Six, C.; Leitner, W. *J. Am. Chem. Soc.*, ***2001***, *123*, 9000.

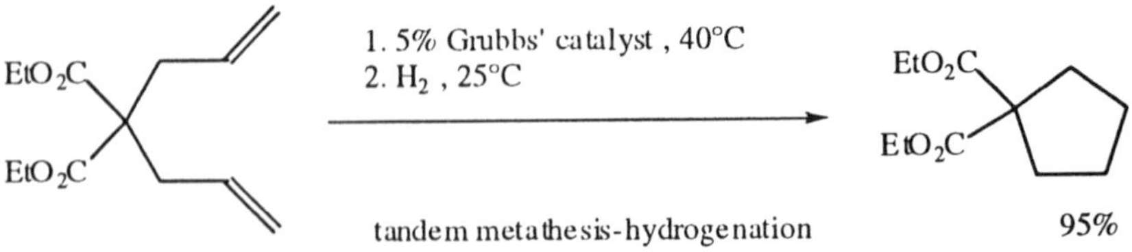

EtO_2C

EtO_2C

1. 5% Grubbs' catalyst, 40°C
2. H_2, 25°C

EtO_2C

EtO_2C

tandem metathesis-hydrogenation

95%

Louie, J.; Bielawski, C.W.; Grubbs, R.H. *J. Am. Chem. Soc.*, ***2001***, *123*, 11312.

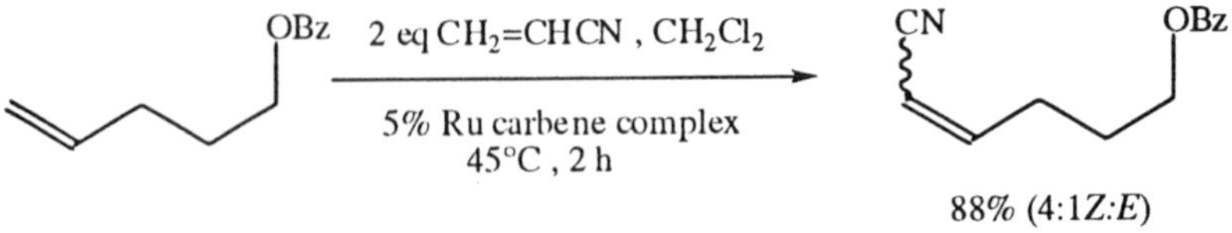

OBz

2 eq $CH_2{=}CHCN$, CH_2Cl_2

5% Ru carbene complex
45°C, 2 h

CN OBz

88% (4:1 *Z*:*E*)

Randl, S.; Gessler, S.; Wakamaatsu, H.; Blechert, S. *Synlett*, ***2001***, 430.

CH_2=CHCN + OBz-alkene; 5% Ru catalyst, CH_2Cl_2, 45°C, 12 h → NC-alkene-OBz

98%

Randal, S.; Buschmann, N.; Connon, S.J.; Blechert, S. *Synlett*, ***2001***, 1547.

CO_2Bn, K_2CO_3, DMF

cat $[PdCl(C_3H_5)]_2$ 130°C, 20h
tetraphosphine ligand

Br → CO_2Bn

quant

Feuerstein, M.; Doucet, H.; Santelli, M. *Synlett*, ***2001***, 1980.

CO/H_2(100 atm), toluene

cat $Rh_4(CO)_{12}$, 100°C

N, Ph → N, Ph

80%

Settambolo, R.; Caizzo, A.; Lazzaroni, R. *Tetraheron Lett.*, ***2001***, *42*, 4045.

Ph—I

CH_2=CHCN, [bmim]PF_6

NEt_3, 100°C, 10% Pd/C

Ph, CN

92%

ionic liquid medium

Hagiwara, H.; Shimizu, Y.; Hoshi, T.; Suzuki, T.; Ando, M.; Ohkubo, K.; Yokoyama, C. *Tetrahedron Lett.*, ***2001***, *42*, 4349.

CH_2=CH_2

$Cl_2(PCy_3)_2$Ru=CHPh
DCM, reflux, 3 h

67%

Chandra, K.L.; Saravanan, P.; Singh, V.K. *Tetrahedron Lett.*, ***2001***, *42*, 5309.

1. *t*-BuLi, TMEDA, 0°C
hexane

2. MeI

N, Si(*i*-Pr)$_3$ → Me, N, Si(*i*-Pr)$_3$

83%

Matsuzuno, M.; Fukuda, T.; Iwao, M. *Tetrahedron Lett.*, ***2001***, *42*, 7621.

TBSO$(CH_2)_4$ (terminal alkene) → CH_2=$CHSO_2Ph$, toluene , 50% / cat Grubbs' catalyst II → TBSO$(CH_2)_4$CH=CHSO_2Ph 86%

Grela, K.; Bieniek, M. *Tetrahedron Lett.*, ***2001***, *42*, 6425.

cycloheptatriene → CH_2=CHCO_2R , THF , MgO / hn (Pyrex) , 10% cycloheptatriene–$Cr(CO)_3$ → bicyclic CO_2R product (H, H) 89% (8:1 de)

R = (–)-8-phenylmenthyl

Rigby, J.H.; Mann, L.W.; Myers, B.J. *Tetrahedron Lett.*, ***2001***, *42*, 8773.

CH_2=CHCO_2Bn → PhBr , TBAB , $NaHCO_3$ / 130°C , Pd-benzthiazole complex, ionic medium → Ph_2C=CHCO_2Bn 82%

Calò, V.; Nacci, A.; Monopoli, A.; Lopez, L.; di Cosmo, A. *Tetrahedron*, ***2001***, *57*, 6071.

3,4-(MeO)$_2C_6H_3$CH(OH)C(=CH_2)CO_2Me → P_2O_5 , DCM , 1 h → 5,6-dimethoxyindene-2-CO_2Me 25%

Basavaiah, D.; Bakthadoss, M.; Reddy, G.J. *Synthesis*, ***2001***, 919.

2,3-dimethylbutadiene → methyl vinyl ketone , 10% $Yb(OTf)_3$ / DCM , rt , 16 h → 4-acetyl-1,2-dimethylcyclohexene 86%

Fang, X.; Warner, B.P.; Watkin, J.G. *Synth. Commun.*, ***2000***, *30*, 2669.

OHC—N(allyl)(CH_2C(Br)=CH_2) → CH_2=CHCO_2t-Bu / cat $Pd(OAc)_2$, dppe / Ag_2CO_3 , MeCN / 85°C , 1 d → OHC—N bicyclic CO_2t-Bu product 54%

Bhat, L.; Steinig, A.G.; Appelbe, R.; de Meijere, A. *Eur. J. Org. Chem.*, ***2001***, 1673.

Ph
Ph

OTs , dioxane , 40°C , 1 d

5% [Rh(nbd)(MeCN)$_2$] PF$_6$

Ph
Ph

51%

Tsukada, N.; Sato, T.; Inoue, Y. *Chem. Commun.*, ***2001***, 237.

C_3H_7
C_3H_7

PhB(OH)$_2$, 10% ni(cod)$_2$, 40% PPh$_3$

H_2O , NMP , 80°C , 5 h

C_3H_7 C_3H_7
Ph
C_3H_7 C_3H_7

77%

Shirakawa, E.; Takahashi, G.; Tsuchimoto, T.; Kawakami, Y. *Chem. Commun.*, ***2001***, 2688.

C
HO

cat Pd(PPh$_3$)$_4$, HCOOH

DMF , 30°C , 12 h

HO HO

HO OH

\+

(1 : 1) 42%

Oh, C.H.; Yoo, H.S.; Jung, S.H. *Chem. Lett.*, ***2001***, 1288.

Ph SiMe$_2$OSiMe$_3$

HSiMe$_2$OSiMe$_3$, 60°C

0.3 RhI(PPh$_3$)$_3$

Ph SiMe$_2$OSiMe$_3$

Mori, A.; Takahisa, E.; Nishihara, Y.; Hiyama, T. *Can. J. Chem.*, ***2001***, *79*, 1522.

CO$_2$*t*-Bu , DCM , 25°C

3% CoBr$_2$(dppe) , 9% ZnI$_2$
Bu$_4$NBH$_4$, 16 h

CO$_2$*t*-Bu

97%

Hilt, G.; de Mesnil, F.-X.; Lüers, S. *Angew. Chem. Int. Ed.*, ***2001***, *40*, 387.

Ph–I

CH$_2$=CHCO$_2$Et , sonication , 2 h

[bmim] Br , cat Pd(OAc)$_2$

Ph CO$_2$Et

87%

Deshmukh, R.R.; Rajagopal, R.; Srinivasan, K.V. *Chem. Commun.*, ***2001***, 1544.

Mo carbene complex

PhH, 5 min

63% conversion

Aeilts, S.L.; Cefalo, D.R.; Bonditatebus Jr., P.J.; Houser, J.H.; Hoveyda, A.H.; Schrock, R.R. *Angew. Chem. Int. Ed.*, ***2001***, *40*, 1452.

OCO_2Et

2.5% $Pd_2(dba)_3$•dba, 1 d
10% PPh_3, DMF, 23°C

2 eq H_2O

CO_2Me MeO_2C MeO_2C CO_2Me

63%

Castaño, A.M.; Méndez, M.; Ruano, M.; Echavarren, A.M. *J. Org. Chem.*, ***2001***, *66*, 585.

OTf

CH_2=CHOBu, microwaves

5 min

OBu

66% (90:10 α:β)

Vallin, K.S.A.; Larhed, M.; Johansson, K.; Hallberg, A. *J. Org. Chem.*, ***2000***, *65*, 4537.

10% $(2,6\text{-}Me_2C_6H_3O)_4Ti$
25% cyclohexylMgCl

THF, 0°C → rt, 1 d

85%

Okamoto, S.; Livinghouse, T. *J. Am. Chem. Soc.*, ***2000***, *122*, 1223.

CO_2Me

$PhNH_2$, AcOH, reflux, 5 h

Pd–Cu–montmorillonite K10

Ph CO_2Me

66%

Waterlot, C.; Couturier, D.; Rigo, B. *Tetrahedron Lett.*, ***2000***, *41*, 317.

C_6H_{13}

Grubbs' catalyst

DCM, 20°C, 20 h

C_6H_{13} C_6H_{13}

70%

Ahmed, M.; Arnauld, T.; Barrett, A.G.M.; Braddock, D.C.; Flack, K.; Procopiou, P.A. *Org. Lett.*, ***2000***, *2*, 551.

Tol
S
O
$Pd(PPh_3)_4$, Ag_2CO_3
MeCN, 50°C
N
I
O
S
Tol
N
74%

Segorbe, M.M.; Adrio, J.; Carretero, J.C. *Tetrahedron Lett.*, ***2000***, *41*, 1983.

$BzO(H_2C)_7$
2 eq CH_2=$CHCO_2Me$
Mes–N N–Mes
Cy_3P–Ru= CH–CH=CMe_2
Cy_3P Cl Cl
CO_2Me
$(CH_2)_7OBz$
91% (4.5:1 *E*:*Z*)

Chatterjee, A.K.; Morgan, J.P.; Scholl, M.; Grubbs, R.H. *J. Am. Chem. Soc.*, ***2000***, *122*, 3783.

OH
220°C , toluene
OH
H
72%

Warrington, J.M.; Yap, G.P.A.; Barriault, L. *Org. Lett.*, ***2000***, *2*, 663.

C_6H_{13}
1. $PhMe_2SiCu$, THF , –40°C
2. O–P(O)(OPh)(OPh) , THF , –40°C
C_6H_{13}
$SiMe_2Ph$
75%

Liepins, V.; Karlström, A.S.E.; Bäckvall, J.-E. *Org. Lett.*, ***2000***, *2*, 1237.

Ph
O
DCE , 0.1 M , rt , 2 h
5% $[Rh(dppb)Cl]_2$/$AgSbF_6$
Ph
O
84%

Cao, P.; Wang, B.; Zhang, X. *J. Am. Chem. Soc.*, ***2000***, *122*, 6490.

SPh PhS Ph $Cp_2Ti[P(OEt)_3]_2$ THF → Ph 71%

Fujiwara, T.; Kato, Y.; Takeda, T. *Tetrahedron,* ***2000***, *56*, 4859.

Me_2N $SiMe_3$ EtO $Cr(CO)_5$ 3 eq PhC≡CH THF, 55°C, 4 d → Ph NMe_2 Ph EtO + Ph NMe_2 Ph EtO

(1.8 : 1) 62%

Schirmer, H.; Flynn, B.L.; de Meijere, A. *Tetrahedron,* ***2000***, *56*, 4977.

$Cr(CO)_2(\eta^6\text{-}C_6H_6)$ P Ph Ph; CO_2Et, Bu_2O, 150°C → CO_2Et H H 92%

Rigby, J.H.; Kondratenko, M.A.; Fiedler, C. *Org. Lett.,* ***2000***, *2*, 3917.

PhC≡CH, cat $CoBr_2$(dppe) cat Bu_4NBH_4, cat ZnI_2 DCM, rt → Ph 94%

Hilt, G.; du Mesnil, F.-X. *Tetrahedron Lett.,* ***2000***, *41*, 6757.

Ph ethylene, $[Ni(MeCN)_6][BF_4]_2$ $AlEt_2Cl$, PPh_3, 1 h → Ph 87%

Fassina, V.; Ramminger, C.; Sefeerin, M.; Monteiro, A.L. *Tetrahedron,* ***2000***, *56*, 7403.

DCM , 0.1 M , rt , 12 h

2.5% Oi-Pr PCy_3 Cl Ru Cl PCy_3

>99%

Yao, Q. *Angew. Chem. Int. Ed.*, ***2000***, *39*, 3896.

DCM , 45°C , 18 h

5% Mes N Cl Ru Cl Mes PCy_3 PhHC

quant

Schürer, S.C.; Gessler, S.; Buschmann, N.; Blechert, S. *Angew. Chem. Int. Ed.*, ***2000***, *39*, 3898.

LiOTf , MeCN , 6 h

alternative to $LiClO_4$

>99%

Augé, J.; Gil, R.; Kalsey, S.; Lubin-Germain, N. *Synlett*, ***2000***, 877.

, $AlMe_3$

2,6-di-*t*-Bu-phenol , heat

(4 : 1) 52%

Bertozzi, F.; Olsson, R.; Frejd, T. *Org. Lett.*, ***2000***, *2*, 1283.

PhI , cat–I , DMF , 140°C , 2 h

95%

cat–I = $PdCl_2$ + $P(OPh_3)_4Br$ on intercalated clay

Varma, R.S.; Naicker, K.P.; Liesen, P.J. *Tetrahedron Lett.*, ***1999***, *40*, 2075.

HN ; 20% Grubbs' catalyst, p-TSA, 6 d ; HN ; 92%

Wright, D.L.; Schulte II, J.P.; Page, M.A. *Org. Lett.*, ***2000***, *2*, 1847.

PhI ; CO_2Me, $Pd_2(dba)_3$, NEt_3 ; D-100, MeCN ; Ph ; CO_2Me ; 89%

D-100 = perflorinated solvent - mainly *n*-perfluorooctane

Moineau, J.; Pozzi, G.; Quici, S.; Sinou, D. *Tetrahedron Lett.*, ***1999***, *40*, 7683.

O ; OMe ; PhI, $Pd(OAc)_2$, PPh_3 ; supercritical CO_2, 75-85°C 1600 psi ; O ; Ph ; OMe ; 84%

Shezad, N.; Oakes, R.S.; Clifford, A.A.; Rayner, C.M. *Tetrahedron Lett.*, ***1999***, *40*, 2221.

$OSiEt_3$; O ; 10% $Cl_2(PCy_2)_2Ru{=}CHPh$; CH_2Cl_2, reflux ; $OSiEt_3$; O ; 87%

Edwards, S.D.; Lewis, T.; Taylor, R.J.K. *Tetrahedron Lett.*, ***1999***, *40*, 4267.

H ; O ; O ; H ; 30% $Cl_2(PCy_2)_3Ru{=}CHPh$; CH_2Cl_2, 50°C ; H ; O ; O ; H ; 65%

Paquette, L.A.; Méndez-Andino, J. *Tetrahedron Lett.*, ***1999***, *40*, 4301.

CO_2Me ; PhCl, 1.5% $Pd_2(dba)_3$; 6% $P(t\text{-}Bu)_3$, 1.1 eq Cs_2CO_3 diamine, 100°C, 42 h ; CO_2Me ; Ph ; 76%

Littke, A.F.; Fu, G.C. *J. Org. Chem.*, ***1999***, *64*, 10.

MeO_2C, Ph — (Me$_4$–Phen)OPd(Me)Cl , DCE; NaBAr$_4$, 25°C , 5 min; Ar = 3,5-$C_6H_3(CF_3)_2$ → $SiMe_2Ph$, MeO_2C, Ph, Me

86% (2:1 isomer ratio)

Widenhoefer, R.A.; Vadehra, A. *Tetrahedron Lett.*, ***1999***, *40*, 8499.

TBDMSO, $SiMe_3$, OMe — e^- , 1M $LiClO_4$ MeOH/THF; 2,6-lutidine → TBDMSO, $CH(OMe)_2$

83%

Frey, D.A.; Reddy, S.H.K.; Moeller, K.D. *J. Org. Chem.*, ***1999***, *64*, 2805.

MeO_2C, MeO_2C, *t*-Bu — 1% $RhCl(PPh_3)_3$ toluene , 0.1 M; reflux → *t*-Bu, H, MeO_2C, MeO_2C, H

96%

Wender, P.A.; Glorius, F.; Husfeld, C.O.; Langkopf, E.; Love, J.A.
J. Am. Chem. Soc., ***1999***, *121*, 5348.

OTBS — 2 eq $PhCH=CH_2$, 5% Mo catalyst PhH , 22°C , Ar → OTBS, Ph

57% (>98% *trans*, 96% ee)

La, D.S.; Ford, J.G.; Satately, E.S.; Bonitatebus, P.J.; Schrock, R.R.; Hoveyda, A.H. *J. Am. Chem. Soc.*, ***1999***, *121*, 11603.

C_8H_{17} — OAc; 5% Grubbs' catalyst II → C_8H_{17}, OAc

60% (2.3:1 *E:Z*)

Chatterjee, A.K.; Grubbs, R.H. *Org. Lett.*, ***1999***, *1*, 1751.

Br / OMe → (CH_2=$CHCO_2Et$, $Pd(OAc)_2$, [bmim] (PF_6), NEt_3, PCl_3; 140°C, 1 d; in ionic liquid) → CO_2Et / MeO 98%

Carmichael, A.J.; Earle, M.J.; Holbrey, J.D.; McCormac, P.B.; Seddon, K.R. *Org. Lett.*, ***1999***, *1*, 997.

—Br → (5 eq [cyclopentene], Na_2CO_3; cat $Pd_2(dba)_3$•dba/4 PCy_3) → Ph—

99% (92:8 of other isomers)

Hartung, C.G.; Köhler, K.; Beller, M. *Org. Lett.*, ***1999***, *1*, 709.

MeO_2C / MeO_2C → ($[(\eta^3C_3H_7)Pd(OEt)PCy_3]^+$ $^-B(3,5\text{-di}CF_3\text{-}C_6H_3$; $HSiEt_3$, DCM) → MeO_2C / MeO_2C 86%

Widenhoefer, R.A.; Perch, N.S. *Org. Lett.*, ***1999***, *1*, 1103.

Ph CO_2Et → (PhH, benzoquinone, AcOH; 10% $Pd(OAc)_2$, *t*-BuOOH, sealed tube, Ac_2O, 90°C) → Ph / Ph CO_2Et 72%

Jia, C.; Lu, W.; Kitamura, T.; Fujiwara, Y. *Org. Lett.*, ***1999***, *1*, 2097.

CO_2Me → (PhI, Pd/C, NEt_3, 80°C, 2 d; supercritical CO_2, 100 bar) → Ph CO_2Me 65%

Cacchi, S.; Fabrizi, G.; Gasparrini, F.; Villani, C. *Synlett*, ***1999***, 345.

PhI → (CO_2Me, NEt_3, toluene, 100°C; 1% [Pd catalyst from guanidinium phosphine supported on glass beads], 16 h) → Ph CO_2Me 87%

Leese, M.P.; Williams, J.M.J. *Synlett*, ***1999***, 1645.

SPh

PhS

Ph

3 $Cp_2Ti[P(OEt)_3]_2$

THF, 12 h

Ph

71%

Fujiwara, T.; Takeda, T. *Synlett*, ***1999***, 354.

Ph

2 eq Ph

5% Grubbs' catalyst

Ph Ph

82% (4:1 *E:Z*)

Blanco, O.M.; Castedo, L. *Synlett*, ***1999***, 557.

2.5% $[p\text{-cymeme})RuCl_2]_2$, 5.5% PCy_3

DCM

82%

Fürstner, A.; Ackermann, L. *Chem. Commun.*, ***1999***, 95.

REVIEWS:

"Recent Chemistry Of Benzocyclobutenes," Mehta, G.; Kotha, S. *Tetrahedron*, ***2001***, *57*, 625.

"The Transannular Diels-Alder Strategy: Application To Total Synthesis," Marsault, D.; Toró, A.; Nowak, P.; Deslongchamps, P. *Tetrahedron*, ***2001***, *57*, 4243.

"The Preparation Of Mono-, 1, 1-Di, *Trans*-1, 2-Di- And Trisubstituted Ethylenes By Benzotriazole Methodology," Katritzky, A.R.; Toader, D. *Synlett*, ***2001***, 458.

"Diels-Alder Reaction On Solid Supports," Yli-Kauhaluoma, J. *Tetrahedron*, ***2001***, *57*, 7053.

"Advances In The Heck Chemistry Of Aryl Bromides And Chlorides," Whitcombe, N.J.; Hii, K.K.; Gibson, S.E. *Tetrahedron*, ***2001***, *57*, 7449.

"Cycloadditions Under Microwave Irradiation Conditions; Methods And Applications," De La Hoz, A.; Díaz-Ortis, A.; Moreno, A.; Lang, F. *Eur. J. Org. Chem.*, ***2000***, 3659.

"Recent Advances In Lewis Acid-Catalyzed Diels-Alder Reactions In Aqueous Media," Fringuelli, F.; Piermatti, O.; Pizzo, F.; Vaccaro, L. *Eur. J. Org. Chem.*, ***2001***, 439.

"Formation Of Cumulenes, Triple-Bonded, And Related Compounds By Flash Vacuum Pyrolysis Of Five-Membered Heterocycles," Yranzo, G.I.; Elguero, J.; Flammang, R.; Wentrup, C. *Eur. J. Org. Chem., **2001***, 2209.

"Synthesis Of Heterocycles Using The Intramolecular Heck Reaction Involving A 'formal' Anti-Elimination Process," Ikeda, M.; El Bialy, S.A.A.; Yakura, T. *Heterocycles, **1999**, 51*, 1957.

SECTION 210: ALKENES FROM MISCELLANEOUS COMPOUNDS

PhCH=N—NHTs → PhCH=CHPh

$B(OMe)_3$, *t*-BuOLi, THF; reflux, 6 h

82% (48:52 *trans:cis*)

Kabalka, G.W.; Wu, Z.; Ju, Y. *Tetrahedron Lett., **2001**, 42*, 4759.

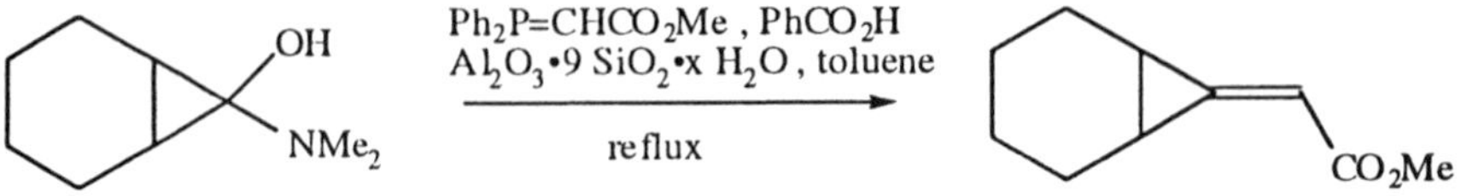

OH, NMe_2 → CO_2Me

$Ph_3P{=}CHCO_2Me$, $PhCO_2H$; $Al_2O_3{\bullet}9\ SiO_2{\bullet}x\ H_2O$, toluene; reflux

82%

Chowdhury, M.A.; Senboku, H.; Tokudda, M.; Masuda, Y.; Chiba, T. *Tetrahedron Lett., **2001**, 42*, 7075.

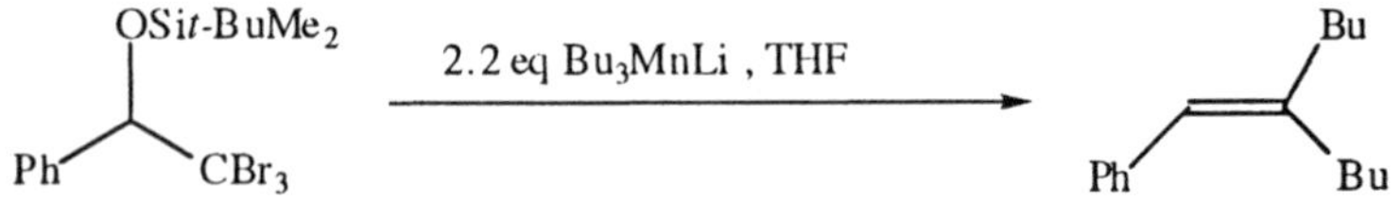

$OSi\textit{t}\text{-}BuMe_2$, Ph, CBr_3 → Ph, Bu, Bu

2.2 eq Bu_3MnLi, THF

58%

Kakiya, H.; Shinokubo, H.; Oshima, K. *Tetrahedron, **2001**, 57*, 10063.

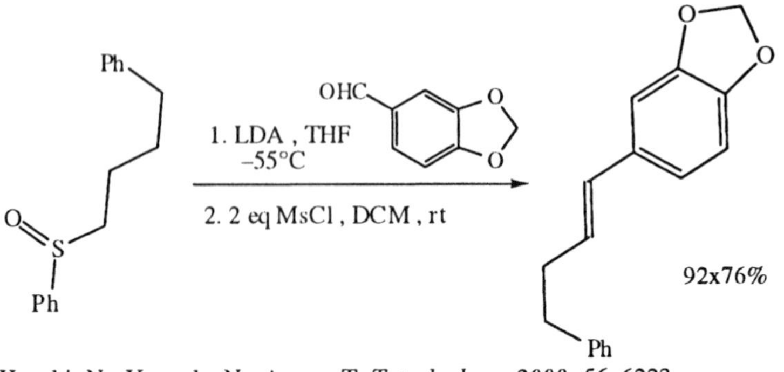

Ph, O, S, Ph → Ph

1. LDA, THF, –55°C; OHC, O, O
2. 2 eq MsCl, DCM, rt

92x76%

Satoh, T.; Hanaki, N.; Yamada, N.; Asano, T. *Tetrahedron, **2000**, 56*, 6223.

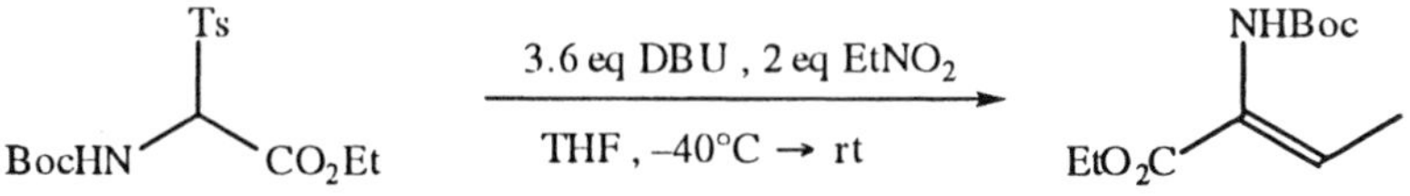

Ts, BocHN, CO_2Et → 3.6 eq DBU, 2 eq $EtNO_2$ / THF, –40°C → rt → NHBoc, EtO_2C

78% (97:3 *Z*:*E*)

Nagano, T.; Kinoshita, H. *Bull. Chem. Soc. Jpn.*, ***2000***, *73*, 1605.

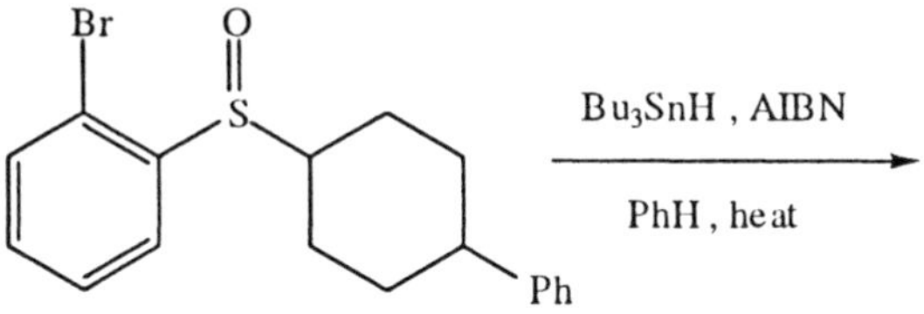

Br, O, S, Ph → Bu_3SnH, AIBN / PhH, heat → Ph 93%

Imboden, C.; Villar, F.; Renaud, P. *Org. Lett.*, ***1999***, *1*, 873.

CHAPTER 15

PREPARATION OF OXIDES

This chapter contains reactions that prepare the oxides of nitrogen, sulfur and selenium. Included are *N*-oxides, nitroso and nitro compounds, nitrile oxides, sulfoxides, selenoxides and sulfones. Oximes are considered to be amines and appear in those sections. Preparation of sulfonic acid derivatives are found in Chapter Two and the preparation of sulfonate esters in Chapter Ten.

SECTION 211: OXIDES FROM ALKYNES

Bu—≡ → 1. $Cp_2Zr(H)Cl$ 2. Me_2Zn 3. CH_2I_2 4. $PhCH{=}N\text{-}P(O)Ph_2$ → HN-P(O)Ph_2 / Ph / Bu cyclopropane product 71% 85:15

Wipf, P.; Kendall, C. *Org. Lett.*, ***2001***, *3*, 2773.

Bu—≡ → 1. $Cp_2Zr(H)Cl$ 2. Me_2Zn 3. $PhHC{=}N{-}P(O){-}Ph_2$, CH_2Cl_2, reflux → Bu, H, Ph, HN-P(=O)Ph Ph cyclopropane product 74%

Wipf, P.; Kendall, C.; Stephenson, C.R.J. *J. Am. Chem. Soc.*, ***2001***, *123*, 5122.

SECTION 212: OXIDES FROM ACID DERIVATIVES

$PhSO_3H$ → Nafion-H, PhH, reflux, 20 h → $PhSO_2Ph$ 48%

Olah, G.A.; Mathew, T.; Prakash, G.K.S. *Chem. Commun.*, ***2001***, 1696.

$PhSO_2Cl$ → BnBr, Sm/$HgCl_2$, THF, 50°C, 2.5 h → $PhSO_2Bn$ 75%

Zhang, J.; Zhang, Y. *J. Chem. Res. (S)*, ***2001***, 516.

1. $CH_2{=}CHCH_2Br$, $AlCl_3$–Fe, H_2O, 4 h
2. 2M HCl

p-$CH_3C_6H_4SO_2Cl$ → p-$CH_3C_6H_4SO_2CH_2CH{=}CH_2$ 85%

Saikia, P.; Laskar, D.D.; Prajapati, D.; Sandhu, J.S. *Chem. Lett.*, ***2001***, 512.

p-$CH_3C_6H_4SO_2Cl$ —PhH, Zn, 70 sec, microwaves→ p-$CH_3C_6H_4SO_2Ph$ 72%

Bandgar, B.P.; Kasture, S.P. *Synth. Commun.*, ***2000***, *31*, 1065.

SECTION 213: OXIDES FROM ALCOHOLS AND THIOLS

$PhCH_2OH$ —$NaNO_2$–AcOH–HCl, 6 h→ $PhCH_2NO_2$ 85%

Baruah, A.; Kalita, B.; Barua, N.C. *Synlett*, ***2000***, 1064.

3 PhOH —$POCl_3$, NaOH, H_2O, 2 min, microwaves→ $(PhO)_3P{=}O$ 90%

Sagar, A.D.; Shinde, N.A.; Bandgar, B.P. *Org. Prep. Proceed. Int.*, ***2000***, *32*, 269.

PhSH → $PhSO_2Me$ 75%

1. MeLi, THF, –78°C
2. –78°C, (Ph, H-oxaziridine with N-SO_2Ph)
3. workup
4. MeI, cat Bu_4NBr, H_2O/toluene, acetone, 85°C, 1 d

Sandrinelli, F.; Perrio, S.; Beslin, P. *Org. Lett.* ***1999***, *1*, 1177.

SECTION 214: OXIDES FROM ALDEHYDES

PhCHO —i-$PrNO_2$, 3 eq Zn, 0°C; AcOH, EtOH, 2 d→ $PhCH{=}N^+(O^-)$-i-Pr 61%

Gautheron-Chapoulaud, V.; Pandya, S.U.; Cividino, P.; Masson, G.; Py, S.; Vallée, Y. *Synlett*, ***2001***, 1281.

SECTION 215: OXIDES FROM ALKYLS, METHYLENES AND ARYLS

CH_3 NO_2 , O_3 , MeCN zeolite , 70°C NO_2 + O_2N NO_2 NO_2 85% 12%

Peng, X.; Suzuki, H. *Org. Lett.* ***2001***, *3*, 34331.

O $SOCl_2$, $CHCl_3$ montmorillonite K10 clay S Cl 61%

Karade, N.N.; Kate, S.S.; Adude, R.N. *Synlett,* ***2001***, 1573.

O N_2O_5 , Fe(acac)$_3$ O NO_2 98% (50:44:6 *o:m:p*)

Bak, R.R.; Smallridge, A.J. *Tetrahedron Lett.,* ***2001***, *42*, 6767.

NO_2 Montmorillonite KSF Bi(NO_3) THF , 10 min >99%

Samajdar, S.; Becker, F.F.; Banik, B.K. *Tetrahedron Lett.,* ***2000***, *41*, 8017.

Ph NO_x , I_2 , CCl_4 ,–5°C Ph CN O_2N CN 80%

Navaarro-Ocaño, A.; Barzana, E.; López-González, D.; Jiménez-Estrada, M. *Org. Prep. Proceed. Int.,* ***1999***, *31*, 117.

SECTION 216: OXIDES FROM AMIDES

Boc–N(H)–CH_2CO_2Me → [$(F_3C)_2$-dioxirane, CH_2Cl_2; 5 h, –20°C → 0°C] → Boc–N(OH)–CH_2CO_2Me 82%

Detomaso, A.; Cursi, R. *Tetrahedron Lett.*, ***2001***, *42*, 755.

SECTION 217: OXIDES FROM AMINES

Ph_3P → [PhIO, montmorillonite K10, MeCN, 30 min, rt] → $Ph_3P{=}O$ 96%

Mielniczak, G.; Łopusiński, A. *Synlett*, ***2001***, 505.

$Ph_3P{\rightarrow}BH_3$ → [2-butanone, *t*-BuOOH, 80°C] → $Ph_3P{=}O$ 94%

Uziel, J.; Darcel, C.; Moulin, D.; Bauduin, C.; Jugé, S. *Tetrahedron Asymm.*, ***2001***, *12*, 1441.

$PhNH_2$ → [H_2O_2, MeOH, 30 min; [Ti(O*i*-Pr)$_4$, 50% H_2O_2]] → $PhNO_2$ 98%

Dewkar, G.K.; Mikalje, M.D.; Ali, I.S.; Paraskar, A.S.; Jagtap, H.; Sudalai, A. *Angew. Chem. Int. Ed.*, ***2001***, *40*, 405.

Me_2NH → [$NaNO_2$, wet SiO_2, DCM, rt; $Mg(HSO_4)_2$, 1 h] → $Me_2N{-}NO$ quant

Zolfigol, M.A.; Ghaemi, E.; Madrakian, E.; Kiany-Borazjani, M. *Synth. Commun.*, ***2000***, *30*, 2057.

With $KHSO_5$ on wet silica gel. Zolfigol, M.A.; Bagherzadeh, M.; Chaghamarani, A.G.; Keypour, H.; Salehzadeh, S. *Synth. Commun.*, ***2001***, *31*, 1161.

2,5-Dibromo-4-nitroaniline (Br, NH_2, Br, O_2N) → [generate HOF *in situ*; aq MeCN, –20°C] → 1,4-dibromo-2,5-dinitrobenzene (Br, NO_2, Br, O_2N) 87%

Dirk, S.M.; Mickelson, E.T.; Henderson, J.C.; Tour, J.M. *Org. Lett.*, ***2000***, *2*, 3405.

4-Chloropyridine (Cl, N) → [urea–H_2O_2, HCO_2H; rt, 4 h] → 4-chloropyridine N-oxide (Cl, $N^+{-}O^-$) 90%

Balicki, R.; Goliński, J. *Synth. Commun.*, ***2000***, *30*, 1529.

Ph–CH$_2$–N(OH)–CH$_2$–Ph → (salen)Mn(III) complex; UHP, MeOH, 0°C → Ph–CH=N$^+$(O$^-$)–CH$_2$–Ph 81%

Cicchi, S.; Cardona, F.; Brandi, A.; Corsi, M.; Goti, A. *Tetrahedron Lett.*, ***1999***, *40*, 1989.

N-hydroxypyrrolidine → 5% aq NaClO, 3 h; 0°C → rt → 1-pyrroline N-oxide 68%

Cicchi, S.; Corsi, M.; Goti, A. *J. Org. Chem.*, ***1999***, *64*, 7243.

PPh_3 → N_2O (140 bar), 100°C, 3 h → $O{=}PPh_3$ quant

Poh, S.; Hernandez, R.; Inagaki, M.; Jessop, P.G. *Org. Lett.*, ***1999***, *1*, 583.

N-methylmorpholine → O_2, Me_2CCH_2CHO, DCE; Fe_2O_3, 50°C, 3 h → N-methylmorpholine N-oxide 70%

Wang, F.; Zhang, H.; Song, G.; Lu, X. *Synth. Commun.*, ***1999***, *29*, 11.

Et_2NH → $NaNO_2$, $C_2H_2O_4$•2 H_2O, DCM; rt, 2 h → Et_2N—NO 95%

Zolfigol, M.A. *Synth. Commun.*, ***1999***, *29*, 905.

SECTION 218: OXIDES FROM ESTERS

NO ADDITIONAL EXAMPLES

SECTION 219: OXIDES FROM ETHERS, EPOXIDES AND THIOETHERS

Bn–S–Bn → 15% (CeCl(tpp), H_2O_2, EtOH; rt → Bn–SO_2–Bn quant

Marques, A.; Marin, M.; Ruasse, M.-F. *J. Org. Chem.*, ***2001***, *66*, 7588.

Bu–S–Bu → 3% H_2O_2, $Me(C_8H_{17})_3N$ HSO_4; 10% Na_2WO_4, $PhCO_3H$; 50°C, 1 h → Bu–S(=O)$_2$–Bu 95%

Sato, K.; Hyodo, M.; Aoki, M.; Zheng, X.-Q.; Noyori, R. *Tetrahedron*, ***2001***, *57*, 2469.

2% Ti salen catalyst, urea–H_2O_2, 0°C, MeOH, 1 d — 78% (96% ee)

Saito, B.; Katsuki, T. *Tetrahedron Lett.*, ***2001***, *42*, 3873.

$(CF_3CO)_2O/H_2O_2$, MeCN — 75%

Nenajdenko, V.G.; Gavryushin, A.E.; Balenkova, E.S. *Tetrahedron Lett.*, ***2001***, *42*, 4397.

$Fe(NO_3)_3$–$FeBr_3$, MeCN, air, 25°C — 92%

Martín, S.E.; Rossi, L.I. *Tetrahedron Lett.*, ***2001***, *42*, 7147.

MB-Bentonite composite, 2 h, $^1O_2$, 0.005 M — quant

Habibi, M.H.; Tangestaninejad, S.; Mirkhani, V.; Yadollahi, B. *Tetrahedron*, ***2001***, *57*, 8333.

OOH, Ti(*Oi*-Pr)$_4$, *R*-BINOL, 1 d, H_2O, toluene, 0°C — 67% (80% ee)

Massa, A.; Lattanzi, A.; Siniscalchi, F.R.; Screttri, A. *Tetrahedron Asymm.*, ***2001***, *12*, 2775.

silica acetate/N_2O_4, DCM, rt, 5 min — 97%

Firouzabadi, H.; Iranpoor, N.; Heydari, R. *Synth. Commun.*, ***2001***, *31*, 2037.

horseradish peroxidase, O_2 — 77% ee

Ozaki, S.-i.; Watanabe, S.; Hayasaka, S.; Konuma, M. *Chem. Commun.*, ***2001***, 1654.

hexamethylene triamine–Br_2, rt, $CHCl_3$•H_2O, < 2 min — 91%

Shaabani, A.; Teimouri, M.B.; Safaei, H.R. *Synth. Commun.*, ***2000***, *30*, 265.

Ph–S–Ph → aq HIO_3, steam bath, 30 min → Ph–S(=O)–Ph 92%

Shirini, F.; Zolfigol, M.A.; Lakouraj, M.M.; Azadbar, M.R.
Russ. J. Org. Chem., ***2001***, *37*, 1340.

Ph–S–Me → Me_3SiCl, KO_2, MeCN, –15°C, 5 h → Ph–S(=O)–Me 93%

Chen, Y.-J.; Huang, Y.-P. *Tetrahedron Lett.*, ***2000***, *41*, 5233.

Bu–S–Bu → (Ph)(Cl)Se-oxaselenolane, DCM, rt, 5 h → Bu–S(=O)–Bu 95%

Zhang, J.; Koizumi, T. *Synth. Commun.*, ***2000***, *30*, 979.

$C_{22}H_{24}$–S–Me → $NaIO_4$, TBAB, MeOH, rt, 1 d → $C_{22}H_{24}$–S(=O)–Me 88%

Yamamoto, T.; Hayakawa, T.; Yoshino, M.; Hata, Si.; Hirayama, Y.
Org. Prep. Proceed. Int., ***2000***, *32*, 192.

Me–C_6H_4–S–Me → 0.5 $PhIO_2$, toluene/H_2O, 2 eq MeOPh tartrate, 20% CTAB, rt → Me–C_6H_4–S(=O)–Me quant, 53% ee *S*

Tohma, H.; Takizawa, S.; Watanabe, H.; Fukuoka, Y.; Maegawa, T.; Kita, Y.
J. Org. Chem., ***1999***, *64*, 3519.

Ph–S–Me → MnO_2, cat H_2SO_4, SiO_2, 40°C, 1 h → Ph–S(=O)–Me 80%

Firouzabadi, H.; Abbasi, M. *Synth. Commun.*, ***1999***, *129*, 1485.

Bu—S—Bu → urea–H_2O_2, TFAA, MeCN, rt, 2 h → Bu—SO_2—Bu 95%

Balicki, R. *Synth. Commun.*, ***1999***, *29*, 2235.

SECTION 220: OXIDES FROM HALIDES AND SULFONATES

Ph–CH$_2$–Cl → [$PO(OEt)_3$, Al_2O_3 / microwaves , 10 min] → Ph–CH$_2$–P(=O)(OEt)$_2$ 85%

Kaboudin, B.; Balakrishna, M.S. *Synth. Commun.*, ***2001***, *31*, 2773.

MeO–C$_6$H$_4$–Br → [PhCHO , $Ni(dppe)Br_2/Zn$ / THF , 75°C] → MeO–C$_6$H$_4$–CH(OH)Ph 91%

Kim, S.; Yoon, J.-Y.; Lim, C.J. *Synlett*, ***2000***, 1151.

SECTION 221: OXIDES FROM HYDRIDES

Methyl 1-naphthoate (CO_2Me) → [NO_2 , O_3 , DCM / −10°C] → 5-nitro (NO_2) + 8-nitro (NO_2, CO_2Me) (23 : 77) 76%

Nose, M.; Suzuki, H.; Suzuki, H. *J. Org. Chem.*, ***2001***, *66*, 4356.

Phenol (OH) → [HNO_3 , $CHCl_3$, 3 h / $(NH_4)_2SO_4$–$NiSO_4$•6 H_2O] → nitrophenol (OH, NO_2)

85% *para*/15% *ortho*

Tasneem, Ali M.M.; Rajanna, K.C.; Saiparakash, P.K. *Synth. Commun.*, ***2001***, *31*, 1123.

4-Fluorophenol (OH, F) → [[$KHSO_4$•2 $KHSO_5$•K_2SO_4] / $NaNO_2$, wet SiO_2 , DCM , rt] → 4-fluoro-2-nitrophenol (OH, NO_2, F) 75%

Zolfigol, M.A.; Bagherzadeh, M.; Madrakian, E.; Ghaemi, E.; Taqian-Nasab, A. *J. Chem. Res. (S)*, ***2001***, 140.

N-hydroxyphthalimide , 60°C
HNO_3 , $PhCF_3$, 15 h

NO_2

56%

Isozaki, S.; Nishiwaki, Y.; Sakaguchi, S.; Ishii, Y. *Chem. Commun.*, ***2001***, 1352.

NO_2 , cat N-hydroxyphthalimide
air , 70°C , 14 h

NO_2

70%

Sakaguchi, S.; Nishiwaki, Y.; Kitamura, T.; Ishii, Y. *Angew. Chem. Int. Ed.*, ***2001***, *40*, 222.

OTs

SiO_2 , $AlCl_3$–$ZnCl_2$
microwaves , 8 min

OH

Ts

92%

Moghaddam, F.M.; Dakamin, M.G. *Tetrahedron Lett.*, ***2000***, *41*, 3479.

OMe

ONO_2 , TfOH
[emim] (OTf)

emim = 1-ethyl-3-methylimidazolium

OMe

95% (67.1:37.9 *o:p*)

Laali, K.K.; Gettwert, V.J. *J. Org. Chem.*, ***2001***, *66*, 35.

$PhSO_2Cl$, microwaves
$FeCl_3$, 45 sec , 170°C

SO_2Ph

86% (77:23 3,4-diMe:2,3-diMe)

Marquié, J.; Laporterie, A.; Dubac, J.; Roques, N.; Desmurs, J.-R. *J. Org. Chem.*, ***2001***, *66*, 421.

Cl

NO_2

1. Pd catalyst
2. Me_2Al complex
3. aq HCl , 90°C , PhH

Me

NO_2

66%

Blum, J.; Berlin, O.; Milstein, D.; Ben-David, Y.; Wassermann, B.C.; Schutte, S.; Schumann, H. *Synthesis*, ***2000***, 571.

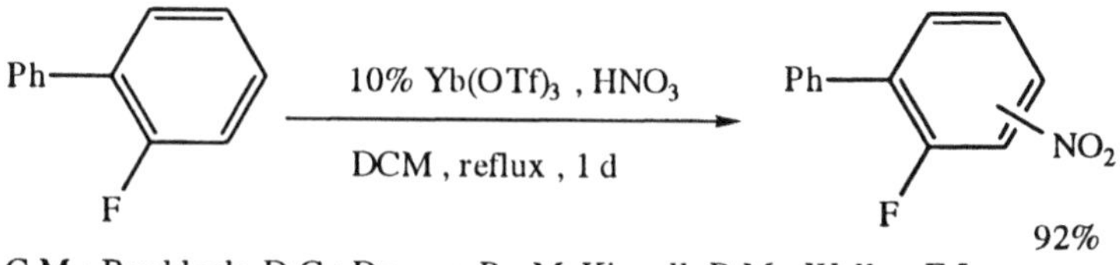

Ph—(2-fluorophenyl) → 10% $Yb(OTf)_3$, HNO_3 / DCM , reflux , 1 d → Ph—(2-fluoro, NO_2-phenyl) 92%

Barrett, A.G.M.; Braddock, D.C.; Ducray, R.; McKinnell, R.M.; Waller, F.J. *Synlett*, ***2000***, 57.

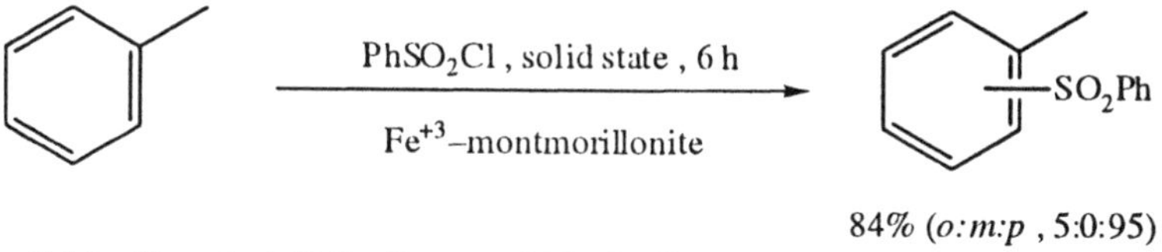

Toluene → $PhSO_2Cl$, solid state , 6 h / Fe^{+3}–montmorillonite → tolyl–SO_2Ph

84% (*o:m:p* , 5:0:95)

Choudary, B.M.; Chowdari, N.S.; Kantam, M.L. *J. Chem. Soc., Perkin Trans. 1*, ***2000***, 2689

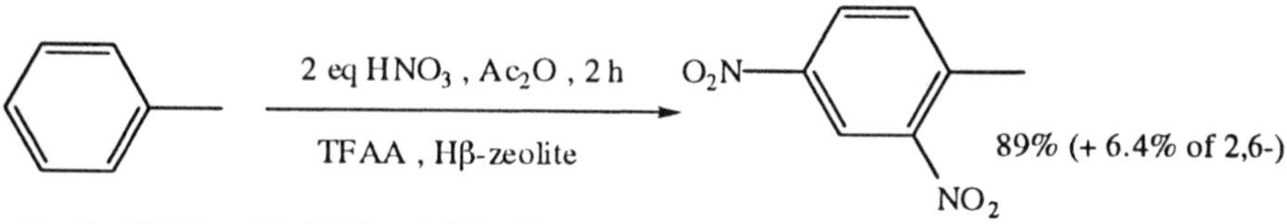

Toluene → 2 eq HNO_3 , Ac_2O , 2 h / TFAA , Hβ-zeolite → O_2N—(2-NO_2-toluene) 89% (+ 6.4% of 2,6-)

Smith, K.; Gibbins, T.; Millar, R.W.; Clardige, R.P. *J. Chem. Soc., Perkin Trans. 1*, ***2000***, 2753.

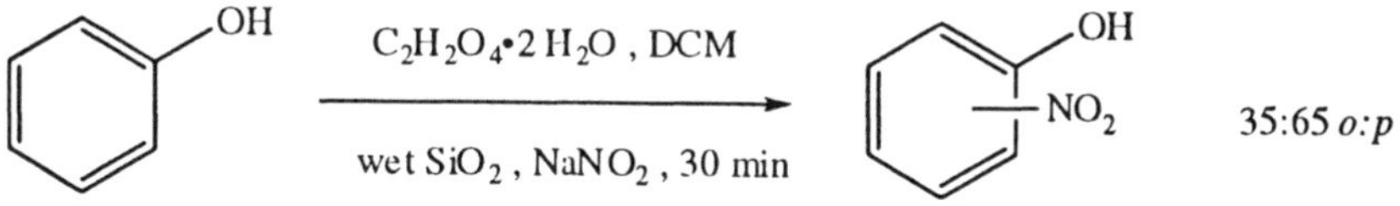

Phenol (OH) → $C_2H_2O_4$•2 H_2O , DCM / wet SiO_2 , $NaNO_2$, 30 min → OH, NO_2 35:65 *o:p*

Zolfigol, M.A.; Ghaemi, E.; Madrakian, E. *Synth. Commun.*, ***2000***, *30*, 1689.

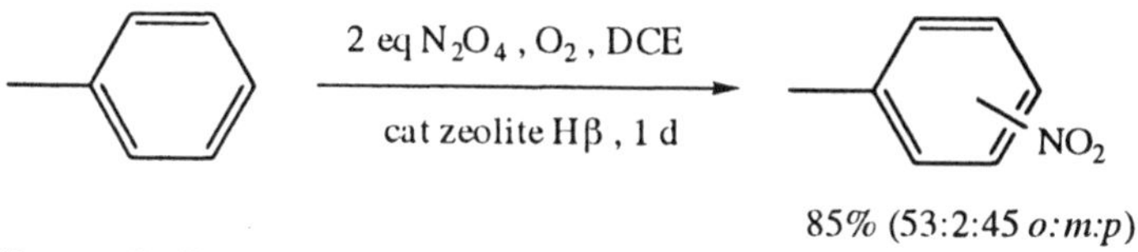

Toluene → 2 eq N_2O_4 , O_2 , DCE / cat zeolite Hβ , 1 d → tolyl–NO_2

85% (53:2:45 *o:m:p*)

Smith, K.; Almeer, S.; Black, S.J. *Chem. Commun.*, ***2000***, 1571.

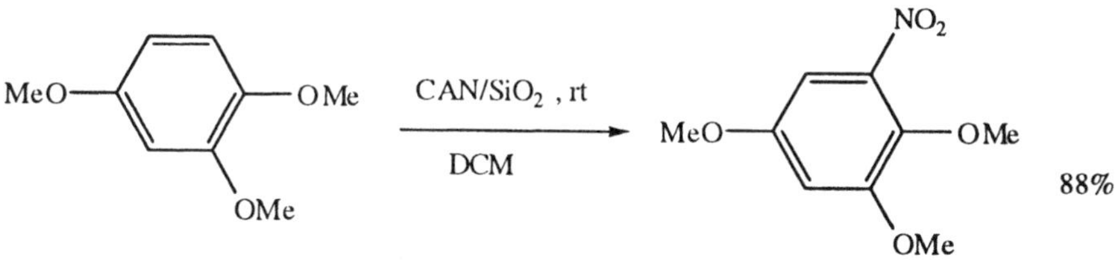

MeO—(OMe, OMe) → CAN/SiO_2 , rt / DCM → MeO—(NO_2, OMe, OMe) 88%

Grenier, J.-L.; Catteau, J.-P.; Cotelle, P. *Synth. Commun.*, ***1999***, *29*, 1201.

2 eq PhH —($SOCl_2$, TfOH, 1 d)→ Ph–S(=O)–Ph 92%

Olah, G.A.; Marinez, E.R.; Prakash, G.K.S. *Synlett*, ***1999***, 1397.

PhOH —(EtOAc, reflux; NO^+–18-crown-6 $H(NO_3)_2$)→ 2,4-dinitrophenol 82%

Iranpoor, N.; Firouzabadi, H.; Heydari, R. *Synth. Commun.*, ***1999***, *29*, 3295.

naphthalene —(70% H_2SO_4/SiO_2, HNO_3; DCM, 25°C, 1 d)→ 1-nitronaphthalene 78%

Smith, A.C.; Narvaez, L.D.; Akins, B.G.; Langford, M.M.; Gary, T.; Geisleer, V.J.; Khan, F.A. *Synth. Commun.*, ***1999***, *29*, 4187.

PhEt —($AcONO_2$–Clay)→ NO_2-C$_6$H$_4$-Et 96% (44:53 *o:p*)

Rodrigues, J.A.R.; Filho, A.P.O.; Moran, P.J.S. *Synth. Commun.*, ***1999***, *29*, 2169.

SECTION 222: OXIDES FROM KETONES

NO ADDITIONAL EXAMPLES

SECTION 223: OXIDES FROM NITRILES

NO ADDITIONAL EXAMPLES

SECTION 224: OXIDES FROM ALKENES

$CH_2=CH-C_6H_{13}$ —(NaH_2PO_2, MeOH, rt; BEt_3, 2 h)→ H(NaO)P(=O)–CH_2CH_2–C_6H_{13} 80%

Deprèle, S.; Montchamp, J.-L. *J. Org. Chem.*, ***2001***, *66*, 6745.

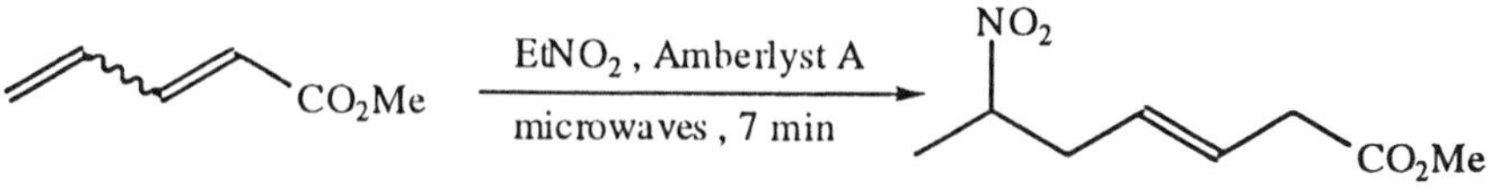

EtNO$_2$, Amberlyst A; microwaves, 7 min — CO_2Me → NO_2 ... CO_2Me

85%

Ballini, R.; Bosica, G.; Fiorini, D. *Tetrahedron Lett.*, ***2001***, *42*, 8471.

SECTION 225: OXIDES FROM MISCELLANEOUS COMPOUNDS

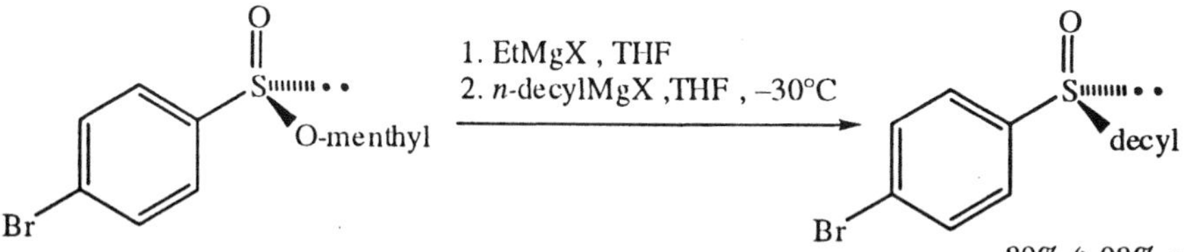

1. EtMgX, THF
2. *n*-decylMgX, THF, –30°C

Br, O, S, O-menthyl → Br, O, S, decyl

89% (>98% ee, *S*)

Capozzi, M.A.M.; Cardellicchio, C.; Naso, F.; Spina, G.; Tortorella, P. *J. Org. Chem.*, ***2001***, *66*, 5933.

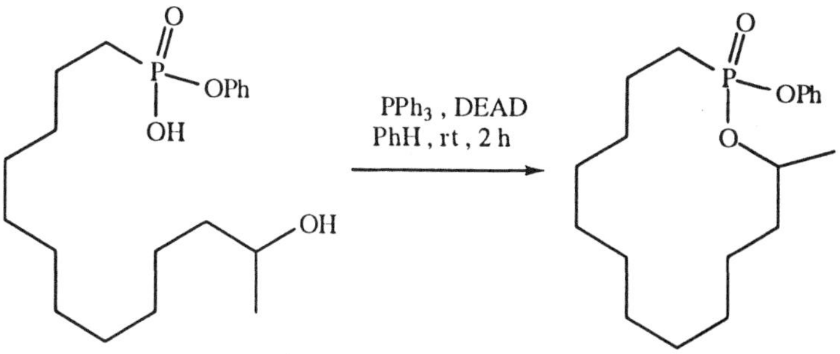

PPh$_3$, DEAD
PhH, rt, 2 h

O, P, OPh, OH, OH → O, P, OPh, O

82%

Pungente, M.D.; Weiler, L. *Org.Lett.*, ***2001***, *3*, 643.

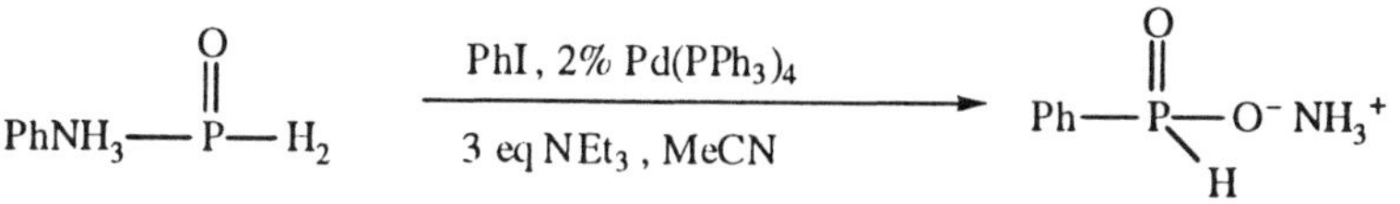

$PhNH_3$—P(O)—H_2 — PhI, 2% Pd(PPh$_3$)$_4$; 3 eq NEt$_3$, MeCN → Ph—P(O)(H)—O^- NH_3^+ 97%

Montchamp, J.-L.; Dumond, Y.R. *J. Am. Chem. Soc.*, ***2001***, *123*, 510.

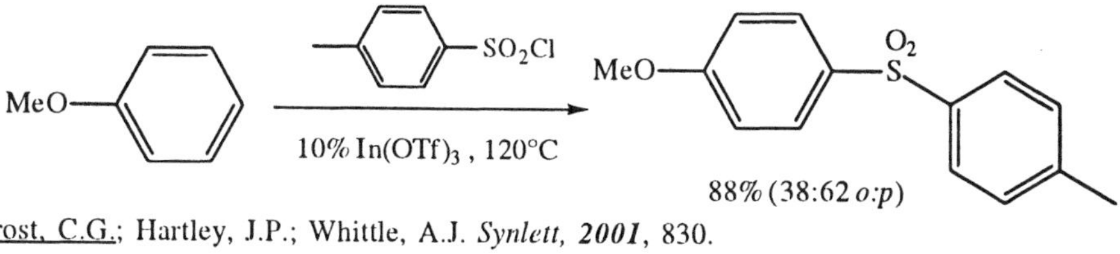

MeO—C$_6$H$_5$ — $CH_3C_6H_4SO_2Cl$; 10% In(OTf)$_3$, 120°C → MeO—C$_6$H$_4$—S(O_2)—C$_6$H$_4$—CH$_3$

88% (38:62 *o:p*)

Frost, C.G.; Hartley, J.P.; Whittle, A.J. *Synlett*, ***2001***, 830.

Ph–CH=CH–CH$_2$–OSO_3^- $PyrH^+$ → MeCN, 2% H_2O, 2.4 eq Br_2, 30 min; $AgNO_3$, 0°C → cyclic sulfate (Ph, Br) 41%

Steinmann, J.E.; Phillips, J.H.; Sanders, W.J.; Kiessling, L.L. *Org. Lett.*, ***2001***, *3*, 3557.

Et–P(O)(OEt)$_2$ → 1. BBr_3, toluene-hexane, –30°C → –70°C; 2. MeOH, 20°C → Et–P(O)(OH)$_2$ 95%

Gauvry, N.; Mortier, J. *Synthesis*, ***2001***, 553.

H–P(O)(OC_3H_7)$_2$ → Ph_2I^+ BF_4^-, DMF; cat Pd(PPh_3)$_4$, K_2CO_3, 90°C, 3 h → Ph–P(O)(OC_3H_7)$_2$ 98%

Zhou, T.; Chen, Z.-C. *Synth. Commun.*, ***2001***, *31*, 3289.

Ph–S(O)–CH$_2$CH$_2$–TMS → $SOCl_2$, DCM; –78°C → rt → Ph–S(O)–Cl 90%

Schwan, A.L.; Strickler, R.R.; Dunn-Dufault, R.; Brillon, D. *Eur. J. Org. Chem.*, ***2001***, 1643.

PhB(OH)$_2$ → NH_4NO_3, TFAA, MeCN → $PhNO_2$ 78%

Salzbrunn, S.; Simon, J.; Surya Prakash, G.K.; Petasis, N.A.; Olah, G.A. *Synlett*, ***2000***, 1485.

Ph$_2$P(O)–CH$_2$CH$_2$–OMe → PhMgBr, 5 h → Ph$_3$P=O 65%

Cardellicchio, C.; Fracchiolla, G.; Naso, F.; Tortorella, P.; Holody, W.; Pietrusiewicz, K.M. *Tetrahedron Lett.*, ***1999***, *40*, 5773.

PhMgBr → (EtO)$_2$P(O)–CN, DCM, 0°C; 30 min → Ph–P(O)(OEt)$_2$ 81%

Guzman, A.; Alfaro, R.; Díaz, E. *Synth. Commun.*, ***1999***, *29*, 2967.

CHAPTER 16

PREPARATION OF DIFUNCTIONAL COMPOUNDS

SECTION 300: ALKYNE - ALKYNE

Ph—≡—TIPS → (6 eq $Cu(OAc)_2$, TBAF, 4 h; Py/ether) Ph—≡—≡—Ph 68%

Hueft, M.A.; Collins, S.K.; Yap, G.P.A.; Fallis, A.G. *Org. Lett.*, ***2001***, *3*, 2883.

→ ($Pd(Ph_3)_4$, CuI, Et_2NH; MeCN, 90°C, 1.5 h) 43%

Dai, W.-M.; Wu, A. *Tetrahedron Lett.*, ***2001***, *42*, 81.

C_8H_{17}—≡ → ($CuCl_2$, KF/Al_2O_3; microwaves, 2 min) C_8H_{17}—≡—≡—C_8H_{17} 61%

Kabalka, G.W.; Wang, L.; Pagni, R.M. *Synlett*, ***2001***, 108.

Cp_2Zr → (1. CuCl; 2. Bn—≡—Br, 20°C, 6 h) 36%

Liu, Y.; Xi, C.; Hara, R.; Nakajima, K.; Yamazaki, A.; Kotora, M.; Takahashi, T. *J. Org. Chem.*, ***2000***, *65*, 6951.

Ph—≡—$SiMe_3$ —(CuCl , DMF , 60°C; 6 h)→ Ph—≡—≡—Ph >99%

Nishihara, Y.; Ikegashira, K.; Hirabayashi, K.; Ando, J.-i.; Mori, A.; Hiyama, T. *J. Org. Chem.*, ***2000***, *65*, 1780.

Ph–CH=CBr$_2$ —(2.5% $Pd_2(dba)_3$, 15% TFP; 20% CuI, 2.5 eq TEA; DMF , 80°C)→ Ph—≡—≡—Ph 71%

TFP = tris-(2-furyl)phosphine

Shen, W.; Thomas, S.A. *Org. Lett.*, ***2000***, *2*, 2857.

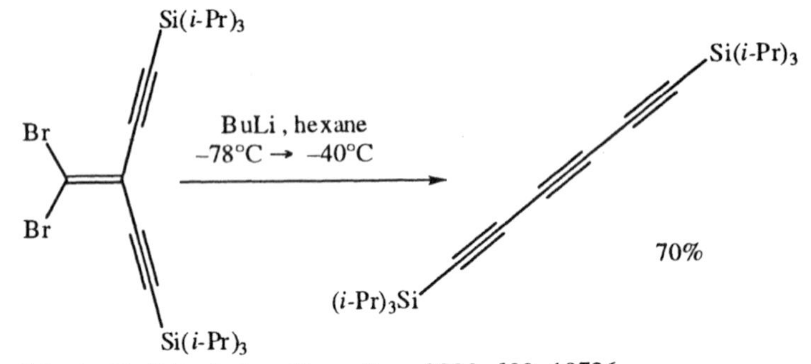

Eisler, S.; Tykwinski, R.R. *J. Am. Chem. Soc.*, ***2000***, *122*, 10736.

Ph—≡ —($CuCl_2$, NaOAc , 3 h; supercritical CO_2)→ Ph—≡—≡—Ph quant

Li, J.; Jiang, H. *Chem. Commun.*, ***1999***, 2369.

REVIEWS:

"Acetylenic Coupling: A Powerful Tool In Molecular Construction," Siemsen, P.; Livingston, R.C.; Diederich, F. *Angew. Chem. Int. Ed.*, ***2000***, *39*, 2632.

SECTION 301: ALKYNE - ACID DERIVATIVES

C_7H_{15}—≡ —(e^- , DMF , CO_2)→ C_7H_{15}—≡—CO_2H 90%

Köster, F.; Dinhus, E.; Duñach, E. *Eur. J. Org. Chem.*, ***2001***, 2507.

SECTION 302: ALKYNE - ALCOHOL, THIOL

MeO SnBu$_3$ TMS — i-PrCHO , DCM → HO TMS OMe + HO TMS OMe (97 : 3) 96%

Savall, B.M.; Powell, N.A.; Roush, W.R. *Org. Lett.*, ***2001***, *3*, 3057.

CHO — TMSO (alkyne), toluene , 60°C; 22% Ph HO NMe$_2$, 50% NEt$_3$; 20% Zn(OTf)$_2$, 5 h → OH OTMS 77% (98% ee)

Anand, N.K.; Carreira,E.M. *J. Am. Chem. Soc.*, ***2001***, *123*, 9687.

EtO$_2$C—CHO — 1. Me$_3$Si(Me)C=C=CH$_2$, cat Sc complex; 2. K$_2$CO$_3$, EtOH → EtO$_2$C OH Me 95% (98% ee)

Evans, D.A.; Sweeney, Z.K.; Rovis, T.; Tedron, J.S. *J. Am. Chem. Soc.*, ***2001***, *123*, 12095.

Ph—≡—C(Me)$_2$OAc — 1. Cp$_2$TiCl$_2$, Mg ,THF; 2. PhCHO; 3. 10% HCl → Ph—≡—C(Me)$_2$CH(OH)Ph 72%

Yang, F.; Zhao, G.; Ding, Y. *Tetrahederon Lett.*, ***2001***, *42*, 2839.

Cl CHO — PhC≡CH , ZnMe$_2$, toluene; 10% chiral amino-alcohol → Cl Ph OH >95% (88% ee)

Lu, G.; Li, X.; Zhou, Z.; Chan, W.L.; Chan, A.S.C. *Tetrahedron Asymm.*, ***2001***, *12*, 2147.

re-LGADH (a reductase)

$SiMe_3$ → $SiMe_3$

92% ee

Schubert, T.; Hummel, W.; Kula, M.-R.; Müller, M. *Eur. J. Org. Chem.*, **2001**, 4181.

HCCH , toluene , $Zn(OTf)_2$

N-Me-ephedrine , i-Pr_2NEt , 7 d

CHO → HO

76% (98% ee)

Sasaki, H.; Boyall, D.; Carreira, E.M. *Helv. Chim. Acta*, **2001**, *84*, 964.

acetone , t-BuOK , 20 min

solvent free

Ph → Ph—OH 94%

Miyamoto, H.; Yasaka, S.; Tanaka, K. *Bull. Chem. Soc. Jpn.*, **2001**, *74*, 185.

PhC≡CH , toluene , 23°C

$Zn(OTf)_2$, NEt_3 , 2 h

(+)-N-methylephedrine

i-PrCHO → OH, i-Pr, Ph

95% (90% ee)

Frantz, D.E.; Fässler, R.; Carreira, E.M. *J. Am. Chem. Soc.*, **2000**, *122*, 1806.

2° alcohol dehydrogenase from

Thermoanaeobacter ethanolicus

O → OH

30% (60% ee , *S*)

Heiss, C.; Phillips, R.S. *J. Chem. Soc., Perkin Trans. 1*, **2000**, 2821.

OH , toluene , 23°C

(–)-N-methylephedrine , NEt_3

$Zn(OTf)_2$

CHO → HO, OH

97% (98% ee)

Boyall, D.; López, F.; Sasaki, H.; Frantz, D.; Carreira, E.M. *Org. Lett.*, **2000**, *2*, 4233.

5-nonanone , 4 eq SmI_2

THF

C_8H_{17}—I → C_8H_{17}—C(Bu)(Bu)OH

84%

Kunishima, M.; Nakata, D.; Tanaka, S.; Hioki, K.; Tani, S. *Tetrahedron*, **2000**, *56*, 9927.

PhC≡CLi , toluene , 1% Me_3Ga

-78°C , 0.1 h

BnO OH Ph

90%

Ooi, T.; Morikawa, J.; Ichikawa, H.; Maruoka, K. *Tetrahedron Lett.*, ***1999***, *40*, 5881.

PhCHO

PhC≡CH , $ZnMe_2$, 10% chiral amino alchol

toluene/THF , –20°C

HO Ph Ph

70% (68% ee, *S*)

Li, Z.; Upadhyay, V.; DeCamp, A.E.; DiMichele, L.; Reider, P.J. *Synthesis*, ***1999***, 1457.

OCOPh O Bn

SmI_2 , $Pd(PPh_3)_4$, THF

OH Bn

87%

Aurrecoechea, J.M.; Fañanás, R.; Arrate, M.; Gorgojo, J.M.; Aurrekoetxea, N. *J. Org. Chem.*, ***1999***, *64*, 1893.

C_3H_7 O

1. $(Cy)_2BH$, 0°C → rt

2. MeOH , 1 h

O C_3H_7 HO C_3H_7

62%

Yu, S.; Li, N.-S.; Kabalka, G.W. *J. Org. Chem.*, ***1999***, *64*, 5822.

O

PhC≡CH , $CsOH\cdot H_2O$

THF , rt , 1 h

OH Ph

88%

Tzalis, D.; Knochel, P. *Angew. Chem. Int. Ed.*, ***1999***, *38*, 1463.

SECTION 303: ALKYNE - ALDEHYDE

$C_{12}H_{25}$—≡—CH_2OH

$TiCl_4/NEt_3$, CH_2Cl_2

0°C , 1 h

$C_{12}H_{25}$—≡—CHO

90%

Han, Z.P.; Shinokubo, H.; Oshima, K. *Synlett*, ***2001***, 1421.

SECTION 304: ALKYNE - AMIDE

Bu(Ts)N–CHO → (1. PPh_3, CCl_4, THF, 60°C; 2. BuLi, THF, –78°C → –30°C) → Bu(Ts)N–C≡CH 99x86%

Brückner, D. *Synlett*, ***2000***, 1402.

BnNHTs → (1. BuLi, toluene, 0°C; 2. Ph(OTf)–I–C≡CH) → BnN(Ts)–C≡CH 90%

Witulski, B.; Göβmann, M. *Synlett*, ***2000***, 1793.

REVIEWS:

"Recent Advances in the Chemistry of Ynamines and Ynamides," Zificsak, C.A.; Mulder, J.A.; Hsung, R.P.; Rameshkumar, C.; Wei, L.-L. *Tetrahedron*, ***2001***, *57*, 7575.

SECTION 305: ALKYNE - AMINE

Ph–CH=N–Bn → ($SiMe_3$–C≡CH; 5% $[IrCl(cod)]_2$, THF, 23°C, 1 d) → Ph–CH(NHBn)–C≡C–$SiMe_3$ 76%

Fischer, C.; Carreira, E.M. *Org. Lett.*, ***2001***, *3*, 4319.

C_8H_{17}–C≡CH → (Bn_2NH, $(CHO)_n$, microwaves; CuI, Al_2O_3) → C_8H_{17}–C≡C–CH_2NBn_2 90%

Kabalka, G.W.; Wang, L.; Pagni, R.M. *Synlett*, ***2001***, 676.

C_5H_{11}–CH=N^+(–O^-)Bn → ($Me_3SiC≡CH$, 10% $Zn(OTf)_2$, 25% i-Pr_2NEt, CH_2Cl_2; 25°C, 12 h) → C_5H_{11}–CH(N(OH)Bn)–C≡C–$SiMe_3$ 62%

Frantz, D.E.; Fässler, R.; Carreira, E.M. *J. Am. Chem. Soc.*, ***1999***, *121*, 11245.

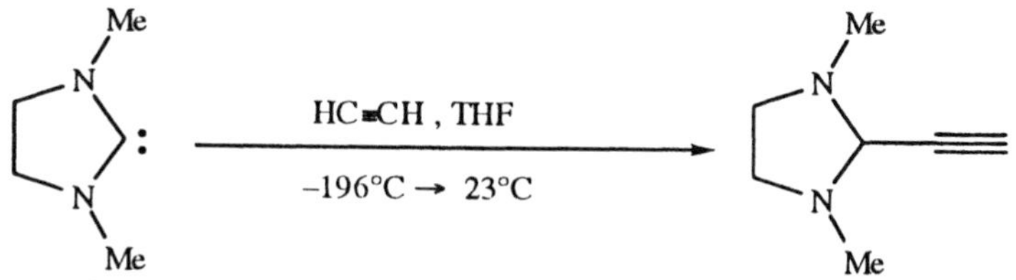

93%

Arduengo III, A.J.; Calabrese, J.C.; Davidson, F.; Diasa, H.V.R.; Goerlich, J.R.; Krafczyk, R.; Marshall, W.J.; Tamm, M.; Schmutzler, R. *Helv. Chim. Acta,* ***1999***, *82*, 2348.

REVIEWS:

"Recent Advances in the Chemistry of Ynamines and Ynamides," Zificsak, C.A.; Mulder, J.A.; Hsung, R.P.; Rameshkumar, C.; Wei, L.-L. *Tetrahedron,* ***2001***, *57*, 7575.

SECTION 306: ALKYNE - ESTER

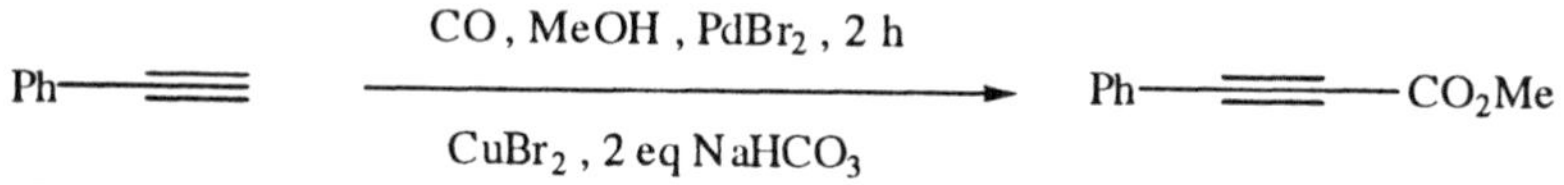

80%

Li, J.; Jiang, H.; Chen, M. *Synth. Commun.,* ***2001***, *31*, 199.

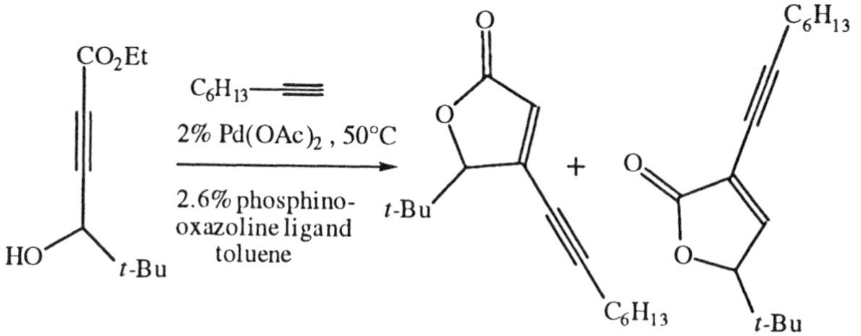

(83 : 17) 90%

Lücking, U.; Pfaltz, A. *Synlett,* ***2000***, 1261.

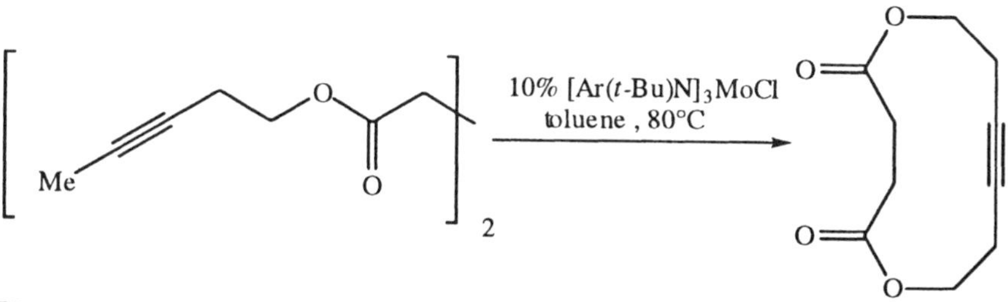

91%

Fürstner, A.; Mathes, C.; Lehmann, C.W. *J. Am. Chem. Soc.,* ***1999***, *121*, 9453.

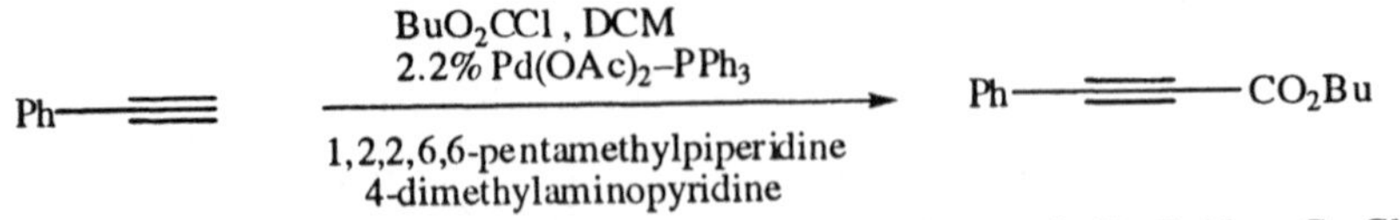

Ph—≡ → (BuO_2CCl, DCM; 2.2% $Pd(OAc)_2$–PPh_3; 1,2,2,6,6-pentamethylpiperidine; 4-dimethylaminopyridine) → Ph—≡—CO_2Bu 89%

Böttcher, A.; Becker, H.; Brunner, M.; Preiss, T.; Henkelmann, J.; De Bakker, C.; Gleiter, R. *J. Chem. Soc., Perkin Trans. 1*, ***1999***, 3555.

SECTION 307: ALKYNE - ETHER, EPOXIDE, THIOETHER

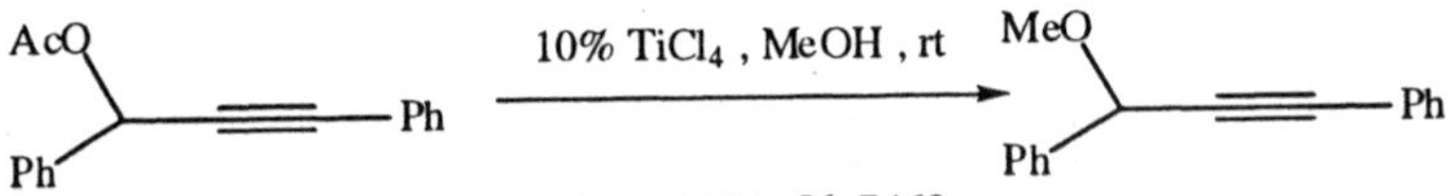

AcO / Ph → (10% $TiCl_4$, MeOH, rt) → MeO / Ph 95%

Mahrwald, R.; Quint, S. *Tetrahedron*, ***2000***, *56*, 7463.

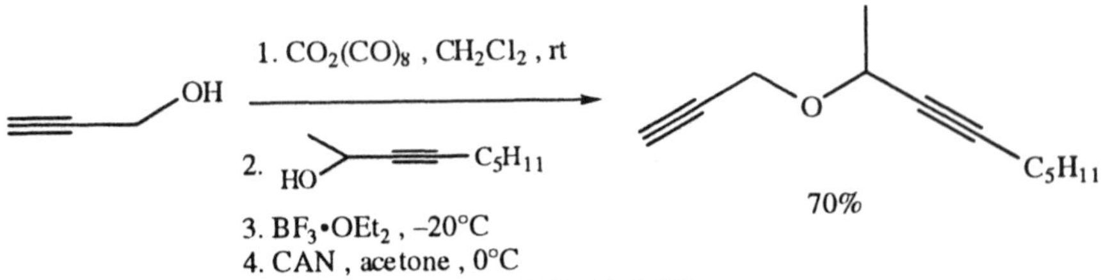

OH → (1. $Co_2(CO)_8$, CH_2Cl_2, rt; 2. HO—≡—C_5H_{11}; 3. BF_3•OEt_2, –20°C; 4. CAN, acetone, 0°C) → O, C_5H_{11} 70%

Diaz, D.D.; Martin, V.S. *Tetrahedron Lett.*, ***2000***, *41*, 9993.

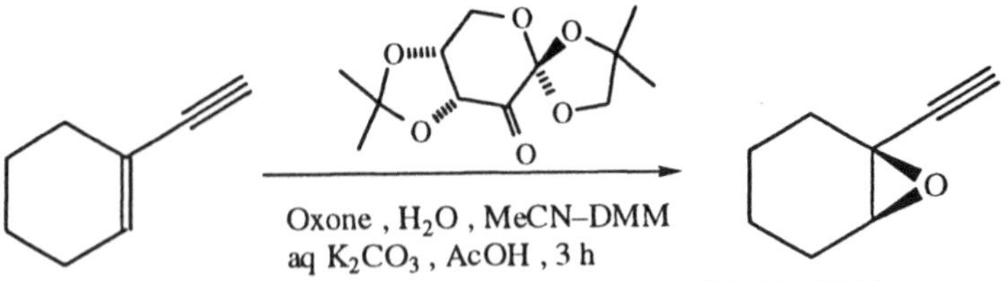

O, O, O, O, O → (Oxone, H_2O, MeCN–DMM; aq K_2CO_3, AcOH, 3 h) → O 78% (93% ee, *R,R*)

Wang, Z.-X.; Cao, G.-A.; Shi, Y. *J. Org. Chem.*, ***1999***, *64*, 7646.

SECTION 308: ALKYNE - HALIDE

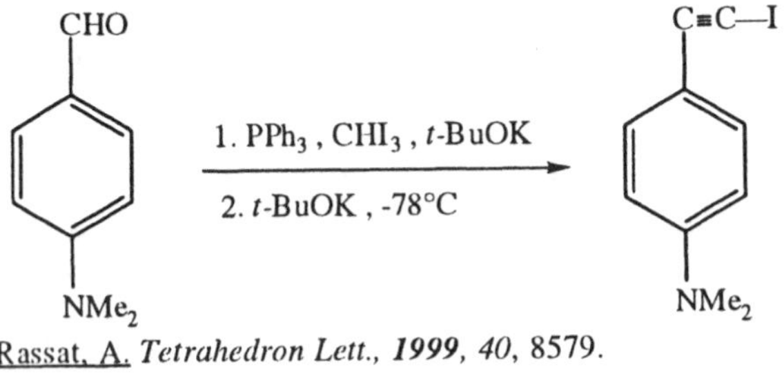

CHO / NMe_2 → (1. PPh_3, CHI_3, *t*-BuOK; 2. *t*-BuOK, -78°C) → C≡C—I / NMe_2 80x80%

Michel, P.; Rassat, A. *Tetrahedron Lett.*, ***1999***, *40*, 8579.

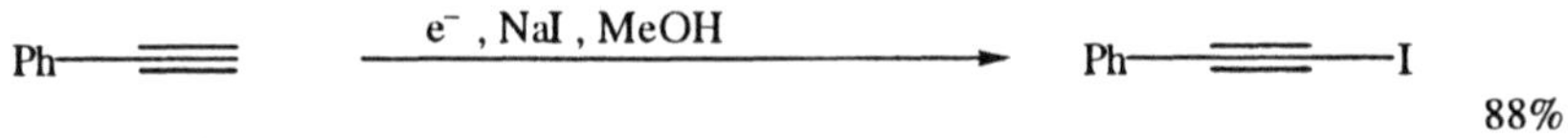

Ph—≡ → (e^-, NaI, MeOH) → Ph—≡—I 88%

Nishiguchi, I.; Kanabe, O.; Itoh, K.; Maekawa, H. *Synlett*, ***2000***, 89.

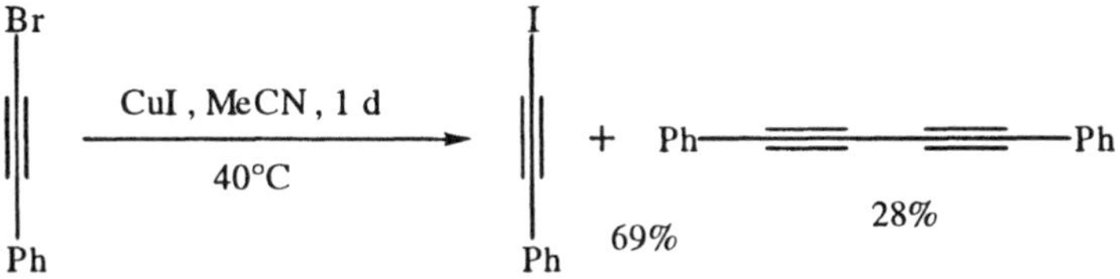

Br—≡—Ph → (CuI, MeCN, 1 d, 40°C) → I—≡—Ph 69% + Ph—≡—≡—Ph 28%

Abe, H.; Suzuki, H. *Bull. Chem. Soc. Jpn.*, ***1999***, *72*, 787.

SECTION 309: ALKYNE - KETONE

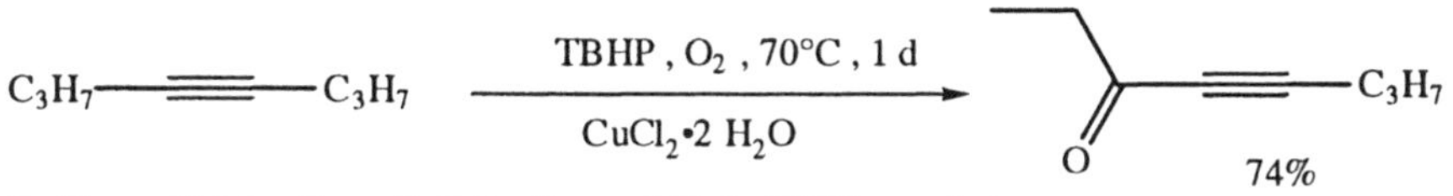

C_3H_7—≡—C_3H_7 → (TBHP, O_2, 70°C, 1 d; $CuCl_2$•2 H_2O) → 74%

Li, P.; Fong, W.M.; Chao, L.C.F.; Fung, S.H.C.; Williams, I.D.
J. Org. Chem., ***2001***, *66*, 4087.

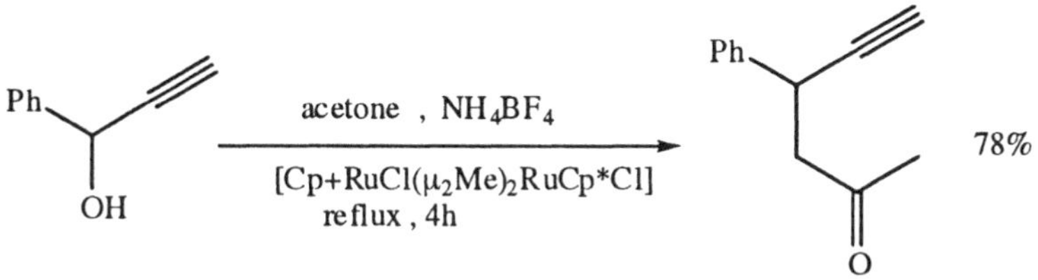

(acetone, NH_4BF_4; [Cp+RuCl(μ_2Me)$_2$RuCp*Cl], reflux, 4h) 78%

Cp*=$N_5C_5Me_5$

Nishibayashi, Y.; Wakiji, I.; Ishii, Y.; Uemura, S.; Hidai, M.
J. Am. Chem. Soc., ***2001***, *123*, 3393.

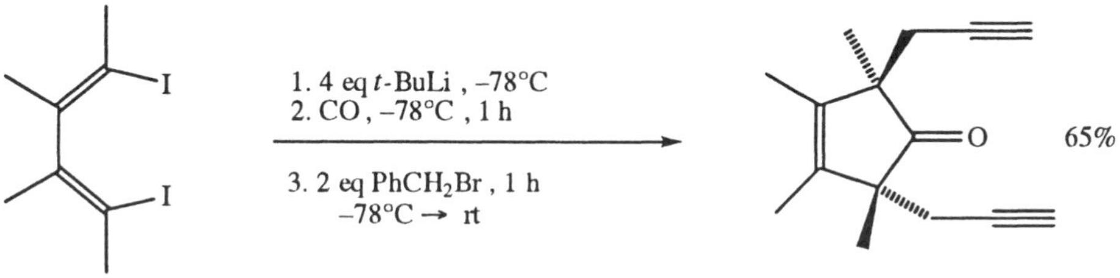

(1. 4 eq *t*-BuLi, –78°C; 2. CO, –78°C, 1 h; 3. 2 eq $PhCH_2Br$, 1 h, –78°C → rt) 65%

Song, Q.; Hen, J.; Jin, X.; Xi, Z. *J. Am. Chem. Soc.*, ***2001***, *123*, 10419.

Ph—≡—Cu → (polymer—SeCOMe; DMF, 80°C) → Ph—≡—C(O)Me 84%

Qian, H.; Shao, L.-X.; Huang, X. *Synlett*, ***2001***, 1571.

Ph—≡ → (PhCOCl, KF/Al_2O_3; cat $Pd(PPh_3)_2Cl_2$–CuI, NEt_3, microwaves, 4 min) → Ph—≡—C(O)Ph 91%

Wang, J.-x.; Wei, B.; Huang, D.; Hu, Y.; Bai, L. *Synth. Commun.*, ***2001***, *31*, 3337.
with CuI/NEt_3, microwaves. See Wang, J.-X.; Wei, B.; Hu, Y.; Liu, Z.; Fu, Y. *Synth. Commun.*, ***2001***, *31*, 3527; Wang, J.-X.; Wei, B.; Hu, Y.; Liu, Z.; Kang, L. *J. Chem. Res. (S)*, ***2001***, 146.

$C_8H_{17}C{\equiv}C$-H → (3 eq methyl vinyl ketone; 0.05 $[RuCl_2(p\text{-cymene})]_2$, 0.2 pyrrolidine, PhH, 60°C, 12 h) → C_8H_{17}—≡—$CH_2CH_2C(O)CH_3$ 91%

Chang, S.; Na, Y.; Choi, E.; Kim, S. *Org. Lett.*, ***2001***, *3*, 2089.

PhCH=CHC(O)Ph → (1. chiral binaphthyl borepin B—≡—C_6H_{13} (Ph, Ph), DCM, $BF_3{\bullet}OEt_2$; 2. aq NH_4Cl) → PhCH(C≡CC_6H_{13})CH_2C(O)Ph

88% (86% ee)

Chong, J.M.; Shen, L.; Taylor, N.J. *J. Am. Chem. Soc.*, ***2000***, *122*, 1822.

Ph_2Sb—≡—Ph → (BzCl, DCM, heat, 3 h; 3% $BnPdCl(PPh_3)_2$) → PhC(O)—≡—Ph

82%

Kakusawa, N.; Yamaguchi, K.; Kurita, J.; Tsuchiya, T. *Tetrahedron Lett.*, ***2000***, *41*, 4143.

Tol-COCl , 5% CuI

NEt_3 , rt , 30 h

83%

Chowdhury, C.; Kundu, N.G. *Tetrahedron*, ***1999***, *55*, 7011.

3% CuI , THF , CO , rt

80%

Sun, A.-M.; Huang, X. *Tetrahedron*, ***1999***, *55*, 13201.

SECTION 310: ALKYNE - NITRILE

NO ADDITIONAL EXAMPLES

SECTION 311: ALKYNE - ALKENE

, THF , 1 h

5% $Pd(PPh_3)_4$, 40°C

65%

Radhakrishnan, U.; Stang, P.J. *Org. Lett.*, ***2001***, *3*, 859.

Bu–C≡C–Ag , 0.5 $Pd(PPh_3)_4$

DMF , 5 min

98%

Dillinger, S.; Bertus, P.; Pale, P. *Org. Lett.*, ***2001***, *3*, 1661.

5% $[(\pi\text{-allyl})PdCl]_2$, $NHEt_2$
10% TDMPP , THF , rt , 2 h

70%

Rubina, M.; Gevorgyan, V. *J. Am. Chem. Soc.*, ***2001***, *123*, 11107.

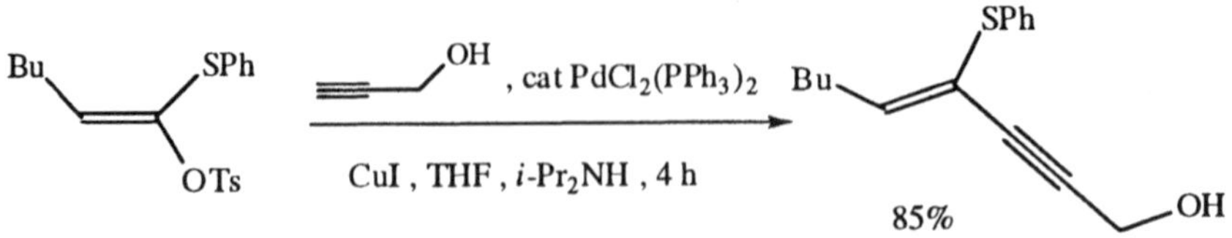

Bu
SPh
OTs
OH
, cat PdCl2(PPh3)2
CuI , THF , i-Pr2NH , 4 h
SPh
Bu
OH
85%

Braga, A.L.; Emmerich, D.J.; Silveira, C.C.; Martins, T.L.C.; Rodrigues, O.E.D.
Synlett, ***2001***, 369.

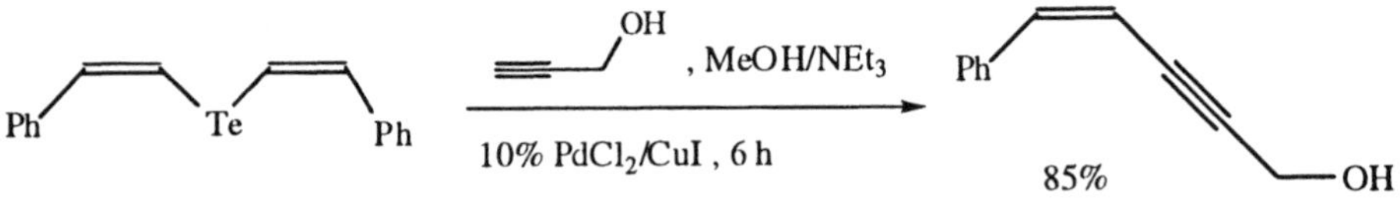

Ph
Te
Ph
OH
, MeOH/NEt3
10% PdCl2/CuI , 6 h
Ph
85%
OH

Zeni, G.; Menezes, P.H.; Moro, A.V.; Braga, A.L.; Silveira, C.C.; Stefani, H.A.
Synlett, ***2001***, 1473.

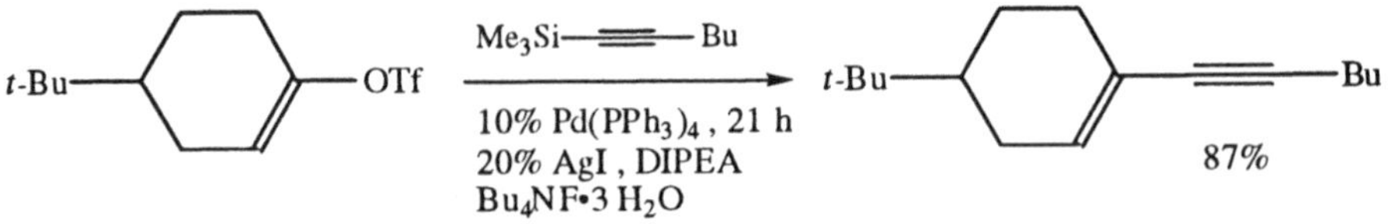

t-Bu
OTf
Me3Si—≡—Bu
10% Pd(PPh3)4 , 21 h
20% AgI , DIPEA
Bu4NF•3 H2O
t-Bu
Bu
87%

Bertus, P.; Halbes, U.; Pale, P. *Eur. J. Org. Chem.*, ***2001***, 4391.

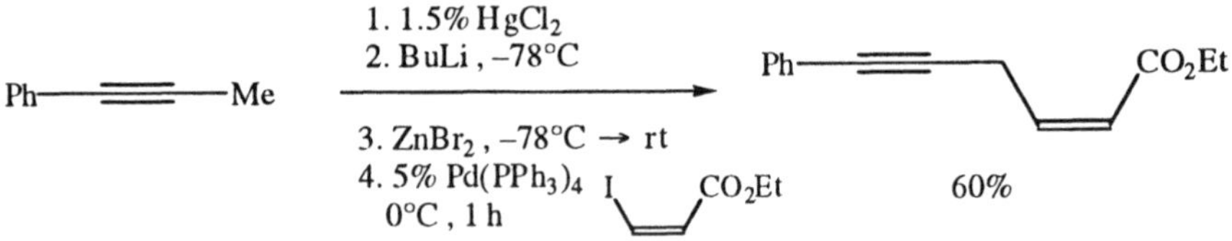

Ph
Me
1. 1.5% HgCl2
2. BuLi , –78°C
3. ZnBr2 , –78°C → rt
4. 5% Pd(PPh3)4
0°C , 1 h
I
CO2Et
Ph
CO2Et
60%

Ma, S.; Zhang, A.; Yu, Y.; Xia, W. *J. Org. Chem.*, ***2000***, *65*, 2287.

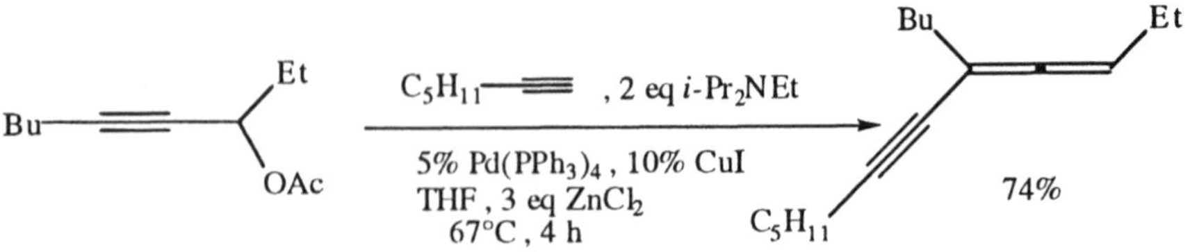

Bu
Et
OAc
C5H11
, 2 eq i-Pr2NEt
5% Pd(PPh3)4 , 10% CuI
THF , 3 eq ZnCl2
67°C , 4 h
Bu
Et
C5H11
74%

Condon-Gueugnot, S.; Linstrumelle, G. *Tetrahedron*, ***2000***, *56*, 1851.

(CH2)10
0.025M , NH4BF4 , MeOH , 60°C
cat thiolate-bridged diruthenium catalyst
(CH2)10
69%

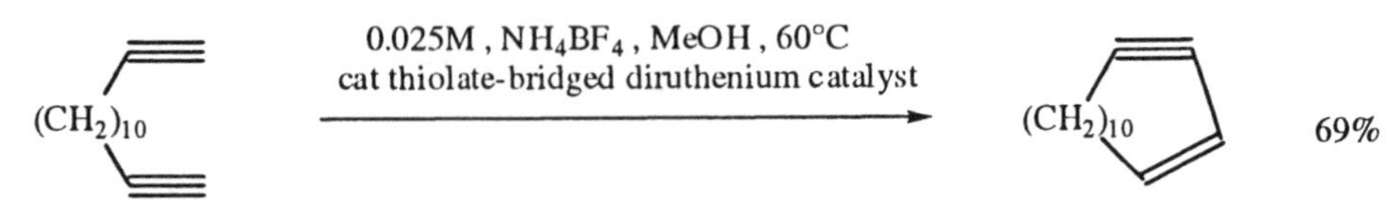

Nishibayashi, Y.; Yamanashi, M.; Wakiji, I.; Hidai, M.
Angew. Chem. Int. Ed., ***2000***, *39*, 2905.

1. 20% $PdCl_2$, 20% CuI, MeOH, rt
2. NEt_3, 4 h, C_5H_{11}—≡

Ph TeBu → Ph ... C_5H_{11} 85%

Zeni, G.; Comasseto, J.V. *Tetrahedron Lett.*, ***1999***, *40*, 4619.

TfO, Bu, I⁺Ph OTf⁻, 3% CuI

2% $PdCl_2(PPh_3)_2$, K_2CO_3
2 eq NEt_3, aq DMF, 30°C, 3 h

C_8H_{17}—≡ → TfO, Bu, C_8H_{17} 77%

Pirguliyev, N.Sh.; Brel, V.K.; Zefirov, N.S.; Stang, P.J. *Tetrahedron*, ***1999***, *55*, 12377.

REVIEWS:

"Cycloaromatization of Open and Masked 1,3-Hexadiene-5-ynes – Mechanistic and Synthetic Aspects," Zimmerman, G. *Eur. J. Org. Chem.*, ***2001***, 457.

SECTION 312: CARBOXYLIC ACID - CARBOXYLIC ACID

CO_2, e⁻, 0.1 M Et_4NClO_4, DMF

Ph → Ph, HO_2C, CO_2H 66%

Senboku, H.; Komatsu, H.; Fujimura, Y.; Tokuda, M. *Synlett*, ***2001***, 418.

$RuCl_3/HIO_5$, H_2O, CCl_4
MeCN, rt

→ CO_2H, CO_2H 86%

Griffith, W.P.; Shoair, A.G.; Suriaatmaja, M. *Synth. Commun.*, ***2000***, *30*, 3091.

SECTION 313: CARBOXYLIC ACID - ALCOHOL, THIOL

1. AD-mix
2. 0.25 TEMPO, 0.02 NaOCl
2 eq $NaOCl_2$

Ph → OH, Ph, CO_2H quant x 60% (>98% ee)

Aladro, F.J.; Guerra, F.M.; Moreno-Dorado, F.J.; Bustamante, J.M.; Jorge, Z.D.; Massanet, G.M. *Tetrahedron Lett.*, ***2000***, *41*, 3209.

CO_2H → $CO/N_2/O_2$ (200/800/100 psi), THF, H_2O, Pd–C, $CuCl_2$, 18 h → CO_2H, OH + OH, CO_2H

(22% 24%) 59% conversion

Shen, C.; Garcia-Zayas, E.A.; Sen, A. *J. Am. Chem. Soc.*, ***2000***, *122*, 4029.

Cl, CO_2H → 1. 2 eq LDA, THF, –78°C; 2. Li, 5% DIBB, *t*-BuCHO, –78°C → OH, *t*-Bu, CO_2H 52%

Pastro, I.M.; Yus, M. *Tetrahedron Lett.*, ***2000***, *41*, 5335.

SECTION 314: CARBOXYLIC ACID - ALDEHYDE

NO ADDITIONAL EXAMPLES

SECTION 315: CARBOXYLIC ACID - AMIDE

O, Me, S, *t*-Bu, N, Ph → 1. Li, O, $AlMe_3$, toluene, 0°C; 2. cat $RuCl_3•H_2O$, $NaIO_4$, CH_2Cl_2, $MeCN/H_2O$ → Me, Ph, O_2, S, *t*-Bu, N, H, CO_2H

80x62% (99:1 dr)

Borg, G.; Chino, M.; Ellman, J.A. *Tetrahedron Lett.*, ***2001***, *42*, 1433.

CHO → 60 eq CO, 120°C, Pd/C; N-methylpyrrolidine, LiBr, 1% H_2SO_4, acetamide → O, CO_2H, N, H 98%

Beller, M.; Moradi, W.A.; Eckert, M.; Neumann, H. *Tetrahedron Lett.*, ***1999***, *40*, 4523.

MeCN → 1. H_2SO_4, H_2O; 2. MeO–⟨ ⟩–CHO, 0.25% $PdBr_2$/2 PPh_3 → MeO, NHAc, CO_2H 76%

Beller, M.; Eckert, M.; Moradi, W.A. *Synlett*, ***1999***, 108.

SECTION 316: CARBOXYLIC ACID - AMINE

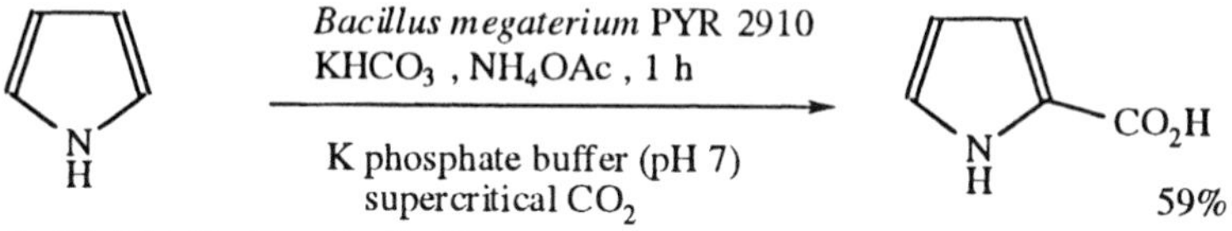

Bacillus megaterium PYR 2910
$KHCO_3$, NH_4OAc , 1 h
K phosphate buffer (pH 7)
supercritical CO_2

CO_2H
N H
59%

Matsuda, T.; Ohaishi, Y.; Harada, T.; Yanagihara, R.; Nagasawa, T.; Nakamura, K. *Chem. Commun.*, ***2001***, 2194.

REVIEWS:

"Recent Developments in Catalytic Asymmetric Strecker-Type Reactions," Yet, L. *Angew. Chem. Int. Ed.*, ***2001***, *40*, 875.

Related Methods:
Section 315 (Carboxylic Acid - Amide).
Section 344 (Amide - Ester).
Section 351 (Amine - Ester).

SECTION 317: CARBOXYLIC ACID - ESTER

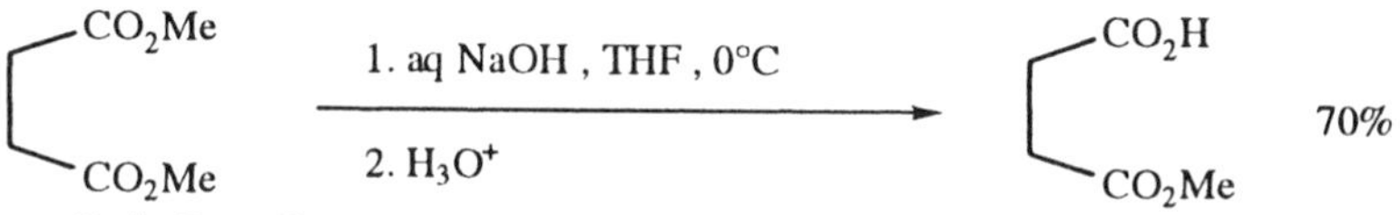

CO_2Me CO_2Me
1. aq NaOH , THF , 0°C
2. H_3O^+
CO_2H CO_2Me
70%

Niwayama, S. *J. Org. Chem.*, ***2000***, *65*, 5834.

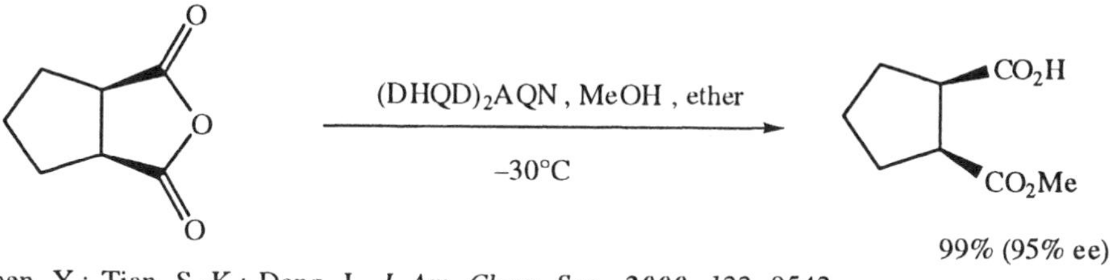

O O O
$(DHQD)_2AQN$, MeOH , ether
–30°C
CO_2H CO_2Me
99% (95% ee)

Chen, Y.; Tian, S.-K.; Deng, L. *J. Am. Chem. Soc.*, ***2000***, *122*, 9542.

$HO_2C(CH_2)_3CO_2H$ → $HO_2C(CH_2)_3CO_2Bu$
HCO_2Bu–octane , Dowex 50 WX2
70°C , 3 h
95%

Nishiguchi, T.; Ishii, Y.; Fujisaki, S. *J. Chem. Soc., Perkin Trans. 1*, ***1999***, 3023.

SECTION 318: CARBOXYLIC ACID - ETHER, EPOXIDE, THIOETHER

NO ADDITIONAL EXAMPLES

SECTION 319: CARBOXYLIC ACID - HALIDE, SULFONATE

NO ADDITIONAL EXAMPLES

SECTION 320: CARBOXYLIC ACID - KETONE

OH, NO_2, Ph → [$Cu(OAc)_2 \cdot H_2O$, air, 90°C; 3% aq AcOH:MeOH] → O, Ph, CO_2H 61%

Nikalje, M.D.; Ali, I..; Dewkar, G.K.; Sudalai, A. *Tetrahedron Lett.*, ***2000***, ***41***, 959.

Also via: Section 360 (Ketone - Ester).

SECTION 321: CARBOXYLIC ACID - NITRILE

NO ADDITIONAL EXAMPLES

Also via: Section 361 (Nitrile - Ester).

SECTION 322: CARBOXYLIC ACID - ALKENE

PhCHO → [1. Me_3Si, OTMS, OTMS, DMF, KOH, rt; 2. aq NH_4Cl] → Ph, CO_2H 96%

Lensen, N.; Mouelhi, S.; Bellassoued, M. *Synth. Commun.*, ***2001***, ***31***, 1007.

PhCHO → [$CH_2(CO_2H_2$, Bentonite; microwaves, 5 min] → Ph, CO_2H, CO_2H 79%

Loupy, A.; Song, S.-J.; Sohn, S.-M.; Lee, Y.-M.; Kwon, T.W. *J. Chem. Soc., Perkin Trans, 1*, ***2001***, 1220.

1. H_2N-CH$_2$CH$_2$-(piperazinyl)NH, rt, 5 h
2. Ph(Me)CHBr
3. H_3O^+

62% (57:43 *RR:RS*)

Brun, E.M.; Gil, S.; Parra, M. *Tetrahedron Asymm.*, ***2001***, *12*, 915.

1. 1 atm CO, 2 DMU, $Ni(cod)_2$, THF, 0°C
2. 2 eq PhZnCl, 0°C, 2h

69%

Takimoto, M.; Shimizu, K.; Mori, M. *Org. Lett.*, ***2001***, *3*, 3345.

PhCHO → ($CH_2(CO_2H)_2$, SiO_2, 4 min; microwaves) → $PhCH{=}CHCO_2H$ 83%

Kumar, H.M.S.; Reddy, B.V.S.; Reddy, P.T.; Srinivas, D.; Yadav, J.S.
Org. Prep. Proceed. Int. ***2000***, *32*, 81.

1M H_2SO_4, reflux, 69 h

89%

Zhu, S.; Pardi, S.; Cohen, T. *Tetrahedron Lett.*, ***2000***, *41*, 9589.

EtZnBr, 5% $PdCl_2(MeCN)_2$, DMF, 25°C, 12 h

90%

Abarbri, M.; Parvain, J.-L.; Kitamura, M.; Noyori, R.; Duchêne, A.
J. Org. Chem., ***2000***, *65*, 7475.

1. ClCH=CHCl, hv
2. 25% Zn, toluene
3. Ac_2O, 85°C

85x80%

Gauvry, N.; Comoy, C.; Lescop, C.; Huet, F. *Synthesis*, ***1999***, 574.

PhCHO → ($CH_2(CO_2H)_2$, piperidine; microwaves, 15 sec) → $PhCH{=}CHCO_2H$ quant

Pellón, R.F.; Mamposo, T.; González, E.; Calderón, O. *Synth. Commun.*, ***2000***, *30*, 3769.

Ph—≡ $\xrightarrow[\text{2 eq DBU , 0°C}]{\text{1 atm } CO_2 \text{ , Ni(cod)}_2}$ Ph–CH=CH–CO_2H 85%

Saito, S.; Nakagawa, S.; Koizumi, T.; Hirayama, K.; Yamamoto, Y.
J. Org. Chem., ***1999***, *64*, 3975.

catechol $\xrightarrow[\text{AcOH , 0°C , 3 d}]{\text{32\% aq } MeCO_3H}$ (Z,Z)-muconic acid + (E,Z)-muconic acid (5 : 2) 70%

McKague, A.B. *Synth. Commun.*, ***1999***, *29*, 1463.

PhCHO $\xrightarrow[\text{microwaves}]{CH_2(CO_2H)_2 \text{ , Py/Pip , 2 min}}$ $PhCH{=}CHCO_2H$ 97%

Mitra, A.K.; De, A.; Karchaudhuri, N. *Synth. Commun.*, ***1999***, *29*, 573.

Also via: Section 313 (Alcohol - Carboxylic Acids).
Section 349 (Amide - Alkene).
Section 362 (Ester - Alkene).
Section 376 (Nitrile - Alkene).

SECTION 323: ALCOHOL, THIOL - ALCOHOL, THIOL

ASYMMETRIC DIOLS

Ph–CH=CH–Me $\xrightarrow[\text{K}_3\text{Fe(CN)}_6 \text{ , aq acetone , } K_2CO_3 \quad \text{2. } K_3Fe(CN)_6 \text{ , } K_2CO_3 \text{ , 20°C}]{\text{1. 5\% microencapsulated } OsO_4 \quad \text{5\% (DHQD)}_2\text{PHAL , 30°C}}$ Ph(H)C(OH)–C(OH)(H)Me

85% (78% ee)

Kobayashi, S.; Ishida, T.; Akiyama, R. *Org. Lett.*, ***2001***, *3*, 2649.

Ph–CH=CH–Ph $\xrightarrow[\text{(DHQD)}_2\text{PHAL , 0°C} \quad t\text{-BuOH}/H_2O]{1.5\ H_2O_2 \text{ , 2 eq TEAA} \quad \text{2\% } OsO_4 \text{ , 5\% flavin}}$ Ph–CH(OH)–CH(OH)–Ph

89% (90% ee)

Jonsson, S.Y.; Adolfsson,H.; Bäckvall, J.-E. *Org. Lett.*, ***2001***, *3*, 3463.

Ph, Me (alkene) → BnSePh , $K_2OsO_2(OH)_4$, K_2CO_3 / $(DHQD)_2PHAL$, aq *t*-BuOH , 1 d → OH, OH, Me, Ph

96% (97% ee)

Krief, A.; Castillo-Colaux, C. *Synlett, **2001***, 501.

(epoxide) O → *RR*-oligomer-salen catalyst / H_2O , MeCN/DCM , THF / NMP , 100°C → OH, OH

98% (94% ee)

Ready, J.M.; Jacobsen, E.N. *J. Am. Chem. Soc., **2001**, 123*, 2719.

Bu, Bu → 3 eq $K_3[Fe(CN)_3]$ / 3 eq K_2CO_3 , 25°C / 0.4 $K_2[OSO_2(OH)_4]$ / 1% PHAL , H_2O/*t*-BuOH → OH, Bu, Bu, OH

96% (93% ee)

Mehltretter, G.M.; Döbler, C.; Sundermeier, U.; Beller, M. *Tetrahedron Lett., **2000**, 41*, 8083.

O, OH → CHO , DMSO , rt / 30% *L*-proline → O, OH, OH

60% (>20:1 dr; >99% ee)

Notz, W.; List, B. *J. Am. Chem. Soc., **2000**, 122*, 7386.

Ph → $K_2OsO_2(OH)_4$, $(DHQD)_2PHAL$ / PhSe(O)Bn , K_2CO_3 , aq *t*-BuOOH → Ph, OH, OH

96% (97% ee)

Krief, A.; Colaux-Castillo, C. *Tetrahedron Lett., **1999**, 40*, 4189.

O, Ph, Ph, O → cat Ru(chiral bis amine) / DMF , HCOOH/NEt$_3$ / 1 d , 30-60°C → OH, Ph, Ph, OH

quant (90.4:9.6 *dl:meso*)
>99% ee , *RR*

Murata, K.; Okano, K.; Miyagi, M.; Iwane, H.; Noyori, R.; Ikariya, T.
*Org. Lett., **1999**, 1*, 1119.

NON-ASYMMETRIC DIOLS

PhCHO → (3 eq Sm , 2M HCl/THF; 20°C , 1 d) → Ph–CH(OH)–CH(OH)–Ph 80% (48:52 *dl:meso*)

Talukdar, S.; Fang, J.-M. *J. Org. Chem.*, ***2001***, *66*, 330.

CHO / CHO → ($Cp_2Ti(Ph)Cl$, Zn , Me_3SiCl; THF (0.05M) , rt) → OH / OH

60% (99:1 *trans:cis*)

Yamamoto, Y.; Hattori, R.; Miwa, T.; Nakagai, Y.-i.; Kubota, T.; Yamamoto, C.; Okamoto, Y.; Itoh, K. *J. Org. Chem.*, ***2001***, *66*, 3865.

O, O → (Cp_2TiCl_2 , Mn , THF; rt (deoxygenated)) → OH, HO 41%

Barrero, A.F.; Cuerva, J.M.; Herrador, M.M.; Valdivia, M.V. *J. Org. Chem.*, ***2001***, *66*, 4074.

O, OMe, OH → (Ph CHO , THF; 1% Zn-linked BINOL , –30°C) → Ph, OH, O, OMe, OH

94% (89 : 1 *syn:anti*)
92% ee 89% ee

Kumagai, N.; Matsunaga, S.; Yoshikawa, N.; Ohshima, T.; Shibasaki, M. *Org. Lett.*, ***2001***, *3*, 1539.

Ph CO$_2$Me → (0.1% K_2OsO_4 , Cinchona alkaloid in ordered inorganic support; $K_3Fe(CN)_6$, K_2CO_3 , *t*-BuOH/H_2O) → Ph, OH, CO_2Me, OH

54% conversion

Motorina, I.; Crudden, C.M. *Org. Lett.*, ***2001***, *3*, 2325.

PhCHO → 10% Ti complex, 3 eq Mn, 25°C; 1.5 eq TMS Cl, MeCN → (1,2-diphenyl-1,2-ethanediol) 94% (96:4 *dl:meso*)

Bensari, A.; Renaud, J.-L.; Riant, O. *Org. Lett.*, ***2001***, *3*, 3863.

(E)-Bu-CH=CH-Bu → cat OsO_4, cat NMM, 2% TEAA; cat flavin → (diol) 95%

Jonsson, S.Y.; Färnegårdh, K.; Bäckvall, J.-E. *J. Am. Chem. Soc.*, ***2001***, *123*, 1365.

PhCHO → 1. 3% $Ce(Oi\text{-}Pr)_3$, Mn, TMSCl; 2. 2N HCl → (diol) + (diol)

(80 : 20) 96%

Groth, U.; Jeske, M. *Synlett*, ***2001***, 129.

(1-phenyl-1,3-butanedione) → chiral Ru complex, HCO_2H, NEt_3, CH_2Cl_2, 50°C, 6 h → (diol) + (diol)

(58 : 42) 79%
83% ee

Cossy, J.; Eustache, F.; Dalko, P.I. *Tetrahedron Lett.*, ***2001***, *42*, 5005.

(hydroxy ketone) → 1. $TiCl_4$; 2. BH_3–Py; 3. H_3O^+; 4. H_2O_2 → (diol) + (diol)

(24 : 76) 90%

Bartoli, G.; Bosco, M.; Marcantoni, E.; Massaccesi, M.; Rinaldi, S.; Sambri, L. *Eur. J. Org. Chem.*, ***2001***, 4679.

OsO_4, $K_3Fe(CN)_6$, K_2CO_3
polymer-bound DHQD-PHAL
t-BuOH, H_2O, 0°C, 18 h

92% (99% ee)

Kuang, Q,-Q.; Zhang, S.-Y.; Wei, L.-L. *Tetrahedron Lett.*, ***2001***, *42*, 5925.

$CeCl_3$•7 H_2O–$(COOH)_2$
MeCN, 2 h

CO_2Et HO OH CO_2Et

98%

Xiao, X.; Bai, D. *Synlett*, ***2001***, 535.

PhCHO

$TiBr_2$/Cu, DCM, 6 h
t-BuCN, –27°C

(96 : 4) 95%

Mukaiyama, T.; Yoshimura, N.; Igarashi, K.; Kagayama, A. *Tetrahedron*, ***2001***, *57*, 2495.

1. LiH, THF, –30°C
2. $TiCl_4$, DCM, –30°C
3. MeMgCl, THF, –78°C
4. H_3O^+

(93 : 7) 88%

Bartoli, G.; Bosco, M.; Di Maartino, E.; Marcantoni, E.; Sambri, L. *Eur. J. Org. Chem.*, ***2001***, 2901.

Bu_3Sn CO_2Et

$SnCl_2$, MeCN, rt, 3 h

86%

Yasuda, M.; Okamoto, K.; Sako, T.; Baba, A. *Chem. Commun.*, ***2001***, 157.

PhCHO → 1. Sm , TMSCl; 2. NH_4Cl → PhCH(OH)CH(OH)Ph 87%

Yu, M.; Zhang, Y. *Org. Prep. Proceed. Int.*, ***2001***, *33*, 187.

PhC(O)CH$_3$ → $TiCl_2$:Zn , *t*-BuCN , DCE , 0°C → PhC(CH$_3$)(OH)CH(OH)CH(CH$_3$)Ph

94% (71:29 *dl:meso*)

Kagayama, A.; Igarashi, K.; Mukaiyama, T. *Can. J. Chem.*, ***2000***, *78*, 657.

2-hydroxytetrahydrofuran → butadiene , 2.4 eq BEt_3; cat Ni(acac)$_2$, THF , rt , 1 d → CH$_2$=CHCH$_2$CH$_2$CH(OH)(CH$_2$)$_4$OH

63%

Kimura, M.; Ezoe, A.; Tanaka, S.; Tamaru, Y. *Angew. Chem. Int. Ed.*, ***2001***, *40*, 3600.

silacyclopentane diol (Si Ph, Ph; Bu; HO, OH) → *t*-BuOOH , KH , DMF → Bu–CH(OH)–CH(OH)–CH(OH)–CH$_2$OH 78%

Liu, D.; Kozmin, S.A. *Angew. Chem. Int. Ed.*, ***2001***, *40*, 4757.

PhCHO → cat $InCl_3$, Mg , TMSCl; THF , 2 h → PhCH(OH)CH(OH)Ph 67% (68:32 *dl:meso*)

Mori, K.; Ohtaka, S.; Uemura, S. *Bull. Chem. Soc. Jpn.*, ***2001***, *74*, 1497.

PhCHO → Zn , aq NH_4Cl , THF , rt → PhCH(OH)CH(OH)Ph 45%

Hekmatshoar, R.; Yavari, I.; Beheshtiha, Y.S.; Heravi, M.M. *Monat. Chem.*, ***2001***, *132*, 689.

PhCHO , InI , DMI

$Pd(PPh_3)_4$, rt , 5 h

62% (22:78 *syn:anti*) 33% (29:71 *E:Z*)

Araki, S.; Kameda, K.; Tanaka, J.; Hirashita, T.; Yamamura, H.; Kawai, M. *J. Org. Chem.*, ***2001***, *66*, 7919.

$CeCl_3{\bullet}7\,H_2O{-}(COOH)_2$

MeCN , 2 h

98%

Xiao, X.; Bai, D. *Synlett*, ***2001***, 535.

polymer supported dicyanoketene acetal

H_2O

94%

Masaki, Y.; Yamada, T.; Tanaka, N. *Synlett*, ***2001***, 1311.

1. $TiCl_4$
2. 0.1 Py
3. BH_3–Py
4. 1M HCl

87% (97:3 *syn:anti*)

Bartoli, G.; Bosco, M.; Bellucci, M.C.; Dalpozzo, R.; Marcantoni, E.; Sambri, L. *Org. Lett.*, ***2000***, *2*, 45.

cat TBAF , THF , –78°C

(96 : 4) 87%

Miura, K.; Nakagawa, T.; Suda, S.; Hosomi, A. *Chem. Lett.*, ***2000***, 150.

1. Li, 5% DTBB, THF, –78°C
2. t-BuCHO, –78°C → –20°C
3. PhCHO, –78°C
4. H_2O, –78°C → 20°C

Cl, OPh; OH, OH, t-Bu, Ph

68%

Foubelo, F.; Saleh, S.A.; Yus, M. *J. Org. Chem.*, ***2000***, *65*, 3478.

$PhCH_2OH$ ← Al/FeF_2, H_2O — PhCHO — Al/LiF, 5 d, H_2O →

OH, Ph, Ph, OH

76% (1.1:1 *meso:dl*)

Li, L.-H.; Chan, T.H. *Org. Lett.*, ***2000***, *2*, 1129.

O, Ph; Br, Sm, rt; MeOH, 3 h

OH, Ph, Ph, OH

69% (58:42 *dl:meso*)

Ghatak, A.; Becker, F.F.; Banik, B.K. *Tetrahedron Lett.*, ***2000***, *41*, 3793.

2 PhCHO

1. $TiCl_4$–Zn
2. $TiCl_4$–Zn/TMEDA

OH, Ph, Ph, OH

77%

Li, T.; Cui, W.; Liu, J.; Zhao, J.; Wang, Z. *Chem. Commun.*, ***2000***, 139.

OH, O, Ph

SmI_2, MeOH/THF, 0°C, 12 h

OH, OH, Ph

95% (99:1 *anti:syn*)

Keck, G.E.; Wager, C.A. *Org. Lett.*, ***2000***, *2*, 2307.

2 PhCHO

1.5 $TiCl_4$, 2.2 Bu_4NI
DCM/hexane, –78°C → rt

OH, Ph, Ph, OH

99% (93:7 *dl:meso*)

Tsuritani, T.; Ito, S.; Shinokubo, H.; Oshima, K. *J. Org. Chem.*, ***2000***, *65*, 5066.

PhCHO —(TiI_4, EtCN)→ Ph–CH(OH)–CH(OH)–Ph 90% (>99:1 *dl:meso*)

Hayakawa, R.; Shimizu, M. *Chem. Lett.*, ***2000***, 724.

$CH_3C(O)C_6H_{13}$ —(2BuI, 2 eq SmI_2)→ C_6H_{13}(Me)C(OH)–C(OH)(Me)C_6H_{13} + HO–C(Me)(C_6H_{13})Bu

+ 8 eq HMPA	<1	:	91
+ 8 eq LiBr	98	:	<1

Miller, R.S.; Sealy, J.M.; Shabangi, M.; Kuhlman, M.L.; Fuchs, J.R.; Flowers II, R.A. *J. Am. Chem. Soc.*, ***2000***, *122*, 7718.

4-Cl-C_6H_4CHO —($TiCl_2$, 2 eq (S)-2-(methoxymethyl)pyrrolidine; THF, 8 h)→ 1,2-bis(4-chlorophenyl)-1,2-ethanediol 59% (73:27 *dl:meso*)

Enders, D.; Ullrich, E.C. *Tetrahedron Asymm.*, ***2000***, *11*, 3861.

Cl–$(CH_2)_3$–OPh —(1. Li, DTAB, THF, -78°C; 2. *t*-BuCHO, 20°C; 3. PhCHO, -78°C; 4. H_2O)→ *t*-Bu–CH(OH)–$(CH_2)_3$–CH(OH)–Ph 68%

Foubelo, F.; Yus, M. *Tetrahedron Lett.*, ***1999***, *40*, 743.

PhCH=CH–C(O)Ph —($LiAlH_4$, O_2)→ Ph–CH(OH)–CH_2–CH(OH)–Ph 90%

Csákÿ, A.G.; Máximo, N.; Plumet, J.; Rámila, A. *Tetrahedron Lett.*, ***1999***, *40*, 6427.

PhC(O)Bu —(Yb-Me_3SiCl, THF; 67°C, 20 h)→ Ph(Bu)C(OH)–C(OH)(Bu)Ph 66% (85:15 *dl:meso*)

Ogawa, A.; Takeuchi, H.; Hirao, T. *Tetrahedron Lett.*, ***1999***, *40*, 7113.

O

Ar Ar

2.2 eq $Ni(cod)_2$, DMF

3 h, rt → 60°C

HO Ar Ar OH

$Ar = 3,4\text{-di}(OC_6H_{13})C_6H_3$

97%

Reisch, H.A.; Enkelmann, V.; Scherf, U. *J. Org. Chem.*, ***1999***, *64*, 655.

Bu

Bu

1% OsO_4, mcpba

N-methylmorpholine

OH OH

Bu

Bu

90% (>99% cis)

Bergstad, K.; Piet, J.J.N.; Bäckvall, J.-E. *J. Org. Chem.*, ***1999***, *64*, 2545.

Ph

O

HO

Ph

$(i\text{-PrO})_2TiBH_4$, CH_2Cl_2

–20°C, 5 min

Ph HO HO Ph + Ph HO HO Ph

(100 : 0) 95%

Ravikumar, K.S.; Sinha, S.; Chandrasekaran, S. *J. Org. Chem.*, ***1999***, *64*, 5841.

O

CHO

Cp_2TiCl, THF, rt

OH H OH

75%

Fernández-Mateos, A.; de la Nava, E.M.; Coca, G.P.; Silvo, A.R.; González, R.R. *Org. Lett.*, ***1999***, *1*, 607.

O OH

1. C_6H_{13} $SiMe_2Cl$

NEt_3

2. $TiCl_4$, toluene, –78°C

OH C_6H_{13}

OH

32x69%

Fujita, K.; Inoue, A.; Shinokubo, H.; Oshima, K. *Org. Lett.*, ***1999***, *1*, 917.

1. *i*-PrCHO , 15% LiO*i*-Pr , ether , rt

2. NaOH , MeOH

74% (96:4 dr)

Mascarenhas, C.M.; Duffey, M.O.; Liu, S.-Y.; Morken, J.P. *Org. lett.*, ***1999***, *1*, 1427.

3% $Cp_2Ti(Ph)Cl$, Zn , Me_3SiCl

THF , rt

40% (99:1 *trans:cis*)

Yamamoto, Y.; Hattori, R.; Itoh, K. *Chem. Commun.*, ***1999***, 825.

PhCHO

1. cat $VOCl_3$, Me_3SiCl/Al

2. H_2O

70% (90:10 *dl:meso*)

Hirao, T.; Hatano, B.; Imamoto, Y.; Ogawa, A. *J. Org. Chem.*, ***1999***, *64*, 7665.

REVIEWS:

"Synthesis and Chemistry of Dithiols," Elgemeie, G.H.; Sayed, S.H. *Synthesis*, ***2001***, 1747.

Also via: Section 327 (Alcohol - Ester). Section 357 (Ester - Ester).

SECTION 324: ALCOHOL, THIOL - ALDEHYDE

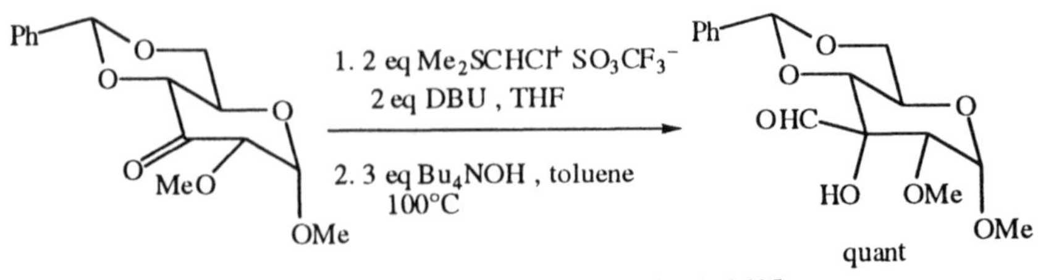

Sato, K.-i.; Sekiguchi, T.; Akai, S. *Tetrahedron Lett.*, ***2001***, *42*, 3625.

$C_3H_7CH_2CHO$ → (PhCHO, 2.5 $Ti(OBu)_4$; *t*-BuOK, THF, 0°C) → OHC–CH(C_3H_7)–CH(OH)Ph

97% (57:43 dr)

Han, Z.; Yorimitsu, H.; Shinokubo, H.; Oshima, K. *Tetrahedron Lett.*, ***2000***, *41*, 4415.

Related Methods: Section 330 (Alcohol - Ketone).

SECTION 325: ALCOHOL, THIOL - AMIDE

5% $Pd(PPh_3)_4$, InI, *i*-PrCHO; THF-HMPA, H_2O, rt

MTS = 2,4,6-trimethylphenylsulfonyl

NHMTS OH 62%

Ohno, H.; Hamaguchi, H.; Tanaka, T. *J. Org. Chem.*, ***2001***, *66*, 1867.

OSO_2NMe_2 → ($AlCl_3$, 140°C; 1 h) → OH, SO_2NMe_2

88% (*o/p* = 0.5)

Benson, G.A.; Maughan, P.J.; Shelly, D.P.; Spillane, W.J. *Tetrahedron Lett.*, ***2001***, *42*, 8729.

O, O → ($BnNH_2$–Dibal, THF; rt, 30 min) → HO, NHBn, O

98%

Huang, P.-Q.; Zheng, X.; Deng, X.-M. *Tetrahedron Lett.*, ***2001***, *42*, 9039.

NBn_2, Se → (1. LDA, THF, –78°C; 2. *i*-PrCHO, –78°C) → NBn_2, Se, HO, *i*-Pr + NBn_2, Se, HO, *i*-Pr

(94 : 6) 87%

Murai, T.; Suzuki, A.; Kato, S. *J. Chem. Soc., Perkin Trans. 1*, ***2001***, 2711.

Boc, N, CHO

1. MeMgBr
2. $MgBr_2$

Br, OH, NHBoc

65%

Righi, G.; Pietrantonio, S.; Bonini, C. *Tetrahedron*, ***2001***, *57*, 10039.

Et, O, O

NH_3 , 50°C

OH, NH_2, Et, O

87%

Taylor, S.K.; Ide, N.D.; Silver, M.E.; Stephan, M.L. *Synth. Commun.*, ***2001***, *31*, 2391.

O, H_2N, O, Ph

4% $K_2Os(OH)_2O_4$, NaOH , rt
t-OBuCl , PrOH , H_2O

i-Pr_2NEt

O, HN, O, Ph, OH

59%

Donohoe, T.J.; Johnson, P.D.; Helliwell, M.; Keenan, M. *Chem. Commun.*, ***2001***, 2078.

Ph, CO_2H

cat [$K_2OsO_2(OH)_4$]
chloramine-T•3 H_2O

aq $NaHCO_3$, rt

$NHSO_2R$, CO_2H, Ph, OH + OH, CO_2H, Ph, $NHSO_2R$

(1.6 : 1) 96%

Fokin, V.V.; Sharpless, K.B. *Angew. Chem. Int. Ed.*, ***2001***, *40*, 3455.

O, N, C_7H_{15}, O

1. (cyclohexanone)=O , 20°C , 10 min

2 eq SmI_2 , THF
2. H_3O^+

O, OH, HN, C_7H_{15}, O

69%

Farcas, S.; Namy, J.-L. *Tetrahedron Lett.*, ***2000***, *41*, 7299.

hν , naphthalene , 20°C

92% ee

Giese, B.; Barbosa, F.; Stähelin, C.; Sauer, S.; Wettstein, P.; Wyss, C. *Pure Appl. Chem.*, ***2000***, *72*, 1623.

SmI_2 , cat $Fe(dbm)_3$

THF

77%

Ha, D.-C.; Yun, C.-S.; Lee, Y. *J. Org. Chem.*, ***2000***, *65*, 621.

PhCHO , $TiCl_4$
TMEDA , DCM

0°C

(92 : 8) 62%

Crimmins, M.T.; Chaudhary, K. *Org. Lett.*, ***2000***, *2*, 775.

$C_3H_7CONHBr$, 1.3 LiOH
4.4% K_2OsO_4•2 H_2O , 4°C

4% $(DHQ)_2PHAL$
t-BuOH–H_2O

94% (95% ee)

Demko, Z.P.; Bartosch, M.; Sharpless, K.B. *Org. Lett.*, ***2000***, *2*, 2221.

$NaBH_4$, $NiCl_2$•6 H_2O , hexane

moist Al_2O_3

90%

Kim, S.H.; Han, E.-H. *Tetrahedron Lett.*, ***2000***, *41*, 6479.

Bn BocHN CHO → 1. 2 eq TFA, *t*-BuNC, 4 eq Py, DCM, 0°C 2. SiO_2 → Bn BocHN CONH*t*-Bu OH 78%

Semple, J.E.; Owens, T.D.; Nguyen, K.; Levy, O.E. *Org. Lett.*, ***2000***, *2*, 2769.

Ph O NHAc O N H CO_2Me → hv, PhH → HO Ph CO_2Me AcHN N H O 75% (dr, >97:3)

Griesbeck, A.G.; Heckroth, H.; Schmickler, H. *Tetrahedron Lett.*, ***1999***, *40*, 3137.

OH CO_2Me → 1.2 TsNNaCl, aq MeCN 1% $K_2OsO_2(OH)_4$ → MeO_2C OH HO NHTs + MeO_2C OH HO NHTs (88 : 12) 88% both rac

Pringle, W.; Sharpless, K.B. *Tetrahedron Lett.*, ***1999***, *40*, 5151.

O N H O Ph → 1. 1.5eq LDA, THF, -78°C 2. 1.5eq MoOPD, -78°C → HO O N H O Ph + HO O N H O Ph (90 : 10) 78%

Hara, O.; Takizwa, J.-i.; Yamatake, T.; Makino, K.; Hamada, Y. *Tetrahedron Lett.*, ***1999***, *40*, 7787.

N Bz OH → 1. TFAA, THF 2. heat, NEt_3 3. 2.5M NaOH → OH N Bz 63%

Cossy, J.; Dumas, C.; Pardo, D.G. *Eur. J. Org. Chem.*, ***1999***, 1693.

SECTION 326: ALCOHOL, THIOL - AMINE

N—Bu

1. LDA, THF
2. PhCHO, –78°C
3. $NaBH_4$, THF/MeOH

NHBu OH Ph + NHBu OH Ph

56% (*syn/anti* = 2.4)

Veenstra, S.J.; Kinderman, S.S. *Synlett*, ***2001***, 1109.

O N Ph Ph

1. $ZrCl_4$, PhH, reflux
2. 10% HCl

OH NH_2 70%

Kikugawa, Y.; Tsuji, C.; Miyazawa, E.; Sakamoto, T. *Tetrahedron Lett.*, ***2001***, *42*, 2337.

PhCHO

$PhNMe_2$, hv, 1 d

Me OH N Ph Ph 79%

Kim, S.S.; Mah, Y.J.; Kim, A.R. *Tetrahedron Lett.*, ***2001***, *42*, 8315.

CHO NBn_2

1. $BrZnCH_2CO_2t$-Bu THF, 0°C, 2 h
2. aq NH_4Cl

OH CO_2t-Bu NBn_2 + OH CO_2t-Bu NBn_2

(78 : 22) 69%

Andrés, J.M.; Pedrosa, R.; Pérez, A.; Pérez-Encabo, A. *Tetrahedron*, ***2001***, *57*, 8521.

O

$BnNH_2$, $LiClO_4$
MeCN, reflux

OH NHBn 95%

de Parrodi, C.A.; Vázquez, V.; Quintero, L.; Juaristi, E. *Synth. Commun.*, ***2001***, *31*, 3295.

CO_2Et CN

1. $MeCO_2Et$, $MgN(i\text{-}Pr)_2$ ether, 0°C
2. $MgN(i\text{-}Pr)_2$

OH CO_2Et N H 79%

Kobayashi, K.; Nakalshima, T.; Mano, M.; Morikawa, O.; Konishi, H. *Chem. Lett.*, ***2001***, 602.

α-$Zr(O_3PMe)_{1.2}(O_3PC_6H_4SO_3H)_{0.8}$

$PhNH_2$, neat, 40°C, 20 h

OH NHPh 92%

Curini, M.; Epifano, F.; Marcotullio, M.C.; Rosati, O. *Eur. J. Org. Chem.*, **2001**, 4149.

Me Br Br N Bn Ph O

MeLi, ether, –90°C

Me Br Ph N OH Bn 64%

Barid, M.S.; Huber, F.A.M.; Tverezovsky, V.V.; Bolesov, I.G. *Tetrahedron*, **2001**, *57*, 1593.

$PhNH_2$, $CeCl_3$•7 H_2O–NaI

DCM, rt

OH NHPh 75%

Reddy, L.R.; Reddy, M.A.; Bhanumathi, N.; Rao, K.R. *Synthesis*, **2001**, 831.

$PhNH_2$, H_2O

β-cyclodextrin

OH NHPh 82%

Reddy, L.R.; Reddy, M.A.; Bhanumathi, N.; Rao, K.R. *Synlett*, **2000**, 339.

O NMe_2

8 atm H_2, *t*-BuOK, *i*-PrOH, rt, 4 h

$RuCl_2$(diphosphine)1,2-diamine) complex

OH NMe_2

99% (92% ee)

Ohkuma, T.; Ishii, D.; Takeno, H.; Noyori, R. *J. Am. Chem. Soc.*, **2000**, *122*, 6510.

Ph Boc—N Ph

1. BuLi, (–)-sparteine, toluene, –78°C
2. Et_2AlCl
3. PhCHO

OH Ph Ph N Boc Ph

85% (90:10 *E*:*Z*, 97:3 er)

Whisler, M.C.; Vaillancourt, L.; Beak, P. *Org. Lett.*, **2000**, *2*, 2655.

1.1 eq $PhNH_2$, reflux , 4 h

hexafluoro-2-propanol

OH

NHPh

84%

Das, U.; Crousse, B.; Kesavan, V.; Bonnet-Delpon, D.; Bégue, J.P. *J. Org. Chem.*, ***2000***, *65*, 6749.

C_4F_9

NH_3 , H_2O , MeOH

1d , rt

OH OH H N C_4F_9 C_4F_9

57%

Charrada, B.; Hedhli, A.; Baklouti, A. *Tetrahedron Lett.*, ***2000***, *41*, 7347.

PhCH=N–Ph

benzophenone , $TiCl_4$, Sm

THF , 60°C , 3 h

Ph OH NHPh Ph Ph

82%

Ma, Y.; Zhang, Y. *Org. Prep. Proceed. Int.*, ***2000***, *32*, 567.

1. 2.2 eq SmI_2 , 1% NiI_2 , THF
2. *i*-PrN=CHPh
3. H_3O^+

OH Ph NH*i*-Pr

80%

Machrouhi, F.; Namy, J.-L. *Tetrahedron Lett.*, ***1999***, *40*, 1315.

NH_4OH , microwaves

8 min

NH_2 OH

77%

Lindström, U.M.; Olofsson, B.; Somfai, P. *Tetrahedron Lett.*, ***1999***, *40*, 9273.

S

O NH

80°C , 4 h

SH N O

90%

Dong, Q.; Fang, X.; Schroeder, J.D.; Garvey, D.S. *Synthesis*, ***1999***, 1106.

$PhNH_2$, $Sn(OTf)_2$, ether

OH NHPh

94%

Sekar, G.; Singh, V.K. *J. Org. Chem.*, ***1999***, *64*, 287.

OHC NOMe N CO_2Bn → Bu_3SnH, AIBN, PhH reflux → HO NHOMe N CO_2Bn 54%

Naito, T.; Nakagawa, K.; Nakamura, T.; Kasei, A.; Ninomiya, I.; Kiguchi, T. *J. Org. Chem.*, ***1999***, *64*, 2003.

SECTION 327: ALCOHOL, THIOL - ESTER

O O Et I O → MesLi, THF −78°C, 1 h → O O HO Et 70%

Kondo, Y.; Asai, M.; Miura, T.; Uchiyama, M.; Sakamoto, T. *Org. Lett.*, ***2001***, *3*, 13.

OH *p*-Tol → I(OH)OTs, $CHCl_3$, reflux → OH OTs *p*-Tol 67%

Abe, S.; Sakuratani, K.; Togo, H. *J. Org. Chem.*, ***2001***, *66*, 6174.

→ CO_2Me, O_2/N_2; cat NHPI, PhCN, 75°C; cat $Co(acac)_3$, 14 h → OH CO_2Me 66%

Hara, T.; Iwahama, T.; Sakaguchi, S.; Ishii, Y. *J. Org. Chem.*, ***2001***, *66*, 6425.

CHO OBn → 1. CH_2=$CHCO_2Me$, 7.5% Py-box 2.5% $[(cod)IrCl]_2$, Et_2SiH, 25°C; 2. H_3O^+ → OH CO_2Me OBn 50% (>95:5 *syn:anti*)

Zhao, C.-X.; Duffey, M.O.; Taylor, S.J.; Morken, J.P. *Org. Lett.*, ***2001***, *3*, 1829.

PhCHO —(LDA, EtOAc, THF)→ OH, Ph, CO_2Et >99%

Huerta, F.F.; Bäckvall, J.-E. *Org. Lett.*, ***2001***, *3*, 1209.

Cl_2Me

1. Cl_2MeSiH, 2.7 DuPhos
1.25% $[RhCl(CO_2)]_2$
2. PhCHO, –46°C
3. H_3O^+

OH, Ph, Cl_2Me

94% (>60:1 *syn:anti*)

Zhao, C.-X.; Bass, J.; Morken, J.P. *Org. Lett.*, ***2001***, *3*, 2839.

Br, OAc

1. In, THF, 0°C → rt
2. PhCHO, 0°C, 4 h

OH, Ph, OAc

86% (85:15 *syn:anti*)

Lombardo, M.; Girotti, R.; Morganti, S.; Trombini, C. *Org. Lett.*, ***2001***, *3*, 2981.

O, O, OMe

1. $EtZnCH_2I$
2. *t*-BuCHO

t-Bu, OH, O, CO_2Me + *t*-Bu, OH, O, CO_2Me

(20 : 1) 85%

Lai, S.; Zercher, C.K.; Jasinski, J.P.; Reid, S.N.; Staples, R.J. *Org. Lett.*, ***2001***, *3*, 4169.

SPh, O

PhCHO, $TiCl_4$•NBu_3

OH, O, Ph, SPh

99% (86:14 *syn:anti*)

Tanabe, Y.; Matusmoto, N.; Funakoshi, S.; Manta, N. *Synlett*, ***2001***, 1959.

OH, O

2 eq *i*-PrCHO, 10% Me_3Al
CH_2Cl_2, rt, 22 h

i-$PrCO_2$, O

80%

Simpura, I.; Nevalainen, V. *Tetrahedron Lett.*, ***2001***, *42*, 3905.

CO_2H (adamantane) + HO–CH(Me)CH$_2$CH$_2$CH$_2$–OH → (2% $HfCl_4(thf)_2$, toluene, 16 h)

R = adamantyl (95 : 4) 79%

Ishihara, K.; Nakayama, M.; Ohara, S.; Yamamoto, H. *Synlett*, ***2001***, 1117.

PhCHO → (OTMS / OMe silyl ketene acetal; chiral Zr catalyst, toluene, C_3H_7OH, 0°C) → OH, CO_2Me 94% (78% ee)

Kobayashi, S.; Ishitani, H.; Yamashita, Y.; Ueno, M.; Shimizu, H. *Tetrahedron*, ***2001***, *57*, 861.

PhCHO → (i-PrCHO, 2% $Y_5O(Oi\text{-}Pr)_{13}$; 13% salen ligand, DCM, MS 4Å) → O_2Ci-Pr, HO, Ph + HO, i-PrCO_2, Ph

(>15 : 1) 70%

Mascarenhas, C.M.; Miller, S.P.; White, P.S.; Morken, J.P. *Angew. Chem. Int. Ed.*, ***2001***, *40*, 601.

Me–(1,3-dioxolane) → (Me, CO_2Me acrylate, O_2, 0.1% $Co(OAc)_2$; 5% N-hydroxyphthalimide, 25°C, 15 h) → O, O, Me, Me, OH, CO_2Me

84%

Hirano, K.; Iwahama, T.; Sakaguchi, S.; Ishii, Y. *Chem. Communn.*, ***2000***, 2457.

(CH$_3$)$_2$CH–CHO → (1. Zn, $BrCH_2CO_2Et$, BF_2•OEt_2; 2. Bz_2O_2) → OH, CO_2Et 92%

Chattopadhyay, A.; Salaskar, A. *Synthesis*, ***2000***, 561.

CH_2Cl_2 , $Et_3N(C_6F_{13}CH_2Cl_4)_2SnO$; TsCl, 50 min , rt — 75 X 80%

Bucher, B.; Curran, D.P. *Tetrahedron Lett.*, **2000**, *41*, 9617.

Fe^{3+}–Montmorillonite , TsOH ; DCE , reflux — 76%

Choudary, B.M.; Chowdari, N.S.; Kantam, M.L. *Tetrahedron*, **2000**, *56*, 7291.

PhCHO → OMe / OTMS , H_2O , rt , 1 d → CO_2Me — 20%

Loh, T.-P.; Feng, L.-C.; Wei, L.-L. *Tetrahedron*, **2000**, *56*, 7309.

OTMS / OMe ; PhCHO , THF , rt , 3 h ; 20% PPh_3 → MeO, Ph — 70%

Matsukawa, S.; Okano, N.; Imamoto, T. *Tetrahedron Lett.*, **2000**, *41*, 103.

Me_3SiO / t-BuS ; $C_8H_{17}CHO$, 100 atm ; supercritical CHF_3 ; 0.2 [R-BINOL–Ti(Oi-Pr)$_2$Cl$_2$] → t-BuS, C_8H_{17}

46% (94:6 *S*:*R*)

Mikami, K.; Matsukawa, S.; Kayaki, Y.; Ikariya, T. *Tetrahedron Lett.*, **2000**, *41*, 1931.

PhCHO → CH_2=$CHCO_2Ph$, Et_2MeSiH ; 2.5% [RhCl(cod)]$_2$, 6.5% *R*-BINAP → Ph, CO_2Ph

72% (3.4:1 *syn:anti*)
87%ee 34% ee

Taylor, S.J.; Duffey, M.O.; Morken, J.P. *J. Am. Chem. Soc.*, **2000**, *122*, 4528.

cat Bu_2SnO , *p*-TsCl ; NEt_3 , 45 min → OTs, OH — 97%

Martinelli, M.J.; Vaidyanathan, R.; Khau, V.V. *Tetrahedron Lett.*, **2000**, *41*, 3773.

PhSCH_2CO_2t-Bu → 1. (Ph,Ph)-bis(SO_2Ar) borolane B–Br; 2. DIPEA, DCM; 3. PhCHO, –78°C → Ph–CH(OH)–CH(SPh)–CO_2t-Bu

Ar = 3,5-di-CF_3-C_6H_3

97% (98:2 *anti:syn*) 90% ee

Corey, E.J.; Choi, S. *Tetrahedron Lett.*, ***2000***, *41*, 2769.

PhCHO → (1-methoxy-1-trimethylsilyloxypropene, $OSiMe_3$, OMe), 0.5 PrOH; 10% BINOL-Zr catalyst, toluene, 0°C → Ph–CH(OH)–CH(CH_3)–CO_2Me

(8:92 *syn:anti*) 61% 93% ee

Ishitani, H.; Yamashita, Y.; Shimizu, H.; Kobayashi, S. *J. Am. Chem. Soc.*, ***2000***, *122*, 5403.

BrCH_2CO_2Et → PhCHO, Et_2Zn, THF; cat $RhCl(PPh_3)_3$, 0°C → Ph–CH(OH)–CH_2–CO_2Et

82%

Kanai, K.; Wakabayashi, H.; Honda, T. *Org. Lett.*, ***2000***, *2*, 2549.

CHO ... CO_2Et → 1. SmI_2, THF, MeOH, 0°C; 2. aq NaCl, citric acid → OH, CO_2Et (cyclobutanol)

65%

Johnston, D.; McCusker, C.F.; Muir, K.; Procter, D.J. *J. Chem. Soc., Perkin Trans. 1*, ***2000***, 681.

F_3COC, MeO_2C, N, Ph, CO_2Me → allyl bromide; In, DMF → HO, CO_2Me, F_3COC, N, CO_2Me, Ph + HO, CO_2Me, F_3COC, N, CO_2Me, Ph

(85 : 15) 70%

Loh, T.-P.; Huang, J.-M.; Xu, K.-C.; Goh, S.-H.; Vittal, J.J. *Tetrahedron Lett.*, ***2000***, *41*, 6511.

OSiMe$_3$ OMe

cat $Me\text{-}Re(O)_2(O_2)$, H_2O_2

Py, MeCN, AcOH

HO—CO_2Me 71%

Stanković, S.; Espenson, J.H. *J. Org. Chem.*, ***2000***, *65*, 5528.

O Ph CHO

10% $Cu(OTf)_2$, DCE, *i*-PrOH

O N N O Ph Ph

OH Ph O*i*-Pr O

57% (28% ee, *R*)

Russell, A.E.; Miller, S.P.; Morken, J.P. *J. Org. Chem.*, ***2000***, *65*, 8381.

O

1. Cl_2CHCO_2*i*-Pr, *i*-PrOK, *i*-PrOH
2. $NaBH_3CN$, *i*-PrOH, 65°C

CO_2*i*-Pr OH 69%

Grison, C.; Coutrot, F.; Comoy, C.; Lemilbeau, C.; Coutrot, P.
Tetrahedron Lett., ***2000***, *41*, 6571.

O

Me_3Si, EtO_2CCHO

$TiCl_4$, CH_2Cl_2

HO CO_2Et O 89%

Ghosh, A.K.; Kawahama, R. *Tetrahedron Lett.*, ***1999***, *40*, 1083.

Me CO_2Et Ph CO_2Et

LiAlH(O*t*-Bu)$_3$, THF

Me OH Ph CO_2Et

rt, 4 d 85%
reflux, 2 h 76%

Ayers, T.A. *Tetrahedron Lett.*, ***1999***, *40*, 5467.

O

5% $Co_2(CO)_8$, CO, 65°C

10% OH N, MeOH/THF

OH CO_2Me

95% (>99% ee)

Hinterding, K.; Jacobsen, E.N. *J. Org. Chem.*, ***1999***, *64*, 2164.

O ; CO_2Et — Bakers yeast, DMSO / aq EtOH, 2.5 h → OH ; CO_2Et

67% (98% ee)

Hayakawa, R.; Nozawa, K.; Kimura, K.; Shimizu, M. *Tetrahedron*, ***1999***, *55*, 7519.

O_2CPh ; O_2CPh — lipase PS-C, THF, pH 7 / rt, 7 h → OH ; O_2CPh 88%

Nair, R.V.; Shukla, M.R.; Patil, P.N.; Salunkhe, M.M. *Synlett*, ***1999***, *29*, 1671.

Also via: Section 313 (Alcohol - Carboxylic Acid).

SECTION 328: ALCOHOL, THIOL - ETHER, EPOXIDE, THIOETHER

HO ; OH — 15% $Pd(OCOCF_3)_2$ / 18% (M.S.S.)iPr-SPRIX / 4 eq *p*-benzoquinone / DCM, rt, 21 h → O ; OH

70% (70% ee)

Arai, M.A.; Kuraishi, M.; Arai, T.; Sasai, H. *J. Am. Chem. Soc.*, ***2001***, *123*, 2907.

Ph ; OC_8H_{17} ; OH — N-methylquinolinium BF_4^- (N+, Me) / NaOAc, DCE, *t*-BuPh, hv → HO ; O

74%

Kumar, V.S.; Floreancig, P.E. *J. Am. Chem. Soc.*, ***2001***, *123*, 3842.

OH ; Ph — PhCHO, 5h / DCM, heat / Montmorillonite KSF → OH ; Ph ; O ; Ph 90%

Yadav, J.S.; Reddy, B.V.S.; Kumar, G.M.; Murthy, Ch.V.S.R. *Tetraahedron Lett.*, ***2001***, *42*, 89.

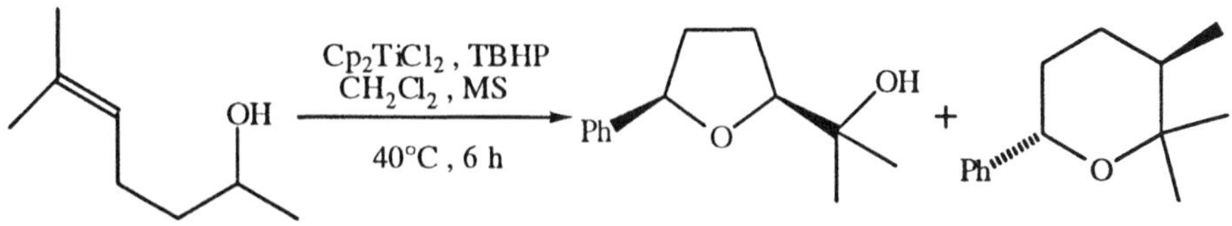

Cp_2TiCl_2 , TBHP
CH_2Cl_2 , MS
40°C , 6 h
OH
Ph
O
OH
+
Ph
O

(93 : 7) 66%

Lattanzi, A.; Sala, G.D.; Russo, M.; Screttri, A. *Synlett,* ***2001***, 1479.

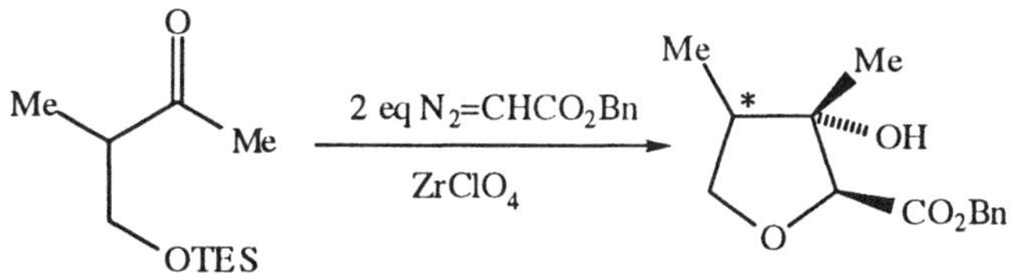

O
Me
Me
OTES
2 eq N_2=$CHCO_2Bn$
$ZrClO_4$
Me
Me
*
OH
CO_2Bn
O

62% (* 5:1 dr)

Angle, S.R.; Chann, K. *J. Org. Chem.,* ***2001***, *42*, 1819.

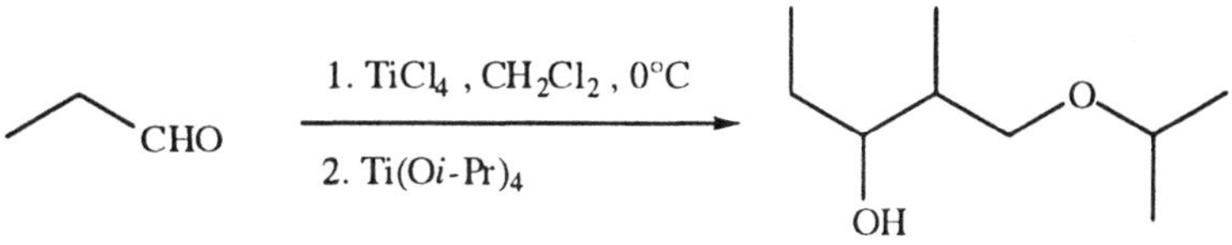

CHO
1. $TiCl_4$, CH_2Cl_2 , 0°C
2. Ti(O*i*-Pr)$_4$
O
OH

52% (56:44)

Aremo, N.; Hase, T. *Tetrahedron Lett.,* ***2001***, *42*, 3637.

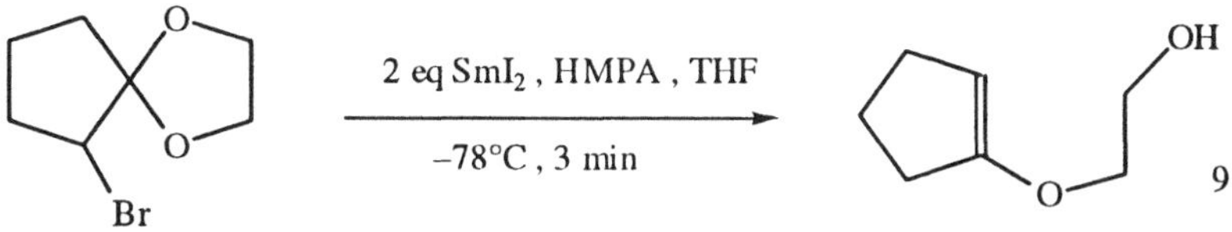

O
O
Br
2 eq SmI_2 , HMPA , THF
–78°C , 3 min
OH
O

91%

Park, H.S.; Kim, S.H.; Park, M.Y.; Kim, Y.H. *Tetrahedron Lett.,* ***2001***, *42*, 3729.

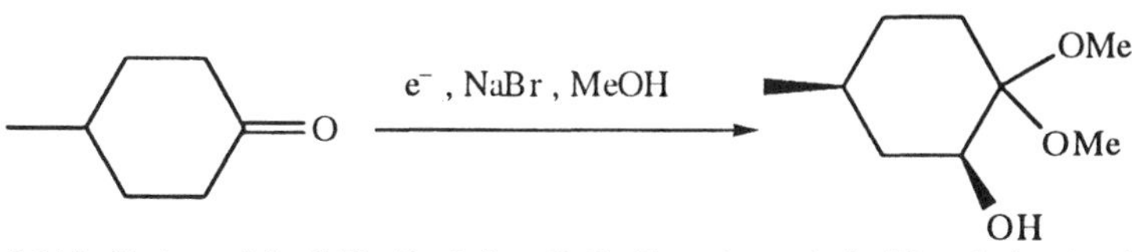

O
e^- , NaBr , MeOH
OMe
OMe
OH

73%

Elinson, M.N.; Feducovich, S.K.; Dmitriev, D.E.; Dorofeev, A.S.; Vereshchagin, A.N.; Nikishin, G.I. *Tetrahedron Lett.,* ***2001***, *42*, 5557.

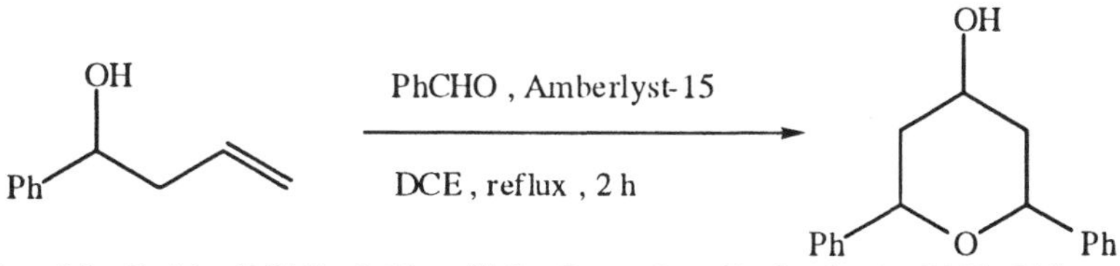

OH
Ph
PhCHO , Amberlyst-15
DCE , reflux , 2 h
OH
Ph
O
Ph

88%

Yadav, J.S.; Reddy, B.V.S.; Sekhar, K.C.; Gunasekar, D. *Synthesis,* ***2001***, 885.

Bu; HO; HO; Ph; $PhSeCl$, K_2CO_3, 1 d; CH_2Cl_2, −78°C → 25°C; OH; Ph; O; Bu; SePh; 60%; +; OH; Ph; O; Bu; SePh; 5%

Grutladauria, M.; Aprile, C.; Riela, S.; Noto, R. *Tetrahedron Lett.*, ***2001***, *42*, 2213.

O; O; 10% Cp_2TiCl_2, THF; collidine•HCl, Zn; 65 h; HO; O; H; 68% (58:42 dr)

Gansäuer, A.; Pierobon, M.; Bluhm, H. *Synthesis*, ***2001***, 2500.

Ph; Ph; Si; Bu; HO; OH; *t*-BuOOH, KH, DMF; Bu; OH; OH; OH; OH; 78%

Denmark, S.E.; Ghosh, S.K. *Angew. Chem. Int. Ed.*, ***2001***, *40*, 4759.

O; $BiCl_3$, MeOH, rt; 0°C, 30 min; OMe; OH; 97%

Mohammadpoor-Baltork, I.; Tangestaninejad, S.; Aliyan, H.; Mirkhani, V. *Synth. Commun.*, ***2000***, *30*, 2365.

Ph; O; $Fe(ClO_4)_3$, MeOH, rt, 1 h; Ph; OH; OMe; 87%

Salehi, P.; Seddighi, B.; Irandoost, M.; Behbahani, I.K. *Synth. Commun.*, ***2000***, *30*, 2967.

O; 5% $[Rh(CO)_2Cl]_2$, 10 eq MeOH; THF, rt; OH; OMe; 94%

Fagnou, K.; Lautens, M. *Org. Lett.*, ***2000***, *2*, 2319.

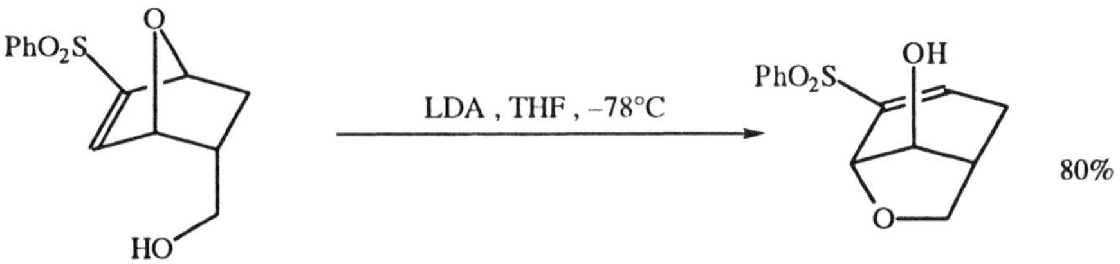

PhO_2S ... HO — LDA, THF, –78°C → PhO_2S ... OH ... O — 80%

Aceña, J.L.; Arjona, O.; Mañas, R.; Plumet, J. *Tetrahedron Lett.*, ***2000***, ***41***, 2549.

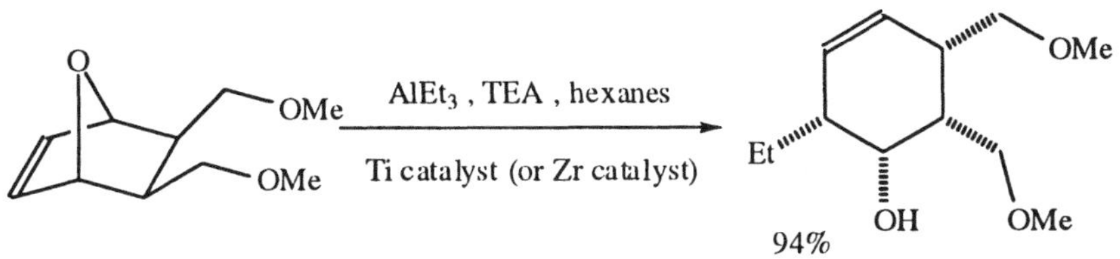

OMe, OMe — $AlEt_3$, TEA, hexanes / Ti catalyst (or Zr catalyst) → Et ... OH ... OMe ... OMe — 94%

Millward, D.B.; Sammis, G.; Waymouth, R.M. *J. Org. Chem.*, ***2000***, ***65***, 3902.

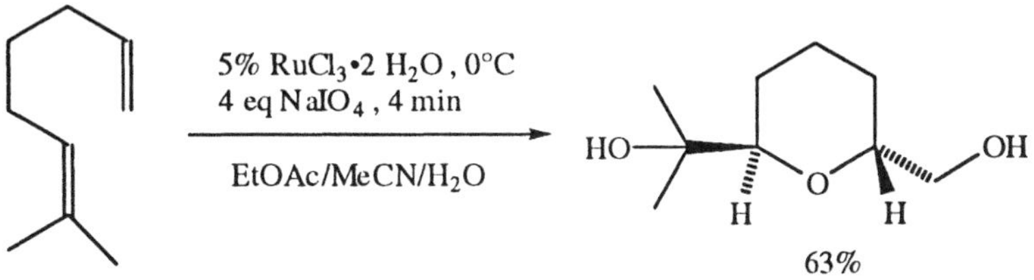

5% $RuCl_3$•2 H_2O, 0°C
4 eq $NaIO_4$, 4 min
$EtOAc/MeCN/H_2O$ → HO ... H O H ... OH — 63%

Niimi, T.; Uchida, T.; Irie, R.; Katsuki, T. *Tetrahedron Lett.*, ***2000***, ***41***, 3647.

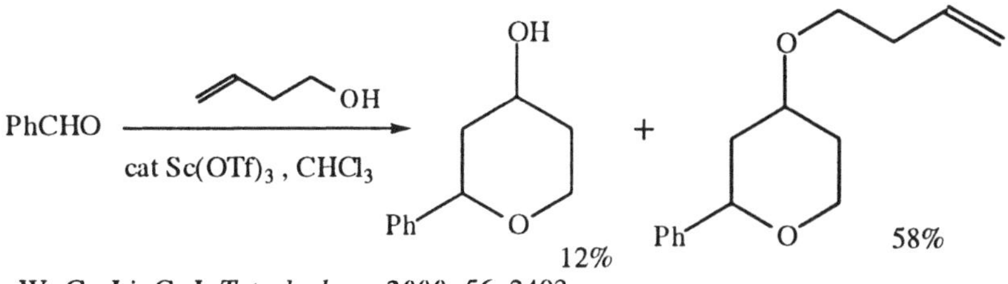

PhCHO — (3-buten-1-ol) OH / cat $Sc(OTf)_3$, $CHCl_3$ → OH, Ph O — 12% + O, Ph O — 58%

Zhang, W.-C.; Li, C.-J. *Tetrahedron*, ***2000***, ***56***, 2403.

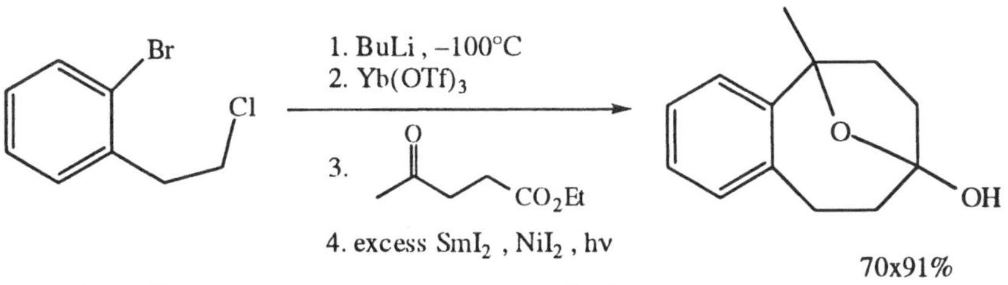

Br, Cl — 1. BuLi, –100°C 2. $Yb(OTf)_3$ 3. (O) CO_2Et 4. excess SmI_2, NiI_2, hν → O, OH — 70x91%

Molander, G.A.; Köllner, C. *J. Org. Chem.*, ***2000***, ***65***, 8333.

1% $VO(Oi\text{-}Pr)_3$, t-BuOOH , 0°C

1.5% chiral hydroxamic acid , 6 h

96% (95% ee)

Hoshino, Y.; Yamamoto, H. *J. Am. Chem. Soc.*, ***2000***, ***12***, 10452.

e^- , NaI–NaOH , MeOH

82%

Elinson, M.N.; Feducovich, S.K.; Dorofeev, A.S.; Vereshchagin, A.N.; Nikishin, G.I. *Tetrahedron,* ***2000***, ***56***, 9999.

1. $(Bu_3Sn)_2O$ toluene

2. $Zn(OTf)_2$

85%

Matsumura, R.; Suzuki, T.; Sato, K.; Inotsume, T.; Hagiwara, H.; Hoshi, T.; Kamat, V.P.; Ando, M. *Tetrahedron Lett.*, ***2000***, ***41***, 7697. Also see Matsumura, R.; Suzuki, T.; Sato, K.; Oku, K.-i.; Hagiwara, H.; Hoshi, T.; Ando, M.; Kamat, V.P. *Tetrahedron Lett.*, ***2000***, ***41***, 770.

0.2 $BF_3 \cdot OEt_2$

1.5 eq $BnO_2CCH=N_2$

THF , -78°C

65%

Liu, Q.; Burton, D.J. *Tetrahedron Lett.*, ***2000***, ***41***, 8045.

10-15 eq Oxone , 70°C

aq acetone , Na_2CO_3

18h

75%

Antonioletti, R.; Righi, G.; Oliveri, L.; Bovicelli, P. *Tetrahedron Lett.*, ***2000***, ***41***, 10127.

Et_3SiO ... CHO; allyl-$SiMe_3$, $BF_3 \cdot OEt_2$; CH_2Cl_2, -45°C; 2,6-di-*t*-butyl-4-methylpyridine; OH, O, $SiMe_3$; 56%

<u>Angle, S.R.</u>; El-Said, N.A. *J. Am. Chem. Soc.*, ***1999***, *121*, 10211.

O, Cl; 1. 2 eq SmI_2, 1% NiI_2, THF, 0°C; I, Cl; 2. hν, rt; OH, O; 71%

<u>Molander, G.A.</u>; Machrouchi, F. *J. Org. Chem.*, ***1999***, *64*, 4119.

O; 1. O, *t*-Bu, Ph, Ph, $O)_3Al$; toluene; 2. LDA, THF; 3. H^+; OH, O, O; 98%

Saito, S.; Yamazaki, S.; Shiozawa, M.; <u>Yamamoto, H.</u> *Synlett*, ***1999***, 581.

O, O, O, O; $BF_3 \cdot OEt_2$, THF; OH, O, O, O; 85%

Fan, J.-F.; <u>Wu, Y.</u>; Wu, Y.-L. *J. Chem. Soc., Perkin Trans. 1*, ***1999***, 1189.

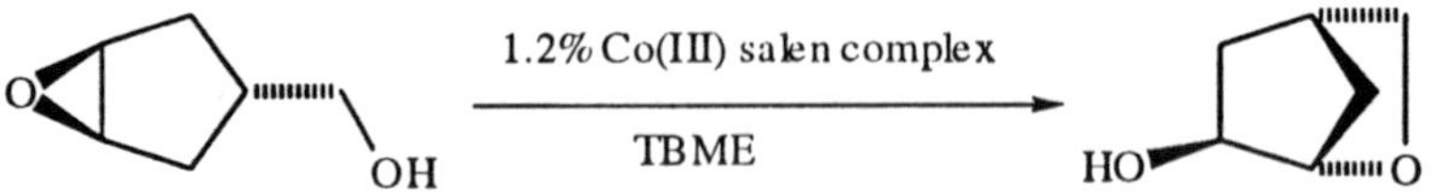

1.2% Co(III) salen complex

TBME

96% (98% ee)

Wu, M.H.; Hansen, K.B.; Jacobsen, E.N. *Angew. Chem. Int. Ed.*, ***1999***, *38*, 2012.

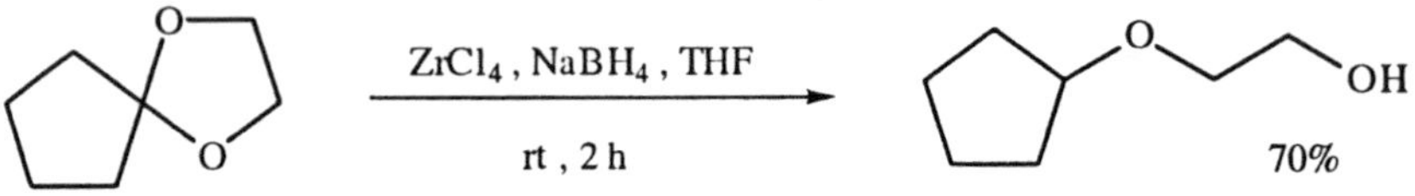

$ZrCl_4$, $NaBH_4$, THF

rt, 2 h

70%

Chary, K.P.; Laxmi, Y.R.S.; Iyengar, D.S. *Synth. Commun.*, ***1999***, *29*, 1257.

REVIEWS:

"Free Radical Chemistry of Three-Membered Heterocycles," Li, J.J. *Tetrahedron*, ***2001***, *57*, 1.

SECTION 329: ALCOHOL, THIOL - HALIDE, SULFONATE

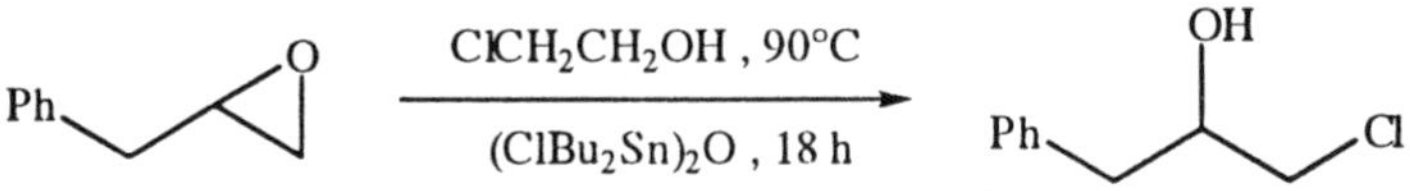

$ClCH_2CH_2OH$, 90°C

$(ClBu_2Sn)_2O$, 18 h

93%

Salomon, C.J. *Synlett*, ***2001***, 65.

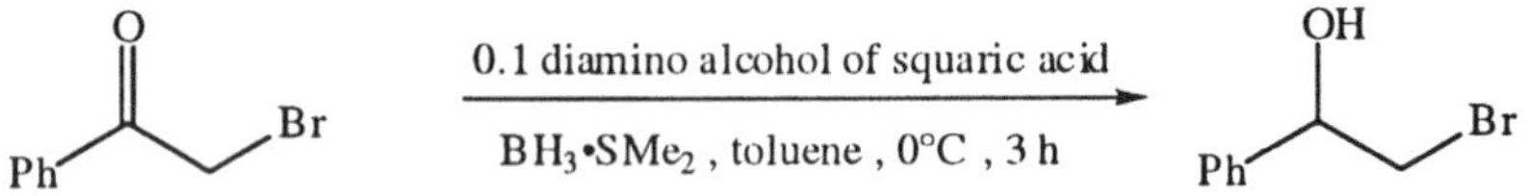

0.1 diamino alcohol of squaric acid

$BH_3 \cdot SMe_2$, toluene, 0°C, 3 h

85-99% (99% ee, *S*)

Zhou, H.; Lü, S.; Xie, R.; Chan, A.S.C.; Yang, T.-K. *Tetrahedron Lett.*, ***2001***, *42*, 1107.

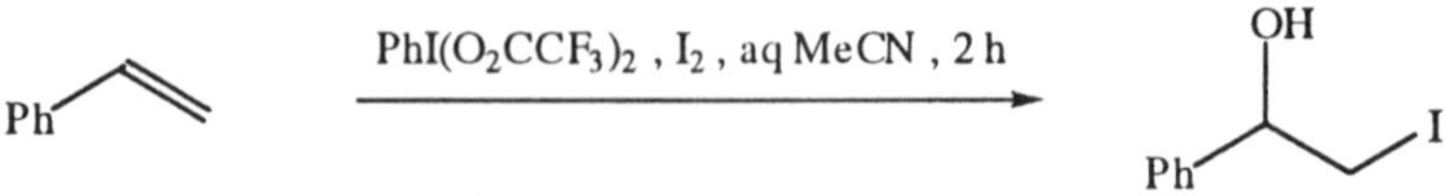

$PhI(O_2CCF_3)_2$, I_2, aq MeCN, 2 h

95%

DeCorso, A.R.; Panunzi, B.; Tingoli, M. *Tetrahedron Lett.*, ***2001***, *42*, 7245.

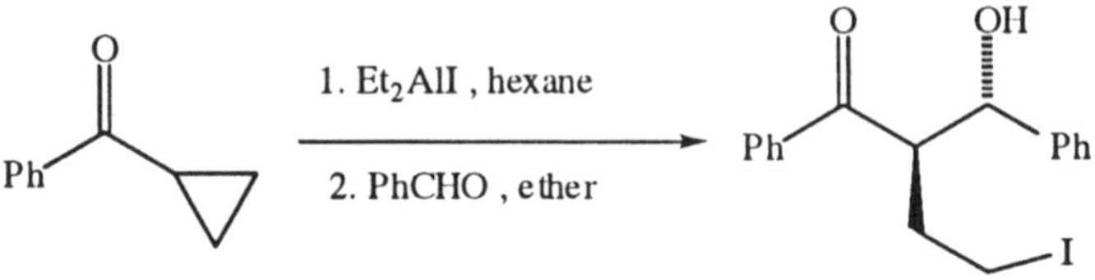

1. Et_2AlI, hexane

2. PhCHO, ether

89% (>99:1 *anti:syn*)

Han, Z.; Uehira, S.; Tsuritani, T.; Shinokubo, H.; Oshima, K. *Tetrahedron*, ***2001***, *57*, 987.

I_2, DCM, rt; cat Py-containing macrocycle — 80%

Sharghi, H.; Niknam, K.; Pooyan, M. *Tetrahedron, **2001**, 57*, 6057.

TiCl$_4$, acetone, 1 h; DCM, 10°C — 71%

Shi, M.; Jiang, J.-K.; Chui, S.-C. *Tetrahedron, **2001**, 57*, 7343.

BH_3•SMe_2, toluene, 110°C; 30% chiral phosphoramide, 90 min — 94% (82% ee, *S*)

Basavaiah, D.; Reddy, G.J.; Chandrashekar, V. *Tetrahedron Asymm.*, ***2001**, 12*, 685.

1. SiCl$_4$, DCM, −78°C, 4 h; 10% chiral diazaphosphonamide; 2. KF, KH_2PO_4 — 77% (>99% ee)

Reymond, S.; Legrand, O.; Brunel, J.M.; Buono, G. *Eur. J. Org. Chem.*, ***2001***, 2819.

F_3C–CO–CO_2Et, 10% chiral Cu complex; ether, 0°C → rt — 80% (83% ee)

Zhuang, W.; Gathergood, N.; Hazell, R.G.; Jørgensen, K.A. *J. Org. Chem.*, ***2001**, 66*, 1009.

1. TiCl$_4$–Bu$_4$NBr, DCM; 2. *i*-PrCHO; 3. H_3O^+ — 60% (>99:1 *syn:anti*)

Han, Z.; Uehira, S.; Shinokubo, H.; Oshima, K. *J. Org. Chem.*, ***2001**, 66*, 7854.

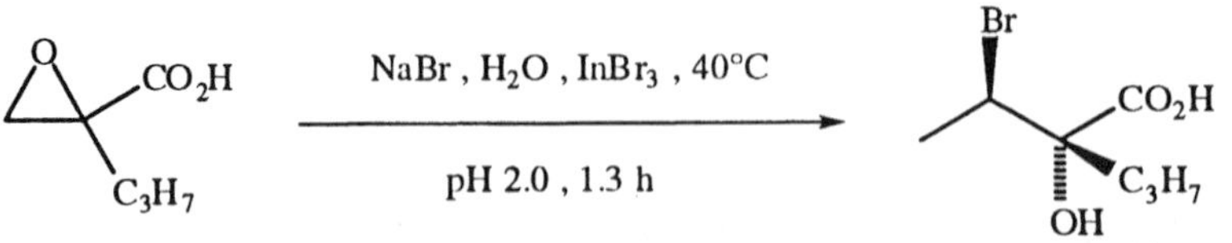
CO2H
C3H7
NaBr , H2O , InBr3 , 40°C
pH 2.0 , 1.3 h
Br
CO2H
C3H7
OH

78%

Amantini, D.; Fringuelli, F.; Pizzo, F.; Vaccaro, L. *J. Org. Chem.*, **2001**, *66*, 4463.

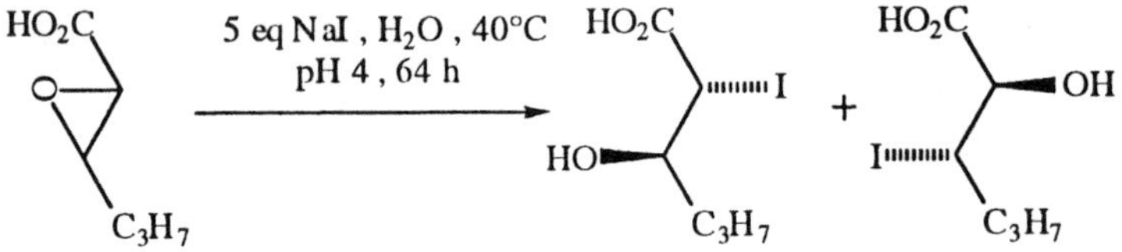
HO2C
C3H7
5 eq NaI , H2O , 40°C
pH 4 , 64 h
HO2C
I
HO
C3H7
+
HO2C
OH
I
C3H7

no catalyst	(99 : 1)	99%
+ $InCl_3$, pH 1.5 30 min	(2 : 98)	>99%

Fringuelli, F.; Pizzo, F.; Vaccaro, L. *J. Org. Chem.*, **2001**, *66*, 4719.

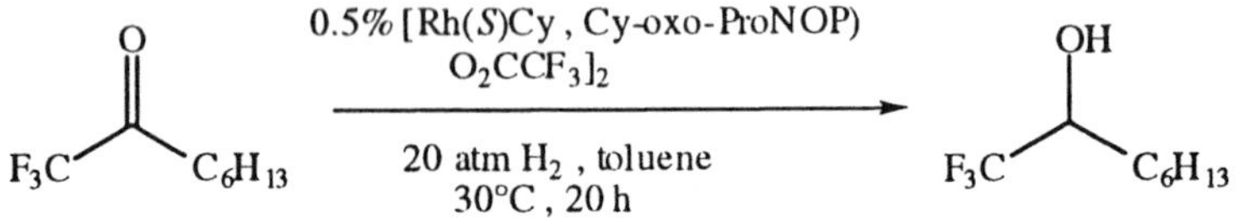
O
F3C
C6H13
0.5% [Rh((S)Cy , Cy-oxo-ProNOP)
O2CCF3]2
20 atm H2 , toluene
30°C , 20 h
OH
F3C
C6H13

98% (97% ee , *R*)

Juroki, Y.; Sakamaki, Y.; Iseki, K. *Org. Lett.*, **2001**, *3*, 457.

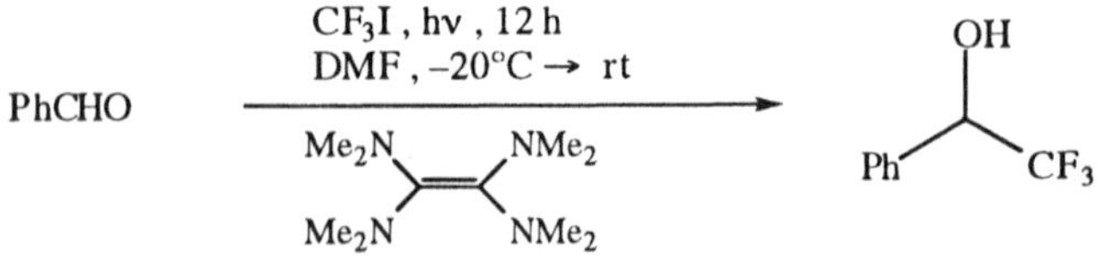
PhCHO
CF3I , hv , 12 h
DMF , –20°C → rt
Me2N
NMe2
Me2N
NMe2
OH
Ph
CF3

78%

Wright, D.L.; Usher, L.C.; Estrella-Jimenez, M. *Org. Lett.*, **2001**, *3*, 4275.

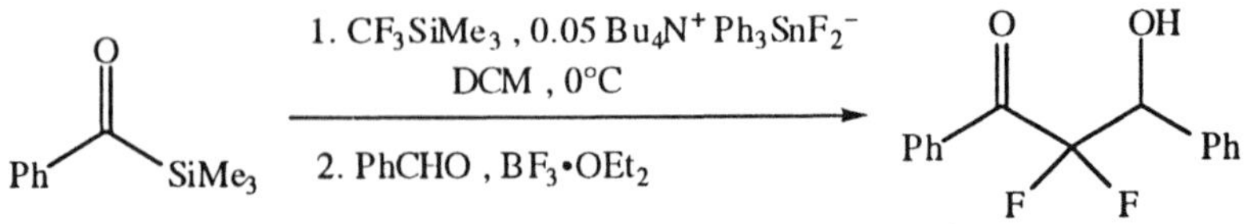
O
Ph
SiMe3
1. CF3SiMe3 , 0.05 Bu4N+ Ph3SnF2−
DCM , 0°C
2. PhCHO , BF3•OEt2
O
OH
Ph
Ph
F
F

78%

Lefebvre, O.; Brigaud, T.; Portella, C. *J. Org. Chem.*, **2001**, *66*, 1941.

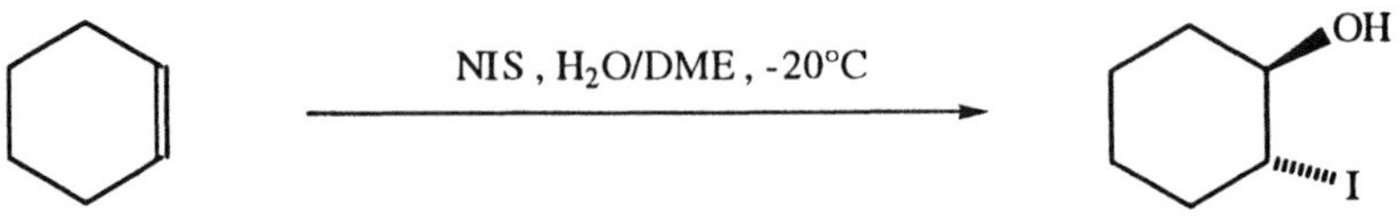
NIS , H2O/DME , -20°C
OH
I

95%

Smietana, M.; Gouverneur, V.; Mioskowski, C. *Tetrahedron Lett.*, **2000**, *41*, 193.

1. 10% $(SnO)_n$, 2 eq BTSP , 2 d
2 eq TMSCl , CH_2Cl_2
2. HCl/MeOH
92%

BTSP = $Me_3SiOOSiMe_3$

Sakurada, I.; Yamasaki, S.; Göttlich, R.; Iida, T.; Kanai, M.; Shibasaki, M. *J. Am. Chem. Soc.*, ***2000***, *122*, 1245.

Ph_3PH^+ Br^- , DCM
–90°C , 43 min
76%

Afonso, C.A.M.; Vieira, N.M.L.; Motherwell, W.B. *Synlett*, ***2000***, 382.

1. CH_2I_2 , toluene , –70°C
MeLi , ether
2. aq NH_4Cl , –70°C → rt
55%

Bessieres, B.; Morin, C. *Synlett*, ***2000***, 1691.

1. TMSCl , DCM , 0°C , 2 d
10% chiral phosphonamide
2. KF , KH_2PO_4
57% (>98% ee)

Reymond, S.; Brunel, J.M.; Buono, G. *Tetrahedron Asymm.*, ***2000***, *11*, 4441.

LiBr, Amberlyst-15 , acetone
rt , 2 h
97%

Antonioletti, R.; Bovicelli, P.; Fazzolari, E.; Righi, G. *Tetrahedron Lett.*, ***2000***, *41*, 9315.

SiF_4 , TBAF , ether
(83 : 17) 78%

Shimizu, M.; Kanemoto, S.; Nakahara, Y. *Heterocycles*, ***2000***, *52*, 117.

I_2 , Mn (salen) catalyst
DCM , rt , 6 h
71%

Sharghi, H.; Naeimi, H. *Bull. Chem. Soc. Jpn.*, ***1999***, *72*, 1525.

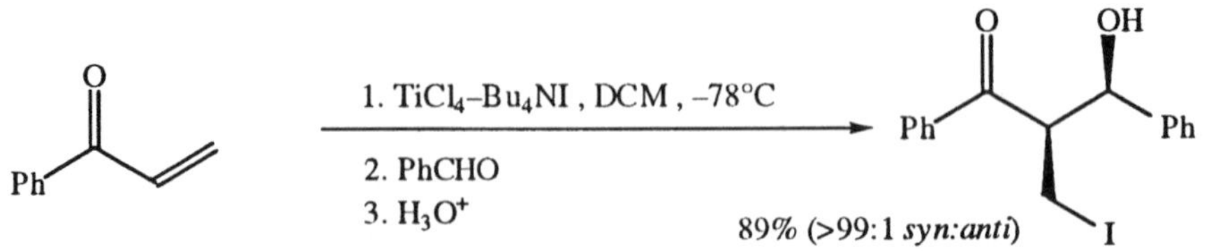

1. $TiCl_4$–Bu_4NI, DCM, –78°C
2. PhCHO
3. H_3O^+

89% (>99:1 *syn:anti*)

Uehira, S.; Han, Z.; Shinokubo, H.; Oshima, K. *Org. Lett.*, ***1999***, *1*, 1383.

SECTION 330: ALCOHOL, THIOL - KETONE

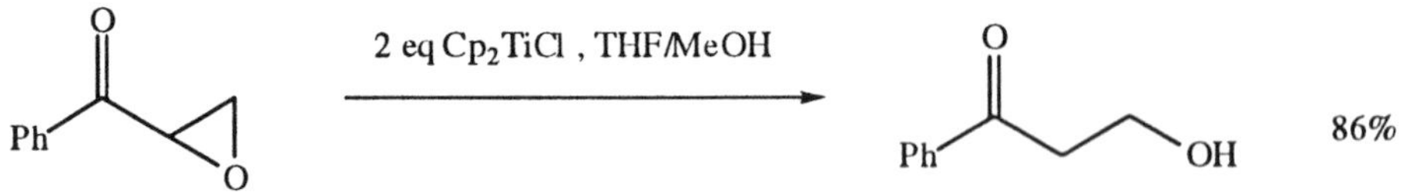

2 eq Cp_2TiCl, THF/MeOH

86%

Hardouin, C.; Chevallier, F.; Rousseau, B.; Doris, E. *J. Org. Chem.*, ***2001***, *66*, 1046.

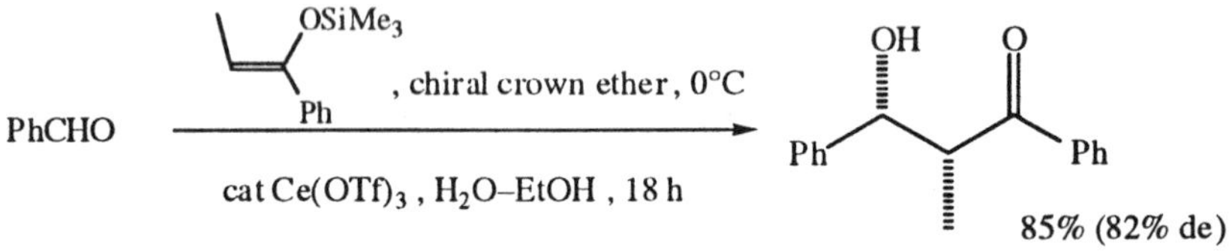

PhCHO

$OSiMe_3$, Ph, chiral crown ether, 0°C

cat $Ce(OTf)_3$, H_2O–EtOH, 18 h

85% (82% de)

Kobayashi, S.; Hamada, T.; Nagayama, S.; Manabe, K. *Org. Lett.*, ***2001***, *3*, 165.

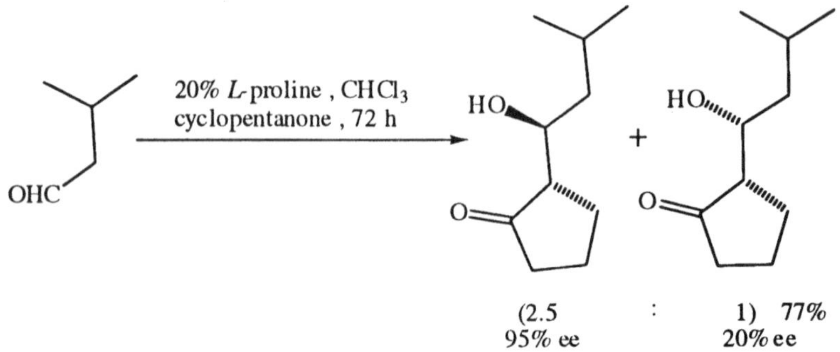

20% *L*-proline, $CHCl_3$
cyclopentanone, 72 h

(2.5 : 1) 77%
95% ee 20% ee

List, B.; Pojarliev, P.; Castello, C. *Org. Lett.*, ***2001***, *3*, 573.

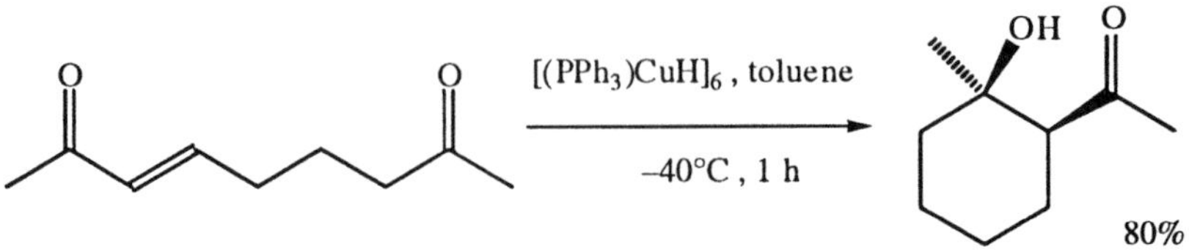

$[(PPh_3)CuH]_6$, toluene
–40°C, 1 h

80%

Chiu, P.; Zeto, C.-P.; Eng, Z.; Cheng, K.-F. *Org. Lett.*, ***2001***, *3*, 1901.

OTES

1. 0.1M SmI_2/THF, 0°C
2. $B(OH)_3$, H_2O

HO OTES O TMS 84%

TMS

Bode, J.W.; Carreira, E.M. *Org. Lett.*, ***2001**, 3*, 1587.

CHO acetone, MS 4Å, 5°C
bis-amine ligand, THF

OH O 62% (87% ee)

Trost, B.M.; Silcoff, E.R.; Ito, H. *Org. Lett.*, ***2001**, 3*, 2497.

O O Ph Ph

$NaBH_4$, EtOH, 100°C, 4 h
aq diamine, $PhB(OH)_2$

5% ketoiminato cobalt complex

OH O Ph Ph

93%
95% anti / 99% ee

Ohtsuka, Y.; Koyashu, K.; Ikeno, T.; Yamada, T. *Org. Lett.*, ***2001**, 3*, 2543.

CHO HO O Ph, THF
10% Zn complex
36 h, -40°C

OH O Ph OH + OH O Ph OH (5 :1) 92%

Yoshikawa, N.; Kumagai, N.; Matsunaga, S.; Moll, G.; Ohshima, T.; Suzuki, T.; Shibasaki, M. *J. Am. Chem. Soc.*, ***2001**, 123*, 2466.

HO O Ph

CHO MS 4Å, THF, 15% Ph_3PS
5% dinuclear Zn catalyst
-35°C, 11 d

OH O Ph OH

94% (90% ee)

Trost, B.M.; Ito, H.; Silcoff, E.R. *J. Am. Chem. Soc.*, ***2001**, 123*, 3367.

CHO

1. OMe $Cr(CO)_5$

100°C , dioxane
2. SiO_2

O H H H OH 66%

Ghorai, B.K.; Herndon, J.W.; Lam, Y.-F. *Org. Lett.*, ***2001***, *3*, 3535.

O Br N_2 OH → Br OH + O Br HO

0.25% $Rh_2(cap)_4$, PhH , reflux , 10 min 64% –
5% $Rh_2(OEt)_4$, PhH , reflux , 10 min – 62%

Hwu, J.R.; Tsay, S.-C.; Lin, L.C.; Chueh, L.L. *J. Am. Chem. Soc.*, ***2001***, *123*, 5104.

OAc $SiMe_3$

1. e^- , MeOH , rt
NaOMe
2. aq HCl

OH O $SiMe_3$ + HO OH O $SiMe_3$

u/ 1.5 eq NaOMe (6.7 : 1) 77%
u/ 2.5 eq NaOMe (1 : 7.3) 66%

Sugoh, K.; Kluniyasu, H.; Sugae, T.; Ohtaka, A.; Takai, Y.; Tanaka, A.; Machino, C.; Kambe, N.; Kurosawa, H. *J. Am. Chem. Soc.*, ***2001***, *123*, 5108.

O Ph CHO

5% $Co(dpm)_2$
$PhSiH_3$, DCE , rt

O OH Ph 70%

Baik, T.-G.; Luis, A.L.; Wang, L.-C.; Krische, M.J. *J. Am. Chem. Soc.*, ***2001***, *123*, 5112.

OTMS Ph

1. PhCHO , cat $HNTf_2$
ether , −78°C
2. 1M HCl , THF

OH O Ph Ph 92%

Ishihara, K.; Hiraiwa, Y.; Yamamoto, H. *Synlett*, ***2001***, 1851.

OH, Ph, Me; $PhCHO$, CH_2Cl_2, 5% $VO(OSiPh)_3$; O, OH, Ph, Ph, Me

86% (80:20 *syn:anti*)

Trost, B.M.; Jonasson, C.; Wuchrer, M. *J. Am. Chem. Soc.*, ***2001***, *123*, 12736.

$OSi(OMe)_3$; PhCHO, 10% chiral phosphine, AgF, BINAP derivative, MeOH; O, OH, Ph

87% (70:30 *syn:anti*), 90% ee

Yanagisawa, A.; Nakatsuka, Y.; Asakawa, K.; Kageyama, H.; Yamamoto, H. *Synlett*, ***2001***, 69.

C_5H_{11}, OH; Cl, CHO; 5% $Ru(PPh_3)_3HCl$, THF, reflux, 35 min; Cl, C_5H_{11}, OH, O

72% (56:44 *syn:anti*)

Uma, R.; Davies, M.; Crévisy, C.; Grée, R. *Tetrahedron Lett.*, ***2001***, *42*, 3069.

O, Ph; *t*-BuCHO, 3% chiral Ca complex, EtCN–THF, 20 h; O, OH, Ph, *t*-Bu

74% (81% ee, *R*)

Suzuki, T.; Yamagiwa, N.; Matsuo, Y.; Sakamoto, S.; Yamaguchi, K.; Shibasaki, M.; Noyori, R. *Tetrahedron Lett.*, ***2001***, *42*, 4669.

Ph, O, EtO_2C, Ph, CO_2Et; 2.2 eq SmI_2, THF, rt, 4 h; O, Ph, HO, CO_2Et, Ph

68%

Liu, Y.; Zhang, Y. *Tetrahedron Lett.*, ***2001***, *42*, 5745.

1. 3 eq SmI_2, THF HMPA, 16 h
2. H_2O^+

OH O H 77% + O H 4%

Kakiuchi, K.; Fujioka, Y.; Yamamura, H.; Tsutsumi, K.; Morimoto, T.; Kurosawa, H. *Tetrahedron Lett.*, ***2001***, *42*, 7595.

OTMS

O, 3 eq $TiCl_4$, DCM
−78°C → −50°C

O OH 46%

Lalić, G.; Petrovski, Ž.; Galonić, D.; Matović, R.; Saičić, R.N. *Tetrahedron*, ***2001***, *57*, 583.

O

MeO_2C–CHO, 5 d
$InCl_3$, untrasound

CO_2Me HO O + O HO CO_2Me

(74 : 26) 78%
66:34 *syn:anti*

Loh, T.-P.; Feng, L.-C.; Wei, L.-L. *Tetrahedron*, ***2001***, *57*, 4231.

O Ph I

PhCHO, TiI_4, EtCN
−95°C → −10°C, 5.5 h

O OH Ph Ph 76%

Shimizu, M.; Kobayashi, F.; Hayakawa, R. *Tetrahedron*, ***2001***, *57*, 9591.

O Ph Ph O

hν, DMPBI, AcOH, THF
DMPBI = 1,3-dimethylbenzimidazoline

O OH Ph Ph 96%

Hasegawa, E.; Chiba, N.; Nakajima, A.; Suzuki, K.; Yoneoka, A.; Iwaya, K. *Synthesis*, ***2001***, 1248.

HO ... O

PhCHO , 10% $Zr(Ot\text{-}Bu)_4$, THF

–20°C , 2 d

OH O Ph

80%

Schneider, C.; Hansch, M. *Chem. Commun.*, ***2001***, 1218.

Ph OTMS , PEG$(OMe)_2$, CO_2

PhCHO

50% $Sc(OTf)_3$, 50°C , 3 h

OH O Ph Ph

72%

Komoto, I.; Kobayashi, S. *Chem. Commun.*, ***2001***, 1842.

O O

PhCHO , $Ti(O\textit{i}\text{-}Pr)_4$-chiral Schiff base

DCE

OH O Ph $CO_2\textit{i}\text{-}Pr$

62% (56% ee)

Hayashi, M.; Yoshimoto, K.; Hirata, N.; Tanaka, K.; Oguni, N.; Harada, K.; Matsushita, A.; Kawachi, Y.; Sasaki, H. *Isr. J. Chem.*, ***2001***, *41*, 241.

$OSi(OMe)_3$

PhCHO , cat *R*-BINAP–AgF
MeOH , –78°C , 1 h

O OH Ph

64% (71:29 *syn:anti*)
71% ee 40% ee

Yanagisawa, A.; Nakatsuka, Y.; Asakawa, K.; Wadamoto, M.; Kageyama, H.; Yamamoto, H. *Bull. Chem. Soc. Jpn.*, ***2001***, *74*, 1477.

O O N MeO Me

1. $TiCl_4$, *i*-Pr_2NEt , DCM , –10°C
2. *i*-PrCHO , DCM , –78°C

HO O O N MeO Me

70% (74:26 *syn,syn:syn,anti*)

Calter, M.A.; Guo, X.; Liao, W. *Org. Lett.*, ***2001***, *3*, 1499.

CHO, NO$_2$; , $TiCl_4$, DCM ; PBu_3 , –78°C , 12 h ; O_2N ; OH ; O ; Cl ; 80%

Shi, M.; Jiang, J.-K.; Cui, S.-C.; Feng, Y.-S. *J. Chem. Soc., Perkin Trans. 1*, ***2001***, 390.

O ; CO_2Et ; 10% Cu-bis-oxazolidine ligand ; ether , 10% Me_2NPh , rt ; EtO_2C ; CO_2Et ; O ; OH ; 79% ee

Juhl, K.; Gathergood, N.; Jørgensen, K.A. *Chem. Commun.*, ***2000***, 2211.

O_2 ; S ; O ; CHO ; PhCl , reflux 1d ; *t*-BuOO*t*-Bu ; OH ; O ; 80%

Motherwell, W.B.; Vázquez, S. *Tetrahedron Lett.*, ***2000***, ***41***, 9667.

O ; 3% Mn(dpm)$_3$; 2% $PhSiH_3$; *i*-PrOH , O_2 ; OH ; O ; 78%

Magnus, P.; Payne, A.H.; Waring, M.J.; Scott, D.A.; Lynch, V. *Tetrahedron Lett.*, ***2000***, ***41***, 9725.

OHC ; NO_2 ; acetone , 30% *S*-proline ; DMSO ; HO ; NO_2 ; O ; 68% (76% ee)

List, B.; Lerner, R.A.; Barbas III, C.F. *J. Am. Chem. Soc.*, ***2000***, ***122***, 2395.

OTMS

1. (epoxide), 3 eq $TiCl_4$, CH_2Cl_2, –50°C

2. H_2O

O

OH

46%

Lalić, G.; Petrovski, Z.; Galonić, D.; Matović, R.; Saičić, R.N. *Tetrahedron Lett.*, ***2000***, *41*, 763.

PhCHO

MeCHO, BFD, ThDP

pH 7 buffer

O

Ph

OH

BFD = benzoylformate decarboxylase
ThDP = thiamin diphosphate

99% (92% ee)

Dünnwald, T.; Demir, A.S.; Siegert, P.; Pohl, M.; Müller, M. *Eur. J. Org. Chem.*, ***2000***, 2161.

Br

N

Ts

$PhB(OH)_2$, Na_2CO_3, 80°C

cat $Pd(PPh_3)_4$, THF–H_2O

Ph

N

Ts

75%

Baldwin, S.W.; Chen, P.; Nikolic, N.; Weinseimer, D.C. *Org. Lett.*, ***2000***, *2*, 1193.

O

Ph CHO

Ni(R), THF, rt, 40 min

O

Ph OH

89%

Barrero, A.F.; Alvarez-Manzaneda, E.J.; Chahboun, R.; Meneses, R. *Synlett*, ***2000***, 197.

O

1. PhCHO, BINOL–$TiO\textit{t}\text{-}Bu_2$

2. *R*-mandelic acid

O HO Ph + O HO Ph

(91 : 5) 71%

Mahrwald, R. *Org. Lett.*, ***2000***, *2*, 4011.

1. 3% $[CuH(PPh_3)]_6$, PMHS , rt
toluene , 39 h

2. $C_5H_{11}CHO$, $TiCl_4$

OH

C_5H_{11}

72%

Lipshutz, B.H.; Chirsman, W.; Noson, K.; Papa, P.; Sclafani, J.A.; Vivian, R.W.; Keith, J.M. *Tetrahedron*, **2000**, *56*, 2779.

Ph

SmI_2 , PhCHO , THF , –45°C

Ph

OH

HO Ph

70%

Mukaiyama, T.; Arai, H.; Shiina, I. *Chem Lett.*, **2000**, 580.

Ph

chiral Ru catalyst , 10°C , 1 d

HCOOH , NEt_3

Ph OH + Ph OH

(89 : 11) 99%

99% ee 12% ee

Koike, T.; Murata, K.; Ikariya, T. *Org. Lett.*, **2000**, *2*, 3833.

Ph Ph

1.2 eq SmI_2 , THF , rt

Ph Ph OH Ph Ph

82%

Zhou, L.; Zhang, Y. *Synth. Commun.*, **2000**, *30*, 597.

HO

cat $ZnBr_2$, DCM , rt

8 h

OH

91%

Tu, Y.Q.; Fan, C.A.; Ren, S.K.; Chan, A.S.C. *J. Chem. Soc., Perkin Trans. 1*, **2000**, 3791.

$OSiCl_3$ —— 1. PhCHO , 10% chiral phosphoramide, DCM , –78°C; 2. aq $NaHCO_3$ ——→ (product, Ph)

92% (12.5/1 *S/R*)

Denmark, S.E.; Stavenger, R.A. *J. Am. Chem. Soc.*, ***2000***, *122*, 8837.

PhCHO —— OTMS (Ph) , 20% $Pb(OTf)_2$, 1 d; 24% crown ether , aq EtOH , 0°C ——→ (product, Ph, Ph)

62% (90:10 *syn:anti*)
55% ee

Nagayama, S.; Kobayashi, S. *J. Am. Chem. Soc.*, ***2000***, *122*, 11531.

PhCHO —— , $TiCl_4$, NEt_3 , 2 d; DCM , –20°C ——→ (product, Ph, Cl) 80%

Shi, M.; Jiang, J.-K.; Feng, Y.-S. *Org. Lett.*, ***2000***, *2*, 2397.

HO (ketone) —— PhCHO , chiral Al catalyst; DCM , rt , 43 h ——→ (product, Ph) 62%

Simpura, I.; Nevalainen, V. *Angew. Chem. Int. Ed.*, ***2000***, *39*, 3422.

C_3H_7CHO —— acetophenone , 5% chiral ligand, 10% Et_2Zn , 15% $Ph_3P{=}S$; THF , MS 4Å ——→ (product, C_3H_7, Ph)

33% (56% ee)

Trost, B.M.; Ito, H. *J. Am. Chem. Soc.*, ***2000***, *122*, 12003.

(enone) —— PhCHO , 2 eq Bu_3SnH; 0.1 CuCl , ether , rt ——→ (product, Ph) 72%

Ooi, T.; Doda, K.; Sakai, D.; Maruoka, K. *Tetrahedron Lett.*, ***1999***, *40*, 2133.

BnO, O, BnO → 1. LDA, THF, –78°C; 2. EtCHO → Et, HO, OBn, O, BnO + Et, HO, OBn, O, BnO

(58 : 42) 80%

Kim, K.S.; Hong, S.D. *Tetrahedron Lett.*, ***2000***, *41*, 5909.

OTMS, Ph → PhCHO, 10% $Sc(O_3SC_{12}H_{25})_3$; 10% HCl, H_2O, 27°C, 3 h → OH, O, Ph, Ph 82%

35% yield without HCl

Manabe, K.; Kobayashi, S. *Tetrahedron Lett.*, ***1999***, *40*, 3773.

OTMS → *i*-PrCHO, $TiCl_4$–NBu_3; 0.05 TMSCl, CH_2Cl_2 → O, OH 87%

Yoshida, Y.; Matsumoto, N.; Hamasaki, R.; Tanabe, Y. *Tetrahedron Lett.*, ***1999***, *40*, 4227.

Ph, CHO → O, Ph, 8% chiral Li catalyst; KHMDS, 16% H_2O, THF, 18 h → OH, O, Ph, Ph 83% (85% ee)

Yoshikawa, N.; Yamada, Y.M.A.; Das, J.; Sasai, H.; Shibasaki, M. *J. Am. Chem. Soc.*, ***1999***, *121*, 4168.

OTMS, *t*-Bu → 1. *R*-BINOL-SEM, $SnCl_4$ CH_2Cl_2, -125°C → -78°C; 2. HF-pyridine, THF → O, *t*-Bu, OH 87x95%

Ishihara, K.; Nakamura, H.; Yamamoto, H. *J. Am. Chem. Soc.*, ***1999***, *121*, 7720.

Ph OH / Ph OH → potassium monoperoxysulfate, K_2CO_3, pH 10.5, MeCN, 0°C → Ph O / Ph OH

89% (45% ee, *R*)

Adam, W.; Saha-Möller, C.R.; Zhao, C.-G. *J. Org. Chem.*, ***1999***, *64*, 7492.

TBDMSO → 1. *t*-BuLi, THF, –78°C; 2. cyclohexanone, THF –78°C → *t*-$BuMe_2Si$ O HO

72%

Stergiades, I.A.; Tius, M.A. *J. Org. Chem.*, ***1999***, *64*, 7547.

Ph O O → H_2O, THF → HO OH O Ph

61%

Howell, A.R.; Ndakala, A.J. *Org. Lett.*, ***1999***, *1*, 825.

O → H OH O O, 20% $InCl_3$, 2 d ultrasound → O OH CO_2H + OH O HO_2C

(84 : 16) 95%
56:44 *syn:anti*

Loh, T.-P.; Wei, L.-L.; Feng, L.-C. *Synlett*, ***1999***, 1059.

PhCHO → $OSiMe_3$ Ph, aq EtOH, –10°C; 20% O N N O *i*-Pr *i*-Pr, 20 h → OH O Ph

74% (3.2 : 1 *syn:anti*)
67% ee

Kobayashi, S.; Nagayama, S.; Busujima, T. *Chem. Lett.*, ***1999***, 71.

PhCHO → (benzoyl formate decarboxylase / aq buffer) → Ph–CO–CH(OH)–Ph 70% (>99% ee)

Demir, A.S.; Dünnwald, T.; Iding, H.; Pohl, M.; Müller, M.
Tetrahedron Asymm., ***1999***, *10*, 4769.

PhCHO → (TMSO(Ph)C=CHCH$_3$, $Cu(OTf)_2$, EtOH / 20% chiral bis oxazoline, 20 h) → PhCH(OH)CH(CH$_3$)COPh

74% (3.2 : 1 *syn:anti*)
67% ee

Kobayashi, S.; Nagayama, S.; Busujima, T. *Tetrahedron,* ***1999***, *55*, 8739.

PhCHO → (CH_2=C(OTMS)Ph, H_2O / activated Montmorillonite K-10 clay) → PhCH(OH)CH$_2$COPh 87%

Loh, T.-P.; Li, X.-R. *Tetrahedron,* ***1999***, *55*, 10789.

2-cyclohexen-1-one → (C_8H_{17}C(O)ZrCpCl / 5% $Pd(OAc)_2$, 10% R-MOP) → 1-(1-hydroxycyclohex-2-enyl)-C(O)C_8H_{17}

88% (66% ee)

Hanzawa, Y.; Tabuchi, N.; Saito, K.; Noguchi, S.; Taguchi, T.
Angew. Chem. Int. Ed., ***1999***, *38*, 2395.

PhCHO → (1-OTMS-cyclopentene, 5M $LiClO_4$ / $MeNO_2$, 4 h) → 2-(hydroxy(phenyl)methyl)cyclopentanone

85%

Sudha, R.; Sankararaman, S. *J. Chem. Soc., Perkin Trans. 1*, ***1999***, 383.

CH_2=C(OAc)CH$_3$ → (PhCHO, Et_2AlOEt, THF, 0°C) → CH$_3$COCH$_2$CH(OH)Ph 74%

Mukaiyama, T.; Shibata, J.; Shimamura, T.; Shiina, I. *Chem. Commun.*, ***1999***, 951.

PhCHO , $(Bu_3Sn)_2$

Bu_2SnCl_2 , aq THF

81% (75:25 *syn:anti*)

Shibata, I.; Kawasaki, M.; Yasuda, M.; Baba, A. *Chem. Lett.*, ***1999***, 689.

Ph CHO , DCM , –23°C

$TiCl_2$•Cu , *t*-BuCN , 3 h

94% (91:9 *syn:anti*)

Mukaiyama, T.; Kagayama, A.; Igarashi, K.; Shiina, I. *Chem. Lett.*, ***1999***, 1157.

$AlCl_3$, PhCl , 106°C

microwaves , 3 min

73% + 23%

Khadilkar, B.M.; Madyar, V.R. *Synth. Commun.*, ***1999***, *29*, 1195.

REVIEWS:

"Diastereoselection in Lewis Acid Mediated Aldol Additions," Mahrwald, R.
Chem. Rev., ***1999***, *95*, 1095

"The Vinylogous Aldol Reaction: A Valuable, Yet Understated Carbon-Carbon Bond Forming Maneuver," Casiraghi, G.; Zanardi, F.; Appendino, G.; Rassu, G.
Chem. Rev., ***2000***, *100*, 1929.

SECTION 331: ALCOHOL, THIOL - NITRILE

PhCHO

3.5 eq HCN , rt , pH 4.5
PVAL-entrapped *R*-oxynitrilane

biphasic , MTBE-hexane

85% (95% ee)

Gröger, H.; Capan, E.; Barthuber, A.; Vorlop, K.-D. *Org. Lett.*, ***2001***, *3*, 1969.

3% $Mn(dpm)_3$, i-PrOH , 1 h

O_2 , $PhSiH_3$

CN OH

54%

Magnus, P.; Scott, D.A.; Fielding, M.R. *Tetrahedron Lett.*, ***2001***, *42*, 4127.

CHO

1. [$InBr/Cl_3CCN$]

2. H_3O^+

OH CN Cl Cl 70%

Nóbrega, J.A.; Gonçalves, S.M.C.; Peppe, C. *Tetrahedron Lett.*, ***2001***, *42*, 4745.

Ph CHO

1. TMSCN , DCM , 36 h , –40°C
Bu_3PO , chiral Al catalyst

2. 2N HCl

OH Ph CN

97% (97% ee)

Hamashima, Y.; Sawada, D.; Nogami, H.; Hanai, M.; Shibasaki, M.
Tetrahedron, ***2001***, *57*, 801.

PhCHO

1. TMSCN , 20% $Ti(Oi\text{-}Pr)_4$, DCM
chiral Schiff base , –40°C

2. H^+

OH Ph Et 98% (70% ee)

Yang, Z.-H.; Wang, L.-X.; Zhou, Z.-H.; Zhou, Q.-L.; Tang, C.-C.
Tetrahedron Asymm., ***2001***, *12*, 1579.

O CN Ph

1. $CeCl_3$, THF

2. $LiBH_4$

OH CN Ph + OH CN Ph

(15 : 85) 75%

Dalpozzo, R.; Bartoli, G.; Bosco, M.; DeNino, A.; Procopio, A.; Sambri, L.; Tagarelli, A.
Eur. J. Org. Chem., ***2001***, 2971.

HO CHO

HCN , pH 5.0 , 1,C , 12 h
R-hydroxynitrile lyase

HO OH CN

72% (86% ee)

Gerrits, P.J.; Marcus, J.; Birikaki, L.; van der Gen, A. *Tetrahedron Asymm.*, ***2001***, *12*, 971.

PhCHO → (1% Li-*R*-(+)-BINOL, TMSCN, 5 min) → Ph–CH(OTMS)–CN 96% (56% ee, S)

Holmes, I.P.; Kagan, H.B. *Tetrahedron Lett.*, ***2000***, *41*, 7453.

C_5H_{11}CH(OBn)CHO → (TMSCN, 5 eq $MgBr_2$•OEt_2; 0°C, 1 h) → HO/CN product + HO/CN diastereomer (OBn, C_5H_{11})

(9 : 1) 95%

Ward, D.E.; Hrapchak, M.J.; Sales, M. *Org. Lett.*, ***2000***, *2*, 57.

3-pentanone (=O) → ($BrCH_2CN$, 2 eq SmI_2; THF) → OH, CN product 87%

Caracoti, A.; Flowers II, R.A. *Tetrahedron Lett.*, ***2000***, *41*, 3039.

PhCHO → (Me_3SiCN, $Ti(Oi\text{-}Pr)_4$, 2 d; cat β-amino-alcohol sulfonato derivatives; DCM, MS 4Å) → Ph–CH(OH)–CN 67%

You, J.-S.; Gau, H.-M.; Choi, M.C.K. *Chem. Commun.*, ***2000***, 1963.

PhCHO → (1. TMSCN, $YCl_3(R\text{-pybox})_2$; 2. 1 M HCl) → Ph–CH(OH)–CN 77% (80% ee)

pybox = bis-(oxazolidinyl)pyridine

Aspinall, H.C.; Greeves, N.; Smith, P.M. *Tetrahedron Lett.*, ***1999***, *40*, 1763.

cyclohexanone (O) → (10% $P(RNCH_2CH_2)_3N$, MeCN; 2.2 eq $MgSO_4$, 25°C, 4 h) → 1-(cyanomethyl)cyclohexanol (OH, CN) 94%

Kisanga, P.; McLeod, D.; D'Sa, B.; Verkade, J. *J. Org. Chem.*, ***1999***, *64*, 3090.

Ph-epoxide (O) → (10% $Ce(OTf)_4$, NaCN; MeCN, 70°C, 1 d) → Ph–CH(OH)–CH$_2$–CN (OH) 79% + Ph–CH(CN)–CH$_2$–OH (CN) 12%

Iranpoor, N.; Shekarriz, M. *Synth. Commun.*, ***1999***, *29*, 2249.

SECTION 332: ALCOHOL, THIOL - ALKENE

Allylic and benzylic hydroxylation (C=C-C-H → C=C-C-OH, etc.) is listed in Section 41 (Alcohols from Hydrides).

CN; 2 eq LDA, THF, –78°C; 4 eq AcOH, 5 min; HO; CN; 83%

Fleming, F.F.; Wang, Q.; Steward, O.W. *J. Org. Chem.*, **2001**, *66*, 2171.

OH; C_5H_{11}; TsBr, AIBN, toluene; 90°C, 3 h; Ts; C_5H_{11}; Br; OH; 75%

Kang, S.-K.; Ko, B.-S.; Ha, Y.-H. *J. Org. Chem.*, **2001**, *66*, 3630.

OTMS; OEt; *i*-PrCHO, THF, rt, 1 d; chiral ammonium fluoride; OH; CO_2Et; 70% (20% ee)

Bluet, G.; Campaagne, J.-M. *J. Org. Chem.*, **2001**, *66*, 4293.

OHC; $(CH_2)_6$; O; O; 1. $(hexyl)_2BH$ 2. Et_2Zn 3. 2% DAIB 4. aq NH_4Cl; HO; $(CH_2)_6$; O; O; 65% (90% ee)

Oppolzer, W.; Radinov, R.N.; El-Sayed, E. *J. Org. Chem.*, **2001**, *66*, 4766.

C_6H_{13}; Br; O; $CrCl_2$, $NiCl_2$, DMF, rt; C_6H_{13}; OH; 70%

Trost, B.M.; Pinkerton, A.B. *J. Org. Chem.*, **2001**, *66*, 7714.

PhCHO → (20% [C_3F_7CO-oxazaborolidinone, Bn], EtCN, –78°C; TMSO-enol ether–I, syringe pump) → Ph-CH(OH)-C(=CH-I)-C(O)CH$_3$

1:2 (*E*:*Z*) 71%

Li, G.; Wei, H.-X.; Phelps, B.S.; Purkiss, D.W.; Kim, S.H. *Org. Lett.*, ***2001***, *3*, 823.

O_2N-C$_6$H$_4$-CHO → (CH_2=CHCN, $TiCl_4$, Bu_4NBr, DCM, 20°C) → O_2N-C$_6$H$_4$-CH(OH)-C(=CH$_2$)-CN

55%

Shi, M.; Feng, Y.-S. *J. Org. Chem.*, ***2001***, *66*, 406.

$(i\text{-Pr})_2NOCO_2$-CH$_2$-C≡C-Me_3Si → (1. BuLi, (–)-sparteine, –78°C, pentane; 2. TiCl(O*i*-Pr)$_3$, –78°C; 3. *i*-PrCHO, –78°C) → $O_2CON(i\text{-Pr})_2$, H, Me_3Si, OH, *i*-Pr

76% (>95% ee)

Schultz-Fademrecht, C.; Wibbeling, B.; Frölich, R.; Hoppe, D. *Org. Lett.*, ***2001***, *3*, 1221.

Methylenecyclohexane → (H-CO-CO_2Et, DCM, –5°C; 2% Pt-complex, C_6F_5OH) → cyclohexenyl-CH$_2$-CH(OH)-CO_2Et

77% conversion (77% ee)

Koh, J.H.; Larsen, A.O.; Gagné, M.R. *Org. Lett.*, ***2001***, *3*, 1233.

Me$_2$Si oxasilacycle (Ph) → (1. Ac-C$_6$H$_4$-I, 2 eq TBAF; 2. 5% Pd(dba)$_2$, rt) → Ph-CH(HO)-CH$_2$-CH=CH-C$_6$H$_4$-Ac

90%

Denmark, S.E.; Yang, S.-M. *Org. Lett.*, ***2001***, *3*, 1749.

Ph, CHO, O; Ph (alkene), , 5% Co complex; $CHCl_3$, −20°C, 2 d; Ph, OH, O, Ph

90% (88% ee)

Kezuka, S.; Ikeno, T.; Yamada, T. *Org. Lett.*, ***2001***, *3*, 1937.

O; excess $Me_2S{=}CH_2$; THF, −10°C → rt; HO

91%

Alcaraz, L.; Cridland, A.; Kinchin, E. *Org. Lett.*, ***2001***, *3*, 4051.

C_6H_{13}; 1. $HB(c\text{-hex})_2$ 2. Et_2Zn 3. chiral N,O-ligand PhCHO; OH, Ph, C_6H_{13}

71% (86% ee, *S*)

Dahmen, S.; Bräse, S. *Org. Lett.*, ***2001***, *3*, 4119.

Br, Ph, OH; 5% $(Ph_3Si)_3V$, PhCHO; DCE, 80°C; Ph, OH, Br, O, Ph

94% (9:1 *Z*:*E*)

Trost, B.M.; Oi, S. *J. Am. Chem. Soc.*, ***2001***, *123*, 1230.

$PhMe_2Si$; $PhCH(OMe)_2$, TMSOTf 2,6-di-*t*-Bu-pyrroline; CH_2Cl_2, −78°C, 1 h; OH, Ph

88%

Braddock, D.C.; Badine, D.M.; Gottschalk, T. *Synlett*, ***2001***, 1909.

O, Bu, OBn; 1. LIDAKOR, THF, −78°C 2. 4 eq BuLi, THF, 0°C → rt; Ph, Bu, OH, OH

70x55%

LIDAKOR, THF, −78°C → 5°C

55%

Thurner, A.; Faigl, E.; Tõke, L.; Mordini, A.; Valacchi, M.; Reginato, G.; Czira, G. *Tetrahdron*, ***2001***, *57*, 8173.

O O

1. (η^2-propene)Ti(O*i*-Pr)$_2$

2. PhCHO

OH Ph O HO

95% (75 : 25 *E*:*Z*)

Okamoto, S.; Teng, X.; Fujii, S.; Takayama, Y.; Sato, F.
J. Am. Chem. Soc., ***2001***, *123*, 3462.

sugar–O O

NO_2 CHO

0.2 DABCO, THF, 2 d

sugar–O O OH NO_2

91% (40% de, *R*)

Krishna, P.R.; Kannan, V.; Ilangovan, A.; Sharma, G.V.M.
Tetrahedron Asymm., ***2001***, *12*, 829.

O

$BiPh_3$, cat $PdCl_2$, rt

aq DMF

HO Ph 74%

Kang, S.-K.; Ryu, H.-C.; Hong, Y.-T.; Kim, M.-S.; Lee, S.-W.; Jung, J.-H.
Synth. Commun., ***2001***, *31*, 2365.

O

1. 2-(lithiomethyl)-1-methylimidazole
+ chiral Li-amide, THF

2. H_3O^+

OH

96% (93% ee)

Lill, S.O.N.; Pettersen, D.; Amedjkouh, M.; Ahlberg, P.
J. Chem. Soc., Perkin Trans. 1, ***2001***, 3054.

OCO_2Me Ph CCl_3

$C_8H_{17}CHO$, $CrCl_2$, DMF

THF, −10°C, 1 d

OH Ph C_8H_{17} Cl

84%

Takai, K.; Kokumai, R.; Nobunaka, T. *Chem. Commun.*, ***2001***, 1128.

CH_3CHO → 1. Br–CH=CH–CO_2Me , In , aq THF , HCl; 2. DBU , PhH

OH

CO_2Me

64x88% (93:7 *E:Z*)

Cha, J.W.; Pae, A.N.; Choi, K.I.I.; Cho, Y.S.; Koh, H.Y.; Lee, E.
J. Chem. Soc., Perkin Trans. 1, ***2001***, 2079.

O

$(CH_2)_4C\equiv CCO_2Et$ → TBAF

CO_2Et

OH

87% (1.5/1)

Wendling, F.; Miesch, M. *Org. Lett.*, ***2001***, *3*, 2689.

CHO → Bn, Me_3Si, C= ; In , DMF , rt

OH C

$SiMe_3$

89%

Lee, P.H.; Bang, K.; Lee, K.; Lee, C.-H.; Chang, S. *Tetrahedron Lett.*, ***2000***, *41*, 7521.

O → Mn , 10% Cp_2TiCl_2 ; 2,4,6-collidine•HCl, THF

HO

75%

Gansäuer, A.; Pierobon, M. *Synlett*, ***2000***, 1357.

O → CHO ; $TiCl_4$, CH_2Cl_2 , 0°C → rt

OH O

68%

Li, G.; Wei, H.-X.; Gao, J.J.; Caputo, T.D. *Tetrahedron Lett.*, ***2000***, *41*, 1.

Ag:KI , AcOH
reflux , 30 min

97%

Sharghi, H.; Aghapour, G. *J. Org. Chem.*, ***2000***, *65*, 2813.

10% PhCHO , 10% $Sn(OTf)_2$

CH_2Cl_2 , 0°C , 2 h

78% (49/1, *E/Z*)

Sumida, S.-i.; Ohga, M.; Mitani, J.; Nokami, J. *J. Am. Chem. Soc.*, ***2000***, *122*, 1310.

CF_3CO_2H , 70°C

20 h

51%

Kim, H.S.; Kim, T.Y.; Lee, K.Y.; Chung, Y.M.; Lee, H.J.; Kim, J.N. *Tetrahedron Lett.*, ***2000***, *41*, 2613.

1.5 eq Et_2Zn , DCM

5% Pd(dppf)Cl_2 , rt

92% (>98% de)

Lautens, M.; Renaud, J.-L.; Hiebert, S. *J. Am. Chem. Soc.*, ***2000***, *122*, 1804.

PhCHO , 5% $PdC_2(PPh_3)_2$

HCl–$SnCl_2$, DMF , rt

>99%

Chang, H.-M.; Cheng, C.-H. *Org. Lett.*, ***2000***, *2*, 3439.

PhCHO → (Br–CH_2–C≡C–CH_2–Br; In, H_2O) → Ph–CH(OH)–C(=CH_2)–CH=CH_2 53%

Lu, W.; Ma, J.; Wang, Y.; Chan, T.H. *Org. Lett.*, ***2000***, *2*, 3469.

1. allyl–BEt_2, ether
2. NaOH, H_2O_2

72%

Zaidlewicz, M.; Krezemimski, M.P. *Org. Lett.*, ***2000***, *2*, 3897.

EtO_2C–CHO, 10% Pd catalyst

DCE/toluene, 60°C, 4 h

OH, CO_2Et

97% (88% ee)

Hao, J.; Hatano, M.; Mikami, K. *Org. Lett.*, ***2000***, *2*, 4059.

Ph–C≡C–Me → (PhCHO, 10% $Ni(cod)_2$; 20% PCy_3, 2 eq BEt_3, toluene, 23°C) → Ph–CH=C(Me)–CH(OH)–Ph 76% (77:23)

Huang, W.-S.; Chan, J.; Jamison, T.F. *Org. Lett.*, ***2000***, *2*, 4221.

PhCHO → (isoprene, cat $Ni(acac)_2$; Me_3B, THF, 2 h) → Ph–CH(OH)–... OH

56% (20:1 *E:Z*)

Kimura, M.; Shibata, K.; Koudahashi, Y.; Tamaru, Y. *Tetrahedron Lett.*, ***2000***, *41*, 6789.

Ph–C(=O)–$SiMe_3$ → (CH_2=CHMgBr, $CeCl_3$, THF; –78°C, 2 h) → Ph–C(OH)($SiMe_3$)–CH=CH_2 60%

Bonini, B.F.; Comes-Franchini, M.; Fochi, M.; Mazzanti, G.; Ricci, A.; Varchi, G. *Synlett*, ***2000***, 1688.

1. *sec*-BuLi, TMEDA, ether, –78°C
2. $Ti(O\textit{i}\text{-}Pr)_4$, –78°C
3. C_3H_7CHO, ether, –78°C → 0°C

71%

Markó, I.E.; Leroy, B. *Tetrahedron Lett.*, ***2000***, *41*, 7225.

1. η^2-C_3H_6–$Ti(O\textit{i}\text{-}Pr)_2$, ether
2. EtCHO, ether/THF
3. H^+

R* = camphor-based auxiliary

95% (97:3 *S*:*R*)

Suzuki, D.; Urabe, H.; Sato, F. *Angew. Chem. Int. Ed.*, ***2000***, *39*, 3290.

sec-BuLi/(–)-sparteine, ether, –90°C, 6 h

85% (62% ee)

Alexakis, A.; Vrancken, E.; Mangeney, P. *J. Chem. Soc., Perkin Trans. 1*, ***2000***, 3354.

Ph–CH2CH2–CHO, SmI_2, THF, –78°C

70%

Youn, S.W.; Park, H.S.; Kim, Y.H. *Chem. Commun.*, ***2000***, 2005.

2 eq Me_2Zn, DCE, L*
5% $Pd(MeCN)_2Cl_2$, heat

L* = *S*-*t*-BuSDIPOF

87% (91% ee)

Lautens, M.; Hiebert, S.; Renaud, J.-L. *Org. Lett.*, ***2000***, *2*, 1971.

PhI → 1. DMF, 10% $Pd(OAc)_2$, tris(2-furyl)phosphine, Ph-allene; 2. In, MeO-C6H4-CHO → 64%

Anwar, U.; Grigg, R.; Rasparini, M.; Savic, V.; Sridharan, V. *Chem. Commun.*, ***2000***, 645.

1.2 (chiral amine), 2 eq LDA, 5 eq DBU, THF, rt, 1 d → OH 78% (95% ee)

Södergren, M.J.; Bertilsson, S.K.; Andersson, P.G. *J. Am. Chem. Soc.*, ***2000***, ***122***, 6610.

CO_2Et → PhCHO, 70% $LiClO_4$; 15% DABCO, ether, 0°C, 20 h → Ph, OH, CO_2Et 81%

Kawamura, M.; Kobayashi, S. *Tetrahedron Lett.*, ***1999***, ***40***, 1539.

$C_7H_{15}CHO$ → 1. $MeCCl_3$, $CrCl_2(Mn)$, THF, rt; 2. LDA → OH, C_7H_{15} 85x96%

Falck, J.R.; Barma, D.K.; Mioskowski, C.; Schlama, T. *Tetrahedron Lett.*, ***1999***, ***40***, 2091.

TBSO, TBSO, O → 2 eq [MeHN, N, Ph, Me /BuLi] → TBSO, TBSO, OH 94% (94% ee)

de Sousa, S.E.; O'Brien, P.; Steffens, H.C. *Tetrahedron Lett.*, ***1999***, ***40***, 8423.

OH → 1. Bu-CH=CH-$B(OH)_2$, THF, MS 4Å, rt; 2. PhMe, 5% BHT, 190°C; 3. Me_3NO, PhH, 80°C → Bu, OH, HO + Bu, OH, HO (25 : 75) 80%

Batey, R.A.; Thadani, A.N.; Lough, A.J. *J. Am. Chem. Soc.*, ***1999***, ***121***, 450.

O
$C_3H_7C{\equiv}CC_3H_7$, 100°C
5% $Pd(OAc)_2$, 2 eq *t*-BuOAc
DMF , 20 h
Br
OH
C_3H_7
C_3H_7
82%

Quan, L.G.; Gevorgyan, V.; Yamamoto, Y. *J. Am. Chem. Soc.*, ***1999***, *121*, 3545.

Ph CHO
$SnBu_3$, 19 h
$BF_3{\bullet}OEt_2$, toluene
OH
Ph
92%

Luo, M.; Iwabuchi, Y.; Hatakeyama, S. *Chem. Commun.*, ***1999***, 267.

PhCHO
$SnBu_3$, $PhCF_3$
S-BINOL-Ti(IV) , *i*-$PrSBEt_2$, –20°C
OH
Ph
81% (98% ee)

Yu, C.-M.; Lee, S.-J.; Jeon, M. *J. Chem. Soc., Perkin Trans. 1*, ***1999***, 3557.

OH
1% Ru(cod)(cot) , PEt_3
hexane , 20°C , 45 h
OH
94% (97% Z)

Sato, T.; Komine, N.; Hirano, M.; Komiya, S. *Chem. Lett.*, ***1999***, 441.

CO_2Me
1. Me_2CuLi/MeMgBr–CuBr–DMS
2. PhCHO
OH
Ph CO_2Me
69% (95/1 *Z/E*)

Wei, H.-X.; Willis, S.; Li, G. *Synth. Commun.*, ***1999***, *29*, 2959.

Also via: Section 302 (Alkyne - Alcohol).

SECTION 333: ALDEHYDE - ALDEHYDE

NO ADDITIONAL EXAMPLES

SECTION 334: ALDEHYDE - AMIDE

NHBt OMe

1. OsO_4/NMO
2. $NaIO_4$ on SiO_2
3. TsOH, toluene

MeO CHO N Bt 73%

Banfield, S.C.; England, D.B.; Kerr, M.A. *Org. Lett.*, ***2001***, *3*, 3325.

O O NAc

$MeNO_2$, rt, 12 h
2 eq $AlCl_3$

CHO N—Ac 69%

Fuchs, J.R.; Funk, R.L. *Org. Lett.*, ***2001***, *3*, 3361.

O Ph N CO_2H

$PhI(OAc)_2$, 0.5 I_2
CH_2Cl_2, 4 h

O Ph NH CHO 66%

Boto, A.; Hernández, R.; Suárez, E. *Tetrahedron Lett.*, ***1999***, *40*, 5945.

SECTION 335: ALDEHYDE - AMINE

NHAc

$POCl_3$–DMF, CTAB
MeCN, reflux, 45 min

CHO N Cl 90%

Ali, M.M.; Tasneem; Rajanna, K.C.; Prakash, P.K.S. *Synlett*, ***2001***, 251.

Me–CH=CH–CHO, 4°C, 72 h; imidazolidinone·HCl, $MeNO_2$–H_2O

(88 : 12) 70%
93% ee

Jen, W.S.; Wiener, J.J.M.; MacMillan, D.W.C. *J. Am. Chem. Soc.*, ***2000***, *122*, 9874.

SECTION 336: ALDEHYDE - ESTER

CO/H_2, supercritical CO_2
cat $P(p\text{-}C_6H_4\text{-}C_6F_{13})_3$
cat $Rh(acac)(CO)_2$, 80°C, 11 h

98% (48% conversion)

Hu, Y.; Chen, W.; Osuna, A.M.B.; Stuart, A.M.; Hope, E.G.; Xiao, J. *Chem. Commun.*, ***2001***, 725.

SECTION 337: ALDEHYDE - ETHER, EPOXIDE, THIOETHER

1. PhS(O)CH(Cl)Li
2. 5 PhSH, 7 *t*-BuOK
t-BuOH, THF, rt, 10 min

98x73%

Satoh, T.; Kubota, K.-i. *Tetrahedron Lett.*, ***2000***, *41*, 2121.

SECTION 338: ALDEHYDE - HALIDE, SULFONATE

1. IPy_2BF_4, DCM, hv, rt, 12 h
2. H_3O^+

91%

Barluenga, J.; González-Bobes, F.; Anathouju, S.R.; García-Martin, M.A.; González, J.M. *Angew. Chem. Int. Ed.*, ***2001***, *40*, 3389.

1% 2,6-lutidine•HCl, 35% HCl
DCM, 65°C

59%

Bellesia, F.; De Buyck, L.; Ghelfi, F.; Libertini, E.; Pagnoni, U.M.; Roncaglia, F. *Tetrahedron*, ***2000***, *56*, 7507.

SECTION 339: ALDEHYDE - KETONE

$C_8H_{17}CH_2CHO$, MeCN, reflux

10% Et_2NSiMe_3

C_8H_{17} CHO 66%

Hagiwara, H.; Komatsubara, N.; Ono, H.; Okabe, T.; Hoshi, T.; Suzuki, T.; Ando, M.; Kato, M. *J. Chem. Soc., Perkin Trans. 1*, ***2001***, 316.

1. PBu_3, THF, –40°C
2. SmI_2
3. 1M HCl

Ph Cl → Ph CHO 80%

Maeda, H.; Huang, Y.; Hino, N.; Yamauchi, Y.; Ohmori, H. *Chem. Commun.*, ***2000***, 2307.

1. LDA, ether, –78°C
2. $HCO_2CH_2CF_3$, –78°C
3. H^+, –78°C

O OH 75%

Zayla, G.H. *Org. Lett.*, ***1999***, *1*, 989.

SECTION 340: ALDEHYDE - NITRILE

NO ADDITIONAL EXAMPLES

REVIEWS:

"Cyanohydrins in Nature and the Laboratory: Biology, Preparations and Synthetic Applications," Gregory, R.J.H. *Chem. Rev.*, ***1999***, *99*, 3649.

SECTION 341: ALDEHYDE - ALKENE

For the oxidation of allylic alcohols to alkene aldehydes, also see Section 48 (Aldehydes from Alcohols).

HO (hexa-1,5-dien-3-ol) → CH_2=CHCHO , 4% Ru catalyst / DCM , 25°C , 12 h → OHC, HO, CHO 70%

Bouzbouz, S.; Cossy, J. *Org. Lett.*, ***2001***, *3*, 1451.

CHO (cyclohexanecarboxaldehyde) → Ph⁄⁄OH / 10% $Pd(OAc)_2$, 20% PPh_3 2.4 eq BEt_3 , THF , NEt_3 LiCl → OHC ... Ph 92% + HO, Ph 4%

Kimura, M.; Horino, Y.; Mukai, R.; Tanaka, S.; Tamaru, Y. *J. Am. Chem. Soc.*, ***2001***, *123*, 10401.

C_6H_{13}—≡ → Me_2PhSiH , 40 atm CO , 40°C , 1 d / 1% $[Rh^+(cod)(\eta^6\text{-}C_6H_5B^-Ph_3)]$ → OHC, $SiMe_2Ph$, C_6H_{13} 78%

Okazaki, H.; Kawanami, Y.; Yamamoto, K. *Chem. Lett.*, ***2001***, 650.

O (furyl) N–NHTs → 2.5 eq PhMgBr → Ph, CHO 75%

Chandrasekhar, S.; Reddy, M.V.; Reddy, K.S.; Ramarao, C. *Tetrahedron Lett.*, ***2000***, *41*, 2667.

O, Bu → DMF , 120°C , 1 h → Bu, CHO 63%

Parsons, P.J.; Thomson, P.; Taylor, A.; Sparks, T. *Org. Lett.*, ***2000***, *2*, 571.

cat $[Ir(cod)Cl]_2$, PCy_3

toluene, Cs_2CO_3

CHO

O

68%

Higashino, T.; Sakaguchi, S.; Ishii, Y. *Org. Lett.*, ***2000***, *2*, 4193.

PhCHO

CH_2=CHOAc, $Ba(OH)_2$

THF, heat, 10 h

Ph CHO 78%

Mahata, P.K.; Barun, O.; Ila, H.; Junjappa, H. *Synlett*, ***2000***, 1345.

TMSO

$Rh(CO)_2(acac)$, $PhMe_2SiH$

PhH, CO, 90°C, 14 h

$SiMe_2Ph$

CHO

65%

Fukuta, Y.; Matsuda, I.; Itoh, K. *Tetrahedron Lett.*, ***1999***, *40*, 4703.

$(cod)\overset{+}{Rh}Ph\ \overset{-}{B}Ph_3$, 60°C

$P(OPh)_3$, CH_2Cl_2, 2 d

6 atm CO, 6 atm H_2

52%

van den Hoven, B.G.; Alper, H. *J. Org. Chem.*, ***1999***, *64*, 3961.

O

Ph CHO

1. $TiCl_4$, DCM, NEt_3

2. H_2O

CHO

Ph Ph

78%

Bharathi, P.; Periasamy, M. *Org. Lett.*, ***1999***, *1*, 857.

Bu CHO

0.2 pyrrolidine, 0.1 $PhCO_2H$

toluene, 20°C, 14 h

Bu CHO

Bu

79%

Ishikawa, T.; Uedo, E.; Okada, S.; Saito, S. *Synlett*, ***1999***, 450.

REVIEWS:

"The Silyloxy-Cope Rearrangement of Syn-Aldol Products: Evolution of a Powerful Synthetic Strategy," Schneider, C. *Synlett*, ***2001***, 1079.

Also via β-Hydroxy aldehydes: Section 324 (Alcohols - Aldehyde).

SECTION 342: AMIDE - AMIDE

1. 2-NsNCl$_2$/MeCN
2. aq Na_2SO_3

NHAc, CO$_2$Me, Ph, NHNs-2 74%

Li, G.; Kim, S.H.; Wei, H.-X. *Tetrahedron Lett.*, ***2000***, *41*, 8699.

8 eq SmI$_2$, THF
16 eq *R*-BINOL
32 eq TMEDA, 4 h
-78°C, 30 min

70% (71% ee) 20%

Kikukawa, T.; Hanamoto, T.; Inanaga, J. *Tetrahedron Lett.*, ***1999***, *40*, 7497.

Also via Dicarboxylic Acids: Section 312 (Carboxylic Acid - Carboxylic Acid)
Diamines Section 350 (Amines - Amines)

SECTION 343: AMIDE - AMINE

t-BuCHO, NaCN, AcOH
H$_2$O, 70°C, 1 d

93% (>99:1)

Boesten, W.H.J.; Seerden, J.-P.G.; de Lange, B.; Dielemans, H.J.A.; Elsenberg, H.L.M.; Kaptein, B.; Moudy, H.M.; Kellogg, R.M.; Broxterman, Q.B. *Org. Lett.*, ***2001***, *3*, 1121.

PhNHMe, DCM, rt
10% Ni(ClO$_4$)$_2$•6 H$_2$O
bis-oxazoline ligand

76% (88% ee)

Zhuang, W.; Hazell, R.G.; Jørgensen, K.A. *Chem. Commun.*, ***2001***, 1240.

1. Et_2AlCl , MeCN
2. AcCl

Et N EtO_2C CO_2Et N Ac

EtO_2C N H

62% (75:25 *anti:syn*)

Shimizu, M.; Niwa, Y. *Tetrahedron Lett.*, ***2001***, *42*, 2829.

N–Ts

PhNHMe , acetone , rt

β-cyclodextrin , H_2O , 1 d

NHTs NHTs

86%

Reddy, M.A.; Reddy, L.R.; Bhanumathi, N.; Rao, K.R. *Chem. Lett.*, ***2001***, 246.

C_5H_{11} O N

NMe NH

ether , –78°C

C_5H_{11} NH O N

86% *trans*

Ishihara, H.; Ito, Y.N.; Katsuki, T. *Chem. Lett.*, ***2001***, 980.

N O

2 eq NH NH , 1 d

BuLi , THF , -78°C → rt

HN N O

82%

Hori, K.; Sugihara, H.; Ito, Y.N.; Katsuki, T. *Tetrahedron Lett.*, ***1999***, *40*, 5207.

NH_2

PhI , Pd/C , NEt_3 , MS 4Å
800 psi CO , 100 gpi H_2

120°C

HN H N Ph O

87%

Lin, Y.-S.; Alper, H. *Angew. Chem. Int. Ed.*, ***2001***, *40*, 779.

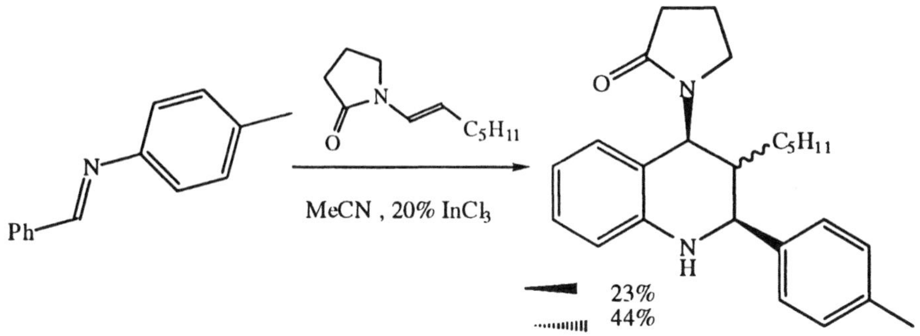

MeCN , 20% $InCl_3$

23%
44%

Hadden, M.; Nieuwenhuyzen, M.; Potts, D.; Stevenson, P.J.; Thompson, N. *Tetrahedron*, ***2001***, *57*, 5615.

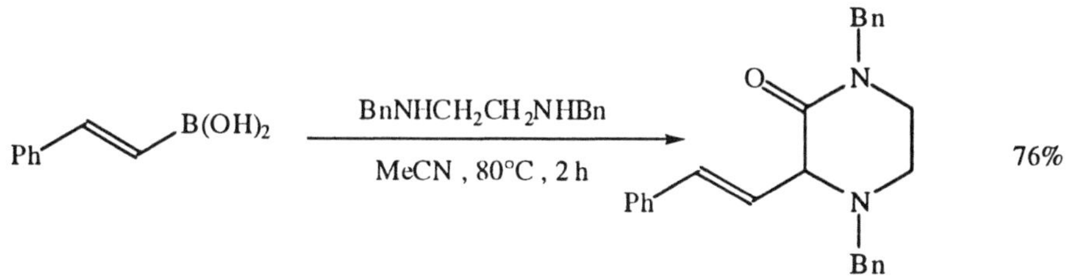

$BnNHCH_2CH_2NHBn$
MeCN , 80°C , 2 h

76%

Petasis, N.A.; Patel, Z.D. *Tetrahedron Lett.*, ***2000***, *41*, 9607.

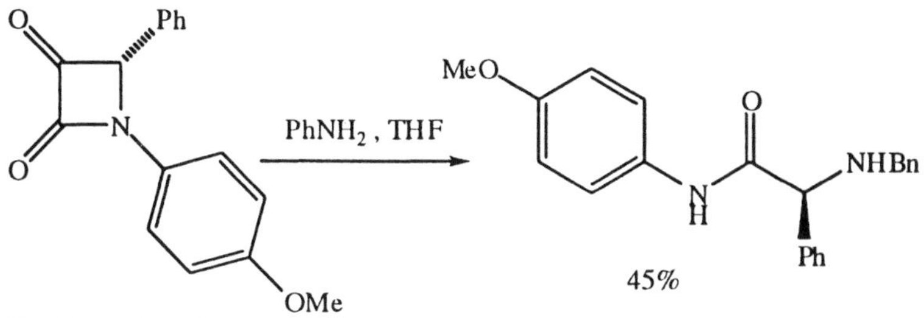

$PhNH_2$, THF

45%

Alcaide, B.; Almendros, P.; Aragoncillo, C. *Chem. Commun.*, ***2000***, 757.

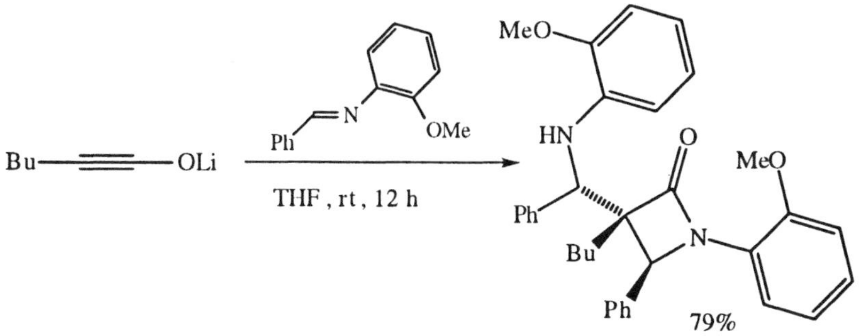

THF , rt , 12 h

79%

Shindo, M.; Oya, S.; Murakami, R.; Sato, Y.; Shishido, K. *Tetrahedron Lett.*, ***2000***, *41*, 5943.

1. $Pd(OAc)_2$, DIPEA, DMF, PhI

2. 10% TFA, CH_2Cl_2

84%

Wang, Y.; Huang, T.-N. *Tetrahedron Lett.*, ***1999***, *40*, 5837.

10% $TiCl_4(thf)_2$, *i*-Pr_2NEt
CH_2Cl_2, 25°C

NHPhth

77% (>99:1 *syn:anti*)

Yoon, T.P.; Dong, Vy.M.; MacMillan, D.W.C. *J. Am. Chem. Soc.*, ***1999***, *121*, 9726.

1. TTMSS, AIBN
80°C, PhH, 5 h

2. H_2O

TTMSS = tris-trimethylsilylsilane

83%

Kizil, M.; Patro, B.; Callaghan, O.; Murphy, J.A.; Hursthouse, M.B.; Hibbs, D. *J. Org. Chem.*, ***1999***, *64*, 7856.

SECTION 344: AMIDE - ESTER

1. *t*-BuNC, THF
$(EtO)_2P(O)CH_2CO_2H$

2. LiBr, NEt_3, THF

87%

Beck, B.; Magnin-Lachaux, M.; Herdtweck, E.; Dömling, A. *Org. Lett.*, ***2001***, *3*, 2875.

OTMS
N–PMP
Ph_3SiO OMe
NHPMP
CO_2Me
Ph
Ph
aq HBF_4 , sodium dodecylsulfate
$OSiPh_3$

90% (90:10 *syn:anti*)

Akiyama, T.; Takaya, J.; Kagoshima, H. *Tetrahedron Lett., **2001**, 42*, 4025.

O
O
N
Cbz
CO_2Me
MeOH , $NaHCO_3$
reflux , 10 min
MeO_2C
CO_2Me
NHCbz
88%

Allevi, P.; Cighetti, G.; Anatasia, M. *Tetrahedron Lett., **2001**, 42*, 5319.

Ph $N(Boc)_2$
KN(cyclohexyl)$_2$
THF , –78°C
CO_2t-Bu
Ph $N(Boc)_2$
92%

Kise, N.; Ozaki, H.; Terui, H.; Ohya, K.; Ueda, N. *Tetrahedron Lett., **2001**, 42*, 7637.

CO_2Me
N
Ts
THF , –78°C → 0°C
$IZnCu(CN)CH_2CH_2CO_2Me$–2 LiCl
CO_2Me
CO_2Me
NHTs
99%

Oishi, S.; Tamamura, H.; Yamashita, M.; Odagaki, Y.; Hamanaka, N.; Otaka, A.; Fujii, N. *J. Chem. Soc., Perkin Trans. 1, **2001***, 2445.

CO_2Me
TsNBrNa , MeCN , rt
cat $Pd(MeCN)_2Cl_2$
Ts
N
CO_2Me
60%

Antunes, A.M.M.; Marto, S.J.L.; Branco, P.S.; Prabhakar, S.; Lobo, A.M. *Chem. Commun., **2001***, 405.

Ph NHAc
CO_2Me
H_2 , Rh (*R,R*)-MeDuPHS) , *i*-PrOH
[bmim] PF_6
ionic liquid medium
Ph NHAc
CO_2Me
quant (99% ee , *R*)

Guernik, S.; Wolfson, A.; Hrshowitz, M.; Greenspoon, N.; Geresh, S. *Chem. Commun., **2001***, 2314.

EtO–C(=O)–CH=N–Ts → (p-TolCHO, EtCN, 3 eq TiI_4; −78°C → −20°C) → EtO–C(=O)–CH(NHTs)–CH(OH)–p-Tol

74% (77:23 *anti:syn*)

Shimizu, M.; Takeuchi, Y.; Sahara, T. *Chem. Lett.*, ***2001***, 1196.

EtO_2C–CH=N–CO_2Et → ($PhNMe_2$, THF, −78°C; 5% *R*-Tol-BINAP/$CuPF_6$) → EtO_2C–CH($NHCO_2Et$)–C_6H_4–NMe_2

75% (96% ee)

Saaby, S.; Fang, X.; Gathergood, N.; Jørgensen, K.A. *Angew. Chem. Int. Ed.*, ***2000***, ***39***, 4114.

t-BuO–C(=O)–Me → (1. LDA, THF; 2. oxaziridine with N–C(=O)NEt_2 and 4-cyanophenyl) → CO_2t-Bu–CH_2CH_2–C(=O)NEt_2 (39%) + CO_2t-Bu–CH_2–CH(OH)–C_6H_4–NC (20%)

Armstrong, A.; Atkin, M.A.; Swallow, S. *Tetrahedron Lett.*, ***2000***, ***41***, 2247.

PhO_2S–CH($NHCO_2Bn$)–CH_2–$CH(CH_3)_2$ → (Br–CH(CH_3)–CO_2Et; Zn–Cu, DCM, rt) → EtO_2C–CH(CH_3)–CH($NHCO_2Bn$)–CH_2–$CH(CH_3)_2$

82% (28:72 *antisyn*)

Mecozzi, T.; Petrini, M. *Tetrahedron Lett.*, ***2000***, ***41***, 2709.

N(BFTC)(CNAP)–CH(CO_2Me)–CH_2–$CH(CH_3)_2$ → (H_2, Pd–C, EtOAc–EtOH; 30 min) → N(BFTC)(H)–CH(CO_2Me)–CH_2–$CH(CH_3)_2$

93%

Papageorgiou, E.A.; Gaunt, M.J.; Yu, J.-q.; Spencer, J.B. *Org. Lett.*, ***2000***, ***2***, 1049.

MeCHO , urea , $InCl_3$ / THF , 0°C — 75%

Ranu, B.C.; Hajra, A.; Jana, U. *J. Org. Chem.*, ***2000***, *65*, 6270.

Me_2N–Boc — 1. *sec*-BuLi, THF 2. CuCN•2 LiCl 3. CH_2=$CHCO_2Me$ — 83%

Dieter, R.K.; Lu, K.; Velu, S.E. *J. Org. Chem.*, ***2000***, *65*, 8715.

NBS , NaOAc , AcOH / PhCl , reflux , 12 h — 85%

Cho, S.-D.; Kim, H.-J.; Ahn, C.; Falck, J.R.; Shin, D.-S. *Tetrahedron Lett.*, ***1999***, *40*, 8215.

5% *S*-Tol-BINAP-$CuPF_6$ / CH_2Cl_2 — 34% (58% ee, *R*)

Fang, X.; Johannsen, M.; Hao, S.; Gathergood, N.; Hazell, R.G.; Jørgensen, K.A. *J. Org. Chem.*, ***1999***, *64*, 4844.

$PySO_3NHBoc$ / NaH, THF — 76%

Hanessian, S.; Johnstone, S. *J.Org. Chem.*, ***1999***, *64*, 5896.

$EtCH_2CO_2Me$, LDA-THF / Ti(O*i*-Pr)$_4$, –70°C → 0°C — (8 : 92) 90%

Kise, N.; Ueda, N. *J. Org. Chem.*, ***1999***, *64*, 7511.

CAN, MeOH, O_2; rt, 15 min — 70%

Nair, V.; Sheeba, V. *J. Org. Chem.*, ***1999***, *64*, 6898.

Bu_3SnH, PhH, AIBN — 70%

Benati, L.; Nanni, D.; Sangiorgi, G.; Spagnolo, P. *J. Org. Chem.*, ***1999***, *64*, 7836.

F_3COCHN CO_2t-Bu; OCO_2Et; 4.5% PPh_3, 2.5 LHMDS; 1.1 $ZnCl_2$, 1% $[(allyl)PdCl]_2$; $-78°C \rightarrow$ rt — 74% (96% ds)

Kazmaier, U.; Zumpe, F.L. *Angew. Chem. Int. Ed.*, ***1999***, *38*, 1468.

Related Methods:

Section 315 (Carboxylic Acid - Amide)
Section 316 (Carboxylic Acid - Amine)
Section 351 (Amine - Ester)

SECTION 345: AMIDE - ETHER, EPOXIDE, THIOETHER

2 eq chiral Lewis acid; i-Pr_2NEt, $-2°C$, 1 d — 88% (83% ee)

Yoon, T.P.; MacMillan, D.W.C. *J. Am. Chem. Soc.*, ***2001***, *123*, 2911.

, toluene; 5% $Dy(OTf)_3$, 4°C, 2 d — 65% (92:8 *endo:exo*)

Batey, R.A.; Powell, D.A.; Acton, A.; Oough, A.J. *Tetrahedron Lett.*, ***2001***, *42*, 7935.

, $Sc(OTf)_3$, DCM

1.5 h

72%

Yadav, J.S.; Reddy, B.V.S.; Pandey, S.K.; Srihari, P.; Prathap, I. *Tetrahedron Lett.*, ***2001***, *42*, 9085.

1. LiBr
2. EtOH , NEt_3 , reflux , 1 h

84x84%

Boukhris, S.; Souizi, A. *Tetahedron Lett.*, ***2000***, *41*, 2559.

EtOH , 10% BF_3•OEt_2

rt , 3 h

99%

Prasad, B.A.B.; Sera, G.; Singh, V.K. *Tetrahedron Lett.*, ***2000***, *41*, 4677.

$MeTi(O$*i*$\text{-}Pr)_3$, THF , rt

C_3H_7MgBr

37% (1:3 *E:Z*)

Kordes, M.; Winsel, H.; De Meijere, A. *Eur. J. Org. Chem.*, ***2000***, 3235.

EtSH , 10% $Hf(OTf)_4$ 0°C
12% chiral oxazolidinone

CH_2Cl_2 , MS 4Å , 15 h

90% (67% ee)

Kobayashi, S.; Ogawa, C.; Kawamura, M.; Sugiura, M. *Synlett*, ***2000***, 983.

SECTION 346: AMIDE - HALIDE, SULFONATE

$InCl_3$, MeCN , rt

8.5 h

78%

Yadav, J.S.; Subba Reddy, B.V.; Kumar, G.M. *Synlett*, ***2001***, 1417.

CF_2Br O N Ts → 30% $Pd(PPh_3)_4$, PPh_3 THF , fluorescent light → F F Br O N Ts 78%

Nagashima, H.; Isono, Y.; Iwamatsu, S.-i. *J. Org. Chem.*, ***2001***, *66*, 315.

Ph N SO_2Tol → $Bu_4N(Ph_3Si)$difluorosilicate, THF , 0°C → CF_3 Ph $NHSO_2Tol$ 90%

Prakash, G.K.S.; Mandal, M.; Olah, G.A. *Synlett*, ***2001***, 77.

Ph CO_2Me → 1. $TsNCl_2$, MeCN , rt , 22 h Pd(II)–(1,10-phenanthroline) 2. aq Na_2SO_3 → Cl Ph CO_2Me NHTs 81% (>95% *anti*)

Wei, H.-X.; Kim, S.H.; Li, G. *Tetrahedron*, ***2001***, *57*, 3869.

Ph CO_2Me → 1. *o*-NsNClNa , MeCN 10% $Cu(OTf)_2$ 2. aq Na_2SO_3 → Cl Ph CO_2Me NHNs 74% (>95:1 *anti:syn*)

Li, G.; Wei, H.-X.; Kim, S.H. *Tetrahedron*, ***2001***, *57*, 8407.

O $MeO_2C(H_2C)_2$ → 1. Ph NH_2 . PhH , 18 h Me_3Al/toluene , 0°C → 80°C 2. aq HCl , 0°C → O O N CH_2CH_2Ph 71%

Tietze, L.F.; Steck, P.L. *Eur. J. Org. Chem.*, ***2001***, 4353.

Et $NHSO_2Ph$ → 3 eq I_2 , MeCN , $NaHCO_3$ 10 min → I Et N SO_2Ph 84%

Jones, A.D.; Knight, D.W.; Hibbs, D.E. *J. Chem. Soc., Perkin Trans. 1*, ***2001***, 1182.

TMSCl, 10% $FeCl_2$, EtOH, 0°C → rt

(91 : 9) 72%

Bach, T.; Schlummer, B.; Harms, K. *Chem. Eur. J.*, ***2001***, *7*, 2581.

Oxone, NaCl, 45°C, wet Al_2O_3, 4 h, $CHCl_3$

Ph NHAc → Ph N(Cl) Ac 93%

Curini, M.; Epifano, F.; Marcotullio, M.C.; Rosati, O.; Tsadjout, A. *Synlett*, ***2000***, 813.

MeO_2C / MeO_2C — BuC≡CH, DCE, 1% Cp*Ru(cod)Cl, rt, 15 min → MeO_2C / MeO_2C ... Bu 89%

Li, G.; Wei, H.-X.; Kim, S.H. *Org. Lett.*, ***2000***, *2*, 2249.

SO_2Me — cat BEt_3, O_2, H_2O, 8 h → N—SO_2Me, I 77%

Wakabayashi, K.; Yorimitsu, H.; Shinokubo, H.; Oshima, K. *Bull. Chem. Soc. Jpn.*, ***2000***, *73*, 2377.

PMB — $Mn_2(CO)_{10}$, 0.11 M, DCM, hv → PMB 76%

Gilbert, B.C.; Kalz, W.; Lindsay, C.I.; McGrail, P.T.; Parsons, A.F.; Whittaker, D.T.E. *J. Chem. Soc., Perkin Trans. 1*, ***2000***, 1187.

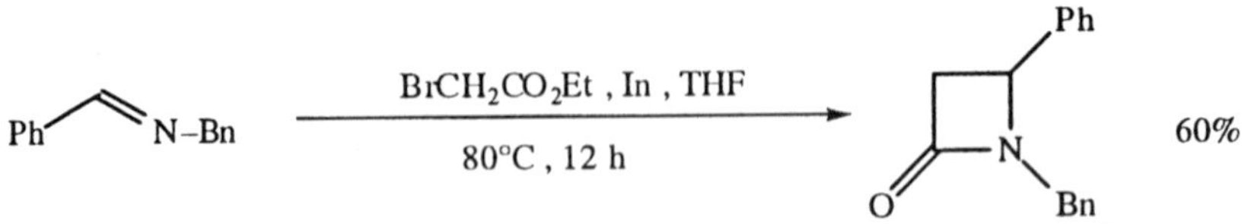
Ph
N–Bn
BrCH2CO2Et , In , THF
80°C , 12 h
Ph
N
O
Bn
60%

Banik, B.K.; Ghatak, A.; Becker, F.F. *J. Chem. Soc., Perkin Trans. 1*, ***2000***, 2179.

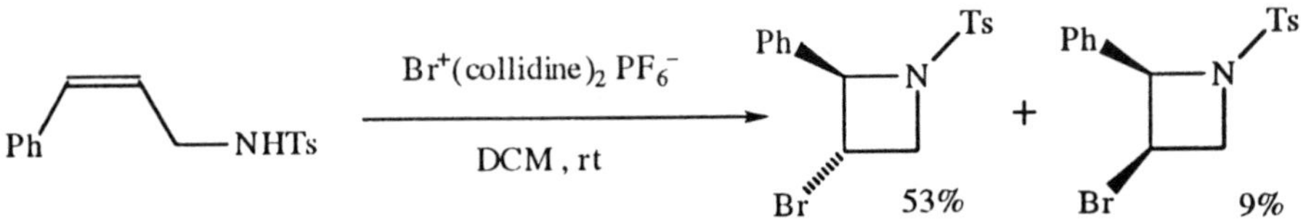
Ph
NHTs
Br+(collidine)2 PF6−
DCM , rt
Ph
Ts
N
Br
53%
+
Ph
Ts
N
Br
9%

Robin, S.; Rousseau, G. *Eur. J. Org. Chem.*, ***2000***, 3007.

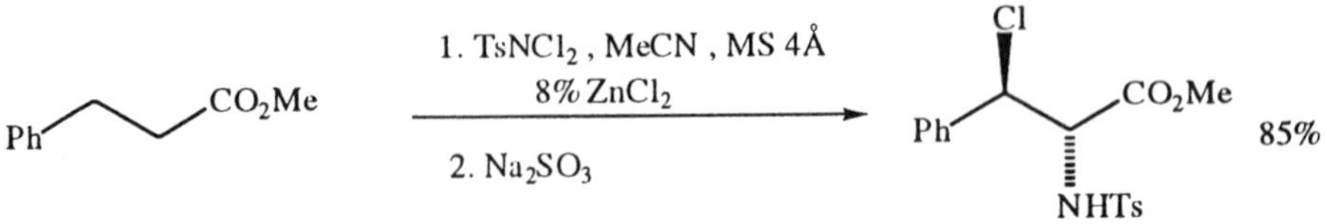
Ph
CO2Me
1. TsNCl2 , MeCN , MS 4Å
8% ZnCl2
2. Na2SO3
Cl
Ph
CO2Me
NHTs
85%

Li, G.; Wei, H.-X.; Kim, S.H.; Neighbors, M. *Org. Lett.*, ***1999***, *1*, 395.

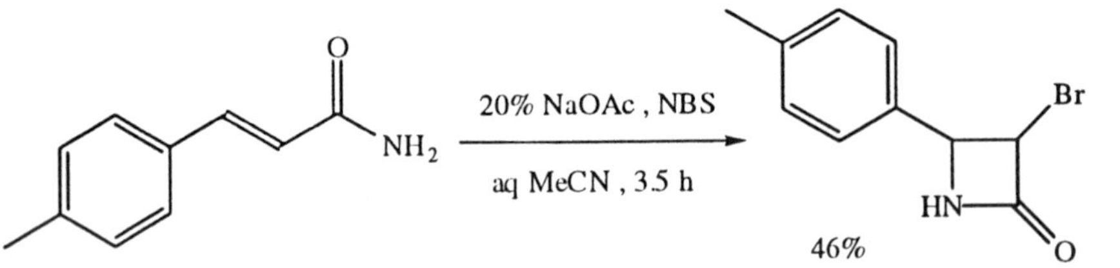
O
NH2
20% NaOAc , NBS
aq MeCN , 3.5 h
Br
HN
O
46%

Naskar, D.; Roy, S. *J. Chem. Soc., Perkin Trans. 1*, ***1999***, 2435.

SECTION 347: AMIDE - KETONE

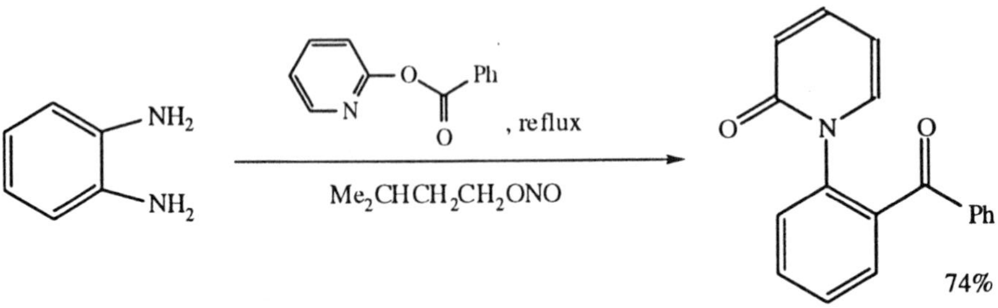
NH2
NH2
N
O
Ph
O
, reflux
Me2CHCH2CH2ONO
O
N
O
Ph
74%

Rayabarapu, D.K.; Majumdar, K.K.; Sambaiah, T.; Cheng, C.-H.
J. Org. Chem., ***2001***, *66*, 3646.

PhI, K_2CO_3, MeCN
5% $Pd(PPh_3)_4$, 20 atm CO
90°C, 6 h

24% + 57%

Kang, S.-K.; Kim, K.J. *Org. Lett.*, ***2001***, *3*, 511.

1. C_3H_7CHO, $MgSO_4$, rt
DCM, K_2CO_3
2. $(EtO_2C)_2O$, EtOH, rt
3. BF_3•OEt_2, DTBMP
DCM, 0°C

72x87% (87% ee)

Kametani, A.; Overman, L.E. *Org. Lett.*, ***2001***, *3*, 1229.

, NEt_3
, CH_2Cl_2
35°C, 30 min

86%

Murry, J.A.; Frantz, D.E.; Soheili, A.; Tillyer, R.; Grabowski, E.J.J.; Reider, P.J. *J. Am. Chem. Soc.*, ***2001***, *123*, 9696.

$AcCo(CO)_4$, NEt_3, THF
rt, overnight

77% (26:1 *trans:cis*)

Bates, R.W.; Satcharoen, V. *Synlett*, ***2001***, 532.

10% $(MeCN)_2PdCl_2$, DCM
1.5 eq $CbzNH_2$, 14 h

84%

Gaunt, M.J.; Spencer, J.B. *Org. Lett.*, ***2001***, *3*, 25.

MeO, MeO, NHTs → (PhC(O)CH(Cl)SPh; $SnCl_4$, DCM, 5 h; −78°C → rt) MeO, MeO, N, Ts, Ph, O 73%

Silveira, C.C.; Bernardi, C.R.; Braga, A.L.; Kaufman, T.S. *Tetrahedron Lett.*, ***2001***, *42*, 8947.

O, O, NBu_2 → ($EtZnCH_2I$) O, NBu_2, O 70%

Hilgenkamp, R.; Zercher, C.K. *Tetrahedron*, ***2001***, *57*, 8793.

O, N_2, Me, N, Me → (cat $Cu(acac)_2$, PhH; reflux) O, Me, N, Me + O, Me, N, Me (1 : 1) 70%

Clark, J.S.; Hodgson, P.B.; Goldsmith, M.D.; Street, L.J. *J. Chem. Soc., Perkin Trans. 1*, ***2001***, 3312.

O → (PhCH=NTs, 20% DMAP; MeOH, 40°C, 1 d) O, NHTs, Ph 50%

Shi, M.; Xu, Y.-M. *Chem. Commun.*, ***2001***, 1876.

Cl, Cl, O, N, O → (2.2 eq Bu_3SnH, AIBN; toluene, heat) O, N, O 58%

Parsons, A.F.; Williams, D.A.J. *Tetrahedron*, ***2000***, *56*, 7217.

TiI_4 , MeCN, 0°C → rt, 22 h

88%

Shimizu, M.; Sahara, T.; Hayakawa, R. *Chem. Lett.*, ***2001***, 792.

N—Boc , Et_2Zn

toluene, 0°C → rt

63% (92% *bis-endo*)

Paparin, J.-L.; Crévisy, C.; Grée, R. *Tetrahedron Lett.*, ***2000***, *41*, 2343.

DBU, toluene, 110°C

1 d

57% (95:5)

Alcaide, B.; Almendros, P.; Salgado, N.R. *J. Org. Chem.*, ***2000***, *65*, 3310.

$Boc(Me)NCH_2SnBu_3$, CuCN

THF, 5 eq TMSCl

98%

Dieter, R.K.; Alexander, C.W.; Nice, L.E. *Tetrahedron*, ***2000***, *56*, 2767.

1. TMSOTf
2. Ph–CH=CH–C(OTMS)=CH₂
3. hν, O_2
4. TBAF

70% (80% de)

Ishimaru, K.; Kojima, T. *J. Chem. Soc., Perkin Trans. 1*, ***2000***, 2105.

PhSe O O N O O $(TMS)_3SiH$, O_2 BEt_3 O H O N O O + O H O N O O

(>19 : 1) 92%

Evans, P.A.; Managan, T.; Rheingold, A.L. *J. Am. Chem. Soc.*, ***2000***, *122*, 11009.

O C_3H_7 Ph N H Bt N O $ZnBr_2$, DCM Bt = benzotriazole Ph N H O C_3H_7 O 50%

Katritzky, A.R.; Fang, Y.; Silina, A. *J. Org. Chem.*, ***1999***, *64*, 7622.

OTMS chiral nitrido-manganese (V)-salen complex TFAA, Py, DCM, 6 h −78°C → rt O $NHCOCF_3$ 83% (75% ee)

Svenstrup, N.; Bøgevig, A.; Hazell, R.G.; Jørgensen, K.A.
J. Chem. Soc., Perkin Trans. 1, ***1999***, 1559.

NHTs O CHN_2 0.1 $Cu(acac)_2$, PhH reflux O N Ts Me 52%

Wang, J.; Hou, Y.; Wu, P. *J. Chem. Soc., Perkin Trans. 1*, ***1999***, 2277.

NH_2 1. Cl_3CCHO, $NH_2OH{\bullet}HCl$ microwaves 2. H_2SO_4, microwaves O O N H 89x75%

Janeshwara, G.K.; Bedekar, A.V.; Deshpande, V.H. *Synth. Commun.*, ***1999***, *29*, 3627.

SECTION 348: AMIDE - NITRILE

NO ADDITIONAL EXAMPLES

SECTION 349: AMIDE - ALKENE

Me_3Si—≡—O^-Li^+ → 1. N-Ts (aziridine), THF, −78°C → 20°C, 12 h; 2. t-BuCHO, −78°C → 20°C, 6 h → t-Bu, O, N, Ts 71%

Iwamoto, K.; Kojima, M.; Chatani, N.; Murai, S. *J. Org. Chem.*, ***2001***, *66*, 169.

BnN → CO_2Me, 120°C, 10 d → O, N, Bn + O, N, Bn

(74 : 26) 89%
98:2 *cis:trans*

Jabin, I.; Revial, G.; Monnier-Benoit, N.; Netchitaïlo, P. *J. Org. Chem.*, ***2001***, *66*, 256.

N, Trs → 1. t-BuLi; 2. $Me_2Cu(CN)Li_2$; 3. H_2O → NHTrs 71%

Neipp, C.E.; Humphrey, J.M.; Martin, S.F. *J. Org. Chem.*, ***2001***, *66*, 531.

Ph, B, O, O → Boc, N, TfO, 5% $PdCl_2(PPh_3)_2$; THF, 2M aq Na_2CO_3 → Boc, N, Ph 76%

Occhiato, E.G.; Trabocchi, A.; Guarna, A. *J. Org. Chem.*, ***2001***, *66*, 2459.

Ph–CH=N–SO_2Ph → (crotyl bromide, In) → Ph–CH($NHSO_2Ph$)–CH(CH_3)–CH=CH_2

H_2O 98% (39:61 *syn:anti*)
THF/H_2O 95% (79:21 *syn:anti*)

Lu, W.; Chan, T.H. *J. Org. Chem.*, ***2001***, *66*, 3467.

SMe, $SiMe_3$, N, CO_2Et → 1. $TiCl_4$, DCM −78°C → −20°C; 2. $KHCO_3$ → MeS, N, O

83%

Duncan, D.; Livinghouse, T. *J. Org. Chem.*, ***2001***, *66*, 5237.

CO_2Me → PhCHO, $TsNH_2$, 2-propanol, 1 d / DABCO, 2% $La(OTf)_3$, MS 4Å → NHTs, Ph, CO_2Me 76%

Balan, D.; Adolfsson, H. *J. Org. Chem.*, ***2001***, *66*, 6498.

1-Iodonaphthalene → Ts–N(Bn)–C(=CH_2)$SnBu_3$, 8% $Pd_2(dba)_3$ / 14% $AsPh_3$, 5 eq CuCl, THF, 50°C → Ts, N, Bn 83%

Minière, S.; Cintrat, J.-C. *J. Org. Chem.*, ***2001***, *66*, 7385.

Ts, N → 10% Grubbs' catalyst, DCM / CH_2=CH_2, rt, 1 d → Ts, N 56%

Kitamura, T.; Mori, M. *Org. Lett.*, ***2001***, *3*, 1161.

toluene , 145°C

3 d

(3 : 1) 87%

Greig, I.R.; Tozer, M.J.; Wright, P.T. *Org. Lett.*, **2001**, *3*, 369.

EtO_2CCH_2CN , toluene , 80°C

cat $Rh_4(CO)_{12}$, 4 h

85%

Takaya, H.; Kojima, S.; Murahashi, S.-i. *Org. Lett.*, **2001**, *3*, 421.

, THF , 1 h

65%

Solin, N.; Narayan, S.; Szabó, K.J. *Org. Lett.*, **2001**, *3*, 909.

1. 2.2 eq MgBr
2. $ClCO_2Ph$
3. Grubbs' catalyst DCM , reflux

80x98%

Pearson, W.H.; Aponick, A. *Org. Lett.*, **2001**, *3*, 1327.

PhN=C=O

5% Cp*Ru(cod)Cl
DCE , reflux , 1 h

62%

Yamamoto, Y.; Takagishi, H.; Itoh, K. *Org. Lett.*, **2001**, *3*, 2117.

C_4H_9 NTs — cat $Ru_3(CO)_{12}$, 1.5 NEt_3, dioxane, 9 h, 20 atm CO → C_4H_9 N Ts O 85%

Kang, S.-K.; Kim, K.-J.; Yu, C.-M.; Hwang, J.-W.; Do, Y.-K. *Org. Lett.*, ***2001***, *3*, 2851.

Ph N Boc Br — 2 eq KHMDS, ether, rt → N Boc Ph 63%

Green, M.P.; Prodger, J.; Sherlock, A.E.; Hayes, C.J. *Org. Lett.*, ***2001***, *3*, 3377.

O N O N Me — O Cl, $Yb(OTf)_3$, CH_2Cl_2, 23°C, $EtNi\text{-}Pr_2$ → O N O O N O

97% (98 : 2 *syn:anti*)

Dong, V.M.; MacMillan, D.W.C. *J. Am. Chem. Soc.*, ***2001***, *123*, 2448.

O N O Me — 1. C_3H_7I, 2.5 eq SmI_2, THF, 1% NiI_2, 20°C, 10 min; 2. 0.5 M HCl → O N Me C_3H_7 95%

Farcas, S.; Namy, J.-L. *Tetrahedron Lett.*, ***2001***, *42*, 879.

Ts N C_8H_{17} I — In, MeOH, reflux, 4 h → NHTs C_8H_{17} 95%

Yadav, J.S.; Bandyopadhyay, A.; Reddy, B.V.S. *Synlett*, ***2001***, 1608.

cat $PdCl_2(PPh_3)_2$, NMP
$ZnCl_2$, Na_2CO_3, 95°C, 1 d

53%

Iyer, S.; Ramesh, C.; Kulkarni, G.M. *Synlett, **2001**,* 1241.

, THF, 65°C

10% [($Pd(OAc)_2/P(Oi\text{-}Pr)_3$/BuLi/toluene]

82%

Hedley, S.J.; Moran, W.J.; Prenzel, A.H.G.P.; Price, D.A.; Harrity, J.P.A. *Synlett*, ***2001***, 1596.

MeOLi, MeOH

0°C, 2.5 h

62%

Bellesia, F.; De Buyck, L.; Colucci, M.V.; Ghelfi, F.; Laureyn, I.; Libertini, E.; Mucci, A.; Pagnoni, U.M.; Pinetti, A.; Rogge, T.M.; Stevens, C.V. *Tetrahedron Lett.*, ***2001**, 42*, 4573.

Dess-Martin periodinane

2 eq Py, DCM, rt, 30 min

92%

Yu, C.; Hu, L. *Tetrahedron Lett., **2001**, 42*, 5167.

Te^{-2}, H_2O/toluene

Adogen 464, 0°C → rt

74%

Chao, B.; Dittmer, D.C. *Tetrahedron Lett., **2001**, 42*, 5789.

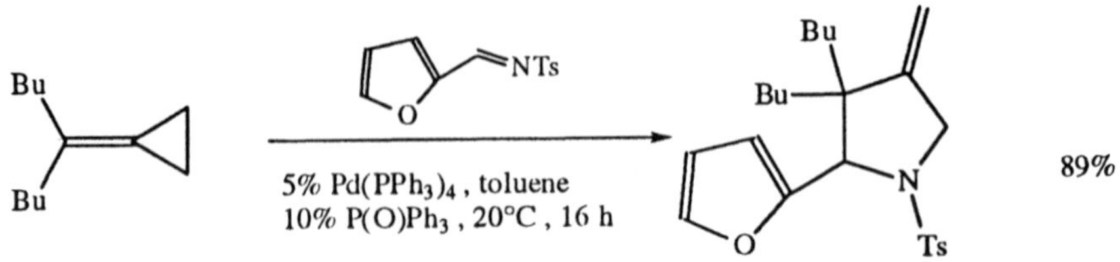

Bu, Bu; NTs; O; 5% $Pd(PPh_3)_4$, toluene; 10% $P(O)Ph_3$, 20°C, 16 h; Bu, Bu, N, Ts, O; 89%

Oh, B.H.; Nakamura, I.; Saito, S.; Yamamoto, Y. *Tetrahedron Lett.*, ***2001***, *42*, 6203.

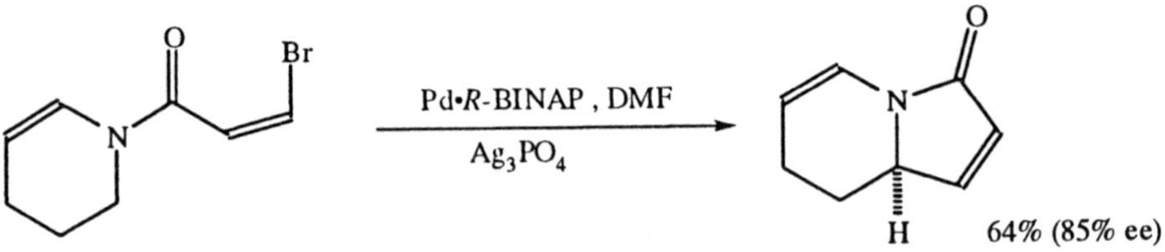

O, Br, N; Pd•*R*-BINAP, DMF; Ag_3PO_4; O, N, H; 64% (85% ee)

Kiewel, K.; Tallant, M.; Sulikowski, G.A. *Tetrahedron Lett.*, ***2001***, *42*, 6621.

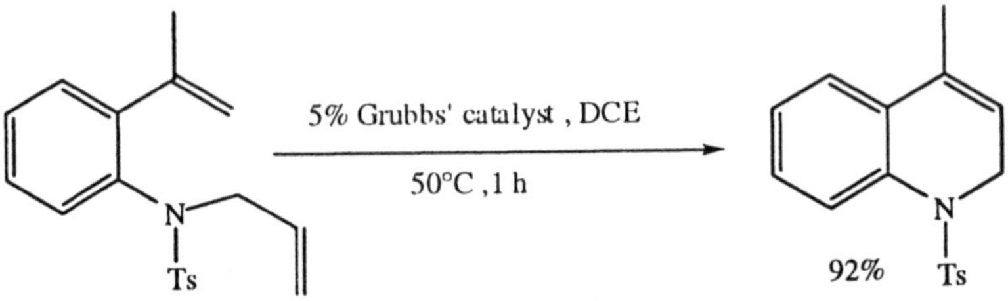

N, Ts; 5% Grubbs' catalyst, DCE; 50°C, 1 h; 92%; N, Ts

Arisawa, M.; Theeraladanon, C.; Nishida, A.; Nakagawa, M. *Tetrahedron Lett.*, ***2001***, *42*, 8029.

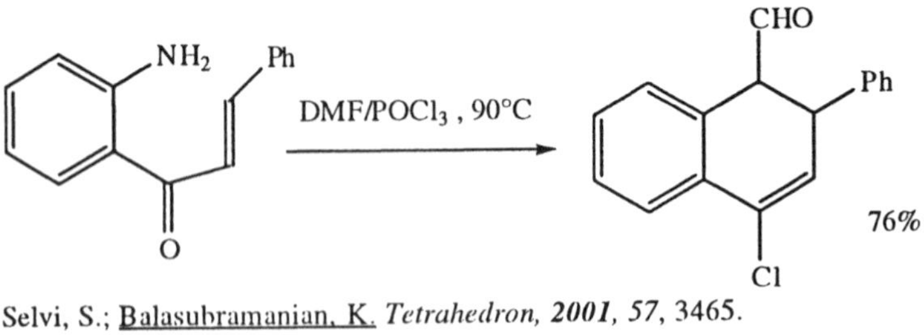

NH_2, Ph, O; DMF/$POCl_3$, 90°C; CHO, Ph, Cl; 76%

Akila, S.; Selvi, S.; Balasubramanian, K. *Tetrahedron*, ***2001***, *57*, 3465.

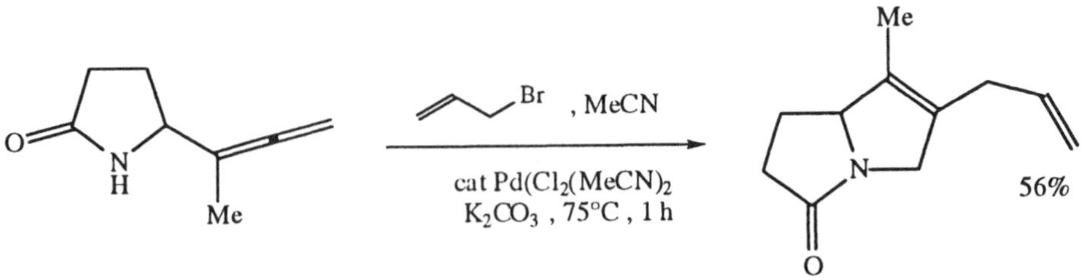

O, N, H, Me; Br, MeCN; cat $Pd(Cl_2(MeCN)_2$; K_2CO_3, 75°C, 1 h; Me, N, O; 56%

Karstens, W.F.J.; Klomp, D.; Rutjes, F.P.J.T.; Hiemstra, H. *Tetrahedron*, ***2001***, *57*, 5123.

OMe

1. $ClSO_2NCO$, Na_2CO_3

2. KOH , Na_2SO_4

NHCO$_2$Me + NHCO$_2$Me

(1 : 1) 70%

Kim, J.D.; Lee, M.H.; Han, G.; Park, H.; Zee, O.P.; Jung, Y.H. *Tetrahedron*, ***2001***, *57*, 8257.

$PhCH_2OH$ → (Me–C(O)–NMe$_2$; (1,10-phenantroline)NiCl$_2$, PhCl , NaH , rt) → Ph–CH=CH–C(O)NMe$_2$ 80%

Ren, H.-J.; Wang, Y.-G. *Synth. Commun.*, ***2001***, *31*, 1201.

2 eq CO , MeOH , 0.5 O_2

cat PdI_2 , KI , 70°C , 15 h

NH$_2$ MeO$_2$C O N H 53%

Gabriele, B.; Salerno, G.; Veltri, L.; Costa, M.; Massera, C. *Eur. J. Org. Chem.*, ***2001***, 4607.

CH_2=C=CH_2 , K_2CO_3

cat Pd complex

I N Ts NHBn Bn N N Ts 54%

Grigg, R.; Köppen, I.; Rasparini, M.; Sridharan, V. *Chem. Commun.*, ***2001***, 964.

OH O Ph NMe$_2$ Cl → (SmI_2 , 25°C) → O Ph NMe$_2$ 90%

Concellón, J.M.; Pérez-Andrés, J.A.; Rodríguez-Solla, H. *Chem. Eur. J*, ***2001***, *7*, 3062.

PhMgBr

THF, reflux

NHTS

NBr_2

Ph

$NHBr_2$

75%

Chandrasekhar, S.; Reddy, M.V.; Rajaiah, G. *Tetrahedron Lett.*, ***2000***, *41*, 10131.

$Pd(OAc)_2$, AcOH

LiBr

Bn—N

CHO

Br

CHO

Bn

87% (>98:2 *Z:E*)

Xie, X.; Lu, X. *Synlett*, ***2000***, 707.

cat $Ru_3(CO)_{12}$, CO

$CH_2=CH_2$

Ph

93%

Berger, D.; Imhof, W. *Tetrahedron*, ***2000***, *56*, 2015.

OTMS

1. Cp_2ZrCl_2, 2 eq BuLi

THF, –78°C → rt

2. EtCHO

Ph

Et

52% (85:15 *E:Z*)

Gandon, V.; Bertus, P.; Szymoniak, J. *Tetrahedron*, ***2000***, *56*, 4467.

Me———OLi

Ph

OMe

THF, rt, 12 h

MeO

Ph

88% (>99:1 *E:Z*)

Shindo, U.; Oya, S.; Murakami, R.; Satoo, Y.; Shishido, K. *Tetrahedron Lett.*, ***2000***, *41*, 5947.

OMe
MeO
MeO MeO
O
1. BuLi , LiCl , THF , –78°C
2. –78°C ,
N
O
i-Pr
3. HCl , HFIP , 0°C
HO
O
i-Pr
67% (67% ee)

Harrington, P.E.; Tius, M.A. *Org. Lett.*, ***2000***, *2*, 2447.

Bz
N
Bn
$SiMe_2Ph$
LDA , THF-HMPA
–78°C → –40°C
H
N
Bz
Ph
$SiMe_2Ph$
84% (20:1)

Anderson, J.C.; Flaherty, A.; Swarbrick, M.E. *J. Org. Chem.*, ***2000***, *65*, 9152.

O
N
O
Br
$SiMe_2Ph$
, BEt_3 , 0°C
$Yb(OTf)_3$, air , THF
+
(65 : 35) 82%

Porter, N.A.; Zhang, G.; Reed, A.D. *Tetrahedron Lett.*, ***2000***, *41*, 5773.

I
N
Bn
O
Bu_3Sn Ph , 10% $Pd(OAc)_2$
20% PPh_3 , toluene , 90°C
Ph
O
N
Bn
89%

Fretwell, P.; Grigg, R.; Sansano, J.M.; Sridharan, V.; Sukirthalingm, S.; Wilson, D.; Redpath, J. *Tetrahedron*, ***2000***, *56*, 7525.

4 eq $MeCu(CN)Li \cdot LiBr \cdot 2 Li$
THF , 3 h , -78°C

MTS = 2,4,6-trimethylbenzenesulfonyl

H, H, N, MTS, H, H, NHMTS, H, Me

93%

Ohno, H.; Toda, A.; Miwa, Y.; Taga, T.; Fujii, N.; Ibuka, T. *Tetrahedron Lett.*, ***1999***, *40*, 349.

$[Pd(MeCN)_4][BF_4]_2$, rt
CH_2Cl_2 , 72 h

Ar = 4-CF_3 phenyl

Ph, Ar, N, O, O, Ar, N, Ph

71% (83% ee)

Jiang, Y.; Longmire, J.M.; Zhang, X. *Tetrahedron Lett.*, ***1999***, *40*, 1449.

Bu

t-$BuMe_2SiCl$, 95°C m 12 h
CO , $Rh_2(CO)_{12}$, PhH
pyrrolidine

$SiMe_2t$-Bu, N, Bu, O

90% (97:3 *Z*:*E*)

Matsuda, I.; Takeuchi, K.; Itoh, K. *Tetrahedron Lett.*, ***1999***, *40*, 2553.

5% Pd ferrocenyl oxazoline
CH_2Cl_2 , 23°C , 3 d

Ph, Ar—N, C_3H_7, O, Ar, N, Ph, C_3H_7

67% (91% ee , *R*)

Donde, Y.; Overman, L.E. *J. Am. Chem. Soc.*, ***1999***, *121*, 2933.

1. (–)-sparteine, BuLi
toluene , 1 h
2. (Z)-$MeCH{=}C(CO_2Et)_2$ [structure]

Ph, N, Ar, Boc, EtO_2C, EtO_2C, Ph, NArBoc

Ar = 4-OMe-C_6H_4

93%

Curtis, M.D.; Beak, P. *J. Org. Chem.*, ***1999***, *64*, 2996.

1. In , THF reflux , 18 h
2. DCC , cat. 4-PPY CH_2Cl_2 , 30 min , rt

37%

4-PPY = 4-pyrrolidinopyridine

Choudhury, P.K.; Foubelo, F.; Yus, M. *J. Org. Chem.*, ***1999***, *64*, 3376.

MeCN , reflux

83%

Bennett, D.M.; Okamoto, I.; Danheiser, R.L. *Org. Lett.*, ***1999***, *1*, 641.

$Pd(PPh_3)_4$, K_2CO_3

PhI, DMF , 80°C 40 h

(10 : 90) 34%

Rutjes, F.P.J.T.; Tjen, K.C.M.F.; Wolf, L.B.; Karstens, W.F.J.; Schoemaker, H.E.; Hiemstra, H. *Org. Lett.*, ***1999***, *1*, 717.

$BnNH_2$, toluene , 110°C , 2 h

82%

Sugihara, T.; Okada, Y.; Yamaguchi, M.; Nishizawa, M. *Synlett*, ***1999***, 768.

1. 3 eq *t*-BuLi , 6 eq HMPA THF , –78°C , 16 h
2. NH_4Cl

75%

Ahmed, A.; Clayden, J.; Yasin, S.A. *Chem. Commun.*, ***1999***, 231.

REVIEWS:

"1,3-Dipolar Cycloadditions of Five- and Six-Membered Cyclic Nitrones to α,β-Unsaturated Acid Derivatives," de March, P.; Figueredo, M.; Font, J. *Heterocycles*, ***1999***, *59*, 1213.

Also via Alkenyl Acids: Section 322 (Carboxylic Acid -Alkene)

SECTION 350: AMINE - AMINE

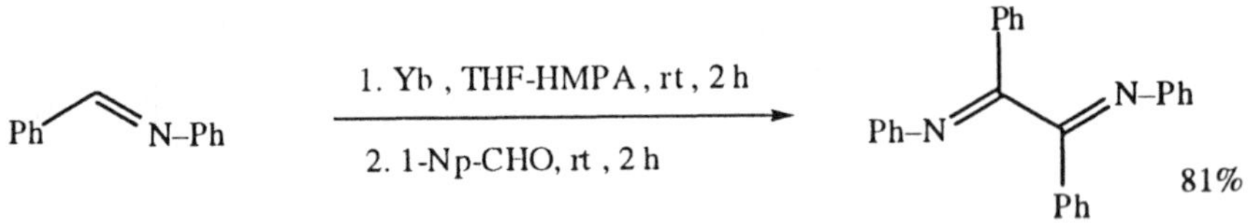

Ph
N–Ph
1. Yb , THF-HMPA , rt , 2 h
2. 1-Np-CHO, rt , 2 h
Ph
Ph–N
N–Ph
Ph
81%

Jin, W.; Makioka, Y.; Kitamua, T.; Fujiwara, Y. *J. Org. Chem.*, ***2001***, *66*, 514.

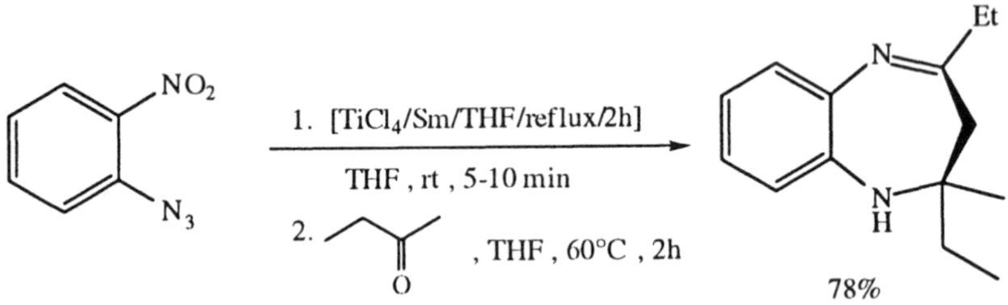

NO2
N3
1. [TiCl4/Sm/THF/reflux/2h]
THF , rt , 5-10 min
2.
O
, THF , 60°C , 2h
Et
N
N
H
78%

Zhang, W.; Zhang, Y.; Chen, X. *Tetrahedron Lett.*, ***2001***, *42*, 73.

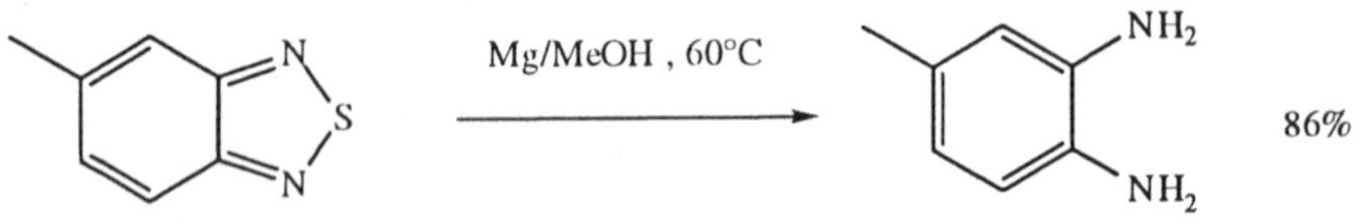

N
S
N
Mg/MeOH , 60°C
NH2
NH2
86%

Prashad, M.; Liu, Y.; Repič, O. *Tetrahedron Lett.*, ***2001***, *42*, 2277.

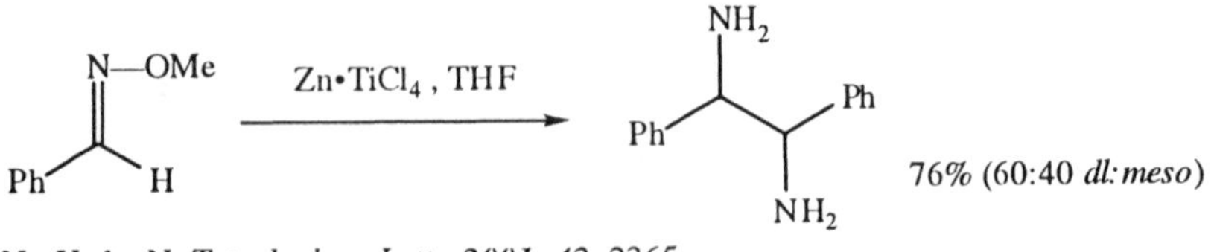

N—OMe
Ph
H
Zn•TiCl4 , THF
NH2
Ph
Ph
NH2
76% (60:40 dl:meso)

Kise, N.; Ueda, N. *Tetrahedron Lett.*, ***2001***, *42*, 2365.

PhCHO → (Me_3Si_2NH, NaH, Li, ether; 5M $LiClO_4$/ether, ultrasound, 5 h) → PhCH(NH_2)CH(NH_2)Ph 42% (76:24 *meso:dl*)

Mojtahedi, M.M.; Saidi, M.R.; Shirzi, J.S.; Bolourtchian, M. *Synth. Commun.*, ***2001***, *31*, 3587.

PhCH(OEt)NEt_2 → ([TiI_4, 2 eq Zn], THF, 3 h) → PhCH(NEt_2)CH(NEt_2)Ph 95% (45:55 *dl:meso*)

Yoshimura, N.; Mukaiyama, T. *Chem. Lett.*, ***2001***, 1334.

PhC(O)CH_3 → (Cp_2TiF_2, $PhMeSiH_2$, toluene; rt, 30 min) → PhCH(NMe_2)CH(NMe_2)Ph 84% (52:48 *meso:dl*)

Selvakumar, K.; Harrod, J.F. *Angew. Chem. Int. Ed.*, ***2001***, *40*, 2129.

PhCH(Cl)CH(morpholino)CO_2Et → (t-$BuNH_2$) → PhCH(NH*t*-Bu)CH(morpholino)CO_2Et + PhCH(morpholino)CH(NH*t*-Bu)CO_2Et

(95 : 5) 84%

Chung, T.-H.; Sharpless, K.B. *Org. Lett.*, ***2000***, *2*, 3555.

PhCH=N-*t*-Bu → (hν, Pyrex, 3 h; *i*-PrOH, acetone) → *t*-BuHN–CH(Ph)–CH(Ph)–NH*t*-Bu + *t*-BuHN–CH(Ph)–CH(Ph)–NH*t*-Bu

(45 : 55) >95%

Campos, P.J.; Arranz, J.; Rodríguez, M.A. *Tetrahedron*, ***2000***, *56*, 7285.

2 PhHC=N+(Ph)O⁻ → (In, EtOH, reflux; 12 h) → PhCH(NHPh)CH(NHPh)Ph 99%

Jeevanandam, A.; Cartwright, C.; Ling, Y.-C. *Synth. Commun.*, ***2000***, *30*, 3153.

N-Ph cyclohexene aziridine → ($PhNHMe$, $Sn(OTf)_2$; 10 min) → trans-2-(NMePh)-cyclohexyl-NHPh 82%

Sekar, G.; Singh, V.K. *J. Org. Chem.*, ***1999***, *64*, 2537.

6,7-dimethoxy-1-methyl-3,4-dihydroisoquinoline → (Rh chiral diamine complex, HCOOH, NEt_3, MeCN) → 6,7-dimethoxy-1-methyl-1,2,3,4-tetrahydroisoquinoline 96% (89% ee)

Mao, J.; Baker, D.C. *Org. Lett.*, ***1999***, *1*, 841.

→ ($ClTi(Oi\text{-}Pr)_3$, THF, rt; $c\text{-}C_5H_9MgCl$) → (72 : 28) 83%

Cao, B.; Xiao, D.; Joullié, M.M. *Org. Lett.*, ***1999***, *1*, 1799.

trans-1,2-diaminocyclohexane → (1-bromonaphthalene, *t*-BuONa; $Pd_2(dba)_3$/dppf, THF) → 2-NH(1-naphthyl)-cyclohexyl-NH_2

Frost, C.G.; Mendonça, P. *Tetrahedron Asymm.*, ***1999***, *10*, 1831.

SECTION 351: AMINE - ESTER

N_2=C(CO_2Me)$_2$ → (Ph_2NH, cat $Rh_2(OAc)_4$; PhH, 1 h) → Ph_2N–CH(CO_2Me)$_2$ 84%

Yang, M.; Wang, X.; Li, H.; Livant, P. *J. Org. Chem.*, ***2001***, *66*, 6729.

Ph_2C=N–CH_2–CO_2t-Bu → (1% chiral ammonium salt, 2.4 eq BnBr, aq KOH, rt; Triton-X-100) → Ph_2C=N–CH(Bn)–CO_2t-Bu

92% (85% ee)

Okino, T.; Takemoto, Y. *Org. Lett.*, ***2001***, *3*, 1515.

Ph_2C=N–CH_2–CO_2t-Bu → (PhCH=CHCH$_2$OAc, 3h; $[(Pd(C_3H_5)Cl]_2$, $P(OPh)_3$, chiral phase transfer catalyst, 50% aq KOH, toluene . 0°C) → Ph_2C=N–CH(CH_2CH=CHPh)–CO_2t-Bu

89% (96% ee)

Nakoji, M.; Kanayama, T.; Okino, T.; Takemoto, Y. *Org. Lett.*, ***2001***, *3*, 3329.

Ph–CH(Me)–NH–C(=O)–CH_2–N=CPh_2 → (1. BnBr, KOH, TBAB, Soxhlet; 2. H^+; 3. AcCl, MeOH; 4. HN=CPh_2) → MeO_2C–CH(Bn)–N=CPh_2

68:32 (*S*:*R*)

Kim, H.J.; Lee, S.-k.; Park. Y.S. *Synlett*, ***2001***, 613.

Ph_2C=N–CH_2–CO_2t-Bu → (3% Cinchona alkaloid, MeI; 50% KOH, toluene–$CHCl_3$, –20°C, 1 d) → Ph_2C=N–CH(Me)–CO_2t-Bu 65% (90% ee, *S*)

Park, H.-g.; Jeong, B.-s.; Yoo, M.-s.; Park, M.-k.; Huh, H.; Jew, S.-s. *Tetrahedron Lett.*, ***2001***, *42*, 4645.

Dimethyl pyridine-2,3-dicarboxylate → (e^-, Et_4NOTs, AcOH; 10°C) → dimethyl 1,2-dihydropyridine-2,3-dicarboxylate 89%

Kita, Y.; Maekawa, H.; Yamasaki, Y.; Nishiguchi, I. *Tetrahedron*, ***2001***, *57*, 2095.

N—PAN, Ph, H; 3 OTMS, SEt, $AlCl_3$, 17h; MS 4Å, DCM, –78°C → rt; EtS, O, NHPAN, O, Ph, SEt; 69%

Shimizu, M.; Ogawa, T.; Nishi, T. *Tetrahedron Lett.*, ***2001***, *42*, 5463.

CO_2Me, I; NEt_3, toluene, reflux; CO_2Me, Et_2N; 96%

Maw, G.; Thirsk, C.; Whiting, A. *Tetrahedron Lett.*, ***2001***, *42*, 8387.

Ph, N, Ph, CO_2t-Bu; 10% chiral phase transfer catalyst, 50% aq KOH; DCM, BnBr; Ph, N, CO_2t-Bu, Ph, Bn; 73% (86% ee)

Lygo, B.; Crosby, J.; Lowden, T.R.; Peterson, J.A.; Wainwright, P.G. *Tetrahedron*, ***2001***, *57*, 2403.

CHO, N; 1. PhNHTMS, 5M $LiClO_4$/ether; 2. $BrZnCH_2CO_2Et$; 3. H_2O; NHPh, CO_2Et, N; 86%

Saidi, M.R.; Azizi, N.; Zali-Boinee, H. *Tetrahedron*, ***2001***, *57*, 6829.

CN, Bn, NHBn; 1. $SnCl_4$; 2. N_2CHCO_2Et; Bn, N, Bn, CO_2Et; 51%

Lee, K.-D.; Suh, J.-M.; Park, J.-H.; Ha, H.-J.; Choi, H.G.; Park, C.S.; Chang, J.W.; Lee, W.K.; Dong, Y.; Yun, H. *Tetrahedron*, ***2001***, *57*, 8267.

CO_2Me, *i*-Pr, CO_2Me; TMSNHOTMS, DCM; cat Cu(bis-oxazoline), –10°C; TMSOHN, CO_2Me, *i*-Pr, CO_2Me; 73% (76% ee)

Cardillo, G.; Gentilucci, L.; Gianotti, M.; Kim, H.; Perciaccante, R.; Tolomelli, A. *Tetrahedron Asymm.*, ***2001***, *12*, 2395.

NH_2, S, CO_2Me → (MeCO-CH=CH-CH$_2$Cl; NEt_3 , PhH) → N-substituted 2-methylpyrrole, S, CO_2Me 80%

Demir, A.S.; Akhmedov, İ.M.; Şwşenoğlu, Ö.; Alptürk, O.; Apaydun, S.; Gerçek, Z.; İbrahimzade, N. *J. Chem. Soc., Perkin Trans. 1*, ***2001***, 1162.

BnO–N=CH–CO_2Et → (t-BuI , BEt_3 , O_2 , DCM; BF_3•OEt_2) → t-Bu–CH(NHOBn)–CO_2Et 85%

Halland, N.; Jørgensen, K.A. *J. Chem. Soc., Perkin Trans. 1*, ***2001***, 1290.

Ph_2C=N–CH_2–CO_2t-Bu → (EtI , 50% KOH , –20°C; 5% chiral Cinchona alkaloid; toluene-$CHCl_3$, 10 h) → Ph_2C=N–CH(Et)–CO_2t-Bu

50% (92% ee)

Jew, S.-s.; Jeong, B.-S.; Yoo, M.-S.; Huh, H.; Park, H.-g. *Chem. Commun.*, ***2001***, 1244.

PhCH=N–p-An → (Me_2C=C(OTMS)(OEt), Amberlyst 15-DRY; EtOH , rt) → PhCH(NHp-An)–CMe_2–CO_2Et

83%

Shimizu, M.; Itohara, S.; Hase, E. *Chem. Commun.*, ***2001***, 2318.

PhCH=N–(o-OH-phenyl) → (Me_2C=C(OMs)(OMe), toluene; cat [Zr(Oi-Pr)$_4$/i-PrOH/S-VAPOL - NMI]; –45°C , 20 h) → (o-OH-phenyl)NH–CH(Ph)–CMe_2–CO_2Me

92% (91% ee)

Xue, S.; Yu, S.; Deng, Y.; Wulff, W.D. *Angew. Chem. Int. Ed.*, ***2001***, *40*, 2271.

MeO_2C–CH=N–OBn → (5 eq i-Pr–I , 5 eq BEt_3; 25°C , DCM , cat AIBN) → i-Pr–CH(CO_2Me)–NHOBn 65% + Et–CH(CO_2Me)–NHOBn 17%

Miyabe, H.; Ueda, M.; Yoshioka, N.; Yamakawa, K.; Naito, T. *Tetrahedron*, ***2000***, *56*, 241.

1. 1% chiral phase transfer catalyst, Ph2C=NCH2CO2t-Bu, toluene, 50% aq KOH
2. 1M citric acid, THF, rt

80% (90% ee)

Ooi, T.; Kameda, M.; Tannai, H.; Maruoka, K. *Tetrahedron Lett.*, **2000**, *41*, 8339.

MeO_2CCHO, $InCl_3$, TsOH, 2,4-dichloroaniline, $MgSO_4$, DCM, reflux

76%

Huang, T.; Li, C.-J. *Tetrahedron Lett.*, **2000**, *41*, 9747.

(MeO)2P(O)H, MeONa/MeOH

37%

Rulev, A.Yu.; Larina, L.I.; Voronkov, M.G. *Tetrahedron Lett.*, **2000**, *41*, 10211.

EtI, BEt_3, toluene, 20°C

(18 : 1) 53%

Miyabe, H.; Fujii, K.; Goto, T.; Naito, T. *Org. Lett.*, **2000**, *2*, 4071.

2 eq $BuNH_2$, DMF, 5 h

55%

de Saint-Fuscien, C.; Tarrade, A.; Dauban, P.; Dodd, R.H. *Tetrahedron Lett.*, **2000**, *41*, 6393.

Ph–CH=N–CH(CH_3)–CO_2Me

1. NaOH , BnBr , Cu-salen complex toluene , 18 h
2. MeOH , AcCl , toluene , rt
3. silica gel

Bn, H_2N, CO_2Me

91% (81% ee)

Belokon', Y.N.; Davies, R.G.; North, M. *Tetrahedron Lett.*, ***2000***, *41*, 7245.

O, CO_2Et

2 eq $EtNH_2$, H_2O

NHEt, CO_2Et

75%

Stefani, H.A.; Costa, I.M.; Silva, D.de O. *Synthesis*, ***2000***, 1526.

N+, CO_2H, Cl^-

EtO_2C, CO_2Et , MnO_2

toluene , NEt_3 , 90°C , 2 h

CO_2Et, N, CO_2Et

92%

Zhang, L.; Liang, F.; Sun, L.; Hu, Y.; Hu, H. *Synthesis*, ***2000***, 1733.

O^-, PhHC=N+, Ph

OTMS, OEt , 10% $La(OTf)_3$

DCM , 4 h

OTMS, N—Ph, Ph, CO_2Et

97%

Qian, C.; Wang, L. *Tetrahedron*, ***2000***, *56*, 7193.

NO_2

O, CO_2Et

$B_{10}H_{14}$, Pd/C

N H, CO_2Et

92%

Bae, J.W.; Cho, Y.J.; Lee, S.H.; Yoon, C.-O.M.; Yoon, C.M. *Chem. Commun.*, ***2000***, 1857.

MeO, N, Ph

$BrCH_2CO_2Me$, N_2 , rt

$ZnCu/CH_2Cl_2$, 1 h

OMe, NH, Ph, CO_2Me

89%

Adrian Jr., J.C.; Barkin, J.L.; Hassib, L. *Tetrahedron Lett.*, ***1999***, *40*, 2457.

CO_2Me + N-Me pyrrolidine → hv (350 nm), MeCN, sensitizer → 87%

sensitizer = 4,4'-dimethoxybenzophenone

Bertand, S.; Glapski, C.; Hoffmann, N.; Pete, J.-P. *Tetrahedron Lett.*, ***1999***, *40*, 3169.

3.5 eq $CsOH/H_2O$, MeI; Ru-polypyridyl complex; CH_2Cl_2, -78°C, 7 h — 82%

Ph, Ph, N, CO_2t-Bu, Me

Tzalis, D.; Knochel, P. *Tetrahedron Lett.*, ***1999***, *40*, 3685.

Cu (salen) complex, BnBr; toluene, solid KOH

Ph, N, CO_2i-Pr

40% (85% ee, *R*)

Belokon, Y.N.; North, M.; Kublitski, V.S.; Ikonnikov, N.S.; Kraskik, P.E.; Maleev, V.I. *Tetrahedron Lett.*, ***1999***, *40*, 6105.

I, NH_2 — AcCl, 300 psi CO, THF; 0.02 $Pd(OAc)_2$, 3 eq i-Pr_2NEt; 100°C, 1 d — O, O, N, Me

99%

Larksarp., C.; Alper, H. *Org. Lett.*, ***1999***, *1*, 1619.

I^+Ph BF_4^- — $CHCl_3$, 60°C, 34 h — I — 93%

Juhl, K.; Hazell, R.G.; Jørgensen, K.A. *J. Chem. Soc., Perkin Trans. 1*, ***1999***, 2293.

Ph, N–Ph — OEt, $OSiMe_3$; SmI_3, EtCN — NHBn, Ph, CO_2Et

9% (93:7 *anti:syn*)

Hayakawa, R.; Shimizu, M. *Chem. Lett.*, ***1999***, 591.

Related Methods:

Section 315 (Carboxylic Acid - Amide)
Section 316 (Carboxylic Acid - Amine)
Section 344 (Amide - Ester)

SECTION 352: AMINE - ETHER, EPOXIDE, THIOETHER

, $InCl_3$, DCM-toluene

, $TiCl_2$

rt, 6 h

(1 : 0.67) 70%

14% ee 5% ee

Sundararajan, G.; Prabagaran, N.; Varghese, B. *Org. Lett.*, ***2001***, *3*, 1973.

PhCH=N–Ph, $BiCl_3$

MeCN, rt, 15 min

95% 30 : 7

Reddy, B.V.S.; Srinivas, R.; Yadav, J.S.; Ramalingam, T. *Synth. Commun.*, ***2001***, *31*, 1075.

$(CHO)_n$, Bu_2NH, neat

CuI/Al_2O_3, microwaves

66%

Kabalka, G.W.; Wang, L.; Pagni, R.M. *Tetrahedron Lett.*, ***2001***, *42*, 6049.

$LiBF_4$, MeCN, rt, 2.5 h

90%

Yadav, J.S.; Reddy, B.V.S.; Madhuri, Ch.; Sabitha, G.; Jagannadh, B.; Kumar, S.K.; Kunwar, A.C. *Tetrahedron Lett.*, ***2001***, *42*, 6381.

Ph_3CS—NO , heat , PhH; *i*-PrOH — 72%

Cavero, M.; Motherwell, W.B.; Potier, P. *Tetrahedron Lett.*, ***2001***, *42*, 4377.

pyrrolidinone , $K_2Pd(SCN)_4$; 70°C , 16 h — 52%

Cheng, X.; Hii, K.K. *Tetrahedron,* ***2001***, *57*, 5445.

PhSH , 5% $ZnCl_2$, DCM — 82%

Wu, J.; Hou, X.-L.; Dai, L.-X. *J. Chem. Soc., Perkin Trans. 1,* ***2001***, 1314.

e^- , $LiClO_4$, DMF — 67%

Batanero, B.; Vago, M.; Barba, F. *Heterocycles,* ***2000***, *53*, 1337.

, 5% SmI_2 , THF; rt , 5 h — 93%

Nishitani, T.; Shiraishi, H.; Sakaguchi, S.; Ishii, Y. *Tetrahedron Lett.*, ***2000***,*41*, 3389.

1. $TiCl_4$, DIPEA , –80°C
2. PhCH=NAr , DCM , –80°C
3. 1M HCl

Ar = 2-chlorophenyl (93 : 7) 83%

Adrian Jr., J.C.; Barkin, J.L.; Fox, R.J.; Chick, J.E.; Hunter, A.D.; Nicklow, R.A. *J. Org. Chem.*, ***2000***, *65*, 6264.

Ni-Pr; Me; Me; Cl; OTMS → 1. MeOH, heat, 2 h; 2. aq K_2CO_3 → O; Me; OMe; Me; NHi-Pr 89x91%

De Kimpe, N.; Aelterman, W.; De Geyter, K.; De Clercq, J.-P. *J. Org. Chem.*, ***1999***, *64*, 5138.

N–OH → 1. 2 eq PhSSPh, DCM; 1.5 eq TEMPO, i-Pr_2NEt; 2. Cl; SOCl; Cl → N; SPh 75%

Lin, X.; Stien, D.; Weinreb, S.M. *Org. Lett.*, ***1999***, *1*, 637.

REVIEWS:

"2-Aminothiophenes by the Gewald Reaction," Sabnis, R.W.; Rangnekar, D.W.; Sonawane, N.D. *J. Heterocyclic Chem.*, ***1999***, *36*, 333.

SECTION 353: AMINE - HALIDE, SULFONATE

C_6H_{13} → Ts; N; Cl; PhH, rt, 3 h → Ts—N; Cl; C_6H_{13} 60% (63:37)

Tsuritani, T.; Shinokubo, H.; Oshima, K. *Org. Lett.*, ***2001***, *3*, 2709.

Ph → 1. t-BuO_2CNBr_2, DCE, reflux; 2. 12% aq Na_2SO_3, 10°C; 3. HCl/DCE, 0°C → rt → Br; Ph; NH_2•HCl 83%

Klepacz, A.; Zwierzak, A. *Tetrahedron Lett.*, ***2001***, *42*, 4539.

Cl; N; Bu → 10% SmI_2, THF/PhH; 50°C, 12 h → Cl; N; Bu 54%

Göttlich, R.; Noack, M. *Tetrahedron Lett.*, ***2001***, *42*, 7771.

Cl
N
Bu

1. CuCl, THF, 50°C, 12 h
2. conc. NH_3, H_2O

Cl
N
Bu

58%

Göttlich, R. *Synthesis*, ***2000***, 1526.

Bu
N
Cl
t-Bu

$TiCl_3 \cdot BF_3$, –78°C
deoxygenated DCM

Cl
Bu—N
t-Bu
+
Cl
Bu—N
t-Bu

(13 : 1) 93%

Hemmerling, M.; Sjöholm, Å.; Sonfai, P. *Tetrahedron Asymm.*, ***1999***, *10*, 4091.

SECTION 354: AMINE - KETONE

SePh
N
Me
O

MeO_2C CO_2Me
Bu_6Sn_2, hν, PhH

CO_2Me
CO_2Me
O
N
Me

45%

Bennasar, M.-L.; Roca, T.; Griera, R.; Bosch, J. *J. Org. Chem.*, ***2001***, *66*, 7547.

O

Bn_2NH, $CeCl_3 \cdot 7\ H_2O$, NaI
SiO_2, 4.5 h

NBn_2
O

91%

Bartoli, G.; Bosco, M.; Marcantoni, E.; Petrini, M.; Sambri, L.; Torregiani, E. *J. Org. Chem.*, ***2001***, *66*, 9052.

PhCHO

H_2N—OMe, acetone
DMSO, rt, 1 d, 20%
S
N
H
CO_2H

O
OMe
Ph

45% (86% ee)

Notz, W.; Sakthivel, K.; Bui, T.; Zhong, G.; Barbas III, C.F. *Tetrahedron Lett.*, ***2001***, *42*, 199.

$C_8H_{17}C(O)ZrClCp_2$ → PhCH=NPh, $Yb(OTf)_3$ / TMSCl, THF, rt, 1 d → $C_8H_{17}COCH(NHPh)NHPh$ 63%

Kakuuchi, A.; Taguchi, T.; Hanzawa, Y. *Tetrahedron Lett.*, ***2001***, *42*, 1547.

PhCOCl, SiO_2, 4 min / microwaves 91%

Loghmani-Khouzani, H.; Sadeghi, M.M.; Safari, J.; Minaeifar, A. *Tetrahedron Lett.*, ***2001***, *42*, 4363.

1. cyclobutanone, CH_2Cl_2, *p*-TsOH / 2. BF_3•OEt_2 / 3. 15% HCl in acetone 70%

Ciblat, S.; Canet, J.-L.; Troin, Y. *Tetrahedron Lett.*, ***2001***, *42*, 4815.

methyl vinyl ketone, 10% $InCl_3$ / DCM, rt, 2 h 85%

Yadav, J.S.; Abraham, S.; Reddy, B.V.S.; Sabitha, G. *Tetrahedron Lett.*, ***2001***, *42*, 8063.

PhCN → 1. Et_3BHLi, THF, rt / 2. OTMS diene, OMe, rt, 5 h → HN, Ph, O 90%

Kawęcki, R. *Synthesis*, ***2001***, 828.

PhS diene → PhCOCHO / $MeNH_2$•HCl → PhS, Ph, N-Me 72%

Chou, S.-S.P.; Hung, C.C. *Synth. Commun.*, ***2001***, *31*, 1097.

1. CHO (crotonaldehyde), DCM
2. p-TSA, toluene, heat
3. H^+, acetone

49% overall

Glasson, S.R.; Canet, J.-L.; Troin, Y. *Tetrahedron Lett.*, ***2000***, *41*, 9797.

1. CH_2=CHMgBr, THF, rt
2. H_2O

77%

Gomtsyan, A. *Org. Lett.*, ***2000***, *2*, 11.

C_3H_7–CH=CH–CH=NH_2^+

EtOAc, toluene, 2 d
150°C, sealed tube

64% (75:25)

Sklenicka, H.M.; Hsung, R.P.; Wei, L.-L.; McLaughlin, M.J.; Gerasyuto, A.I.; Degin, S.J. *Org. Lett.*, ***2000***, *2*, 1161.

$HNEt_2$, HCHO, EtOH, 10 min

microwaves

60%

Gadhwal, S.; Baruah, M.; Prajapati, D.; Sandhu, J.S. *Synlett*, ***2000***, 341.

OMe (2-methoxypropene), PhCHO

$Cu(DS)_2$, H_2O, 0°C

65%

Kobayashi, S.; Ueno, M.; Suzuki, R.; Ishitani, H.; Kim, H.-S.; Wataya, Y. *J. Org. Chem.*, ***1999***, *64*, 6833.

N H
1. Et_2AlCl , DCM , 0°C
2. AcCl , DCM , 0°C
Ac N H

86%

Okauchi, T.; Itonaga, M.; Minami, T.; Owa, T.; Kitoh, K.; Yoshino, H. *Org. Lett.*, ***2000***, *2*, 1485.

O Ph
MeO⌒NEt$_2$, La(OTf)$_3$•n H_2O
a chiral *bis*-naphthoxide
O Ph NEt$_2$

53% (30% ee)

Yamasaki, S.; Iida, T.; Shibasaki, M. *Tetrahedron Lett.*, ***1999***, *40*, 307.

N Ph
1. 2 eq $CH_2{=}NMe_2^+$ Cl^-
CH_2Cl_2 , -30°C , 5 h
2. HCl
O Ph NMe_2 NMe_2

70%

Arend, M.; Risch, N. *Tetrahedron Lett.*, ***1999***, *40*, 6205.

CHO Bz
$PhNH_2$, THF , 0°C
Bu_2SnClH–HMPA , 2 h
N–Ph Bz

5%

Suwa, T.; Shibata, I.; Nishino, K.; Baba, A. *Org. Lett.*, ***1999***, *1*, 1579.

PhCHO
acetophenone , $PhNH_2$, 23°C , 12 h
10% dodecylbenzenesulfonic acid in water
NHPh O Ph Ph

69%

Manabe, K.; Kobayashi, S. *Org. Lett.*, ***1999***, *1*, 1965.

Ph N-Ph
OTMS Ph , 0.1 HBF_4
aq MeOH , 30 min
NHPh O Ph Ph

96%

Akiyama, T.; Takaya, J.; Kagishima, H. *Synlett*, ***1999***, 1045.

OTMS, Ph, PhCHO, MeCN; 2 eq BF_3•OEt_2, 10 eq H_2O, 0°C

NHPh O Ph Ph

97%

Akiyama, T.; Takaya, J.; Kagoshima, H. *Chem. Lett.*, ***1999***, 947.

OEt OH S

$MeNH_2$, NEt_3, MeCN, rt

OEt O NHMe

84%

Furukawa, I.; Fujisawa, H.; Abe, T.; Ohta, T. *Synth. Commun.*, ***1999***, *29*, 599.

OMe NH MeO OH

20% HCl; dioxane

HO MeO N H O

26%

Tanaka, H.; Doi, M.; Shimizu, H.; Etoh, H. *Heterocycles*, ***1999***, *51*, 2415.

SECTION 355: AMINE - NITRILE

CN

1. Chloramine-T•3 H_2O, dark, 8 h; $AgNO_3$, THF, rt
2. NaOH, THF, 3 h; sealed tube, reflux; CN

CN N CN

42x56%

Kumar, K.A.; Rai, K.M.L.; Umesha, K.B. *Tetrahedron*, ***2001***, *57*, 6993.

CN

O NH; clay, EDC, reflux

O N CN

82%

Shaikh, N.S.; Deshpande, V.H.; Bedekar, A.V. *Tetrahedron*, ***2001***, *57*, 9045.

NC, OMe (4-methoxyphenyl isocyanide) → OCO$_2$Me (allyl methyl carbonate), toluene; TMSN$_3$, cat $Pd_2(dba)_3$•$CHCl_3$, dppe, 60°C, 1 h → MeO–C$_6$H$_4$–N(allyl)CN 99%

Kamijo, S.; Jin, T.; Yamamoto, Y. *J. Am .Chem. Soc.*, ***2001***, *123*, 9453.

O_2N–C$_6$H$_4$–CHO → OH (1-naphthol), H_2O; $CH_2(CN)_2$, cetyltrimethylammonium chloride → NH_2, CN, NO_2 92%

Ballini, R.; Bosica, G.; Conforti, M.L.; Maggi, R.; Mazzacani, A.; Righi, P.; Sartori, G. *Tetrahedron,* ***2001***, *57*, 1395.

Ph, =N–Bn → 1. 10% Sc(*R*-BINOL)$_2$Li, TMSCN toluene, –20°C, 9 h; 2. H_2O → Ph, CN, NHBn

>95% (88% ee)

Chavarot, M.; Byrne, J.J.; Chavant, P.Y.; Vallée, Y. *Tetrahedron Asymm.*, ***2001***, *12*, 1147.

Ph, N, Ph → Bu_3SnCN, 0.1 Zr(O*t*-Bu)$_4$ 0.3 *R*-6-Br-BINOL, 12 h; toluene, PhH, –65°C → 0°C → NHPh, Ph, CN 92% (91% ee)

Ishitani, H.; Komiyama, S.; Hasegawa, Y.; Kobayashi, S. *J. Am. Chem. Soc.*, ***2000***, *122*, 762.

i-PrCHO → PhNHOH, Me_3SiCN, rt 5M $LiClO_4$, ether; 15 min → CN, *i*-Pr, N, Ph, OH 97%

Heydari, A.; Lavijani, H.; Emami, J.; Karami, B. *Tetrahedron Lett.*, ***2000***, *41*, 2471.

NC, Bn—N; CH_2=CHCN , MeCN, hv (254 nm) , rt; Bn, N, CN, CN, 52%; + Bn, N, CN, CN, 26%

Ishii, K.; Shimada, Y.; Sugiyama, S.; Noji, M. *J. Chem. Soc., Perkin Trans. 1*, ***2000***, 3022.

Ph, N, $CHPh_2$; 10% chiral Schiff base , toluene, 10% $Ti(Oi\text{-}Pr)_4$, 2 eq TMSCN; 1.5 *i*-PrOH (add over 210 h); CN, Ph, $NHCHPh_2$; 82% (>99% ee)

Krueger, C.A.; Kuntz, K.W.; Dzierba, C.D.; Wirschun, W.G.; Gleason, J.D.; Snapper, M.L.; Hoveyda, A.H. *J. Am. Chem. Soc.*, ***1999***, ***121***, 4284.

$N\text{-}CHPh_2$, Ph, H; 1. Ph, N, N, N, H, Ph; 2. HCN , 0.2 M in toluene, −40°C , 20 h; $NHCHPh_2$, H, Ph, CN; 96% (86% ee)

Corey, E.J.; Grogan, M.J. *Org. Lett.*, ***1999***, ***1***, 157.

SECTION 356: AMINE - ALKENE

N, Ph, I; 2 eq Ph—≡—Et , 100°C, 5% $Pd(OAc)_2$, LiCl , 36 h, *i*-Pr_2NEt , Bu_4NCl; Ph, N, Et; 75%

Roesch, K.R.; Larock, R.C. *J. Org. Chem.*, ***2001***, ***66***, 412.

NHBoc; PhOTf , $Pd(OAc)_2$, dppf; MeCN , K_2CO_3 , 909°C; Ph, NHBoc; 70%

Olofsson, K.; Sahlin, H.; Larhed, M.; Hallberg, A. *J. Org. Chem.*, ***2001***, ***66***, 544.

Ph–CH=CH–CH$_2$OH → (i-BuNH$_2$, MnO$_2$, DCM; MS 4Å, reflux, 1 d) → Ph–CH=CH–CH=N–i-Bu >95%

Blackburn, L.; Taylor, R.J.K. *Org. Lett.*, ***2001***, *3*, 1637.

H

Bu ... N–t-Bu → (30% CuI, NEt$_3$; DMA, 110°C) → Bu, N, t-Bu 86%

Kel'in, A.; Sromek, A.W.; Gevorgyan, V. *J. Am. Chem. Soc.*, ***2001***, *123*, 2074.

O—menthyl

$(OC)_5Cr$

Ph

Ph–CH=N–Ph → (20% GaCl$_3$, DCE, reflux, 3 h) → Ph, N, Ph, Ph, menthyl—O 60% (76:24 *trans:cis*)

Kagoshima, H.; Okamura, T.; Akiyama, T. *J. Am. Chem. Soc.*, ***2001***, *123*, 7182.

C_3H_7–CH=CH–CH$_2$ONHBn → (BuLi, THF, 0°C) → C_3H_7, N(Bn)OH 82%

Ishikawa, T.; Kawakami, M.; Fukui, M.; Yamashita, A.; Urano, J.; Saito, S.
J. Am. Chem. Soc., ***2001***, *123*, 7734.

OH, CHO → (Ph–CH=CH–B(OH)$_2$; BnNHMe, EtOH, rt) → Bn, N, Me, OH, Ph 81%

Petasis, N.A.; Boral, S. *Tetrahedron Lett.*, ***2001***, *42*, 539.

PhCHO → (CH$_3$CH=CHCH$_2$SiMe$_3$; CbzNH$_2$, BF$_3$•OEt$_2$, MeCN) → NHCbz + NHCbz

(84 : 16) 62%

Billet, M.; Klotz, P.; Mann, A. *Tetrahedron Lett.*, ***2001***, *42*, 631.

OMe, Me, Me, N, Et → $(CH_2{=}CH)_2CuMgBr$, THF / $BF_3{\bullet}OEt_2$, –78°C → OMe, Me, Me, NHEt 45%

Penkett, C.S.; Simpson, I.D. *Tetrahedron Lett.*, ***2001***, *42*, 1179.

BnHN, Bu → cat $PdCl_2$, KCl , DMA / 100°C , 6 h → Bu, N, Bn 53%

Liu, L.T.; Huang, H.-L.; Wang, C.-L.J. *Tetrahedron Lett.*, ***2001***, *42*, 1329.

CF_3, CO_2Me, N, SO_2Me → $Cl_2(PCy_3)_2Ru{=}CHPh$ / CH_2Cl_2 , rt → CF_3, CO_2Me, N, SO_2Me 77%

Osipov, S.N.; Kobelíkova, N.M.; Shchetnikov, G.T.; Kolomiets, A.F.; Bruneau, C.; Dixneuf, P.H. *Synlett*, ***2001***, 621.

$OCOC_6F_5$, N, Ph → 10% $Pd(PPh_3)_4$, DMF / 5 eq NEt_3 , MS 4Å / 110°C , 30 min → N, Ph 82%

Kitamura, M.; Zaman, S.; Narasaka, K. *Synlett*, ***2001***, 974.

Ph, CHO → NH_2NHBz , $Sn(CH_2CH{=}CH_2)_4$ / 5% $Sc(OTf)_3$, H_2O/THF / 30°C , 20 h → NHNHBz, Ph 85%

Kobayashi, S.; Hamada, T.; Manabe, K. *Synlett*, ***2001***, 1140.

Me_3Si—C≡C—C(O)Me; AcOH, PhMe, 50°C — 79%

Bagley, M.C.; Dale, J.W.; Bower, J. *Synlett*, ***2001***, 1149.
In toluene, ethyl ester, catalyzed by 15% $ZnBr_2$ (90%). Bagley, M.C.; Dale, J.W.; Hughes, D.D.; Ohnesorge, M.; Philips, N.G.; Bower, J. *Synlett*, ***2001***, 1523.

toluene, pressure tube, cat $Pd(OAc)_2$, dppf; *t*-BuONa, Ar, 90°C, 15 h — 81%

Sung, J.J.; Yee, N.K. *Tetrahedron Lett.*, ***2001***, *42*, 2937.

0.2 $TsNH_2$, DMF, 1 h; 2 eq K_2CO_3, 90°C — 76%

Kim, J.N.; Lee, H.F.; Lee, K.Y.; Kim, H.S. *Tetrahedron Lett.*, ***2001***, *42*, 3737.

$PhNH_2$; , $PhNH_2$•HCl, 9 h; cat $[Ru_3(CO)_{12}]$, 140°C, neat — 95%

Tokunaga, M.; Ota, M.; Haga, M.-a.; Wakatsuki, Y. *Tetrahedron Lett.*, ***2001***, *42*, 3865.

, DCE, rt; 10% $SmI_2(thf)_2$, 1 d — 61%

Collin, J.; Jaber, N.; Lannon, M.I. *Tetrahedron Lett.*, ***2001***, *42*, 7405.

DNP–NHOH, $FeCl_2/FeCl_3$; DNP = 2,4-dinitrophenyl — 75%

Singh, S.; Nicholas, K.M. *Synth. Commun.*, ***2001***, *31*, 3087.

C_6H_{13} O N O Ph I — In, MeOH, reflux, 4 h → NHC_6H_{13} H Ph 87%

Yadav, J.S.; Bandyopadhyay, A.; Reddy, B.V.S. *Tetrahedron Lett.*, ***2001***, *42*, 6385.

Ts N Ph + OMe TMSO — 0.3 $YbCl_3$, toluene; 75°C, 6 h → OMe N TMSO Ph 55%

Ray, C.A.; Risberg, E.; Somfai, P. *Tetrahedron Lett.*, ***2001***, *42*, 9289.

N–NMe_2 Ph — Li MeO; THF, rt, 16 h → NMe_2 N Ph MeO 53% + Ph NMe_2 N MeO 22%

Breuil-Desvergnes, V.; Goré, J. *Tetrahedron*, ***2001***, *57*, 1951.

Me Ph Me—N Li — 1. 2 eq Me MgBr; 2. 3 eq $ZnBr_2$; 3. H_3O^+ → Me Me N Me Ph

62% (88/12 dr)

Rezaei, H.; Marek, I.; Normant, J.F. *Tetrahedron*, ***2001***, *57*, 2477.

$PhNHNH_2$ — acetophenone; N—Bu/$AlCl_3$, 185°C, 1 h; ionic liquid medium → Ph N H 73%

Rebeiro, G.L.; Khadilkar, B.M. *Synthesis*, ***2001***, 370.

O

CO_2Et , silica agel

PhCHO

urea , microwaves , 3 min

EtO_2C CO_2Et

N H

90%

Yadav, J.S.; Redy, B.V.S.; Reddy, P.T. *Synth. Commun.*, ***2001***, *31*, 425.

Cl

, $PhNH_2$,H_2O

Bieliaca clay , 40°C , 5 h

Cl

Cl

HN

96%

Sartori, G.; Bigi, F.; Maggi, R.; Mazzacani, A.; Oppici, G. *Eur. J. Org. Chem.*, ***2001***, 2513.

Ph

piperidine , THF , reflux , 20 h

cat Rh(cod) BF_4 , 2 PPh_3

Ph N

55%

Tillack, A.; Trauthwein, H.; Hartung, C.G.; Eichberger, M.; Pitter, S.; Jansen, A.; Beller, M. *Monat. Chem.*, ***2000***, *141*, 1327.

SPh

N SPh

Bn

$[Cp_2Ti / P(OEt)_3]_2$, 0.03 M

THF

N

Bn

61%

Fujiwara, T.; Kato, Y.; Takeda, T. *Heterocycles*, ***2000***, *52*, 147.

O

O

Me_2N $OSiMe_3$

2h

NMe_2

55%

Kardon, F.; Mörtl, M.; Knausz, D. *Tetrahedron Lett.*, ***2000***, *41*, 8937.

Ph

N O

1. 1.5 eq Cp_2TiMe_2
1.1M toluene , 110°C

2. flash chromatography
SiO_2

PhHN

O

63x80%

Tehrani, K.A.; De Kimpe, N. *Tetrahedron Lett.*, ***2000***, *41*, 1975.

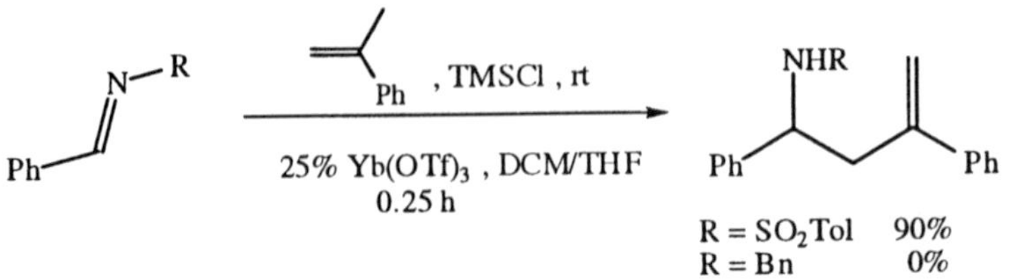
N—R
Ph
Ph
, TMSCl , rt
25% Yb(OTf)$_3$, DCM/THF
0.25 h
NHR
Ph
Ph
R = SO_2Tol 90%
R = Bn 0%

Yamanaka, M.; Nishida, A.; Nakagawa, M. *Org. Lett.*, ***2000***, *2*, 159.

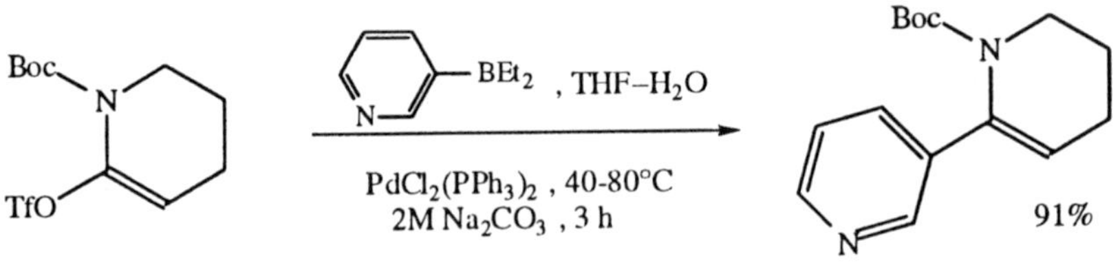
Boc
N
TfO
BEt$_2$, THF–H_2O
N
$PdCl_2(PPh_3)_2$, 40-80°C
2M Na_2CO_3 , 3 h
Boc
N
N
91%

Occhiato, E.G.; Trabocchi, A.; Guarna, A. *Org. Lett.*, ***2000***, *2*, 1241.

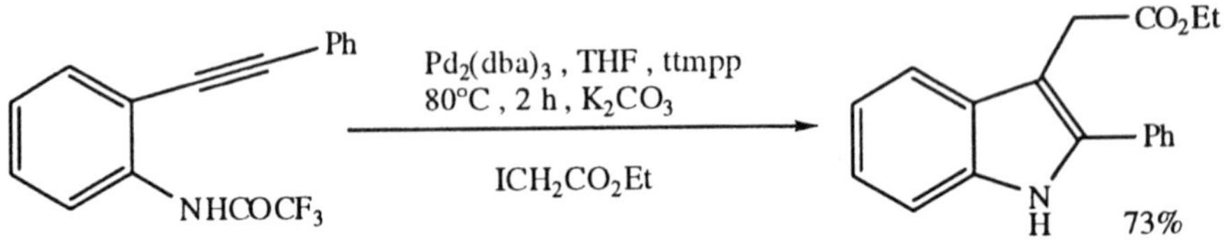
Ph
$NHCOCF_3$
$Pd_2(dba)_3$, THF , ttmpp
80°C , 2 h , K_2CO_3
ICH_2CO_2Et
CO_2Et
Ph
N
H
73%

ttmpp = tris(2,4,6-trimethoxyphenyl)phosphine

Arcadi, A.; Cacchi, S.; Fabrizi, G.; Marinelli, F. *Synlett*, ***2000***, 394.

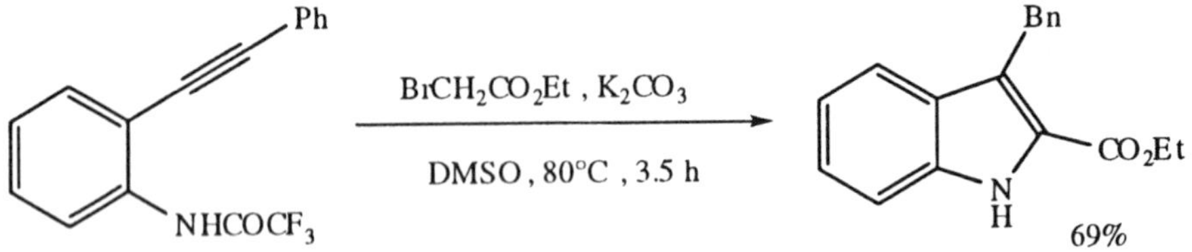
Ph
$NHCOCF_3$
$BrCH_2CO_2Et$, K_2CO_3
DMSO , 80°C , 3.5 h
Bn
CO_2Et
N
H
69%

Arcadi, A.; Cacchi, S.; Fabrizi, G.; Marinelli, *Synlett*, ***2000***, 647.

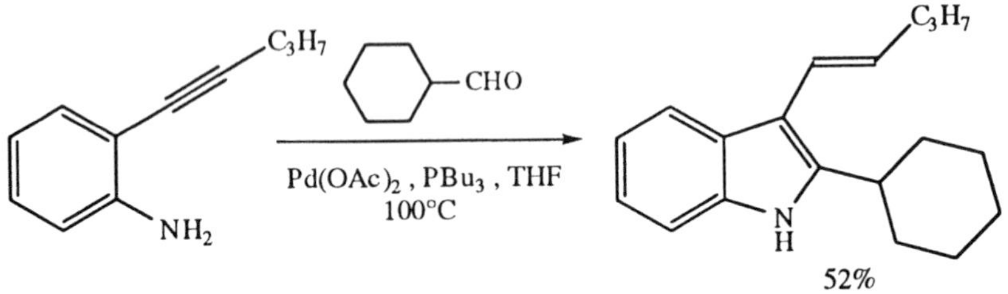
C_3H_7
NH_2
CHO
$Pd(OAc)_2$, PBu_3 , THF
100°C
C_3H_7
N
H
52%

Takeda, A.; Kamijo, S.; Yamamoto, Y. *J. Am. Chem. Soc.*, ***2000***, *122*, 5662.

Ph, H_2N ; , CuI , TMG ; $PdCl_2(PPh_3)_2$, DMF , 20 h ; MsHN, I ; Ph, H_2N, N, SO_2Me

73% (95% ee)

Messina, F.; Botta, M.; Corelli, F.; Villani, C. *Tetrahedron Asymm.*, ***2000***, *11*, 1681.

CO_2i-Pr, i-PrO_2C, N ; 1. 3.5 eq Na NH_3/THF , –78°C 2. MeI ; 3. NH_4Cl ; CO_2i-Pr, i-PrO_2C, N H

99%

Donohoe, T.J.; McRiner, A.J.; Sheldrake, P. *Org. Lett.*, ***2000***, *2*, 3861.

I, NHCHO ; 1. TMS—≡ , CuI $PdCl_2(PPh_3)_2$; 2. $POCl_3$, i-Pr_2NH 3. 2.2 eq Bu_3SnH , AIBN 4. H_3O^+ ; TMS, N H

93x82x82%

Ranier, J.D.; Kennedy, A.R. *J. Org. Chem.*, ***2000***, *65*, 6213.

OTf ; NMe_2 , MeCN ; $Pd(OAc)_2$, dppf , K_2CO_3 sealed tube , 80°C ; NMe_2

40%

Olofsson, K.; Larhed, M.; Hallberg, A. *J. Org. Chem.*, ***2000***, *65*, 7235.

Bn—N—H ; O , 15% $TiCl_4$; 10% $CpRu(MeCN)_3 PF_6$ DMF , 60°C ; O, Bn, N

Trost, B.M.; Pinkerton, A.B.; Kremzow, D. *J. Am. Chem. Soc.*, ***2000***, *122*, 12007.

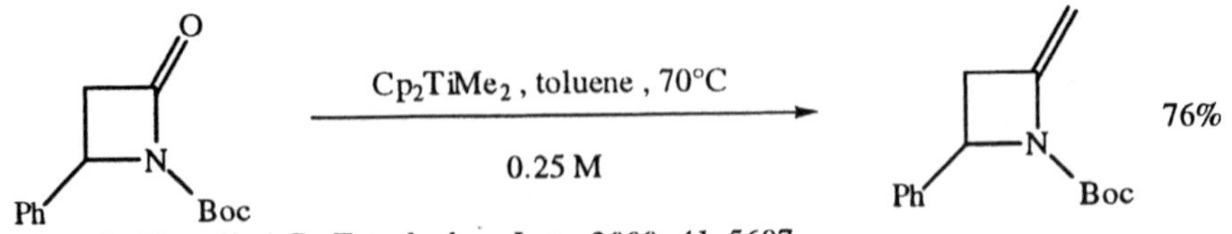
O
N
Ph
Boc
Cp2TiMe2 , toluene , 70°C
0.25 M
N
Ph
Boc
76%

Martínez, I.; Howell, A.R. *Tetrahedron Lett.*, ***2000***, *41*, 5607.

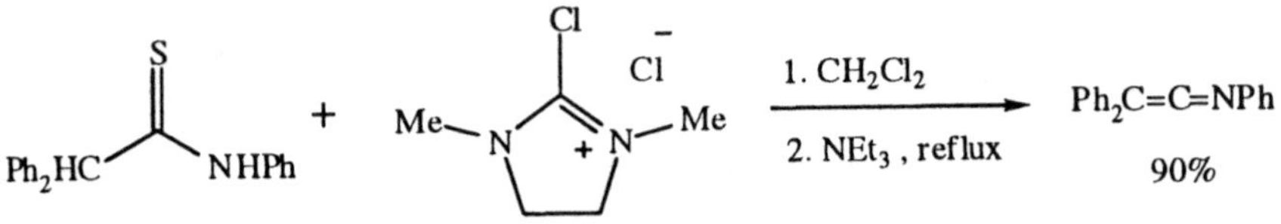
S
Ph2HC
NHPh
+
Cl
Cl
Me
N
+
N
Me
1. CH2Cl2
2. NEt3 , reflux
Ph2C=C=NPh
90%

Shimizu, M.; Gama,Y.; Takagi, T.; Shibakami, M.; Shibuya, I. *Synthesis*, ***2000***, 517.

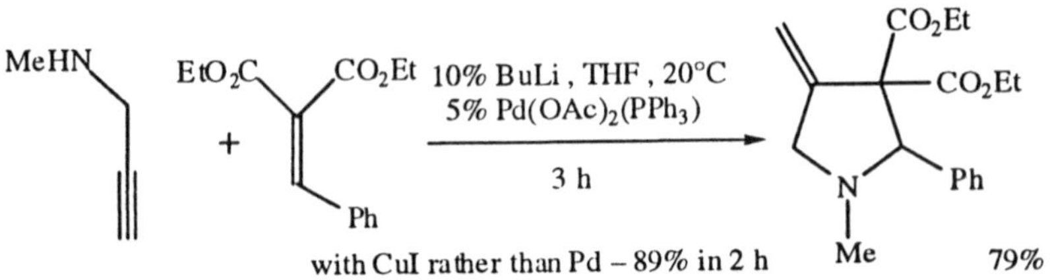
MeHN
EtO2C
CO2Et
+
Ph
10% BuLi , THF , 20°C
5% Pd(OAc)2(PPh3)
3 h
CO2Et
CO2Et
N
Ph
Me
with CuI rather than Pd – 89% in 2 h
79%

Clique, B.; Monteiro, N.; Balme, G. *Tetrahedron Lett.*, ***1999***, *40*, 1301.

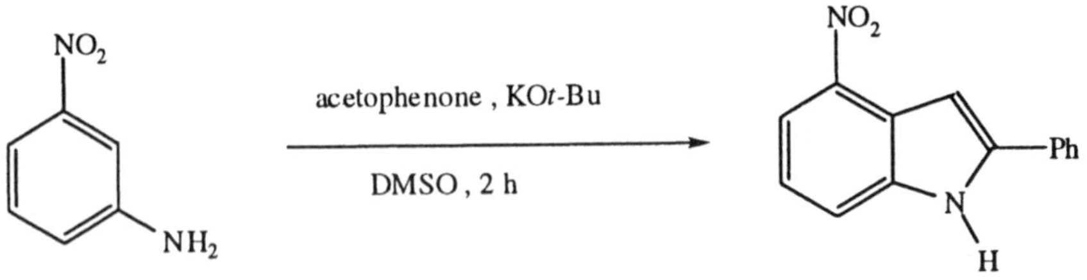
NO2
NH2
acetophenone , KOt-Bu
DMSO , 2 h
NO2
Ph
N
H

Moskalev, N.; Makosza, M. *Tetrahedron Lett.*, ***1999***, *40*, 5395.

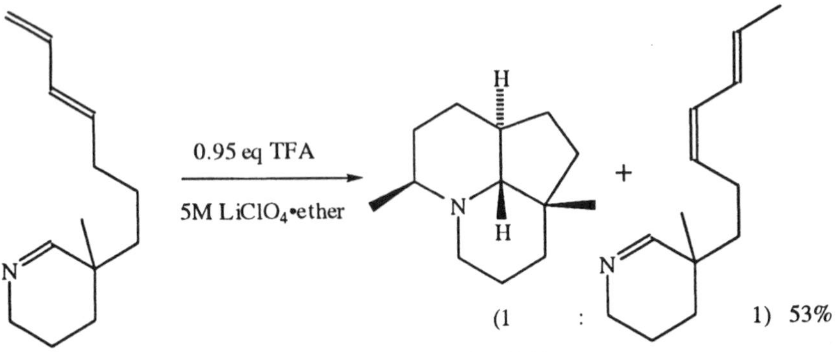
N
0.95 eq TFA
5M LiClO4•ether
H
N
H
+
N
(1 : 1) 53%

Grieco, P.A.; Kaufman, M.D. *J. Org. Chem.*, ***1999***, *64*, 6041.

PhHN—≡ → (i-Pr_2NH, $(HCHO)_{aq}$, CuI; dioxane, reflux, 11 h) → N-Ph 75%

Jayaprakash, K.; Venkatachalam, C.S.; Balasubramanian, K.K. *Tetrahedron Lett.*, ***1999***, *40*, 6493.

Me, H, H, NHMts → (cat $Pd(PPh_3)_4$, 4 PhI, 4 K_2CO_3, dioxane; reflux) → Me, Ph, N—Mts + Me, Ph, N—Mts (82 : 18) 80%

Ohno, H.; Toda, A.; Miwa, Y.; Taga, T.; Osawa, E.; Tamaoka, Y.; Fujii, N.; Ibuka, T. *J. Org. Chem.*, ***1999***, *64*, 2992.

Ph—≡— → (Bn_2NH, dioxane, 100°C; 5% $Pd(PPh_3)_4$, 10% $PhCO_2H$) → Ph, NBn_2 98%

Kadota, I.; Shibuya, A.; Lutete, L.M.; Yamamoto, Y. *J. Org. Chem.*, ***1999***, *64*, 4570.

Cl, NHBn → (1. BuC≡CTs, DCM, rt, 5 h; 2. 2 eq LDA, THF, –78°C; 3. basic alumina) → Ts, N, Bu, Bn 94%

Back, T.G.; Nakajima, K. *Org. Lett.*, ***1999***, *1*, 261.

NC, OAc → (2 eq $C_3H_7NH_2$, THF; rt, 8 h) → C_3H_7HN, CN 87% (84:16 *E*:*Z*)

Hbaïeb, S.; Latiri, Z.; Amri, H. *Synth. Commun.*, ***1999***, *29*, 981.

REVIEWS:

"Addition of Carbon-Centered Radicals to Imines and Related Compounds," Friestad, G.K. *Tetrahedron*, ***2001***, *57*, 5461.

"Recent Developments in Imino-Diels-Alder Reactions," Buonora, P.; Olsen, J.-C.; Oh, T. *Tetrahedron*, ***2001***, *57*, 6099.

"Generation and Reactivity of α-Metalated Vinyl Ethers," Friesen, R.W. *J. Chem. Soc., Perkin Trans. 1*, ***2001***, 1969.

SECTION 357: ESTER - ESTER

5% Grubbs' catalyst 2, DCM
3 mM, 40°C

81%

Lee, C.W.; Grubbs, R.H. *J. Org. Chem.*, ***2001***, *66*, 7155.

CHO

2 eq Ac_2O
0.2% $FeCl_3$
0°C, 30 min

OAc
OAc

67%

Trost, B.M.; Lee, C.B. *J. Am. Chem. Soc.*, ***2001***, *123*, 3671.

OTMS
t-BuS

EtO_2C

10% bis(oxazoline)Cu(II) complex
DCM, −78°C

t-BuS
CO_2Et

73% (99% ee, 99:1 *syn:anti*)

Evans, D.A.; Scheidt, K.A.; Johnson, J.N.; Willis, M.C. *J. Am. Chem. Soc.*, ***2001***, *123*, 4480.

Ac_2O, zeolite HY, 10 h

OAc
OAc

85%

Ramesh, P.; Reddy, V.L.N.; Venugopal, D.; Subrahmanyam, M.; Venkateswarlu, Y. *Synth. Commun.*, ***2001***, *31*, 2599.

OMe
$^-$OTs
Et OH

Ph (styrene) → OTs, OTs, Ph — −30°C — 65% ee

Hirt, U.H.; Schuster, M.F.H.; French, A.N.; Wiest, O.G.; Wirth, T.
Eur. J. Org. Chem., ***2001***, 1569.

PhCHO — 3% $VOCl_3$, 2 eq Zn , 2 eq Ac_2O / 80°C → OAc, Ph, Ph, OAc — 85% (80:20 *dl:meso*)

Hirao, T.; Takeuchi, H.; Ogawa, A.; Sakurai, H. *Synlett*, ***2000***, 1658.

CHO, Br — $PhCH_2CO_2H$, NEt_3 / cat $PdCl_2$/4 PPh_3 , CO , MeCN → O, Ph, O, O, O — 81%

Cho, C.S.; Back, D.Y.; Kim, H.-Y.; Shim, S.C.; Oh, D.H. *Synth. Commun.*, ***2000***, *39*, 1139.

REVIEWS:

"1, 2-Diacetals: A New Opportunity for Organic Synthesis," Ley, S.V.; Baeschlin, D.K.; Dixon, D.J.; Foster, A.C.; Ince, S.J.; Priepke, H.W.M.; Reynolds, D.J. *Chem. Rev.*, ***2001***, *101*, 53.

Also via Dicarboxylic Acids: Section 312 (Carboxylic Acids - Carboxylic Acids)
Hydroxy-esters Section 327 (Alcohol - Ester)
Diols Section 323 (Alcohol - Alcohol)

SECTION 358: ESTER - ETHER, EPOXIDE, THIOETHER

Bu—≡ — $PhSCO_2Me$ / $Pd(PCy_3)_2$, 110°C toluene , octane → Bu, PhS, CO_2Me — 96%

Hua, R.; Takeda, H.; Onozawa, S.-y.; Abe, Y.; Tanaka, M.
J. Am. Chem. Soc., ***2001***, *123*, 2899.

CO_2Me N$_2$ Oi-Pr → 1% Ru catalyst, hexane / –5°C, 72 h → CO_2Me O

98% (98% ee)

Davies, H.M.L.; Grazini, M.V.A.; Aouad, E. *Org. Lett.*, ***2001***, *3*, 1475.

HO OH → Dess Martin periodinane / CH_2Cl_2, rt, 35 min → O OAc

92%

Roels, J.; Metz, P. *Synlett*, ***2001***, 789.

HO_2C → PhSeSePh, DDQ / MeCN, 30°C → O O SePh

95%

Tiecco, M.; Testaferri, L.; Temnperini, A.; Bagnoli, L.; Marini, F.; Santi, C. *Synlett*, ***2001***, 1767.

CO_2Et EtO_2C → DCM/hexane, 4 h / [PhSH/BuLi], –50°C → EtO_2C PhS CO_2Et + EtO_2C SPh CO_2Et

(40 : 60) 91%

Kamimura, A.; Kawahara, F.; Omata, Y.; Murakami, N.; Morita, R.; Otake, H.; Mitsudera, H.; Shirai, M.; Kakehi, A. *Tetrahedron Lett.*, ***2001***, *42*, 8497.

O O O → 5% Grubbs' catalyst II 0.05 M, DCM, 40°C / CH_2=CHPh → O O O

81%

Choi, T.-L.; Grubbs, R.H. *Chem. Commun.*, ***2001***, 2648.

HN O O → PhSH, C_3H_7OH, C_3H_7ONa / 6 h → PhS NH_2

98%

Ishibashi, H.; Uegaki, M.; Sakai, M.; Takeda, Y. *Tetrahedron*, ***2001***, *57*, 2115.

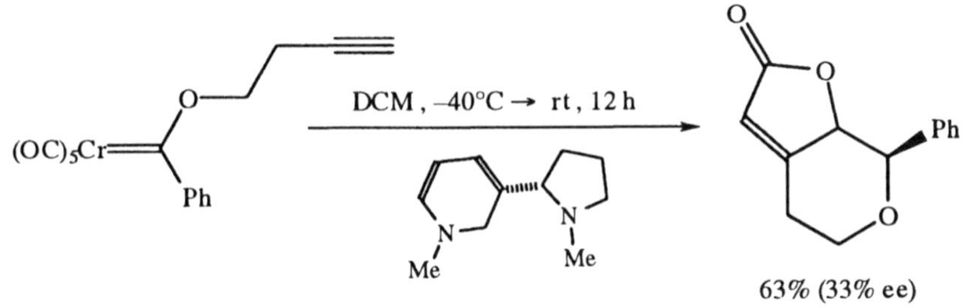

O

$(OC)_5Cr$ Ph

DCM, –40°C → rt, 12 h

N Me N Me

O O Ph O

63% (33% ee)

Rudler, H.; Parlier, A.; Certal, V.; Frison, J.-C. *Tetrahedron Lett.*, ***2001***, *42*, 5235.

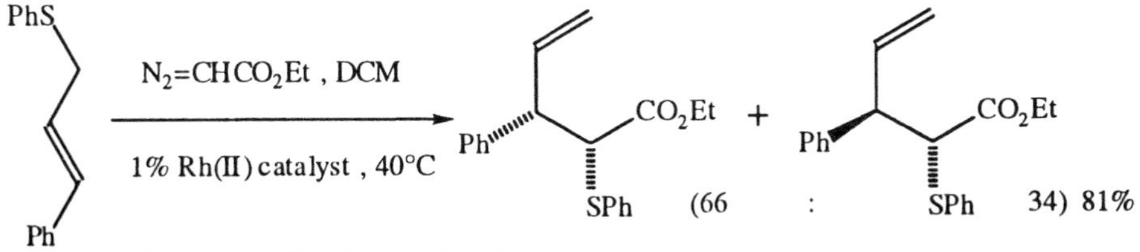

PhS Ph

N_2=CHCO$_2$Et, DCM

1% Rh(II) catalyst, 40°C

Ph CO_2Et SPh + Ph CO_2Et SPh (66 : 34) 81%

Kitagaki, S.; Yanamoto, Y.; Okubo, H.; Nakajima, M.; Hashimoto, S. *Heterocycles*, ***2001***, *54*, 623.

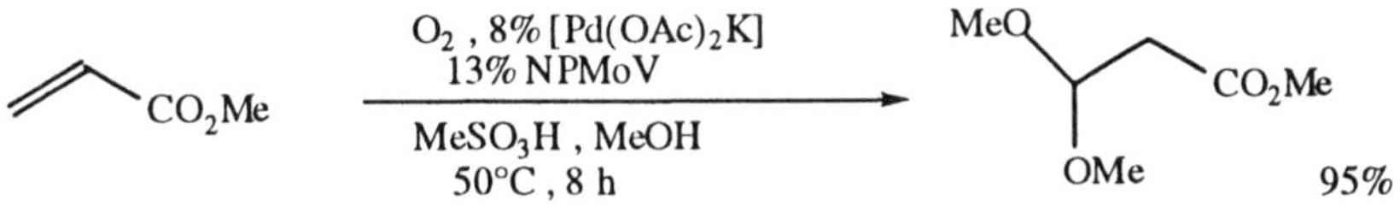

CO_2Me

O_2, 8% [Pd(OAc)$_2$K]
13% NPMoV

$MeSO_3H$, MeOH
50°C, 8 h

MeO OMe CO_2Me 95%

NPMoV = molybdovanadophosphate

Kishi, A.; Sakaguchi, S.; Ishii, Y. *Org. Lett.*, ***2000***, *2*, 523.

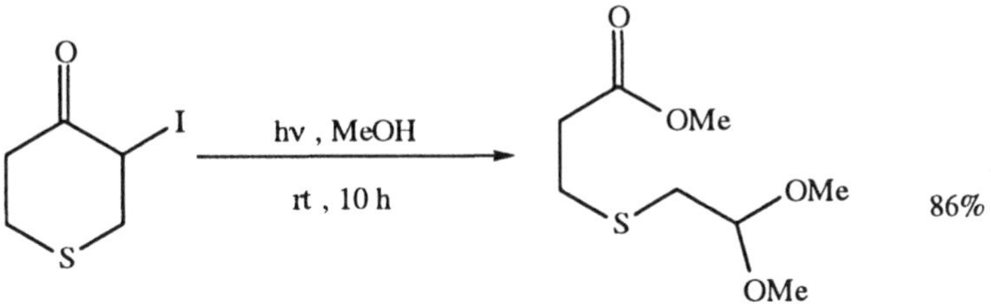

O I S

hv, MeOH

rt, 10 h

O OMe S OMe OMe 86%

Ji, S.-J.; Horiuchi, C.A. *Bull. Chem. Soc. Jpn.*, ***2000***, *73*, 1645.

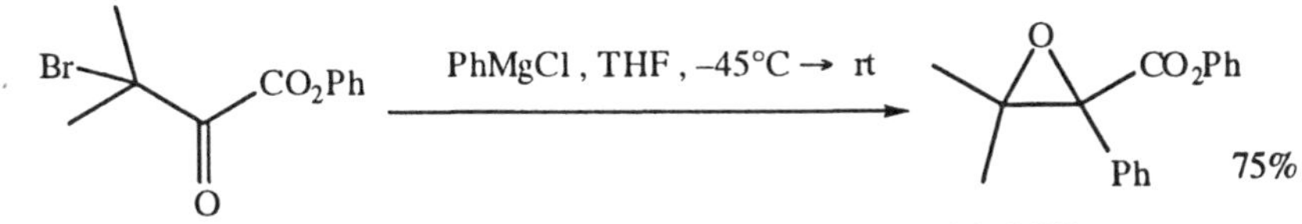

Br CO_2Ph O

PhMgCl, THF, –45°C → rt

O CO_2Ph Ph 75%

Jung, M.E.; Mengel, W.; Newton, T.W. *Synth. Commun.*, ***1999***, *29*, 3659.

PhSSPh → 1. Cp_2TiCl_2/*i*-BuMgBr, THF, –10°C → 0°C; 2. $BrCH_2CO_2$–Et, THF–HMPA → $PhSCH_2CO_2Et$ 80%

Huang, X.; Zheng, W.-X. *Synth. Commun.*, ***1999***, *29*, 1297.

CH_2=CHOEt, 10% BF_3•OEt_3

OMe, CF_3, N, HN, OEt

56% (>98:2 *cis:trans*)

Crousse, B.; Bégué, J.-P.; Bonnet-Delpon, D. *J. Org. Chem.*, ***2000***, *65*, 5009.

$Ce(SO_4)_4$•4 H_2O; O_2, MeOH, 50°C, 8 h

O, Cl, MeO, MeO, OMe, O 65%

He, L.; Horiuchi, C.A. *Bull. Chem. Soc., Jpn.*, ***1999***, *72*, 2515.

SECTION 359: ESTER - HALIDE, SULFONATE

BrF_3, $CDCl_3$; 0°C

N–OMe, O, O, F, F, OMe, N, O, O 50%

Rozen, S.; Ben-David, I. *J. Org. Chem.*, ***2001***, *66*, 496.

10% ammonium salts, THF, –78°C → rt, 3 h

Ph, O, Cl, Cl, O, Cl, Cl, Cl, Cl, Cl, Ph, O, C_6Cl_5

81% (99% ee)

Wack, H.; Taggi, A.E.; Hafez, A.M.; Drury III, W.J.; Lectka, T. *J. Am. Chem. Soc.*, ***2001***, *123*, 1531.

$FeCl_3$, NaI , MeCN

reflux

76%

Chavan, S.P.; Sharma, A.K. *Tetrahedron Lett.*, ***2001***, *42*, 4923.

dilauroyl peroxide

PhH , reflux

89% (5.4:1 dr)

Ollivier, C.; Bark, T.; Renaud, P. *Synthesis*, ***2000***, 1598.

I_2 , aq MeCN , 16 h

reflux

98% (91:9 *trans:cis*)

Ha, H.-J.; Lee, S.-Y.; Park, Y.-S. *Synth. Commun.*, ***2000***, *30*, 3645.

SECTION 360: ESTER - KETONE

10% $YbCl_3$
DCM , rt

10% TsOH
$MeNO_2$, rt

84% (96% ee)

77% (97% ee)

Zhu, Y.; Shu, L.; Tu, Y.; Shi, Y. *J. Org. Chem.*, ***2001***, *66*, 1818.

Candida antarctica lipase B
10 torr , 40°C

95%

Córdova, A.; Janda, K.D. *J. Org. Chem.*, ***2001***, *66*, 1906.

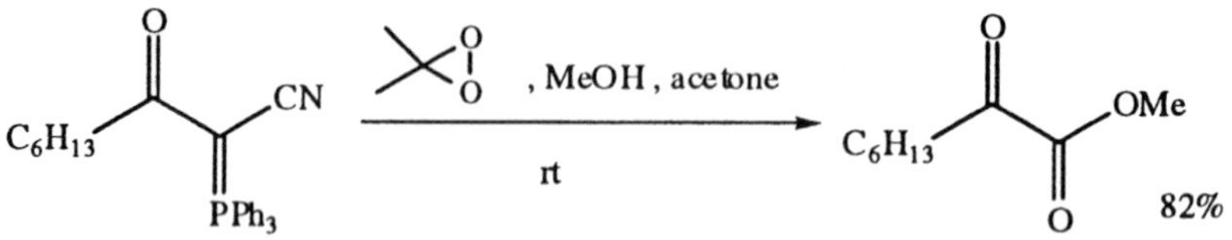

O; C_6H_{13}; CN; PPh_3 → O, O (dimethyldioxirane), MeOH , acetone; rt → O; C_6H_{13}; OMe; O

82%

Wong, M.-K.; Yu, C.-W.; Yuen, W.-H.; Yang, D. *J. Org. Chem.*, ***2001***, *66*, 3606.

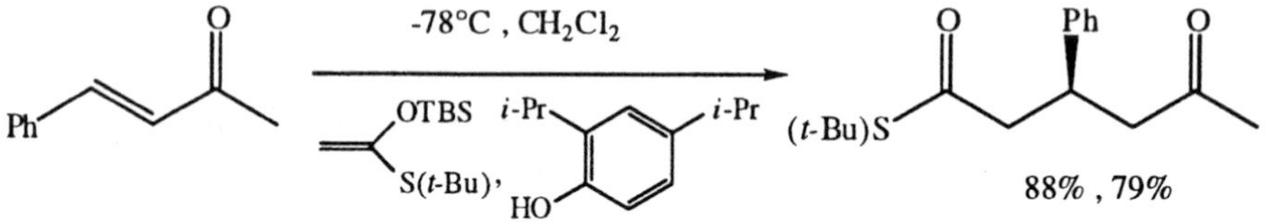

O; Ph → $-78°C$, CH_2Cl_2; OTBS, S(*t*-Bu), *i*-Pr, *i*-Pr, HO → O; Ph; O; (*t*-Bu)S

88% , 79%

Harada, T.; Iwai, H.; Yakatsuki, H.; Fujita, K.; Kubo, M.; Oku, A. *Org. Lett.*, ***2001***, *3*, 2101.

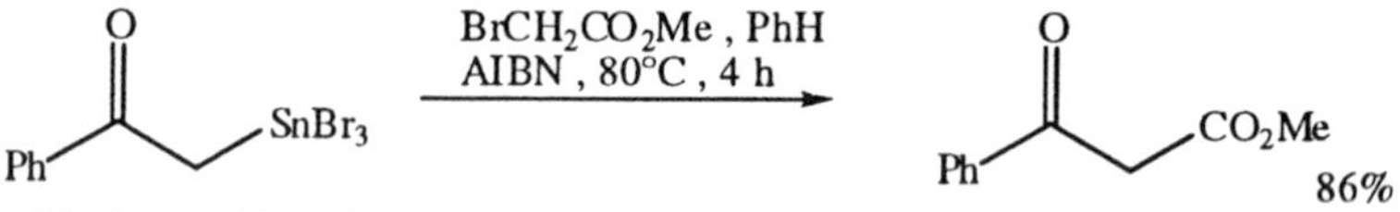

O; Ph; $SnBr_3$ → $BrCH_2CO_2Me$, PhH; AIBN , 80°C , 4 h → O; Ph; CO_2Me

86%

Miura, K.; Fukisawa, N.; Saito, H.; Wang, D.; Hosomi, A. *Org. Lett.*, ***2001***, *3*, 2591.

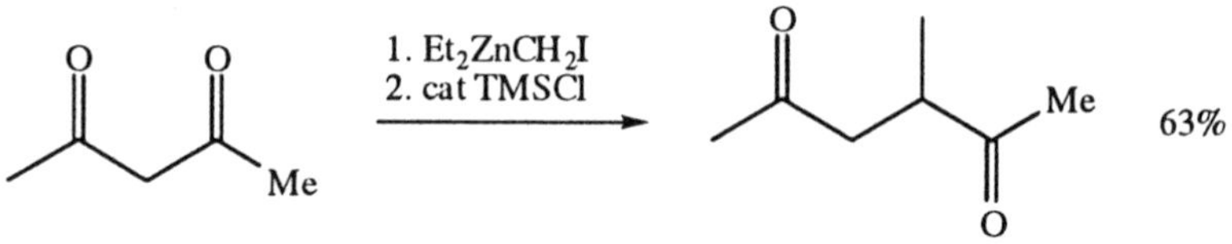

O; O; Me → 1. Et_2ZnCH_2I; 2. cat TMSCl → O; Me; O

63%

Hilgenkamp, R.; Zercher, C.K. *Org. Lett.*, ***2001***, *3*, 3037.

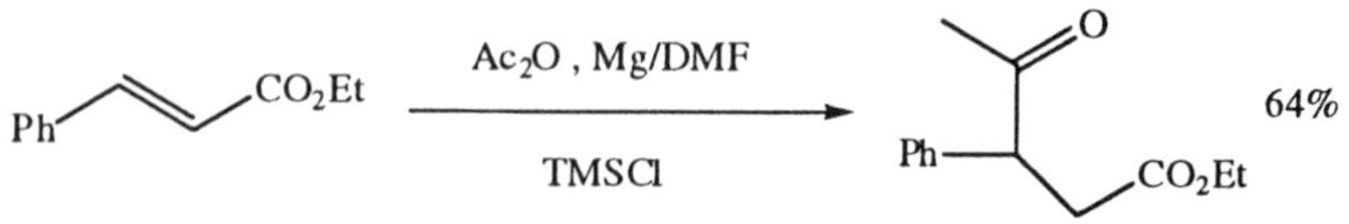

Ph; CO_2Et → Ac_2O , Mg/DMF; TMSCl → O; Ph; CO_2Et

64%

Ohno, T.; Sakai, M.; Ishino, Y.; Shibata, T.; Maekawa, H.; Nishiguchi, I. *Org. Lett.*, ***2001***, *3*, 3439.

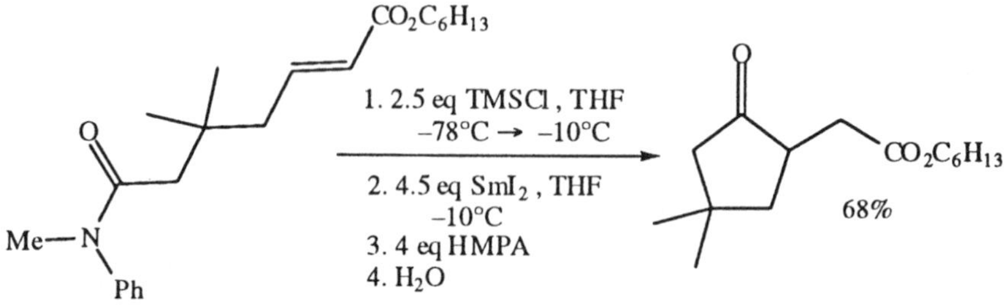

$CO_2C_6H_{13}$; O; Me—N; Ph → 1. 2.5 eq TMSCl , THF; −78°C → −10°C; 2. 4.5 eq SmI_2 , THF; −10°C; 3. 4 eq HMPA; 4. H_2O → O; $CO_2C_6H_{13}$

68%

McDonald, C.E.; Galka, A.M.; Green, A.I.; Keane, J.M.; Kowalchick, J.E.; Micklitsch, C.M.; Wisnoski, D.D. *Tetrahedron Lett.*, ***2001***, *42*, 163.

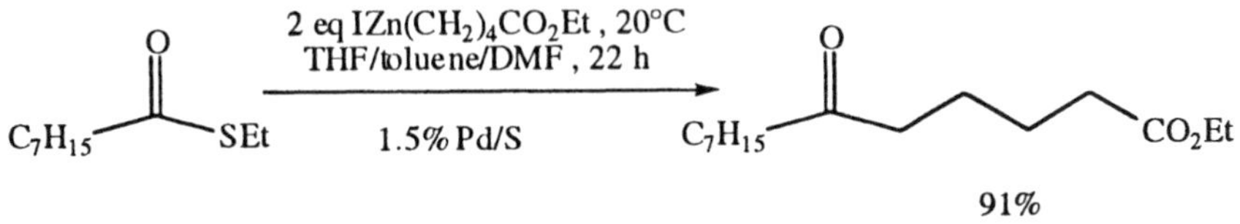

2 eq $IZn(CH_2)_4CO_2Et$, 20°C
THF/toluene/DMF , 22 h
1.5% Pd/S

C_7H_{15} SEt → C_7H_{15} CO_2Et

91%

Shimizu, T.; Seki, M. *Tetrahedron Lett.*, ***2001***, *42*, 429.

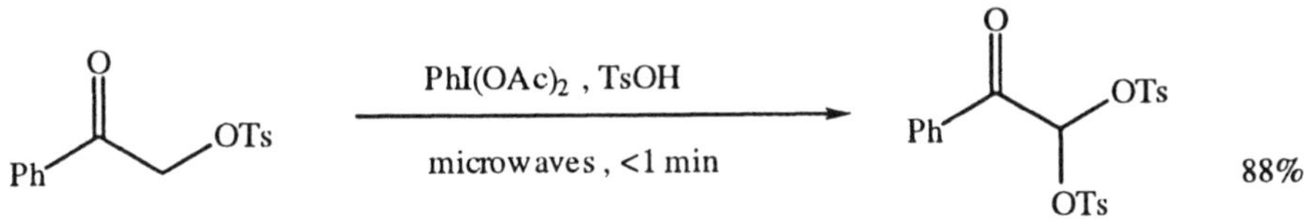

$PhI(OAc)_2$, TsOH
microwaves , <1 min

Ph OTs → Ph OTs OTs 88%

Lee, J.C.; Choi, J.-H. *Synlett*, ***2001***, 234.

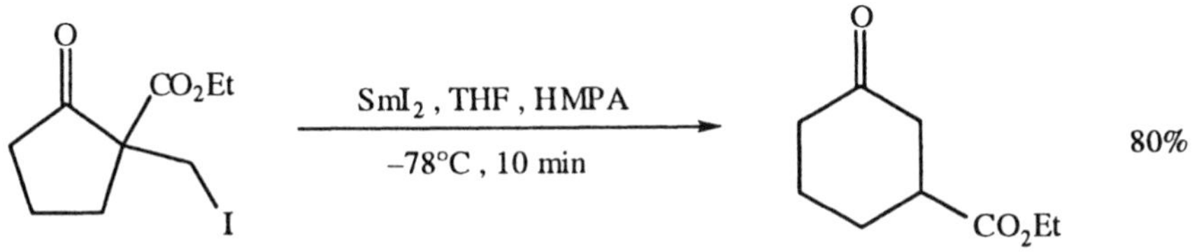

SmI_2 , THF , HMPA
–78°C , 10 min

CO_2Et I → CO_2Et 80%

Chung, S.H.; Cho, M.S.; Choi, J.Y.; Kwon, D.W.; Kim, Y.H. *Synlett*, ***2001***, 1266.

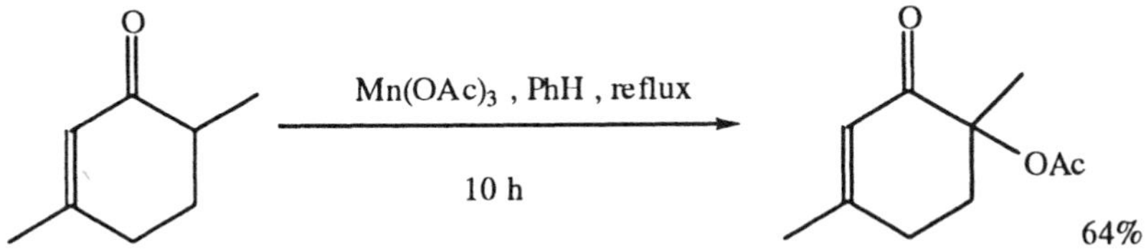

$Mn(OAc)_3$, PhH , reflux
10 h

OAc 64%

Tanyeli, C.; Sezen, B.; Iyigün, Ç.; Elmali, O. *Tetrahedron Lett.*, ***2001***, *42*, 6397.

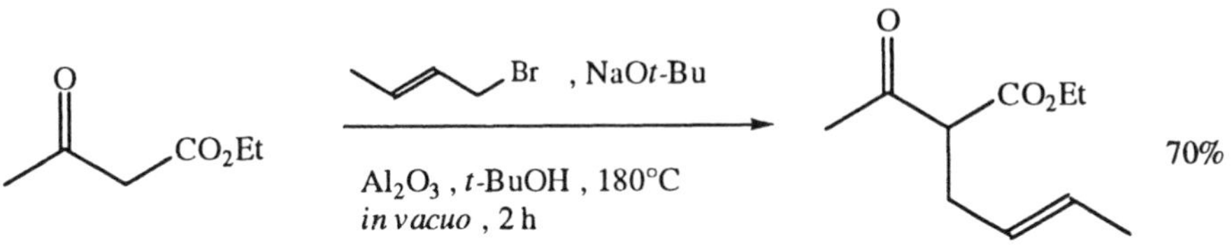

Br , NaO*t*-Bu
Al_2O_3 , *t*-BuOH , 180°C
in vacuo , 2 h

CO_2Et → CO_2Et 70%

Bhar, S.; Chaudhuri, S.K.; Sahu, S.G.; Panja, C. *Tetrahedron*, ***2001***, *57*, 9011.

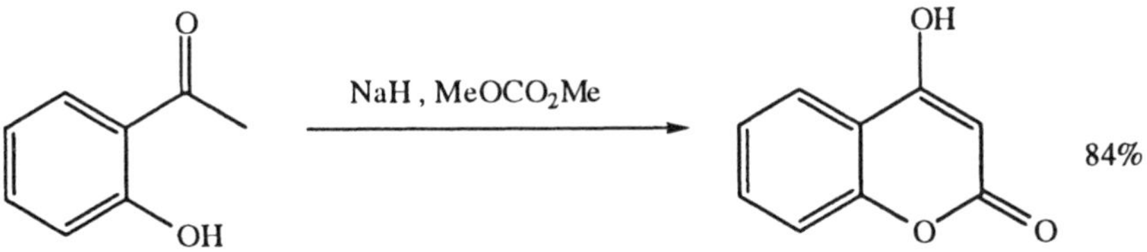

NaH , $MeOCO_2Me$

OH → OH O O 84%

Jung, J.-C.; Jung, Y.-J.; Park, O.-S. *Synth. Commun.*, ***2001***, *31*, 1195.

$i\text{-Pr–CO}_2\text{Ph}$ → ($ZrCl_4$, $i\text{-Pr}_2\text{NEt}$, DCM; –20°C, 3 h) → i-Pr–C(=O)–C(H_3C)(CH_3)–CO_2Ph 72%

Tanabe, Y.; Hamasaki, R.; Funakoshi, S. *Chem. Commun.*, ***2001***, 1674.

1. CH_2=$CHCO_2Me$, THF, 6 h; 10% $RhCl(PPh_3)_3$, 135°C
2. aq HCl

Ph–C(=O)–CH_2CH_2–CO_2Me 73%

Willis, M.C.; Sapmaz, S. *Chem. Commun.*, ***2001***, 2558.

$Mn(OAc)_3$; PhH, 8h

61%

Tanyeli, C.; Sezen, B. *Tetrahedron Lett.*, ***2000***, ***41***, 7973.

OCOPh, CN → (CO_2Me; LHMDS) → Ph, CO_2Me 65%

Kraus, G.A.; Dneprovskaia, E. *Tetrahedron Lett.*, ***2000***, ***41***, 21.

OTMS OTMS, OEt → (, 2 eq $TiCl_4$; DCM, –78°C) → CO_2Et, HO 63%

Langer, P.; Köhler, V. *Org. Lett.*, ***2000***, ***2***, 1597.

OH, Ph, CO_2Et → (O_3, MeOH, DMS; –78°C) → Ph–C(=O)–CO_2Et 77%

Mahmood, S.J.; McLaughlin, M.; Hossain, M.M. *Synth. Commun.*, ***1999***, ***29***, 2967.

CO_2Me

Ph—≡—Ph → CO , $Rh_6(CO)_{16}$, MeOH; 130°C , dioxane → Ph, O; 88%

Yoneda, E.; Kaneko, T.; Zhang, S.-W.; Onitsuka, K.; Takahashi, S. *Tetrahedron Lett.*, ***1999***, *40*, 7811.

Ph; O → DCE , 0.1 M , rt , 2 h; 5% $[Rh(dppb)Cl]_2/AgSbF_6$ → Ph; O; 84%

Ley, S.V.; Thomas, A.W.; Finch, H. *J. Chem. Soc., Perkin Trans. 1*, ***1999***, 669.

C_6H_{13}; O; OMs → 1. $Zn(OAc)_2$, aq AcOH 80°C , AcOH; 2. Ac_2O , Py → C_6H_{13}; H; OAc; O; 70%

Nagasawa, K.; Hori, N.; Koshino, H.; Nakata, T. *Heterocycles*, ***1999***, *50*, 919.

Also via Ketoacids	Section 320 (Carboxylic Acid - Ketone)
Hydroxyketones	Section 330 (Alcohol - Ketone)

SECTION 361: ESTER - NITRILE

O → TMSCN , THF , 4 h; 5% LiOMe , 22°C → OTMS; CN + CN; OTMS

(96 : 4) >99%

Wilkinson, H.S.; Grover, P.T.; Vanddenbossche, C.P.; Bakale, R.P.; Bhongle, N.N.; Wald, S.A.; Senanayake, C.H. *Org. Lett.*, ***2001***, *3*, 553.

CN / CO_2H — DCC, t-BuOH → CN / CO_2t-Bu 75%

Nahmany, M.; Melman, A. *Org. Lett.*, ***2001***, *3*, 3733.

EtO_2C—CN, –24°C, 4 h; modified Cinchona alkaloid → NC, OCO_2Et 76% (95% ee)

Tian, S.-K.; Deng, L. *J. Am. Chem. Soc.*, ***2001***, *123*, 6195.

OAc / Ph / OAc — Me_3SiCN, $TiCl_4$, DCM; –78°C → rt, 2 h → OAc / Ph / CN 92%

Sandberg, M.; Sydnes, L.K. *Org. Lett.*, ***2000***, *2*, 687.

O — 5 eq MeO_2C–CN, THF; 20 eq DIPA, rt → MeO_2CO, CN 94%

Berthiaume, D.; Poirier, D. *Tetrahedron*, ***2000***, *56*, 5995.

CHO — cat Cp_2•Sn(thf)$_2$, toluene, rt, 3 h; HO, CN, AcO → CN, O, O 86%

Kawasaki, Y.; Fujii, A.; Nakano, Y.; Sakaguchi, S.; Ishii, Y. *J. Org. Chem.*, ***1999***, *64*, 4214.

O — 10 eq MeO_2C–CN, THF; 20 eq DIPA, rt, 18 h → CN, OCO_2Me 65%

Poirier, D.; Berthiaume, D.; Boivin, R.P. *Synlett*, ***1999***, 1423.

SECTION 362: ESTER - ALKENE

This section contains syntheses of enol esters and esters of unsaturated acids as well as ester molecules bearing a remote alkenyl unit.

PhSH , CO , $Pd(OAc)_2$/*RR*-DIOP

400 psi , DCM , 110°C

PhS O

71% (89% ee)

Xiao, W.-J.; Alper, H. *J. Org. Chem.*, ***2001***, *66*, 6229.

Me

O O OAc

5% $Pd(OAc)_2$, 6 eq bpy

AcOH , 60°C , 10 h

Me AcO O O

87%

Zhang, Q.; Lu, X.; Han, X. *J. Org. Chem.*, ***2001***, *66*, 7676.

OH

$Me_3SiC{\equiv}CH$, 25% $SnCl_4$, 3 h

50% BuLi , PhCl , 105°C

OH

90%

Kobayashi, K.; Yamaguchi, M. *Org. Lett.*, ***2001***, *3*, 241.

OH *i*-Pr OH

$PdCl_2(PPh_3)_2$, dppp

CsOAc , MeCN , CO

55°C , 12 h

O O O *i*-Pr

53%

Hu, Y.; Yang, Z. *Org. Lett.*, ***2001***, *3*, 1387.

Cl

NaOMe , $MeCH(CO_2Me)_2$

5% $Pd_2(dba)_3$•$CHCl_3$, THF

DPE phos , 70°C , 3 h

=C= Me CO_2Me CO_2Me

94%

Ogasawara, M.; Ikeda, H.; Nagano, T.; Hayashi, T. *Org. Lett.*, ***2001***, *3*, 2615.

CO_2Me Ph Br

1. DABCO, H_2O/THF rt , 15 min

2. $NaBH_4$, rt , 15 min

Ph CO_2Me

83%

Basavaiah, D.; Kumaragurubaran, N. *Tetrahedron Lett.*, ***2001***, *42*, 477.

MeO_2C $C_6H_4(pBr)$ + MeO_2C $C_6H_4(pBr)$

N_2=C(CO_2Me)($C_6H_4(p$-Br)); 23°C, Rh_2(S-DOSP)$_4$

(75 : 25) 67%

86% ee 66% ee

Davies, H.M.L.; Ren, P.; Jin, Q. *Org. Lett.*, ***2001***, *3*, 3587.

TMSO, OMe; PhCHO; 10% CuF•*S*-tol-BINAP; Ph; COMe; HO; Ph

(86 : 14) 85%

Bluet, G.; Bazán-Tejeda, B.; Campagne, J.-M. *Org. Lett.*, ***2001***, *3*, 3807.

Ph, Br; HC(NHAc)(CO_2Et)$_2$, DCM; *t*-BuOK; 10% Pd(dba)$_2$ / *R*-BINAP; NHAc, CO_2Et, CO_2Et, Ph

98% (75% ee *R*)

Ogasawara, M.; Ikeda, H.; Nagano, T.; Hayashi, T. *J. Am. Chem. Soc.*, ***2001***, *123*, 2089.

BuO; Ru enoic carbene complex; CH_2Cl_2, reflux, 3 h; BuO, OBu

87%

Choi, T.-L.; Lee, C.W.; Chatterjee, A.K.; Grubbs, R.H. *J. Am. Chem. Soc.*, ***2001***, *123*, 10417.

Ph—≡—SMe; 1. TsOH, CH_2Cl_2, rt, 17 h; 2. SiO_2; Ph, SMe, OTs

82%

Braga, A.L.; Emmerich, D.J.; Silveira, C.C.; Martins, T.L.C.; Rodrigues, O.E.D. *Synlett*, ***2001***, 371.

CO/H_2, PhH, 90°C, 3 h; 0.25% $Rh_4(CO)_{12}$

SiMe₃ → SiMe₃, O, O; 92%

Fukuta, Y.; Matsuda, I.; Itoh, K. *Tetrahedron Lett.*, ***2001***, *42*, 1301.

C_3H_7 —≡ → PhSH, cat $Pd(OAc)_2$/dppp; *p*-TsOH, THF, 110°C, 6 h; 300 psi CO → C_3H_7, SPh, O; 86%

El Ali, B.; Tijani, J.; El-Ghanam, A.; Fettouhi, M. *Tetrahedron Lett.*, ***2001***, *42*, 1567.

OH → cat $Ru_3(CO)_{12}$, 5 atm CO; NEt_3, rt → O, O; 71%

Yoneda, E.; Zhang, S.-W.; Onitsuka, K.; Takahashi, S. *Tetrahedron Lett.*, ***2001***, *42*, 5459.

OH, Ph, CO_2Me → basic Al_2O_3, $NaBH_4$, 10min; microwaves, 165°C → CO_2Me, Ph

90% (95:5 *E:Z*)

Ravichandran, S. *Synth. Commu.*, ***2001***, *31*, 2055.

OH, **Ph**, **CO_2Me** → TMSOTf, PhH; reflux, 30 min → **CO_2Me**, **Ph**, **Ph**

86% (97:3 *Z:E*)

Ravichandran, S. *Synth. Commun.*, ***2001***, *31*, 2345.

Ph—≡ → CO, MeOH, $CuBr_2$; $PdBr_2$ → Ph, MeO_2C, CO_2Me; 76% (77:23 *Z:E*)

Li, J.; Jiang, H.; Chen, M. *Synth. Commu.*, ***2001***, *31*, 3131.

Ph, CO_2H — 4% CuCl, MeOH / reflux, 2 h → Ph, O, O — 95%

Ma, S.; Yu, Z.; Wu, S. *Tetrahedron,* ***2001****, 57*, 1585.

OAc, Ph, CO_2Me — $TiCl_3$, $LiAlH_4$, THF / reflux, 1 h → Ph, CO_2Me — 46% (88:12 *E:Z*)

Shadakshari, U.; Nayak, S.K. *Tetrahedron,* ***2001****, 57*, 4599.

CO_2R^* — hν, DCM, –40°C, $HOCH_2CH_2NMe_2$ → CO_2R^* — 81% (83% de)

R* = Ph

Bargiggia, F.; Piva, O. *Tetrahedron Asymm.,* ***2001****, 12*, 1389.

PhCHO — $BrCH_2CO_2Et$, toluene, K_2CO_3 / 2% PEG-supported TeBu, cat $P(OPh)_3$ → Ph, CO_2Et — 98% (>99:1 *E:Z*)

Huang, Z.-Z.; Ye, S.; Xia, W.; Tang, Y. *Chem. Commun.,* ***2001***, 1384.

TMSO, EtS, Ph — 1. ≡—$SiMe_3$, $GaCl_3$, methylcyclohexane; 2. THF, 10M HCl → O, EtS, Ph — 85%

Arisawa, M.; Miyagawa, C.; Yoshimura, S.; Kido, Y.; Yanaguchi, M.
Chem. Lett., ***2001***, 1080.

2 ≡—C(Me)₂—OH — $MeCO_2H$, isoprene, 40°C / 5% [CpRuCl(cod)], 20 h → O_2CMe, HO — 57%

La Paih, J.; Dérien, S.; Bruneau, C.; Demerseman, B.; Toupet, L.; Dixneuf, P.H.
Angew. Chem. Int. Ed., ***2001****, 40*, 2912.

e^-, $Bu_4N\,ClO_4$

Ph, Br, O, Ph, Ph, O, O, +, Ph, O_2CPh, $PhCO_2$, Ph

Pb cathode/MeCN	65%	24%
Sn cathode/MeCN	0%	78%

Kise, N.; Ueda, N. *Bull. Chem. Soc. Jpn.*, ***2001***, *74*, 755.

MeS, C=O, Cl, O, OH, MeS, O, O

xylene , reflux , 8 h

57%

Lieb, F.; Benet-Buchholz, J.; Fäcke, T.; Fischer, R.; Graff, A.; Lefebvre, I.M.; Stetter, J. *Tetrahedron,* ***2001***, *57*, 4133.

CO_2Et

LiBr, AcOH , 80°C

O, O

67%

Ma, S.; Li, L.; Wei, Q.; Xie, H.; Wang, G.; Shi, Z.; Zhang, J. *Pure. Appl. Chem.*, ***2000***, *72*, 1739.

cat Cu–Na–HSZ–320 zeolite

t-Bu, O, O, O, O, Ph , 80°C , 21h

O, O, Ph

60%

Carloni, S.; Frullanti, B.; Maggi, R.; Mazzacani, A.; Bigi, F.; Sartori, G. *Tetrahedron Lett.*, ***2000***, *41*, 8947.

OMe, $(OC)_5Cr$, OMe, MeO

OH

THF , heat , 3.25 h

O, O, OMe, Ac

69%

Good, G.M.; Kemp, M.I.; Kerr, W.J. *Tetrahedron Lett.*, ***2000***, *41*, 9323.

Ph—≡—Ph → 1. Cp_2ZrEt_2 2. $ClCO_2Et$ 3. HCl → Ph, Ph, CO_2Et 70%

Takahashi, T.; Xi, C.; Ura, Y.; Nakajima, K. *J. Am. Chem. Soc.*, ***2000***, *122*, 3228.

CO_2H, I → 5 eq, 10% $Pd(OAc)_2$; 5 eq $NaHCO_3$, Bu_4NCl DMF, 60°C → O, O 70%

Gagnier, S.V.; Larock, R.C. *J. Org. Chem.*, ***2000***, *65*, 1525.

OH → 1% $Ru_3(CO)_{12}$, 100°C 10 atm CO, dioxane; NEt_3, 8 h → O, O 99%

Yoneda, E.; Kaneko, T.; Zhang, S.-W.; Onitsuka, K.; Takahashi, S. *Org. Lett.*, ***2000***, *2*, 441.

O → polymer-supported triflating reagent; *i*-Pr_2NEt → OTf 95%

Wentworth, A.D.; Wentworth Jr. P.; Mansoor, U.F.; Janda, K.D. *Org. Lett.*, ***2000***, *2*, 477.

→ PhSH, $Pd(OAc)_2$, PPh_3; 400 psi CO, DCM 110°C, 60 h → PhS, O 83%

Xiao, W.-J.; Vasapollo, G.; Alper, H. *J. Org. Chem.*, ***2000***, *65*, 4138.

C_3H_7, CO_2H → EtI, $Pd(PPh_3)_4$, K_2CO_3; MeCN, 70°C, 5.5 h → Et, C_3H_7, O, O 76%

Rossi, R.; Bellina, F.; Biagetti, M.; Catanese, A.; Mannina, L. *Tetrahedron Lett.*, ***2000***, *41*, 5281.

Ph—≡—Ph , DMF , Py

5% $Pd(OAc)_2$, Bu_4NCl , 2 d

59%

Kadnikov, D.V.; Larock, R.C. *Org. Lett.*, ***2000***, *2*, 3643.

CO_2Me

, Et_2MeSiH , DCM

slow addition , 25°C , 1 h

OCO_2Me MeO_2C Ph

93%

Muraoka, T.; Matsuda, I.; Itoh, K. *J. Am. Chem. Soc.*, ***2000***, *122*, 9552.

1. 2 eq SmI_2 , THF , rt
2. 1M HCl

C_7H_{15} CO_2Et Cl Me

75%

Concellón, J.M.; Pérez-Andrés, J.A.; Rodríguez-Solla, H. *Angew. Chem. Int. Ed.*, ***2000***, *39*, 2773.

1. $Cp_2Zr(H)Cl$
2. CO
3. PhSeSePh

76%

Zhong, P.; Xiong, Z.-X.; Huang, X. *Synth. Commun.*, ***2000***, *30*, 887.

$NaBH_4$, 10% $Pd(OAc)_2$,
20% PPh_3 , THF , reflux

51%

Grigg, R.; Savic, V. *Chem. Commun.*, ***2000***, 2381.

1. $Cp_2Zr(H)Cl$
2. CO
3. PhSCl

80%

Zhong, P.; Xiong, Z.-X.; Huang, X. *Synth. Commun.*, ***2000***, *30*, 2793.

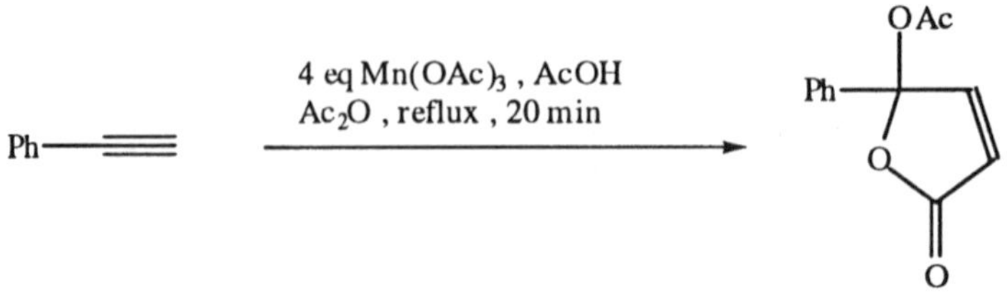

Ph—≡ → 4 eq $Mn(OAc)_3$, AcOH; Ac_2O, reflux, 20 min → (OAc, Ph, O, O) 78%

Montevecchi, P.C.; Navacchia, M.L. *Tetrahedron*, ***2000***, *56*, 9339.

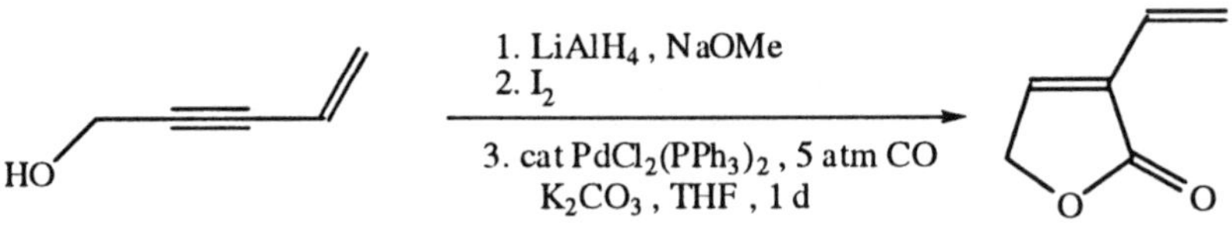

HO → 1. $LiAlH_4$, NaOMe; 2. I_2; 3. cat $PdCl_2(PPh_3)_2$, 5 atm CO; K_2CO_3, THF, 1 d → (O, O) 76%

Liao, B.; Negishi, E.-i. *Heterocycles*, ***2000***, *52*, 1241.

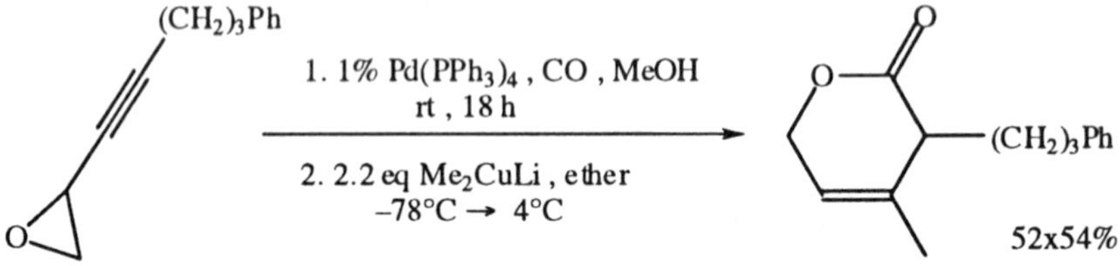

$(CH_2)_3Ph$, O → 1. 1% $Pd(PPh_3)_4$, CO, MeOH; rt, 18 h; 2. 2.2 eq Me_2CuLi, ether; –78°C → 4°C → (O, O, $(CH_2)_3Ph$) 52x54%

Knight, J.G.; Ainge, S.W.; Baxter, C.A.; Eastman, T.P.; Harwood, S.J. *J. Chem. Soc., Perkin Trans. 1*, ***2000***, 3188.

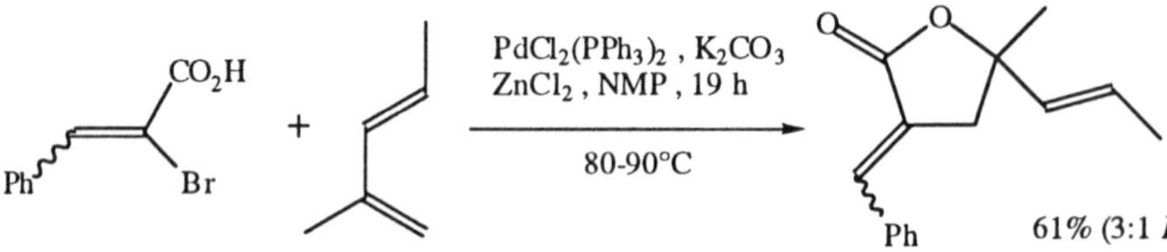

(CO_2H, Ph, Br) + → $PdCl_2(PPh_3)_2$, K_2CO_3; $ZnCl_2$, NMP, 19 h; 80-90°C → (O, O, Ph) 61% (3:1 *E:Z*)

Iyer, S.; Ramesh, C. *Tetrahedron Lett.*, ***1999***, *40*, 4719.

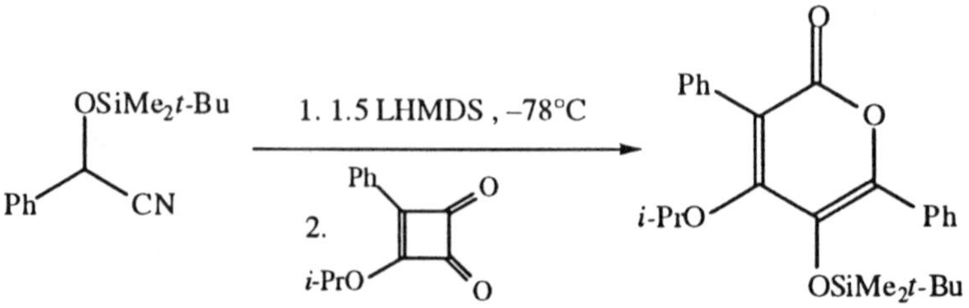

($OSiMe_2t$-Bu, Ph, CN) → 1. 1.5 LHMDS, –78°C; 2. (Ph, O, *i*-PrO, O) → (O, Ph, O, *i*-PrO, Ph, $OSiMe_2t$-Bu) 71%

Mingo, P.; Zhang, S.; Liebeskind, L.S. *J. Org. Chem.*, ***1999***, *64*, 2145.

PhSH , CO , THF , 6 h
cat $Pd(OAc)_2$, dppp
110°C
SPh
O
76%

Xiao, W.-J.; Vasapollo, G.; Alper, H. *J. Org. Chem.*, ***1999***, *64*, 2080.

Ph
CO , MeOH , $PdCl_2$, 3 $CuCl_2$
Cl
CO_2Me
Ph
58%

Li, J.; Jiang, H.; Feng, A.; Jia, L. *J. Org. Chem.*, ***1999***, *64*, 5984.

CO_2Me
I
Me—≡—*t*-Bu , DMF , 100°C
2% $Pd(OAc)_2$, 4% P(*o*-Tol)$_3$
0.3% NEt_3 , NaOAc , 1 d
O
O
t-Bu
72% Me

Larock, R.C.; Doty, M.J.; Han, X. *J. Org. Chem.*, ***1999***, *64*, 8770.

C_7H_{15}
CO_2H
4 eq $CuCl_2$, MeCN
25°C , 50 h
Cl
C_7H_{15}
O
O
52%

Ma, S.; Wu, S. *J. Org. Chem.*, ***1999***, *64*, 9314.

C_8H_{17}
I^+Ph BF_4^-
Me_2NCHO , 50°C , 6 h
OCHO
C_8H_{17}
83%

Ochiai, M.; Yamamoto, S.; Sato, K. *Chem. Commun.*, ***1999***, 1363.

2 Bu
PhCOOH , dioxane , rt
5% RuCl(cod)(C_5Me_5) , 15 h
Ph
OBz
Ph
98%

Le Paih, J.; Dérien, S.; Dixneuf, P.H. *Chem. Commun.*, ***1999***, 1437.

PhCHO
EtO_2CCH_2CN , 5 min
microwaves
Ph
CO_2Et
CN
97%

Mitra, A.K.; De, A.; Karchaudhuri, N. *Synth. Commun.*, ***1999***, *29*, 2731.

OTBS OAc

1. O_3

2. OMe OLi

OTBS O O

89%

Keck, G.E.; Li, X.-Y.; Knutson, C.E. *Org. Lett.*, ***1999***, *1*, 411.

Ph OTf

DMF, 40°C, $KHCO_3$
5% $Pd(OAc)_2$

CO_2Me OH

Ph O O

74%

Arcadi, A.; Cacchi, S.; Fabrizi, G.; Marinelli, F.; Pace, P. *Eur. J. Org. Chem.*, ***1999***, 3305.

Ph

CO/MeOH, $PdCl_2$, $CrCl_2$

NaOAc, PhH

MeO_2C CO_2Me Ph

64%

Li, J.; Jiang, H.; Jia, L. *Synth. Commun.*, ***1999***, *29*, 3733.

OH

PhI, CO_2, cat $PdCl_2(PPh_3)_2$

DMAC, 100°C, K_2CO_3, 8 h

Ph O O O

99%

Uemura, K.; Shiraishi, D.; Noziri, M.; Inoue, Y. *Bull. Chem. Soc. Jpn.*, ***1999***, *72*, 1063.

Related Methods: Section 60A (Protection of Aldehydes).
Section 180A (Protection of Ketones).

Also via Acetylenic Esters: Section 306 (Alkyne - Ester).
Alkenyl Acids: Section 322 (Carboxylic Acid - Alkene).
β-Hydroxy-esters: Section 327 (Alcohol - Ester).

SECTION 363: ETHER, EPOXIDE, THIOETHER - ETHER, EPOXIDE, THIOETHER

See Section 60A (Protection of Aldehydes) and Section 180A (Protection of Ketones) for reactions involving formation of Acetals and Ketals.

HO (dithiane substrate) → PVC anode, Pt wire cathod; e^-, 0.1 M Et_4NOTs, THF; 30% MeOH, 2,6-lutidine → S, S, O, OMe; 94%

Sun, Y.; Liu, B.; Kao, J.; Andre d'Avignon, D.; Moeller, K.D. *Org. Lett.*, ***2001***. *3*, 1729.

$Tl(NO_3)_3$, MeOH, 0°C; 5 min → $CH(OMe)_2$; 87%

Ferraz, H.M.C.; Silva Jr., L.; Vieira, T.O. *Tetrahedron,* ***2001***, *57*, 1709.

Ph, SEt, SEt → [Ti–In], 2 eq Zn, DCM, rt, 1 h → SEt, Ph, Ph, SEt; 95% (49:51 *dl:meso*)

Yoshimura, N.; Igarashi, K.; Funasaka, S.; Mukaiyama, T. *Chem. Lett.*, ***2001***, 640.

O, O → Et_2NSiMe_3, 2 eq MeI; toluene, 90°C → O, $OSiMe_3$; 76%

Ohshita, J.; Iwata, A.; Tang, H.; Yamamoto, Y.; Matsui, C.; Kunai, A.
Chem. Lett., ***2001***, 740.

OEt, O, Br → 1. Bu_3SnH, toluene, BEt_3, –78°C; 2. air, –78°C, 3 h; 3. 1M NaOH, rt, 12 h → O, OEt; 65% (>98% ds)

Villar, F.; Equey, O.; Renaud, P. *Org. Lett.*, ***2000***, *2*, 1061.

OH, CHO → 2 eq $Me_2C(OMe)_2$, I_2, rt; DCM, 30 min → O, OMe, Me, OMe; 90%

Yadav, J.S.; Reddy, B.V.S.; Hashim, S.R. *J. Chem. Soc., Perkin Trans. 1,* ***2000***, 3082.

PhCHO → ($Mn/TMSCl/Cp_2TiCl_2$, THF; 0°C, 1 d) → PhCH(OTMS)CH(OTMS)Ph 73% (13:1 dr)

Dunlap, M.S.; Nicholas, K.M. *Synth. Commun.*, ***1999***, *29*, 1097.

SECTION 364: ETHER, EPOXIDE, THIOETHER - HALIDE, SULFONATE

PhCHO, $InCl_3$, DCM; rt, overnight 85%

Yang, X.-F.; Mague, J.T.; Li, C.-J. *J. Org. Chem.*, ***2001***, *66*, 739.

1. vinylMgBr, THF, 0°C → rt; 2. piperidine (NH) 95%

Gomtsyan, A.; Koenig, R.J.; Lee, C.-H. *J. Org. Chem.*, ***2001***, *66*, 3613.

BF_3•OEt_2, AcOH; hexane, –78°C 61%

Jaber, J.J.; Mitsui, K.; Rychnovsky, S.D. *J. Org. Chem.*, ***2001***, *66*, 4679.

CAN, LiBr, MeCN; propargyl alcohol, rt 87%

Roy, S.C.; Guin, C.; Rana, K.K.; Maiti, G. *Synlett*, ***2001***, 226.

, $InCl_3$, $CHCl_3$

rt , 5 h

95% (3:1 *cis:trans*)

Li, J.; Li, C.-J. *Tetrahedron Lett.*, ***2001***, *42*, 793.

$Br^+(coll)_2\ PF_6^-$

CH_2Cl_2

78%

Albert, S.; Robin, S.; Rousseau, G. *Tetrahedron Lett.*, ***2001***, *42*, 2477.

I_2 , HgO ,

58%

Talybov, G.M.; Mekhtieva, V.Z.; Karaev, S.F. *Russ. J. Org. Chem.*, ***2001***, *37*, 600.

$SiCl_4$, *i*-Pr_2NEt
CH_2Cl_2 , -85°C

5% fluorinated arene ligand

88% (94% ee)

Tao, B.; Lo, M.M.-C.; Fu, G.C. *J. Am. Chem. Soc.*, ***2001***, *123*, 353.

10% CsF , 0°C , glyme , 5h

86%

Billard, T.; Langlois, B.R.; Blond, G. *Tetrahedron Lett.*, ***2000***, *41*, 8777.

$NaIO_4$ / $NaHSO_3$

MeCN

61%

Okimoto, Y.; Kikuchi, D.; Sakaguchi, S.; Ishii, Y. *Tetrahedron Lett.*, ***2000***, *41*, 10223.

MeOH, dioxane, rt, 0.75 I_2

0.25 $Ce(OTf)_2$

I

OMe

80%

Iranpoor, N.; Shekarriz, M. *Tetrahedron Lett.*, ***2000***, *56*, 5209.

OH

Ph

$PhCHO$, $InCl_3$

Cl

Ph O Ph

81%

Yang, J.; Viswanathan, G.S.; Li, C.-J. *Tetrahedron Lett.*, ***1999***, *40*, 1627.

O

Et_2NSiMe_3, 2 eq MeI

80-90°C, 5 h

I

$OSiMe_3$

81%

Ohshita, J.; Iwata, A.; Kanetani, F.; Kunai, A.; Yamamoto, Y.; Matui, C. *J. Org. Chem.*, ***1999***, *64*, 8024.

SECTION 365: ETHER, EPOXIDE, THIOETHER - KETONE

OTMS

MeS

AcO

O

, cat TMSOTf

MeCN, –30°C → rt, 10 min

O

SMe

94%

Matsugi, M.; Murata, K.; Gotanda, K.; Nambu, H.; Anilkumar, G.; Matsumoto, K.; Kita, Y. *J. Org. Chem.*, ***2001***, *66*, 2434.

OTIPS

NH_2

1. C_3H_7CHO, $MgSO_4$, rt
DCM, K_2CO_3
2. $(EtO_2C)_2O$, EtOH, rt

3. $BF_3 \bullet OEt_2$, DTBMP
DCM, 0°C

O

H

N

EtO_2C

H

C_3H_7

72x87% (87% ee)

Cohen, F.; MacMillan, D.W.C.; Overman, L.E.; Romero, D. *Org. Lett.*, ***2001***, *3*, 1225.

PhSH , THF , LiAl-poly2a

30 min

95% (41% ee)

SPh

Sundararajan, G.; Prabagaran, N. *Org. Lett.*, ***2001***, *3*, 389.

chiral phase transfer agent , PhMe

aq NaOCl , rt

98% (86% ee)

Lygo, B.; To, D.C.M. *Tetrahedron Lett.*, ***2001***, *42*, 1343.

OTMS

Ph

$1\%\ Sc(OTf)_3$

57%

Yoshimatsu, M.; Kuribayashi, M.; Koike, T. *Synlett*, ***2001***, 1799.

2 eq OTMS *t*-Bu , –78°C

$3\ eq\ Et_2AlCl$, DCM

86% (61:39 *tanscis*)

Schmitt, A.; Reißig, H.-U. *Eur. J. Org. Chem.*, ***2001***, 1169.

1. MeOTf , DCM , 0°C
2,6-di-*t*-Bu-Py

2. aq $NaHCO_3$

81%

Armstrong, A.; Draffen, A.G. *J. Chem. Soc., Perkin Trans. 1*, ***2001***, 2861.

1. O_2, Mn(II) abietate PhH, rt
2. aq KOH, rt

73%

Kulinkovich, O.G.; Astashko, D.A.; Tyvorskii, V.I.; Ilyina, N.A. *Synthesis*, ***2001***, 1453.

OTBS, OMOM, Ph, H

1. BCl_3, DCM, –78°C
2. pH 7 buffer

85%

Gasparski, C.M.; Herrinton, P.M.; Overman, L.E.; Wolfe, J.P. *Tetrahedron Lett.*, ***2000***, *41*, 9431.

$PhCH_2CO_2H$

1. HO, OH (resorcinol), BF_3•OEt_2, 85°C, 90 min
2. BF_3•OEt_2, DMF/PCl_5 rt, 1 h

80%

Balasubramanian, S.; Nair, M.G. *Synth. Commun.*, ***2000***, *30*, 469.

OTMS

CO/H_2 (80 bar), 90°C, 2 d
$[Rh(cod)Cl]_2$, DCM

64% (1:1 dr)

Hollmann, C.; Eilbracht, P. *Tetrahedron*, ***2000***, *56*, 1685.

Ph

CAN, MeOH, rt

OMe, Ph 44%

Nair, V.; Nair, L.G.; Panicker, S.B.; Sheeba, V.; Augustine, A. *Chem. Lett.*, ***2000***, 584.

Cl, O, N_2

5% $Sc(OTf)_3$, 10% MeOH
CH_2Cl_2

Cl, O, OMe 93%

Pansare, S.V.; Jain, R.P.; Bhattacharyya, A. *Tetrahedron Lett.*, ***1999***, *40*, 5255.

1. $CH_2=C(Al\textit{i}\text{-}Bu_2)CO_2Et$
2. K_2CO_3, acetone, rt

Br, O, CO_2Et, O

65x75%

Ramachandran, P.V.; Krzeminski, M.P. *Tetrahedron Lett.*, ***1999***, *40*, 7879.

OTMS

$Me_2C(OMe)SPh$

$SnCl_4$

O, OMe

62%

with $TiCl_4$, RSPh is obtained in 86%

Braga, A.L.; Dronelles, L.; Silveira, C.C.; Wessjohann, L.A. *Synthesis*, ***1999***, 562.

CHO

Cl, O, Ph, LiOH–H_2O

10% chiral phase transfer agent
Bu_2O, 4°C, 60 h

O, O, Ph

80% (53% ee)

Arai, S.; Shirai, Y.; Ishida, T.; Shioiri, T. *Tetrahedron*, ***1999***, *55*, 6375.

REVIEWS:

"Epoxy Ketones as Versatile Building Blocks in Organic Synthesis," Lauret, C. *Tetrahedron Asymm.*, ***2001***, *12*, 2359.

SECTION 366: ETHER, EPOXIDE, THIOETHER - NITRILE

O

10% Ti complex, TMSCN

THF, –5°C, 36 h

TMSO, CN

88% (76% ee)

Hamashima, Y.; Kanai, M.; Shibasaki, M. *Tetrahedron Lett.*, ***2001***, *42*, 691.

O, Ph, OAc

TMS–CN, $InBr_3$, CH_2Cl_2

rt, 3 h

OH, Ph, OAc, CN

97%

Bandini, M.; Cozzi, P.G.; Melchiorre, P.; Umani-Ronchi, A. *Tetrahedron Lett.*, ***2001***, *42*, 2041.

bis-Ti-salen complex (0.5%)
Me_3SiCN , DCM , 1 d

OTMS, Ph, CN

quant (66% ee)

Belokon, Y.N.; Green, B.; Ikonnikov, N.S.; North, M.; Persons, T.; Tararov, V.I. *Tetrahedron*, ***2001***, *57*, 771.

PhCHO

1. [DCM , Et_2AlCl , rt , 2 h
carbohydrate phosphine oxide ligand]
−78°C
2. TMSCN , −60°C

OTMS, Ph, CN

96% (80% ee, *S*)

Kanai, M.; Hamashima, Y.; Shibasaki, M. *Tetrahedron Lett.*, ***2000***, *41*, 2405.

TMSCN , $CHCl_3$, −45°C , 4 d

OTMS, CN

90% (91% ee)

Schaus, S.E.; Jacobsen, E.N. *Org. Lett.*, ***2000***, *2*, 1001.

TMSCN , 10% Ti(O*i*-Pr)$_4$, THF
10% chiral phosphine oxide
−30°C , 36 h

TMSO, CN, Ph

85% (92% ee)

Hamashima, Y.; Kanai, M.; Shibasaki, M. *J. Am. Chem. Soc.*, ***2000***, *122*, 7412.

Cl, CN

PhCHO , phase transfer catalyst
NaOH

CN, O, Ph

46% (2.1:1 *trans:cis*)

Makosza, M.; Przyborowski, J.; Klajn, R.; Kwast, A. *Synlett*, ***2000***, 1773.

SePh, O, O, CO_2Me

CH_2=CHCN , Bu_3SnH
AIBN , PhH , heat

O, O, CN, CO_2Me 53% + O, O, CN, CO_2Me 15%

Nimkar, K.S.; Mash, E.A. *Tetrahedron*, ***2000***, *56*, 5793.

1. allyl ethyl carbonate; 10 eq allyl alcohol; 5% $Pd(PPh_3)_4$, rt, 1d
2. 5% Grubbs' catalyst, DCM; CH_2=CH_2 atmosphere, 1 d

Ph, CN, CN → NC, CN, Ph, O — 81 X 94%

Xie, R.L.; Hauske, J.R. *Tetrahedron Lett.*, ***2000***, *41*, 10167.

Ph, Me, O — Me_3SiCN, 3 M $LiClO_4$/ether; 0.1 mPa → CN, Ph, Me, $OSiMe_3$ — 98%

Jenner, G. *Tetrahedron Lett.*, ***1999***, *40*, 491.

PhCHO — TMSCN, 0.1% chiral Ti (salen) complex; CH_2Cl_2, rt, 1 d → OTMS, Ph, CN — 86% ee

Belokon', Y.N.; Caveda-Cepas, S.; Green, B.; Ikonnikov, N.S.; Khrustalev, V.N.; Larichev, V.S.; Moscalenko, M.A.; North, M.; Orizu, C.; Tararov, V.I.; Tasinazzo, M.; Timofeeva, G.I.; Yashkina, L.V. *J. Am. Chem. Soc.*, ***1999***, *121*, 3968.

HO, Et, CO_2Me — 1. $PhSeOSO_3^-$, MeCN; 2. $(NH_4)_2Se_2O_8$ → MeO_2C, O, Et — 90%

Tiecco, M.; Testaferri, L.; Santi, C. *Eur. J. Org. Chem.*, ***1999***, 797.

PhCHO — TMSCN, $Zr(KPO_4)_2$, DCM; reflux → CN, Ph, OTMS — 98%

Curini, M.; Epifano, F.; Marcotullio, M.C.; Rosati, O.; Rossi, M. *Synlett*, ***1999***, 315.

Ph, O — Me_3SiCN, 1 d; 1% chiral Ti salen complex → OTMS, Ph, CN — quant (62% ee)

Belokon, Y.N.; Green, B.; Ikonnikov, N.S.; North, M.; Tarapov, V.I. *Tetrahedron Lett.*, ***1999***, *40*, 8147.

SECTION 367: ETHER, EPOXIDE, THIOETHER - ALKENE

Enol ethers are found in this section as well as alkenyl ethers.

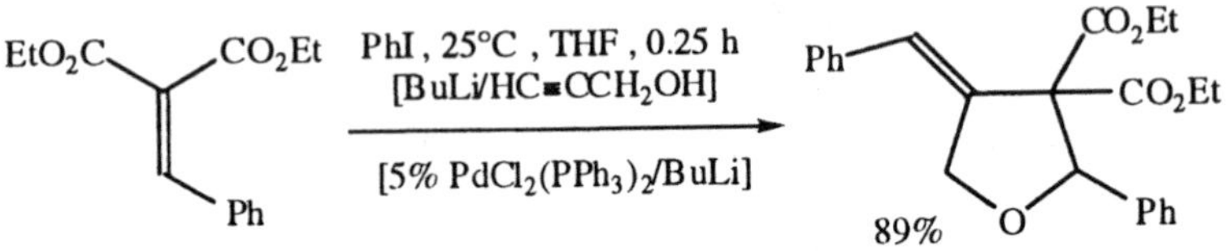

EtO_2C CO_2Et Ph → PhI, 25°C, THF, 0.25 h; [BuLi/HC≡CCH$_2$OH]; [5% PdCl$_2$(PPh$_3$)$_2$/BuLi] → Ph CO_2Et CO_2Et O Ph 89%

Bottex, M.; Cavicchioli, M.; Hartmann, B.; Monteiro, N.; Balme, G. *J. Org. Chem.*, ***2001***, *66*, 175.

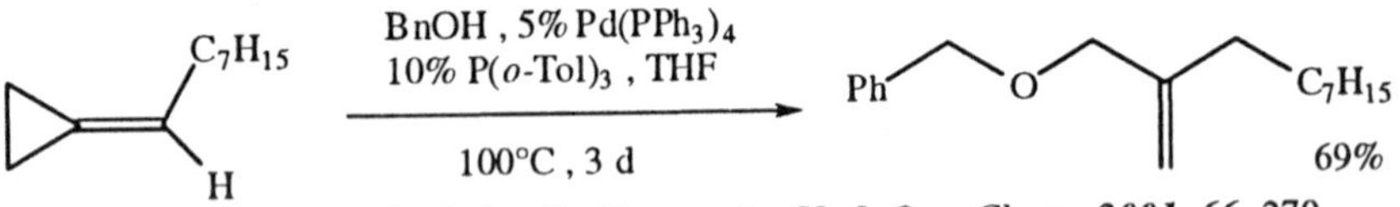

C_7H_{15} H → BnOH, 5% Pd(PPh$_3$)$_4$; 10% P(*o*-Tol)$_3$, THF; 100°C, 3 d → Ph O C_7H_{15} 69%

Camacho, D.H.; Nakamura, I.; Saito, S.; Yamamoto, Y. *J. Org. Chem.*, ***2001***, *66*, 270.

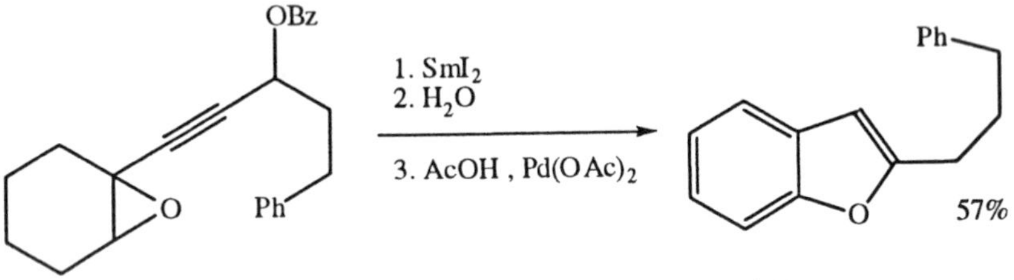

OBz O Ph → 1. SmI$_2$; 2. H$_2$O; 3. AcOH, Pd(OAc)$_2$ → Ph O 57%

Aurrecoechea, J.M.; Pérez, E.; Solay, M. *J. Org. Chem.*, ***2001***, *66*, 564.

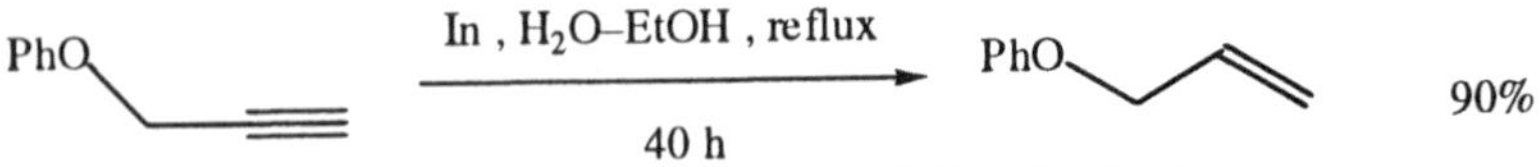

PhO → In, H$_2$O–EtOH, reflux; 40 h → PhO 90%

Ranu, B.C.; Dutta, J.; Guchhait, S.K. *J. Org. Chem.*, ***2001***, *66*, 5624.

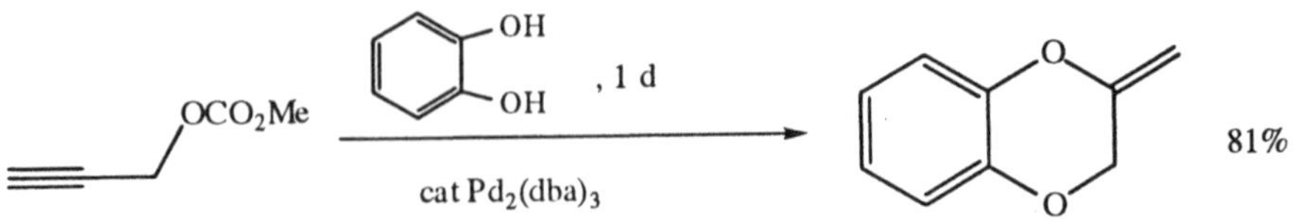

OCO$_2$Me → (catechol: OH, OH), 1 d; cat Pd$_2$(dba)$_3$ → O O 81%

Labrosse, J.-R.; Lhoste, P.; Sinou, D. *J. Org. Chem.*, ***2001***, *66*, 6634.

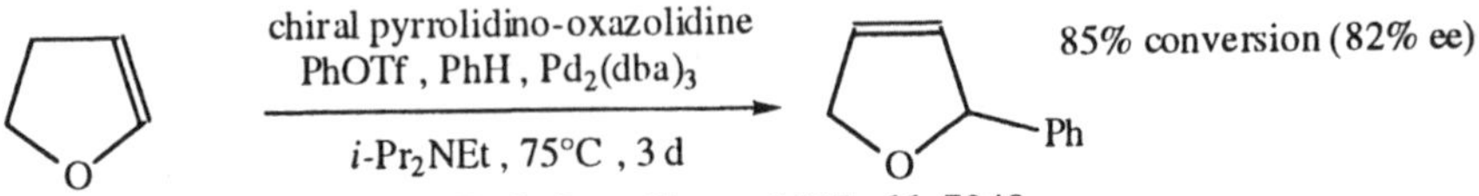

O → chiral pyrrolidino-oxazolidine; PhOTf, PhH, Pd$_2$(dba)$_3$; *i*-Pr$_2$NEt, 75°C, 3 d → O Ph 85% conversion (82% ee)

Gilbertson, S.R.; Xie, D.; Fu, Z. *J. Org. Chem.*, ***2001***, *66*, 7240.

Ph–C6H4–OH → Ph–C6H4–O–CH=CH2 93%
$(CH_2{=}CH)_4Sn$, $Cu(OAc)_2$
O_2 , MeCN , rt , 22 h

Blouin, M.; Frenette, R. *J. Org. Chem.*, ***2001***, *66*, 9043.

OTf , chiral P,N-ligand
$Pd_2(dba)_3$, *i*-Pr_2NEt , PhH
70°C , 22 h

quant (94% ee)

Gilbertson, S.R.; Ru, Z. *Org. Lett.*, ***2001***, *3*, 161.

OTMS OTMS
Bu_4NBr , BSA
105°C , 4 h

BSA = (bis-trimethylsilyl)acetamide

(90 : 10) 90%

Smietana, M.; Mioskowski, C. *Org. Lett.*, ***2001***, *3*, 1037.

Grubbs' catalyst , PhH
rt , 2 h

65%

Heck, M.-P.; Baylon, C.; Nolan, S.P.; Mioskowski, C. *Org. Lett.*, ***2001***, *3*, 1989.

t-Bu, Me, H, H, CO_2Et, OH
10% $AgCl_3$
CH_2Cl_2
t-Bu, *t*-Bu, Me, CO_2Et

74%

Hoffmann-Röder, A.; Krause, N. *Org. Lett.*, ***2001***, *3*, 2537.

1. PPh_3 , TBSOTf
THF
2. In , Br

OTBS

65%

Lee, P.H.; Lee, K.; Kim, S. *Org. Lett.*, ***2001***, *3*, 3205.

MeO, OMe, O, O, $SiPh_2tBu$ → $TiCl_2$ / DCM, –30°C, 3 h → CO_2Me, CO_2Me 96%, O, $ButPh_2Si$ 96%

Yadav, V.K.; Balamuragan, R. *Org. Lett.*, ***2001***, *3*, 2717.

PhCHO → 1. MeO_2C, PhO_2S, $=N_2$, DMAD, 25°C / 2. NEt_3, 65°C → MeO_2C, CO_2Me, MeO_2C, O, Ph 42%

Johnson, T.; Cheshire, D.R.; Stocks, M.J.; Thurston, V.T. *Synlett*, ***2001***, 646.

O, Ph → $C_8H_{17}CHO$, $CrCl_2$, H_2O / Me_3SiCl, THF, 25°C, 1 d → C_8H_{17}, O, Ph 80%

Takai, K.; Morita, R.; Sakamoto, S. *Synlett*, ***2001***, 1614.

PhO_2S, SO_2Ph → 1. Ph_2CHOH, LHMDS / THF, –78°C / 2. Na/Hg, NaH_2PO_4, MeOH → Ph_2HCO 93%

Cabianca, E.; Chéry, F.; Rollin, P.; Cossu, S.; De Lucchi, O. *Synlett*, ***2001***, 1962.

MeO, O, OMe → CH_2=CHOEt, $MgBr_2$•OEt_2 / ether, rt, 15 min → O, OEt, OEt 90%

Malanga, C.; Mannucci, S. *Tetrahedron Lett.*, ***2001***, *42*, 2023.

MeO, OMe, Ph → 2 eq *i*-Bu_3Al, DCE, 5 h → OMe, Ph 78%

Cabrera, G.; Fiaschi, R.; Napolitano, E. *Tetrahedron Lett.*, ***2001***, *42*, 5867.

Ph—≡—Me → BnOH, 10% $Pd(PPh_3)_4$ / 10% $PhCO_2H$, dioxane, 100°C → Ph, OBn 96%

Kadota, I.; Lutete, L.M.; Shibuya, A.; Yamamoto, Y. *Tetrahedron Lett.*, ***2001***, *42*, 6207.

AcO CN

1. SmI_2
2. H_2O
3. PhI, cat $Pd(PPh_3)_4$
NEt_3, 60°C

O Ph CN 53%

Aurrecoechea, J.M.; Pérez, E. *Tetraheron Lett.*, ***2001***, *42*, 3839.

OH CO_2Me cat $Pd(MeCN)_2Cl_2$ 0°C, 0.5 h O CO_2Me CO_2Me 95%

Kato, K.; Nishimura, A.; Yamamoto, Y.; Akita, H. *Tetrahedron Lett.*, ***2001***, *42*, 4203.

C_6H_{13} 0.5 PhSeSePh, 3 h $NaBH_4$/EtOH, reflux PhSe 75% C_6H_{13}

Dabdoub, M.J.; Baroni, A.C.M.; Lenardão, E.J.; Gianeti, T.R.; Hurtado, G.R. *Tetrahedron*, ***2001***, *57*, 4271.

OH CHO CO_2Me DABCO, H_2O, rt, 2 h O CO_2Me 74%

Ravichandran, S. *Synth. Commun.*, ***2001***, *31*, 1233.

SPh PhS CO_2Et Ph $Cp_2Ti[P(OEt_3]_2$, 2 h reflux OEt Ph 62%

Rahim, Md.A.; Sasaki, H.; Saito, J.; Fujiwara, T.; Takeda, T. *Chem. Commun.*, ***2001***, 625.

OEt OEt

1. LiC-KOR, THF, –95°C
2. BBu_3, –45°C
3. H_2O, –95°C

Bu OEt 86%

Tivola, P.B.; Deagostino, A.; Prandi, C.; Venturello, P. *Chem. Commun.*, ***2001***, 1536.

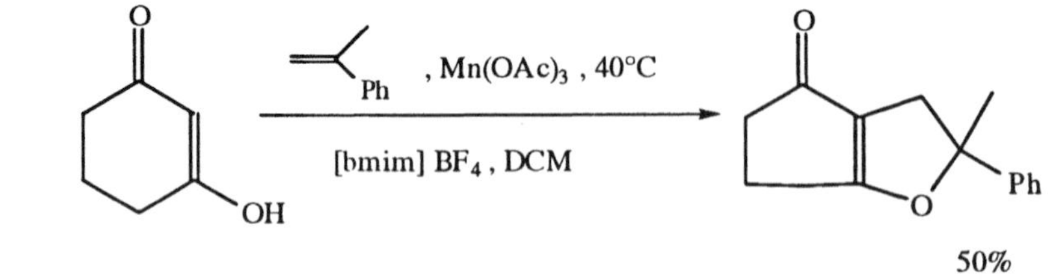

, $Mn(OAc)_3$, 40°C

[bmim] BF_4 , DCM

50%

Bar, G.; Parsons, A.F.; Thomas, C.B. *Chem. Commun.*, ***2001***, 1350.

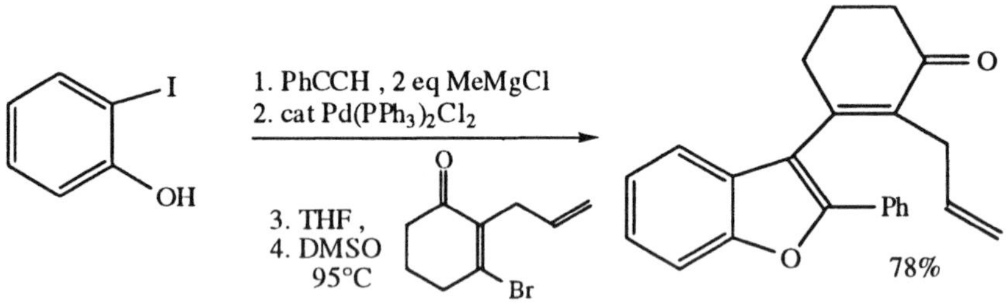

1. PhCCH , 2 eq MeMgCl
2. cat $Pd(PPh_3)_2Cl_2$

3. THF ,
4. DMSO
95°C

78%

Chaplin, J.H.; Flynn, B.L. *Chem. Commun.*, ***2001***, 1594.

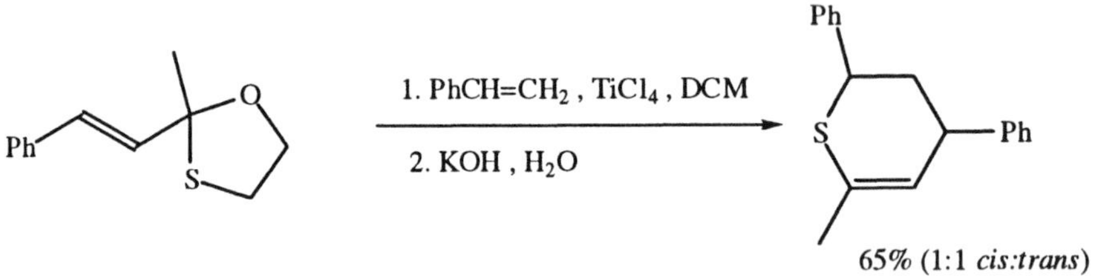

1. $PhCH{=}CH_2$, $TiCl_4$, DCM

2. KOH , H_2O

65% (1:1 *cis:trans*)

Kervredo, S.; Loiseau, M.; Lizzani-Cuvelier, L.; Duñach, E. *Chem. Commun.*, ***2001***, 2284.

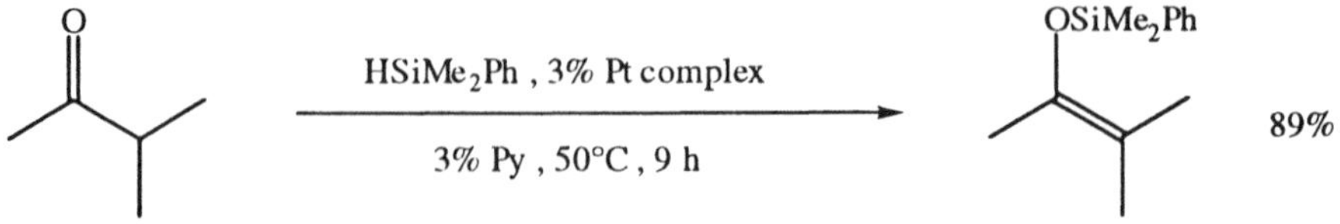

$HSiMe_2Ph$, 3% Pt complex

3% Py , 50°C , 9 h

89%

Ozawa, F.; Yamamoto, S.; Kawagishi, S.; Hiraoka, M.; Ikeda, S.; Minami, T.; Ito, S.; Yoshifuji, M. *Chem. Lett.*, ***2001***, 972.

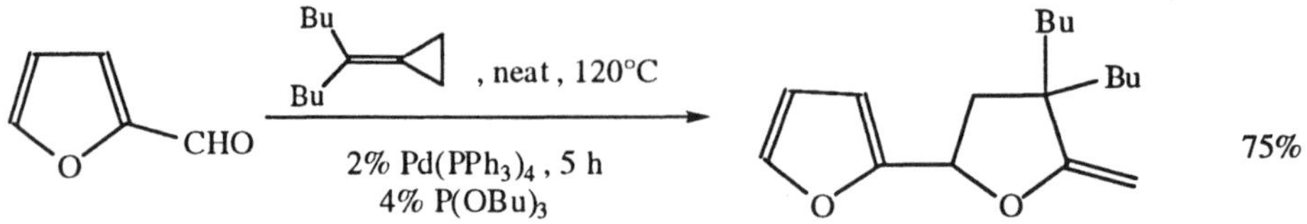

, neat , 120°C

2% $Pd(PPh_3)_4$, 5 h
4% $P(OBu)_3$

75%

Aggarwal, V.K.; Alonso, E.; Hynd, G.; Lydon, K.M.; Palmer, M.J.; Porcelloni, M.; Studley, J.R. *Angew. Chem. Int. Ed.*, ***2001***, *40*, 1430.

MeO_2CO HO; HO–C6H4–OMe; 5% $[Pd_2(dba)_3]\cdot CHCl_3$, 2 h; 20% dppe , dioxane , 50°C; O O O; 85%; OMe

Yoshida, M.; Ihara, M. *Angew. Chem. Int. Ed.*, ***2001***, *40*, 616.

OTMS; TMS; MeO–C6H4–$CH(OMe)_2$; cat $SnCl_2$–AcCl , DCM , rt , 4 h; MeO; O; 53%

Oriyama, T.; Ishiwata, A.; Suzuki, T. *Bull. Chem. Soc. Jpn.*, ***2001***, *74*, 569.

O; OTMS OTMS; OEt; , 2 eq $TiCl_4$; DCM , MS 4Å , –78°C → 20°C; H; O; CO_2Et; 70% (>98:2 *E*:*Z*)

Langer, P.; Eckardt, T. *Angew. Chem. Int. Ed.*, ***2000***, *39*, 4343.

Ph—≡; 1. $Cp_2Zr(H)Cl$,THF , rt , 30 min; 2. Se , rt , 30 min; 3. BnBr , 50°C , 1 h; Ph; Se—Bn; 72%

Huang, X.; Wang, J.-H. *Synth. Commun.*, ***2000***, *30*, 307.

Ph—≡; 1. $ZrCp_2(H)Cl$, DCM; 2. Ph_2Te_2; Ph; TePh; 67%

Huang, X.; Liang, C.-G. *Synth. Commun.*, ***2000***, *30*, 1737.

O; Ph; CO_2Et; $BnCH_2C≡CH$, toluene; CuI , DBU , aq THF; EtO_2C; Ph; O; 91%

Arcadi, A.; Cerichelli, G.; Chiarini, M.; De Giuseppe, S.; Marinelli, F. *Tetrahedron Lett.*, ***2000***, *41*, 9195.

Bu; C=C; Me; OH → 5 eq allyl bromide (Br), DMA; rt, 5% $PdCl_2$, 10.5 h → Bu; Me; 76%

Ma, S.; Gao, W. *Tetrahedron Lett.*, ***2000***, ***41***, 8933.

O; +IPh; O → $MeOCH_2C\equiv CH$, 15 min; $Rh_2(OAc)_3$, reflux → O; O; CH_2OMe; 47%

Gogonas, E.P.; Hadjiarapoglou, L.P. *Tetrahedron Lett.*, ***2000***, ***41***, 9299.

t-Bu; H; C; CO_2Et → 1. LDA; 2. Cp_2TiCl_2; 3. DMDU; 4. HCl, $CHCl_3$ → t-Bu; Me; O; CO_2Me

Krause, N.; Laux, M.; Hoffmann-Röder, A. *Tetrahedron Lett.*, ***2000***, ***41***, 9613.

$PhCO_2Me$ → 2 eq $Me_3P{=}CHCN$, toluene; 100°C, 2 d → Ph; MeO; CN

89% (0.9/1, *E/Z*)

Tsunoda, T.; Takagi, H.; Takaba, D.; Kaku, H.; Itô, S. *Tetrahedron Lett.*, ***2000***, ***41***, 235.

Ph; OH → PdI_2–thiourea, CBr_4; Cs_2CO_3, CO, MeOH; 40°C → CO_2Me; Ph; O; 80%

Nan, Y.; Miao, H.; Yang, Z. *Org. Lett.*, ***2000***, *2*, 297.

O; PhO_2S; Ph; Ph → $HC\equiv CCH_2OH$, THF; $Pd(PPh_3)_2Cl_2$, t-BuOK; reflux → O; Ph; O; Ph; 57%

Monteiro, N.; Balme, G. *J. Org. Chem.*, ***2000***, ***65***, 3223.

$2.1\ Ce(NBu_4)_2(NO_3)_6$, 25°C

MeCN–DCM , MS 4Å

TMS; O; i-Bu; $SnBu_3$; O; i-Bu

88%

Chen, C.; Mariano, P.S. *J. Org. Chem.*, ***2000***, *65*, 3252.

Amberlyst-15

$CHCl_3$, –20°C , 1 h

HO; OH; O

quant

Young, J.-j.; Jung, L.-j.; Cheng, K.-m. *Tetrahedron Lett.*, ***2000***, *41*, 3411.

$0.1\ Ag_2CO_3$, PhH

80°C , 1 h

OH; O; O; O

99%

Pale, P.; Chuche, J. *Eur. J. Org. Chem.*, ***2000***, 1019.

Ph–CH=CH–$SnBu_3$

LiCl , $Pd(PPh_3)_4$

O; N; Ph; OTf; O; N; Ph; Ph

75%

Schaus, J.V.; Panek, J.S. *Org. Lett.*, ***2000***, *2*, 469.

Ph–C≡C–CH(Ph)–OCO_2Me

$Pd_2(dba)_3$, *R*-MeO-Biphep
25°C

OH; OH; Ph; Ph

99% (87% ee)

Labrosse, J.-R.; Lhoste, P.; Sinou, D. *Org. Lett.*, ***2000***, *2*, 527.

$Me_3SiCH=N_2$, DCE , 84°C
5% $FeBr_2PPh_3$, 2 h

PhS; Ph; Ph; $SiMe_3$; SPh

88% (85:15 dr)

Carter, D.S.; Van Vranken, D.L. *Org. Lett.*, ***2000***, *2*, 1303.

TBSO SiMe$_3$ → 1. $Co_2(CO)_8$ 2. toluene, heat, 12 h → OTBS SiMe$_3$ 70%

Dolaine, R.; Gleason, J.L. *Org. Lett.*, ***2000***, *2*, 1753.

N_2 O O OTBS → Rh_2(O-oct)$_4$, hexanes reflux → O O O OTBS 83%

Davies, H.M.L.; Calvo, R.L.; Townsend, R.J.; Ren, P.; Churchill, R.M. *J. Org. Chem.*, ***2000***, *65*, 4261.

O → PhOTf, 5% $Pd(dba)_3$•dba, PhH, 5 d; 15% chiral P-containing oxazoline ligand; *i*-Pr_2NEt, 70°C → O Ph 81% (96% ee)

Hashimoto, Y.; Horie, Y.; Hayashi, M.; Saigo, K. *Tetrahedron Asymm.*, ***2000***, *11*, 2205.

I O → PhCHO, CH_2=C=CH_2; DMF, In, 10% $Pd(OAc)_2$; 20% tris(2-formyl)phosphine → OH Ph O 52%

Anwar, U.; Grigg, R.; Sridharan, V. *Chem. Commun.*, ***2000***, 933.

CHO OH → (O), DABCO → O O 81%

Kaye, P.T.; Nocanda, X.W. *J. Chem. Soc., Perkin Trans. 1*, ***2000***, 1331.

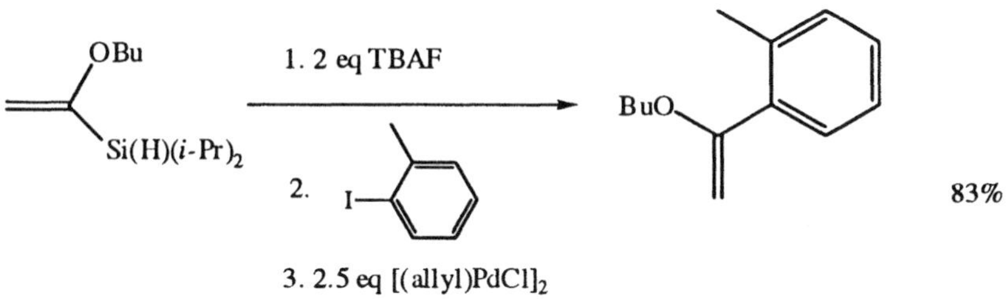

OBu, Si(H)(*i*-Pr)$_2$ → 1. 2 eq TBAF; 2. I-(2-methylphenyl); 3. 2.5 eq [(allyl)PdCl]$_2$ → BuO 83%

Denmark, S.E.; Neuville, L. *Org. Lett.*, ***2000***, *2*, 3221.

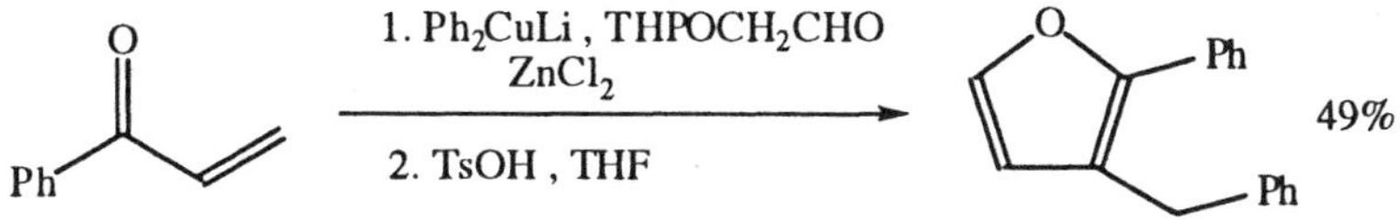

Ph, O → 1. Ph_2CuLi, $THPOCH_2CHO$, $ZnCl_2$; 2. TsOH, THF → O, Ph, Ph 49%

Méndez-Andino, J.; Paquette, L.A. *Org. Lett.*, ***2000***, *2*, 4095.

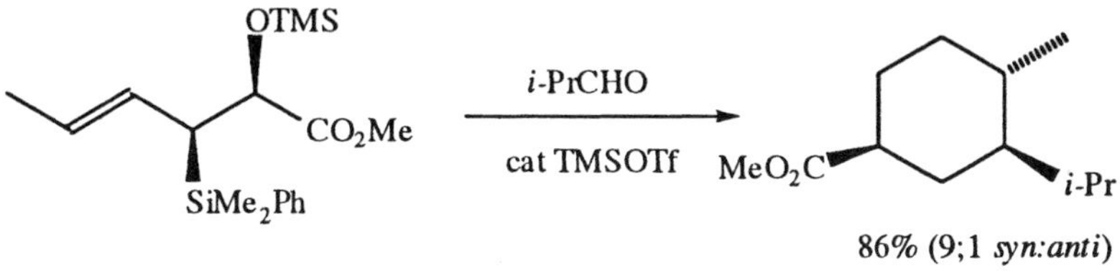

OTMS, CO_2Me, $SiMe_2Ph$ → *i*-PrCHO, cat TMSOTf → MeO_2C, *i*-Pr 86% (9;1 *syn:anti*)

Huang, H.; Panek, J.S. *J. Am. Chem. Soc.*, ***2000***, *122*, 9836.

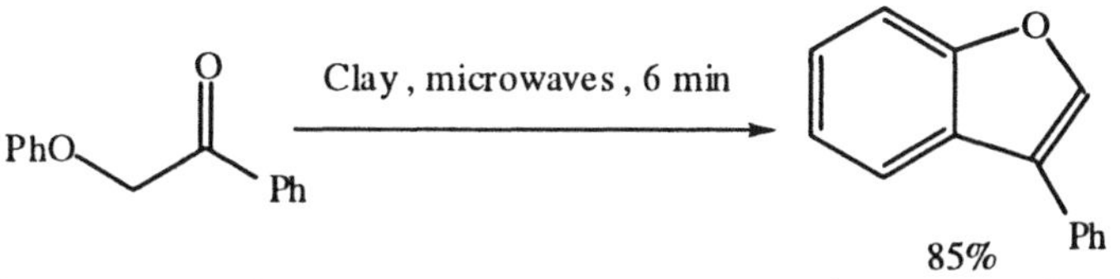

PhO, O, Ph → Clay, microwaves, 6 min → O, Ph 85%

Meshram, H.M.; Sekhar, K.C.; Ganesh, Y.S.S.; Yadav, J.S. *Synlett*, ***2000***, 1273.

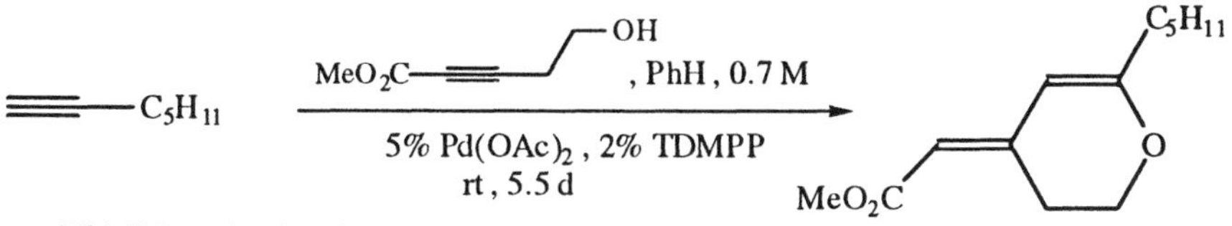

C_5H_{11} → MeO_2C–≡–CH$_2$CH$_2$–OH, PhH, 0.7 M; 5% Pd(OAc)$_2$, 2% TDMPP, rt, 5.5 d → C_5H_{11}, O, MeO_2C 61%

TDMPP = tris-(2,6-dimethoxyphenyl)-phosphine

Trost, B.M.; Frontier, A.J. *J. Am. Chem. Soc.*, ***2000***, *122*, 11727.

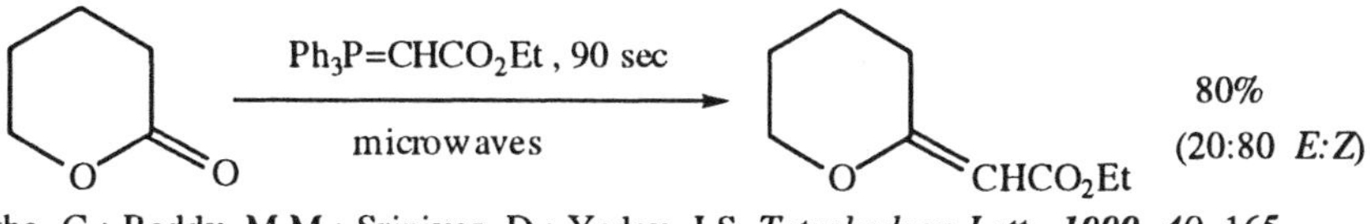

O, O → Ph_3P=$CHCO_2Et$, 90 sec, microwaves → O, $CHCO_2Et$ 80% (20:80 *E:Z*)

Sabitha, G.; Reddy, M.M.; Srinivas, D.; Yadav, J.S. *Tetrahedron Lett.*, ***1999***, *49*, 165.

PhO_2S PhO_2S Ph — $PtCl_2$, MeOH, 2 d → PhO_2S PhO_2S OMe Ph 72%

Méndez, M.; Muñoz, M.P.; Echavarren, A.M. *J. Am. Chem. Soc.*, ***2000***, *122*, 11549.

EtCHO, BF_3•OEt_2 → O Et 70% (70:30 *cis:trans*)

Regás, D.; Afonso, M.M.; Galindo, A.; Palenzuela, J.A. *Tetrahedron Lett.*, ***2000***, *41*, 6781.

O Ph — Mg, TMSCl, DMF, 15–25°C → OTMS Ph + OTMS Ph

63% (11.8, *E/Z*) 31%

Ishino, Y.; Kita, Y.; Maekawa, H.; Ohno, T.; Yamasaki, Y.; Miyata, Y.; Nishiguchi, I. *Tetrahedron Lett.*, ***1999***, *40*, 1349.

N Ph CHO — Et_3SiH, PBu_3, $Ni(cod)_2$, 45°C → Ph N $OSiEt_3$ 85%

Tang, X.-Q.; Montgomery, J. *J. Am. Chem. Soc.*, ***1999***, *121*, 6098.

OH — allyl alcohol (OH), 5% $RuCl(cod)C_5Me_5$, 30 min, rt → O OH 80%

at 80°C for 1 h, obtain 80% of allyl ether derivative

Dérien, S.; Ropartz, L.; Le Paih, J.; Dixneuf, P.H. *J. Org. Chem.*, ***1999***, *64*, 3524.

Ph–(oxetan-2-one) $\xrightarrow[75\text{-}80°C]{Cp_2TiMe_2\text{ , toluene}}$ Ph–(2-methyleneoxetane) 76%

Dollinger, L.M.; Ndakala, A.J.; Hashemzadeh, M.; Wang, G.; Wang, Y.; Martinez, I.; Arcari, J.T.; Galluzzo, D.J.; Howell, A.R. *J. Org. Chem.*, ***1999***, *64*, 7074.

HO $\xrightarrow[25°C\text{ , 18 h}]{PdI_2\text{ , 2 eq KI , DMA}}$ 92%

Gabriele, B.; Salerno, G.; Lauria, E. *J. Org. Chem.*, ***1999***, *64*, 7687.

ON–N(–)–C(=O)NH$_2$ $\xrightarrow{HCO_3^-\text{ , MeOH}}$ Ph–CH=CH–CH$_2$OMe + Ph–CH(OMe)–CH=CH$_2$

25% 75%

Wiberg, K.B.; Österle, C.G. *J. Org. Chem.*, ***1999***, *64*, 7756.

OH $\xrightarrow[MeCN\text{ , 60°C , 3 h}]{Ph_2I^+\ BF_4^-\text{ , }K_2CO_3}$ Ph, O 76%

Kang, S.-K.; Baik, T.-G.; Kulak, A.N. *Synlett*, ***1999***, 324.

C_7H_{15}C(=O)OEt $\xrightarrow[8\text{ eq TMEDA , THF, 25°C}]{CH_2\text{-}(ZnI)_2\text{ , 4 eq }TiCl_2\text{ , 4 h}}$ C_7H_{15}C(=CH$_2$)OEt 75%

Matsubara, S.; Ukai, K.; Mizuno, T.; Utimoto, K. *Chem. Lett.*, ***1999***, 825.

Ph–C≡CH $\xrightarrow[2.\ PhSCl]{1.\ Cp_2Zr(H)Cl\text{ , THF}}$ Ph–CH=CH–SPh 85%

Huang, X.; Zhong, P.; Guo, W.-r. *Org. Prep. Proceed. Int.*, ***1999***, *31*, 201.

Ph_3P O $\xrightarrow{(CH_2O)_n\text{ , toluene , reflux , 1 h}}$ O, O 90%

Okuma, K.; Tanaka, Y.; Shuzui, I.; Shioji, K. *Heterocycles*, ***1999***, *50*, 125.

Ph–CO–CH$_2$–CO–CO$_2$Et → 1. HO_2C–C$_6$H$_4$–SO_2N_3, NEt_3, MeCN, 25°C; 2. HO–CH$_2$CH$_2$–OH; 3. p-TsOH, PhH, 80°C → 2-CO$_2$Et-3-Ph-5,6-dihydro-1,4-dioxin 22%

Hilagenkamp, R.; Brogan, J.B.; Zercher, C.K. *Heterocycles*, ***1999***, *51*, 1073.

Related Methods: Section 180A (Protection of Ketones)

SECTION 368: HALIDE, SULFONATE - HALIDE, SULFONATE

Halocyclopropanations are found in Section 74F (Alkyls from Alkenes).

(Z)-2-pentene → Br_2, [bmim] Br, ionic liquid medium → 2,3-dibromopentane (Me, Br, Br, Et) + diastereomer (>99 : 1) 95%

Chiappe, C.; Capraro, D.; Conte, V.; Pieraccini, D. *Org. Lett.*, ***2001***, *3*, 1061.

cyclohexene → e^-, Et_4NI, NEt_3–THF, CH_2Cl_2, rt → 1-fluoro-2-iodocyclohexane (F, I) 90%

Kobayashi, S.; Sawaguchi, M.; Ayuba, S.; Fukuhara, T.; Hara, S. *Synlett*, ***2001***, 1938.

PhCHO → PhCH=CH$_2$, BCl_3, CH_2Cl_2 → Ph–CHCl–CH$_2$–CHCl–Ph 90% (53:47 *RR(SS):RS(SR)*)

Kabalka, G.W.; Wu, Z.; Hu, Y. *Tetrahedron Lett.*, ***2001***, *42*, 5793.

Bu–CH=CH$_2$ → CAN, KBr, H_2O, DCM, rt, 20min → Bu–CHBr–CH$_2$Br 82%

Nair, V.; Panicker, S.B.; Augustine, A.; George, T.G.; Thomas, S.; Vairamani, M. *Tetrahedron*, ***2001***, *57*, 7417.

SECTION 369: HALIDE, SULFONATE - KETONE

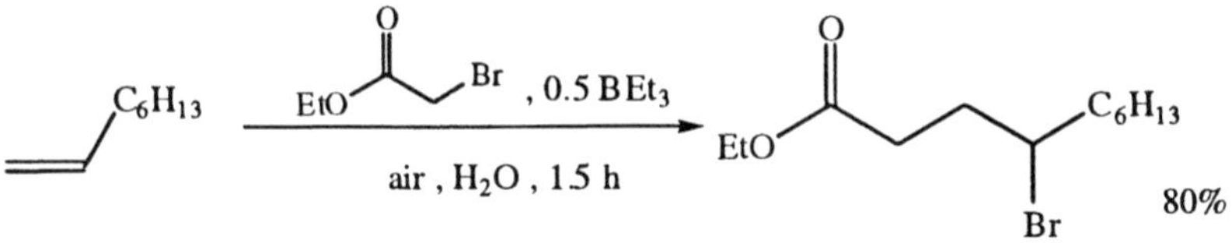

C_6H_{13} — EtO(C=O)CH₂Br, 0.5 BEt_3; air, H_2O, 1.5 h → EtO(C=O)CH₂CH₂CH₂CH(Br)C_6H_{13} 80%

Yorimitsu, H.; Shinokubo, H.; Matsubara, S.; Oshima, K.; Omoto, K.; Fujimoto, H. *J. Org. Chem.*, ***2001***, *66*, 7776.

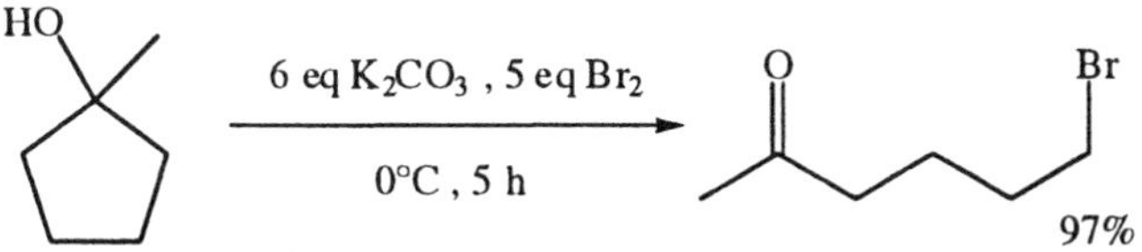

6 eq K_2CO_3, 5 eq Br_2; 0°C, 5 h — 97%

Zhang, W.-C.; Li, C.-J. *J. Org. Chem.*, ***2000***, *65*, 5831.

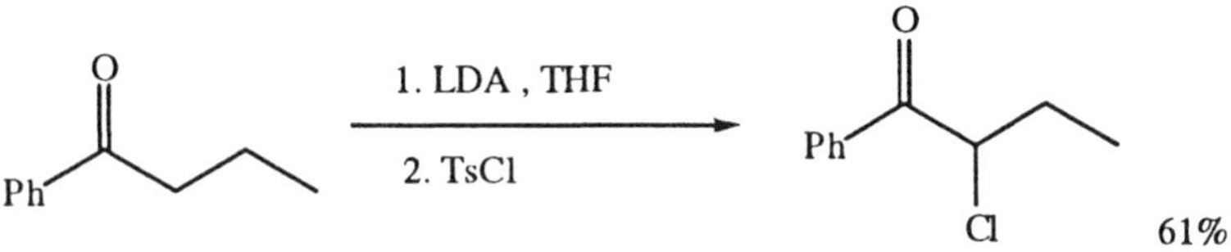

1. LDA, THF; 2. TsCl — Ph, Cl — 61%

Brummond, K.M.; Gesenberg, K.D. *Tetrahedron Lett.*, ***1999***, *40*, 2231.

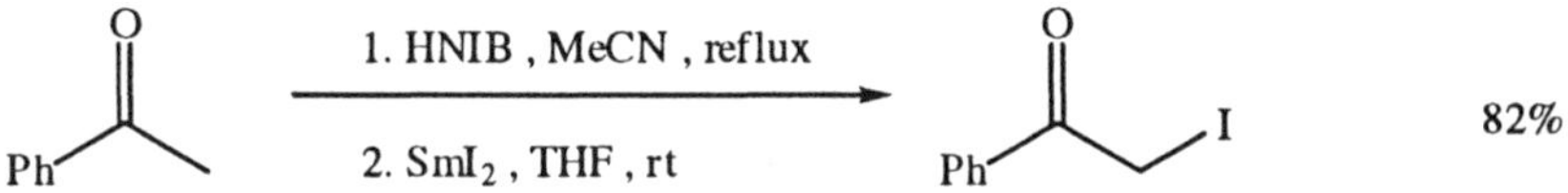

1. HNIB, MeCN, reflux; 2. SmI_2, THF, rt — Ph, I — 82%

HNIB = (hydroxy-*p*-nitrobenzenesulfonyloxy)benzene

Lee, J.C.; Jin, Y.S. *Synth. Commun.*, ***1999***, *29*, 2769.

SECTION 370: HALIDE, SULFONATE - NITRILE

NO ADDITIONAL EXAMPLES

SECTION 371: HALIDE, SULFONATE - ALKENE

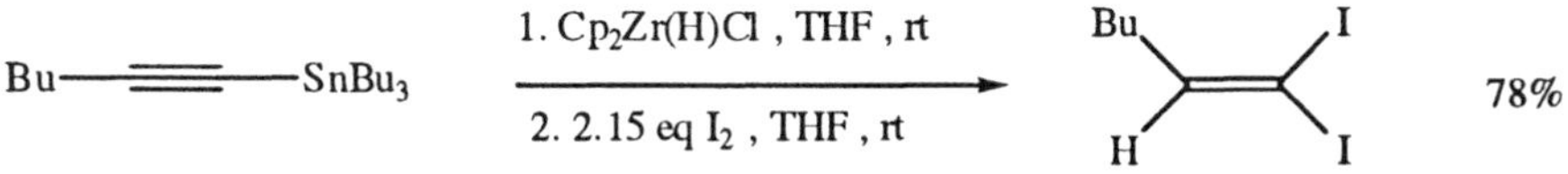

Bu—≡—$SnBu_3$ — 1. $Cp_2Zr(H)Cl$, THF, rt; 2. 2.15 eq I_2, THF, rt → Bu, I, H, I — 78%

Dabdoub, M.J.; Dabdoub, V.B.; Baroni, A.C.M. *J. Am. Chem. Soc.*, ***2001***, *123*, 9694.

C_7H_{15} OAc C_5H_{11} — 1.5% $Pd(OAc)_2$, 40°C; 2.5 eq LiBr, AcOH, 25 h; acetone → C_7H_{15} Br C_5H_{11} 85%

Horváth, A.; Bäckvall, J.-E. *J. Org. Chem.*, ***2001***, *66*, 8120.

BrO ... $SiMe_3$ — 1. 2.5 eq $Cp_2Zr(H)Cl$, THF, 53°C; 2. I_2, DCM, rt; 3. MeZnCl, $Pd(PPh_3)_4$•THF; 4. I_2, DCM → BrO ... Me I 50x95x75%

Arefolov, A.; Langille, N.F.; Panek, J.S. *Org. Lett.*, ***2001***, *3*, 3281.

H F F I — 1. Zn, DMAC; 2. PhI, CuBr, rt, cat $Pd(PPh_3)_4$, 4 h, DMAC → H F F Ph 76 X 74%

Liu, Q.; Burton, D.J. *Tetrahedron Lett.*, ***2000***, *41*, 8045.

O Ph — AcCl, $ZnCl_2/SiO_2$, DCE; 30°C, 1 h → Cl Ph Me 91%

Kodomari, M.; Nagaoka, T.; Furusawa, Y. *Tetrahedron Lett.*, ***2001***, *42*, 3105.

Bu—≡—SPh — TMSBr, CH_2Cl_2, –40°C; MeOH, 1 h → Bu SPh Br 99%

Su, M.; Yu, W.; Jin, Z. *Tetrahedron Lett.*, ***2001***, *42*, 3771.

O NH_2 — NBS, K-10 Montmorillonite; MeOH → O Br NH_2 60%

Braibante, M.E.F.; Braibante, H.T.S.; Rosso, G.b.; da Roza, J.K. *Synthesis*, ***2001***, 1935.

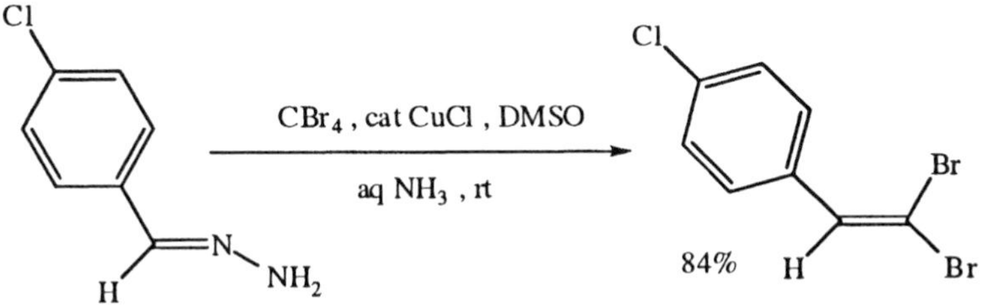

$\xrightarrow[\text{aq } NH_3\text{ , rt}]{CBr_4\text{ , cat CuCl , DMSO}}$ 84%

Shastin, A.V.; Korotchenko, N.; Nenajdenko, V.G.; Balenkova, E.S. *Synthesis*, ***2001***, 2081.

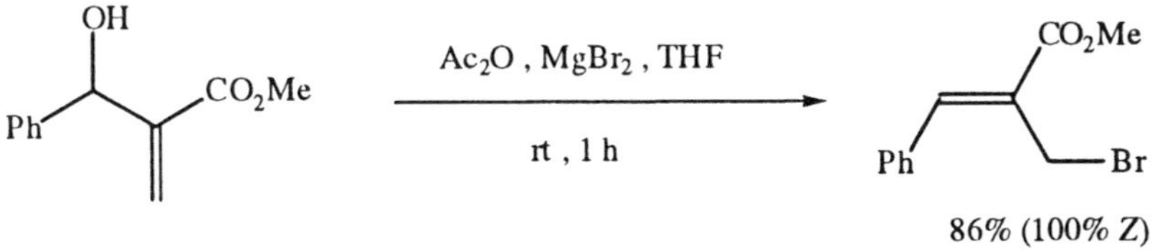

$\xrightarrow[\text{rt , 1 h}]{Ac_2O\text{ , }MgBr_2\text{ , THF}}$ 86% (100% Z)

Ravichandran, S.; *Synth. Commun*, ***2001***, ***31***, 2059.

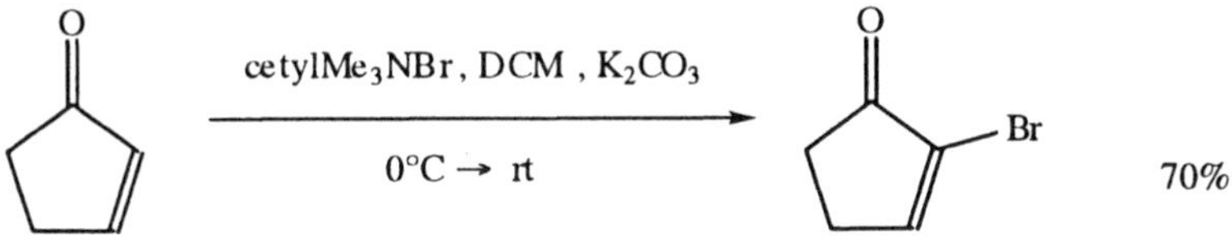

$\xrightarrow[0°C \to \text{rt}]{\text{cetylMe}_3\text{NBr, DCM , }K_2CO_3}$ 70%

Bose, G.; Barua, P.M.B.; Chaudhuri, M.K.; Kalita, D.; Khan, A.T. *Chem. Lett.*, ***2001***, 290.

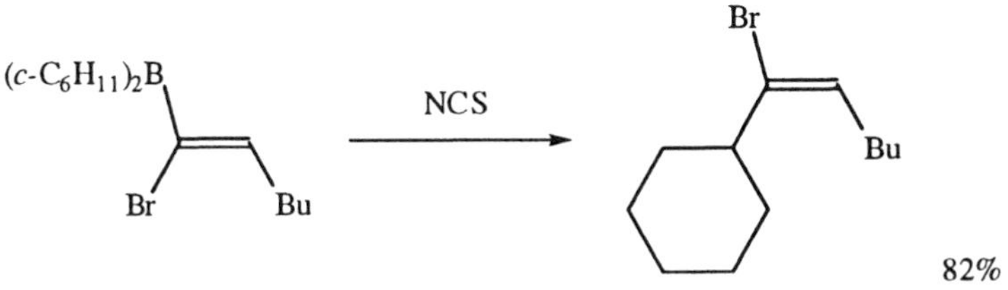

$(c\text{-}C_6H_{11})_2B$ $\xrightarrow{\text{NCS}}$ 82%

Hoshi, M.; Shirakawa, K. *Tetrahedron Lett.*, ***2000***, ***41***, 2595.

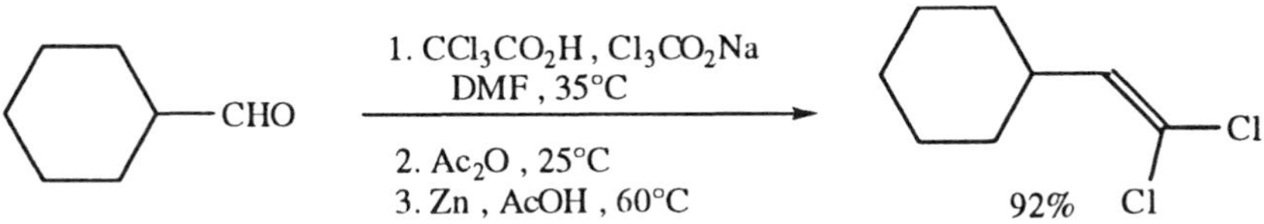

1. CCl_3CO_2H , Cl_3CO_2Na DMF , 35°C
2. Ac_2O , 25°C
3. Zn , AcOH , 60°C

92%

Wang, Z.; Campagna, S.; Xu, G.; Pierce, M.E.; Fortunak, J.M.; Confalone, P.N. *Tetrahedron Lett.*, ***2000***, ***41***, 4007.

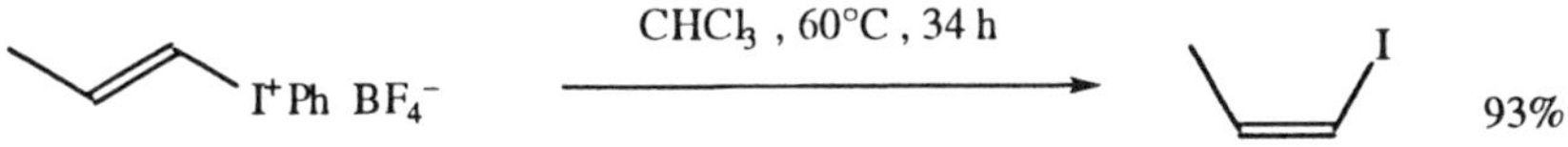

$I^+Ph\ BF_4^-$ $\xrightarrow{CHCl_3\text{ , 60°C , 34 h}}$ 93%

Okuyama, T.; Fujita, M.; Gronheid, R.; Lodder, G. *Tetrahedron Lett.*, ***2000***, ***41***, 5125.

TMSBr , MeOH , DCM , –40°C

OMe

OMe

Br

89%

Yu, W.; Jin, Z. *J. Am. Chem. Soc.*, ***2000***, *122*, 9840.

Ph Me O

Me Cl O

CF_3COOH

Ph Me Cl

83%

+ CH_3COOH

Moughamir, K.; Mezgueldi, B.; Atmani, A.; Mestdagh, H.; Rolando, C. *Tetrahedron Lett.*, ***1999***, *40*, 59.

CO_2Bn

CO_2Bn

1. $SnCl_2$, NEt_3 , CH_2Cl_2 , rt
2. I_2
3. aq HCl

I

CO_2Bn

CO_2Bn

84%

Kitagawa, O.; Suzuki, T.; Fujiwara, H.; Taguchi, T. *Tetrahedron Lett.*, ***1999***, *40*, 2549.

O

1. DMD , acetone , rt
2. NaBr , Amberlyst-15 rt , 12 h

O Br

96%

Righi, G.; Bovicelli, P.; Sperandio, A. *Tetrahedron Lett.*, ***1999***, *40*, 5889.

OAc

C_7H_{15} I I

2 eq SmI_2

C_7H_{15} I

88% (94:6 *Z*:*E*)

Concellón, J.M.; Bernad, P.L.; Pérez-Andrés, J.A. *Angew. Chem. Int. Ed.*, ***1999***, *38*, 2384.

I

O O

1. dilauryl peroxide , PhH , reflux
2. 5 eq $EtSO_2N_3$, PhCl
dilauryl peroxide , 100°C

O O

N_3

80% (74:26 *exo*:*endo*)

Huang, X.; Wang, J.-H.; Yang, D.-Y. *J. Chem. Soc., Perkin Trans. 1*, ***1999***, 673.

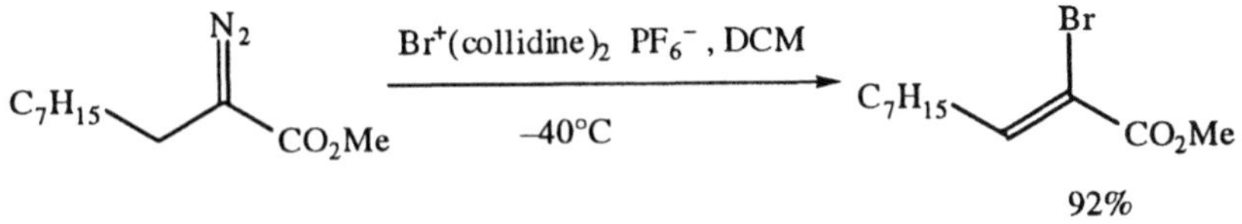
N2
C7H15
CO2Me
Br+(collidine)2 PF6−, DCM
−40°C
Br
C7H15
CO2Me
92%

<u>Rousseau, G.</u>; Marie, J.-X. *Synth. Commun.*, ***1999***, *29*, 3705.

REVIEWS:

"An Efficient New Methodology For The Synthesis Of 1-Functionalized-2-Halo-2-Alkenes Via Hydrohalogentation Reaction Of Electron-Deficient Allenes," <u>Ma, S.</u>; Li, L. *Synlett*, ***2001***, 1206.

SECTION 372: KETONE - KETONE

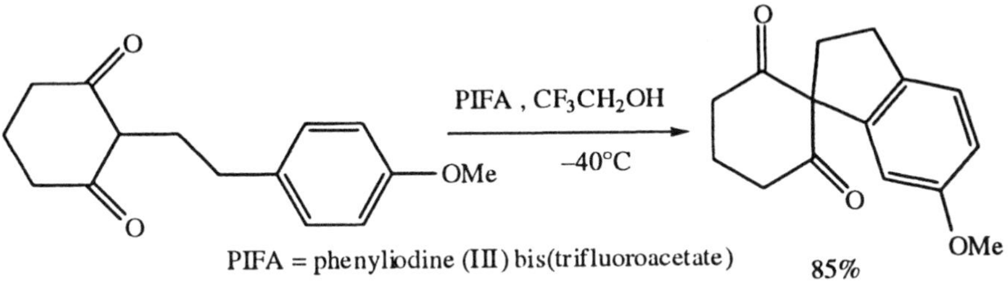
O
O
OMe
PIFA, CF3CH2OH
−40°C
O
O
OMe
PIFA = phenyliodine (III) bis(trifluoroacetate)
85%

Arisawa, M.; Ramesh, N.G.; Nakajima, M.; Tohma, H.; <u>Kita, Y.</u> *J. Org. Chem.*, ***2001***, *66*, 59.

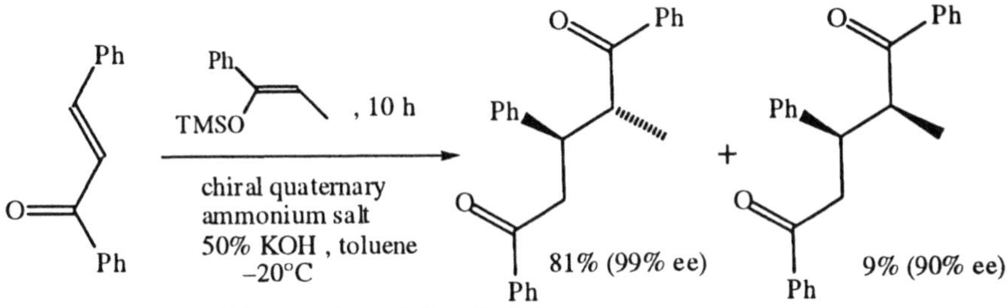
Ph
O
Ph
Ph
TMSO
, 10 h
chiral quaternary
ammonium salt
50% KOH, toluene
−20°C
O
Ph
Ph
O
Ph
81% (99% ee)
+
O
Ph
Ph
O
Ph
9% (90% ee)

Zhang, F.-Y.; <u>Corey, E.J.</u> *Org. Lett.*, ***2001***, *3*, 639.

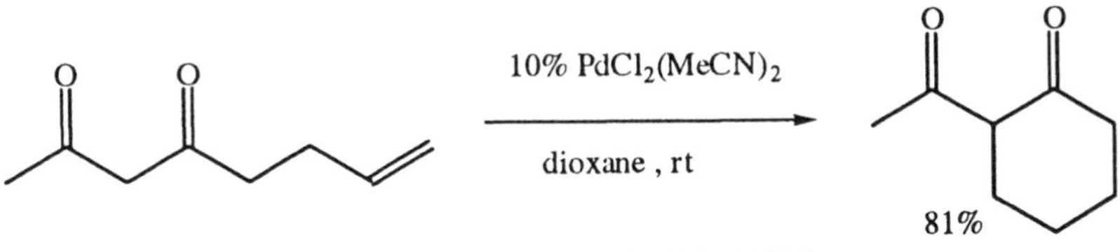
O
O
10% PdCl2(MeCN)2
dioxane, rt
O
O
81%

Pei, T.; <u>Widenhoefer, R.A.</u> *J. Am. Chem. Soc.*, ***2001***, *123*, 11290.

1. OH / Ph alkyne, 2% $PdCl_2(PPh_3)_2$, 1% CuI, NEt_3, heat
2. NC–C6H4–CH, 20% HO thiazolium salt N+ Me I^-

88%

Braun, R.U.; Zeitler, K.; Müller, T.J.T. *Org. Lett.*, ***2001***, *3*, 3297.

1.5 eq PCC, DCM / rt, 10 h

84%

Ren, S.-K.; Wang, F.; Dou, H.-N.; Fan, C.-A.; He, L.; Song, Z.-L.; Xia, W.-J.; Li, D.-R.; Jia, Y.-X.; Li, X.; Tu, Y.Q. *Synthesis*, ***2001***, 2384.

OTMS → , $VOCl_3$, MS 4Å / Ph, OTMS, EtOH, DCM, –75°C → –30°C

63%

Kurihara, M.; Hayashi, T.; Miyata, N. *Chem. Lett.*, ***2001***, 1324.

Ph—≡—Ph → $(Cu^*)RuO_2(CO_2CF_3)$, MeCN / 0.2 TFA, rt

98%

Che, C.-M.; Yu, W.-Y.; Chan, P.-M.; Cheng, W.-C.; Peng, S.-M.; Lau, K.-C.; Li, W.-K. *J. Am. Chem. Soc.*, ***2000***, *122*, 11380.

PhCOCN → In, ultrasound / DMF, rt, 8h

62%

Baek, H.; Lee, S.J.; Yoo, B.W.; Ko, J.J.; Kim, S.H.; Kim. J.H. *Tetrahedron Lett.*, ***2000***, *41*, 8097.

acetone , $ZnCl_2$•Et_2NH

t-BuOH , PhH , rt , 3 d

Br Br O O O 64%

Nevar, N.M.; Kel'in, A.V.; Kulinkovich, O.G. *Synthesis*, ***2000***, 1259.

REVIEWS:

"Chemistry of 2-Acylcycloalkane-1,3-diones," Rubinov, D.B.; Rubinova, I.L.; Akhrem, A.A. *Chem. Rev.*, ***1999***, *99*, 1047.

SECTION 373: KETONE - NITRILE

CN

1. O_3 , MeOH , Me_2S
2. TsOH

O CN 80x40%

Fleming, F.F.; Huang, A.; Sharief, V.A.; Pu, Y. *J. Org. Chem.*, ***1999***, *64*, 2830.

SECTION 374: KETONE - ALKENE

For the oxidation of allylic alcohols to alkene ketones, see Section 168 (Ketones from Alcohols and Phenols)

For the oxidation of allylic methylene groups ($C{=}C{-}CH_2 \rightarrow C{=}C{-}C{=}O$), see Section 170 (Ketones from Alkyls and Methylenes).

For the alkylation of alkene ketones, also see Section 177 (Ketones from Ketones) and for conjugate alkylations see Section 74E (Alkyls form Alkenes).

PhCHO

O CO_2Me , $Cu(OTf)_2$

CH_2Cl_2 , reflux , 40 h

Ph MeO_2C O 76%

Dujardin, G.; Leconte, S.; Bénard, A.; Brown, E. *Synlett*, ***2001***, 147.

Ph, CO_2i-Pr, O, C_3H_7

0.1 $Yb(OTf)_3$, 40°C
DCM, MS 3Å, 96 h

C_3H_7, O, CO_2i-Pr, Ph

99% (34:66 *syn:anti*)

Hiersemann, M.; Abraham, L. *Org. Lett.*, ***2001***, *3*, 49.

Bu, O, O

1. Ph, MeO, $Cr(CO)_5$
2. Ce(IV)
3. PhBr, dioxane
$PdCl_2(dppf)$

O, Ph, Ph, O

66x98%

Davies, M.W.; Johnson, C.N.; Harrity, J.P.A. *J. Org. Chem.*, ***2001***, *66*, 3525.

Bu, O^- Li^+

1. EtO_2C, Ph, O, –78°C, 5 h
2. PhH, SiO_2, reflux

O, Bu, Ph

74%

Shindo, M.; Sato, Y.; Shishido, K. *J. Org. Chem.*, ***2001***, *66*, 7818.

NH_2, N, OMe

1.1 eq NBS, H_2O, THF
H_2SO_4, rt, 5 min

O, N, O

80%

Kim, D.W.; Choi, H.Y.; Lee, K.-J.; Chi, D.Y. *Org. Lett.*, ***2001***, *3*, 445.

TMSO, CO_2Et

1. TMSO—≡, $GaCl_3$
2. H^+

O, CO_2Et

74%

Arisawa, M.; Akamatsu, K.; Yamaguchi, M. *Org. Lett.*, ***2001***, *3*, 789.

$2.5\%\ Co_2(CO)_8$, 30 atm CO
DCM , 130°C , 18 h

TIPS

TIPS

75%

Son, S.U.; Yoon, Y.A.; Choi, D.S.; Park, J.K.; Kim, B.M.; Chung, Y.K. *Org. Lett.*, ***2001***, *3*, 1065.

BnO_2C Ph
BnO_2C Ph

5% $IrCl(CO)(PPh_3)_2$
CO , xylene , 120°C

BnO_2C Ph
BnO_2C O Ph

86%

Shibata, T.; Yamashita, K.; Ishida, H.; Takagi, K. *Org. Lett.*, ***2001***, *3*, 1217.

O Ph O

i-$PrNO_2$, DBU , rt
MeCN

O Ph

77%

Ballini, R.; Bosica, G.; Fiorini, D.; Gil, M.V.; Petrini, M. *Org. Lett.*, ***2001***, *3*, 1265.

O Br

5% Pd catalyst , 130°C
2 eq Nu_4NOAc , DMF
MeCN , H_2O, 16 h

O

72%

Högenauer, K.; Mulzer, J. *Org. Lett.*, ***2001***, *3*, 1495.

Se Ph N

C≡CLi
TMEDA , ether
MeI

O SeMe Ph

50%

Murai, T.; Mutoh, Y.; Kato, S. *Org. Lett.*, ***2001***, *3*, 1993.

1. cat $[RhCl(CO)(dppp)]_2$, $\equiv$—$CH_2CNa(CO_2Me)_2$, MeCN, 30°C
2. CO atmosphere, reflux, 1 d

MeO_2CO ... MeO_2C CO_2Me + MeO_2C CO_2Me

(7 : 1) 88 X 87%

Evans, P.A.; Robinson, J.E. *J. Am. Chem. Soc.*, ***2001***, *123*, 4609.

$Cr(CO)_5$, OMe

1. (dimethylfulvene), MeCN, 80°C
2. 1N HCl, DCM

80%

Barluenga, J.; Martínez, S.; Suárez-Sobrino, A.L.; Tomás, M. *J. Am. Chem. Soc.*, ***2001***, *123*, 11113.

$C_{10}H_{21}$—≡—CHO → cat $[Rh(dppe)_2]_2(BF_4)_2$, MeCN, 100°C → $C_{10}H_{21}$ (cyclopentenone)

67%

Tanaka, K.; Fu, G.C. *J. Am. Chem. Soc.*, ***2001***, *123*, 11492.

$SiMe_3$, OTf

cat $[(\pi\text{-}C_3H_5)PdCl]_2$, dppe, CsF, (allyl OAc), CO, 1 atm, MeCN, 80°C, 4 h

80%

Chatani, N.; Kamitani, A.; Oshita, M.; Fukumoto, Y.; Murai, S. *J. Am. Chem. Soc.*, ***2001***, *123*, 12686.

Ts—N → 5% Co catalyst, CO, 70°C, THF, 1 d → Ts—N ... =O

84%

Comely, A.C.; Gibson, S.E.; Stevenazzi, A.; Hales, N.J. *Tetrahedron Lett.*, ***2001***, *42*, 1183.

1. (ethyl vinyl ketone)
4 eq Li naphth , –78°C
2. BnBr , 20 h
3. p-TsOH , PhH , reflux

99%

Tai, C.-L.; Ly, T.W.; Wu, J.-D.; Shia, K.-S.; Liu, H.-J. *Synlett*, ***2001***, 214.

EtO_2C CO_2Et

NaH , THF , 55°C

PPh_3 CO_2Et

EtO_2C EtO_2C OEt

42% (>98:2 *trans:cis*)

Langer, P.; Kracke, B. *Synlett*, ***2001***, 1790.

OTBS CO_2Me

$Pd(OAc)_2$, MeCN , 100 min

rt

CO_2Me

79%

Toyota, M.; Majo, V.J.; Ihara, M. *Tetrahedron Lett.*, ***2001***, *42*, 1555.

2.2 eq $Co_2(CO)_8$, DMSO
CH_2=CH_2 , 110°C , 4 d

16%

Rausch, B.J.; Gleiter, R. *Tetrahedron Lett.*, ***2001***, *42*, 1651.

C_8H_{17} $ZrClCp_2$

Bn Ph

cat $PdCl_2(PPh_3)_2$, rt
THF-ether , 12 h

C_8H_{17} Bn Ph

1:2.4 *E:Z*

Hanzawa, Y.; Kakuuchi, A.; Yabe, M.; Narita, K.; Tabuchi, N.; Taguchi, T. *Tetrahedron Lett.*, ***2001***, *42*, 1737.

OH

1. $Co_2(CO)_8$, MS 4Å toluene

2. Me_3N–O , rt

OH

60% (1:1 *cis:trans*)

Blanco-Urgeiti, J.; Casarrubios, L.; Domínguez, G.; Pérez-Castells, J. *Tetrahedron Lett.*, ***2001***, *42*, 3315.

HO

, DMF , 60°C , 30 h

10% $[CpRu(MeCN)_3]$ PF_6

71%

Yoshida, M.; Sugimotot, K.; Ihara, M. *Tetrahedron Lett.*, ***2001***, *42*, 3877.

1. proline , DMF , rt , 6 h

2. 45 h ,

3. H_2O

4. PhH , reflux , *p*-TsOH , 2 h

70% (76% ee)

Rajagopal, D.; Narayanan, R.; Swaminathan, S. *Tetrahedron Lett.*, ***2001***, *42*, 4887.

PhCHO —acetophenone , MeOH , 18 h / natural phosphate , $NaNO_3$→ PhCH=CHBz 98%

Sebti, S.; Solhy, A.; Tahir, R.; Boulaajaj, S.; Mayoral, J.A.; Fraile, J.M.; Kossir, A.; Oumimoun, H. *Tetrahedron Lett.*, ***2001***, *42*, 7953.

$Co_2(CO)_8$, NMO , 2 h

NMO , DCM , rt

H

71%

Ishizaki, M.; Iwahara, K.; Niimi, Y.; Satoh, H.; Hoshino, O. *Tetrahedron*, ***2001***, *57*, 2729.

Ph

1. $Co_2(CO)_3$, DCM

2. NMO , 0°C → rt

Ph O

O

O 73%

Lovely, C.J.; Seshadri, H. *Synth. Commun.*, ***2001***, *31*, 2479.

O

$Ni(cod)_2$, TMSCl
CO , toluene

OHC O

O

56%

OHC

García-Gómez, G.; Moretó, J.M. *Eur. J. Org. Chem.*, ***2001***, 1359.

O

1.5 eq $AlCl_3$, ICl , 0°C

DCM , 45 min

I

I

O

83%

Sha, C.-K.; Lee, F.-C.; Lin, H.-H. *Chem. Commun.*, ***2001***, 39.

O

CH_2=C=CH_2 , 5% $Ni(cod)_2$

Ph $SnMe_3$

toluene , 1.5 h

O $SnMe_3$

Ph

64%

Shirakawa, E.; Nakao, Y.; Hiyama, T. *Chem. Commun.*, ***2001***, 263.

O

, Grubbs' catalyst II

DCM , 45°C , 6 h

O

O

86%

Randl, S.; Connon, S.J.; Blechert, S. *Chem. Commun.*, ***2001***, 1796.

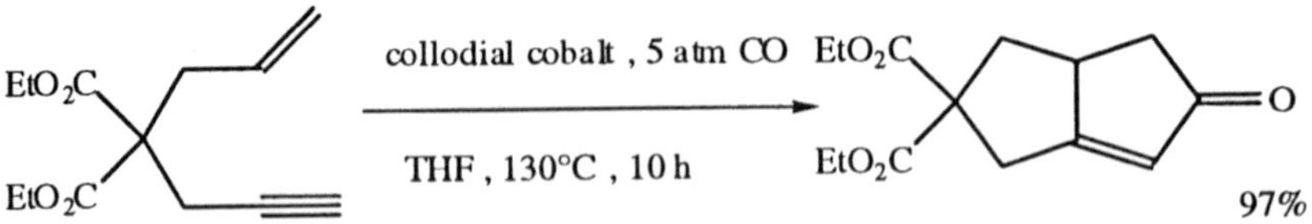
EtO2C
EtO2C
collodial cobalt , 5 atm CO
THF , 130°C , 10 h
EtO2C
EtO2C
O
97%

Kim, S.-W.; Son, S.U.; Lee, S.S.; Hyeon, T.; Chung, Y.K. *Chem. Commun.*, ***2001***, 2212.

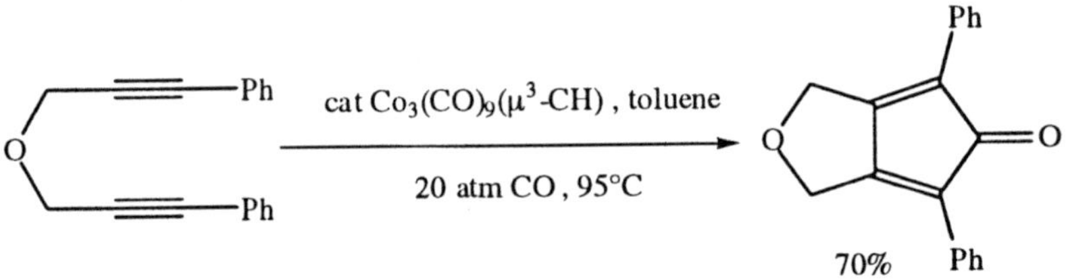
Ph
O
Ph
cat Co3(CO)9(μ3-CH) , toluene
20 atm CO , 95°C
Ph
O
O
70%
Ph

Sugihara, T.; Wakabayashi, A.; Takao, H.; Imagawa, H.; Nishizawa, M. *Chem. Commun.*, ***2001***, 2456.

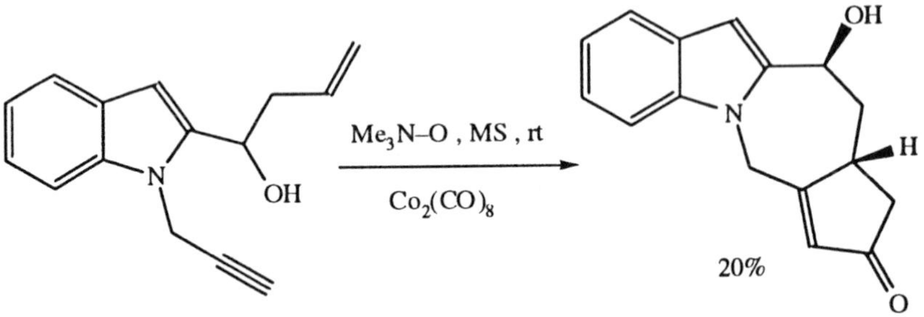
N
OH
Me3N–O , MS , rt
Co2(CO)8
OH
N
H
20%
O

Pérez-Serrano, L.; Casarrubios, L.; Domínguez, G.; Pérez-Castells, J. *Chem. Commun.*, ***2001***, 2602.

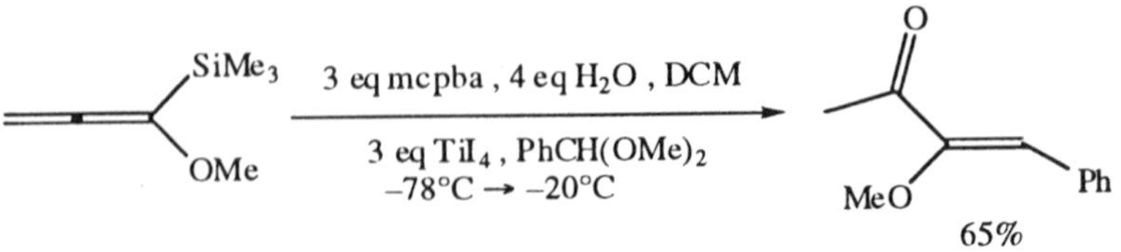
SiMe3
OMe
3 eq mcpba , 4 eq H2O , DCM
3 eq TiI4 , PhCH(OMe)2
–78°C → –20°C
O
MeO
Ph
65%

Hayakawa, R.; Makino, H.; Shimizu, M. *Chem. Lett.*, ***2001***, 756.

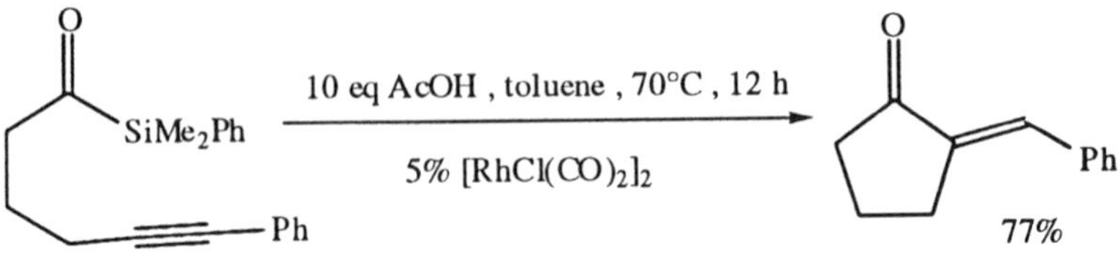
O
SiMe2Ph
Ph
10 eq AcOH , toluene , 70°C , 12 h
5% [RhCl(CO)2]2
O
Ph
77%

Yamane, M.; Amemiya, T.; Narasaka, K. *Chem. Lett.*, ***2001***, 1210.

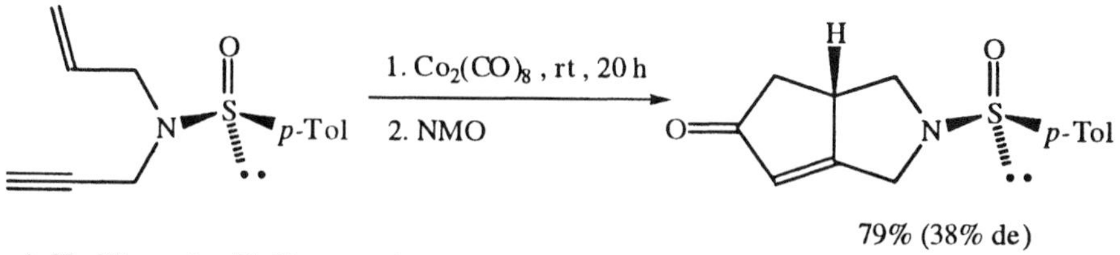

1. $Co_2(CO)_8$, rt, 20 h
2. NMO

79% (38% de)

Hiroi, K.; Watanabe, T. *Heterocycles*, ***2001***, *54*, 73.

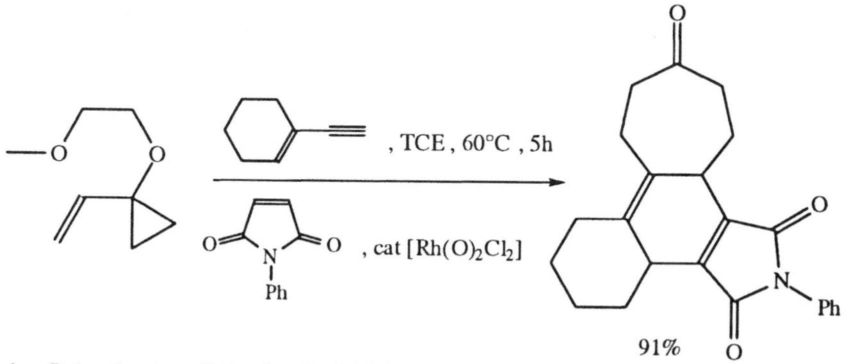

, TCE, 60°C, 5h

, cat $[Rh(O)_2Cl_2]$

91%

Wender, P.A.; Gamber, G.G.; Scanio, M.J.C. *Angew. Chem. Int. Ed.*, ***2001***, *40*, 3895.

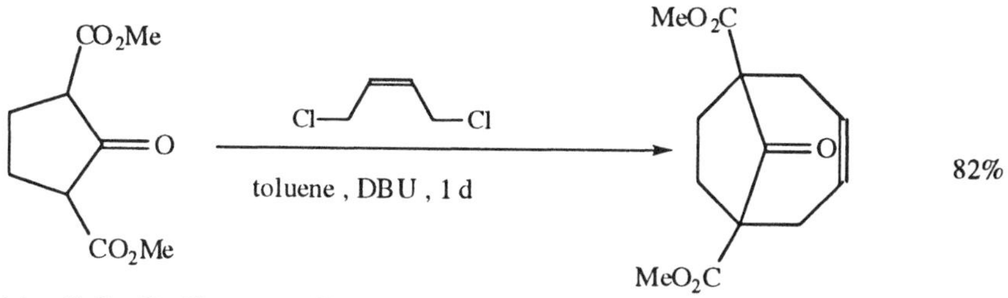

toluene, DBU, 1 d

82%

Lavoisier-Gallo, T.; Charonnet, E.; Pons, J.-M.; Rajzman, M.; Faure, R.; Ridriguez, J. *Chem. Eur. J.*, ***2001***, *7*, 1056.

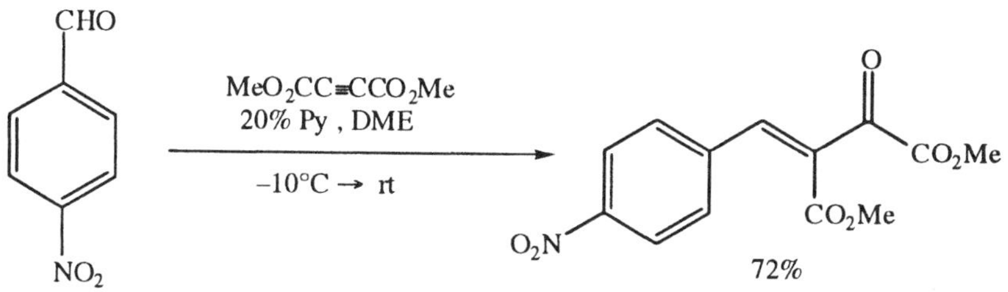

$MeO_2CC{\equiv}CCO_2Me$
20% Py, DME
−10°C → rt

72%

Nair, V.; Sreekanth, A.R.; Vinod, A.U. *Org. Lett.*, ***2001***, *3*, 3495.

, LiBr, DMF
10% $CpRu(NCMe)_3{}^+PF_6^-$
15% $SnCl_4 \cdot 5\ H_2O$, 60°C
C_6H_{13}
MeO–C6H4–CHO
MeO
OH
O
C_6H_{13}
Br
52%
+
O
C_6H_{13}
Br
20%

Trost, B.M.; Pinkerton, A.B. *J. Am. Chem. Soc.*, **2000**, *122*, 8081.

PhOTF, chiral Pd catalyst
i-Pr_2NEt, THF, 50°C, 2 d
Ph
87% (85% ee)

Ogasawara, M.; Yoshida, K.; Hayashi, T. *Heterocycles*, **2000**, *52*, 195.

EtO_2C
EtO_2C
cat Co, charcoal, 18 h
20 atm CO, 130°C, THF
EtO_2C
EtO_2C
O
98%

Son, S.U.; Lee, S.i.; Chung, Y.K. *Angew. Chem. Int. Ed.*, **2000**, *39*, 4158.

$Cr(CO)_5$
OMe
PhC≡CH, THF
reflux
MeO
Ph
O
71%

Barluenga, J.; Aznar, F.; Palomero, M.A. *Angew. Chem. Int. Ed.*, **2000**, *39*, 4346.

OH
Me
Me
O_2, cat dicopper complex
mushroom tyrosinase, 6 h
phosphate buffer (pH 6.8)
Me Me
O O
Me Me
85%

Gupta, R.; Mukherjee, R. *Tetrahedron Lett.*, **2000**, *41*, 7763.

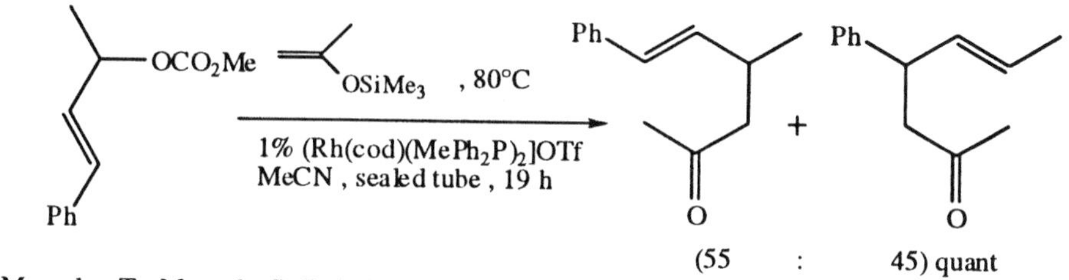

OCO_2Me $OSiMe_3$, 80°C

1% $(Rh(cod)(MePh_2P)_2]OTf$
MeCN, sealed tube, 19 h

Ph   Ph   O   +   O

(55 : 45) quant

Muraoka, T.; Matsuda, I.; Itoh, K. *Tetrahedron Lett.*, ***2000***, *41*, 8807.

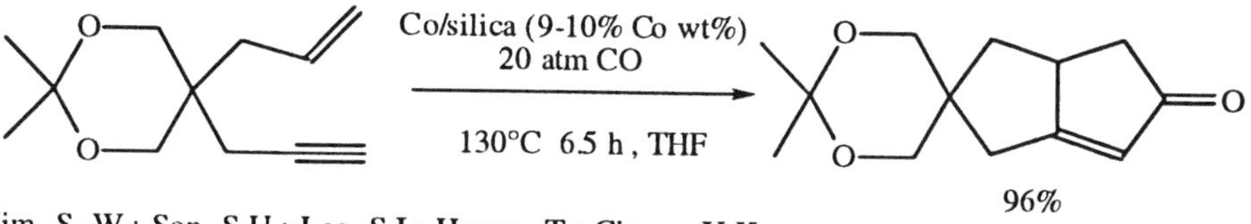

Co/silica (9-10% Co wt%)
20 atm CO
130°C 6.5 h, THF

96%

Kim, S.-W.; Son, S.U.; Lee, S.I.; Hyeon, T.; Chung, Y.K.
J. Am. Chem. Soc., ***2000***, *122*, 1550.

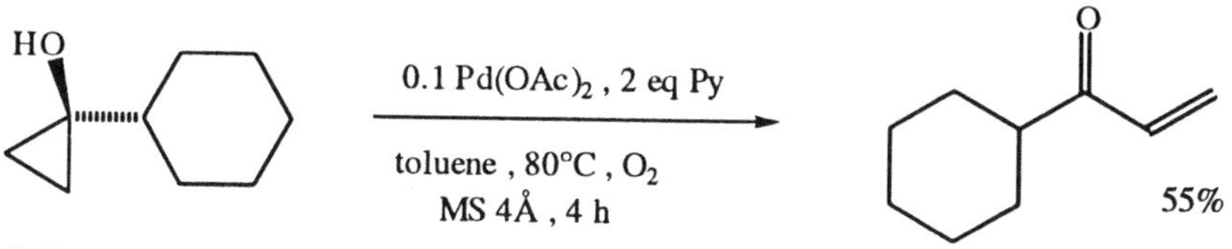

HO

0.1 $Pd(OAc)_2$, 2 eq Py
toluene, 80°C, O_2
MS 4Å, 4 h

55%

Park, S.-B.; Cha, J.K. *Org. Lett.*, ***2000***, *2*, 147.

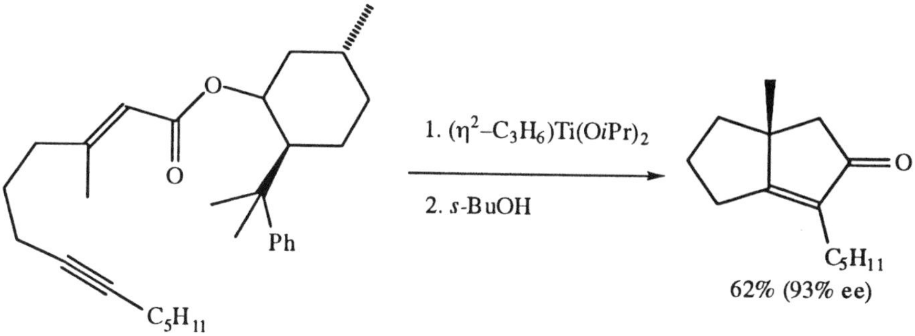

1. $(\eta^2\text{-}C_3H_6)Ti(OiPr)_2$
2. *s*-BuOH

C_5H_{11} Ph   O

C_5H_{11}
62% (93% ee)

Urabe, H.; Hideura, D.; Sato, F. *Org. Lett.*, ***2000***, *2*, 381.

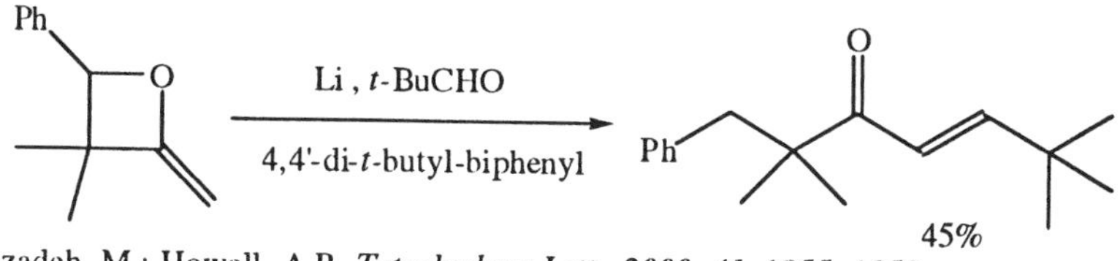

Ph   O

Li, *t*-BuCHO
4,4'-di-*t*-butyl-biphenyl

Ph   O

45%

Hashemzadeh, M.; Howell, A.R. *Tetrahedron Lett.*, ***2000***, *41*, 1855, 1859.

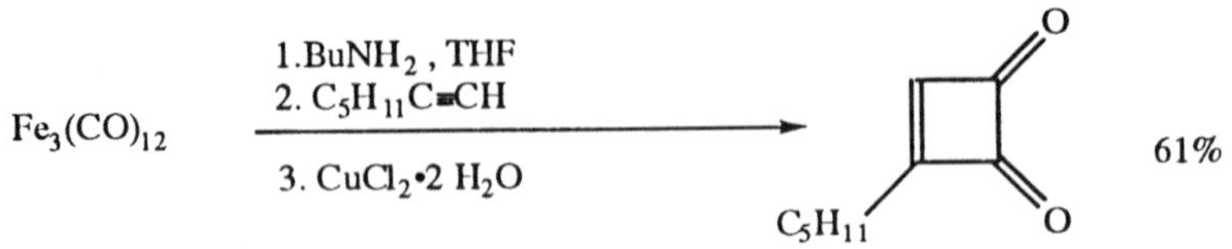

$Fe_3(CO)_{12}$ → 1. $BuNH_2$, THF; 2. $C_5H_{11}C{\equiv}CH$; 3. $CuCl_2{\bullet}2\ H_2O$ → C_5H_{11}, O, O — 61%

Rameshkumar, C.; Periasamy, M. *Tetrahedron Lett.*, ***2000***, *41*, 2719.

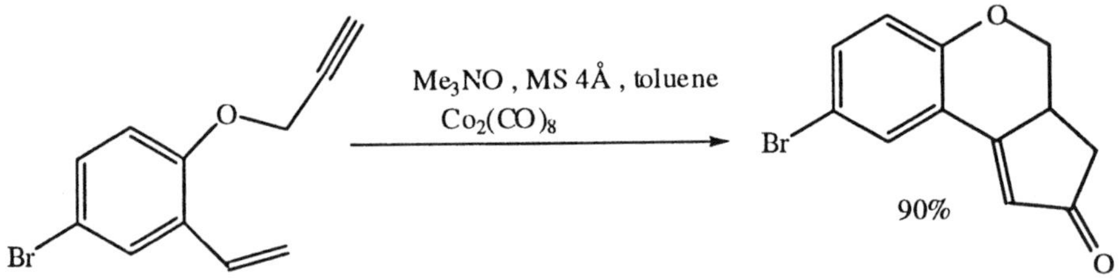

Br, O → Me_3NO, MS 4Å, toluene; $Co_2(CO)_8$ → Br, O, O — 90%

Pérez-Serrano, L.; Blanco-Urgoiti, J.; Casarrubios, L.; Domínguez, G.; Pérez-Castells, J. *J. Org. Chem.*, ***2000***, *65*, 3513.

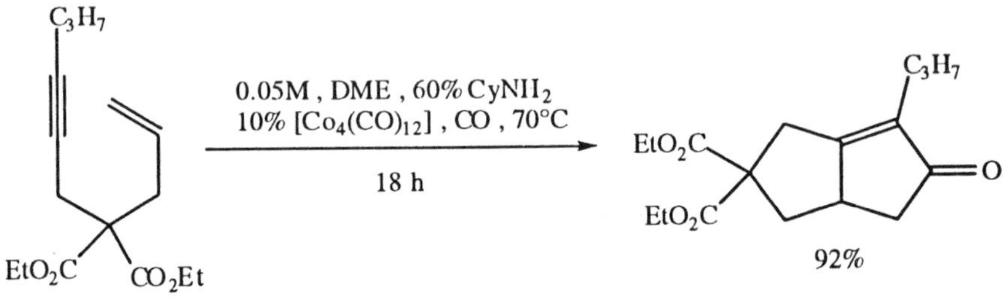

C_3H_7, EtO_2C, CO_2Et → 0.05M, DME, 60% $CyNH_2$; 10% $[Co_4(CO)_{12}]$, CO, 70°C; 18 h → C_3H_7, EtO_2C, EtO_2C, O — 92%

Krafft, M.E.; Boñaga, L.V.R. *Angew. Chem. Int. Ed.*, ***2000***, *39*, 3676.

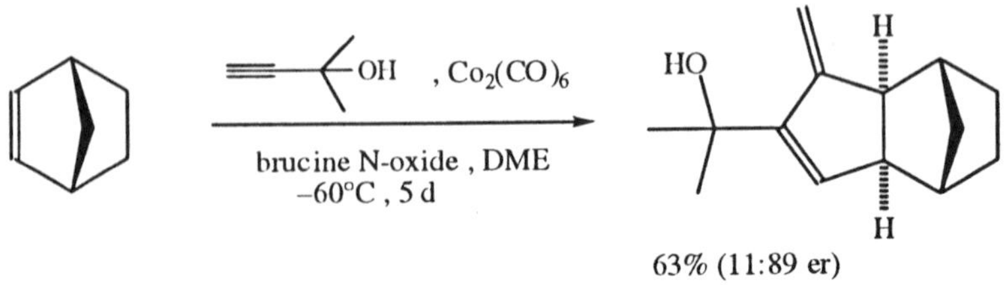

→ OH, $Co_2(CO)_6$; brucine N-oxide, DME; −60°C, 5 d → HO, H, H — 63% (11:89 er)

Kerr, W.J.; Lindsay, D.M.; Rankin, E.M.; Scott, J.S.; Watson, S.P. *Tetrahedron Lett.*, ***2000***, *41*, 3229.

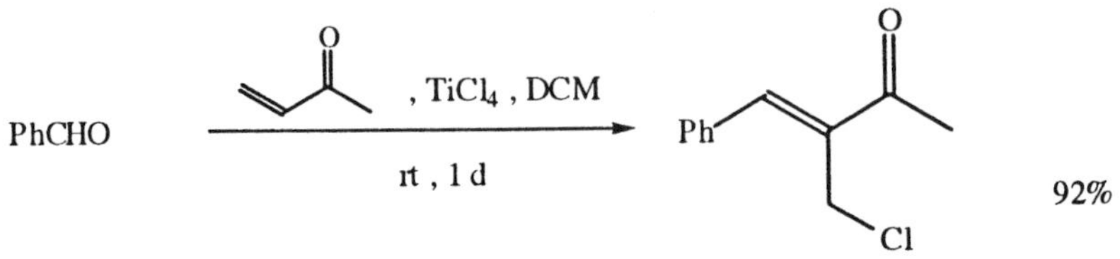

PhCHO → O, $TiCl_4$, DCM; rt, 1 d → Ph, O, Cl — 92%

Li, G.; Gao, J.; We, H.-X.; Enright, M. *Org. Lett.*, ***2000***, *2*, 617.

, 3 atm CO , THF

$[RuCl_2(CO)_3]_2$, NEt_3 , 120°C , 5 h

80% (100% *exo*)

Morisaki, Y.; Kondo, T.; Mitsudo, T.-a. *Org. Lett.*, ***2000***, *2*, 949.

1. $CF_3(CF_2)_5$–C_6H_4–SeCl
2. H_2O_2
3. $Na_2S_2O_5$
4. fluorous extraction

86%

Crich, D.; Barba, G.R. *Org. Lett.*, ***2000***, *2*, 989.

HC≡CCH_2OMe , 0.5M DCE

0.5% $[Rh(CO)_2Cl]_2$, 80°C , H^+

92%

Wender, P.A.; Dyckman, A.J.; Husfeld, C.O.; Scanio, M.J.C. *Org. Lett.*, ***2000***, *2*, 1609.

$CeCl_3$•7 H_2O , NaI

MeCN , 9 h

89%

Bartoli, G.; Bellucci, M.C.; Petrini, M.; Marcantoni, E.; Sambri, L.; Torregiani, E. *Org. Lett.*, ***2000***, *2*, 1791.

2 eq PhC≡CH , 30 atm CO

5% $Co_2(CO)_8$, 130°C
DCM , 18 h

68%

Son, S.U.; Choi, D.S.; Chung, Y.K.; Lee, S.-G. *Org. Lett.*, ***2000***, *2*, 2097.

OH

Pd(dba)$_2$, MeCN , 50°C

20 h

Ph

O

98%

Okumoto, H.; Jinnai, T.; Shimizu, H.; Harada, Y.; Mishima, H.; Suzuki, A. *Synlett.*, ***2000***, 629.

EtO_2C

EtO_2C

0.02 DME , 35% $Co_2(CO)_8$
1.5 eq $CyNH_2$, 60°C

N_2 atmosphere

EtO_2C

EtO_2C

O

80%

Okumoto, H.; Nishihara, S.; Yamamoto, S.; Hino, H.; Nozawa, A.; Suzuki, A. *Synlett*, ***2000***, 991.

C_3H_7—≡—SeEt

1. Cp_2ZrCl , DCM

2. BzCl , 15% CuBr–SMe_2

O

Ph

C_3H_7

SeEt

82%

Suwa, T.; Sugiyama, E.; Shibata, I.; Baba, A. *Synthesis*, ***2000***, 789.

OH

160°C , Ph_2O , 1 h

H

H

O

76%

Ovaska, T.V.; Roses, J.B. *Org. Lett.*, ***2000***, 2, 2361.

Me

MeO_2C

MeO_2C

3% $[RhCl(CO)_2]$, 9% *S*-BINAP
12% AgOTf , 2 atm CO , THF

30°C , 20 h

Me

MeO_2C

MeO_2C

O

91% (62% ee) H

Jeong, N.; Sung, B.S.; Choi, Y.K. *J. Am. Chem. Soc.*, ***2000***, *122*, 6771.

O

PhCHO , In/$InCl_3$, rt

aq THF

Ph

O

69%

Kang, S.; Jang, T.-S.; Keum, G.; Kang, S.B.; Han, S.-Y.; Kim, Y. *Org. Lett.*, ***2000***, 2, 3615.

$Co_2(CO)_8$, $Bu_3P{=}S$, PhH; 1 atm CO, 70°C

EtO_2C EtO_2C → EtO_2C EtO_2C =O 90%

Hayashi, M.; Hashimoto, Y.; Yamamoto, Y.; Usuki, J.; Saigo, K. *Angew. Chem. Int. Ed.*, ***2000***, *39*, 631.

Ph— ; 15 atm CO, 110 atm $CH_2{=}CH_2$, 85°C; 3% $Co_4(CO)_{11}[P(OPh_3)\}$, 41 h → O, Ph 80%

Jeong, N.; Hwang, S.H. *Angew. Chem. Int. Ed.*, ***2000***, *39*, 636.

2 eq *o*-iodoxybenzoic acid; 65°C, 10 h → O + O

(1 : 2) 76%

Nicolaou, K.C.; Zhong, Y.-L.; Baran, P.S. *J. Am. Chem. Soc.*, ***2000***, *122*, 7596.

OH, Ph; 2 eq PhI , 80°C , 12 h , DMF; 10% $Pd(OAc)_2$, 20% PPh3; 2 eq Nu_4NCl , 2 eq *i*-Pr_2NEt → O, Ph, Ph

Larock, R.C.; Reddy, Ch.K. *Org. Lett.*, ***2000***, *2*, 3325.

Bu, Bu, Bu, Bu; 1. 4 eq *t*-BuLi , –78°C; 2. CO_2 , –78°C → –30°C → O, Bu, Bu, Bu, Bu 70%

Xi, Z.; Song, Q. *J. Org. Chem.*, ***2000***, *65*, 9157.

O, Ph, $SnMe_3$; , 50°C , 0.2 h; 5% $Ni(cod)_2$, toluene → Ph, O, $SnMe_3$ 72%

Shirakawa, E.; Nakao, Y.; Yoshida, H.; Hiyama, T. *J. Am. Chem. Soc.*, ***2000***, *122*, 9030.

, 30 atm CO , 110°C

5% $Co_2(CO)_8$, DCM , 18 h

80%

Son, S.U.; Chung, Y.K.; Lee, S.-G. *J. Org. Chem.*, ***2000***, *65*, 6142.

PhC≡CH , $Co_2(CO)_6$, toluene
pulegone phosphine–BH_3 complex

50°C

99% (99% ee)

Verdaguer, X.; Moyano, A.; Pericàs, M.A.; Rivera, A.; Maestro, M.A.; Mahía, J. *J. Am. Chem. Soc.*, ***2000***, *122*, 10242.

1. , solvent free , KOH

rt , 3 h
2. 3M HCl

25%

Miyamoto, H.; Kanetaka, S.; Tanaka, K.; Yoshizawa, K.; Toyota, S.; Toda, F. *Chem. Lett.*, ***2000***, 888.

1. LDA , THF , –78°C

2. Ph(Cl)S=N*t*-Bu , THF , –78°C

95%

Mukaiyama, T.; Matsuo, J.-i.; Kitagawa, H. *Chem. Lett.*, ***2000***, 1250.

, ultrasound

NEt_3 , toluene , 15 min

99%

Ford, J.G.; Kerr, W.J.; Kirk, G.G.; Lindsay, D.M.; Middlemiss, D. *Synlett*, ***2000***, 1415.

Ph O

1. $TMSC(Li)N_2$, THF
–78°C → rt
2. 1N HCl, 0°C

82%

Sakai, A.; Aoyama, T.; Shioiri, T. *Tetrahedron Lett.*, ***2000***, *41*, 6859.

HO $Co_2(CO)_6$

, DCM, rt, 2.5 h

OH H H quant

Brown, D.S.; Campbell, E.; Kerr, W.J.; Lindsay, D.M.; Morrison, A.J.; Pike, K.G.; Watson, S.P. *Synlett*, ***2000***, 1573.

Et—≡—Et

1. BuLi, $CHCl_3$, THF, –78°C
2. HCl

Et Et 90%

Netland, K.A.; Gundrsen, L.-L.; Rise, F. *Synth. Commun.*, ***2000***, *30*, 1767.

OH OMe OH

MnO_2, bentonite
microwaves, 15 min

O OMe O 92%

Gómez-Lara, J.; Gutiérrez-Pérez, R.; Penieres-Carrillo, G.; López-Cortés, J.G.; Escudero-Salas, A.; Alvarez-Toledano, C. *Synth. Commun.*, ***2000***, *30*, 2713.

MeO

1. *sec*-BuLi, THF, -78°C
2. CuI, DMF
3. TMSCl
4. (methyl vinyl ketone)
5. TBAF, 3 H_2O, THF

O OMe

95% (65:35 *Z*:*E*)

Berrien, J.-F.; Raymond, M.-N.; Moskowitz, H.; Mayrargue, J. *Tetrahedron Lett.*, ***1999***, *40*, 1313.

1. Bu(2-Th)CuCNLi$_2$, THF
BF$_3$•OEt-2
2.

65%

Araújo, M.A.; Barrientos-Astigarraga, R.E.; Ellensohn, R.M.; Comasseto, J.V. *Tetrahedron Lett.*, ***1999***, *40*, 5115.

Bu$_2$CuLi , THF , –78°C

80% (1:1.2 *cis:trans*)

Lee, P.H.; Park, J.; Lee, K.; Kim, H.-C. *Tetrahedron Lett.*, ***1999***, *40*, 7109.

10% Pd(OAc)$_2$
Py , MS 3Å , toluene
O$_2$, 80°C , 2 d

60%

Nishimura, T.; Ohe, K.; Uemura, S. *J. Am. Chem. Soc.*, ***1999***, *121*, 2645.

10% CpRu(cod)Cl , DMF
15% CeCl$_3$•7 H$_2$O , 60°C
10% EtC≡CCH$_2$CH$_2$OH

55%

Trost, B.M.; Pinkerton, A.B. *J. Am. Chem. Soc.*, ***1999***, *121*, 4068.

Me$_2$AlCl

93%

Cunico, R.F.; Zaporowski, L.F.; Rogers, M. *J. Org. Chem.*, ***1999***, *64*, 9307.

PhCHO

acetone , microwaves
aq NaOH , 15 min

80%

Kad, G.L.; Kaur, K.P.; Singh, V.; Singh, J. *Synth. Commun.*, ***1999***, *29*, 2583.

TMSO, Ph — 1. $Me_3SiC{\equiv}CH$, $GaCl_3$; 2. H^+ → Ph (ketone) 75%

Yamaguchi, M.; Tsukagoshi, T.; Arisawa, M. *J. Am. Chem. Soc.*, ***1999***, *121*, 4074.

cat [Rh{(*RR*)Me-DuPHOS})cod)]PF_6, 5 atm CO, DME, 55-60°C, 6-20 h → O

99% (78% ee, *SS*)

Murakami, M.; Itami, K.; Ito, Y. *J. Am. Chem. Soc.*, ***1999***, *121*, 4130.

C_3H_7, H, H, O, Me — 1. (cyclopentenone), hν, pentane; 2. 200-240°C, BHT, PhH, sealed tube, 2-4 h → C_3H_7, O, O, Me 91x41%

Randall, M.L.; Lo, P.C.-K.; Bonitatebus Jr., P.J.; Snapper, M.L. *J. Am. Chem. Soc.*, ***1999***, *121*, 4534.

HO, HO — 1. $Co_2(CO)_8$; 2. NMO, CH_2Cl_2 → OH, HO, O, H

(H : H, 5:95) 88%

Mukai, C.; Kim, J.S.; Sonobe, H.; Hanaoka, M. *J. Org. Chem.*, ***1999***, *64*, 6822.

2% Al 38 C2, buffer (pH 7.4)
MeCN, 2 d

>95% (42% ee)

List, B.; Lerner, R.A.; Barbas III, C.F. *Org. Lett.*, ***1999***, *1*, 59.

1. $CO_2(CO)_8$, toluene
MS 4Å, −10°C
2. TMANO

90%

Pérez-Serrano, L.; Casarrubios, L.; Domínguez, G.; Pérez-Castells, J. *Org. Lett.*, ***1999***, *1*, 1187.

$Co_2(CO)_8$

74%

Ishizaki, M.; Iwahara, K.; Kyoumura, K.; Hoshino, O. *Synlett*, ***1999***, 587.

$Co_2(CO)_6$

3.5 eq PhSMe, DCE
83°C, 10 min

98%

Sugihara, T.; Yamada, M.; Yamaguchi, M.; Nishizawa, M. *Synlett*, ***1999***, 771.

4 eq CoF_3, dioxane, rt
4 eq H_2O, rt

87%

Tomatsu, A.; Takemura, S.; Hashimoto, K.; Nakata, M. *Synlett*, ***1999***, 1474.

SECTION 375: NITRILE - NITRILE

PBu_3 , 300 mPa , 50°C , 1 d

quant (77:23 *trans:cis*)

Jenner, G. *Tetrahedron Lett.*, ***2000***, *41*, 3091.

SECTION 376: NITRILE - ALKENE

10 eq Zn , MeOH; 0°C → rt

40%

Chavan, S.P.; Shrma, A.K.; Ethiraj, K.S. *Synlett*, ***2001***, 857.

Me—≡—NEt_2 → Me_3SiCN , DCM , 20°C; 15 h → 59%

Lukashev, N.V.; Kazantsev, A.V.; Borisenko, A.A.; Beletskaya, I.P. *Tetrahedron*, ***2001***, *57*, 10309.

, cat Bu_4NHSO_4; 50% aq NaOH , cyclohexane ether , 55°C , 5.5 h → 71%

Jończyk, A.; Gierczak, A.H. *Synthesis*, ***2001***, 93.

PhCHO → $CH_2(CN)_2$, Na_2CO_3 , MS 4Å; 70°C → 97%

Siebenhaar, B.; Casagrande, B.; Studer, M.; Blaser, H.-U. *Can. J. Chem.*, ***2001***, *79*, 566.

PhCHO → $CH_2(CN)_2$, TEBA , neat → $PhCH{=}C(CN)_2$ 88%

Bose, D.S.; Narsaiah, A.V. *J. Chem. Res. (S)*, ***2001***, 36.

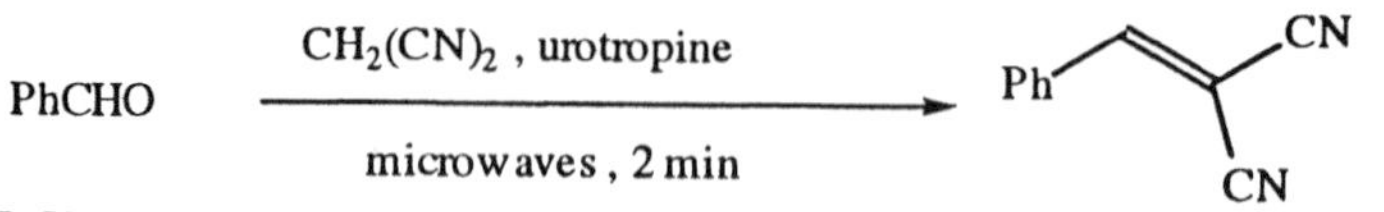

PhCHO → ($CH_2(CN)_2$, urotropine; microwaves, 2 min) → Ph–CH=C(CN)CN 87%

Wang, J.-X.; Wei, B.; Hu, Y.; Liu, Z.; Kang, L. *J. Chem. Res. (S)*, ***2001***, 146

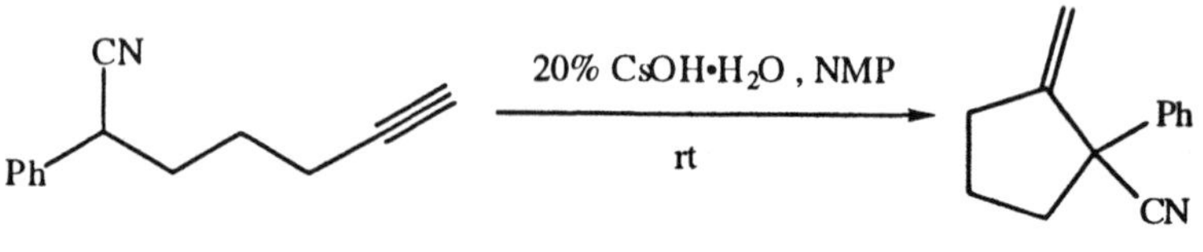

CN / Ph → (20% $CsOH{\bullet}H_2O$, NMP; rt) → Ph / CN 80%

Koradin, C.; Rodríguez, A.; Knochel, P. *Synlett*, ***2000***, 1452.

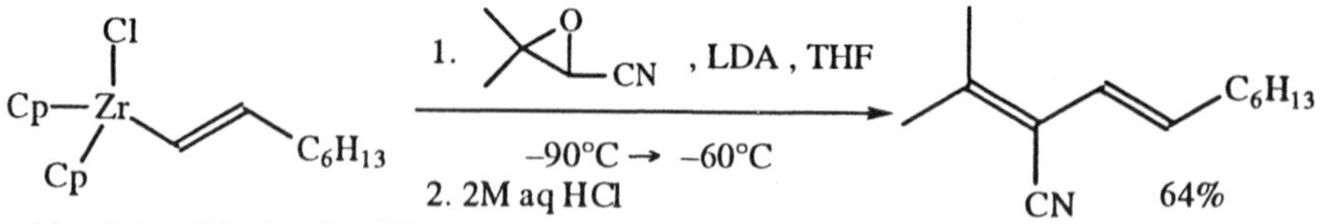

Cl / Cp—Zr / Cp / C_6H_{13} → (1. O / CN, LDA, THF; −90°C → −60°C; 2. 2M aq HCl) → C_6H_{13} / CN 64%

Kasatkin, A.N.; Whitby, R.J. *Tetrahedron Lett.*, ***2000***, *41*, 6201.

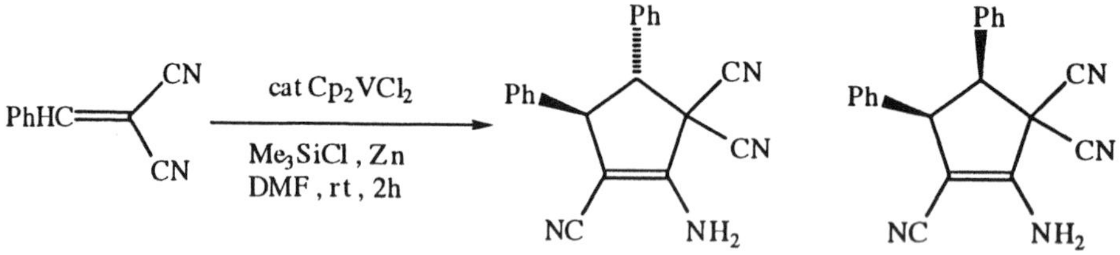

PhHC=C(CN)CN → (cat Cp_2VCl_2; Me_3SiCl, Zn; DMF, rt, 2h) → Ph / Ph / CN / CN / NC / NH_2 + Ph / Ph / CN / CN / NC / NH_2 (57 : 43) 72%

Zhou, L.; Hirao, T. *Tetrahedron Lett.*, ***2000***, *41*, 8517.

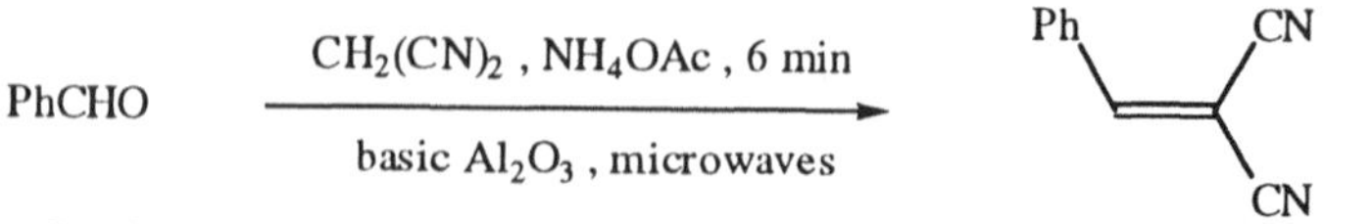

PhCHO → ($CH_2(CN)_2$, NH_4OAc, 6 min; basic Al_2O_3, microwaves) → Ph / CN / CN 80%

Balalaie, S.; Namati, N. *Synth. Commun.*, ***2000***, *30*, 869.

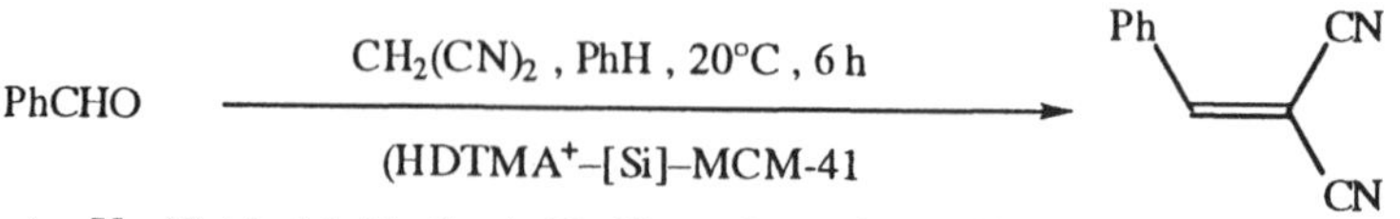

PhCHO → ($CH_2(CN)_2$, PhH, 20°C, 6 h; $(HDTMA^+$–[Si]–MCM-41) → Ph / CN / CN 94%

Kubota, Y.; Nishizaki, Y.; Sugi, Y. *Chem. Lett.*, ***2000***, 998.

SECTION 377: ALKENE - ALKENE

CH_2=CHCN , Pd(OAc)$_2$
t-amyl alcohol , 83°C
K_2CO_3 , 2 h

OEt OEt CN 34%

Shvo, Y.; Arisha, A.H.I. *J. Org. Chem.*, ***2001***, *66*, 4921.

H$^+$, PhH (degassed)
sealed tube , 240°C
1.5 eq BHT

H CO$_2$Me TBSO 64%

Deak, H.L.; Stokes, S.S.; Snapper, M.L. *J. Am. Chem. Soc.*, ***2001***, *123*, 5152.

TMS HO SPh
−78°C
Tf_2O , DCM , 2,6-lutidine

SPh + SPh
(10 : 1) 65%

Harmata, M.; Bohnert, G.; Barnes, C.L. *Tetrahedron Lett.*, ***2001***, *42*, 149.

OH
7% [CpRu(NCMe$_3$)$_3$]$^+$ PF$_6^-$
acetone/THF , H_2O ,−20°C , 4 h

OH O 77%

Trost, B.M.; Rudd, M.T. *J. Am. Chem. Soc.*, ***2001***, *123*, 8862.

MeO$_2$C MeO$_2$C
$PtCl_2$, dioxane
70°C , 20 h

MeO$_2$C MeO$_2$C
quant

Méndez, M.; Muñoz, M.P.; Nevado, C.; Cárdenas, D.J.; Echavarren, A.M. *J. Am. Chem. Soc.*, ***2001***, *123*, 10511.

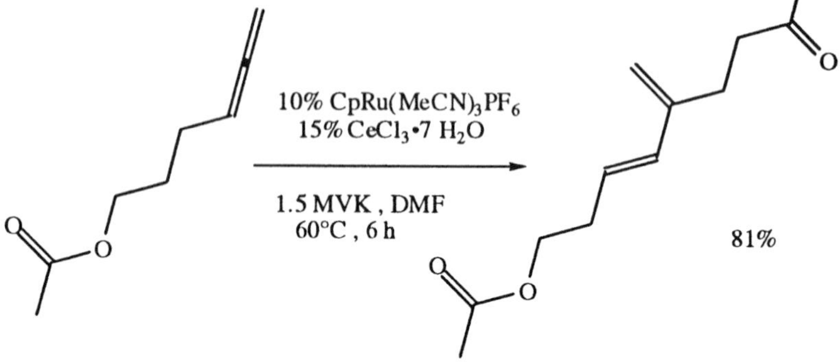

10% $CpRu(MeCN)_3PF_6$
15% $CeCl_3 \cdot 7\ H_2O$

1.5 MVK, DMF
60°C, 6 h

81%

Trost, B.M.; Pinkerton, A.B.; Seidel, M. *J. Am. Chem. Soc.*, ***2001***, *123*, 12466

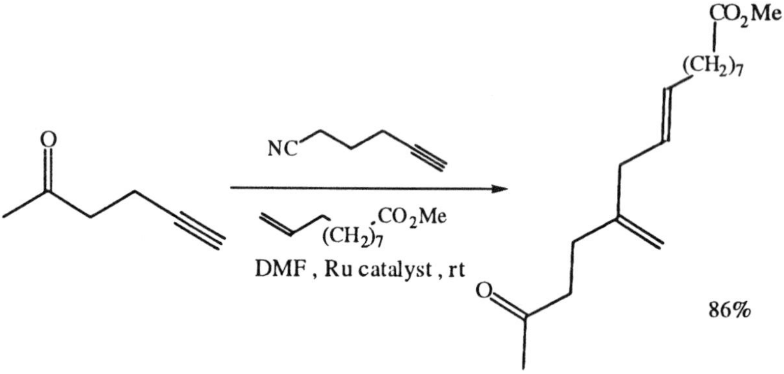

CO_2Me
$(CH_2)_7$

O

NC

CO_2Me
$(CH_2)_7$
DMF, Ru catalyst, rt

O

86%

Trost, B.M.; Pinkerton, A.B.; Toste, F.D.; Sperrle, M. *J. Am. Chem. Soc.* ***2001***, *123*, 12504.

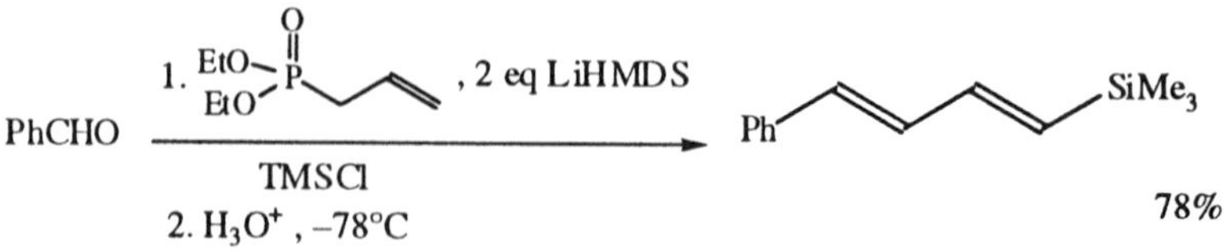

1. EtO, EtO, P, O, 2 eq LiHMDS
PhCHO
TMSCl
2. H_3O^+, −78°C

Ph SiMe$_3$

78%

Lee, B.S.; Gil, J.M.; Oh, D.Y. *Tetrahedron Lett.*, ***2001***, *42*, 2345.

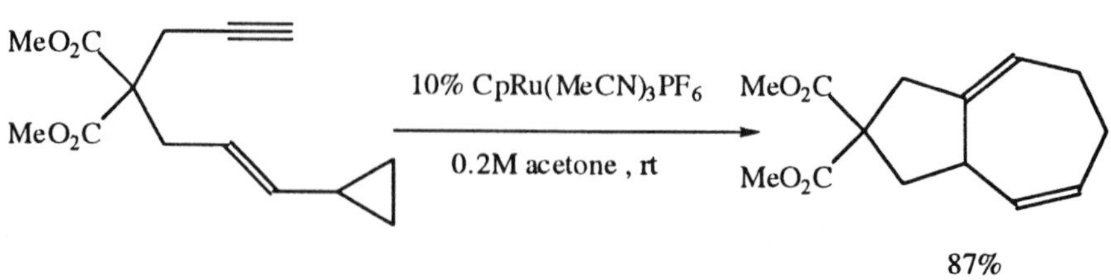

MeO_2C
MeO_2C

10% $CpRu(MeCN)_3PF_6$
0.2M acetone, rt

MeO_2C
MeO_2C

87%

Trost, B.M.; Toste, F.D.; Shen, H. *J. Am. Chem. Soc.*, ***2000***, *122*, 2379.

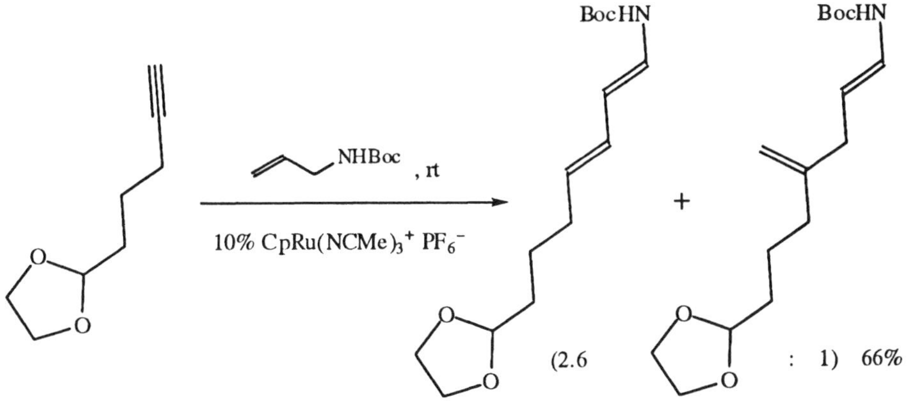

BocHN BocHN

NHBoc, rt

10% $CpRu(NCMe)_3^+ PF_6^-$

+

(2.6 : 1) 66%

Trost, B.M.; Surivet, J.-P. *Angew. Chem. Int. Ed.*, ***2001***, *40*, 1468.

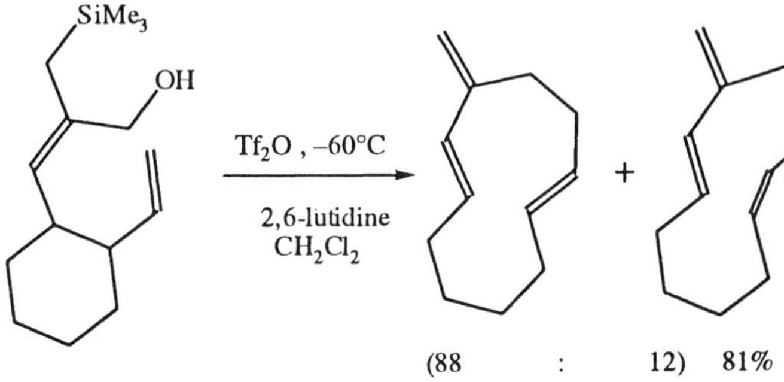

$SiMe_3$

OH

Tf_2O, –60°C

2,6-lutidine
CH_2Cl_2

+

(88 : 12) 81%

Suzuki, H.; Monda, A.; Kuroda, C. *Tetrahedron Lett.*, ***2001***, *42*, 1915.

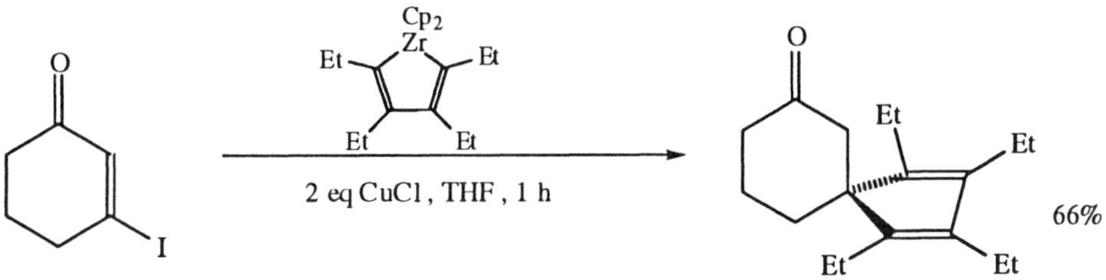

Cp_2 Zr, Et, Et, Et, Et

2 eq CuCl, THF, 1 h

66%

Xi, C.; Kotora, M.; Nakajima, K.; Takahashi, T. *J. Org. Chem.*, ***2000***, *65*, 945.

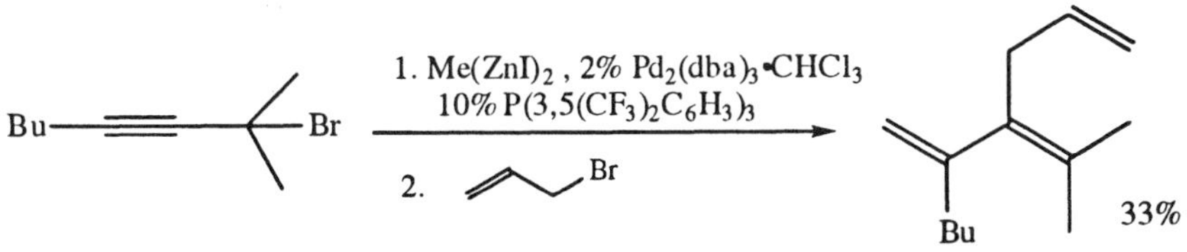

Bu—≡—C(Me)₂—Br

1. $Me(ZnI)_2$, 2% $Pd_2(dba)_3 \cdot CHCl_3$
10% $P(3,5(CF_3)_2C_6H_3)_3$

2. allyl bromide

Bu

33%

Matsubara, S.; Ukai, K.; Toda, N.; Utimoto, K.; Oshima, K. *Synlett*, ***2000***, 995.

Ph–CH=CH–$SiMeF_2$ —(CuCl, DMF, air; rt, 3 h)→ Ph–CH=CH–CH=CH–Ph 98%

Nishihara, Y.; Ikegashira, K.; Toriyama, F.; Mori, A.; Hiyama, T.
Bull. Chem. Soc. Jpn., ***2000***, *73*, 985.

2 $CH_2=C=CH–CO_2Et$ —(10% $Ni(PPh_3)_4$, toluene; –15,C → –10°C, 30 min)→ 1,2-bis(CO_2Et-methylene)cyclobutane 71%

Saito, S.; Hirayama, K.; Kabuto, C.; Yamamoto, Y. *J. Am. Chem. Soc.*, ***2000***, ***122***, 10776.

Me_3Si–CH=C($SnBr_3$)–CH_2–C(=O)–$OSnBr_3$ —(1. I_2 ether, rt, 1 h; 2. Bu_3Sn–CH=CH–C_5H_{11}, DMF, $PdCl_2(MeCN)_2$, 60°C, 6h; 3. SiO_2)→ Me_3Si–CH=CH–CH=CH–C_5H_{11} 64%

Lunot, S.A.; Thibonnet, J.; Duchêne, A.; Parrain, J.-L.; Abarbri, M.
Tetrahedron Lett., ***2000***, *41*, 8893.

TBDMSO–C(Me)(C≡C–CO_2Me)–CH_2CH_2CH=CMe_2 —(10% $CpRu(MeCN)_3\ BF_6$; 0.1 M acetone, rt, 4 h)→ TBDMSO, Me-substituted cycloheptene with =CH–CO_2Me 67%

Trost, B.M.; Toste, F.D. *J. Am. Chem. Soc.*, ***1999***, *121*, 9728.

Me_3Si–CH_2–CH(Bt)–CH=CH_2 —(1. BuLi, –78°C; 2. $C_6H_{13}Br$; 3. heat)→ C_6H_{13}–C(=CH_2)–CH=CH_2 82c70%

B t = benzotriazyl

Katritzky, A.R.; Serdyuk, L.; Toader, D.; Wang, X. *J. Org. Chem.*, ***1999***, *64*, 1888.

PhCHO —(HC≡C–CH_2–$SiMe_3$, BuLi, $ZnCl_2$; Ni(cod)$_2$)→ Ph–CH=CH–C(Bu)=CH_2 58%

Cunico, R.F.; Zaporowski, L.F.; Rogers, M. *J. Org. Chem.*, ***1999***, *64*, 9307.

HO H H 10% $Rh(PPh_3)_3OTf$ toluene, 110°C 0.01M MeO_2C CO_2Me HO H MeO_2C CO_2Me 96%

Wender, P.A.; Dyckman, A.J. *Org. Lett.*, ***1999***, *1*, 2089.

SECTION 378: OXIDES - ALKYNES

SO_2Ph Me Ph—≡—Me, DCM 10% Mo[N(*t*-Bu)(Ar)]$_3$ Ar = 3,5-dimethylphenyl SO_2Ph Ph 71%

Fürstner, A.; Mathes, C. *Org. Lett.*, ***2001***, *3*, 221.

Ph BuTe P OEt OEt O PhC≡CH, MeOH, NEt_3 20% $PdCl_2$, 20% CuI rt, 8 h Ph Ph P OEt OEt O 73%

Braga, A.L.; de Andrade, L.H.; Silveira, C.C.; Moro, A.V.; Zeni, G. *Tetrahedron Lett.*, ***2001***, *42*, 8563.

O Ph 1. NaH 2. NaH, $(EtO)_2P(O)H$ THF, CCl_4 O EtO—P EtO O Ph 62% (>95% *Z*:*E*)

Kabalka, G.W.; Yang, K.; Wang, Z. *Synth. Commun.*, ***2001***, *31*, 511.

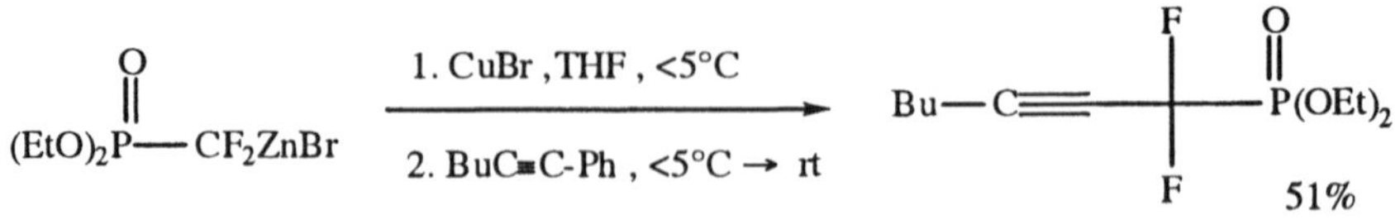

$(EtO)_2P(O)—CF_2ZnBr$ → 1. CuBr, THF, <5°C; 2. BuC≡C-Ph, <5°C → rt → $Bu—C≡C—CF_2—P(O)(OEt)_2$ 51%

Zhang, X.; Burton, D.J. *Tetrahedron Lett.*, ***2000***, *41*, 7791.

REVIEWS:

"The Chemistry of Acetylenes and Allenic Sulfones," Back, T.G. *Tetrahedron*, ***2001***, *57*, 5263.

SECTION 379: OXIDES - ACID DERIVATIVES

NO ADDITIONAL EXAMPLES

SECTION 380: OXIDES - ALCOHOLS, THIOLS

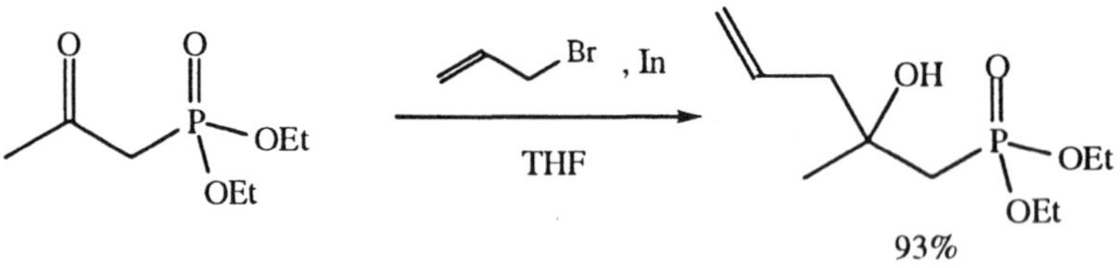

Allyl bromide, In; THF → 93%

Ranu, B.C.; Samanta, S.; Hajra, A. *J. Org. Chem.*, ***2001***, *66*, 7519.

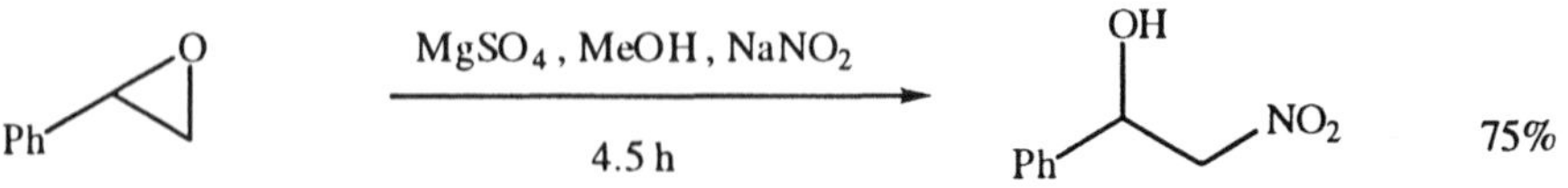

$MgSO_4$, MeOH, $NaNO_2$; 4.5 h → 75%

Kalita, B.; Barua, N.C.; Bezbarua, M.; Bez, G. *Synlett*, ***2001***, 1411.

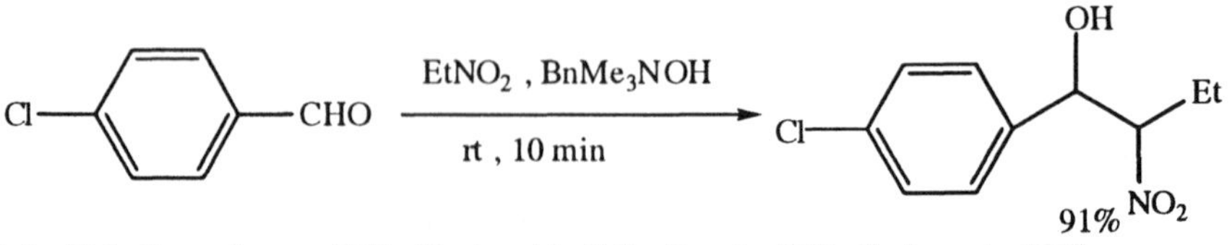

$EtNO_2$, $BnMe_3NOH$; rt, 10 min → 91%

Bulbule, V.J.; Jnaneshwara, G.K.; Deshmukh, R.R.; Borate, H.B.; Seshpande, V.H. *Synth. Commun.*, ***2001***, *3*, 3623.

PhO$_2$S ... O → 1. BF$_3$•OEt$_2$ 2. PhH → SO$_2$Ph, OH, Ph 47% (80:20 *trans:cis*)

Brandänge, S.; Bäckvall, J.-E.; Leijonmarck, H. *J. Chem. Soc., Perkin Trans. 1*, ***2001***, 2051.

O, CO$_2$Et → MeNO$_2$, 20% NEt$_3$, Cu(OTf)$_3$; cat *S*-*t*-Bu-bis-oxazoline ligand → HO, NO$_2$, CO$_2$Et >95% (92% ee)

Christensen, C.; Juhl, K.; Jørgensen, K.A. *Chem. Commun.*, ***2001***, 2222.

O → 1. PhSH, HFIP 2. H$_2$O$_2$ → OH, S, Ph, O 84% (65:35 *trans:cis*)

HFIP = hexafluoroisopropanol

Kesavan, V.; Bonnet-Delpon, D.; Bégué, J.-P. *Tetrahedron Lett.*, ***2000***, ***41***, 2895.

OHC, Ph → EtNO$_2$, THF, 0°C; 10% LiAlH$_4$, 6 h → Ph, O$_2$N, OH, Me + Ph, O$_2$N, OH, Me (2.1 : 1) 91%

Youn, W.; Kim, Y.H. *Synlett*, ***2000***, 880.

O, S, Tol, Ph, O → Dibal, THF, –78°C → O, S, Tol, Ph, OH 98% (86:14 *S,S:S,R*)

Nakamura, S.; Kuroyanagi, M.; Watanabe, Y.; Toru, T. *J. Chem. Soc., Perkin Trans. 1*, ***2000***, 3143.

NO$_2$ → BuCHO, KOH (powder); no solvent → NO$_2$, Bu, OH 65%

Ballini, R.; Bosica, G.; Parrini, M. *Chem. Lett.*, ***1999***, 1105.

cat

OH

NO_2 95%

Kisanga, P.B.; Verkade, J.G. *J. Org. Chem.*, ***1999***, *64*, 4298.

PhCHO $\xrightarrow{EtNO_2\text{ , Mg-Al hydrotalcite}}$

87% (3.25:1 *threo:erythro*)

Bulbule, V.J.; Deshpande, V.H.; Velu, S.; Sudalai, A.; Sivasankar, S.; Sathe, V.T. *Tetrahedron*, ***1999***, *55*, 9325.

REVIEWS:

"The Henry Reaction; Recent examples," Luzzio, F.A. *Tetrahedron*, ***2001***, *57*, 915.

SECTION 381 OXIDES - ALDEHYDES

NO ADDITIONAL EXAMPLES

SECTION 382: OXIDES - AMIDES

cat $Rh_2(OAc)_4$

DCE, 12 h reflux

(49 : 51) 95%

Yoon, C.H.; Zaworotko, M.J.; Moulton, B.; Jung, K.W. *Org. Lett.*, ***2001***, *3*, 3539.

TMSO N⁺ O⁻ Et + PMP N CO_2Et —— 2% Cu catalyst , –100°C / THF ——> O_2N NHPMP CO_2Et Et

94% (25:1 *erythro:threo*)
95% ee *erythro*

Knudsen, K.R.; Risgaard, T.; Nishiwaki, N.; Gothelf, K.V.; Jørgensen, K.A. *J. Am. Chem. Soc.*, ***2001***, *123*, 5843.

O P OEt OEt —— $NsONHCO_2Et$, CaO , CH_2Cl_2 / rt , 1 d ——> O P OEt OEt N EtO_2C 40%

Fazio, A.; Loreto, M.A.; Tardella, P.A. *Tetrahedron Lett.*, ***2001***, *42*, 2185.

PhCHO —— 1. Cbz–NH_2 , $MeOPCl_2$ / 2. EtOH , NEt_3 ——> O P OMe OEt CbzHN Ph + O P OMe OH CbzHN Ph

47% 39%

Xu, J.; Fun, N. *J. Chem. Soc., Perkin Trans. 1*, ***2001***, 1223.

O P OEt OEt N —— stage II *B. bassiana* / pH 5.0 , 20°C , 7 d ——> O P OEt OEt N OH 43%

Hemenway, M.S.; Olivo, H.F. *J. Org. Chem.*, ***1999***, *64*, 6312.

SECTION 383: OXIDES - AMINES

O S *t*-Bu N OTBS —— EtMgBr , THF ——> O S *t*-Bu N H Et OTBS

91% (91:9 dr)

Tang, T.P.; Volkman, S.K.; Ellman, J.A. *J. Org. Chem.*, ***2001***, *66*, 8772.

PhCHO → [PhCH(Me)NH$_2$, HOP(OEt)$_2$; montmorillonite KSF, microwaves, 3 min] → PhCH(NHCH(Me)Ph)P(O)(OEt)$_2$ 85%

Yadav, J.S.; Subba Reddy, B.V.; Madan, Ch. *Synlett*, **2001**, 1131.

PhNHOH → [i-PrCHO, 5 M $LiClO_4$/ether, $(MeO)_2PO(SiMe_3)$, rt, 15 min] → i-PrCH(N(OSiMe$_3$)Ph)P(O)(OMe)$_2$ 97%

Heydari, A.; Zarei, M.; Alijanianzadeh, R.; Tavakol, H. *Tetrahedron Lett.*, **2001**, *42*, 3629.

(EtO)$_2$P(O)C≡CH → [Et_2NH, EtOH, reflux; cat CuI, 30 min] → (EtO)$_2$P(O)CH=CHNEt$_2$ 42%

Panarina, A.E.; Dugadina, A.V.; Zakharov, V.I.; Ionin, B.I. *Tetrahedron Lett.*, **2001**, *42*, 4365.

PhCH=N–Ts → [$MeNO_2$, 5% Yb(Oi-Pr)$_3$; i-PrOK, THF, rt] → PhCH(NHTs)CH$_2$NO$_2$ quant

Qian, C.; Gao, F.; Chen, R. *Tetrahedron Lett.*, **2001**, *42*, 4673.

4-MeC$_6$H$_4$CHO → [$PhNH_2$, (EtO)$_2$P–OH, 10% $TaCl_5$–SiO_2, DCE; rt, 22 h] → 4-MeC$_6$H$_4$CH(NHPh)P(O)(OEt)$_2$ 92%

Chandrasekhar, S.; Prakash, S.J.; Jagadeshwar, V.; Narsihmulu, Ch. *Tetrahedron Lett.*, **2001**, *42*, 5561.

PhCHO → [$PhNH_2$, Al_2O_3, (EtO)$_2$P(O)H; microwaves, 6 min] → PhCH(NHPh)P(O)(OEt)$_2$ 87%

Kaboudin, B.; Nazari, R. *Tetrahedron Lett.*, **2001**, *42*, 8211.

NO_2 rt, 1 h; BnNHOH•HCl, EtOH, NEt_3 → NO_2, OH, N, Bn 93%

O'Neil, I.A.; Clator, E.; Southern, J.M.; Bickley, J.F.; Tapolczay, D.J. *Tetrahedron Lett.*, ***2001***, *42*, 8251.

Ph, H, N, Bn; $HOP(OEt)_2$, $ZrCl_4$, MeCN; rt, 3.5 h → NHBn, OEt, OEt, Ph, P, O 87%

Yadav, J.S.; Reddy, B.V.S.; Raj, K.S.; Reddy, B.; Prasad, A.R. *Synthesis*, ***2001***, 2277.

$C_5H_{11}CHO$ 1. H, P, OEt, OEt, O, Al_2O_3, NH_4O_2CH microwaves, 3 min 2. *p*-TsOH, H_2O 3. NaOH → NH_2, C_5H_{11}, P, OEt, OEt, O 66%

Kaboudin, B. *Chem. Lett.*, ***2001***, 880.

OMe, N, EtO_2C; $C_3H_7NO_2$, cat $Cu(OTf)_2$; chiral bis-oxazoline ligand → OMe, HN, NO_2, EtO_2C, Et

92 : 8 *erythro:threo*
91%ee 85% ee

Nishiwaki, N.; Knudsen, K.R.; Gothelf, K.V.; Jørgensen, K.A. *Angew. Chem. Int. Ed.*, ***2001***, *40*, 2992.

O, NHBn; 1. NaH 2. SPh, PPh_3I^+ 3. mcpba 4. *t*-BuOK, *t*-BuOH → SO_2Ph, N, Ph 85 X 93%

Burley, I.; Bilic, B.; Hewson, A.T.; Newton, J.R.A. *Tetrahedron Lett.*, ***2000***, *41*, 8969.

PhCHO → ($PhNH_2$, HOP(OEt)$_2$, $InCl_3$; THF, ultrasound, 5 h) → O=P(OEt)(OEt)–CH(Ph)(NHPh) 93%

Ranu, B.C.; Hajra, A.; Jana, U. *Org. Lett., 1999, 1,* 1141.

SECTION 384: OXIDES - ESTERS

PhC(=O)–P(=O)(OEt)$_2$ → (PhCHO, 0.2 Sm, THF–HMPA; rt, 3 h) → PhCH(OBz)–P(=O)(OEt)$_2$ 85%

Takaki, K.; Itono, Y.; Nagafuji, A.; Naito, Y.; Shishido, T.; Takehira, K.; Makioka, Y.; Tankguchi, Y.; Fujiwara, H. *J. Org. Chem.* ***2000***, *65*, 475.

$CH_3C(=O)Cl$ → (1. P(OMe)$_3$, 0°C → 70°C; 2. Bz_2O, DBU, THF, 0°C → rt; 3. chiral bis-phosphine, 25°C, Rh(cod)OTf, 4 atm H_2, MeOH, 2 d) → (MeO)$_2$P(=O)–CH(OBz)CH$_3$ 100% conversion, 84% ee

Burk, M.J.; Stammers, T.; Straub, J.A. *Org. Lett., **1999**, 1,* 387.

SECTION 385: OXIDES - ETHERS, EPOXIDES, THIOETHERS

HOCH$_2$CH$_2$C(S(O)Ph)=C=CH$_2$ → (*t*-BuOK, *t*-BuOH; rt, 5 min) → 3-(phenylsulfinyl)-2-methyl-4,5-dihydrofuran quant

Mukai, C.; Yamashita, H.; Hanaoka, M. *Org. Lett., **2001**, 3,* 3385.

SECTION 386: OXIDES - HALIDES, SULFONATES

$(F_3C)_2HCO$–P(=O)(H)–$OCH(CF_3)_2$ → (NIS, MeCN) → $(F_3C)_2HCO$–P(=O)(I)–$OCH(CF_3)_2$ 51%

Timperley, C.M.; Waters, M.J. *Chem. Commun., **2001***, 797.

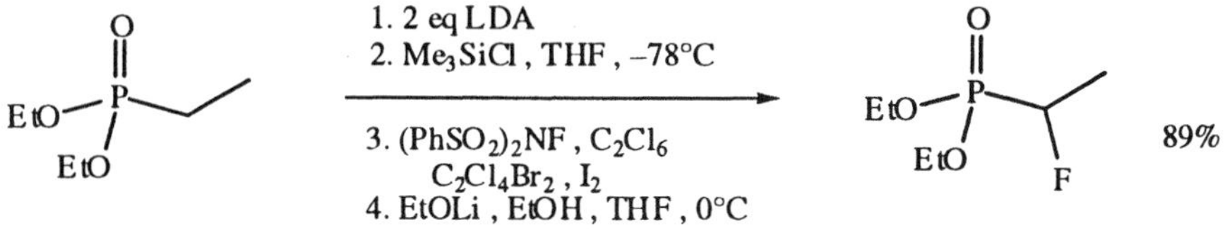

1. 2 eq LDA
2. Me_3SiCl, THF, –78°C
3. $(PhSO_2)_2NF$, C_2Cl_6
$C_2Cl_4Br_2$, I_2
4. EtOLi, EtOH, THF, 0°C

EtO, EtO, O, P, F

89%

Iorga, B.; Eymery, F.; Savignac, P. *Synthesis*, ***2000***, 576.

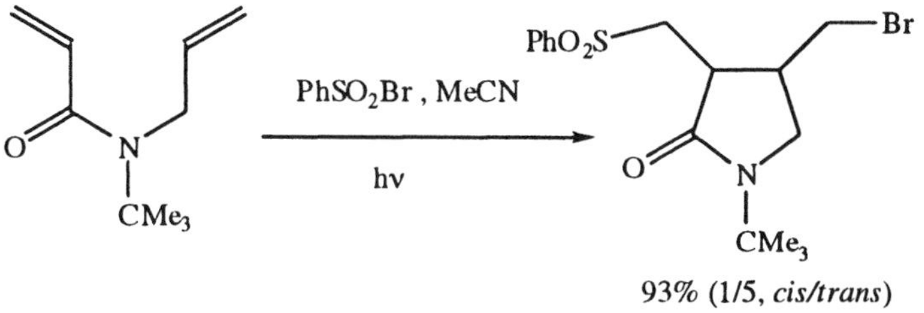

$PhSO_2Br$, MeCN

hv

O, N, CMe_3, PhO_2S, Br

93% (1/5, *cis/trans*)

Wang, C.; Russell, G.A. *J. Org. Chem.*, ***1999***, *64*, 2346.

SECTION 387: OXIDES - KETONES

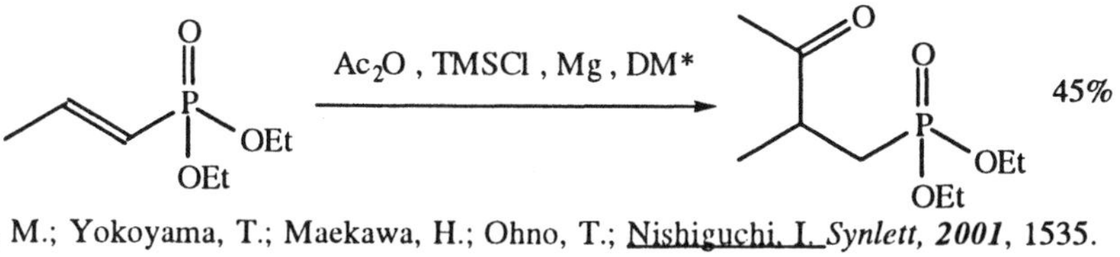

Ac_2O, TMSCl, Mg, DM*

O, P, OEt, OEt

45%

Kyoda, M.; Yokoyama, T.; Maekawa, H.; Ohno, T.; Nishiguchi, I. *Synlett*, ***2001***, 1535.

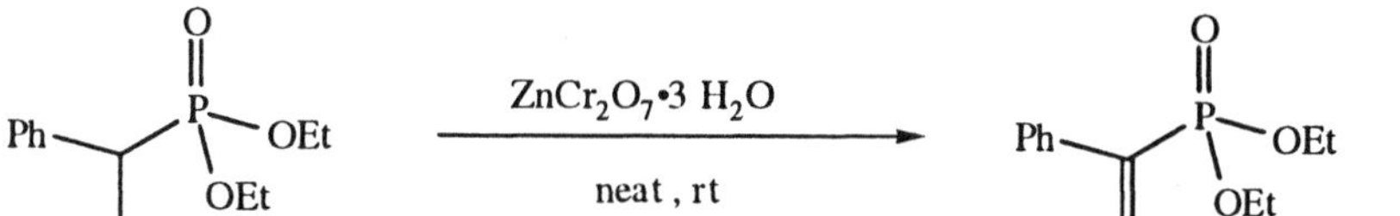

$ZnCr_2O_7•3\ H_2O$

neat, rt

Ph, O, P, OEt, OEt, OH

94%

Firouzabadi, H.; Iranpoor, N.; Sabhani, S.; Sardarian, A.-R. *Tetrahedron Lett.*, ***2001***, *42*, 4369.

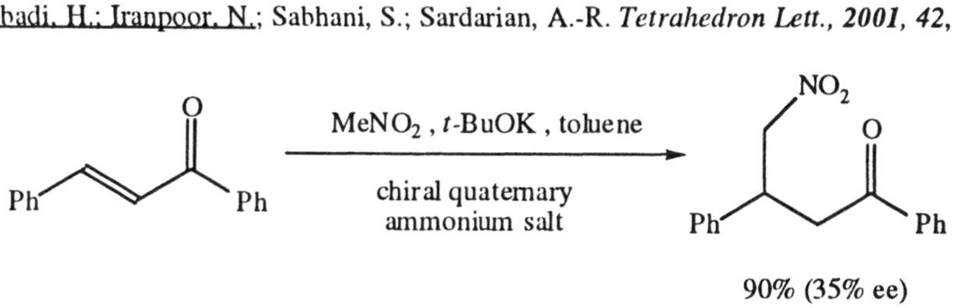

$MeNO_2$, *t*-BuOK, toluene

chiral quaternary ammonium salt

Ph, O, Ph, NO_2

90% (35% ee)

Kim, D.Y.; Huh, S.C. *Tetrahedron*, ***2001***, *57*, 8933.

Bu–CH(OH)–P(O)(OEt)(OEt) → (Al_2O_3/CrO_3, microwaves, 2 min) → Bu–C(O)–P(O)(OEt)(OEt) 75%

Kaboudin, B.; Nazari, R. *Synth. Commun.*, ***2001***, *31*, 2245.

(EtO)(EtO)P(O)–$CH_2CH=CH_2$ →

1. O_2NEt, 4 eq ClPhNCO, NEt_3, THF
2. LDA, THF
3. AcOH
4. $TiCl_3$, aq HCl, DMP

→ (EtO)(EtO)P(O)–$CH_2CH=CH$–C(O)–CH_3 88x64%

Lee, S.Y.; Lee, B.S.; Lee, C.-W.; Oh, D.Y. *J. Org. Chem.*, ***2000***, *65*, 256.

Ph–CH=CH–NO_2 → (CH_2=C(Ph)$OSnBu_3$, Bu_4NBr, THF, –45°C, 4 h) → Ph–C(O)–CH_2–CH(Ph)–CH_2–NO_2

94% (0% without TBAB)

Yasuda, M.; Ohigashi, N.; Baba, A. *Chem. Lett.*, ***2000***, 1266.

Ph–C(O)–CH=CH–Ph → ($MeNO_2$, Mg–Al–O*t*-Bu, hydrotalcite catalyst) → Ph–C(O)–CH_2–CH(Ph)–CH_2–NO_2 93%

Choudary, B.M.; Kantam, M.L.; Kavita, B.; Reddy, Ch.V.; Figueras, F. *Tetrahedron*, ***2000***, *56*, 9357.

Ph–CH=CH–C(O)–Ph → ($MeNO_2$, KF/PN, 2 h) → Ph–CH(CH_2–O_2N)–CH_2–C(O)–CH_2–Ph

PN = "natural mineral phosphate"

98%

Loupy, A.; Régnier, S. *Tetrahedron Lett.*, ***1999***, *40*, 6221.

SECTION 388: OXIDES - NITRILES

Ph–CH=N–$OCOC_6F_5$ → (20% heterobimetallic complex, $EtNO_2$, CH_2Cl_2, –40°C, 2 d) → Ph–CH(NHP(O)Ph_2)–CH(Et)–NO_2

77% (6:1 *anti:syn*)

Yamada, K.i.; Moll, G.; Shibasaki, M. *Synlett*, ***2001***, 980.

Et_2AlCN , THF
65°C , 15 min

Tol
CN O
H Et Tol

80% (>98% de)

Ruano, J.L.G.; García, M.C.; Laso, N.M.; Castro, A.M.M.; Ramos, J.H.R. *Angew. Chem. Int. Ed.*, ***2001***, *40*, 2507.

SECTION 389: OXIDES - ALKENES

Ph
Ph

1. Cl–C_6H_4–SO_2Na , AcOH , H_2O
PhTeTePh , 80°C
2. AcOH

O_2 S Cl PhTe

85%

Huang, X.; Liang, C.-G.; Xu, Q.; He, Q.-W. *J. Org. Chem.*, ***2001***, *66*, 74.

C_6H_{13}

Ph–P(O)(Ph)–H , toluene , 25°C , 1 h
3% $Rh(I)(PPh_3)_3$

C_6H_{13} Ph Ph O

quant

Han, L.-B.; Zhao, C.-Q.; Tanaka, M. *J. Org. Chem.*, ***2001***, *66*, 5929.

Ph S

3 eq H_2O_2

Ph S=O + Ph S O O

$MeReO_3$, 0°C , 1 h	70%	14%
Na_2WO_2 , rt , 1 h	-	85%

Choi, S.; Yang, J.-D.; Ji, M.; Choi, H.; Kee, M.; Ahn, K.-H.; Byeon, S.-H.; Baik, W.; Ko, S. *J. Org. Chem.*, ***2001***, *66*, 8192.

Ph

$PhSO_2Na$, NaI , CAN
MeCN , rt , 45 min

Ph SO_2Ph 76%

Nair, V.; Augustine, A.; George, T.G.; Nair, L.G. *Tetrahedron Lett.*, ***2001***, *42*, 6763.

1. C_5H_{11}–Ar(OMOM)$_2$–)$_2Cu(CN)Li_2$, $BF_3•OEt_2$
2. EtMgBr
3. $(EtO)_2P(O)Cl$

OTES; O; P—OEt; OEt; MOMO; MOMO; C_5H_{11}; TESO

William, A.D.; Kobayashi, Y. *Org. Lett.*, ***2001***, *3*, 2017.

cat $PdMe_2$(dppb)
dioxane , 12 h
100°C

76%

Mirzaei, F.; Han, L.-B.; Tanaka, M. *Tetrahedron Lett.*, ***2001***, *42*, 297.

PhCHO → ($MeNO_2$, 90°C; MCM41-NH_2 silica catalyst) → Ph–CH=CH–NO_2 98%

Demicheli, G.; Maggi, R.; Mazzacani, A.; Righi, P.; Sartori, G.; Bigi, F. *Tetraehedron Lett.*, ***2001***, *42*, 2401.

SO_2CHBr_2

cyclopentanone , THF
Sm/SmI_2/cat $CrCl_3$

SO_2

65%

Liu, Y.; Wu, H.; Zhang, Y. *Synth. Commun.*, ***2001***, *31*, 47.

Bu—≡—$SiMe_3$

1. $Cp_2Zr(H)Cl$, THF , rt
2. PhSOCl ,THF , 40°C

Bu; $SiMe_3$; S—Ph; O

83%

Zhong, P.; Guo, M.-P.; Huang, X. *Synth. Commun.*, ***2001***, *31*, 615.

Ph—≡ → Ph–CH=CH–P(O)(pinacolato); 2% [RhBr(PPh$_3$)], H–P(O)(pinacolato), acetone, 25°C, 20 h; 93%

Zhao, C.-Q.; Han, L.-B.; Goto, M.; Tanaka, M. *Angew. Chem. Int. Ed.*, ***2001***, *40*, 1929.

Ph–C≡C–S(O)Ph → (Z)-Ph–CH=CH–S(O)Ph; 1. $Cp_2Zr(H)Cl$, THF, rt 2. H_2O; 78%

Zhong, P.; Huang, X.; Ping-Guo, M. *Tetrahedron*, ***2000***, *56*, 8921.

PhO_2S–CH$_2$–Me → PhO_2S(Me)C=CHPh; 1. 2 eq BuLi, THF, 0°C 2. $(EtO)_2POCl$, THF, 0°C 3. PhCHO; 84%

Lee, J.W.; Lee, C.-W.; Jung, J.H.; Oh, D.Y. *Synth. Commun.*, ***2000***, *30*, 279.

2-nitrobenzaldehyde (CHO, NO_2) → 2-NO_2-C$_6$H$_4$–CH=CH–NO_2; $MeNO_2$, 25°C, 10 min, gel entrapped base catalyst; 96%

Bandgar, B.P.; Uppalla, L.S. *Synth. Commun.*, ***2000***, *30*, 2071.

C_6H_{13}—≡ → C_6H_{13}(PhS)C=CH–P(O)(OPh)$_2$; $(PhO)_2P(O)SPh$, toluene, 5 h, 3% $Pd(PPh_3)_4$, 100°C; 92%

Han, L.-B.; Tanaka, M. *Chem. Lett.*, ***1999***, 863.

REVIEWS:

"α-Phosphonovinyl Carbanions in Organic Synthesis," Minami, T.; Okauchi, T.; Kouno, R. *Synthesis*, ***2001***, 349.

SECTION 390: OXIDES - OXIDES

N

$(EtO)_2P(=O)H$, $Pd(PPh_3)_4$

O
EtO—P
EtO
OEt
P
OEt
O
N

89%

Allen Jr., A.; Manke, D.R.; Lin, W. *Tetrahedron Lett.*, **2000**, *41*, 151.

Ph
NO_2

1. chiral TADDOL phosphite, Et_2Zn
TMEDA, THF, –78°C
2. TMSCl, MeCN, reflux
3. DCM, H_2O, rt

O
HO—P
HO
NO_2
Ph

89x87% (92% ee)

Enders, D.; Tedeschi, L.; Bats, J.W. *Angew. Chem. Int. Ed.*, **2000**, *39*, 4605.